Illustrated Handbook of Succulent Plants

Series Editors: U. Eggli, H. E. K. Hartmann

Springer

Berlin
Heidelberg
New York
Barcelona
Hong Kong
London
Milan
Paris
Singapore
Tokyo

Illustrated Handbook of Succulent Plants
(Eggli/Hartmann Eds.)

Illustrated Handbook of Succulent Plants: Monocotyledons
Ed. by Urs Eggli (2001)

Illustrated Handbook of Succulent Plants: Dicotyledons
Ed. by Urs Eggli (2002)

Illustrated Handbook of Succulent Plants: Aizoaceae A–E
Ed. by H. E. K. Hartmann (2001)

Illustrated Handbook of Succulent Plants: Aizoaceae F–Z
Ed. by H. E. K. Hartmann (2001)

Illustrated Handbook of Succulent Plants: Asclepiadaceae
Ed. by F. Albers and U. Meve (2002)

Illustrated Handbook of Succulent Plants: Crassulaceae
Ed. by Urs Eggli (2003)

Urs Eggli (Ed.)

Illustrated Handbook of Succulent Plants: Monocotyledons

With 227 Colour Photos, Printed in 32 Colour Plates

Springer

DR. URS EGGLI

Sukkulenten-Sammlung Zürich
Mythenquai 88
8002 Zürich
Switzerland

e-mail: urs.eggli@gla.stzh.ch

Compiled in cooperation with the International Organization for Succulent Plant Study (IOS)

With contributions by:
S. Arroyo-Leuenberger, M. B. Bayer, J. Bogner, U. Eggli, P. I. Forster, D. R. Hunt, E. van Jaarsveld,
N. L. Meyer, L. E. Newton, G. D. Rowley, G. F. Smith, J. Thiede, C. C. Walker

ISBN 3-540-41692-7 Springer-Verlag Berlin Heidelberg New York

Library of Congress Cataloging-in-Publication Data

Illustrated handbook of succulent plants. Monocotyledons / Eggli (ed.)
   p. cm.
  Includes bibliographical references (p.).
  ISBN 3540416927 (alk. paper)
  1. Monocotyledons--Classification. I. Eggli, Urs.

  QK495.A14 I45 2001
  584'.01'2--dc21

                2001034208

Springer-Verlag Berlin Heidelberg New York
a member of BertelsmannSpringer Science + Business Media GmbH
© Springer-Verlag Berlin Heidelberg 2001
Printed in Germany

Data conversion: Colour plates by Büro Stasch, Bayreuth, Germany (*stasch@stasch.com*)
Cover design: *design & production* GmbH, 69121 Heidelberg, Germany
Typesetting: Camera-ready by Urs Eggli
SPIN: 10765911   31/3130   – 5 4 3 2 1 0 – Printed on acid-free paper

Preface

Handbooks devoted to succulent plants (including cacti) have a long-standing tradition, and the demand for updated editions is a good indication of the high degree of interest that exists in this fascinating group of plants.

While first handbooks devoted to the family *Cactaceae* were already published in the 19th century, the first handbook dealing with the so-called 'other' succulents was authored by Hermann Jacobsen and published in 1954 - 1955, then called "Handbuch der sukkulenten Pflanzen". A revised and enlarged English edition was made available in 1959 and was repeatedly reprinted in the following years.

A major step towards a compact handbook including short descriptions, full synonymy and hundreds of illustrations was the publication of the first edition of Hermann Jacobsen's celebrated "Sukkulentenlexikon" in 1970, followed by the English edition in 1975, and a revised German edition in 1981, finished by Klaus Hesselbarth following Hermann Jacobsen's death in August 1978.

The demand for updated compact information on succulent plants has not diminished since then, and contrary to Hermann Jacobsen's opinion (preface to the German edition 1970) that the number of papers dealing with succulent plants was on the decline, the interest in succulent plants was growing, both among hobbyists and among botanists. The results are numerous new taxa in many families, and many monographs of previously little-known groups have been published in the last 30 years. Accordingly, a need for a "New Lexicon" was beginning to be felt, both for 'other' succulents and for cacti.

After some informal initial discussions held among various botanists and interested specialist collectors, and faced with the desire of Gustav Fischer Verlag Jena, the publisher of both Backeberg's "Kakteenlexikon" and Jacobsen's "Sukkulentenlexikon", to prepare revised editions, a meeting was called for during the 1990-congress of the International Organization for Succulent Plant Study (IOS) in Zürich. There was a general feeling that the situation presented a unique occasion to produce a set of authoritative volumes embracing *all* succulent plants, and produced according to a rigid set of common standards adopted to suit all families involved. It was planned that the project, informally termed the "New IOS Succulent Plant Lexicon", should consist of three volumes, one devoted to the *Cactaceae*, one to the *Aizoaceae*, and one to the remaining succulent taxa from various families.

For various reasons, the Lexicon project took much longer to be completed. The *Cactaceae* part had not become available in time for inclusion in the present handbook edition. In the case of the *Aizoaceae*, time permitted a much more complete treatment of the family, which will be published in two volumes. Moreover, it soon became apparent that a single volume will not be sufficient to cover the vast array of the "remaining succulents". Accordingly, they will be split into four volumes (Monocotyledons, Dicotyledons, *Asclepiadaceae* and *Crassulaceae*). It is hoped that these volumes will − together with their companions − be a useful successor to Jacobsen's celebrated Lexicon.

Already Jacobsen counted on the help of various contributors for his Lexicon, and this is even more the case with the present publication. The Illustrated Handbook of Succulent Plants has sought the help of numerous specialists to provide the most up-to-date contributions on the various families and genera. The present volume derives from the close and most enjoyable collaboration between many specialists. In contrast to Jacobsen's Lexicon, the descriptions have been vastly expanded to make them as diagnostic as possible. Numerous additional data items, such as typification, etymology, etc., have been incorporated, and it is hoped that this will add to the usefulness of the "New Lexicon" when compared with its predecessor – for users from both the botanical, the horticultural and hobby communities.

Thanks to the availability of computers and sophisticated database technology, it has been possible to considerably reduce the time needed to adapt the final manuscripts of the volumes on the "remaining succulents" for publication. The whole text of the volumes on Monocotyledons, Dicotyledons, *Asclepiadaceae* and *Crassulaceae*, including the complete nomenclature and all literature references is stored in a database, and the final version for the printing process was directly generated from this database. With this technology, the reference section as well as the index to taxa could be compiled automatically, which guarantees maximum completeness and reliability, as well as standardization. With the help of this "database publishing", it was possible to get this volume published within four months after the completion of all manuscripts and proof-reading.

Succulent plant botany will not end with the publication of this volume. In fact, the contrary is expected: It has always been the experience that the publication of a major revision or monograph hastens new research within the groups treated. This, it is hoped, will also be the result of the present publication. The compilation of it has exposed many instances where our knowledge is feeble or even non-existent. This shows that additional research is urgently called for. It is therefore the hope of the editor – and this is somewhat in contrast to the hopes of Hermann Jacobsen upon publication of his Lexicon – that succulent plant research will be enhanced by the publication of these volumes, and that a revised and updated edition will become possible in the not too distant future.

Acknowledgments

A special word of thanks goes to all the contributors of this volume. Without their help, the project to compile and edit this handbook would not have been the pleasure that it was.

The editor is most grateful to the authors for their patience with regards to the delayed publication of the handbook. The economics of book production have changed considerably during the time when the manuscripts were completed. The publisher of the original handbooks, Gustav Fischer Verlag Jena, has lamentably disappeared. After considerable uncertainties, the project has found a firm base in the publishing houses of Springer-Verlag, Heidelberg (for the English edition) and Ulmer-Verlag, Stuttgart (for the concurrent German edition). It is my pleasure to thank Dr. Dieter Czeschlik and Dr. Jutta Lindenborn from Springer-Verlag, and Roland Ulmer from Ulmer-Verlag, for their willingness to step in when everything appeared to become lost.

The compilation of the volumes treating the "remaining succulents" has been undertaken using computing infrastructure provided by the Municipal Succulent Collection Zürich (ZSS), and the permission to do so is heartily acknowledged with a deep feeling of gratitude, both to its former director Diedrich J. Supthut as well as to the present director Dr. Thomas Bolliger, and to the authorities of the City Parks Department.

Numerous persons have helped during the compilation and editing process. Foremost is Auli Lenzi (ZSS) who is thanked for her helping hand to keep the project papers organized. Gordon Rowley (Reading) has been of vital help supplying biographical information and literature in numerous cases.

Finally, I would like to thank my family, wife and son, as well as all my colleagues at ZSS and elsewhere for their patience when the handbook project kept me busy. A special word of thanks is due to Dr. Johanna Schlüter of Fischer Verlag, Jena, who so ably accompanied this project during the first three quarters of its long history.

Zürich, May 2001   U. EGGLI

Contents

Note: ♦ = family or genus completely covered; ◊ = family or genus only partially covered; ~ = family or genus only mentioned; numbers in brackets = numbers of species / infraspecific taxa covered

List of Abbreviations and Symbols

∅	diameter	**Gl**	gland
±	more or less, circa	Gr.	Greek
>	greater than	**Gy**	gynostegium
≥	greater or equal than	**Ha**	hair(s)
<	less than	Herb. / herb.	herbarium
≤	less or equal than	holo	holotype
×	hybrid indicator	Hort. / hort.	of gardens
		I	illustration
Anth	anthers	ICBN	International Code of
Ar	areole		Botanical Nomenclature,
Art.	article (of ICBN)		Tokyo Edition
Ax	axil	illeg.	illegitimate
BG	Botanical Garden	in sched.	"in schedis", i.e. in the
Bo	body		herbarium
BPH	Botanico-Periodicorum-	incl.	including
	Huntianum (1968)	**Inf**	inflorescence
BPH/S	idem., Supplementum (1991)	ING	Index Nominorum
Br	branch		Genericorum
Bra	bract	**Int**	internode
Bri	bristles	inval.	invalid
C	central	**ITep**	inner tepals
Ca	carpels	KG, KGW	Karoo Garden (Worcester,
Cal	calyx		RSA)
Cap	capitulum	l.c.	"loco citato", i.e. at the place
Ci	interstaminal corona		cited
Cl	corolla	**L**	leaf, leaves
Cn	corona	lecto	lectotype
Cs	staminal corona	**Lit**	literature
cv.	cultivar	ms.	manuscript
Cy	cyathium	Mt. / Mts.	Mount / Mountains
D	distribution	N	North, northern
E	East, eastern	NBG	National Botanical Garden
e.g.	for example		(RSA)
epi	epitype	NE	North-East, north-eastern
esp.	especially	**Nec**	nectary
Etym	Etymology	neo	neotype
excl.	excluding	**NGl**	nectar gland
f.	figure	nom.	"nomen", i.e. botanical name
fa.	forma	**NSc**	nectar scale
fig.	figure	NW	North-West, north-western
Fil	filaments	**OTep**	outer tepals
Fl	flower	**Ov**	ovary
fl.	"floruit", i.e. living	p.	page
Fr	fruit	p.a.	"per annum", i.e. per year
		p.p.	"pro parte", i.e. partly

Pc	pericarpel
Ped	pedicel
Per	perianth
pers. comm.	personal comment
Pet	petals
Phy	phyllaries
pl.	plate
Poll	pollinia
pp.	pages
R	roots
Rec	receptacle
Ri	ribs
Ros	rosette
RSA	Republic of South Africa
S	South, southern
s.a.	"sine anno", i.e. without year
s.l.	"sensu lato", i.e. in a wide sense
s.n.	"sine numero", i.e. without (collection) number
s.str.	"sensu stricto", i.e. in a strict sense
Sc	scale
SE	South-East, south-eastern
Se	seeds
Sect.	section
SEM	Scanning Electron Micrograph
Sep	sepals
Ser.	series
Sp	spines
SpS	spine shield
ssp.	subspecies
St	stamens
Sti	stigma
Sty	style
Subgen.	subgenus
Subsect.	subsection
Subser.	subseries
SUG	Stellenbosch University Gardens (RSA)
SW	South-West, south-western
syn	syntype
T	(nomenclatural) type
t.	plate
Tep	tepals
TL2	Taxonomic Literature, ed. 2 (1976-2000)
tt.	plates
Tu	tubercles
unpubl.	unpublished
USA	United States of America
var.	variety
W	West, western

Abbreviations for frequently cited publications

BJS	Botanische Jahrbücher für Systematik
BMI	Bulletin of Miscellaneous Information [Kew]
BT	Bothalia
CBM	Curtis's Botanical Magazine
CSJA	Cactus and Succulent Journal (US)
CUSNH	Contributions of the US National Herbarium
EJ	Euphorbia Journal
FC	Flora Capensis (ed. Harvey & al., 1860 - 1925)
FPA	Flowering Plants of Africa
FPSA	Flowering Plants of South Africa
FTA	Flora of Tropical Africa (ed. Oliver & al., 1868 - 1937)
HIP	Hooker's Icones Plantarum
JLSB	Journal of the Linnean Society, Botany
JSAB	Journal of South African Botany
KB	Kew Bulletin
KuaS	Kakteen und andere Sukkulenten
NPF2	Die natürlichen Pflanzenfamilien Ed. 2 (ed. Engler & Prantl, 1924 - 1959)
PSRV	Prodromus Systematis Naturalis Regni Vegetabilis (De Candolle, 1824 - 1873)
RSN	Repertorium Specierum Novarum Regni Vegetabilis
SAJB	South African Journal of Botany
WDSE	Succulent Euphorbieae (South Africa) (White & al., 1941)

List of Contributors

Arroyo-Leuenberger, Dr. Silvia; Albrechtstrasse 115, 12167 Berlin, Germany

Bayer, Martin Bruce; Hope Street 16, Cape Town 8001, Republic of South Africa

Bogner, Josef; Augsburgerstrasse 43a, D-86368 Gersthofen, Germany

Eggli, Dr. Urs; Sukkulenten-Sammlung Zürich, Mythenquai 88, 8002 Zürich, Switzerland

Forster, Dr. Paul I.; Queensland Herbarium, Brisbane Botanic Gardens, Mt. Coot-tha Road, Toowong, Queensland 4066, Australia

Hunt, Dr. David R.; The Manse, Chapel Lane, Milborne Port, Sherborne, DT9 5DL, England

Jaarsveld, Ernst J. van; Kirstenbosch National Botanic Garden, Private Bag X7, Claremont 7735, Republic of South Africa

Meyer, Nicole; address as for G. Smith, below

Newton, Prof. Len E.; Department of Botany, Kenyatta University, P. O. Box 43844, Nairobi, Kenya

Rowley, Gordon D.; 1 Ramsbury Drive, Earley, Reading, Berksh. RG6 7RT, England

Smith, Prof. Dr. Gideon F.; National Botanical Institute, Private Bag X101, Pretoria 0001, Republic of South Africa

Thiede, Dr. Joachim; Botanisches Institut und Botanischer Garten, Universität Hamburg, Ohnhorststrasse 18, 22609 Hamburg, Germany

Walker, Dr. Colin C.; Department of Biological Sciences, The Open University, Milton Keynes MK7 6AA, England

Illustration Credits

B. Descoings: XII.h, XV.c, XX.a

U. Eggli: I.e, II.a, II.e, II.f, III.a, III.e, III.g, IV.c, IV.f, V.a, V.b, VI.a, VI.b, VI.c, VI.e, VIII.b, VIII.c, VIII.d, VIII.e, VIII.f, IX.a, IX.b, IX.c, IX.e, IX.f, X.a, X.b, X.c, X.f, X.g, XI.b, XI.d, XI.e, XI.h, XII.a, XII.d, XIII.b, XIII.c, XIII.g, XIV.f, XVI.a, XVI.b, XVIII.a, XIX.b, XIX.f, XX.c, XX.d, XX.e, XX.f, XXI.a, XXII.c, XXII.e, XXIII.f, XXIV.b, XXIV.c, XXIV.d, XXIV.f, XXV.a, XXV.b, XXV.c, XXV.d, XXVI.e, XXVI.f, XXVI.g, XXVI.h, XXVII.a, XXVII.b, XXVII.c, XXVII.d, XXVII.e, XXVII.f, XXVII.g, XXIX.b, XXIX.c, XXIX.d, XXIX.e, XXX.a, XXX.d, XXXI.a, XXXI.b, XXXI.c, XXXI.d, XXXI.f, XXXI.g, XXXI.h, XXXII.b, XXXII.c, XXXII.d, XXXII.e, XXXII.f, XXXII.g, XXXII.h

J. Etter & M. Kristen: I.d, II.g, III.c, III.d, III.f, IV.b, V.c, V.d, V.e, VI.g, VII.b, VII.d, VII.e, XXXII.a

P. Forster: XXVII.h

M. Gilbert: XII.c, XIV.c

K. Grantham: XIII.f

M. Grubenmann: XIX.a

E. van Jaarsveld: XX.g, XX.h, XXI.b, XXI.c, XXI.d, XXI.e, XXI.f, XXI.g, XXII.a, XXII.b, XXII.d, XXII.f, XXIII.a, XXIII.b, XXIII.c, XXIII.d, XXIII.e, XXIII.g, XXIII.h, XXIV.a, XXIV.e, XXV.e, XXV.f, XXVI.c, XXX.b, XXX.c, XXX.e, XXX.f, XXX.g, XXXI.e

B. Jonkers: XXVIII.a

R. Lucas: XXVIII.d

A. Miller: XXVIII.c

L. Newton: X.h, XI.c, XI.g, XII.b, XII.g, XIII.a, XIII.d, XIII.h, XIV.b, XIV.d, XIV.h, XV.a, XV.b, XV.d, XV.e, XV.f, XV.g, XVI.c, XVI.d, XVI.e, XVI.g, XVI.h, XVII.a, XVII.b, XVII.c, XVII.d, XVII.f, XVII.h, XVIII.c, XVIII.d, XIX.d, XX.b, XXVIII.b, XXVIII.e, XXVIII.f, XXIX.a

G. Rowley: XVI.f

G. Smith: X.e, XI.a, XI.f, XII.e, XIV.a, XIV.e, XIV.g, XVII.e, XVII.g, XVIII.b, XVIII.e

D. Supthut: X.d, XII.f, XIII.e, XIX.c, XIX.e

J. Thiede: I.a, I.b, I.c, I.f, II.b, II.c, II.d, III.b, IV.a, IV.d, IV.e, IV.g, V.f, V.g, VI.d, VI.f, VII.a, VII.c, VII.f, VII.g

J. Trager: VIII.a, IX.d, XXVI.a, XXVI.b, XXVI.d, XXVIII.g, XXVIII.h

Introduction

What is a succulent ?

It is probably impossible to define what constitutes a *succulent plant* – at least in view of the several competing definitions. For the purpose of this handbook, a pragmatic approach has been selected, and apart from the multitude of unambiguous succulents, many borderline cases are included as well, especially if the species in question are encountered in cultivation together with other succulents, and if they are native to more or less semi-arid regions and consequently show some degree of xerophytic adaptation. This, then, includes most of the caudex and pachycaul plants now popular in cultivation.

Other borderline cases included are a number of bulbous and rhizomatous monocotyledons, where examples from several genera are covered, as well as several weakly developed leaf succulents from the *Gesneriaceae* (e.g. *Columnea*).

On the other hand, purely halophytic succulents (such as *Salicornia*) are omitted from these pages since they are as a whole neither adapted to climatically dry conditions nor encountered in collections devoted to succulent plants.

Finally, some families with undoubted claim to (xerophytic) succulence have been excluded from this set of volumes. This notably is the case for the *Cactaceae*, which will be treated elsewhere. In addition, the families *Bromeliaceae* and *Orchidaceae* are also excluded. Both count with a considerable number of mostly leaf succulents, but for both, vast specialist literature and numerous specialist societies are in existence, and this effort does not need to be duplicated here. For all these excluded families, however, a family description is included in the present volumes for the sake of completeness.

How to use this handbook

Since all information is presented in strictly alphabetical sequence of families, genera and species (except that monocotyledons and dicotyledons are treated separately, and that the families *Aizoaceae*, *Asclepiadaceae* and *Crassulaceae* occupy their own volumes), it is easy to find the entry for a given species as long as its family placement is known.

An alternative way is to use the taxonomic cross-reference index supplied at the end of the volume. This index contains all the names treated in the volume and for accepted names indicates the page where a treatment can be found, or in the case of synonyms gives the name of the accepted taxon and a page reference as above. For names merely mentioned in the text, the index gives the page reference and the name under which information can be found.

If a completely unknown plant is to be identified, the handbook supplies keys to the genera with succulent representatives for each family. Please note that these keys are designed to work for the succulent taxa treated, and do not necessarily include the total variation encountered in a genus. If the family is not known, the reader is referred to general botanical books that include keys to plant families. Rowley (1980) and Eggli (1994) provided keys for flowering and non-flowering succulents, and Geesink & al. (1981) produced a well-known book of keys to all flowering plants worldwide.

Scope of information presented

Families

The family names adopted are always conforming to the standard form (ending in *-aceae*); alternative names (such as *Compositae* for *Asteraceae*) are not used.

Within each family, the genera are treated in alphabetical sequence, and the same applies to the sequence of species within genera. Some genera of minimal importance or with borderline succulence are only mentioned or at the most described, but no individual species are treated.

The following families are included as a whole, i.e. with all their component species: *Agavaceae, Aloaceae* and *Doryanthaceae* in the present volume covering the Monocotyledons, and the *Didiereaceae, Fouquieriaceae* and *Nolanaceae* in the Dicotyledons volume. The *Aizoaceae* and *Crassulaceae* are covered in their entirety in separate volumes within this series, and the succulent taxa of the *Asclepiadaceae* likewise occupy a separate volume of the Handbook.

The family description characterizes the family as a whole, which often includes much more variation than that observed amongst its succulent representatives.

This is followed by notes on the distribution, classification and economic importance of the family, and the occurrence of succulence if this is not a general feature of the family as a whole. Also, a key to genera with succulents is included, and special terminology used for genera and species descriptions is discussed.

The family concept adopted more or less follows Mabberley (1987), except for the monocotyledons, where Dahlgren & al. (1985) is used as a base, with a number of small modifications.

Genera and species

The entries for genera and species follow the same layout. Names of authors are given in full, with initials added where necessary according to Brummitt & Powell (1992). The literature reference of the original description or combination is followed by information on typification (where available, see below). In the case of genera, important literature is then cited. This is followed by information on geographical distribution (including notes on ecology where available) and an explanation of the etymology for generic names.

The main part of the entry is made up by the diagnostic description of the taxon, followed by a discussion of its variability, circumscription and/or application where necessary. It should be noted that these descriptions reflect major variability only, but do not include all the reported minor variations.

For larger genera, an outline of the accepted formal or informal classification is also given, with individual taxa or groups numbered in sequence. These sequence numbers are then given at the start of each taxon description to indicate its placement within the genus.

If recent conflicting classifications are available for a given group, this is shortly discussed and the classification adopted is indicated.

Minor spelling variants of epithets are not indicated; instead, the 'corrected' spelling is used throughout for accepted names and synonyms.

Infraspecific taxa

Infraspecific taxa are given in strict alphabetic order of rank and name (i.e. ranks in the sequence cv., fa., ssp., var.). This is due to the strict alphabetical sorting used when the output for the handbook was generated from a computerized database. It also means that the typical infraspecific taxon (i.e. the one repeating the species name) is not treated first as in many handbooks, but in its appropriate alphabetical sequence.

Cultivars, hybrids

Cultivars (rank abbreviated as cv.) are not included on an exhaustive base. Cultivars not associated with a species are enumerated first, i.e. between the generic entry and the first species. Cultivars associated with a species name are included under that species, either as an entry of their own (and in the same form as subspecies etc.), or, in the case of cultivars of minor importance, in the form of a short mention in the species discussion. Cultivar nomenclature follows the guidelines of the ICBN.

Formally named hybrid genera are either included as 'genera' of their own, or dealt with in the discussion of their parent genera. The same applies to formally named hybrid species (incl. those named as cultivars). Hybrids only known with their hybrid formula are either discussed in the generic entry, or mentioned under one or the other of their parent species. No attempt has been made, however, to include all the numerous formally named hybrids.

Descriptions

The descriptions are as compact, concise and diagnostic as possible. Characters that do not vary for the group concerned are not repeated from the family or genus descriptions. In the case of genera further subdivided, information already presented in the group definitions is also not normally repeated in the descriptions of individual taxa.

Measurements

All measurements are given in metric units. Measurements without further qualifications *always refer to the long axis* of the organ described (i.e. length, height etc.); two measurements united with the ×-sign stand for length × width.

Terminology

Special terms used in descriptions are explained when first used; other botanical terminology is not further explained, and the readers are referred to the numerous botanical glossaries, of which Stearn (1992) is cited by way of a most important and useful example.

Typification

This information is included for convenience when readily available, but is lacking in numerous cases. The type citations include the country and major administrative unit where the type was collected, the collector and collection number, and the herbaria where material is said or known to be deposited. The herbarium acronyms conform to *Index Herbariorum*, Ed. 8 (Holmgren & al. 1990). Where more than one herbarium acronym is given, the first relates to the holotype, the others to isotypes. Additional information on typification is sometimes added, especially in the case of lecto- or neotypes.

Nomenclatural status of names

For all taxa treated, every attempt has been made to use only valid and legitimate names, but this was not achievable in a small number of cases. In the synonym lists, the nomenclatural status (invalid, illegitimate, rejected) is indicated by citing the ICBN articles violated (following the numbering in the "Tokyo" Code). Spelling variants are considered as invalidly published according to ICBN Art. 61.

Synonymies

The synonymies given for genera and especially species are as exhaustive as possible and include all names recognized as synonyms. The first synonym(s) − if applicable − is / are the basionym and / or later combination(s) for the accepted name of the entry. All combinations of the same basionym are given in sequence of publication and are united with the ≡-sign to indicate that these are homotypic (nomenclatural) synonyms. Please note that the ≡-sign is only used for *combinations* based on the same basionym and does not indicate other homotypic synonyms (e.g. *nomina nova*). All other synonyms are headed with 'incl.' to indicate that they are, with the exception of *nomina nova* based on the same type, taxonomic synonyms (= heterotypic synonyms). Again, groups of combinations based on the same basionym are united with ≡-signs. Basionyms are given first and in chronological order.

Geographical names

Country names are listed roughly in a North to South and West to East sequence. Every attempt has been made to standardize geographical names (of countries, administrative units, regions, etc.) as far as possible, but there is a surprising amount of change relating to such names. This is specifically the case for the names of the RSA provinces, which have changed considerably during 1995, especially affecting the former Cape Province, which has been split up into 4 units (North-West Province, Northern Cape, Western Cape, Eastern Cape). We have tried as best as we could to provide the modern names in the distribution information included in the handbook, but it has been impossible to adjust all the data for type localities, where the name "Cape Prov." is still used in some cases. This results in some inconsistencies, but it is hoped that these are tolerable under the present circumstances.

Some difficulties were also encountered in a few cases where countries have been amalgamated (as in the case of the former North Yemen and South Yemen) or divided (e.g. Eritrea, formerly part of Ethiopia). Full consistency in all these cases cannot be guaranteed.

In order to save space, geographical directions such as North, South, etc., are *always* abbreviated (N, S, etc.). Please note that *SW Africa* indicates 'southwestern Africa' and *not* the former Southwest-Africa (now Namibia). Similarly, *S Africa* indicates 'southern Africa' and *not* the Republic of South Africa, for which the abbreviation RSA is always used.

Literature references

Literature references are given for all accepted names. Normally, the publication is cited with a full abbreviation according to the standards defined in Eggli (1985) and Eggli (1998) for specialist succulent plant periodicals, or BPH (Lawrence & al. 1968) and BPH/S (Bridson & Smith 1991) for other periodicals, while TL2 of Stafleu & Cowan (1976-1988) and supplements (Stafleu & Mennega 1992-2000) are followed for book abbreviations (in both causes with some minor exceptions to conserve uniformity).

A number of frequently used titles of journals and books are further abbreviated to a short acronym, and a list of these acronyms follows the list of Abbreviations and symbols (page XI).

In the running text, literature is cited in the usual way (author and year, sometimes supplemented by a page reference), and full details can be found in the list of references at the end of the volume.

Illustrations

An attempt has been made to cite one readily accessible illustration for each species or infraspecific taxon when no illustrations are included in the literature reference for the accepted name. If the name used in the cited publication differs from the accepted name in the handbook, it is indicated (genus name abbreviated to first letter if identical, specific or infraspecific epithet omitted if identical to the accepted name).

Illustrations in the illustration section of this volume are given in bold print to distinguish them from other material cited.

Indication of authorships

For families, authorship is indicated at the end of the entry. For genera, authorship is given as a subheading after the genus heading. If more than one author has contributed species entries for a genus, each entry has its own indication of authorship as far as its authorship differs from the authorship given for the genus as a whole. It is thus possible to identify the author(s) of any entry in the handbook.

Agavaceae

Small to large perennial **Ros** plants, monocarpic (**Ros** unbranched) or polycarpic (**Ros** branched), herbs, shrubs or trees, terrestrial, very rarely epiphytic; stems none or short (then not rarely caespitose), or ± arborescent (then mostly ± branching), in part with secondary thickening growth (*Agave, Furcraea, Yucca*), partly with spreading or thick and upright rhizomes (*Yucca* p.p., *Agave* Subgen. *Manfreda*); **L** spirally arranged in **Ros**, dorsiventral, lanceolate, linear or subulate, often broadest near the base and gradually tapering towards the tip, thick and succulent, tough and fibrous or ± thin and weak (*Agave* Subgen. *Manfreda*), tip either a hard pungent **Sp** (*Yucceae, Agave* s.str.) or a ± soft (more rarely hard) point (*Beschorneria, Furcraea, Agave* Subgen. *Manfreda*), margins entire, with horny marginal teeth (these often on prominences), or filiferous; **Inf** terminal, mostly large, 0.5 - 13 m, with few or numerous **Bra**, mostly complex much-branched panicles with cymose lateral part-**Inf** consisting of monochasial units; scape (peduncle) mostly present, more rarely short to nearly none; peduncular **Bra** mostly ± similar to rosette **L**, diminishing in size upwards; floral **Bra** present; **Ped** normally present; **Fl** mostly bisexual, 3-merous, actinomorphic or slightly zygomorphic, generally in the **Ax** of well-developed **Bra**, pendent or ± upright (*Agave*), anthesis diurnal and / or nocturnal; **Tep** mostly whitish to yellow or greenish, rarely ± reddish; **Per** tube none (then **Tep** ± spreading) or **Tep** connivent to form a tube-like structure (*Beschorneria, Hesperaloe* p.p.), or **Per** tube present and of variable length (*Agave*), tubular or campanulate and sometimes abruptly widened and urceolate in the upper part; **St** 3 + 3, inserted at or somewhat above the base of the free **Tep** or within the **Per** tube; **Fil** mostly long and slender-filiform, rarely short, sometimes widened basally (*Furcraea*) or apically (*Yucca*), mostly glabrous, sometimes puberulent (*Yucca* p.p.); **Anth** dorsifixed, 2-thecous, sagittate to hastate (*Yucceae*) or oblong to linear (*Agaveae*), long or short, dehiscing introrsely with longitudinal slits; pollen primarily sulcate, rarely bisulcate (*Agave* Subgen. *Manfreda* p.p.), mainly released in monads, rarely also in tetrads (*Beschorneria, Furcraea*); **Ov** superior (*Yucceae*) or inferior (*Agaveae*), 3-locular, each locule with several to many ovules, often (generally?) with septal **Nec**; **Sty** rather long and simple or apically with 3 free and short branches (*Yucca*); **Sti** mostly single and either capitate or 3-lobed or on 3 **Sty** branches (*Yucca*), surface dry or wet; **Fr** mostly loculicidal capsules, rarely septicidal capsules or berries (*Yucca* p.p.), with several to many **Se**; **Se** mostly flat and plate-like crescent-shaped or semicircular, but sometimes less compressed (*Yucca*), black (due to phytomelans), storage tissue a perisperm (*Yucca*) or a (helobial or

nuclear) endosperm. – *Cytology:* x = 30, generally with 25 short and 5 long chromosome pairs.

Distribution: S Canada, N, C and S USA, Mexico (= distribution centre), C America to Panama, Caribbean Region, Colombia, Venezuela, Peru, Bolivia; widely cultivated throughout the world in suitable climates and often also naturalized.

Literature: Krause (1930); Pax & Hoffmann (1930); Hutchinson (1934); Dahlgren & al. (1985); Eguiarte & al. (1994); Bogler & Simpson (1995); Bogler & Simpson (1996); Verhoek-Williams (1998).

The *Agavaceae* are mostly adapted to arid conditions; the majority are xeromorphic and ± succulent rosette plants of desertic regions, and the family is therefore here covered in its entirety. They exhibit a high water-use efficiency connected with the common occurence of the water-conserving CAM-mode of photosynthesis (present in all species of *Agave* so far studied and in some *Yucca* and *Hesperaloe*; see esp. Nobel (1988)). More specialized flower pollinators are common (bumblebees and carpenter bees, moths, hawkmoths, bats, hummingbirds; see Verhoek-Williams (1998)).

The bimodal karyotype (McKelvey 1933) represents a major distinguishing feature of the *Agavaceae* and similar karyotypes are otherwise only found in *Hesperocallis* (n = 24; *Hesperocallidaceae*) and *Hosta* (with n = 30 as in the *Agavaceae*; *Funkiaceae*) according to Tamura (1995) and Verhoek-Williams (1998). The *Hosta*-karyotype differs from that of the *Agavaceae* in being less pronouncedly bimodal with 4 long pairs, 2 - 3 medium-short pairs and 23 - 25 short pairs (Tamura 1995), but relations to the *Agavaceae* are also supported by embryological (Cave 1948) and serological data (Chupov & Kutiavina 1981). Molecular and morphological phylogenetic studies leave somewhat uncertain whether *Hosta* represents the sister-group of *Agavaceae* (see Eguiarte & al. (1994), Bogler & Simpson (1995), Eguiarte (1995), Hernández Sandoval (1995) and Bogler & Simpson (1996)). Alternatively, *Hosta* holds an unresolved basal position within the *Agavaceae* (Bogler & Simpson 1995) or is even nested within (i.e., part of) the *Agavaceae* (Eguiarte & al. 1994). Based on the latter data, *Hosta* is included within the *Agavaceae* in the most recent molecular consensus classification (APG [Angiosperm Phylogeny Group] 1998), which is not followed here due to contrasting data and since *Hosta* differs from the *Agavaceae* in its temperate distribution (centred in Asia), non-xerophytic habit and the less bimodal karyotype; it is therefore best regarded as basal to the *Agavaceae*.

The infrafamilial classification presented below and the genera accepted in the subsequent treatments reflect the recent phylogenetic studies and thus deviates from the most recent overview treatment of the family by Verhoek-Williams (1998):

[1] Tribe *Yucceae* Bartlett 1830 (incl. subfam. *Yuccoideae* Kosteletzky 1831; incl. *Yuccaceae* J. Agardh 1858): **L** margins mostly filiferous, rarely serrulate; **Fl** pendent, actinomorphic; **Tep** free; **Ov** superior; **Anth** sagittate to hastate; embryo erect. – Recent phylogenetic studies are inconclusive as to whether tribe *Yucceae* is the sister group of tribe *Agaveae* (see Clary & Simpson (1995) and Bogler & Simpson (1996)), or whether the monophyletic tribe *Agaveae* is nested within tribe *Yucceae* rendering the latter paraphyletic (see Bogler & Simpson (1995) and Hernández Sandoval (1995)).

[2] Tribe *Agaveae* (incl. tribe *Poliantheae* Hutchinson 1934): **L** margins entire and often with teeth, or rarely filiferous; **Fl** pendent or ± upright, actinomorphic or slightly zygomorphic; **Tep** free or ± fused to form a tube; **Ov** inferior; **Anth** oblong to linear; embryo curved.

Key to the genera

1 **L** margins almost always filiferous, rarely serrulate but never toothed; **Fl** pendent, actinomorphic; **Tep** free; **Anth** sagittate to hastate; **Ov** superior: **2**
– **L** margins entire and with or without (often strong) teeth, rarely filiferous; **Fl** pendent or ± upright, actinomorphic or slightly zygomorphic; **Tep** free or usually forming a ± long tube; **Anth** oblong to linear; **Ov** inferior: **4**
2 **L** margins filiferous, **L** epidermis without papillate cells; **Fl** often white and wax-like; **Fil** free, apically swollen and outcurved, sometimes puberulent; **Sty** thickened with 3 short branches; **Sti** 3 with 2 lobes each (S Canada to Guatemala):   **Yucca**
– **Ros** stemless; **L** margins filiferous or finely serrulate, **L** epidermis with papillate cells over the veins; **Fil** adnate to the base or lower parts of the **Tep**, not swollen, straight, glabrous; **Sty** short and slender; **Sti** distinctly capitate and fringed with papillae: **3**
3 Habit ± grass-like; **Ros** few-leaved; **L** margins filiferous; **Tep** connivent and **Fl** therefore narrowly campanulate, whitish to reddish (S USA (Texas), N and C Mexico):   **Hesperaloe**
– Habit *Yucca*-like; **Ros** many-leaved; **L** margins finely serrulate; **Tep** openly spreading, whitish (SW USA, NW Mexico):   **Hesperoyucca**
4 **L** margins entire or toothed; **Fl** pendent, actinomorphic; **Tep** ± free; **Fil** filiform or basally swollen; **Sty** swollen and with 3 basal ridges, distally abruptly narrowed: **5**
– **L** margins entire, with or without teeth, or filiferous; **Fl** ± upright, actinomorphic or slightly zygomorphic; **Tep** basally fused to form a ± long tube; **Fil** filiform; **Sty** basally not swollen

(S USA to Colombia and Venezuela, Caribbean region):   **Agave**
5 Plants polycarpic; **L** margins entire or minutely denticulate; **Tep** free but connivent and forming a tube-like structure, reddish to yellowish; **Fil** filiform (Mexico):   **Beschorneria**
– Plants monocarpic; **L** margins mostly toothed; **Tep** openly spreading, whitish to greenish; **Fil** basally swollen (Mexico to Bolivia, Caribbean region):   **Furcraea**

In earlier classifications not recognizing the *Agavaceae* as a separate family, the 2 tribes here recognized were placed in widely separated positions, emphasizing the ovary position in a broadly circumscribed *Amaryllidaceae* with inferior ovaries (tribe *Agaveae*) and an equally broadly circumscribed *Liliaceae* with superior ovaries (tribe *Yucceae*). Hutchinson (1934) resurrected the *Agavaceae* to include 6 tribes of xerophytic genera from the Americas, Africa and Asia. Huber (1969) and Dahlgren & al. (1985) segregated the Old World genera and the New World *Nolina*-group as several separate families, leaving the *Agavaceae* as presently circumscribed.

Many species of *Agavaceae* are of considerable importance for man ('man-*Agave*-symbiosis'): Many species of *Agave* and *Yucca* had multiple uses in the former indigenous civilizations in the USA and Mexico. A a number of species of *Agave* and *Furcraea* are important sources of fibres and are cultivated in large-scale plantations in suitable climates around the world. Several species of *Agave* are used to manufacture distilled alcoholic beverages and are the base of the Mexican pulque industry. Almost all taxa have considerable horticultural potential, and many are frequently planted as ornamentals in suitable climates. Especially in mediterranean climates, numerous species have become naturalized and today form an important feature of the landscape, together with *Opuntia* ("Prickly Pear", *Cactaceae*), which has a similar New World origin.

[J. Thiede]

AGAVE

J. Thiede

Agave Linné (Spec. Pl. [ed. 1], 323, 1753). **T:** *Agave americana* Linné [Lectotype, designated by Britton & P. Wilson, Sci. Survey Puerto Rico, 5: 156, 1923 (fide ING).]. – **Lit:** Trelease (1913); Trelease (1915b); Berger (1915); Hummelinck (1936); Hummelinck (1938); Gentry (1972); Verhoek-Williams (1975); Gentry (1982); Piña Luján (1985); Piña Luján (1986); McVaugh (1989); Hummelinck (1993); Lott & García-Mendoza (1994). **D:** S USA, Mexico, C America to Panama, whole Caribbean region, Colombia, Venezuela; cultivated worldwide in tropical and subtropical to frost-free temperate climates and often naturalized. **Etym:** Gr. 'Agave', daughter of Kadmos and sister of Semele in Gr. mythology, also the mother of Pentheus, which she murdered in an outburst of fury; also Gr. 'agavos', stately, noble, illustrious; for the stately nature of many species, but also for the ferocious leaf margin teeth present in many species.

Incl. *Polianthes* Linné (1753). **T:** *Polianthes tuberosa* Linné [Typification by inference, only element included.].

Incl. *Pothos* Adanson (1763) (*nom. illeg.*, Art. 53.1). **T:** not typified.

Incl. *Tuberosa* Heister *ex* Fabricius (1769) (*nom. illeg.*, Art. 52.1). **T:** *Polianthes tuberosa* Linné.

Incl. *Bonapartea* Willdenow (1814). **T:** *Bonapartea juncea* Willdenow.

Incl. *Littaea* Tagliabue (1816). **T:** *Littaea geminiflora* Tagliabue.

Incl. *Bravoa* Llave & Lexarza (1824). **T:** *Bravoa geminiflora* Lexarza [Typification by inference, only element included.].

Incl. *Coetocapnia* Link & Otto (1828). **T:** *Coetocapnia geminiflora* Link & Otto [Typification by inference, only element included.].

Incl. *Robynsia* Draparnaud (1841) (*nomen rejiciendum*, Art. 56.1). **T:** *Robynsia geminiflora* Draparnaud [Typification by inference, only element included.].

Incl. *Ghiesbreghtia* Roezl (1861) (*nom. inval.*, Art. 32.1c?).

Incl. *Manfreda* Salisbury (1866). **T:** *Agave virginica* Linné.

Incl. *Allibertia* Marion (1882). **T:** *Allibertia intermedia* Marion [Typification by inference, only element included.].

Incl. *Prochnyanthes* S. Watson (1887). **T:** *Prochnyanthes viridescens* S. Watson.

Incl. *Leichtlinia* H. Ross (1893). **T:** *Agave protuberans* Engelmann *ex* Baker [Typification by inference, only element included.].

Incl. *Delpinoa* Ross (1897). **T:** *Delpinoa gracillima* Ross [Typification by inference, only element included.].

Incl. *Pseudobravoa* Rose (1899). **T:** *Bravoa densiflora* B. L. Robinson & Fernald.

Incl. *Runyonia* Rose (1922). **T:** *Runyonia longiflora* Rose [Typification by inference, only element included.].

Perennial xerophytic **Ros** plants, mostly **L** succulents, mono- or polycarpic, terrestrial (very rarely epiphytic); **R** tough and fibrous, sometimes fusiformly thickened (Subgen. *Manfreda*), shallowly radiating; **Ros** acaulescent or stems short, rarely elongated (*A. pedunculifera*), mostly thick, solitary or branched, sometimes rhizomatous (esp. in Subgen. *Manfreda*); **Ros** small to very large; **L** mostly long-lived, predominantly succulent and xeromorphic, more rarely ± soft and annual (Subgen. *Manfreda* p.p.), ± thick and fibrous, linear to lanceolate to ovate, usually rich in steroidal sapogenins, tip a terminal ± strongly developed **Sp** or a soft point (Subgen. *Manfreda*); **L** margins entire, minutely to strongly toothed (then partly on prominences of the **L** margin), or filiferous; **Inf** paniculate, small to up to 12 m tall (= complete **Inf**, i.e. scape and floriferous part); part-**Inf** cymose, consisting of monochasial units, either ± short-stalked and mostly with few (rarely only paired) **Fl** (**Inf** then wrongly termed 'spikes' by Gentry (1982); Subgen. *Littaea*) or long-stalked, often several times compound, with many ± densely arranged **Fl** (**Inf** then wrongly termed 'panicles' and part-**Inf** termed 'umbels' by Gentry (1982); Subgen. *Agave*), or without a stalk and paired to single **Fl** (Subgen. *Manfreda*); **Inf** sometimes bulbilliferous, esp. in species of anthropogenic origin; peduncular **Bra** ± similar to **Ros** leaves, diminishing in size upwards; **Ped** long to short or nearly none (esp. Subgen. *Manfreda*); **Fl** diurnal and / or nocturnal, actinomorphic or (slightly) zygomorphic (Subgen. *Manfreda* p.p., rarely in Subgen. *Littaea*), generally proterandrous; **Per** tubular to campanulate, usually yellow or greenish to brownish, more rarely reddish, **Tep** basally usually fused and forming a **Per** tube, much varying in length, lobes of varying length (wrongly termed 'tepals' by Gentry (1982)); **St** exserted; **Fil** filiform, normally ± long, rarely ± short, inserted in the tube or at mouth of the tube or at the **Tep** base; **Anth** versatile, oblong; pollen released in monads, sulcate (rarely bisulcate); **Ov** inferior, thick-walled, 3-locular, with numerous axile ovules in 2 rows per locule, often constricted above to a ± conspicuous neck; septal **Nec** present; **Sty** elongate, filiform, tubular, not yet fully expanded at anthesis; **Sti** 3-lobed, papillate-glandular; **Fr** dehiscent loculicidal capsules, apically often beaked (i.e. narrowly elongated); **Se** flattened, black. – *Cytology:* x = 30, with multiples present.

The large genus *Agave* is traditionally subdivided into the subgenera *Littaea* (flowers 'spicate' in pairs or clusters or more rarely racemose in small distinct clusters) and *Agave* (with flowers paniculate in large 'umbellate' clusters on lateral peduncles) (e.g. Gentry (1982: 61)), implying fundamental differences between both groups ('spicate' vs. 'paniculate-umbellate'). Morphologically, however, the inflorescences in (all?) *Agave* species generally represent complex panicles with cymose-monochasial part-inflorescences (Dahlgren & al. 1985). Inflorescences of *Agave* differ in the degree of branching (= number of flowers) of the part-inflorescences only and are therefore only gradually rather than fundamentally different. The few-flowered part-inflorescences characterizing Subgen. *Littaea* apparently represent the plesiomorphic condition, rendering *Littaea* paraphyletic, see the phylogenies presented by Bogler & Simpson (1996). This character is retained in Subgen. *Manfreda*, whereas the long-stalked, much-branched and many-flowered part-inflorescences of Subgen. *Agave* apparently represent an apomorphic feature. This might reveal monophyly of the latter, but a multiple origin of this inflorescence type cannot be ruled out.

Polyploidy (up to 8x) is common in both subgenera, but esp. so in Subgen. *Agave* (Pinkava & Baker 1985).

In all phylogenies recently published (see Bogler & Simpson l.c., Clary & Simpson (1995) and Hernández Sandoval (1995)), the genera *Manfreda, Prochnyanthes* and *Polianthes*, which are recognized in most recent treatments (e.g. McVaugh (1989), Verhoek-Williams (1978)) form a well-supported monophyletic clade together with the genus *Agave*. As far as these phylogenies permit a higher level of resolution, these 3 former genera (= subtribe *Poliantheae* in the sense of Verhoek-Williams (1975)) together form a monophyletic clade in which they are successively nested within *Agave*, rendering *Agave* in its traditional circumscription paraphyletic. Moreover, the generic separation of *Manfreda, Prochnyanthes* and *Polianthes* within the framework of the traditional classification is based on rather weak (and sometimes inconsistent) character differences, which is also noted by most authors. An example is *Agave polianthiflora*, which shows the same flower type that is used as generic character to delimit *Polianthes*. The deviating features of Subgen. *Manfreda* merely appear to result from an ecological shift towards more humid habitats connected with a change of growth form (rhizome vs. leaves as storage organ) together with a possible shift of pollinators connected with the often rather long perianth tubes. Consequently, the 3 genera of subtribe *Poliantheae* were included within a more broadly defined genus *Agave* (Thiede & Eggli 1999), and are together treated as Subgen. *Manfreda*.

The infrageneric classification of *Agave* into species groups follows Gentry (1982) for continental taxa and Trelease (1913) for the Caribbean species, albeit some of these groups appear to represent artificial assemblages only (see e.g. Ullrich (1990d)). The Caribbean species (exclusively belonging to Subgen. *Agave*) are very insufficiently known and have only been studied for regional floristic works after the last general treatment by Trelease (1913). Detailed habitat studies will most certainly result in a reduction of names and recognize some currently upheld species as mere island forms of more widespread species. Since groups represent rank-less informal taxa, no attempt has been made to clarify the complicated infrageneric nomenclature of the genus, and all untypified names were generally not further considered.

[1] Subgen. *Littaea* (Tagliabue) Baker 1888: **Inf** with ± short-stalked part-**Inf** mostly with few (rarely only paired) **Fl** ('spicate' **Inf**):
[A] 'Weakly armed group'; **L** margins entire, serrulate, filiferous or with weak teeth.

[a] **L** margins firm, not filiferous, **L** surface without white marks left by the central bud, striate, not soft, **L** margins finely serrulate; **Tep** tube well-developed; **Ov** without neck:
[1a] Group *Striatae* Baker 1888 (incl. Sect. *Chonanthagave* A. Berger 1915): Plants perennial forming large clumps; **L** narrow, linear, hard, not softly succulent, striate, margins finely serrulate; **Fl** geminate, with deep **Tep** tube; **Ov** half-inferior, without neck, merging into a well-developed **Tep** tube; **Fil** inserted in the middle of the tube, frequently at 2 levels. – 5 species in NE, C and S Mexico.
This group, esp. *A. dasylirioides*, holds the most basal position in *Agave* based on both molecular phylogenetic data (Bogler & Simpson 1995, 1996) as well as its plesiomorphic morphological features (Gentry 1982: 241-242).

[b] **L** not striate, margins firm, not filiferous, softly succulent, surface without white marks left by the central bud, margins smooth or irregularly serrulate; **Tep** tube short or none; **Ov** with a neck:
[1b] Group *Serrulatae* Baker 1877 (incl. Group *Amolae* Gentry 1982; incl. Sect. *Nizandensae* (B. Ullrich) B. Ullrich 1991 ≡ Ser. *Nizandensae* B. Ullrich 1991; incl. Sect. *Yuccaefoliae* (A. Terracciano) B. Ullrich 1996 ≡ Subser. *Yuccaefoliae* A. Terracciano 1885): **L** soft, margins entire, unarmed, terminal **Sp** present; **Inf** with densely arranged **Fl**; **Tep** tube shallow to medium-sized; **Fil** inserted at the mouth of the **Tep** tube. – 8 species in N, C and S Mexico (mainly Sierra Madre Occidental).
Gentry's name Group *Amolae* has to be replaced by the earlier name *Serrulatae* Baker, described to include *A. pruinosa* Lemaire, which represents a synonym of *A. attenuata* (see Ullrich (1990h), where the name *Serrulatae* is wrongly applied).

[1c] Group *Choritepalae* Gentry 1982: **L** without a terminal **Sp** (except *A. guiengolensis*); **Inf** short, dense; **Tep** tube lacking or **Tep** arising from a discoid **Rec**; **Fil** insertion not elevated. – 3 species from N Mexico and the Isthmus of Tehuantepec.
Regarded as artificial by Ullrich (1990b).

[c] **L** margins filiferous (i.e. decomposing into white threads), **L** surface with white marks left by the central bud:
[1d] Group *Filiferae* Gentry 1982: **Ros** mostly medium-sized, 30 - 90 cm tall; **L** narrow, unarmed, filiferous; **Fl** geminate, campanulate, 30 - 55 mm, **Tep** tube usually much shorter than the lobes. – 5 species in N and C Mexico (mainly Sierra Madre Occidental).

[1e] Group *Parviflorae* Gentry 1982: **Ros** small; **L** short (very rarely exceeding 30 cm), filiferous; **Fl** small; **Per** with a tube 6 - 30 mm long, usually much exceeding the short free lobes. – 4 species from the N Sierra Madre Occidental (Mexico) and adjacent Arizona (USA).

[B] 'Strongly armed group': **L** margins generally with large teeth (except some species of Group *Polycephalae* and Group *Marginatae*):

[a] Plants mainly polycarpic; stems branched; **Tep** tube deep, lobes 1 - 2× as long as the tube:

[1f] Group *Polycephalae* Gentry 1982: **L** broad, softly succulent, fleshy, margins usually with closely set teeth; **Tep** tube grooved, lobes 1 - 2× as long as the tube; **Ov** 3-angled. – 6 species from E-C to S Mexico. A wholly tropical and mostly mesophytic group.

[b] Plants mainly monocarpic; stems simple; **Ros** often surculose; **Tep** tube shallow, lobes 2 - 6× as long as the tube (except *A. pelona*):

[1g] Group *Marginatae* Gentry 1982: **L** with conspicuous horny margins and with conspicuous marginal teeth; **Fl** small with small **Tep** tube, lobes 5 - 6× as long as the shallow cup-like tube, frequently involute around the **Fil**. – 18 species from the S USA (S New Mexico, S Texas), Mexico and Guatemala.

[1h] Group *Urceolatae* Gentry 1982: **L** with or without horny margins; **Ros** in small tight clusters; **Tep** lobes 2 - 3× as long as the urceolate **Tep** tube. – 1 species (*A. utahensis*) in the USA (NW Arizona, S Utah, Nevada, S California).

[2] Subgen. *Agave*: **Inf** with long-stalked, often several times compound part-**Inf** with many ± densely arranged **Fl** ('umbellate' part-**Inf**).

[A] 'Large-sized group': Plants medium-sized to (very) large; **Ros** ± 1 - 2.5 m tall; **L** ± 1 - 2.5 m. – Continental Species:

[a] **L** generally not ensiform but lanceolate to ovate, much less than 10× longer than wide; **Fr** ± oblong:

[2a] Group *Agave* (incl. Group *Americanae* Baker 1888 (*nom. illeg.*): **Ros** medium-sized to large; stems short, (freely) surculose; **L** light glaucous-grey to pale green, marginal teeth well-developed; **Inf** axis with smaller chartaceous **Bra**; **Inf** linear to long-oval in outline, part-**Inf** several times compound, not crowded; **Fl** yellow, rather slender; **Tep** tube furrowed, **Tep** not wilting until after anthesis; **Ov** shorter than the **Tep**. – 8 mainly cultivated species.

[2b] Group *Salmianae* A. Berger 1915: **Ros** large, massive; stems short, thick, usually closely and freely surculose; **L** light green

to green, mostly very large, very thick towards the base; **Inf** axis with large appressed imbricate fleshy **Bra**; **Inf** pyramidal to ovoid in outline, part-**Inf** widely branching, several times compound; **Fl** large, succulent; **Tep** tube broad, thick-walled, lobes longer than the tube, dimorphic, becoming incurved when wilting after anthesis; **Fil** often inserted at 2 levels. – 5 mainly cultivated species.

[2c] Group *Crenatae* A. Berger 1915 (≡ Sect. *Crenatae* (A. Berger) B. Ullrich 1993): Plants solitary, unbranched, rarely surculose; stems short; **L** longer, 70 - 150 cm, green to yellow-green, young pruinose, usually with clearly visible marks left by the central bud, **L** margins deeply crenate and undulate, marginal teeth large, irregular in size and spacing, frequently with small interstitial teeth; **Inf** tall, narrow; **Tep** red or purple in bud, yellow at anthesis; **Tep** lobes 2 - 4× as long as the tube. – 7 species mainly from the Mexican Sierra Madre Occidental and the Trans-Mexican Volcanic Belt.

[2d] Group *Campaniflorae* Trelease 1912: Stems short; **Ros** rather open, mostly solitary (except *A. capensis*); **L** rather softly succulent, 70 - 150 cm, green, margins not horny, marginal teeth moderate, uniform in size and spacing; **Fl** campanulate; **Tep** tube thin-walled, deep, broad (15 - 22 mm), lobes short, slightly exceeding the tube in length; **Fil** deeply attached. – 3 species from Baja California (Mexico).

[2e] Group *Umbelliflorae* Trelease 1912: Stems commonly branching from **L** axils, frequently developing long branching stems resulting in fragmented clones; **L** broad and short (< 70 cm), mostly bright green; **Inf** stout, compact, part-**Inf** large, ± globular ('umbellate'), subtended by large succulent sheathing **Bra**; **Fl** large, fleshy. – 2 species in the USA (SW California) and Mexico (N Baja California).

[b] **L** ensiform, linear, patulous, 10 - 20× as long as wide; **Tep** lobes drying reflexed on the tube; **Fr** broadly ovoid:

[2f] Group *Viviparae* Trelease 1913 (incl. Group *Sisalanae* Trelease 1913; incl. Group *Rigidae* A. Berger 1915): Stems short to elongate; **Ros** surculose; **L** linear, ensiform, narrow, rigid, usually patulous, 10 - 20× as long as wide, margins nearly straight; **Inf** small, open; **Fl** greenish-yellow, weak; **Fr** sometimes replaced by bulbils. – 12 species, mainly in Mexico, 1 in the S USA (Florida), 1 also in C America, some cultivated only.

Ullrich (1990d) abandoned Group *Sisala-*

nae sensu Gentry (1982) as an artificial assemblage and rearranged its species in other groups. He placed *A. sisalana* (the type species of Group *Sisalanae*) in Gentry's Group *Rigidae*. The latter group, however, must be renamed with the oldest name, Group *Viviparae*, whose type species *A. vivipara* turned out to be the correct name for the taxon previously named *A. angustifolia* (the type species of Gentry's *Rigidae*). The remaining species of the *Sisalanae* sensu Gentry were reclassified as follows by Ullrich (l.c.): *A. desmetiana* = [2?, unclear]; *A. kewensis* sensu Gentry (= *A. grijalvensis*) = [2j]; *A. neglecta* = [2a]; *A. sisalana* = [2f]; *A. weberi* = [2a]. The former predominantly Caribbean Group *Viviparae* is here renamed as Group *Vicinae* (see there).

[B] 'Small-sized group': Plants mostly small to medium-sized; **Ros** ± 0.4 - 1 m tall; **L** ± 0.2 - 1 m. – Continental species:

[a] Plants generally surculose; flowering period spring to summer; **Tep** tube deep, lobes short, strongly dimorphic, outer lobes conspicuously larger; **Fil** inserted at 2 levels:

[2g] Group *Applanatae* Trelease 1912 (incl. Group *Ditepalae* Gentry 1982): Plants small to large, single or sparingly surculose; **L** mostly light glaucous, marginal teeth well-developed; **Inf** open; **Bra** scarious, reflexing, persistent; **Tep** long, leathery, unequal, generally reddish in bud, yellow at anthesis, **Tep** tube long, deep, as long or longer than the lobes, lobes dimorphic, short, usually red-tipped; **Fr** long oblong. – 12 species in N, C and S Mexico and the S USA (Arizona, New Mexico).

[b] As in [a], but **Tep** tube shallow or deep, lobes much longer than the tube, subequal; **Fil** generally inserted all on the same level:

[2h] Group *Deserticolae* Trelease 1912: Plants small to medium-sized; **Ros** solitary; stems none or short; **L** glaucous-grey to greenish, greyish, rough with papillae, marginal teeth firm or weak and easily detached; **Inf** narrow, part-**Inf** short; **Fl** small; **Tep** tube very short, open, lobes (1-) 3 - 5× as long as the tube. – 10 species from the USA (SE California, Arizona) and Mexico (mainly Baja California, also Sonora) (Sonoran Desert Region).
Without close relationship to any other Group (Gentry 1982: 354).

[2i] Group *Parryanae* Gentry 1982: Plants small to medium-sized; **Ros** compact, suckering sparingly or prolifically with vigorous rhizomes; **L** short, broad, closely imbricate, glaucous-grey to green; marginal teeth conspicuously larger towards the

L tip; **Inf** scape strong; **Tep** long, slender, red to purplish in bud, **Tep** tube well-developed, shorter than the lobes; **Fr** rather small, strong-walled, ovoid to oblong. – 5 species from the S USA (California to SW Texas) and Mexico (Central Mexican Plateau to Guanajuato).

[c] Plants not surculose; flowering period winter to early spring:

[2j] Group *Marmoratae* A. Berger 1915: **L** grey, scabrous, crenate, terminal **Sp** small; **Fl** small, bright yellow, tube small. – 4 species from C and S Mexico and Sonora.

[2k] Group *Costaricenses* Trelease 1913 (incl. Group *Guatemalenses* Trelease 1915 ≡ Sect. *Guatemalenses* (Trelease) B. Ullrich 1992; incl. Group *Scolymoides* A. Berger 1915; incl. Group *Hiemiflorae* Gentry 1982): **Inf** narrow with very short scape; **Fl** in tightly balled clusters. – 12 species from C America (Guatemala to Costa Rica) but esp. from E-C and S Mexico.

[C] Caribbean species:

[2l] Group *Antillanae* Trelease 1913: **Ros** suckering; **L** fleshy, usually curved, usually green, marginal teeth usually rather large, terminal **Sp** usually elongated; **Fl** rather large (40 - 80 mm); panicles several times compound, freely fruiting, sometimes bulbilliferous; **Se** rather large (6 - 9 × 4 - 6 mm). – 14 species mainly from the Greater Antilles.

[2m] Group *Antillares* Trelease 1913: **Ros** suckering; **L** fleshy, usually curved, usually green, marginal teeth small, terminal **Sp** usually elongated; **Fl** rather small (30 - 45 mm); panicle **Br** rather simple; **Se** small (5 - 6 × 4 - 5 mm). – 6 species from Cuba.

[2n] Group *Bahamanae* Trelease 1913: **Ros** rarely suckering; **L** usually grey, terminal **Sp** elongated; **Fl** rather large (40 - 60 mm); panicles several times compound; **Se** rather large (7 - 8 × 4 - 6 mm). – 6 species from the Bahamas.

[2o] Group *Caribaeae* Trelease 1913 (incl.? Ser. *Columbianae* A. Berger 1915): **Ros** solitary and not suckering; **L** fleshy, usually curved, usually green, marginal teeth usually small, terminal **Sp** with stout involutely slitted base, above usually short and oblique; **Fl** rather large (40 - 80 mm); panicles several times compound, freely bulbilliferous, not always fruiting; **Se** rather large (6 - 9 × 4 - 6 mm). – 8 species from the Windward Islands, recently reduced to a single species by Rogers (2000).

[2p] Group *Inaguenses* Trelease 1913: **Ros** freely suckering; **L** hard and straight, grey; **Fl** rather small (35 - 50 mm); **Se** small (5 × 4 mm). – 2 species from the Bahamas.

[2q] Group *Vicinae* Thiede 2001 (introduced here, type *Agave vicina* Trelease; incl. Group *Viviparae sensu* Trelease, misapplied): **Ros** suckering; **L** fleshy, usually curved, usually green, marginal teeth usually small, terminal **Sp** elongated, slender; **Fl** rather large (40 - 80 mm); panicles several times compound, freely bulbilliferous, not always fruiting; **Se** rather large (6 - 9 × 4 - 6 mm). − 7 species from N South America (Colombia, Venezuela) and the Leeward Islands.

Since Trelease's Group *Viviparae* has to replace Gentry's Group *Rigidae* for nomenclatural reasons (see also there), it is here renamed based on *A. vicina*.

[3] Subgen. *Manfreda* (Salisbury) Baker (1877) ≡ *Manfreda* Salisbury 1866 (incl. *Polianthes* Linné 1753; incl. *Bravoa* Llave & Lexarza 1824; incl. *Allibertia* Marion 1882; incl. *Delpinoa* Ross 1887; incl. *Prochnyanthes* S. Watson 1887; incl. *Leichtlinia* Ross 1893; incl. *Pseudobravoa* Rose 1899; incl. *Runyonia* Rose 1922):

[A] Perennials with upright fleshy rhizomes; **R** fleshy and fibrous, arising from the base of the rhizome; **L** chartaceous to fleshy, green for one season or slightly longer, ending in a soft point; marginal teeth, if present, soft; **Inf** 'racemes' or 'spikes' with solitary or paired **Fl** at the nodes. − S USA, Mexico, N C America.

[3a] Group *Manfreda* (≡ *Manfreda* Salisbury; incl. *Manfreda* Subgen. *Eumanfreda* Rose 1899 (*nom. inval.*); incl. *Manfreda* Subgen. *Pseudomanfreda* Rose 1899): Rhizome globose or oblong, large; **Fl** usually solitary at the nodes (paired only in aberrant specimens), scent sweet or unpleasant; **Tep** mostly greenish or brownish (rarely white or pink), tube short to long; **St** and **Sty** long-exserted; **Sti** trigonous or rarely 3-lobed. − S USA, Mexico, N C America.

[3a1] *A. brunnea* Subgroup: **L** succulent, evergreen, not dying back at the end of the growing season, tip with a short soft point; **L** margins with large to small teeth ± ≥ 1 mm, spaced apart from each other (**L** not fleshy and with a long pungent apical point in *A. hauniensis*). − 7 species in the USA (Texas) and N Mexico.

[3a2] *A. scabra* Subgroup: **L** thin to semi-succulent, dying back at the end of the growing season; **L** margins entire or minutely papillate; **Ov** not protruding into the **Tep** tube; **Tep** tube inserted at the tip of the **Ov**, funnel-shaped, narrowed above the **Ov**. − 10 species in C and S Mexico. This subgroup is possibly an artifical paraphyletic hold-all of the least specialized species.

[3a3] *A. guttata* Subgroup: **L** thin to semi-succulent, dying back at the end of the growing season; **L** margins hyaline, usually minutely erose-denticulate and thus rough to the touch; **Ov** protruding into the **Tep** tube; **Tep** tube cylindrical, not narrowed above the **Ov**; **Fr** with a scar from the **Tep** in a ring around the shoulder. − 8 species from C and S Mexico and Guatemala.

[3a4] *A. virginica* Subgroup: **L** thin to semi-succulent, dying back at the end of the growing season; **Tep** lobes erect; **Sty** markedly shorter than the **St**; **Sti** 3-lobed, lobes reflexed at maturity. − Only *A. virginica* from the C and SE USA.

[B] Rhizome small or large; **Fl** usually paired at the nodes (solitary in *A. confertiflora*), scent sweet or absent; **Tep** white (sometimes tinged with green) to reddish, tube long; **St** and **Sty** included; **Sti** 3-lobed, lobes reflexed at maturity.

[3b] Group *Polianthes* (≡ *Polianthes* Linné 1753): Plants medium-sized to small; **L** linear to lanceolate, herbaceous; **Tep** white, pink, red or coral-pink, tube straight or with a wide curve, narrow, gradually widening above. − 14 species from Mexico.

[3b1] *Polianthes* Subgroup (≡ *Polianthes* Subgen. *Polianthes*): **Inf** with 3 - 8 flowering nodes; **Fl** sweet-scented; **Ov** erect or spreading; **Tep** white to pink, tube nearly horizontal towards the mouth, lobes erect or reflexed to revolute; **St** inserted near the mouth of the **Tep** tube. − 11 species from N and C Mexico.

[3b2] *Bravoa* Subgroup (≡ *Bravoa* Llave & Lexarza 1824 ≡ *Polianthes* Subgen. *Bravoa* (Llave & Lexarza) M. Roemer 1847): **Inf** with 8 - 20 or more flowering nodes (only 3 - 5 (-9) in *A. bicolor*); **Fl** unscented; **Ov** horizontal or curved downwards; **Tep** pinkish-red, red, or coral-coloured, tube curved so that the **Fl** are pendent, lobes short, erect or flaring; **St** inserted below the middle of the **Tep** tube. − 3 species from C and S Mexico.

[3c] Group *Prochnyanthes* (≡ *Prochnyanthes* S. Watson 1887): Mature plants large; **L** broad, narrowed basally, chartaceous; **Tep** greenish-white to greenish-red, abruptly curved near the middle, narrow below, abruptly widened above. − Only *A. bulliana* from N-C Mexico.

The genus *Agave* is of considerable importance ethnobotanically, both for the fibre and pulque industry, as well as for horticultural use (see the family description). Many species are cultivated and / or naturalized world-wide esp. in mediterranean climates.

The following names are of unresolved application but are referred to this genus: *Agave abortiva* A. Terracciano (1885); *Agave aloides* Jacobi (1866); *Agave amaniensis* Trelease & Nowell (1933); *Agave americana* Grisebach (1864) (*nom. illeg.*, Art. 53.1); *Agave ×armata* hort. *ex* A. Berger (1915); *Agave aspera* A. Terracciano (1885) (*nom. illeg.*, Art. 53.1); *Agave banlan* Perrotet (1824); *Agave baxteri* Baker (1888); *Agave beaulueriana* Jacobi (1868); *Agave ×beguinii* hort. *ex* A. Berger (1912); *Agave bennetii* hort. *ex* A. Berger (1915); *Agave bernhardii* Jacobi (1868); *Agave bollii* A. Terracciano (1885); *Agave bonnetii* hort. *ex* A. Berger (1915); *Agave brauniana* Jacobi (1866); *Agave bromeliaefolia* Salm-Dyck (1834); *Agave calderonii* Trelease (1923); *Agave chinensis* F. P. Smith (1871); *Agave cinerascens* Jacobi (1864); *Agave collina* Greenman (1897); *Agave concinna* Lemaire in Hort. Vanhoutte (1846); *Agave conduplicata* Jacobi & C. D. Bouché (1867) ≡ *Agave virginica* var. *conduplicata* (Jacobi & C. D. Bouché) A. Terracciano (1885) ≡ *Manfreda conduplicata* (Jacobi & C. D. Bouché) Rose (1903); *Agave cucullata* Lemaire *ex* Jacobi (1865); *Agave cyanophylla* Jacobi (1866); *Agave davillonii* Baker (1892); *Agave deamiana* Trelease (1915); *Agave decaisneana* Jacobi (1868); *Agave demeesteriana* Jacobi (1866); *Agave diacantha* Royle (1855); *Agave drimiaefolia* Hort. Petropol. *ex* Baker (1888) (*nom. inval.*, Art. 34.1c); *Agave echinoides* Jacobi (1868) ≡ *Agave striata* var. *echinoides* (Jacobi) Baker (1877); *Agave ehrenbergiana* Baker (1877); *Agave ehrenbergii* Jacobi (1865); *Agave elizae* A. Berger (1915); *Agave entea* Hartwich (1897); *Agave erosa* A. Berger (1915); *Agave fenzliana* Jacobi (1866); *Agave flaccida* Jacobi (1866) (*nom. illeg.*, Art. 53.1); *Agave fourcroydes* Jacobi (1865) (*nom. illeg.*, Art. 53.1); *Agave fragrantissima* Jacquin (1762); *Agave ×franzosinii* Hort. Hanbury *ex* W. Watson (1889) (*nom. inval.*, Art. 32.1c?); *Agave friderici* A. Berger (1912); *Agave galeottei* Baker (1877); *Agave glabra* Karwinsky in M. Roemer (1847); *Agave glaucescens* Otto in M. Roemer (1847) (*nom. inval.*, Art. 32.1c?); *Agave goeppertiana* Jacobi (1865); *Agave grandibracteata* Ross (1892); *Agave granulosa* Scheidweiler *ex* C. Koch (1861); *Agave guedeneyrii* Houllet (1875); *Agave gutierreziana* Trelease (1920); *Agave haworthiana* M. Roemer (1847); *Agave haynaldii* Todaro (1876); *Agave henriquesii* Baker (1887); *Agave heteracantha* A. Berger (1898) (*nom. illeg.*, Art. 53.1); *Agave hookeri* Baker (1881) (*nom. illeg.*, Art. 53.1); *Agave horizontalis* Jacobi (1868); *Agave horizontinalis* Baker (1887) (*nom. inval.*, Art. 61.1); *Agave horrida* var. *micracantha* Baker (1877); *Agave humboldtiana* Jacobi (1866); *Agave inghamii* [?]

longissima Hort. Whitacker *ex* A. Berger (1915); *Agave ixtli* C. Koch (1860) (*nom. illeg.*, Art. 53.1); *Agave kellermanniana* Trelease (1915); *Agave keratto* Salm-Dyck (1859) (*nom. inval.*, Art. 61.1); *Agave kerratto* Baker (1892) (*nom. inval.*, Art. 61.1); *Agave kewensis* Jacobi (1866); *Agave laticincta* Verschaffelt (1868); *Agave leguayana* hort. *ex* Besaucèle (s.a.) (*nom. illeg.*, Art. 53.1); *Agave leguayana* Baker (1877) (*nom. illeg.*, Art. 53.1); *Agave lemairei* Hort. Verschaffelt *ex* Ill. Hort. (1864); *Agave lempana* Trelease (1925); *Agave lindleyi* Jacobi (1868); *Agave littaeoides* Pampanini (1909); *Agave longisepala* Todaro (1878); *Agave macrantha* Todaro (1879) (*nom. illeg.*, Art. 53.1) ≡ *Agave macracantha* var. *macrantha* (Todaro) A. Terracciano (1885); *Agave maculata* Regel (1856); *Agave maculata* hort. *ex* A. Berger (1915) (*nom. illeg.*, Art. 53.1); *Agave malinezii* C. Koch (1862); *Agave massiliensis* hort. *ex* A. Berger (1912); *Agave maximowicziana* Regel (1890); *Agave milleri* Salm-Dyck (1834) (*nom. illeg.*, Art. 53.1); *Agave minarum* Trelease (1915); *Agave monostachya* Sessé & Moçiño (1894); *Agave ×mortolensis* A. Berger (*pro sp.* (1912); *Agave muelleriana* A. Berger (1915); *Agave multiflora* Todaro (1890); *Agave nigromarginata* Hort. De Smet *ex* Besaucèle (s.a.); *Agave nirvana* Herbin & Robins (1968) (*nom. inval.*, Art. 36.1, 37.1); *Agave nissonii* Baker (1874); *Agave offroyana* De Smet *ex* Jacobi (1865); *Agave ortgiesiana* Todaro (s.a.) (*nom. illeg.*, Art. 53.1) ≡ *Agave filifera* fa. *ortgiesiana* (Todaro) H. Jacobsen (1954) (*nom. inval.*, Art. 33.2); *Agave pallida* Sartorius *ex* Jacobi (1865) (*nom. illeg.*, Art. 53.1); *Agave pampaniniana* A. Berger (1915); *Agave paupera* A. Berger (1915); *Agave pavoliniana* Pampanini (1910); *Agave perlucida* Jacobi (1868); *Agave ×pfersdorfii* hort. *ex* Besaucèle (*pro sp.*) (s.a.); *Agave planera* Fasio (1903) (*nom. inval.*, Art. 32.1c); *Agave polianthoides* Schiede *ex* Schlechtendal (1844); *Agave polianthoides* M. Roemer (1847) (*nom. illeg.*, Art. 53.1); *Agave polyacantha* Haworth (1821); *Agave pringlei* hort. *ex* A. Berger (1912); *Agave prostrata* Martius *ex* Dragendorff (1898) (*nom. inval.*, Art. 32.1c); *Agave pulcherrima* Otto in M. Roemer (1847); *Agave pulverulenta* Verschaffelt (1863); *Agave pumila* Simon *ex* Besaucèle (s.a.) (*nom. illeg.*, Art. 53.1); *Agave purpurea* Souza Novelo (1941) (*nom. inval.*, Art. 36.1); *Agave ragusae* Todaro (1897); *Agave regia* Baker (1877); *Agave richardsii* hort. *ex* G. Nicholson (1884); *Agave rohanii* Jacobi (s.a.); *Agave rohanii* hort. *ex* A. Berger (1915) (*nom. illeg.*, Art. 53.1); *Agave romani* Hort. De Smet *ex* Besaucèle (s.a.); *Agave ×romanii* Hort. De Smet *ex* Baker (1888); *Agave rovelliana* Todaro (1876); *Agave rudis* Lemaire *ex* Jacobi (1865); *Agave saponifera* H. Grothe (1880); *Agave schidigera* var. *ortgiesiana* Baker (1877) ≡ *Agave ortgiesiana* (Baker) Trelease (1914) (*nom. illeg.*, Art. 53.1); *Agave schneideriana* A. Berger (1915); *Agave scolymus* Dietrich (1843) (*nom. illeg.*, Art. 53.1); *Agave scolymus*

Kunth (1850) (*nom. illeg.*, Art. 53.1); *Agave serrulata* Steudel (1841); *Agave silvestris* hort. *ex* A. Berger (1915); *Agave simoni* André (1904); *Agave simonii* hort. *ex* Besaucèle (s.a.); *Agave smithiana* Jacobi (1866); *Agave sordida* A. Berger (1915); *Agave subinermis* M. Roemer (1847); *Agave ×taylorea* Hort. Veitch (1877) (*nom. illeg.*, Art. 52.1); *Agave taylorii* Hort. Williams (1874); *Agave teoxomuliana* Karwinsky *ex* M. Roemer (1847); *Agave terraccianoi* Pax (1893); *Agave thomsoniana* Jacobi (1866); *Agave toeniata* hort. *ex* Besaucèle (s.a.); *Agave toneliana* hort. *ex* Besaucèle (s.a.) (*nom. illeg.*, Art. 53.1); *Agave toneliana* Baker (1881) (*nom. illeg.*, Art. 53.1); *Agave troubetskoyana* Baker (1892); *Agave undulata* Klotzsch (1840) ≡ *Manfreda undulata* (Klotzsch) Rose (1903); *Agave undulata* var. *strictior* Jacobi & C. D. Bouché (1865); *Agave vandervinnenii* Lemaire (1864); *Agave ×villae* Pirotti *ex* Baker (1892) (*nom. inval.*, Art. 61.1); *Agave ×villarum* André (1886); *Agave viridissima* Baker (1877); *Agave vivipara* Salm-Dyck (1859) (*nom. illeg.*, Art. 53.1); *Agave washingtonensis* Rose (1898); *Agave watsonii* J. R. Drummond & C. H. Wright (1907); *Agave weissenburgensis* Wittmach *ex* Baker (1889) (*nom. inval.*, Art. 61.1?); *Agave wiesenbergensis* Wittmack (1885); *Agave wiesenburgensis* Wittmack (1885) (*nom. inval.*, Art. 61.1); *Agave wildringii* Britton (1911); *Agave ×winteriana* A. Berger (1915); *Agave zuccarinii* Otto (1842); *Polianthes americana* Sessé & Moçiño (1888); *Polianthes ensifolia* hort. *ex* Steudel (1840); *Polianthes pygmaea* Jacquin (1793).

A. acicularis Trelease (Mem. Nation. Acad. Sci. 11: 34, t. 52, 1913). **T:** Cuba, Santa Clara (*Britton & al.* 5926 [NY]). – **Lit:** León (1946). **D:** C Cuba.

[2l] **L** lanceolate, ± 100 × 12 cm, slightly greyish, dull, margins slightly concave; marginal teeth gently upcurved, 2 - 3 mm, chestnut-brown, 10 - 15 mm apart, below the middle more distant, larger and reflexed, occasionally with outcurved point; terminal **Sp** stoutly acicular, straight, subtriangularly grooved below the middle, smooth, 2.5 cm, grey-brown, slightly glossy, decurrent; **Inf** 'paniculate', not known to be bulbilliferous; **Ped** scarcely 5 mm; **Fl** 40 - 45 mm; **Ov** fusiform, 25 mm, exceeding the **Tep** in length; **Tep** yellow, tube open, ± 5 mm, lobes 12 - 15 mm; **Fr** apparently becoming almost pear-shaped and a little stipitate or beaked.

A. acklinicola Trelease (Mem. Nation. Acad. Sci. 11: 41, t. 91, 1913). **T** [syn]: Bahamas, Acklin Island (*Brace* 4442 [NY]). – **Lit:** Correll & Correll (1982). **D:** Bahamas (Acklin Island).

[2n] **Ros** solitary; **L** rather narrowly lanceolate, concave, occasionally somewhat canaliculate, to 300 × 15 cm, dull greyish; marginal teeth straight or gently curved, rather acuminately deltoid, often from oblique green prominences or with lenticular

bases, 1 - 1.5 mm, 5 - 10 mm apart; terminal **Sp** conical, somewhat flexuously recurved, involutely grooved to or beyond the middle, smooth, 2 - 2.5 cm, red-brown, becoming grey, glossy, decurrent; **Inf** unknown.

A. aktites Gentry (US Dept. Agric. Handb. 399: 148-150, ill., 1972). **T:** Mexico, Sinaloa (*Gentry* 11470 [US, DES, MEXU]). – **D:** Mexico (Sonora, Sinaloa); sand dunes with coastal thorn forest. **I:** Gentry (1982: 555, 557).

[2f] Stems broadly globose; **Ros** small, 40 - 70 × 60 - 110 cm, surculose; **L** linear, straight, patent, unequal within a **Ros**, broadly clasping at the base, smooth or asperous, 40 - 60 × 2 - 4 cm, bluish glaucous-grey, sometimes with transverse zonal pattern; marginal teeth generally upcurved, with slender flexuous tips, 3 - 5 mm, irregularly spaced, 1 - 3 or 4 - 5 cm apart; terminal **Sp** abruptly subulate, usually broad at the base and flattened above, 1.2 - 2 cm, dark brown to greyish; **Inf** 3 - 4 m, 'paniculate', narrow, part-**Inf** 10 - 15 in the upper ¼ - ⅓ of the **Inf**, short, small; **Fl** 64 - 70 mm; **Ov** 26 - 31 mm, neckless; **Tep** sticking together and not opening properly, quickly wilting, pale greenish, tube 14 - 16 mm, lobes unequal, 21 - 25 mm.

This is the only Mexican *Agave* growing naturally and regularly in the maritime zone of beach dunes, to which habitat it appears to be limited (Gentry 1982: 558).

A. albescens Trelease (Mem. Nation. Acad. Sci. 11: 44, t. 53, 116, 1913). **T:** Cuba (*Britton* 2085 [NY?]). – **Lit:** León (1946). **D:** SE Cuba.

[2m] **Ros** solitary; **L** oblong-lanceolate, conduplicate, flattish, slightly rough, ± 45 × 15 cm, light grey, dull, passing to glaucous and banded, margins nearly straight; marginal teeth usually straight or gently curved, broadly triangular or acuminately deltoid, 2 - 3 mm, ± 1 cm apart; terminal **Sp** conical, somewhat recurved, shallowly grooved below the middle, dull, sometimes roughened except at the tip, 1.5 cm, blackish-chestnut-brown, very shortly decurrent; **Inf** 'paniculate', 5 m; **Ped** slender; **Fl** 30 - 35 mm; **Ov** fusiform, 15 mm, shorter than the **Tep**; **Tep** golden-yellow, tube open, 5 - 6 mm, lobes 12 - 14 mm, rather shorter than the **Ov**.

A small species variable in leaf characters. According to the protologue, it differs from the few other grey-leaved Cuban species in the granular roughening of the leaves and the finally purplish-black colour of the terminal spine and marginal teeth.

A. albomarginata Gentry (Agaves Cont. North Amer., 129-131, ill., 1982). **T:** Ex cult. (*Gentry* 19811 [US, DES]). – **D:** Known from cultivation only.

[1g] Stem subcaulescent; **Ros** open, freely suckering; **L** few, lanceolate-linear, straightly as-

cending, convex below, somewhat keeled towards the base, nearly flat above, 100 - 125 × 4 (near the base) / 2.5 cm (in the middle), greyish-green, margins thin, horny, somewhat friable, white; marginal teeth in the middle of the **L** white like the margin, thin, recurved, 2 - 4 mm, 3 - 5 cm apart, towards the **L** base blunt, 1 - 2 cm apart, distal ⅓ of the **L** toothless; terminal **Sp** subulate, with a rounded groove above, 1.5 cm, grey with dark tip, thinly decurrent; **Inf** 4 - 6 m, 'spicate', slender, laxly flowered, part-**Inf** with 2 or 3 **Fl**; **Ped** short, thick; **Fl** 35 - 40 mm; **Ov** fusiform, 18 - 22 mm, neck thick, grooved; **Tep** pale to greenish-yellow, tube openly spreading, 4 - 5 mm, lobes equal, 13 - 14 mm.

Perhaps better treated as a var. or ssp. of the closely related *A. lechuguilla*, but more extreme in morphological characters than other variants of the taxon (Gentry 1982: 130).

A. americana Linné (Spec. Pl. [ed. 1], 323, 1753). **T:** LINN 443.1. – **D:** USA, Mexico; cultivated worldwide in frost-free climates, and locally naturalized.

A. americana ssp. **americana** – **D:** USA (SE Texas), Mexico; cultivated widely. **I:** Gentry (1982: 276, 280).

Incl. *Agave americana* var. *americana*; **incl.** *Agave virginica* Miller (1768) (*nom. illeg.*, Art. 53.1); **incl.** *Agave ramosa* Moench (1794); **incl.** *Agave spectabilis* Salisbury (1796); **incl.** *Agave theometel* Zuccagni (1809) ≡ *Agave americana* var. *theometel* (Zuccagni) A. Terracciano (1885); **incl.** *Agave milleri* Haworth (1812); **incl.** *Agave variegata* Steudel (1821) (*nom. inval.*, Art. 32.1c); **incl.** *Agave picta* Salm-Dyck (1859) ≡ *Agave longifolia* var. *picta* (Salm-Dyck) Regel (1865) ≡ *Agave mexicana* var. *picta* (Salm-Dyck) Cels (1865) ≡ *Agave milleri* var. *picta* (Salm-Dyck) van Houtte (1868) ≡ *Agave americana* var. *picta* (Salm-Dyck) A. Terracciano (1885) ≡ *Agave ingens* var. *picta* (Salm-Dyck) A. Berger (1912); **incl.** *Agave altissima* Zumaglini (1864); **incl.** *Agave fuerstenbergii* Jacobi (1870); **incl.** *Agave communis* Gaterau (1889); **incl.** *Agave picta* A. Berger (1904) (*nom. illeg.*, Art. 53.1); **incl.** *Agave americana* var. *marginata* Trelease (1908); **incl.** *Agave americana* var. *medio-picta* Trelease (1908); **incl.** *Agave americana* var. *striata* Trelease (1908); **incl.** *Agave americana* var. *marginata alba* Trelease (1908) (*nom. inval.*, Art. 23.1); **incl.** *Agave americana* var. *marginata aurea* Trelease (1908) (*nom. inval.*, Art. 23.1); **incl.** *Agave americana* var. *marginata pallida* Trelease (1908) (*nom. inval.*, Art. 23.1); **incl.** *Agave celsiana* hort. *ex* A. Berger (1911) (*nom. illeg.*, Art. 53.1); **incl.** *Agave ingens* A. Berger (1912); **incl.** *Agave complicata* Trelease *ex* Ochoterena (1913); **incl.** *Agave gracilispina* Engelmann *ex* Trelease (1914); **incl.** *Agave melliflua* Trelease (1914); **incl.** *Agave zonata* Trelease (1914); **incl.** *Agave tingens* A. Berger (1915); **incl.** *Agave felina* Trelease (1920); **incl.** *Agave rasconensis* Trelease (1920); **incl.** *Agave subzonata* Trelease (1920); **incl.** *Agave americana* [?] *nairobensis* Herbin & Robins (1968) (*nom. inval.*, Art. 36.1, 37.1).

[2a] Stem short; **Ros** 1 - 2 × 2 - 3.7 m, freely suckering; **L** lanceolate, narrowed above the thickened base, usually acuminate, some **L** reflexed above the middle of the lamina, plane or guttered, smooth to slightly asperous, mostly 1 - 2 m × 15 - 25 cm, light grey-glaucous to light green, sometimes variegated, margins undulate to crenate; marginal teeth variable, larger 5 - 10 mm, brown to pruinose-grey, 2 - 6 cm apart, from broad low bases, the slender cusps straight to flexuous or curved; terminal **Sp** conical to subulate, mostly 3 - 5 cm, shiny brown to pruinose-grey; **Inf** 5 - 9 m, 'paniculate', slender, straight, long-oval in outline, rather open, part-**Inf** 15 - 35, in the upper ⅓ - ½ of the **Inf**, spreading; **Fl** slender, 7 - 10 cm; **Ov** 3 - 4.5 cm, greenish, neck grooved, tapering to the narrower base; **Tep** yellow, tube funnel-shaped, 8 - 20 mm, lobes unequal, 25 - 35 mm. – *Cytology:* 2n = 60, 120, 180, 240.

The leaves are often reflexed above the middle, and this is a characteristic feature of the species. The flowers exhibit a short tapering ovary that is shorter than the tepals. It is a very polymorphic species cultivated worldwide in many variants, esp. in winter-rainfall climates (Gentry 1982: 278).

A. americana ssp. **protoamericana** Gentry (Agaves Cont. North Amer., 287-290, ills., 1982). **T:** Mexico, Nuevo León (*Gentry & Barclay* 20156 [US, DES, MEXU]). – **D:** USA (Texas); Mexico (Nuevo León, Tamaulipas, San Luis Potosí); open slopes in tropical deciduous and thorn forest, 500 - 1400 m.

[2a] Differs from ssp. *americana*: **L** generally shorter, shortly narrowed above the thick and fleshy base, 0.8 - 1.35 m × 17 - 22 cm, light glaucous-grey to pale green, sometimes cross-zoned; **Inf** generally with fewer (15 - 20) part-**Inf**; **Tep** tube deeply funnel-shaped, longer, 15 - 20 mm.

Regarded as the wild progenitor of the many cultivated *A. americana* ssp. *americana* types. It shows apparent introgression with *A. asperrima* (as *A. scabra*) (Gentry 1982: 289).

A. americana var. **expansa** (Jacobi) Gentry (US Dept. Agric. Handb. 399: 80, 1972). – **Lit:** McVaugh (1989). **D:** Cultivated only (USA [California, Arizona], Mexico [Jalisco]). **I:** Gentry (1982: 276, 283).

≡ *Agave expansa* Jacobi (1868); **incl.** *Agave abrupta* Trelease (1920).

[2a] Differs from ssp. *americana*: Stem forming a short trunk in age, to 60 cm; **L** glaucous-grey, frequently cross-zoned, margins crenate; marginal teeth along the middle of the lamina on several sharply angled low tubercles; **Fl** 7 - 8.5 cm. – *Cytology:* 2n = 119.

Known only as a cultivar (and better to be named as such) used for the pulque industry, first introduced into W Europe where it was described by Jacobi (Gentry 1982: 283).

A. americana var. **oaxacensis** Gentry (Agaves Cont. North Amer., 285-287, ills., 1982). **T:** Mexico, Oaxaca (*Gentry & Arguelles* 12260 [US, DES, MEXU]). – **D:** Cultivated only (Mexico: Oaxaca).

[2a] Differs from ssp. *americana*: **L** not reflexed, spreading, very large, 1.2 - 2 m × 18 - 24 cm, glaucous-white, margins nearly straight; marginal teeth closely set, not mamillate; **Inf** large, to 10 m; **Fl** large, 9.5 - 10.5 cm.

Observed in cultivation only (and thus better treated as a cultivar), esp. in the Oaxaca Valley, but similar plants were also observed elsewhere. Formerly grown for fibre and pulque, but now abandoned (Gentry 1982: 287).

A. angustiarum Trelease (CUSNH 23: 138, 1920). **T:** Mexico, Guerrero (*Trelease* 17+77 [MO]). – **D:** Mexico (Michoacán, México, Morelos, Puebla, Guerrero, Oaxaca); on cliffs, 600 - 1500 m. **I:** Gentry (1982: 134).

[1g] **Ros** subcaulescent, open, solitary; **L** few, linear to lanceolate, straight, thick, firm, long acuminate, plane to concave above, convexly thickened below, 50 - 80 × 6 - 7 cm, green or pruinose-glaucous (both forms in mixed populations), margins horny, continuous; marginal teeth straight or slanted downwards but commonly upcurved, flattened, largest somewhat scattered, mostly 4 - 7 mm, brown to grey, 1 - 3 cm apart, characteristically without teeth below the **L** tip for ¼ - ⅓ of the **L** length; terminal **Sp** acicular, well rounded, narrowly grooved above, with conspicuous median protrusion below, 3 - 4.5 cm, long decurrent to the upper teeth; **Inf** 'spicate', 2 - 4 m, slender; **Fl** 35 - 40 mm; **Ov** 15 mm, neck narrow; **Tep** glaucous greenish-white, tube 4 - 5 mm, lobes 16 mm.

Distinguished by its long narrow leaves that are toothless along the long-tapering apex, and the protruding spine-base. The taxon may be confused with some forms of *A. kerchovei* (Gentry 1982: 134-135).

A. anomala Trelease (Mem. Nation. Acad. Sci. 11: 36, t. 66, 1913). **T:** Cuba (*Shafer* 1409 [MO?]). – **D:** E Cuba (Holguin to Myabe).

[2l] **Ros** unknown; **L** elongate-lanceolate, rather gradually pointed, 75 - 100 × 7.5 cm, green, margins not repand, unarmed or with few and very small teeth towards the base; terminal **Sp** unguiculately recurved, conically subulate, smooth, rather dull, 3 - 10 mm, reddish-brown, shortly decurrent and dorsally immersed into the green tissue; **Inf** unknown, not known to be bulbilliferous; **Ped** ± 10 mm and slender or 40 mm and much stouter; **Fl** 55 - 60 (-70) mm; **Ov** oblong-fusiform, 30 - 40 mm, rather longer than the **Tep**; **Tep** yellow, tube conical, 8 - 10 mm,

lobes 20 × 4 - 5 mm; **Fr** (abnormal) narrowly pear-shaped and oblong, 40 × 15 mm, somewhat stipitate and beaked.

The only native Caribbean species that lacks marginal teeth (Trelease l.c.). The species was described based on the type collection only and has apparently not been recollected since.

A. antillarum Descourtilz (Fl. Méd. Antilles 4: 239, pl. 284, 1827). – **Lit:** Trelease (1913: with ill.); León (1946). **D:** Hispaniola (Haiti, Dominican Republic).

Incl. *Agave vivipara* Lamarck (1783) (*nom. illeg.*, Art. 53.1); **incl.** *Agave dominicensis* Rüse (1893) (*nom. inval.*, Art. 32.1c); **incl.** *Agave americana* Urban (1903) (*nom. illeg.*, Art. 53.1).

This is the earliest named species of the group, but is only inaccurately described. Engelmann (1875) equated it with the short-flowered *Agave* of S Haiti, and he was followed by later authors (Trelease 1913: 31).

A. antillarum var. **antillarum** – **D:** Hispaniola (Haiti, Dominican Republic).

[2l] Stem none; **Ros** solitary; **L** lanceolate, gradually acute, somewhat concave, ± 100 × 8 cm, bright green, margins typically nearly straight; marginal teeth straight or upcurved, narrowly triangular from lenticular bases or acuminately deltoid, 2 - 3 mm, 1 - 2.5 cm apart; terminal **Sp** conical, nearly straight, involutely grooved near the base, smooth, 1.5 - 2 cm, brown, dull, decurrent; **Inf** ± 5 m (?), 'paniculate', narrowly oblong, part-**Inf** with ascending **Br** and densely clustered **Fl**, in the upper ¼ or ⅓ of the **Inf**, not known to be bulbilliferous; **Ped** 5 - 10 mm; **Fl** 40 - 50 mm; **Ov** oblong-fusiform, 25 - 30 mm, longer than the **Tep**; **Tep** deep orange, tube open, scarcely 5 mm, lobes 15 mm; **Fr** narrowly oblong, stipitate and beaked, 40 - 45 × 15 mm.

A. antillarum var. **grammontensis** Trelease (RSN 23: 362, 1927). **T:** Hispaniola, Haiti (*Ekman* 3355 [MO?]). – **D:** Hispaniola (Haiti).

[2l] Differs from var. *antillarum*: **L** glaucous; marginal teeth in the middle of the **L** heavily triangular, 5 mm, 0.5 - 1 cm apart, margins between the teeth concave; **Tep** orange, almost cochineal-red within; **Fr** unknown.

A. apedicellata Thiede & Eggli (KuaS 50(5): 111, 1999). **T:** Mexico, San Luis Potosí (*Parry & Palmer* 867 [US]). – **D:** Mexico (San Luis Potosí); 2450 m, known from the type collection only.

Incl. *Bravoa sessiliflora* Hemsley (1880) ≡ *Polianthes sessiliflora* (Hemsley) Rose (1903).

[3b1] Plants slender; **L** linear, narrow, obtuse, 3 - 5 mm broad; **Inf** 'spicate', almost glabrous, 2 - 3× as long as the **L**; **Bra** broadly ovate, acute or shortly acuminate, small; **Fl** sessile, to 54 mm, mostly geminate; **Tep** white, tube narrow; **Sti** included.

Insufficiently known. Upon transfer to *Agave*, *Polianthes sessiliflora* needed a new name due to *Agave sessiliflora* Hemsley 1880.

A. applanata Koch *ex* Jacobi (Hamburg. Gart.- & Blumenzeit. 20: 550, 1864). **T** [neo]: Mexico, Veracruz (*Trelease* 1 [MO]). – **D:** Mexico (Chihuahua, Durango, Zacatecas, Guanajuato, Querétaro, Hidalgo, México, Puebla, Veracruz, Oaxaca). **I:** Gentry (1982: 421, 424).

    Incl. *Agave schnittspahnii* Jacobi (1865).

    [2g] **Ros** 0.5 - 1 (-2 cult.) × 1 - 2 (-3 cult.) m, solitary; **L** many, linear-lanceolate, very rigid, usually widest at or near the base, 40 - 60 × 7 - 10 cm, mature **L** much longer than earlier stages, margins horny throughout or lacking in the middle of the lamina; marginal teeth nearly straight or frequently curved downwards, very strong, sharp, larger teeth (middle of the lamina) 8 - 15 mm, dark brown becoming light waxy pruinose, mostly 4 - 6 cm apart; terminal **Sp** very strong, flat or broadly hollowed above, 3 - 7 cm, dark reddish-brown becoming greyish with age, decurrent along the margin; **Inf** 4 - 8 m, 'paniculate', narrow, scape rather short, part-**Inf** numerous; **Fl** 55 - 80 mm; **Ov** angularly cylindrical, 35 - 38 mm, greenish; **Tep** yellow, tube 15 - 22 mm, lobes unequal, outer 15 - 22 mm.

This taxon appears to be endemic to Veracruz and adjacent Puebla and is cultivated as a cottage plant elsewhere. It was possibly disseminated to the North by men in (pre-) historic times (Gentry 1982: 424-425).

A. ×arizonica Gentry & J. H. Weber *pro sp.* (CSJA 42(5): 222-225, ills., 1970). **T:** USA, Arizona (*Weber* s.n. [US, ASU, DES]). – **D:** USA (Arizona); open rocky slopes in Chaparral or Juniper Grassland, 1100 - 1450 m. **I:** Gentry (1982: 255-256).

Identified as the possible natural hybrid *A. toumeyana* ssp. *bella* × *A. chrysantha* by Reichenbacher (1985), Pinkava & Baker (1985) and others. Only about 50 - 60 individuals are known (Hodgson 1999).

A. arubensis Hummelinck (Recueil Trav. Bot. Néerl. 33: 236-237, 248, fig. 14-15, pl. 3a, 4, 1936). **T** [syn]: Aruba (*Hummelinck* 17a+b [U]). – **Lit:** Hummelinck (1993: with ills.). **D:** Leeward Islands (Aruba); debris of coral rocks.

    [2q] **Ros** 1.3 - 1.6 m ∅, suckering (?); **L** rather few, broadly lanceolate, slighty S-curved, widest somewhat below or in the middle, usually slightly acuminate, usually guttered and lower face round, 60 - 80 × 13 - 14 cm; marginal teeth usually pointing downwards somewhat below the middle of the **L**, often upcurved at the top, slender-aciculate, 4 - 6 (-7) mm, 8 - 12 per 10 cm, on rather weakly to strongly developed green or hardening tubercles; terminal **Sp** acicular, straight or very slightly up-

curved at the tip, narrowly to broadly grooved below or to the middle, usually rough, covered with many minute tubercles, 2.7 - 3.2 cm, shortly decurrent; **Inf** 3.5 - 5 m, 'paniculate', narrowly oblong, part-**Inf** rather many, in the upper ¼ - ⅔ of the **Inf**, forming few **Fr** but freely bulbilliferous; **Tep** 19 - 21 mm, tube 7 - 8 mm; **Fr** narrowly oblong, long-stipitate and beaked, 3.3 - 4 × 1.2 - 1.5 cm.

According to the protologue hardly different from *A. vicina* (as *A. vivipara*) in vegetative characters, but clearly differing in generative parts (small number of bracts, long tube with filaments inserted low down, form of capsules).

A. asperrima Jacobi (Hamburg. Gart.- & Blumenzeit. 20: 561, 1864). **T** [neo]: USA, Texas (*Gentry & Barclay* 20012 [US, DES]). – **D:** USA (Texas); Mexico (widespread).

Since *A. scabra* Salm-Dyck (1859) is an illegitimate later homonym of *A. scabra* Ortega (1797) but was misapplied to this plant by Gentry (1982: 296), Ullrich (1992f) resurrected *A. asperrima* Jacobi as the valid name for *A. scabra sensu* Gentry and placed the 'true' *A. scabra* Salm-Dyck in the synonymy of *A. parryi*. The taxon is related to *A. americana*, with which it intergrades. It is the most widespread and common *Agave* in the Chihuahuan Desert of N Mexico, with the exception of *A. lechuguilla* (Gentry 1982: 296).

A. asperrima ssp. **asperrima** – **D:** USA (Texas), Mexico (Chihuahua, Coahuila, Nuevo León, Durango, Zacatecas); dry Chihuahuan Desert areas, 1200 - 1900 m. **I:** Gentry (1982: 276, 297-298, as *A. scabra*).

    Incl. *Agave caeciliana* A. Berger (1915).

    [2a] **Ros** rather open, 0.7 - 1 × to nearly 2 m, freely suckering; **L** 30 - 40, lanceolate, rigid, very broad at the base and constricted just above, long-acuminate, convex below, flat above, then deeply guttering through the middle of the lamina, scabrous, generally 60 - 110 × 12 - 16 cm, light green to glaucous-grey, margins sometimes horny along the upper ½; marginal teeth generally deflected below the middle of the lamina, larger teeth 8 - 15 mm, brown to pruinose-grey, on cusps from broadly rounded bases; terminal **Sp** subulate to acicular, very narrowly grooved above, base scabrous, 3.5 - 6 cm, long decurrent on the involute margin; **Inf** mostly 4 - 6 m, 'paniculate', part-**Inf** small, compact, 8 - 12 in the upper ⅓ of the **Inf**; **Fl** 60 - 80 mm; **Ov** slender, 30 - 40 mm, greenish; **Tep** yellow, tube 13 - 20 mm, lobes unequal, 18 - 25 mm. – *Cytology:* 2n = 128 - 186.

A. asperrima ssp. **maderensis** (Gentry) B. Ullrich (Sida 15(2): 254, 1992). **T:** Mexico, Coahuila (*Gentry & Engard* 23251 [DES, MEXU, US]). – **D:** Mexico (Coahuila); local endemic in (limestone) canyons of desert mountains, 1850 - 2000 m. **I:** Gentry (1982: 301, as *A. scabra* ssp.).

≡ *Agave scabra* ssp. *maderensis* Gentry (1982) (*incorrect name*, Art. 11.4).

[2a] Differs from ssp. *asperrima*: Stem short, thick; **Ros** solitary; **L** triangularly linear-lanceolate, relatively smooth, 50 - 60 × 7 - 12 cm, green to yellow-green; marginal teeth slender, larger teeth 5 - 8 mm, mostly 2 - 3 cm apart; **Inf** with ≥ 12 large spreading several times compound part-**Inf**; **Fl** 65 - 70 mm; **Tep** tube 11 - 16 × 12 - 15 mm, lobes linear, 15 - 20 mm.

A. asperrima ssp. **potosiensis** (Gentry) B. Ullrich (Sida 15(2): 254, 1992). **T:** Mexico, San Luis Potosí (*Gentry & al.* 20162 [US, DES, MEXU]). – **D:** Mexico (Nuevo León, San Luis Potosí, Querétaro); plains and hills of the S Chihuahuan Desert. **I:** Gentry (1982: 276, 301, as *A. scabra* ssp.).

≡ *Agave scabra* ssp. *potosiensis* Gentry (1982) (*incorrect name*, Art. 11.4).

[2a] Differs from ssp. *asperrima*: **Ros** more open, spreading, sparingly surculose; **L** broadly lanceolate, tip outcurving and sigmoid, asperous to nearly smooth, 65 - 110 × 14 - 20 cm, glaucous-grey to nearly white, frequently cross-zoned; marginal teeth sometimes on tuberculate elevantions; **Inf** with 10 - 18 small part-**Inf**; **Ov** slender, 32 - 50 mm; **Tep** tube large, 15 - 22 mm.

Best distinguished by its more open and spreading rosettes with broader leaves, flowers with slender ovaries and a large tube, but often difficult to separate due to apparent introgression with *A. americana* ssp. *protoamericana* (Gentry 1982: 301).

A. asperrima ssp. **zarcensis** (Gentry) B. Ullrich (Sida 15(2): 254, 1992). **T:** Mexico, Durango (*Gentry & Arguelles* 22084 [US, DES, MEXU]). – **D:** Mexico (Durango). **I:** Gentry (1982: 276, 303, as *A. scabra* ssp.).

≡ *Agave scabra* ssp. *zarcensis* Gentry (1982) (*incorrect name*, Art. 11.4).

[2a] Differs from ssp. *asperrima*: **Ros** surculose, forming large clumps; **L** linear-ovate, hollowed above, 55 - 60 × 15 - 20 cm, greyish-green; marginal teeth mostly reflexed, moderate, larger 5 - 7 mm, 1 - 2 cm apart; part-**Inf** 8 - 14, in the upper ⅓ of the **Inf**, on sigmoid **Br**; **Fl** 68 - 92 mm; **Ov** 3-angled and 6-grooved, 35 - 50 mm; **Tep** tube deeply furrowed, thickly 12-ridged within.

A highland ecotype within *A. asperrima*, distinguished best by its short broad leaves with moderate teeth, large flowers, 2-level insertion of filaments, and large woody fruits (Gentry 1982: 302).

A. atrovirens Karwinsky *ex* Salm-Dyck (Hort. Dyck., 7: 302, 1834). **T** [neo]: Mexico, Oaxaca (*Gentry* 22377 [US, DES, MEXU]). – **D:** S Mexico.

A. atrovirens var. **atrovirens** – **D:** Mexico (Guerrero, Puebla, Veracruz, Oaxaca); strictly high-montane, 1850 - 3400 m. **I:** Gentry (1982: 468, 471-472); Cact. Suc. Mex. 39(4): front cover, 1994.

Incl. *Agave tehuacanensis* Karwinsky *ex* Otto (1842); **incl.** *Agave latissima* Jacobi (1864); **incl.** *Agave schlechtendalii* Jacobi (1864); **incl.** *Agave coccinea* Roezl *ex* Jacobi (1865) ≡ *Agave americana* var. *coccinea* (Roezl *ex* Jacobi) A. Terracciano (1885); **incl.** *Agave ottonis* Jacobi (1866); **incl.** *Agave canartiana* Jacobi (1868); **incl.** *Agave canartiana* var. *laevior* Jacobi (1868); **incl.** *Agave deflexispina* Jacobi (1870); **incl.** *Agave gracilis* Jacobi (1870); **incl.** *Agave macroculmis* Todaro (1878); **incl.** *Agave coccinea* hort. *ex* A. Berger (1898) (*nom. illeg.*, Art. 53.1).

[2k] **Ros** openly spreading, large to very large, 1.5 - 2 × 3 - 4 m, solitary; **L** lanceolate, thick at the base (to 25 cm), usually narrowed below the middle of the lamina, openly concave, mostly 150 - 200 × 25 - 40 cm, dark to blackish-green to light glaucous or glaucous-variegated, margins ± straight; marginal teeth moderate, regular, bases broad, larger teeth mostly 4 - 7 mm (in the **L** middle), brown to greyish-brown, 1 - 4 cm apart; terminal **Sp** straight or sinuous, strong, broad at the base, widely openly grooved above, 3 - 5 cm, keel rounded below and markedly intruding into the **L** tip; **Inf** 8 - 12 m, 'paniculate', narrow, part-**Inf** congested in the upper ⅓ - ½ of the **Inf**, globose, mostly 18 - 30; **Fl** thickly fleshy, 70 - 100 mm; **Ov** cylindrical, tapering at the base, 30 - 50 mm, neck thick, furrowed, 4 - 7 mm; **Tep** red to purple in bud, when opening yellowish within, tube 11 - 15 mm, lobes unequal, outer 30 - 34 mm.

See Piña Luján (1994) on the type locality.

A. atrovirens var. **mirabilis** (Trelease) Gentry (Agaves Cont. North Amer., 473, ill. (p. 476), 1982). **T:** Mexico, Veracruz (*Trelease* 7 p.p. [MO, DES]). – **D:** Mexico (Puebla, Veracruz); cool montane habitats, 2150 - 2480 m.

≡ *Agave mirabilis* Trelease (1920).

[2k] Differs from var. *atrovirens*: **L** consistently light grey-glaucous.

A. attenuata Salm-Dyck (Hort. Dyck., 3, 1834). **T:** [neo – icono]: Curtis's Bot. Mag. ser. 3, 18: t. 5333, 1862. – **Lit:** Ullrich (1990h: with ills.). **D:** Mexico (Jalisco, Michoacán, México); high rocky outcrops in pine forests, 1900 - 2500 m. **I:** Gentry (1982: 67-68). **Fig. I.a**

Incl. *Agave attenuata* var. *brevifolia* Jacobi (s.a.); **incl.** *Agave attenuata* var. *latifolia* Salm-Dyck (s.a.); **incl.** *Agave elliptica* hort. *ex* Besaucèle (s.a.); **incl.** *Agave virens* hort. *ex* Besaucèle (s.a.); **incl.** *Agave spectabilis* hort. *ex* Besaucèle (s.a.) (*nom. illeg.*, Art. 53.1); **incl.** *Agave compacta* hort. *ex* Besaucèle (s.a.) (*nom. inval.*, Art. 32.1c?); **incl.** *Ghiesbreghtia mollis* Roezl (1861) (*nom. inval.*, Art.

32.1c?); **incl.** *Agave glaucescens* Hooker (1862) (*nom. illeg.*, Art. 53.1); **incl.** *Agave attenuata* var. *compacta* Jacobi (1865); **incl.** *Agave pruinosa* Lemaire *ex* Jacobi (1865); **incl.** *Agave debaryana* Jacobi (1868); **incl.** *Agave ghiesbreghtii* [?] *dentata* Hort. Belg. *ex* Jacobi (1868); **incl.** *Agave ghiesbreghtii* [?] *mollis* Hort. Belg. *ex* Jacobi (1868); **incl.** *Agave kellockii* Jacobi (1868); **incl.** *Agave dentata* hort. *ex* Baker (1877) (*nom. illeg.*, Art. 53.1); **incl.** *Agave attenuata* var. *serrulata* A. Terracciano (1885) ≡ *Agave cernua* var. *serrulata* (A. Terraciano) A. Berger (1915); **incl.** *Agave cernua* A. Berger (1915).

[1b] Stems 1 to several, usually ascending-curved, 0.5 - 1.5 m, becoming naked in age; **L** indeterminate in number, relatively short-lived, ovate-acuminate, softly succulent, broadest in the middle, plane to concave, 50 - 70 × 12 - 16 cm, light glaucous-grey to pale yellowish-green, margins smooth or serrulate; terminal **Sp** absent but **L** tip finely tapered, soon fraying; **Inf** 2 - 3.5 m, 'spicate', densely flowered, part-**Inf** shortly pedicellate 'fascicles' with 3 - 8 **Fl** in the **Ax** of chartaceous **Bra**; **Fl** 35 - 50 mm; **Ov** fusiform, 15 - 25 mm, green, neck constricted; **Tep** greenish-yellow, tube shallowly funnel-shaped, 3 - 5 mm, lobes equal, 16 - 24 mm.

The presence of a flower tube groups this species in Group *Serrulatae* and not in Group *Choritepalae*, which includes the otherwise similar *A. bracteosa* and *A. ellemeetiana*. It is closely related to *A. gilbertii* and esp. to *A. pedunculifera*, which form a broad-leaved group within Group *Serrulatae* (as *Amolae*) (Gentry 1982: 70).

A. aurea Brandegee (Proc. Calif. Acad. Sci., ser. 2, 2: 207, 1889). **T:** Mexico, Baja California (*Brandegee* s.n. [UC]). – **Lit:** Turner & al. (1995). **D:** Mexico (Baja California Sur); lava fields, mostly 300 - 1070 m. **I:** Gentry (1982: 310, 312, 314).

Incl. *Agave campaniflora* Trelease (1912).

[2d] Stem short; **Ros** rather open, 1 - 1.2 × 1.5 - 2 m, solitary; **L** linear to long-lanceolate, widely arching, pliable, guttered, rounded below, thickly fleshy towards the base, (63-) 86 (-110) × (7-) 8.6 (-12) cm, green to somewhat glaucous, margins straight to undulate; marginal teeth moderate, regular, mostly 4 - 7 mm, 1 - 2 cm apart, dark to light brown, on straight or moderately curved cusps from low angular bases; terminal **Sp** subulate, 2.5 - 3.5 cm, dark brown or greyish-red, shortly decurrent or decurrent as dark horny margin through the uppermost 8 - 10 teeth bases; **Inf** 2.5 - 5 m, 'paniculate', part-**Inf** broad, congested, 15 - 25, in the upper ½ of the **Inf**; **Fl** campanulate, 43 - 70 mm; **Ov** 25 - 35 mm, reddish, neck constricted, 6 - 10 mm; **Tep** red to purplish in bud, opening yellow to orange-yellow, tube 8 - 14 mm, lobes 16 - 19 mm.

Easily recognized by the long narrow lanceolate green leaves arching out in open rosettes, broad ra-

ther diffuse reddish lateral part-inflorescences, and bright yellow flowers from reddish buds and ovaries (Gentry 1982: 313).

A. avellanidens Trelease (Annual Rep. Missouri Bot. Gard. 22: 60, 1912). **T:** Mexico, Baja California (*Brandegee* 6 [UC]). – **D:** Mexico (Baja California). **I:** Gentry (1982: 361). **Fig. I.b**

[2h] Stem to 0.5 m; **Ros** 0.6 - 1.2 × 1 - 1.5 m, solitary; **L** many, broadly linear-lanceolate to ovate, thickly fleshy, rigid, little or not narrowed at the base, shortly acuminate, smooth, 40 - 70 × 9 - 14 cm, green, margins straight or undulate, frequently horny; marginal teeth straight or variously curved, variable in size and curvature, flattened, 5 - 15 mm, dusky grey over brown, mostly 1 - 3 cm apart, rather regularly spaced; terminal **Sp** conical, strong, 2.5 - 4.5 cm, brown to greyish, strongly decurrent as horny margin; **Inf** 4 - 6 m, 'paniculate', part-**Inf** dense, large, globose, 25 - 35; **Fl** small, slender, 40 - 70 mm; **Ov** 20 - 40 mm, neck sometimes constricted; **Tep** pale yellow, drying orange-yellow, tube 4 - 6 mm, lobes ± equal, 16 - 24 mm.

Resembling *A. shawii* (Group *Umbelliflorae*) within Group *Deserticolae*, from which it clearly differs in its flowers (Gentry 1982: 363).

A. bahamana Trelease (Mem. Nation. Acad. Sci. 11: 40, t. 84-86, 1913). **T:** Bahamas, Great Harbor Cay (*Britton & Millspaugh* 2340 [NY]). – **Lit:** Correll & Correll (1982). **D:** Bahamas (Berry Islands); open rocky plains and ridges, open coppices and pinelands.

Incl. *Agave sobolifera* Hitchcock (1893) (*nom. illeg.*, Art. 53.1); **incl.** *Agave rigida* Northrop (1902) (*nom. illeg.*, Art. 53.1).

[2n] Acaulescent; **Ros** solitary; **L** rather narrowly lanceolate, concave, occasionally somewhat conduplicate, 2 - 3 m × 15 cm, dull greyish, margins nearly straight; marginal teeth straight or the longer teeth appressed-recurved, triangular, scarcely lenticular at the base, 3 - 5 mm, usually 5 - 10 mm apart, reduced above and below, sometimes on small green tubercles; terminal **Sp** slightly recurved, stoutly conical, usually becoming involutely grooved below the middle, smooth, 1 - 1.5 cm, brownish becoming grey, dull, decurrent; **Inf** to ± 10 m, 'paniculate', ovoid, part-**Inf** on slightly ascending **Br**, ± in the upper ⅓ of the **Inf**; **Ped** ± 10 mm; **Fl** 50 - 60 mm; **Ov** oblong-fusiform, 30 - 35 mm; **Tep** ± 15 × 4 mm, golden-yellow, tube conical, ± 7 mm, lobes 15 mm; **Fr** oblong, shortly stipitate and beaked, 5 × 2.5 cm.

A. barbadensis Trelease (Mem. Nation. Acad. Sci. 11: 28-29, t. 34-38, 65, 107, 1913). – **Lit:** Howard & al. (1979). **D:** Lesser Antilles (Barbados); spontaneous in dune areas, frequently escaped from cultivation.

Incl. *Agave americana* Dillenius (1774) (*nom. il-*

leg., Art. 53.1); **incl.** *Furcraea tuberosa* Drummond (1907) (*nom. illeg.*, Art. 53.1).

[2o] **Ros** moderately surculose; **L** broadly lanceolate, curved outwards, up to 20 cm thick at the base, rather abruptly acute, almost cochleate and conduplicate towards the tip, concave, 150 - 250 × 25 - 30 (at the base) cm, dull dark green, glaucous when young, margins straight; marginal teeth straight or curved, distinct, 2 - 3 mm, usually 10 - 12 mm apart; terminal **Sp** rather unguiculately-conically subulate, involute at the base, smooth, polished near the tip, (0.7-) 1 - 1.5 cm, blackish-brown at the curved tip, decurrent and dorsally intruding into the green tissue; **Inf** 5 - 6 m and more, narrowly oblong, 'paniculate', part-**Inf** on very ascending **Br**, in the upper ⅓ or more of the **Inf**, freely and densely bulbilliferous, not known to produce **Fr**; **Ped** 10 (-20) mm; **Fl** aborting before completely opening, 65 - 75 mm and more; **Ov** oblong-fusiform, 45 - 55 mm; **Tep** yellow, tube conical, ± 15 mm, lobes 20 - 25 mm.

A. bicolor (Solano & García-Mendoza) Thiede & Eggli (KuaS 52: [in press], 2001). **T:** Mexico, Oaxaca (*García-Mendoza & al.* 2403 [MEXU, BRIT, FEZA]). − **D:** Mexico (Oaxaca); grassland and in pine-oak forests, 2300 - 2500 m, flowering July to August.

≡ *Polianthes bicolor* Solano & García-Mendoza (1998).

[3b2] Plants glabrous; rhizome (2-) 3 - 4 (-5) × (1-) 1.5 - 2.5 cm; **R** fleshy; **L** (3-) 4 - 6 (-12), lanceolate, semisucculent, undulate, (5-) 8 - 15 × 0.6 - 1 (-1.4) cm, smooth to papillose, margins undulate, papillose, hyaline; **Inf** 24 - 40 (-54) cm, 'spicate', with 3 - 5 (-9) **Fl**-bearing nodes with paired **Fl**; **Bra** lanceolate, 3 - 5 (-7.5) cm; **Ped** 6 - 13 mm, reddish; **Fl** (20-) 23 - 29 mm; **Ov** 9 - 15 (-19) mm; **Tep** orange-greenish, tube abruptly widened and curved, (1.6-) 2.4 - 3 (-5.5) mm ∅ at the mouth, lobes 2 - 3 (-4) × (1.7-) 2 - 3 (-4) mm, green, apiculate; **Anth** included; **Sty** (10-) 19 - 26 (-32) mm; **Sti** 3-lobed; **Fr** semiglobose, 1.1 × 1.1 cm; **Se** semiglobose, 4.5 × 2.4 mm, black.

According to the protologue closest to *A. duplicata* (as *Polianthes geminiflora*). *A. bicolor* is the S-most species in the *Polianthes* Group.

A. ×blissii (Worsley) Thiede & Eggli (KuaS 50(5): 111, 1999). − **I:** Worsley (1911).

≡ *Polianthes ×blissii* Worsley (1911).

This is the garden hybrid *A. duplicata* (as *Bravoa geminiflora*) × *A. polianthes* (as *Polianthes tuberosa*).

A. boldinghiana Trelease (Mem. Nation. Acad. Sci. 11: 21, t. 11-13, 1913). **T:** Curaçao (*Boldingh* A2 [not indicated]). − **Lit:** Hummelinck (1938). **D:** Leeward Islands (Aruba, Bonaire, Curaçao). **I:** Hummelinck (1993).

[2q] Stem almost none; **Ros** suckering; **L** narrowly oblanceolate, subacuminate, openly concave, 90 - 125 × ± 15 cm, green, passing into somewhat glaucous, margins rather straight; marginal teeth often irregularly upcurved above and recurved below, heavily triangular or from lunate bases, mostly 2 - 5 mm, scarlet becoming chestnut-brown, mostly 1 - 1.5 cm apart; terminal **Sp** acicular, somewhat upcurved-flexuous, grooved and usually involute towards the base, smooth, polished towards the tip, 2.5 - 3 cm, red-brown, shortly decurrent; **Inf** ± 5 m, 'paniculate', narrowly oblong, part-**Inf** few, distant, on ascending **Br**, in the upper ½ or less of the **Inf**, freely bulbilliferous; **Ped** 5 mm; **Fl** 45 mm; **Ov** broadly fusiform, 20 - 25 mm; **Tep** golden-yellow, tube conical, ± 7 mm, lobes 15 mm; **Fr** unknown.

Always distinguishable from all other species in its geographical range, but less well defined compared with certain forms of *A. cocui* and *A. vicina* (as *A. vivipara*) from other regions (Hummelinck 1938).

A. bovicornuta Gentry (Publ. Carnegie Inst. Washington 527: 92, 1942). **T:** Mexico, Sonora (*Gentry* 3672 [DS, ARIZ, DES]). − **D:** Mexico (Sonora, Chihuahua, Sinaloa); rocky open slopes, oak woodland and pine-oak forest, 930 - 1850 m. **I:** Gentry (1982: 325, 330). **Fig. I.c, I.e**

[2c] **Ros** 0.8 - 1 × 1.5 - 2 m, solitary; **L** lanceolate to spatulate, much narrowed towards the base, widest at or above the middle, smooth, 60 - 80 × 14 - 17 cm, yellowish-green to green, younger **L** frequently shining glaucous, with conspicuous imprints from the central bud, margins crenate; marginal teeth dimorphic, larger teeth mostly 8 - 12 mm, flexuous and slender above a broad base, mostly 2 - 4 cm apart, on prominent tubercles, smaller teeth mostly 2 - 5 mm, 1 to several between the larger teeth, all chestnut-brown or dark brown to greyish-brown in age; terminal **Sp** strong; **Inf** 5 - 7 m, 'paniculate', narrow, scape short, part-**Inf** short, compact, 20 - 30 in the upper ½ of the **Inf**; **Fl** small, 55 - 65 mm; **Ov** 30 - 35 mm incl. a neck 4 - 6 mm long, pale green; **Tep** yellow, tube 6 - 8 mm, lobes 18 - 21 mm.

Distinguished within Group *Crenatae* by light to yellowish-green leaves with narrow bases, relatively small flowers, and the low insertion of the filaments in the middle of the perianth tube (Gentry 1982: 330).

A. braceana Trelease (Mem. Nation. Acad. Sci. 11: 40, t. 83, 1913). **T:** Bahamas (*Brace* 1982 [NY]). − **Lit:** Correll & Correll (1982). **D:** Bahamas (Abaco); rocky or sandy soils in pinelands or coastal coppices.

Incl. *Agave mexicana* Dolley & al. (1889) (*nom. illeg.*, Art. 53.1).

[2n] Acaulescent; **Ros** solitary; **L** broadly oblanceolate, nearly flat, to ± 70 × 20 cm, grey, margins between the teeth straight or concave when the

teeth are raised on low green tubercles; marginal teeth straight or the lower teeth gently recurved, triangular, 2 - 3 mm, usually 5 - 10 mm apart; terminal **Sp** conical, straight or gently curved, flat or round-grooved to about the middle or becoming involute, smooth, 1 - 1.5 cm, brownish becoming grey, dull, slightly decurrent; **Inf** to ± 7 m, 'paniculate'; **Ped** ± 10 mm; **Fl** 40 - 45 mm; **Ov** oblong-fusiform, 20 mm; **Tep** 15 - 17 × 3 - 4 mm, golden-yellow, tube conical, ± 7 mm, lobes 15 - 17 mm; **Fr** broadly oblong, shortly stipitate and beaked, 2 - 3.5 × 2 cm.

A. bracteosa S. Watson *ex* Engelmann (Gard. Chron., ser. nov. 18: 776, ills., 1882). **T:** [lecto — icono]: l.c. figs. 138-139. − **D:** Mexico (Coahuila, Nuevo León); scattered on limestone cliffs and rocky slopes of the N Sierra Madre Oriental, 900 - 1700 m. **I:** Gentry (1982: 90, 92). **Fig. I.d**

[1c] **Ros** open, small to medium-sized, forming caespitose mounds by above-ground axillary budding; **L** relatively few, long-lanceolate, arching and recurving, with weak fibres, widest near the base, convex in the basal ⅓, plane above, asperous, 50 - 70 × 3 - 5 cm, yellow-green, margins minutely serrulate; terminal **Sp** absent, leaf tip drying early, friable, yellowish; **Inf** ascending to erect, 1.2 - 1.7 m, 'spicate', densely flowered in the upper ⅓, part-**Inf** with geminate **Fl**; **Fl** 22 - 26 mm; **Ov** fusiform, 12 - 14 mm, virtually neckless; **Tep** white to pale yellow, tube virtually none, reduced to a short **Rec**, lobes 11 mm; **Fil** long-exserted, 50 - 60 mm.

Very distinctive, even within the Group *Choritepalae*, with its unarmed curling leaves and white flowers (Gentry 1982: 91). This prompted Ullrich (1990b) to place it within a reconsidered monotypic Group *Serrulatae* Baker (see also the note for *A. brevispina*). Gentry (l.c.) reports the inflorescences to emerge laterally from upper leaf axils so that one rosette may flower repeatedly. This atypical behaviour may simply be a misinterpretation, with the flowering rosettes merely forming rosettes from upper leaf axils after flowering. − The neotypification by Gentry (1982: 91) is superseded by Ullrich (l.c.).

A. brevipetala Trelease (RSN 23: 362, 1927). **T:** Hispaniola, Haiti (*Ekman* 1604 [MO?]). − **D:** Hispaniola (Haiti: Morne Cabaio, La Selle).

[2?] Acaulescent; **Ros** solitary; **L** broadly lanceolate, 100 cm, green, rather dull; marginal teeth broadly triangular, variously curved, with lenticular bases, 5 - 10 mm but apical and basal teeth smaller, glossy chestnut-brown, 10 - 15 mm apart, teeth of the middle of the **L** on clasping green marginal prominences, margin in-between nearly straight; terminal **Sp** smooth, curved, subterete, involutely narrow-grooved, 20 - 25 × 6 mm, rather glossy chestnut-brown, decurrent for some 10 cm and connecting to the small upper teeth; **Inf** 'paniculate', **Fl** densely clustered at the tips of the part-**Inf**, not known to be bulbilliferous; **Ped** ± 5 mm; **Fl** ± 35 mm; **Ov** thick, oblong, 20 mm, longer than the **Tep**; **Tep** colour not described, drying dark, lobes 10 mm; **Fr** unknown.

Described on the base of the dried type material only and apparently not mentioned by any later author. See also the note for *A. brevispina*.

A. brevispina Trelease (RSN 23: 363, 1927). **T:** Hispaniola, Haiti (*Ekman* 5371 [MO?]). − **D:** Hispaniola (Haiti: Croix-des-Bouquets, Plaine Cul de Sac).

[2?] Acaulescent; **Ros** solitary; **L** broadly lanceolate, 100 × 10 cm, dark green, rather dull; marginal teeth straight or some of the lower teeth recurved, rather narrowly triangular, with lenticular base, 1 - 3 mm, 5 - 15 mm apart, margin in-between nearly straight; terminal **Sp** straight, somewhat flattened, involutely narrowly grooved, slightly granular, 10 × 3 mm, rather dull hazel-brown, tip darker, decurrent for 2× its length; **Inf** 4 m, 'paniculate', part-**Inf** rather slender, shortly few-flowered at the tips and **Fl** densely clustered, not known to be bulbilliferous; **Ped** ± 5 mm; **Fl** ± 40 mm; **Ov** 25 mm, longer than the **Tep**; **Tep** yellow, tube 5 mm, lobes 12 × 5 mm; **Fr** unknown.

Described on the base of the type material only and apparently not mentioned by any later author. This and the aforementioned species do not fit well into any of the groups established by Trelease. They are of uncertain affinities and in need of study. Trelease himself did not give further data in the protologues.

A. brittoniana Trelease (Mem. Nation. Acad. Sci. 11: 44-45, t. 98-99, 1913). **T:** Cuba, Santa Clara (*Britton & al.* 4776 [NY]). − **Lit:** Álvarez de Zayas (1996b). **D:** C Cuba.

A polymorphic species, which occurs abundantly and prolifically at anthropogenic sites (Álvarez de Zayas l.c.).

A. brittoniana ssp. **brachypus** (Trelease) A. Álvarez (Fontqueria 44: 121, 1996). **T:** Cuba (*Britton & al.* 6183 [NY?]). − **D:** C Cuba; xeromorphic scrub and derived secondary formations. **I:** Trelease (1913: t. 99: 1, as var.).

≡ *Agave brittoniana* var. *brachypus* Trelease (1913).

[2m] Differs from ssp. *brittoniana*: **L** tip canaliculate, normally with small denticles at the inner margin; **Inf** somewhat laxer; **Fl** smaller; **Fr** more cylindrical.

The shorter pedicels given as diagnostic in the protologue are not a constant feature (Álvarez de Zayas l.c.).

The basionym could be regarded as a provisional and hence invalid (Art. 34.1b) name, but Trelease's illustration and the adjacent caption leaves no doubt that he fully accepted the taxon. The lectotypifica-tion proposed by Álvarez de Zayas (l.c.) is unnecessary and moreover erroneous, since an ele-ment not originally included by Trelease is selected.

A. brittoniana ssp. **brittoniana** – **D:** C Cuba; evergreen forests to xeromorphic scrub, 100 - 1000 m.

[2m] Stems not rhizomatous; **L** broadly lanceo-late, abruptly acute towards the tip, (70-) 80 - 100 (-110) × (13-) 15 - 20 (-24) cm, green, sometimes somewhat greyish, slightly glossy, margins often concave; marginal teeth variously curved, 1 - 6 (-8) mm, (6-) 8 - 10 (-15) mm apart, slender-cusped from lenticular or heavy bases, which may stand on retrorse green prominences in the lower ⅓ of the **L**; terminal **Sp** unguiculately curved, subconical or in-volutely much thickened below, openly grooved to the middle or involute, smooth, somewhat polished, 1 - 2.5 cm, brown, dotted with white, ± decurrent; **Inf** (4-) 5 - 8 m, 'paniculate', scape very short or nearly none, part-**Inf** ascending, (11-) 15 - 30 cm; **Ped** 5 - 10 mm; **Fl** 25 - 35 (-45) mm; **Ov** fusiform, 15 - 20 (-25) mm; **Tep** yellow, outer face greenish, tube open, 3 - 6 mm, lobes 9 - 14 (-16) × 3 - 5 mm; **Fr** oblong, sometimes nearly cylindrical, basally strongly stipitate, tip slightly beaked, 2.3 - 4 (-4.5) × 1.1 - 1.5 (-1.7) cm.

A. brittoniana ssp. **sancti-spirituensis** A. Álvarez (Fontqueria 44: 125, ill. (p. 122), 1996). **T:** Cuba, Sancti Spíritus (*Jiménez & al.* 69532 [HAJB]). – **D:** C Cuba.

[2m] Differs from ssp. *brittoniana*: **L** much smaller, more broadly oblong and less lanceolate; **Fl** as well as **Tep** and **Anth** larger.

A. brunnea S. Watson (Proc. Amer. Acad. Arts 26: 156, 1891). **T:** Mexico, Coahuila (*Pringle* 2218 [GH, US [photo]]). – **D:** Mexico (SE Chihuahua, W Coahuila); dry hills or desert plains, volcanic or alluvial alkaline soils (sandy or gravelly clay), 1125 - 1400 m, flowering late June to August. **I:** Piña Luján (1985: 28-29).

≡ *Manfreda brunnea* (S. Watson) Rose (1903) ≡ *Polianthes brunnea* (S. Watson) Shinners (1966).

[3a1] Plants robust, reproducing vegetatively by buds from the rhizome below the **L** bases; rhizome usually oblong, ± 2 × 0.9 - 2.5 cm; **R** fleshy; **L** 4 - 8, succulent, recurved, linear-lanceolate to broadly lanceolate, tip acute, with a long point, smooth, to 32 × 1.3 - 2.9 (-3.6) cm, somewhat glaucous, mot-tled with red; margins toothed, teeth cartilaginous, usually large, deltoid or truncate-erose, 0.3 - 1.1 cm apart, with narrow pale band on the **L** margin be-tween the teeth; remains of **L** bases membranous, fraying into fine fibres at the tip, (4.5-) 5.5 - 9.5 cm; **Inf** to 1.3 m, 'spicate', flowering part 6.5 - 29 cm, with 9 - 29 solitary sessile, nearly erect **Fl**; **Ov**

long-ellipsoid, 10 - 20 (-23) mm; **Tep** tube narrowly funnel-shaped, straight, gradually constricted above the **Ov**, (15-) 20 - 32 (-35) mm, outer face yellow-ish-green, inner face brown; **Tep** lobes obtuse, not swollen at the tip; **Sty** exceeding the **Tep** tube for 35 - 53 (-65) mm; **Sti** clavate, trigonous, deeply fur-rowed; **Fr** woody, ellipsoid to oblong, 1.8 - 3.6 × 1.2 - 1.6 cm; **Se** 5 × 3 - 4 mm. – *Cytology:* n = 30.

Easily distinguished from the other species in the *A. brunnea* Subgroup by its long narrow tepal tube with exserted stamens and styles and by the coarse teeth on the leaf margin (Verhoek-Williams 1975: 190).

A. bulliana (Baker) Thiede & Eggli (KuaS 50(5): 112, 1999). **T:** Mexico (*Karwinsky* s.n. [not pre-served]). – **D:** Mexico (Durango, Aguascalientes, Jalisco, Nayarit, Michoacán, Zacatecas); dry rocky slopes or roadcuts, in pine-oak grassland, or in shaded moist ravines, 1150 - 3100 m, flowers late June to early September. **I:** CBM 121: pl. 7427, 1895; McVaugh (1989: fig. 41, as *Prochnyanthes mexicana*).

≡ *Bravoa bulliana* Baker (1884) ≡ *Prochnyanthes bulliana* (Baker) Baker (1895); **incl.** *Polianthes me-xicana* Zuccarini (1837) ≡ *Prochnyanthes mexicana* (Zuccarini) Rose (1903); **incl.** *Prochnyanthes viri-descens* S. Watson (1887).

[3c] Plants large (for Subgen. *Manfreda*), usually single; **R** semifleshy with a wiry core, rhizomes cy-lindrical, 1 - 3 × 1.5 - 2.5 cm; **L** few, (1-) 2 - 4 (-5), chartaceous, thin, fibrous, erect or occasionally curved backwards from about the middle, often twisted, with a distinct midrib, shallowly chan-nelled over the midrib, lamina flat, broadly undu-late, or revolute, linear-lanceolate to oblanceolate, narrowed towards the base, 20 - 47 (-62) × (0.7-) 1.3 - 5.2 (-6.9) cm, light or dark green, dull, often speckled with magenta towards the base, spotted or not, veins slightly prominent on both surfaces, papillate, fibres of old **L** bases 7 - 10 (-12.5) cm, margins very narrow, hyaline, papillate to erose-papillate or papillate-denticulate; **Inf** 0.9 - 2 m and more, 'spicate', flowering part elongate, (9-) 17.5 - 47 (-83.5) cm, with 4 - 22 flowering nodes with paired **Fl**; **Ped** 3 - 46 (-68) mm; **Fl** functionally pendent by an abrupt curve in the **Tep** tube; **Ov** el-lipsoid, 4 - 8 (-12) mm; **Tep** white tinged with grey-green or dull green and red, white or creamy within, tube curved near the middle or at ⅓ from the **Ov**, narrow below, abruptly widened above the bend, (11-) 15 - 27 mm, lobes flaring, broadly delt-oid, (3-) 4 - 9 (-10) mm; **Sty** finally equalling the tube or longer, white; **Fr** 1 - 1.9 × 1 - 1.4 cm; **Se** 2.5 - 3 × 3.5 - 4 mm.

When transferring *Prochnyanthes mexicana* to *Agave*, a new name was necessary to avoid homo-nymy with *A. mexicana* Lamarck 1783. Therefore, the second-oldest synonym *Prochnyanthes bulliana* had to be chosen.

A. ×bundrantii (Howard) Thiede & Eggli (KuaS 50(5): 111, 1999).

≡ *Polianthes ×bundrantii* Howard (1978).

This is the garden hybrid *A. polianthes* (as *Polianthes tuberosa*) × *A. howardii* (as *Polianthes howardii*).

A. cacozela Trelease (Mem. Nation. Acad. Sci. 11: 41, t. 89-91, 1913). **T:** Bahamas, New Providence (*Cunningham* s.n. [MO?]). – **D:** Bahamas (New Providence); rocky margins of salt marshes.

[2n] Acaulescent; **Ros** solitary; **L** lanceolate, deeply concave, typically roughish, 150 - 200 × 20 cm, yellowish-green, somewhat overcast with grey, margins straight or somewhat concave between the teeth; marginal teeth nearly straight or the larger teeth appressed-recurved, narrowly triangular, 2 - 5 mm (middle of the lamina), usually 1 - 1.5 cm apart; terminal **Sp** triquetrously conical, straight or the tip slightly refracted, openly grooved below the middle, smooth, 1.5 - 2 cm, brownish becoming grey, dull, decurrent; **Inf** 6 - 7 m, 'paniculate', dense, ovoid, part-**Inf** on horizontal or slightly ascending **Br**, reportedly sometimes bulbilliferous; **Ped** ± 10 mm; **Fl** 50 - 60 mm; **Ov** oblong-fusiform, 35 - 40 mm; **Tep** ± 20 × 4 - 5 mm, golden-yellow, tube rather open, ± 7 mm; **Fr** narrowly oblong, shortly conical-stipitate, 3.5 - 4.5 × 1.5 cm.

According to the protologue, seedlings of the type collection were decidedly papillate-roughened on both leaf faces.

A. cajalbanensis A. Álvarez (Revista Jard. Bot. Nac. Univ. Habana 1(2/3): 33-39, ill., 1981). **T:** Cuba, Pinar del Río (*Bisse & Álvarez* 32466 [HAJB]). – **D:** W Cuba; steep ultrabasic slopes.

[2l] Stem short; **Ros** solitary; **L** many, oblanceolate in the lower ⅔, straight, fleshy, coriaceous, only slightly concave, 50 - 60 × 8 - 10 cm, grey-green, slightly opaque, margins with asymmetrical slightly recurved prominences, these 3 × 4 - 5 mm, in between margin nearly straight; marginal teeth basally slightly recurving, 2 - 4 mm, dark chestnut-brown to nearly black, 1 - 2 cm apart; terminal **Sp** conical, straight, basally flattened, 1 - 1.5 cm, dark chestnut-brown, not lustrous, not decurrent; **Inf** 3 - 5 m, 'paniculate', part-**Inf** 3-parted, 30 - 40 cm; **Ped** 18 - 25 (-35) mm; **Fl** 40 - 45 mm; **Ov** fusiform, trigonous, basally constricted, 15 - 20 mm; **Tep** orange, slightly yellow, tube 5 - 6 mm, lobes 12 - 15 mm; **Fr** oblong, apically acute or apiculate, 2 - 2.5 × 1.5 cm.

Easily identifiable by its somewhat lobed leaf margins, the recurved marginal teeth, its orange flowers and small fruits. It is closest to *A. grisea*, but the leaves and inflorescences are only ½ as large according to the protologue.

A. calodonta A. Berger (Hort. Mortol. 364, 1912). **T:** Ex cult. La Mortola (*Berger* s.n. [US [lecto?]]).

– **D:** Only known from cultivation. **I:** Gentry (1982: 334).

Incl. *Agave scolymus* A. Berger (1898) (*nom. illeg.*, Art. 53.1).

[2c] **Ros** semiglobose, 1.5 - 1.6 m ∅, solitary; **L** many, spatulate, ± erect, older **L** spreading, fleshy, narrowed towards the base (7.5 - 8 cm), shortly acuminate, basally convex on both faces, upper face shallowly hollowed, upwards markedly thin, ≥ 80 × 20 - 21 (upper ⅓) cm, light green, with light grey bloom, both faces with imprints of the central bud, margins sinuous in the middle of the lamina; marginal teeth irregular, 10 - 13 mm, 2.5 - 3.5 cm apart, with broad horny bases and deltoid cusps hooked forwards or backwards, on broad fleshy prominences, intersinuses with much smaller intermittent teeth, teeth in the lower ½ of the **L** much smaller, straight or reflexed, all teeth light brown; terminal **Sp** 3 - 4 cm, decurrent to the upper 3 - 4 teeth; **Inf** tall, 'paniculate', long pyramidal, scape strong; **Fl** 85 mm (dried); **Ov** narrow, 35 - 40 mm; **Tep** yellow, tube broadly funnel-shaped, ± 10 mm, lobes 35 - 40 mm.

The only plant known up to now flowered in La Mortola in 1897 and died without producing seed or offsets (Berger 1915: 196).

A. cantala Roxburgh (Hort. Bengal., 25, 1814). – **D:** Known from cultivation only.

≡ *Furcraea cantala* (Roxburgh) Voigt (1845); **incl.** *Agave cantula* Roxburgh (1832) (*nom. inval.*, Art. 61.1).

A. cantala var. **acuispina** (Trelease) Gentry (Agaves Cont. North Amer., 569, ill. (p. 555), 1982). **T:** El Salvador, San Miguel Dept. (*Calderón* 2084 [US]). – **D:** Known from cultivation only (S Mexico, Honduras, El Salvador).

≡ *Agave acuispina* Trelease (1925).

[2f] Differs from var. *cantala*: **L** sturdier, shorter, mature **L** 140 - 170 × 6 - 8 cm, margins straight to undulate; terminal **Sp** 3 - 5 mm broad at the base, longer, > 1.5 cm; **Inf** with 20 - 35 part-**Inf**; **Fl** shorter, 57 - 63 mm; **Ov** shorter, 25 - 30 mm, neck short; **Tep** green, lobes subequal, 19 - 21 mm, light greenish-yellow.

A. cantala var. **cantala** – **Lit:** McVaugh (1989). **D:** Cultivated worldwide, esp. in SE Asia; not known from the wild. **I:** Gentry (1982: 570).

Incl. *Agave bulbifera* Salm-Dyck (1834); **incl.** *Agave laxa* Karwinsky *ex* Otto (1842) (*nom. illeg.*, Art. 53.1); **incl.** *Agave vivipara* Dalzell & A. Gibson (1861) (*nom. illeg.*, Art. 53.1); **incl.** *Agave rumphii* Jacobi (1865) (*nom. illeg.*, Art. 53.1); **incl.** *Agave candelabrum* Todaro (1878).

[2f] Stem 30 - 60 cm; **Ros** tall, slender, laxly leafy, 2 - 2.5 m ∅, surculose; **L** linear, long-acuminate, thin, frequently reflexing, roundly keeled below towards the base, rough below, smooth above,

150 - 200 × 7 - 9 cm, light or dark green, margins straight; marginal teeth antrorsely curved, small, larger teeth 3 - 4 mm, brown, mostly 2 - 3 cm apart, reduced or lacking towards the **L** tip; terminal **Sp** very small, 0.5 - 1.5 cm; **Inf** 6 - 8 m, 'paniculate', scape slender, part-**Inf** lax, ± 20, in the upper ½ of the **Inf**, sometimes bulbilliferous; **Fl** slender, 70 - 85 mm; **Ov** fusiform, tapering below to a basal rim, 32 - 42 mm, virtually neckless; **Tep** greenish tinged purple or reddish, tube 14 - 17 mm, lobes subequal, 25 - 28 mm.

Recognizable by its thin long narrow leaves (weak and frequently reflexed above the middle), small teeth and green flowers in broad 'panicles' (Gentry 1982: 569).

A. capensis Gentry (Occas. Pap. Calif. Acad. Sci. 130: 72-73, ills. (pp. 74-76), 1978). **T:** Mexico, Baja California Sur (*Gentry & Fox* 11247 [US]). – **D:** Mexico (Baja California Sur); arid slopes, ± sea-level to ± 3200 m. **I:** Gentry (1982: 310, 317-318).

[2d] Stem short; **Ros** open, small, caespitose by axillary budding, eventually in large clusters 0.6 - 0.8 × 0.8 - 1.2 m; **L** narrowly lanceolate, straight to arching, soft, brittle, succulent, commonly sigmoid towards the tip, convex below, concave above, mostly 30 - 60 × 4 - 7 cm, light glaucous-green, margins undulate, not horny; marginal teeth mildly curved, regular, 4 - 5 mm, reddish-brown to greyish, mostly 1 - 2 cm apart, with short mamillate bases; terminal **Sp** subulate, 1.5 - 3 cm, dark brown, shortly decurrent for 1 - 2 cm; **Inf** mostly 2.5 - 3.5 m, 'paniculate', part-**Inf** small, 15 - 24, in the upper ½ - ⅔ of the **Inf**; **Fl** 50 - 65 mm; **Ov** 25 - 35 mm, green, neck constricted; **Tep** in bud reddish-brown or purplish, opening yellow outside, tube 8 - 14 mm, lobes equal, 13 - 23 mm. – *Cytology:* 2n = 60.

Distinguished within the Group *Campaniflorae* by its small narrow leaves and clustered growth (Gentry 1982: 316).

A. caribaeicola Trelease (Mem. Nation. Acad. Sci. 11: 27, t. 30, 1913). **T:** Martinique (*Hahn* 114 [NY, MO?]). – **Lit:** Howard & al. (1979: with ill.). **D:** Windward Islands (Dominica, Martinique, St. Lucia, St. Vincent, Grenadines, Grenada). **I:** KuaS 48: 98, as *A. unguiculata*.

Incl. *Agave martiana* C. Koch (1860); **incl.** *Agave caribaea* Baker (1888) (*nom. illeg.*, Art. 53.1); **incl.** *Agave grenadina* Trelease (1913); **incl.** *Agave medioxima* Trelease (1913); **incl.** *Agave unguiculata* Trelease (1913); **incl.** *Agave ventum-versa* Trelease (1913).

[2o] **Ros** solitary; **L** lanceolate, ascending, curving and twisted, rather gradually and very concavely acute, 100 - 200 × 15 cm, green and glossy, very slightly glaucous beneath, margins straight; marginal teeth straight, 1 - 3 mm, or those below the middle of the lamina 2× as long and recurved, narrowly triangular, red to chestnut-brown, ± 5 mm

apart, commonly with intermediate smaller intermittent teeth; terminal **Sp** conical and grooved or involute, recurved-mucronate or with oblong-conical involute light brown basal thickening, smooth, 1.5 - 2.5 cm, nearly black, rather dull, decurrent, dorsally intruding into the green **L** tissue; **Inf** 3 - 5 m, 'paniculate', bulbilliferous, not known to produce **Fr**; **Fl** ± 60 mm; **Tep** golden-yellow, tube openly conical, 8 mm, lobes 18 - 20 mm.

Berger (1915: 216) regards *A. martiana* as possibly very close to *A. medioxima* and *A. grenadina* (both here included as synonyms); it would antedate *A. caribaeicola* and its definite identification is open to debate.

A. cerulata Trelease (Annual Rep. Missouri Bot. Gard. 22: 55, 1912). **T:** Mexico, Baja California (*Nelson & Goldman* 7180 [US]). – **Lit:** Turner & al. (1995). **D:** Mexico (Baja California).

A. cerulata ssp. **cerulata** – **D:** Mexico (C and S Baja California, N Baja California Sur). **I:** Gentry (1982: 356, 364).

[2h] **Ros** small, 0.2 - 0.5 m, abundantly surculose; **L** few, narrowly lanceolate to triangular-lanceolate, long-acuminate, mostly 25 - 50 × 4 - 7 cm and 6 - 12× as long as broad, yellow to light green, sometimes cross-zoned, rarely light glaucous-grey, margins nearly straight to mildly undulate; marginal teeth small, weakly attached, 1 - 4 mm, greyish-brown, bordered with a brown ring at the base, irregularly spaced, on low tubercles, sometimes lacking through much of the lamina; terminal **Sp** acicular, 3 - 6 cm, light to dark grey, decurrent only to the uppermost teeth or less; **Inf** 2 - 3.5 m, 'paniculate', part-**Inf** small, 6 - 12; **Fl** mostly 45 - 60 mm; **Ov** fusiform, 22 - 32 mm; **Tep** in bud white waxy glaucous, opening pale yellow, tube broadly funnel-shaped or discoid, with thick **Nec** and bulges opposite the **Fil** insertions, 3 - 5 mm, lobes 16 - 22 mm.

A. cerulata ssp. **dentiens** (Trelease) Gentry (Occas. Pap. Calif. Acad. Sci. 130: 43, 1978). **T:** Mexico, Baja California (*Rose* 16819 [MO, US]). – **D:** Mexico (Baja California: San Esteban Island).

≡ *Agave dentiens* Trelease (1912).

[2h] Differs from ssp. *cerulata*: **Ros** 0.5 - 0.7 × 0.8 - 1.5 m; **L** long-acuminate, 40 - 55 (-70) cm, light glaucous grey, margins nearly straight to mildly undulate; marginal teeth small, weakly attached, friable, 1 - 2 mm, or nearly toothless; terminal **Sp** acicular, 3 - 5 cm, brown to grey; **Inf** broad, part-**Inf** 8 - 18 in the upper ½ of the **Inf**, on 30 - 40 cm long **Br**.

A. cerulata ssp. **nelsonii** (Trelease) Gentry (Occas. Pap. Calif. Acad. Sci. 130: 44, 1978). **T:** Mexico, Baja California (*Nelson & Goldman* 7111 [US]). – **D:** Mexico (C Baja California); igneous highlands. **I:** Gentry (1982: 356, 372-373). **Fig. II.b**

≡ *Agave nelsonii* Trelease (1912); **incl.** *Agave shawii* E. C. Nelson (1911) (*nom. illeg.*, Art. 53.1).

[2h] Differs from ssp. *cerulata*: **Ros** 0.5 - 0.75 m ∅; **L** short-acuminate, 20 - 40 × 6 - 8 cm, mostly 3 - 6× as long as broad, mostly light grey to bluish-glaucous over green, margins undulate with small prominences or nearly straight; marginal teeth firmly attached, larger, 3 - 9 mm, frequently on small tubercles; terminal **Sp** strongly subulate, 2 - 4 cm; **Inf** with 15 - 20 part-**Inf**.

A. cerulata ssp. **subcerulata** Gentry (Occas. Pap. Calif. Acad. Sci. 130: 44-48, ills. (pp. 46-47, 49), 1978). **T:** Mexico, Baja California Sur (*Gentry* 10330 [US, DES, MEXU]). – **D:** Mexico (N Baja California Sur). **I:** Gentry (1982: 356, 373). **Fig. II.c**

[2h] Differs from ssp. *cerulata*: **Ros** 0.15 - 0.3 × 0.3 - 0.5 m; **L** short-acuminate, 15 - 30 × 2.5 - 7 cm, mostly 3 - 6× as long as broad, mostly light grey to bluish-glaucous over green, margins conspicuously crenate with prominent tubercles; marginal teeth well developed, weakly attached, larger, 3 - 8 mm in the middle of the lamina; terminal **Sp** subulate, usually sinuous, 2 - 4 cm.

Resembles *A. subsimplex* from the opposite Sonoran coast (Gentry 1982: 375).

A. chamelensis (E. J. Lott & Verhoek-Williams) Thiede & Eggli (KuaS 50(5): 110, 1999). **T:** Mexico (*Lott & Wendt* 1663 [MICH, BH, CAS, MEXU]). – **D:** Mexico (Jalisco); uncommon along arroyos in tropical (semi-) deciduous forests, 50 - 75 m, flowers in December. **I:** Lott & Verhoek-Williams (1991: as *Manfreda*).

≡ *Manfreda chamelensis* E. J. Lott & Verhoek-Williams (1991).

[3a2] Plants reproducing vegetatively by buds from the rhizome; **R** fleshy; rhizome upright, cylindrical, 3 - 15 × 2 - 3 cm; **L** up to 9, spreading, narrowly channelled, nearly conduplicate near the base, base narrow, tip acute, brittle, herbaceous to somewhat fleshy, veins papillate on both faces, 37 - 77 (-91) × (1-) 1.6 - 4.8 (-6.5) cm, margins with a narrow yellowish cartilaginous band, minutely denticulate, teeth regular; remains of **L** bases membranous, not separating into fibres; **Inf** 0.75 - 1.2 (-2) m, 'spicate', flowering part 10 - 20 cm, with 10 - 25 (-35) sessile **Fl**; mature **Fl** nearly erect; **Ov** oblong to ovate, not protruding into the tube, 5 - 10 mm; **Tep** green, tube funnel-shaped, 6 - 13 mm, lobes oblong, reflexed to tightly revolute, 8 - 11 mm; **Sty** 25 - 35 mm exserted; **Sti** clavate, trigonous; **Fr** globose, 1.2 - 1.6 × 1 - 1.5 cm; **Se** ± cuneiform, 5 - 6 × 4 - 5 mm.

This species appears to be closest to *A. scabra* and *A. jaliscana*. It differs from both by shorter floral bracts, filaments curved near the tip at bud opening, and by the absence of coarse fibrous remains of old leaves, from the first-named by its shorter floral

tube and globose beakless capsules, and from the last-named by its wider leaves and shorter styles and filaments (Lott & Verhoek-Williams 1991). The tropical lowland habitat is untypical and apparently not known for any other species of Subgen. *Manfreda* except *A. littoralis*.

A. chiapensis Jacobi (Hamburg. Gart.- & Blumenzeit. 22: 213, 1866). **T** [neo]: Mexico, Chiapas (*Gentry* 12178 [US, DES, MEXU, MICH]). – **D:** Mexico (Chiapas). **I:** Gentry (1982: 217, 225).

Incl. *Agave chiapensis* var. *major* hort. *ex* A. Berger (1915); **incl.** *Agave teopiscana* Matuda (1974).

[1f] Stem short; **Ros** openly spreading, robust, medium-sized, caespitose; **L** variable, ovate, narrowed near the base, shortly acuminate, rounded below, plane to slightly hollowed above and upcurving, smooth, mostly 30 - 50 × 7 - 16 cm, light shiny grey-green, margins slightly undulate to crenate; larger marginal teeth deltoid, upcurved, 3 - 4 mm, closely spaced, or more remote and subulate, 5 - 10 mm, dark brown to greying; terminal **Sp** subulate, straight to sinuous, strong, 2 - 3.5 cm; **Inf** ± 2 m, 'spicate', with a long scape, **Fl** in the terminal ¼ - ⅓; **Fl** trigonous, fleshy, 60 - 70 mm, obscured in large tufts of broad-based **Bra**; **Ov** grooved to the base, 20 - 30 mm; **Tep** yellow or green flushed with reddish or purple, tube funnel-shaped, 8 - 12 mm, lobes unequal, 30 - 32 mm.

Appears to be closely related to *A. warelliana*, which differs in its closely serrate marginal teeth on a red margin (Gentry 1982: 226).

A. chrysantha Peebles (Proc. Biol. Soc. Wash. 48(4): 139, 1935). **T:** USA, Arizona (*Peebles & Harrison* 5543 [US]). – **Lit:** Turner & al. (1995). **D:** USA (C Arizona); granitic and volcanic mountain slopes, 900 - 1800 m. **I:** Gentry (1982: 421, 426-427).

≡ *Agave palmeri* var. *chrysantha* (Peebles) Little *ex* Benson (1943) ≡ *Agave palmeri* ssp. *chrysantha* (Peebles) B. Ullrich (1992).

[2g] **Ros** small and compact to rather large and open, 0.5 - 1 × 0.8 - 1.8 m, usually solitary; **L** linear-lanceolate to lanceolate, straight, usually only a little narrowed below the middle, widest in the middle, deeply guttered, mostly 40 - 75 × 8 - 10 cm, greyish to yellowish-green, margins nearly straight to repand; larger marginal teeth straight or flexed, 5 - 10 mm, ± 1 - 3 cm apart, smaller towards the base and with small intermittent teeth; terminal **Sp** slender, openly grooved above, 2.5 - 4.5 cm, brown or castaneous to grey in age, decurrent for 5 - 15 cm to the upper teeth; **Inf** 4 - 7 m, 'paniculate', small, narrow, part-**Inf** small, congested, 8 - 18 in the upper ¼ - ⅓ of the **Inf**; **Fl** 40 - 55 mm; **Ov** slender, 22 - 30 mm, neck short, constricted; **Tep** yellow, rarely red-tipped, tube 8 - 13 mm, lobes dimorphic, outer 9 - 15 mm. – *Cytology:* 2n = 60.

Benson (1943) and Ullrich (1992e) both suggest

infraspecific ranks under its closest largely allopatric relative *A. palmeri*, with which it shows introgression where they meet (Gentry 1982: 429, 446). The species hybridizes with *A. murpheyi*, *A. palmeri*, *A. parryi* var. *couesii* and *A. delamateri* (Hodgson 1999).

A. chrysoglossa I. M. Johnston (Proc. Calif. Acad. Sci., Ser. 4, 12: 998-999, 1924). **T:** Mexico, Baja California (*Johnston* 3123 [CAS]). – **Lit:** Turner & al. (1995). **D:** Mexico (Sonora); on often bare rocks in hot coastal and lowland regions. **I:** Gentry (1982: 67, 73). **Fig. II.d**

[1b] Stem short; **Ros** openly spreading, 1 - 1.3 × 2 - 2.4 m, mostly solitary, sometimes suckering profusely; **L** few, linear-lanceolate, straight or slightly curved, deflexed at maturity, convex below, flat above, smooth, 70 - 120 × 4 - 7 cm (wider at the base), light green, margins fragile, 1 mm wide, brown; marginal teeth none; terminal **Sp** acicular, with a short fine groove at the base above, 2 - 4 cm, brown, aging greyish; **Inf** mostly 2 - 4 m, 'spicate', densely flowered in the upper ¾, part-**Inf** with geminate **Fl**; **Ped** bifurcate, 10 - 15 mm; **Fl** 35 - 45 mm; **Ov** slender, 16 - 20 mm incl. a 3 - 5 mm long neck; **Tep** yellow, tube shallow, 4 - 4.5 mm, lobes ± equal, outcurved at anthesis, 14 - 16 mm.

Closely related to *A. vilmoriniana*, but with straight narrow plane leaves and without bulbils in the inflorescence. It represents the xerophytic northern coastal / lowland relative of the Group *Serrulatae* (as *Amolae*) (Gentry 1982: 74-75).

A. cocui Trelease (Mem. Nation. Acad. Sci. 11: 19, t. 5-7, 1913). **T:** Venezuela (*Zuloaga* s.n. [MO?]). – **Lit:** Hummelinck (1936). **D:** Venezuela, Colombia; mainly coastal; also on the Leeward Islands (Aruba, Bonaire, Curaçao, Margarita), probably introduced. **I:** Hummelinck (1993).

Incl. *Agave americana* Humboldt (1808) (*nom. illeg.*, Art. 53.1); **incl.** *Agave cocui* var. *cucutensis* Hummelinck (1936); **incl.** *Agave cocui* var. *laguayrensis* Hummelinck (1936).

[2q] **Ros** solitary; **L** broadly lanceolate, guttered, sharply acute or subacuminate, deeply and sometimes tortuously concave, (80-) 100 - 120 (-140) × ± 30 cm, glaucous, soon green and glossy, margins concave; marginal teeth mostly upcurved above and recurved below, acuminately triangular or from lunate bases on green or at length hardening prominences, 2.5 - 6 mm, reddish chestnut-brown, usually 10 - 20 mm apart; terminal **Sp** triquetrously conical, shallowly grooved below the middle and involute below, smooth, (1-) 1.2 - 2 (-3) cm, red-brown, decurrent and dorsally immersed into the green tissue; **Inf** 5 - 10 m, 'paniculate', narrowly oblong, part-**Inf** on nearly horizontal **Br**; **Fl** 40 - 65 mm; **Ov** 25 - 40 mm; **Tep** yellow, tube openly conical, 3 - 7 mm, lobes ± 18 - 25 mm; **Fr** oblong, little stipitate or beaked, 4 - 5 × 1.7 - 2.5 cm.

Not always clearly separated from *A. vicina* (as *A. vivipara*) (Hummelinck 1938).

A. colimana Gentry (CSJA 40: 212-213, ills., 1968). **T:** Mexico, Colima (*Gentry* 18325 [US, DES, MEXU]). – **Lit:** McVaugh (1989). **D:** Mexico (SW Jalisco, Colima, Michoacán); primarily coastal on rocky sites, or more inland in tropical deciduous forest, 0 - 1000 m. **I:** Gentry (1982: 104, 107-108).

Incl. *Agave ortgiesiana* Roezl (1871); **incl.** *Agave angustissima* var. *ortgiesiana* Trelease (1920).

[1d] Shortly caulescent; **Ros** 0.4 - 0.6 × 1 - 1.2 m, solitary; **L** many, linear, straight, slightly narrowed above the base, widest near the middle, thin and flat above, smooth, 40 - 70 × 1 - 2.5 cm, green, margins narrow, brown, filiferous with fine long brown threads; terminal **Sp** weak, short, 5 - 8 mm, greyish-brown to dark brown, decurrent into the **L** margin; **Inf** 2 - 3 m, 'spicate', slender, flowering from ± 1 m above the base, part-**Inf** not crowded, with geminate **Fl**; **Ped** 10 - 15 mm; **Fl** 40 - 50 mm; **Ov** 14 - 20 mm, greenish-yellow, neck 4 - 7 mm, slightly constricted; **Tep** pale yellow or lavender, tube narrow, 9 - 17 mm, lobes nearly equal, 14 - 19 mm.

Distinctive with its elongate leaves and deep narrow flower tube, but sometimes approaching *A. filifera* ssp. *schidigera* in flower tube length and *A. filifera* ssp. *multifilifera* in leaf characters (Gentry 1982: 103). Gentry (l.c.) and also McVaugh (1989: 135) regard the earlier name *A. ortgiesiana* as invalid but referring to the same species. *A. ortgiesiana* was published in an excerpt of a letter, and if the name must be regarded as valid, as advocated by Ullrich (1991a), *A. colimana* should be formally proposed for conservation.

A. colorata Gentry (Publ. Carnegie Inst. Washington 527: 93, 1942). **T:** Mexico, Sonora (*Gentry* 3050 [CAS, ARIZ]). – **Lit:** Turner & al. (1995). **D:** Mexico (Sonora, N Sinaloa); foothills or coastal regions, open rocky sites in thorn forest. **I:** Gentry (1982: 421, 432); KuaS 44(10): centre page pullout 1993/10. **Fig. I.f**

[2g] **Ros** compact, small to medium-sized, sparingly suckering; **L** few, ovate, shortly acuminate to lanceolate, thick, firm, convex below towards the base, plane to concave above, asperous, 25 - 60 × 12 - 18 cm, light grey, glaucous, frequently cross-zoned and red-tinted, margins prominently crenate or mamillate; marginal teeth straight or flexuous, mostly 5 - 10 mm (in the middle of the lamina), brown to greyish, 1.5 - 3 cm apart, smaller below; terminal **Sp** subulate, straight or flexuous, narrowly grooved above in the upper ½, mostly 3 - 5 cm, brown to grey; **Inf** 2 - 3 m, 'paniculate', narrow, part-**Inf** densely flowered, 15 - 20 in the upper ⅓ - ½ of the **Inf**; **Fl** 50 - 70 mm; **Ov** 25 - 40 mm, pale green, neck short, not constricted; **Tep** reddish in bud, opening yellow, apex usually remaining red-

dish, tube 15 - 20 mm, lobes unequal, outer 12 - 16 mm.

The closest relative, both morphologically and geographically, is *A. shrevei* (Gentry 1982: 431).

A. confertiflora Thiede & Eggli (KuaS 50(5): 111, 1999). **T:** Mexico, Chihuahua (*Hartman* 536 [US?]). – **D:** Mexico (Chihuahua).

Incl. *Bravoa densiflora* B. L. Robinson & Fernald (1895) ≡ *Pseudobravoa densiflora* (B. L. Robinson & Fernald) Rose (1899) ≡ *Polianthes densiflora* (B. L. Robinson & Fernald) Shinners (1966).

[3b2] **R** numerous, spreading, thickened; stem bulb-like, oblong, to 5 × ≥ 2.5 cm; **L** linear, attenuate, 7.5 - 10 × 0.25 cm, scape **L** reduced to **Bra**, 2.5 - 5 cm, with broad scarious and attenuate tips, floral **Bra** similar; **Inf** 'spicate', short, dense; **Fl** solitary, slender, spreading, curved, 43 - 55 mm; **Tep** pulverulent-tomentose on the outer face, dull yellow (dry), tube scarcely widened, throat oblique, lobes erect, ovate, obtuse, with a tuft of short white **Ha** at the tip, 2.5 - 4 mm; **Fr** (immature) ovoid, ≥ 0.5 cm ∅; **Se** unknown.

A hardly known but seemingly very distinct species, apparently known only from the type collection. When *Polianthes densiflora* is transferred to *Agave*, a new name is necessary to avoid homonymy with *A. densiflora* Hooker.

A. congesta Gentry (Agaves Cont. North Amer., 476-479, ills., 1982). **T:** Mexico, Chiapas (*Gentry* 23651 [US, DES, MEXU]). – **D:** Mexico (Chiapas); widely scattered in pine-oak forest, 2150 - 2480 m.

[2k] Stem short; **Ros** compact, to 1 × 1 - 2 m, solitary; **L** lanceolate to lanceolate-spatulate, at first curved-ascending, then horizontally spreading, thick, (shortly) acuminate, plane, (40-) 70 - 120 × 10 - 22 cm, green to yellow-green, sometimes faintly glaucous or pruinose; margins undulate to crenate, variously mamillate; marginal teeth straight to variously curved, moderate to rather large, dark to greyish-brown, usually remote, 3 - 5 cm apart, mostly on cusps, 5 - 10 mm, base broad and low; terminal **Sp** stout, base very broad, widely flatly grooved above, 3 - 7 cm, grey to chestnut-brown, sharply decurrent to the upper teeth; **Inf** 6 - 8 m, 'paniculate', straight, part-**Inf** as congested rounded clusters, 40 - 50 per **Inf**; **Fl** 55 - 70 mm; **Ov** 30 - 40 mm, neck short; **Tep** orange to reddish or purplish, opening yellow, tube deeply funnel-shaped, 10 - 13 mm, lobes unequal, 17 - 25 mm.

The closely related *A. hiemiflora* is distinguished by its smaller rosettes with fewer leaves and less congested flowers with less dimorphic and paler tepals (Gentry 1982: 479).

A. cundinamarcensis A. Berger (Agaven, 222, 1915). **T:** Colombia, Cundinamarca (*Wercklé* s.n. [not indicated]). – **D:** Colombia (Cundinamarca: Río Magdalena drainage).

[2o?] **Ros** solitary; **L** very thick above the base, only 15 cm broad, straightly spreading, then curved upwards and rapidly becoming broader, then again curved outwards, but the last 15 cm again curved upwards, 200 × 45 cm, steel-grey on yellowish-green ground; marginal teeth flat and broad, nearly obtuse, very short, hardly pungent; terminal **Sp** rather short; **Inf** 'paniculate', part-**Inf** sparingly bulbilliferous.

Hardly known and possibly a redescription of the first species described from Colombia, *A. wallisii*. Berger (1915) placed both species in his Ser. *Columbianae*, which he regards as possibly closest to the *Caribaeae*. Neither of these 2 species appears close to the geographically adjacent Colombian-Venezuelan *A. cocui* from Group *Vicinae*.

A. cupreata Trelease & A. Berger (Agaven, 197, 1915). **T:** Mexico, Michoacán / Guerrero (*Langlassé* 867 [B [status?], MEXU, US]). – **D:** Mexico (Michoacán, Guerrero); mountain slopes, 1220 - 1850 m. **I:** Gentry (1982: 325, 336-337). **Fig. II.f**

[2c] **Ros** caulescent, openly spreading, medium-sized, solitary; **L** broadly lanceolate or ovate, thick-fleshy, strongly narrowed at the base, plane to slightly concave above, 40 - 80 × 18 - 20 cm, bright shiny green, margins deeply crenate-mamillate; marginal teeth straight to curved, strongly flattened, dimorphic, larger teeth 10 - 15 mm on prominences, 3 - 6 cm apart, smaller teeth in the intersinuses of the **L** margin, of varying sizes, copper-coloured to grey; terminal **Sp** slender, sinuous, 3 - 5 cm, light brown to greyish, with a sharp border decurrent to the upper teeth; **Inf** 4 - 7 m, 'paniculate', rather broad, part-**Inf** lax, diffuse, 14 - 25 in the upper ½ of the **Inf**; **Fl** 55 - 60 mm; **Ov** 30 - 35 mm, olive-green, neck constricted; **Tep** rufous in bud, open orange-yellow, tube broadly funnel-shaped, 6 - 7 mm, lobes subequal, 20 - 21 mm.

Distinguished by the broad, shiny green leaves with high prominences and conspicuous patterns from the central bud, which are bright copper-coloured in early stages (Gentry 1982: 335).

A. dasylirioides Jacobi & Bouché (Hamburg. Gart.- & Blumenzeit. 21: 344, 1865). **T:** "Guatemala" (*Warszewicz* s.n. [B]). – **Lit:** Ullrich (1990g: with ills.). **D:** Mexico (San Luis Potosí, Morelos); cliffs on mountain slopes, mixed pine and hardwood forest, 1500 - 2200 m. **I:** Gentry (1982: 236, 240-241).

Incl. *Agave dealbata* Lemaire *ex* Jacobi (1865) ≡ *Agave dasylirioides* var. *dealbata* (Lemaire *ex* Jacobi) Baker (1877); **incl.** *Agave intrepida* Greenman (1899).

[1a] **Ros** symmetrical, 0.3 - 0.5 × 0.6 - 1 m, generally solitary; **L** 70 - 100, linear-lanceolate, straightly spreading but pliable, relatively thin, scarcely succulent, plane above, mostly 40 - 60 × 2 - 3 cm, glaucous-green, smoothly striate below and above, margins 1 mm wide, pale yellowish-white,

minutely serrulate; terminal **Sp** acicular, 0.5 - 1.5 cm, reddish-brown; **Inf** 1.5 - 2 m, 'spicate', arching, with **Fl** in the upper ⅓ - ½, part-**Inf** with 1 - 2 **Fl**; **Fl** persistent; **Ov** linear-tapered, 9 - 12 mm, neckless; **Tep** greenish-yellow, tube funnel-shaped, 8 - 12 mm, lobes equal, 9 - 11 mm.

The type of this name was erroneously said to have been collected in Guatemala. Gentry (1982: 241) assumed this species to hold a very basal position in the genus based on its 'primitive' features (leaves serrulate and scarcely succulent, inflorescences relatively simple, ovary incompletely inferior, tepals all equal, lobes nearly of equal length), and this assumption is principally confirmed by molecular data (Bogler & Simpson 1996). See also the comment for *A. petrophila*.

A. datylio Simon *ex* F. A. C. Weber (Bull. Mus. Hist. Nat. (Paris) 8: 224, 1902). **T** [neo]: Mexico, Baja California (*Gentry & Arguelles* 11200 [US, DES, MEXU, MICH]). – **Lit:** Turner & al. (1995). **D:** Mexico (Baja California Sur).

Without close relatives in Group *Vivipara* (as *Rigidae*) (Gentry 1982: 572).

A. datylio var. **datylio** – **D:** Mexico (Baja California Sur: Cape region); granitic sandy soils at lower elevations. **I:** Gentry (1982: 571).

[2f] **Ros** 0.6 - 1 × 1 - 1.5 m, suckering freely, rhizomes frequently elongate; **L** radiately spreading, lanceolate-linear, rather rigid, rounded below, canaliculate above, 50 - 80 × 3 - 4 cm, green to yellowish-green, young somewhat glaucous, margins nearly straight; marginal teeth deltoid, flattened, rather blunt, mostly 3 - 5 mm, dark brown, usually remote or 3 - 6 cm apart, more closely spaced below; terminal **Sp** conical to subulate, scarcely or flatly grooved above, large, 2.5 - 4 cm, dark brown to greyish, shortly decurrent; **Inf** 3 - 5 m, 'paniculate', part-**Inf** small, 8 - 15 in the upper ½ of the **Inf**; **Fl** 40 - 55 mm; **Ov** 20 - 30 mm; **Tep** greenish-yellow, tube funnel-shaped, 5 - 10 mm, lobes 15 - 20 mm.

A. datylio var. **vexans** (Trelease) I. M. Johnston (Proc. Calif. Acad. Sci., Ser. 4, 12: 1001, 1924). – **D:** Mexico (Baja California Sur); sandy soils at lower elevations. **I:** Gentry (1982: 572).

≡ *Agave vexans* Trelease (1912).

[2f] Differs from var. *datylio*: **Ros** smaller; **L** smaller, 30 - 50 cm, more glaucous or yellowish. – *Cytology:* 2n = 174.

This appears to represent the xerophytic ecotype of the species (Gentry 1982: 572).

A. debilis A. Berger (Agaven, 33, 1915). **T** [lecto]: Mexico, Oaxaca (*Pringle* 4745 [US, BM, BR, G, GH, LE, M, MEXU, P]). – **Lit:** McVaugh (1989). **D:** Mexico (D.F., Hidalgo, México, Morelos, Michoacán, Oaxaca, Puebla); moist pine forests or

pine-oak ericaceous woods, 1830 - 3960 m, flowers mid-July to mid-September.

Incl. *Manfreda angustifolia* Rose *in sched.* (s.a.) (*nom. inval.*, Art. 29.1); **incl.** *Manfreda pringlei* Rose (1903); **incl.** *Polianthes debilis* Shinners (1966).

[3a3] Plants of moderate size (for Subgen. *Manfreda*), reproducing vegetatively by horizontal rhizomes with plantlets at the tips; rhizome cylindrical or rarely ovoid, (1-) 2 - 4 × 1.3 - 2 cm; **R** half-fleshy, fibrous, vertical; **L** 2 - 6 (-8), erect-spreading, linear-lanceolate, somewhat succulent, slightly channelled, occasionally gently undulate, tip acute, with medium-sized point, smooth, 14 - 56 × 0.8 - 2.2 (-2.8) cm, dark green, sometimes spotted, at times red-speckled on the lower face near the base, margins narrow, hyaline, sometimes streaked with purplish-red, papillate to erose-denticulate, usually rough to the touch; remains of the **L** bases fibrous, 4 - 9 cm; **Inf** (28-) 91.3 - 135 (-149) cm, 'spicate', flowering part short, crowded, 2.5 - 7.5 (-11) cm, with 5 - 18 sessile **Fl**; **Fl** nearly erect, slightly curved at the junction of **Ov** and **Tep** tube; **Ov** ellipsoid, 9 - 11 (-12) mm; **Tep** tube cylindrical, slightly widened towards the mouth, not constricted above the **Ov**, 9 - 15 × 3 - 5 (in the middle) mm; **Tep** lobes oblong, revolute, 8 - 13 (-16) mm, green or purple; **Sty** exserted for 30 - 48 mm; **Sti** clavate, trigonous, shallowly furrowed; **Fr** globose, 1.5 - 1.8 × 1.3 - 1.7 cm; **Se** 3 - 4 × 4 - 5 mm.

This species shares many floral and fruit characters with *A. guttata*, but is nevertheless easily distinguished by its longer, narrower and more pliable and herbaceous leaves and its distribution. It may represent the moist-forest counterpart of the more xeromorphic *A. guttata* (Verhoek-Williams 1975: 252-253).

A. decipiens Baker (BMI 1892: 183, 1892). **T:** [neo – icono]: Curtis's Bot. Mag. 122: t. 7477, 1896, sub *A. laxifolia*. – **D:** USA (Florida); coastal sands. **I:** Gentry (1982: 555, 574).

Incl. *Agave laxifolia* Baker (1896); **incl.** *Agave spiralis* Brandegee *ex* A. Berger (1912).

[2f] Arborescent, trunk 1 - 3 m, very broad through bulging **L** bases; **Ros** extending down for some distance from the stem tip; **L** narrowly lanceolate, rigidly spreading to recurving, fleshy, narrowed at the thickened base, long-acuminate, concave, mostly 75 - 100 × 7 - 10 cm, green, margins repand; marginal teeth 2 - 3 mm (in the middle of the lamina), dark brown, 1 - 2 cm apart, on low prominences, slender cusps upcurving, with few smaller intermittent teeth; terminal **Sp** conical, ungrooved, 1 - 2 cm, dark brown, not decurrent; **Inf** 3 - 5 m, 'paniculate', part-**Inf** 10 - 12 and more in the upper ½ of the **Inf**, often bulbilliferous; **Fl** 60 - 80 mm, foetid; **Ov** large and thick, 40 - 48 mm, neckless; **Tep** greenish-yellow, tube funnel-shaped, 11 - 13 mm, lobes subequal, 18 - 22 mm.

Geographically isolated from the remainder of the genus (except *A. neglecta*). The taxon is most probably of cultivated origin. It was reported to occur on old Indian village sites in 1933 and may well represent an old pre-Columbian food or fibre plant, comparable to *A. delamateri*. It is said to reach 4 m in height with leaves 2 m long in fertile soil (Gentry 1982: 573).

A. delamateri W. C. Hodgson & Slauson (Haseltonia 3: 130-140, ills., 1995). **T:** USA, Arizona (*Hodgson* 5478 [DES, ASU]). – **D:** USA (C Arizona); open steep slopes, 725 - 1550 m. **I:** Hodgson (1999).

Incl. *Agave repanda* Trelease *ex* Gentry (1982) (*nom. inval.*, Art. 34.1c).

[2g] **Ros** ± 1 × 1 m, caespitose; **L** lanceolate, erect, broadest near or just below the middle, acuminate, inwardly arcuate at the tip, guttered above, mostly 50 - 63 (-74) × 7.5 - 9 cm, bluish-grey glaucous with purple-maroon tinge and green cross-banding, margins straight to repand; marginal teeth variable, usually reflexed, becoming porrect near the **L** base, larger teeth 3.5 - 5 mm, smaller teeth 1 - 1.5 mm, dark glossy brown to grey and pruinose (esp. towards the tip), 1.5 (near the **L** base) - 11 (-40) mm apart; terminal **Sp** 2.8 - 3.5 (-4.9) cm, brownish-grey, decurrent for ⅙ - ⅓ of the **L** length; **Inf** 4.5 - 6 m, 'paniculate', broad, open, part-**Inf** widely spaced, 12 - 17 in the upper ⅗ - ⅝ of the **Inf**, with 14 - 20 **Fl** each; **Fl** long-lived, 47 - 67 mm; **Ov** 21 - 29 mm, neck 1 - 3.5 mm; **Tep** pale cream tinged light green, tube 11 - 16 mm, lobes unequal, 14 - 18 mm.

According to the protologue already collected around 1920 and recognized as distinct by Trelease who used the unpublished name *A. repanda*. It is most closely related to the allopatric *A. fortiflora* and *A. palmeri*, but distinguished esp. by its numerous rhizomatous offsets, easily cut leaves, and 1- (instead of 2-) seriate filaments. It hybridizes with *A. chrysantha* (Hodgson 1999).

A. delamateri is regarded as a pre-Columbian food or fibre plant that originated farther S in Mexico since it occurs in direct or indirect association with archaeological features (cf. protologue). A further such (undescribed) taxon from the Grand Canyon region in Arizona is mentioned by Hodgson (1999).

A. deserti Engelmann (Trans. Acad. Sci. St. Louis 3: 310-311, 370, 1875). **T** [syn]: USA, California (*Emory* s.n. [MO]). – **Lit:** Turner & al. (1995). **D:** SW USA, NW Mexico.

A large and variable complex with hard-to-define limits, and difficult to separate from *A. cerulata* (Gentry 1982: 376). *A. aquariensis* Trelease *ex* Gentry 1970 is either a synonym of *A. deserti* or of *A. subsimplex* (Gentry 1982: 390).

A. deserti ssp. **deserti** – **D:** USA (Arizona), Mexico (Baja California). **I:** Gentry (1982: 356, 377-378). **Fig. II.a, II.e**

Incl. *Agave deserti* Orcutt (1883) (*nom. illeg.*, Art. 53.1); incl. *Agave consociata* Trelease (1912).

[2h] **Ros** mostly 30 - 50 × 40 - 60 cm, sparingly or prolifically suckering; **L** variable, lanceolate to linear-lanceolate, thick, rigid, scarcely narrowed above the broad clasping base, moderately acuminate, convex below, concave above, mostly 25 - 40 × 6 - 8 cm, 4 - 7× as long as broad, grey to bluish-glaucous, often cross-zoned, margins usually straight; marginal teeth usually regularly spaced, loosely attached, smaller teeth 2 - 3 mm, longer teeth 6 - 8 mm, grey, mostly 15 - 30 mm apart, slender-cusped; terminal **Sp** strong, generally 2 - 4 cm, light brown to greyish, decurrent to the 1. or 2. tooth; **Inf** 2.5 - 4 m, 'paniculate', part-**Inf** small, 6 - 15 in the upper ⅕ - ¼ of the **Inf**; **Fl** 40 - 60 mm; **Ov** 22 - 40 mm, neck slightly narrowed, 4 - 6 mm; **Tep** yellow, tube 4 - 6 mm, lined with a nectariferous disk, lobes equal, 14 - 20 mm. – *Cytology:* 2n = 118.

A. deserti ssp. **pringlei** (Engelmann *ex* Orcutt) Gentry (Occas. Pap. Calif. Acad. Sci. 130: 20, 1978). **T:** Mexico, Baja California Sur (*Orcutt* s.n. [K, MEXU]). – **D:** Mexico (Baja California). **I:** Gentry (1982: 378, 381).

≡ *Agave pringlei* Engelmann *ex* Orcutt (1883); incl. *Agave scaberrima* Hort. Peacock *ex* Baker (1888).

[2h] Differs from ssp. *deserti*: **Ros** 40 - 70 × 50 - 80 cm; **L** very long-acuminate, mostly 40 - 70 × 5 - 7 cm, 8 - 12× as long as broad, green to yellowish-green or light glaucous-grey, margins straight; marginal teeth firmly attached; terminal **Sp** acicular, 3 - 4 cm, conspicuously decurrent in a horny margin frequently extending to the middle of the lamina or even below; **Inf** 3 - 6 m; **Tep** tube 5 - 8 mm.

Gentry (1982: 380) wrongly ascribes the basionym name to 'Engelmann *ex* Baker'.

A. deserti ssp. **simplex** Gentry (Occas. Pap. Calif. Acad. Sci. 130: 22, ills. (pp. 23-24), 1978). **T:** USA, Arizona (*Gentry* 23404 [US, ARIZ, DES, MEXU]). – **D:** USA (Arizona, California), Mexico (Sonora); low desert scrub, 350 - 1200 m. **I:** Gentry (1982: 356, 383-384). **Fig. II.g**

[2h] Differs from ssp. *deserti*: **Ros** generally solitary, rarely with 1 - 3 offsets; **L** moderately acuminate, mostly 25 - 40 (-50) × 6.5 - 10 cm, 4 - 7× as long as broad, margins usually straight; marginal teeth weakly attached; terminal **Sp** decurrent as a horny margin only to the 1. or 2. tooth; **Inf** 4 - 6 m; **Tep** in bud pale yellow to ferrugineous, tube 5 - 10 mm.

Hybridizing with *A. schottii* ssp. *schottii*, and possibly with *A. mckelveyana* (Hodgson 1999).

A. desmetiana Jacobi (Hamburg. Gart.- & Blumenzeit. 22: 217, fig. 32, 1866). **T** [neo]: Mexico, Sinaloa (*Gentry* 11569 [US, DES, MEXU]). – **D**: Cultivated only. **I**: Gentry (1982: 621, 623-624).

Incl. *Agave regeliana* Jacobi (1866) ≡ *Agave miradorensis* var. *regeliana* (Jacobi) A. Terracciano (1885) (*incorrect name*, Art. 11.1); **incl.** *Agave ananassoides* Jacobi (1868); **incl.** *Agave miradorensis* Jacobi (1868).

[2?] **Ros** 70 × 90 cm, surculose when young; **L** linear-lanceolate, arching, openly ascending, turgidly brittle, abruptly or gradually narrowed towards the base, 50 - 80 × 7 - 12 cm, dark to glaucous-green, margins smooth, without distinct coloration; marginal teeth none or small, regular, 1 - 2 mm, chestnut-brown, 1 - 2 cm apart or few and irregularly spaced; terminal **Sp** subulate, shortly and broadly grooved above, 2 - 3 cm, dark brown to reddish-brown; **Inf** 2.5 - 3 m, 'paniculate', long, narrow, part-**Inf** congested, 20 - 25 in the upper ½ - ⅔ of the **Inf**; **Fl** 40 - 60 mm; **Ov** shortly stipitate, small, 15 - 26 mm, green, neck very short, not constricted; **Tep** green in bud, open pale yellow, tube 10 - 12 mm, lobes 13 - 15 mm.

Distinguished by its smooth unarmed arching leaves and the short compact inflorescences with small flowers with a very short ovary and a broad tube (Gentry 1982: 623). It was originally introduced from Cuba as *A. anomala*, where it indeed may have originated (Ullrich 1990d). Its systematic position is unclear, since it does not fit well into any group; Gentry's placement in the former Group *Sisalanae* is regarded as artificial (Ullrich l.c.).

A. difformis A. Berger (Agaven, 95-96, 1915). **T**: US, K. – **D**: Mexico (San Luis Potosí, Hidalgo); coarse limestone rocky soils, arid side of the Sierra Madre Oriental, 1560 - 1875 m. **I**: Gentry (1982: 137-138).

[1g] **Ros** subcaulescent, open, rather vigorous, variable, 0.7 - 1 × 1 - 1.5 m, freely suckering; **L** polymorphic, straight or falcate or sinuous, stiffly ascending, thickly convex below, concave above, 50 - 80 × 4 - 6 cm, green to yellow-green, margins straight or undulate, firm or detachable, predominantly light grey; marginal teeth variable, generally 5 - 10 mm, dark brown to grey, 2 - 3 cm apart, rarely double, sometimes with smaller intermittent teeth, or reduced or entirely lacking; terminal **Sp** conical-subulate, stout, 1.5 - 3 cm, dark brown to grey; **Inf** 3.5 - 5 m, 'spicate', slender, scape waxy-glaucous, **Fl** in the upper ⅓ - ½ of the **Inf**; **Fl** 30 - 40 mm; **Ov** 15 - 21 mm, green, neck short; **Tep** light green to yellow and pink, tube 2.5 - 3.5 mm, lobes equal, 15 - 18 mm.

A robust species, within the Group *Marginatae* characterized by its polymorphic long-ensiform leaves (Gentry 1982: 137).

A. dolichantha Thiede & Eggli (KuaS 50(5): 111,

1999). **T**: Mexico, Jalisco (*Rose & Hay* 6290 [US]). – **D**: Mexico (Jalisco). **I**: Cházaro Basáñez & Machuca Núñez (1995: as *Polianthes longiflora*).

Incl. *Polianthes longiflora* Rose (1903).

[3b1] Plants glabrous; basal (= **Ros**) **L** unknown, tips of the bulb-**Sc** coarsely fibrous; **Inf** 'spicate', perhaps 1 m tall, with 2 - 8 **Fl** in 2 - 4 pairs well separated along the axis, or 2 - 4 **Fl** in an apical 'cluster', scape somewhat red-spotted; **Fl** sessile, fragrant at anthesis; **Ov** 9 - 16 (-19) mm; **Tep** pink in bud, at anthesis white or tinged with purple, tube 60 - 100 mm, basal portion erect or nearly so, narrowly tubular, 2 mm ∅, gradually funnel-shaped and dilated in the distal ½ or ⅓, curved outwards at or above the middle, mouth oblique, lobes elliptic, 15 - 20 (-26) mm; **Anth** tips scarcely surpassing the **Tep** tube; **Sty** 80 - 100 mm, lobes not exserted, flat, ± 1.5 mm; **Fr** unknown.

This species was long known only from incomplete flowering specimens offered for sale (McVaugh 1989: 253) until its recent discovery in the wild, see Cedano M. & al. (1993) and Cházaro Basáñez & Machuca Núñez (1995). Placed in *Agave*, a new name was necessary to avoid homonymy with *A. longiflora* (Rose) G. D. Rowley 1977.

A. duplicata Thiede & Eggli (KuaS 50(5): 111, 1999). **T**: Mexico, Michoacán (*Lexarza* s.n. [not preserved?]). – **Lit**: McVaugh (1989: as *Polianthes geminiflora*). **D**: Mexico.

Incl. *Bravoa geminiflora* Lexarza (1824) ≡ *Coetocapnia geminiflora* (Lexarza) Link & Otto (1828) ≡ *Polianthes geminiflora* (Lexarza) Rose (1903); **incl.** *Bravoa coetocapnia* Roemer (1847); **incl.** *Bravoa graminiflora* Hemsley (1884) (*nom. inval.*, Art. 61.1).

When transferring *Polianthes geminiflora* to *Agave*, a new name was necessary to avoid homonymy with *A. geminiflora* (Tagliabue) Ker Gawler 1817.

The hybrid of this species with *A. bulliana* (as *Prochnyanthes bulliana*) is known as *Bravoa* ×*kewensis* hort. and was first recorded 1889.

A. duplicata ssp. **clivicola** (McVaugh) Thiede & Eggli (KuaS 50(5): 111, 1999). **T**: Mexico, Jalisco (*Wilbur* 2133 [MICH]). – **D**: Mexico (Jalisco, Michoacán); mostly on steep shaded slopes, barrancas, gullies, in oak-pine forests, (900-) 1200 - 2150 m, flowers July to September (to October).

≡ *Polianthes geminiflora* var. *clivicola* McVaugh (1989).

[3b2] Differs from ssp. *duplicata*: **L** somewhat lustrous, flaccid, basal **L** mostly (15-) 25 - 30 (-48) × (0.8-) 1.5 - 2.5 (-3.7) cm, margins usually very narrowly revolute, with very thin pale scarious or hyaline edges, smooth and entire, rarely obscurely roughened; **Inf** 0.7 - 1.25 m, flowering part 20 - 40 (-60) cm, with 6 - 16 widely spaced flowering nodes.

A. duplicata ssp. **duplicata** – **D**: Mexico (Nayarit,

Guanajuato, Jalisco, Michoacán, Guerrero, México, Distrito Federal, Oaxaca, Hidalgo); mainly on rocky slopes in oak or pine forests, chiefly in the Trans-Mexican Volcanic Belt, (1000-) 2200 - 2800 m, flowers June to August (to November). **I:** Ic. Pl. Rar. Hort. Reg. Bot. Berol., 35, 1828.

[3b2] **R** fleshy, fascicled; **L** emerging from narrowly ovoid bulbs, 1 - 5 in basal **Ros** (or 1 - 3 additional a few cm above the base of the scape), ascending to prostrate, linear or broadly linear, widest in the middle, acute to long-attenuate at the tip, soft, mostly 15 - 30 (-50) × (0.15-) 0.5 - 1.5 cm, sometimes red-spotted near the base, margins thin, pale and scarious, sometimes revolute, usually somewhat erose, sometimes evenly papillose; **Inf** 0.5 - 0.9 (-1.4) m, 'spicate', flowering part 10 - 20 (-40) cm, with 4 - 12 (-16) widely spaced flowering nodes with paired **Fl** (often only 1 developing); **Ped** 4 - 6 (-8) mm (in **Fr** (5-) 8 - 13 mm), strongly ascending; **Ov** erect, ellipsoid; **Tep** pale red, coral-pink, red, orange-red or scarlet, sometimes green distally, tube at anthesis curving strongly outwards from near the base and **Fl** becoming almost horizontal or decurved, slender-terete basally, widened from near or below the middle, mostly 14 - 20 (-23) mm, lobes short, spreading, 1.5 - 3 × 1.5 - 3 mm (outer larger than inner); **Sty** with 3 flat flaring lobes < 1 mm; **Fr** shortly oblong or almost globose, ± 7 - 10 × 7 - 8 mm; **Se** wedge-shaped, sharply angled, 2.5 - 3 mm.

Rather widespread and variable (McVaugh 1989: 249).

A. duplicata ssp. **graminifolia** (Rose) Thiede & Eggli (KuaS 50(5): 112, 1999). **T:** Mexico, Jalisco (*Rose* 2571 [US]). – **D:** Mexico (S Zacatecas, Aguascalientes, Jalisco, Guanajuato?); grasslands, rocky slopes and grassy openings in oak forests, (1400?-) 2000 - 2250 m, flowers July to September.

≡ *Polianthes graminifolia* Rose (1903) ≡ *Polianthes geminiflora* var. *graminifolia* (Rose) McVaugh (1989).

[3b2] Differs from ssp. *duplicata*: Lower **L** face pectinately hispidulous on the veins and margins with thick erect blunt **Gl**-tipped **Ha** 0.1 - 0.2 mm long; exposed portion of **L** sheaths and to a lesser extent lower stem parts similarly pubescent.

Hardly different from ssp. *duplicata* except by the distinctive indumentum (McVaugh 1989).

A. durangensis Gentry (Agaves Cont. North Amer., 433-436, ills., 1982). **T:** Mexico, Durango (*Gentry & Gilly* 10576 [US]). – **D:** Mexico (S Durango, Zacatecas); rocky slopes and gravelly bajadas in grassland, 1700 - 2600 m.

[2g] Stem short; **Ros** 0.8 - 1.2 × 1.2 - 1.8 m, solitary or caespitose; **L** broadly lanceolate, narrowed above the broad base, widest in the middle, straight to outcurving, asperous, 40 - 90 × 14 - 22 cm, glaucous-grey, pruinose, margins heavily armed, deeply

crenate-mamillate; marginal teeth variously curved, prominent, broadly flattened, 1 - 2 cm, generally 1 - 2 cm apart; terminal **Sp** strong, broadly channelled above, 4 - 6 cm, pruinose-grey over brown; **Inf** 7 - 8 m, 'paniculate', open, scape short, axis zigzag, part-**Inf** sinuously spreading, trifurcate, small, in the upper ¾ of the **Inf**; **Fl** persistently erect, 60 - 80 mm; **Ov** 30 - 45 mm incl. the unconstricted neck; **Tep** yellow, tube cylindrical, 15 - 22 mm; lobes unequal, outer 10 - 12 mm.

Without close relationship to other species of Group *Ditepalae*. Vegetatively, it may be confused with the sympatric *A. scabra* of Group *Agave* (as *Americanae*) (Gentry 1982: 436).

A. dussiana Trelease (Mem. Nation. Acad. Sci. 11: 26-27, t. 28-29, 1913). **T:** Guadeloupe (*Duss* 3961 [NY]). – **D:** Windward Islands (St. Barts, Antigua, Montserrat, Guadeloupe, Martinique, Dominica). **I:** Succulenta 66: 188-189, 1987.

Incl. *Agave montserratensis* Trelease (1913).

[2o] **Ros** acaulescent, solitary; **L** oblong-lanceolate, erect, arching, slightly concave, abruptly or gradually acute, 100 - 175 × 15 cm, slightly greyish deep green, becoming blue-glaucous, then rather glossy, margins straight; marginal teeth curved, slender, or recurved-appressed in the middle of the lamina, somewhat lenticular at their bases, 2 - 3 (-5) mm; terminal **Sp** conically subulate, strongly recurved, base strongly involutely thickened, gradually pointed, 5 - 7 mm, black; **Inf** 5 - 9 m, 'paniculate', part-**Inf** without bulbils, but bulbils occasionally produced in the **Ax** of the lower **Bra** of the scape; **Fl** 60 - 65 mm; **Ov** oblong-fusiform, 30 - 35 mm; **Tep** yellow, tube open, ± 8 mm, lobes 25 mm; **Fr** narrowly to broadly oblong, stipitate at the base, slightly beaked at the tip, (2-) 3 - 4.5 × ± 2 cm.

The occurrence on Dominica was recently reported by Hill & James (1998).

A. ellemeetiana Jacobi (Hamburg. Gart.- & Blumenzeit. 21: 457, 1865). **T** [neo]: Ex cult. (*Anonymus* s.n. [K]). – **D:** Known from cultivation only. **I:** Gentry (1982: 95-96).

[1c] Stem nearly none; **Ros** open, 0.35 - 0.5 × 0.7 - 1 m, surculose; **L** rather few, ovate to oblong, somewhat recurved, reclining at maturity, thickly soft-succulent, widest in the middle, acuminate, plane below beyond the thick base, concave to plane above, smooth, 50 - 70 × 12 - 20 cm, light bright green, margins friable, smooth, sometimes reddish and finely serrulate towards the **L** tip; terminal **Sp** none, but **L** tip shortly acuminate and slightly calloused; **Inf** erect, 3 - 4.5 m, 'spicate', densely flowered from near the base, part-**Inf** usually with 4 **Fl**; **Ped** united in pairs, 15 - 20 mm; **Fl** campanulate, 28 - 40 mm; **Ov** 13 - 20 mm, neck conspicuously elongate; **Tep** pale greenish-yellow, tube very short, 1 - 2 mm, lobes 13 - 15 mm; **Fil** long exserted, 50 - 60 mm.

Introduced into cultivation from Mexico ± 1864 and apparently persisting in cultivation in Europe up to the present (Gentry 1982: 97). Very similar to some forms of *A. pedunculifera* of Group *Serrulatae* (as *Amolae*) (Ullrich 1990b).

A. ensifera Jacobi (Nachtr. Versuch syst. Glied. Agaveen 1: 138, 1868). **T** [neo]: Ex cult. La Mortola (*Berger* s.n. [US 1023791+1023763]). – **D:** Known from cultivation only.

    Incl. *Agave heteracantha* Baker (1877) (*nom. illeg.*, Art. 53.1); **incl.** *Agave lophantha* var. *latifolia* A. Berger (1915) ≡ *Agave univittata* var. *latifolia* (A. Berger) Breitung (1959).

    [1g] **Ros** dense, caespitose; **L** linear-lanceolate, ensiform, leathery-fleshy, strongly convex below and above, 50 - 60 × 4 - 5 (3.8 - 4 near the base) cm, smooth and dark green with a clear light stripe 5 - 7 mm wide, margins with a narrow grey border 0.5 - 1 mm wide; marginal teeth mostly antrorsely curved, 4 - 6 mm, light grey, closely set, 1 - 2 cm apart, interspersed with smaller teeth, altogether 30 - 40 teeth per side; terminal **Sp** short, basal groove above short and opening broadly with decurrent border, 1 - 1.5 cm, brown to grey; **Inf** 2 - 2.5 m, 'spicate', part-**Inf** mostly with **Fl** in pairs; **Ped** 2 - 3 mm; **Fl** 35 - 42 mm; **Ov** 20 - 24 mm, neck constricted, ± 3 mm; **Tep** light yellowish, tube open, short, 2 - 3 mm, lobes subequal, 14 - 17 mm.

Of unknown origin and commonly cultivated along the Mediterranean Riviera at Berger's time; apparently related to the *A. lechuguilla - A. difformis* group (Gentry 1982: 139).

A. evadens Trelease (Mem. Nation. Acad. Sci. 11: 20-21, t. 9-10, 116, 1913). **T:** Trinidad (*Crueger* 1333 [Herb. Urban]). – **Lit:** Hummelinck (1938). **D:** Trinidad.

    Incl. *Agave polyacantha* Baker (1888) (*nom. illeg.*, Art. 53.1); **incl.** *Agave vivipara* Hart (1890) (*nom. illeg.*, Art. 53.1); **incl.** *Agave polyantha* Dodge (1897).

    [2q] **Ros** shortly caulescent, somewhat suckering (?); **L** narrowly oblanceolate, gradually acute, openly concave or somewhat conduplicate, or with inrolled margins above, 70 - 100 cm, margins almost straight; marginal teeth small, 0.5 - 1.5 mm, rather close together; terminal **Sp** conical, straight or somewhat recurved, slightly involute at the base, 1 - 1.4 cm, slightly or not decurrent; **Inf** slender, 'paniculate', part-**Inf** few and lax, on ascending **Br**; **Fl** 47 - 55 mm; **Ov** oblong-fusiform, 25 mm; **Tep** colour not described, tube open, 2 - 3.5 mm, lobes 19 - 25 mm; **Fr** distinctly stipitate, ± 4 cm.

According to the protologue intermediate in foliage between *A. cocui* and *A. boldinghiana* and known to Trelease from photographs and dissociated flowers only. If the name *A. polyantha* Dodge really proves to be conspecific, it would have priority.

A. felgeri Gentry (US Dept. Agric. Handb. 399: 60-62, ills., 1972). **T:** Mexico, Sonora (*Gentry* 11343 [US, DES, MEXU, MICH]). – **Lit:** Ullrich (1991f: with ills.); Turner & al. (1995). **D:** Mexico (Sonora); arid desert lowlands near coasts. **I:** Gentry (1982: 108).

    [1d] **Ros** small, surculose, forming rather closely caespitose groups; **L** rather few, linear to narrowly lanceolate, straight or falcate, widest at the base, convex below, plane above, epidermis rugose or scabrous above, 25 - 35 × 0.7 - 1.5 cm, green to yellow-green, with faint imprints from the central bud, frequently with pale median stripe, margins with weakly filiferous narrow brown border, smooth; terminal **Sp** weak, small, 0.8 - 1.5 cm, grey; **Inf** 1.5 - 2.5 m, 'spicate', flowering in the upper ¼, part-**Inf** with 1 - 2 **Fl**; **Ped** strong, single or geminate, 2 - 5 mm; **Fl** 25 - 30 mm (dry, relaxed); **Ov** 12 - 14 mm; **Tep** yellow (?), tube 2 - 4 mm, lobes about equal, 10 - 12 mm.

Very similar to *A. schottii* from Group *Parviflorae* in vegetative features, but aligned with Group *Filiferae* due to its open shallow flower tube and long lobes (Gentry 1982: 109). Ullrich (l.c.) emphasizes vegetative and geographical criteria and suggests a placement in Group *Parviflorae*.

A. filifera Salm-Dyck (Hort. Dyck., 309, 1834). **T:** [neo – icono]: Ill. Hort. 7(4): t. 243, 1860. – **D:** Mexico.

A. filifera ssp. **filifera** – **D:** Mexico (San Luis Potosí, Aguascalientes, Guanajuato, Hidalgo, Querétaro, Michoacán, México, Veracruz). **I:** Gentry (1982: 104, 111).

    Incl. *Agave filifera* var. *elatior* hort. *ex* Besaucèle (s.a.); **incl.** *Agave filifera* var. *immaculata* hort. *ex* Besaucèle (s.a.); **incl.** *Agave filifera* var. *longifolia* hort. *ex* Besaucèle (s.a.); **incl.** *Agave filifera* var. *mediopicta* hort. *ex* Besaucèle (s.a.); **incl.** *Agave filifera* var. *splendens* hort. *ex* Besaucèle (s.a.); **incl.** *Agave filifera* var. *viridis* hort. *ex* Besaucèle (s.a.); **incl.** *Agave filifera* var. *candida superba* hort. *ex* Besaucèle (s.a.) (*nom. inval.*, Art. 24.2); **incl.** *Agave filamentosa* Salm-Dyck (1859) ≡ *Agave filifera* var. *filamentosa* (Salm-Dyck) Baker (1877); **incl.** *Agave pseudofilifera* Ross & Lanza (1892).

    [1d] **Ros** dense, small, forming large clumps with age; **L** many, lanceolate, straight, broadest in the middle, acuminate, thickened and convex above and below from the base to the middle of the lamina, 15 - 30 × 2 - 4 cm, green, with white impressions from the central bud, smooth, margins finely filiferous; terminal **Sp** flat above, rounded below, 1 - 2 cm, greyish; **Inf** 2 - 2.5 m, 'spicate', tapering, densely flowered in the upper ½, part-**Inf** mostly with **Fl** in pairs; **Ped** thick, short; **Fl** ascending-outcurving, 30 - 35 mm; **Ov** fusiform, 13 - 15 mm, neck furrowed; **Tep** reddish, tube funnel-shaped, furrowed, 5 - 6 mm, lobes equal, 14 mm. – *Cytology:* 2n = 60.

Separable from the closely related ssp. *schidigera* by its caespitose habit, shorter and thicker leaves, and smaller flowers with a shorter tube (Gentry 1982: 110). Ullrich (1992d) reduced *A. schidigera* and *A. multifilifera* to subspecies of *A. filifera*. This reclassification needs further field study.

A. filifera ssp. **microceps** Kimnach (CSJA 67(5): 306-310, ills., 1995). **T:** Mexico, Sinaloa (*Kimnach 1923* [HNT, MEXU, US]). – **D:** Mexico (Sinaloa).

Incl. *Agave filifera* var. *compacta* Trelease (1914).

[1d] Differs from ssp. *filifera*: **Ros** 20 - 30 × 20 - 35 cm, densely caespitose, forming clusters ≥ 1 m ∅; **L** linear to linear-oblanceolate, abruptly widened to 2.5 - 3.5 cm within 2 cm from the base, acuminate, acute, 12 - 20 × 1 - 2 cm, margins with a brownish or purplish-grey band ± 2 mm wide, white-filiferous, youngest **L** sometimes with 1 - 3 white streaks on the upper face; **Inf** 1 - 1.35 m, part-**Inf** in the upper 80 cm, with paired **Fl**; **Fl** at right angles to the axis, 50 - 55 mm; **Ov** 7 - 8 mm; **Tep** greenish-yellow.

According to the protologue similar in its rosettes to ssp. *schidigera* but much smaller and proliferous. It is possibly a redescription of *A. filifera* var. *compacta*, but nevertheless with priority on subspecies level.

A. filifera ssp. **multifilifera** (Gentry) B. Ullrich (Brit. Cact. Succ. J. 10(3): 66, 1992). **T:** Mexico, Chihuahua (*Gentry 8167* [US 2558493 + 2558494, DES, MEXU]). – **D:** Mexico (Sonora, Chihuahua, Sinaloa, Durango); cliffs and rocky sites in pine-oak forests, 1400 - 2200 m. **I:** Gentry (1982: 104). **Fig. III.b**

≡ *Agave multifilifera* Gentry (1972).

[1d] Differs from ssp. *filifera*: Stems short but clearly developed; **Ros** ± 1 × 1.5 m, solitary; mature **L** 200 in number, linear-lanceolate, erectly spreading to declining, firm but pliable, broadest at the base, 50 - 80 × 1.2 - 3.5 cm, light green, margins long and copiously filiferous; terminal **Sp** chestnut-brown to grey with age; **Inf** to 5 m, densely flowered from above the **L**, part-**Inf** with 2 - 3 **Fl**; **Fl** 40 - 43 mm; **Ov** 20 - 21 mm, neck constricted, faintly grooved, 5 mm; **Tep** green with lavender hue in bud, at anthesis pale green with pink tinge on the lobes, lobes subequal, 16 - 17 mm.

One of the most robust taxa in Group *Filiferae*, differing from the other ssp. of *A. filifera* mainly by its larger dimensions. See also the note for ssp. *schidigera*.

A. filifera ssp. **schidigera** (Lemaire) B. Ullrich (Brit. Cact. Succ. J. 10(3): 65, 1992). **T:** [neo – icono]: Ill. Hort. 9(7): t. 330, 1862. – **Lit:** McVaugh (1989); Ullrich (1992d). **D:** Mexico (Chihuahua, Sinaloa, Durango, Zacatecas, San Luis Potosí, Nayarit, Jalisco, Guanajuato, Aguascalientes, Michoacán, Guerrero). **I:** Gentry (1982: 104, 120-121).

≡ *Agave schidigera* Lemaire (1861) ≡ *Agave filifera* var. *schidigera* (Lemaire) A. Terracciano (1885); **incl.** *Agave taylorii* hort. *ex* Besaucèle (s.a.) ≡ *Agave schidigera* var. *taylorii* (Besaucèle) H. Jacobsen (1955); **incl.** *Agave filifera* var. *adornata* Scheidweiler (1861); **incl.** *Agave filifera* var. *pannosa* Scheidweiler (1861); **incl.** *Littaea roezlii* Roezl (1861) (*nom. inval.*, Art. 34.1b); **incl.** *Agave filifera* var. *angustifolia* Lemaire (1865); **incl.** *Agave filifera* var. *ignescens* Lemaire (1865); **incl.** *Agave schidigera* var. *angustifolia* Lemaire (1865); **incl.** *Agave schidigera* var. *ignescens* Lemaire (1865); **incl.** *Agave schidigera* var. *plumosa* Lemaire (1865); **incl.** *Agave vestita* S. Watson (1890); **incl.** *Agave wrightii* J. R. Drummond (1909) (*nom. illeg.*, Art. 53.1); **incl.** *Agave discreptata* J. R. Drummond (1912); **incl.** *Agave perplexans* Trelease (1914).

[1d] Differs from ssp. *filifera*: Stems short; **Ros** symmetrical, solitary, 0.7 - 1 m ∅; **L** sometimes falcate, relatively thin, pliable, widest at or below the middle, 30 - 40 (-50) × 1.5 - 3 (-4) cm, green to greyish-green or yellowish-green, rarely reddish, with imprints from the next younger **L**, margins brown to white, coarsely white-filiferous; terminal **Sp** 0.5 - 1.6 (-2) cm, brown to grey with age, shortly decurrent; **Inf** 2 - 3.5 m, slender, laxly flowered in the upper ½; **Fl** 30 - 45 mm; **Ov** fusiform, 12 - 20 mm; **Tep** green to yellow or flushed with purple, tube narrowly funnel-shaped, 7 - 10 mm, lobes equal, 13 - 20 mm.

Closely related to ssp. *filifera*, but separable by its non-surculose habit, longer, thinner, more pliable leaves and coarse rather than finely filiferous margins (Gentry 1982: 120). Ullrich (1992d) reduced this taxon as well as *A. multifilifera* to subspecific rank under *A. filifera* based on mere literature interpretations, which for corroboration would require a critical field study. Since *A. filifera* ssp. *microceps* is closest to *A.* [*filifera* ssp.] *schidigera*, the concept of Ullrich is nevertheless followed here in order to avoid a new combination for ssp. *microceps* under *A. schidigera*.

A. flexispina Trelease (CUSNH 23: 133, 1920). **T:** Mexico, Durango (*Palmer 330* [US, NY]). – **D:** Mexico (S Chihuahua, Durango, Zacatecas); grassland and oak woodland, 1300 - 2300 m. **I:** Gentry (1982: 437).

[2g] **Ros** open, small, 25 - 35 × 50 - 70 cm, solitary or caespitose; **L** few (± 40 in mature **Ros**), ovate, acuminate, 16 - 30 × 6 - 8 cm, glaucous- to yellowish-green, margins undulate to crenate; marginal teeth mostly retrorse, larger teeth mostly 5 - 8 mm, brown to pruinose, 1 - 1.5 cm apart, on small tubercles, sometimes with small intermittent teeth; terminal **Sp** acicular, usually flexuous, flat to openly grooved near the base, 2.5 - 3.5 cm, brown to pruinose-grey, decurrent to the upper teeth; **Inf** 2.5 - 3.5 m, 'paniculate', slender, rather open, frequently

narrow, part-**Inf** small, 6 - 12, few-flowered; **Fl** 50 - 70 mm; **Ov** cylindrical, slightly angular, 22 - 35 mm, neck obscure; **Tep** greenish-yellow with red tinge, tube cylindrical to urceolate, 13 - 18 mm, lobes unequal, outer 10 - 18 mm.

In appearance like a small *A. shrevei* or *A. palmeri*, but different in its flowers (Gentry 1982: 438).

A. fortiflora Gentry (US Dept. Agric. Handb. 399: 122-126, ills., 1972). **T**: Mexico, Sonora (*Gentry 19808* [US]). – **D**: Mexico (Sonora). **I**: Gentry (1982: 421, 439-440).

[2g] **Ros** open, up to 1 × 1.8 m, mostly solitary; **L** straightly ascending or outcurving and conduplicate, long-acuminate, gradually narrowed above the dilated base, widest in the middle, finely tuberculate-rugose (incl. teeth and **Sp**), to 50 - 100 × 8 - 12 cm, light grey-glaucous, usually cross-zoned, margins straight or teeth in the middle of the lamina on small tubercles; marginal teeth curved downwards or erect, 5 - 10 mm (middle of the lamina), 1 - 3 cm apart, with irregularly arranged smaller intermittent teeth; terminal **Sp** subulate, rounded below, narrowly grooved above, chestnut-brown to light grey, decurrent along the margin to the uppermost teeth; **Inf** 4 - 6 m, 'paniculate', open and ovoid, scape short, part-**Inf** dense, 12 - 18; **Fl** long-lived, erect, 72 - 82 mm; **Ov** 45 - 50 mm, pale green; **Tep** yellow, tube broadly bulging, 11- 13 mm, elliptic in cross-section, lobes 20 - 23 mm.

Distinct by its large, strong and long-lasting flowers and without close relatives in Group *Ditepalae* (Gentry 1982: 439).

A. fourcroydes Lemaire (Ill. Hort. 11(Misc.): 65, 1864). – **D**: Cultivated only; mainly E Mexico. **I**: Gentry (1982: 576).

Incl. *Agave ixtlioides* Lemaire *ex* Jacobi (1866); **incl.** *Agave rigida* var. *longifolia* Engelmann (1875); **incl.** *Agave ixtli* var. *elongata* Baker (1877) ≡ *Agave rigida* var. *elongata* (Baker) Baker (1881) ≡ *Agave elongata* (Baker) A. Berger (1912) (*nom. illeg.*, Art. 53.1); **incl.** *Agave longifolia* hort. *ex* A. Berger (1915); **incl.** *Agave ixtli* hort. *ex* A. Berger (1915) (*nom. illeg.*, Art. 53.1); **incl.** *Agave sullivanii* Trelease (1920).

[2f] Stem thick, 1 - 1.7 m; **Ros** large, suckering; **L** straight, linear, rigid, thickly rounded at the base, acuminate, guttered, 120 - 180 × 8 - 12 cm, margins straight; marginal teeth slender, 3 - 6 mm, dark brown, regularly spaced; terminal **Sp** conical, stout, openly short-grooved above, mostly 2 - 3 cm, dark brown; **Inf** 5 - 6 m, 'paniculate', part-**Inf** 10 - 18 in the upper ½ of the **Inf**, bulbilliferous, never producing **Se**; **Fl** 60 - 70 mm; **Ov** fusiform, roundly-trigonous, 35 - 40 mm, neck briefly constricted; **Tep** greenish-yellow, tube urceolate, 12 - 16 mm, lobes subequal, 16 - 18 mm. – *Cytology:* 2n = 60, ± 140, 150.

Widely cultivated for fibre ("Henequen") esp. in E Mexico (Gentry 1982). The different cultivars subsumed under *A. fourcroydes* show different degrees of similarity with its wild progenitor *A. vivipara* (Colunga-García Marín & al. 1996: as *A. angustifolia*). The isozyme studies of Colunga-García Marín & al. (1999) indicated *A. fourcroydes* to represent a polyphyletic assemblage of different cultivars independently derived from within the variable *A. vivipara*. Consequently, the species name *A. fourcroydes* should be abandoned and the different cultivars be named under its progenitor species (e.g. *A. vivipara* 'Sac Ki').

A. franzosinii Baker (BMI 1892: 3, 1892). **T** [neo]: Ex cult. Huntington (*Gentry 10163+19866* [US, DES]). – **D**: Known from cultivation only. **I**: Gentry (1982: 276, 291).

[2a] **Ros** widely spreading, very large, 2 - 2.7 (-3) × to 4.5 m, freely suckering; **L** lanceolate, spreading, recurved, or sharply reflexed, narrowed at the base, thickened and convex below towards the base, hollowed above, somewhat asperous, 180 - 220 × 22 - 35 cm, light glaucous-grey or bluish-glaucous variously marked with green below the middle of the lamina, margins straight to repand; larger marginal teeth (middle of the lamina) 8 - 10 mm, dark brown, remote, on fleshy prominences; terminal **Sp** 3 - 6 cm, dark brown, decurrent along the inrolled **L** tip; **Inf** 8 - 11.4 m, 'paniculate', broadly cylindrical, to 2.9 m broad, scape short, axis strong, part-**Inf** broadly spreading, several times compound; **Fl** large, 83 - 100 mm; **Ov** 35 - 45 mm, light bright green; **Tep** yellow, soon withering, tube 18 - 22 mm, lobes 30 - 32 mm.

A distinctive species not easily confused with, and obviously related to *A. americana* (Gentry 1982: 291). Howard & al. (1979) use *A. beaulueriana* Jacobi 1869 (here treated amongst the unresolved names) as older valid name for this taxon and consequently list *A. franzosinii* as synonym. Berger (1915: 157) ascribes *A. franzosinii* to Nissen *ex* Ricasoli (Della utilità dei giardini d'acclimazione, 7, 1888) as name only. Sewell (Gard. Chron. ser. 3, 1889: 639) and W. Watson (Bull. Misc. Inf. [Kew] 1889: 301) also used the name before Baker. The correct name and author for this plant therefore needs further study.

A. funkiana K. Koch & C. D. Bouché (Wochenschr. Vereines Beförd. Gartenbaues Königl. Preuss. Staaten 3: 47, 1860). **T**: Mexico, Hidalgo (*Gentry 12273* [US, DES]). – **D**: Mexico (Nuevo León, San Luis Potosí, Hidalgo); 250 - 1800 m. **I**: Gentry (1982: 126, 140).

[1g] **Ros** open, 0.6 - 0.9 × 1.2 - 1.8 m, freely suckering; **L** linear, radiating, firm, straight or somewhat falcate, patulous, base broadly clasping, convexly thickened below, concave above, mostly 60 - 80 × 3.5 - 5.5 cm, yellowish-green to dark

green, frequently with pale median stripe, margins horny, nearly straight, firm, thin, brown to grey; marginal teeth mostly directed downwards, regular, slender, 3 - 5 mm, 1 - 2.5 cm apart, with a few small irregularly arranged intermittent teeth; terminal **Sp** conical-subulate, with a narrow to open groove above, 1 - 3 cm, brown to white; **Inf** slender, 3.5 - 4.5 m, 'spicate', laxly flowered in the upper ½, part-**Inf** with paired **Fl**; **Ped** geminate, ± 1 cm; **Fl** 40 - 45 mm; **Ov** oblong-fusiform, 20 - 24 mm, neck constricted; **Tep** pale glaucous-green, tube 3.5 - 4 mm, lobes 18 - 19 mm.

Obviously related to *A. lophantha*, but differing in its larger size, regular and linear slightly concave leaves with nearly straight fine margins, and numerous regular fine teeth (Gentry 1982: 140). The specimen cited by Gentry (l.c., 189) from Chiapas (*Gentry* 12195) belongs to *A. ghiesbreghtii* (Lott & García-Mendoza 1994).

A. fusca (Ravenna) Thiede & Eggli (KuaS 52: [in press], 2001). **T:** Guatemala (*Ravenna* 325 [Herb. Ravenna]). – **D:** Guatemala (Chimaltenango - Comalapa); sandy plains.
    ≡ *Manfreda fusca* Ravenna (1987).
    [3a3] Rhizome 2 - 2.3 cm ∅; **L** several, sprawling, narrowly lanceolate, canaliculate, lower face carinate, slightly scabrous, 30 - 50 × 2.2 - 3 cm, ash-green; **Inf** scape stiff, **Bra** rather distant; **Fl** ± 20, crowded, single, apparently sessile, with foetid odour; **Ov** oblong, 12 - 17.8 × 6.6 mm, greenish; **Tep** 37 mm, outer face glaucous-green, lobes spreading to reflexed, linear-lanceolate, 14.8 - 16 × 4 - 5.8 mm, tip apiculate-tuberculate, outer face dark brown; **Fil** sparsely glandular-pilose, 39 - 42 mm, dirty greenish-white with diminutive dark streaks; **Sty** reflexed and twisted before the Anth dehisce, almost straight or slightly curved afterwards, to 57 mm; **Sti** capitately 3-lobed; **Fr** unknown.

In the protologue, no differential diagnosis is given nor is the relationship indicated. García-Mendoza & Castañeda Rojas (2000) clearly place the species in the *A. guttata* subgroup of Subgen. *Manfreda*, based on a study of additional specimens at MEXU. It is distinct in its flower colour and foetid odour. The species was erroneously omitted from the treatment of the family for the 'Flora Mesoamericana' (Lott & García-Mendoza 1994).

A. geminiflora (Tagliabue) Ker Gawler (J. Sci. Arts (London) 2: 86-90, 1817). **T:** [lecto – icono]: Bibliot. Ital. Giorn. Lett. 1: 100, fig. – **Lit:** McVaugh (1989). **D:** Mexico (Nayarit); on rocks in oak woodland, 1000 - 1400 m, only known from one small area. **I:** Gentry (1982: 104, 113-114); Kaktusblüte 1995: 41-44.
    ≡ *Littaea geminiflora* Tagliabue (1816); **incl.** *Agave geminiflora* var. *stricta-viridis* hort. *ex* Besaucèle (s.a.); **incl.** *Yucca boscii* Hort. Panorm. *ex* Hor-

nemann (1813); **incl.** *Bonapartea juncea* Willdenow (1814); **incl.** *Yucca boscii* Desfontaines (1815) (*nom. illeg.*, Art. 53.1); **incl.** *Bonapartea flagelliformis* Donnersmark (1820); **incl.** *Agave angustissima* Engelmann (1875); **incl.** *Agave geminiflora* var. *knightiana* J. R. Drummond (1909); **incl.** *Dracaena boscii* Hort. Cels *ex* A. Berger (1915).
    [1d] Stem short; **Ros** dense, 0.7 - 1 m, somewhat broader, solitary; **L** many, linear, eventually arching, flexible, narrow, pliable, abruptly acute, roundly convex below and above, 45 - 60 × 0.6 - 0.8 cm, green, smooth, margins finely filiferous or rarely naked; terminal **Sp** shortly subulate, 5 - 7 mm, greyish; **Inf** long tapering, stout at the base, 4 - 6 m, 'spicate', flowering in the upper ⅔ - ¾, part-**Inf** with mostly geminate **Fl**; **Ped** slender, 5 - 8 mm; **Fl** 40 - 52 mm; **Ov** slender, 16 - 20 mm, neck grooved; **Tep** greenish below, flushed above with red or purple, tube narrowly funnel-shaped, 6 - 11 mm, lobes slightly unequal, 18 - 21 mm.

Distinct within Group *Littaea* by its relatively large simple stem with innumerable, very narrow, pliable, smooth leaves and large 'spikes' with relatively remote long flowers. It is most closely related to *A. ornithobroma* (Gentry 1982: 112). The report of a specimen of this taxon from S Sinaloa by McVaugh (1989: 135) is erroneous, since the collection cited (*Gentry* 18358) is the type collection of *A. ornithobroma*.

A. gentryi B. Ullrich (Succulenta 69(10): 210-214, ills., 1990). **T:** Mexico, Nuevo León (*Gentry* 20159 [DES]). – **D:** Mexico (Coahuila, Nuevo León, Tamaulipas, Durango, Zacatecas, San Luis Potosí, Hidalgo, México, Puebla); on limestone in pine-oak forests or chaparral, 1850 - 2800 m. **I:** Gentry (1982: 597, 599, as *A. macroculmis*).
    [2b] Stem thick, short; **Ros** rigid, medium to large, 0.6 - 1 m ∅, solitary, with rhizomatous offsets; **L** 30 - 45, triangularly long-acuminate, base very broad and thick, concave, 60 - 100 × 17 - 26 cm, dark to light green, sometimes faintly glaucous, margins partly or entirely horny; marginal teeth nearly straight, commonly 8 - 12 mm (middle of the lamina), chestnut-brown to greyish-brown, 2 - 4 cm apart, cusps from well-rounded bases; terminal **Sp** subulate, broadly channelled above for ⅔ of its length, very strong, 4.5 - 6.5 cm, dark brown to greyish, long decurrent; **Inf** 3 - 5 m, 'paniculate', ellipsoid, scape stout, scape **Bra** large and fleshy, with entire margins, closely imbricate at the base of the part-**Inf**, part-**Inf** dense, 10 - 28; **Fl** 70 - 90 mm; **Ov** 35 - 55 mm, green, neck grooved; **Tep** reddish in bud, opening yellow, tube funnel-shaped, 11 - 16 mm, lobes unequal, 20 - 28 mm.

Easily recognizable by the large fleshy scape bracts congested below the inflorescences, and its extremely broad-based rigid long-pointed green leaves (Gentry 1982: 598, as *A. macroculmis*). Gentry's use of the name *A. macroculmis* for this plant must be rejected, since this name is a synonym of *A. atrovirens*. Since the plant was without name, Ullrich (l.c.) described it as new. See also under *A. montana*.

A. ghiesbreghtii Lemaire *ex* Jacobi (Hamburg. Gart.- & Blumenzeit. 20: 545, 1864). **T:** US. – **D:** Mexico (México, Guerrero, Puebla, Oaxaca, Chiapas), Guatemala. **I:** Gentry (1982: 141). **Fig. III.c**

Incl. *Agave inghamii* hort. (s.a.) (*nom. inval.*, Art. 29.1); **incl.** *Agave leguayana* Verschaffelt (1868); **incl.** *Agave roezliana* Baker (1877); **incl.** *Agave roezliana* var. *inghamii* hort. *ex* Baker (1877); **incl.** *Agave purpusorum* A. Berger (1915); **incl.** *Agave roezlii* hort. *ex* A. Berger (1915) (*nom. inval.*, Art. 61.1); **incl.** *Agave huehueteca* Standley & Steyermark (1943).

[1g] **Ros** open, short, copiously suckering; **L** few, broadly lanceolate, ovate or deltoid, straight or upcurving, thick, rigid, narrowed above the base and widest in the middle, apex acuminate, convex below, plane to slightly hollowed or guttered above, 30 - 40 × 7 - 10 cm, or more rarely broadly linear and 35 - 38 × 5.5 - 6 cm, dark to light green, margins horny, relatively narrow, brown; marginal teeth frequently straight, sometimes curved upwards or downwards, larger teeth 5 - 8 (-10) mm, brown to greyish, 1 - 3 cm apart, reduced at the **L** tip; terminal **Sp** subulate, 2 - 4 cm, brown to grey; **Inf** 3 - 4 m, 'spicate', densely flowered; **Fl** 40 - 50 mm; **Ov** cylindrical, 16 - 20 mm, neck constricted; **Tep** greenish-brown to purplish, paler within, tube broadly funnel-shaped, 3 - 5 (-10?) mm, lobes subequal, 15 - 21 mm.

Closely related to *A. kerchovei*, but different by its shorter and broader leaves with more and smaller teeth on narrower and darker horny margins (Gentry 1982: 142).

A. gigantensis Gentry (Occas. Pap. Calif. Acad. Sci. 130: 63-67, ills. (pp. 65-66, 68), 1978). **T:** Mexico, Baja California Sur (*Gentry & McGill* 23320 [US, DES, MEXU, SD]). – **D:** Mexico (Baja California Sur); 600 - 1520 m. **I:** Gentry (1982: 387-388).

[2h] **Ros** rather open, 0.5 - 1 × 0.8 - 1.2 m, solitary; **L** few, broadly lanceolate, plane, rigid, thick, fleshy, markedly narrowed at the base, widest in the middle, acuminate, smooth, 40 - 75 × 11 - 16 cm, green to glaucous-green, turning red to purplish when plants are flowering, margins undulate to prominently mamillate; marginal teeth variously curved, basis thick, frequently 2 - 3 teeth cuspidate, confluent along the upper leaf margins, large, 10 - 20 mm and more, brown to light greyish, up to 6 - 8 cm apart; terminal **Sp** strongly subulate, 3 - 6 cm, grey, long decurrent as pronounced horny margin; **Inf** 4 - 5 m, 'paniculate', rather narrow, part-**Inf** rather small, 15 - 25 in the upper ¼ - ⅓ of the **Inf**; **Fl** slender, 48 - 60 mm; **Ov** slender, fusiform, neck

constricted; **Tep** buds waxy white, open bright pale yellow, tube spreading, 4 - 5 mm, lobes 18 - 25 mm.

A. avellanidens and *A. moranii* appear to be the closest relatives (Gentry 1982: 386).

A. gilbertii A. Berger (Monatsschr. Kakt.-kunde 14: 126, 1904). **T** [lecto]: Ex cult. (*Anonymus* s.n. [K [sub *A. bakeri*]]). – **D:** Known from cultivation only. **I:** Gentry (1982: 72).

Incl. *Agave bakeri* Hooker *fil. ex* W. Watson (1903) (*nom. illeg.*, Art. 53.1).

[1b] **Stem** short; **Ros** solitary, not suckering; **L** many, lanceolate, recurving, coriaceous, narrowed and thickened near the base, broadest in the middle, convex beneath, concave to plane above, 90 - 100 × 10 - 12 cm, glaucous-green, margins thin, brown; marginal teeth none; terminal **Sp** slender, 0.5 - 2 cm; **Inf** to 3 m, 'spicate', cylindrical, densely flowered from near the base; **Fl** 50 - 60 mm; **Ov** fusiform, slender, ± 20 mm, neck slender; **Tep** greenish outside, whitish within, tube 6-furrowed, 11 - 12 mm, lobes reflexed or revolute, 20 mm.

The name *A. bakeri* Hooker *fil. ex* W. Watson (1903) represents an illegitimate later homonym of *A. bakeri* Ross (1894). The valid name for the taxon is the replacement name *A. gilbertii* published by Berger, in contrast to the treatment by Gentry (1982: 71). The species apparently disappeared in cultivation after the original plant died. It has never been recollected (Gentry l.c.).

A. ×glomeruliflora (Engelmann) A. Berger *pro sp.* (Agaven, 95, 1915). **T:** [lecto – icono]: Gard. Chron. ser. nov., 1883: 19, fig. 6. – **D:** USA (Texas), Mexico (Coahuila); grasslands, 620 - 1520 m. **I:** Gentry (1982: 143).

≡ *Agave heteracantha* var. *glomeruliflora* Engelmann (1883) ≡ *Agave lechuguilla* fa. *glomeruliflora* (Engelmann) Trelease (1920); **incl.** *Agave chisosensis* C. H. Müller (1939).

Obviously of hybrid origin with morphological gradations between *A. lechuguilla* and *A. neomexicana* and / or *A. havardiana* (Gentry 1982: 143) and thus only representing an aggregate of different habitually similar natural hybrids. Gentry's selection of a neotype (l.c.) is superseded by his simultaneous selection of a lectotype (l.c.).

A. ×gracilipes Trelease *pro sp.* (Annual Rep. Missouri Bot. Gard. 22: 95, 1912). **T:** USA, Texas (*Mulford* 293+293a [MO, NY]). – **D:** USA (SE New Mexico, W Texas); Mexico (Chihuahua); on limestone among grama grass, 1250 - 1850 m. **I:** Gentry (1982: 522, 527-528, 536).

A very variable complex, identified as the natural hybrid *A. lechuguilla* × *A. neomexicana* by Trelease (l.c.), Gentry (1982: 530), and Pinkava & Baker (1985). Since the latter putative parent is absent from major parts of the area of *A. ×gracilipes*, other parent species must be involved as well (*A. parryi* ?).

A. gracillima A. Berger (Agaven, 33, 288 [erratum], 1915). **T:** Mexico, Durango (*Rose* 2341b [US]). – **Lit:** McVaugh (1989). **D:** Mexico (S Durango, Nayarit); valleys, grassy plains and oak forests in the mountain region, 1100 - 1370 m, flowers mid-August to late November.

Incl. *Manfreda elongata* Rose (1903) ≡ *Polianthes elongata* (Rose) Shinners (1966) (*nom. illeg.*, Art. 53.1); **incl.** *Agave gracilis* A. Berger (1915) (*nom. illeg.*, Art. 53.1); **incl.** *Polianthes rosei* Shinners (1967) (*nom. illeg.*, Art. 52.1).

[3a2] Plants large (for Subgen. *Manfreda*); **L** 4, arching, linear-lanceolate, strongly recurved, deeply channelled, herbaceous, with many closely-spaced veins, 35 - 49.5 (-76) × 2.8 - 3.9 (-5 fide Rose (1903)) cm, margins narrow to medium broad, hyaline, entire; **Inf** 90 - 120 cm (fide Rose l.c.), 'spicate', flowering part half-dense, 17.5 - 31.5 cm, with 20 - 27 sessile **Fl**; mature **Fl** (apparently) almost horizontal; **Ov** ellipsoid, 10 - 14 mm, without neck; **Tep** brownish or yellowish-green, tube cylindrical, bluntly 6-angled, straight, not markedly constricted above and at an angle with the **Ov**, (8-) 11 - 15 mm, lobes oblong, reflexed, 12 - 16 mm; **Fil** curved upwards at maturity; **Sty** curved upwards at maturity, exserted for 28 - 36 mm; **Sti** clavate, trigonous, shallowly furrowed; **Fr** and **Se** unknown.

Distinguished by its long recurving deeply channelled leaves, the long-acuminate floral bracts, and the styles, which are usually much longer than the stamens (Verhoek-Williams 1975: 285).

A. grijalvensis B. Ullrich (KuaS 41(6): 102-108, ills., 1990). **T:** Mexico, Chiapas (*Gentry* 12204 [DES, US]). – **D:** Mexico (Chiapas); calcareous soil. **I:** Gentry (1982: 621, 626, as *A. kewensis*).

[2g] **Ros** open, large, solitary; **L** few, narrowly lanceolate, arching or sprawling, pliable, thickly succulent, guttered, 120 - 180 × 12 - 15 cm, yellowish-green, margins straight or nearly so; marginal teeth straight, small, larger teeth 3 - 4 mm (in the upper ⅓ of the lamina), 1 - 3 cm apart, much reduced below or lacking on the lower ⅓ of the lamina; terminal **Sp** acicular, narrowly grooved above, 3 - 4.5 cm, not decurrent; **Inf** 3 - 5 m, 'paniculate', scape usually short, part-**Inf** 3-branched, 12 - 20 in the upper ½ of the **Inf**; **Fl** slender, 60 - 74 mm; **Ov** slender, cylindrical, 30 - 40 mm, neck slightly constricted; **Tep** yellow, tube 12 - 15 mm, lobes unequal, 18 - 20 mm.

Ullrich (l.c.) rejected Gentry's use of the name *A. kewensis* for this plant, and provided the new name. He provisionally placed the species in Group *Marmoratae* based on vegetative features (instead of Gentry's placement of his "*A. kewensis*" in his Group *Sisalanae*).

A. grisea Trelease (Mem. Nation. Acad. Sci. 11: 34-35, t. 54-56, 1913). **T:** Cuba, Santa Clara (*Grey* 1 [MO?]). – **Lit:** León (1946). **D:** S-C Cuba.

Incl. *Agave grisea* var. *grisea*; **incl.** *Agave grisea* var. *cienfuegosana* Trelease (1913); **incl.** *Agave grisea* var. *obesispina* Trelease (1913).

[21] **Ros** solitary; **L** lanceolate, somewhat concave, 150 - 200 × 10 - 20 (-25) cm, green, passing into glaucous, or grey, rather dull, margins between the teeth from nearly straight to decidedly concave; marginal teeth gently curved, heavily triangular, sometimes wider or sublenticular at the **L** base, 2 - 3 (-5) mm, 15 - 25 (-45) mm apart; terminal **Sp** triquetrously conical or somewhat subulate, slightly curved, flattened or shallowly concave to or beyond the middle or becoming subinvolute, smooth, 1 - 1.5 (-2) cm, reddish-chestnut-brown or brown, decurrent for its length or more; **Inf** 6 - 8 m, 'paniculate', oblong, part-**Inf** in the upper ½ of the **Inf**; **Fl** 40 - 55 mm; **Ov** oblong-fusiform, 20 - 30 mm; **Tep** golden-yellow, tube ± 8 mm, lobes 15 - 18 mm; **Fr** oblong, shortly stipitate and beaked, 4 × 2 cm.

A polymorphic species. Since the 2 varieties merely differ gradually and in addition appear to be connected by transitional forms (Berger 1915: 209), they are provisionally included in the synonymy here.

A. guadalajarana Trelease (CUSNH 23: 123, 1920). **T:** Mexico, Jalisco (*Pringle* 4473 [MO, K, MEXU, NY, US]). – **Lit:** McVaugh (1989). **D:** Mexico (Jalisco); grassy slopes of oak woodland, 1500 - 2000 m. **I:** Gentry (1982: 532).

[2i] **Ros** compact, small, 25 - 35 cm ∅, broader than tall, solitary, rarely suckering; **L** numerous, obovate to oblong, rigid, closely imbricate, obtuse, plane to incurved, 20 - 30 × 8 - 12 cm, inner **L** shiny but glaucous, outer **L** dull grey, margins nearly straight, upper part mamillate; upper marginal teeth 8 - 10 mm, remote, those from the middle of the lamina towards the base much smaller, 3 - 4 mm, reddish-brown to dusty grey, 5 - 10 mm apart; terminal **Sp** subulate, straight to sinuous, flat to shallowly hollowed above, roundly keeled below, 2.5 cm, greyish; **Inf** 4 - 5 m, 'paniculate', scape slender, part-**Inf** small, 15 - 20 in the upper ½ of the **Inf**; **Fl** 60 mm; **Tep** slender, lobes much longer than the tube.

Distinct from all other members of Group *Parryanae* in its mamillate leaf margins and lax inflorescence (Gentry 1982: 531).

A. guerrerensis (Matuda) G. D. Rowley (Repert. Pl. Succ. 26: 4, 1977). **T:** Mexico, Guerrero (*González Medrano & al.* s.n. [MEXU]). – **D:** Mexico (Guerrero); 1250 m. **I:** Cact. Suc. Mex. 20: 47, 1975.

≡ *Manfreda guerrerensis* Matuda (1975).

[3a3] Rhizome fleshy, fasciculate; **L** 2 - 3, base subamplexicaul, lamina obovate-lanceolate, acute, entire, half-fleshy, gradually tapering towards a narrow petiole-like lower part ± 8 - 10 cm long and 2 mm broad, white-pubescent, 40 - 45 × 6 - 8 cm,

glaucous, plain green or spotted with spots towards the base; remains of **L** bases broadly triangular, later fibrous, ± 4.5 cm; **Inf** glabrous, 1.2 - 1.5 m, scape spotted purplish towards the base, with few **Fl**; **Ped** short; **Fl** incl. **Ov** apparently ± 67 mm; **Ov** 3 mm ⌀; **Tep** tube campanulate, lobes erect or spreading, ± 17 mm.

This species, which is only known from the type collection, is doubtfully distinct from *A. debilis* and differs mainly in its much broader leaves only. The sparse flower measurements given in the protologue remain unclear ("length of tube including ovary 5 cm") and need to be corroborated from the type material. Fruit data is lacking in the protologue, though the accompanying illustration shows a fruiting but apparently flowerless specimen. – The typification of this taxon was discussed by González Medrano (1991).

A. guiengola Gentry (Brittonia 12: 98-100, ills., 1960). **T:** Mexico, Oaxaca (*Gentry* 16436 [US, DES, MEXU]). – **Lit:** Ullrich (1991g). **D:** Mexico (Oaxaca: Cerro Guiengola); limestone, 100 - 1000 m, only known from the type locality. **I:** Gentry (1982: 90, 98-99). **Fig. III.d**

[1c] **Ros** open, mostly solitary; **L** few, mature ± 30, ovate to ovate-lanceolate, openly ascending, short-acuminate, nearly plane above but briefly and narrowly channelled apically, light grey or white-glaucous, epidermis finely and densely papillate, margins variously serrate; marginal teeth flattened, blunt, fine or coarse, 1- to 2-cuspidate, dark brown; terminal **Sp** acicular, dark brown, not decurrent or decurrent for ± its own length in a horny margin; **Inf** erect, 1.6 - 2 m, 'spicate', flowering from near the base, part-**Inf** with 2 - 3 **Fl**; **Fl** inconspicuous, 33 - 35 mm; **Tep** pale yellow or yellowish-white.

A very distinct species esp. due to its broad thick white-glaucous leaves. The tubeless flowers place it in Group *Choritepalae*, but it differs from the 2 other species (*A. bracteosa*, *A. ellemeetiana*) geographically as well as in the shape and coloration of the toothed leaves (Gentry 1982: 97) and seed morphology (Ullrich 1991g). Ullrich (l.c.) therefore regards the Group *Choritepalae* as artificial.

A. guttata Jacobi & C. D. Bouché (Hamburg. Gart.- & Blumenzeit. 21: 190, 1865). – **Lit:** McVaugh (1989). **D:** Mexico (San Luis Potosí, Aguascalientes, Jalisco, Zacatecas, Durango); open sun, grassy roadsides, rocky fields, summit of hills, roadcuts, 1220 - 2440 m, flowers mid-July to late August. **I:** Desert Pl. Life 12: 174-175, 1940.
≡ *Manfreda guttata* (Jacobi & C. D. Bouché) Rose (1903) ≡ *Polianthes guttata* (Jacobi & C. D. Bouché) Shinners (1966); **incl.** *Agave protuberans* Engelmann *ex* Baker (1888) ≡ *Leichtlinia protuberans* (Engelmann *ex* Baker) H. Ross (1893); **incl.** *Leichtlinia commutata* H. Ross (1896).

[3a3] Plants medium-sized to large (for Subgen.

Manfreda), reproducing vegetatively by buds from the storage rhizome and by spreading rhizomes producing a plantlet at the tip; rhizome bulbous to oblong, 1.8 - 4 × 1.3 - 3 cm; **R** half-fleshy; **L** 2 - 7 (-13 in cultivation), spreading or erect-spreading, lanceolate to lanceolate-elliptic, often narrowed towards the base, tip obtuse with a short firm point, channelled, undulate, semisucculent, (8-) 14 - 38 × (0.9-) 1.3 - 3.1 cm, glaucous, plain green or spotted with small or large and confluent green or dark brown spots; margins with a narrow white cartilaginous band sometimes streaked with red, minutely denticulate or erose; remains of **L** bases fibrous, fibres 3 - 6 cm; **Inf** (61-) 90 - 156 cm, 'spicate', flowering part crowded, (2.3-) 3 - 14 cm, with (2-) 4 - 25 (-33) **Fl**; **Ped** none or very short in the lower **Fl**; **Fl** nearly erect, with strong scent of cooked potatoes or onions; **Ov** cylindrical, 6 - 12 (-15) mm; **Tep** tube cylindrical to oval in cross section, slightly curved, short, 3 - 12 mm, lobes oblong, tightly revolute to coiled, (6-) 10 - 15 mm, greenish-yellow; **Fil** exceeding the tube by 20 - 29 (-41) mm, pale green, often speckled with brown; **Sty** at maturity longer than the **St** and arched upwards, exceeding the tube, pale greenish-white; **Sti** clavate, 3-lobed; **Fr** ellipsoid, 1.6 - 2.4 × 1 - 1.9 cm; **Se** 3 × 4 mm.

The round storage rhizome and the presence of spreading rhizomes, the denticulate-erose leaf margin, dense inflorescence, and the stubby flowers with exserted stamens and styles are diagnostic for this species (Verhoek-Williams 1975: 238).

A. gypsophila Gentry (Agaves Cont. North Amer., 510-512, ills., 1982). **T:** Mexico, Guerrero (*Floyed & Ryan* 103 [MICH, UC]). – **Lit:** McVaugh (1989); Ullrich (1991b). **D:** Mexico (Jalisco, Colima, Michoacán, Guerrero); calcareous or gypseous rocks, understorey of lowland thorn forest, 300 - 1000 m.

[2j] **Ros** openly spreading, solitary; **L** few (20 - 30 in mature **Ros**), linear-lanceolate, generally arching, weak, brittle and with few fibres, thick, slightly narrowed near the base, deeply convex below, flat above, asperous, 45 - 100 (-110) × 7 - 12 cm, glaucous-grey, margins closely dentate with small prominences; marginal teeth weak, 1 - 2 mm, with small intermittent teeth; terminal **Sp** conical, very small, 0.5 - 1.5 cm, dark brown, not decurrent; **Inf** 2 - 3 m, 'paniculate', part-**Inf** relatively few, widely spreading, few-flowered, in the upper ½ of the **Inf**, in cultivation bulbilliferous; **Fl** 30 - 35 mm (dried, relaxed); **Ov** fusiform, 18 - 20 mm, neck furrowed; **Tep** yellow, tube broadly funnel-shaped, 4 - 5 mm, lobes about equal, 10 - 11 mm.

A 'distinctive oddity' easily distinguished by its linear brittle grey leaves with close-set prominences and small teeth and spines (Gentry 1982: 512). Ullrich (1991b) is the first to record the species from Michoacán.

A. harrisii Trelease (Mem. Nation. Acad. Sci. 11:

34, t. 50-51, 1913). – **Lit:** Adams (1972). **D:** Jamaica; interior limestone plateau, ± 650 m.

[2l] **Ros** solitary; **L** narrowly lanceolate, curved, gradually acute, nearly flat, 100 - 200 × 15 - 35 cm, rather glossy dark green, margins between the teeth straight or concave; marginal teeth straight or curved, narrowly triangular, scarcely 2 mm, 1 - 2 cm apart, often from the tops of green prominences; terminal **Sp** conical, somewhat flexuous or recurved, narrowly channelled towards the base, smooth, 1 - 1.5 cm, reddish-brown, glossy, not decurrent; **Inf** 8 - 10 m, 'paniculate', part-**Inf** with **Br** ± 60 cm; **Ped** rarely > 10 mm; **Fl** 45 - 50 mm; **Ov** fusiform, 25 - 30 mm, distinctly longer than the **Tep**; **Tep** ± 20 mm, yellow, tube open, 7 - 8 mm, lobes erect, 12 - 15 mm; **Fr** narrowly oblong, turbinately narrowed rather than stipitate, shortly beaked, 4.5 - 5 × 1.5 - 2 cm.

A. hauniensis J. B. Petersen (Bot. Tidsskr. 48: 158-159, 1947). **T:** Cult. BG Kobenhaven (*Anonymus* P1875/459 [C]). – **D:** Mexico (Guerrero, México, Morelos); lava fields, rocky slopes in oak woods or in full sun in glades, 700 - 2010 m, flowers in November.
≡ *Manfreda hauniensis* (J. B. Petersen) Verhoek-Williams (1978); **incl.** *Manfreda insignis* Matuda (1966); **incl.** *Manfreda malinaltenangensis* Matuda (1976).

[3a1?] Plants very large (for Subgen. *Manfreda*), reproducing vegetatively by lateral shoots from the storage rhizome, rhizome 7 cm ∅; **L** arching, linear-lanceolate, shallowly channelled, somewhat coriaceous, tip acute, with a long pungent point, (35-) 49 - 77 (-92) × 4 - 10.5 cm, green, slightly glaucous on the lower face, margins cartilaginous, hyaline, sometimes streaked with dark green or purple, irregularly denticulate to denticulate-erose, teeth small and simple to large and bifid to trifid; **Inf** 2 - 3.5 (-3.8) m, 'spicate', flowering part dense, (22.5-) 41.5 cm or more, with 23 - 40 or more **Fl**; **Fl** sessile, erect-spreading, fleshy; **Ov** ellipsoid to ovoid, 12 - 22 mm; **Tep** tube funnel-shaped, nearly straight, 14 - 20 mm, lobes spreading, oblong, cucullate, 20 - 27 (35 - 46 in cult.) mm, yellowish-green or also dark red within; **Fil** spreading, reddish; **Sty** exserted from the tube for 72 - 108 mm, reddish-brown; **Sti** clavate, trigonous; **Fr** ovoid to oblong, rounded below, ± 3 × 1.2 - 2 cm.

Best distinguished within Subgen. *Manfreda* by the very large size of the plants and the long pungent point terminating the leaves (Verhoek-Williams 1975: 229). *Manfreda malinaltenangensis*, which is compared with *M. insignis* in the protologue, falls well within the range of *A. hauniensis* and is thus here placed in its synonymy.

A. havardiana Trelease (Annual Rep. Missouri Bot. Gard. 22: 91, 1912). **T:** USA, Texas (*Havard* s.n. [MO]). – **D:** USA (Texas: Big Bend region),

Mexico (N Chihuahua, N Coahuila); rocky slopes in grassland, frequently on limestone, 1240 - 2000 m. **I:** Gentry (1982: 522, 527, 533). **Fig. III.e**

[2i] **Ros** rather open, 0.5 - 0.8 × 1 - 1.6 m, mostly solitary, suckering sparingly; **L** ovate-acuminate, thick, rigid, broadest at the clasping base, slightly narrowed above the base, widest below the middle, rounded below, concave above, 30 - 60 × 15 - 20 cm, rarely larger, glaucous-grey to light green, occasionally yellowish; uppermost marginal teeth ± straight, other teeth reflexed, numerous, larger teeth towards the **L** tip, mostly 7 - 10 mm, gradually diminishing downwards, 1.5 - 2 cm apart; terminal **Sp** stout, straight to sinuous, roundedly keeled below, broadly grooved above, mostly 3 - 5 (-8 or even 10) cm, dark brown to greyish, long decurrent, sometimes as a complete horny margin; **Inf** 2 - 4 m, 'paniculate', broad, open, part-**Inf** large, 12 - 20; **Fl** 68 - 88 mm; **Ov** 30 - 40 mm, green, neck short, thick; **Tep** yellow, tube deeply funnel-shaped, 14 - 22 mm, lobes unequal, 18 - 24 mm.

Distinguished within Group *Parryanae* by the very broad-based acuminate leaves with reflexed teeth, and tepals forming a deep tube and with relatively short lobes (Gentry 1982: 535).

A. hiemiflora Gentry (Agaves Cont. North Amer., 480-482, ills., 1982). **T:** Guatemala (*Gentry* 23640 [US, DES, MEXU]). – **D:** Mexico (Chiapas), Guatemala.

[2k] **Ros** compact, solitary; **L** 50 - 90, lanceolate, openly spreading, rather softly fleshy, gradually narrowed and thickened towards the base, acuminate, plane to slightly hollowed above, mostly 30 - 55 × 10 - 15 cm, light grey-glaucous to pale green, margins undulate to deeply crenate; larger marginal teeth 5 - 8 mm in the middle of the lamina, 1 - 3 cm apart, light to dark brown, or teeth smaller and on undulate margins, on prominences, slender cusps variously curved up or down; terminal **Sp** slender or thick, sinuous or contorted to straight, openly grooved to flat above, generally 2 - 4 cm; **Inf** 4 - 5 m, 'paniculate', slender, narrow, part-**Inf** small, 20 - 30 in the upper ½ - ⅔ of the **Inf**; **Fl** slender, 50 - 70 mm; **Ov** 25 - 40 mm, neck short, a little constricted; **Tep** in bud sometimes red, opening yellow, tube funnel-shaped, 8 - 13 mm, lobes unequal, 16 - 23 mm.

A highland relative of *A. seemanniana* and *A. congesta* (Gentry 1982: 480).

A. hookeri Jacobi (Hamburg. Gart.- & Blumenzeit. 22(4): 168, 1866). **Nom. illeg.**, Art. 53.1. **T** [neo]: Ex cult. (*Brown* s.n. [K]). – **Lit:** McVaugh (1989). **D:** Mexico (Jalisco, Michoacán, Guerrero); apparently cultivated only or as spontaneous escape (?). **I:** Gentry (1982: 325, 338-339).

[2c] Stem short, thick; **Ros** large, up to 2 m, solitary; **L** lanceolate, arching in age, thickly fleshy, gradually narrowed toward base and tip, generally

concave above, 120 - 175 × 20 - 25 cm, glaucous to green or yellow-green, margins undulate to crenate, esp. in the middle of the lamina, nearly straight below with small teeth; marginal teeth straight or curved, 8 - 12 mm (middle of the lamina), dark brown to greyish-brown, 2 - 5 cm apart, with few smaller intermittent teeth, much reduced and closely spaced towards the base, broadly based on fleshy prominences; terminal **Sp** subulate, 3.5 - 6 cm, edges decurrent as smooth horny **L** border for 15 - 20 cm; **Inf** 7 - 8 m, 'paniculate', part-**Inf** compact, 20 - 40 in the upper ½ of the **Inf**; **Ped** long; **Fl** slender, 63 - 80 mm; **Ov** 34 - 41 mm, neck long, constricted; **Tep** in bud red to pink, opening yellow, lobes red to pink, tube 5 - 8 mm, lobes unequal, 28 - 32 mm.

Recognizable among the species of Group *Crenatae* by its large size, the glaucous leaves with long-decurrent spine-bases, the short flower tube and very long tepal lobes (Gentry 1982: 340). The name unfortunately is an illegitimate homonym of *A. hookeri* Koch 1865 and must be proposed for conservation.

A. horrida Lemaire *ex* Jacobi (Hamburg. Gart.- & Blumenzeit. 22: 64, 1866). **T** [neo]: Mexico, Morelos (*Pringle* 8206 [US, NY]). − **D:** S Mexico.

A. horrida ssp. **horrida** − **D:** Mexico (San Luis Potosí, México, Morelos); volcanic rocks and mountains. **I:** Gentry (1982: 145).

Incl. *Agave horrida* var. *latifrons* hort. *ex* Besaucèle (s.a.); **incl.** *Agave horrida* var. *monstruosa* hort. *ex* Besaucèle (s.a.); **incl.** *Agave horrida* var. *recurvispina* hort. *ex* Besaucèle (s.a.); **incl.** *Agave horrida* var. *viridis* hort. *ex* Besaucèle (s.a.); **incl.** *Agave regelii* hort. *ex* Besaucèle (s.a.); **incl.** *Agave grandidentata* Hort. Belg. *ex* Jacobi (1866); **incl.** *Agave maigretiana* Jacobi (1866); **incl.** *Agave gilbeyi* Hort. Haage & Schmidt (1873) ≡ *Agave horrida* var. *gilbeyi* (Hort. Haage & Schmidt) Baker (1877); **incl.** *Agave horrida* var. *macrodonta* Baker (1877); **incl.** *Agave desmetiana* hort. *ex* Baker (1877) (*nom. illeg.*, Art. 53.1); **incl.** *Agave regeliana* hort. *ex* Baker (1877) (*nom. illeg.*, Art. 53.1); **incl.** *Agave artichaut* Hort. C. Besserer *ex* A. Berger (1915); **incl.** *Agave killischkii* hort. *ex* A. Berger (1915).

[1g] **Ros** compact, small, solitary; **L** 80 - 100 in mature **Ros**, ovate to elliptic-lanceolate, patulous, rigidly thick-fleshy, slightly narrowed above the base, short-acuminate, convex below, plane to hollowed above, generally 18 - 35 × 4 - 7.5 cm, yellowish-green to green, margins thickly horny, straight to sinuous between the teeth; marginal teeth straight to variously curved, broadly flattened at the base, even hooked, large, generally 10 - 15 mm, rarely much smaller, light grey, 5 - 10 mm apart, continuing to near the base of the terminal **Sp**; terminal **Sp** semicircular to subdeltoid in cross-section, very pungent, flattened rather than grooved above, 2.5 -

4 cm; **Inf** 2 - 2.5 m, 'spicate', slender, scape 1 - 1.5 m, part-**Inf** with 1 - 2 **Fl**; **Ped** slender, 4 - 8 mm; **Fl** 35 - 40 mm; **Ov** fusiform, 17 - 20 mm, neck constricted, smooth or slightly grooved; **Tep** dark purple red or yellow, tube shortly funnel-shaped, 3 - 5 mm, lobes equal, 15 - 16 mm.

Distinguished from the closely related *A. ghiesbreghtii* by larger, more numerous marginal teeth continuing nearly to the base of the terminal spine, but sometimes hardly separable (Gentry 1982: 146).

A. horrida ssp. **perotensis** B. Ullrich (Cact. Suc. Mex. 35(4): 80, ill. (p. 96), 1990). **T:** Mexico, Veracruz (*Gentry & al.* 20417 [US, DES, MEXU]). − **D:** Mexico (N Puebla, C Veracruz). **Fig. III.f**

[1g] Differs from ssp. *horrida*: **Ros** small to medium-sized; **L** 25 - 40 × 5 - 8 cm, pale green to green; marginal teeth variable, straight to curved or flexuous, frequently slanted downwards and curved, < 1 cm or 2 - 3 cm apart; terminal **Sp** conical to subulate, 3 - 5 cm, broadly decurrent; **Inf** long, tapering, 3 - 5 m, densely flowered in the upper ⅔ in spiralling sequence, part-**Inf** with geminate **Fl**.

Plants belonging to this taxon were previously misinterpreted as *A. obscura* (see there).

A. howardii (Verhoek-Williams) Thiede & Eggli (KuaS 50(5): 112, 1999). **T:** Mexico, Colima (*Howard & al.* 72-70 [RSA]). − **Lit:** Verhoek-Williams (1976: with ill.); McVaugh (1989). **D:** Mexico (Jalisco, Colima); well-drained soils in partial shade in oak or tropical deciduous forests, 1000 - 1100 m, flowers July to August.

≡ *Polianthes howardii* Verhoek-Williams (1976).

[3b1] Plants glabrous; **R** fleshy; **L** 5 - 6, in a basal **Ros** from a fibrous-coated bulb, erect-spreading, narrowly oblanceolate to linear, tip acute or mucronate, 22 - 27 (-36) × (1-) 1.5 - 2.5 cm, glossy green, lower face sometimes flecked with magenta, margins entire; **Inf** 0.6 - 1.1 m, 'spicate', flowering part 20 - 70 cm, with 13 - 30 widely spaced solitary semi-pendent **Fl**; **Ped** erect, 17 - 29 (-50) mm; **Ov** 9 - 16 (-19) mm; **Tep** glaucous, outer face coral-red at the base, grading to green in the distal ⅓, irregularly streaked with yellow, inner face greenish-yellow, often with maroon stripes in the tube, tube nearly straight, at a slight angle with the **Ov**, mouth slightly oblique, 3 - 5 mm ∅, lobes rounded, slightly flaring, 1.5 - 3 mm; **St** included or **Anth** exserted for 2 mm; **Sty** white, with 3 reflexed lobes; **Fr** globose, 0.8 - 1 cm ∅.

A. hurteri Trelease (Trans. Acad. Sci. St. Louis 23(3): 136, t. 8-10, 1915). **T:** Guatemala (*Trelease* 3 [ILL]). − **D:** NW Guatemala; mountains in the pine-oak forest zone, 1800 - 3300 m. **I:** Gentry (1982: 483-484).

Incl. *Agave samalana* Trelease (1915).

[2k] Stem thick, short; **Ros** rather open, 1 - 1.8 × 2 - 3 m, solitary; **L** numerous, lanceolate, out-

curving to ascending, broadest at or above the middle, acuminate, plane to slightly hollowed above, slightly rough above, more asperous below, mostly 70 - 130 × 15 - 22 cm, light glaucous to pale green and yellow-green, margins ± straight; marginal teeth straight to curved, small to moderate, larger teeth 3 - 8 mm (middle of the lamina), dark brown, 1 - 3 cm apart, rarely smaller and closer or margins quite toothless; terminal **Sp** subulate, broad at the base, openly grooved above, 4 - 6 cm, dark brown to greyish-brown; **Inf** stout, narrow, 5 - 7 m, 'paniculate', scape short, part-**Inf** rounded, 30 - 45 in the upper ⅔ of the **Inf**; **Fl** 55 - 85 mm; **Ov** cylindrical, 30 - 45 mm, neck short; **Tep** greenish-yellow to purple-tinged, tube funnel-shaped to angulate-cylindrical, 9 - 15 mm, lobes unequal, outer 16 - 28 mm.

A variable complex, usually distinguishable from other members of the Group *Guatemalenses* with rounded part-inflorescences by its larger many-leaved rosettes and consistently longer stalked part-inflorescences (Gentry 1982: 485).

A. impressa Gentry (Agaves Cont. North Amer., 146-149, ills., 1982). **T:** Mexico, Sinaloa (*Gentry 23366* [US, ARIZ, DES, MEXU]). – **Lit:** McVaugh (1989). **D:** Mexico (Sinaloa, Nayarit); volcanic rocks in the hot lowlands.

[1g] **Ros** subacaulescent, openly spreading, small to medium-sized, solitary; **L** linear to lanceolate, rigidly spreading, thickly fleshy with viscid adhesive sap and few fibres, convex below, plane to hollowed above, 40 - 60 × 5 - 9 cm, pale yellowish-green, with conspicuous white imprints from the central bud on the upper face, margins horny, continuous, straight to sinuous between the teeth, 2 - 3 mm wide, light to dark grey; marginal teeth straight or slightly curved, flattened, regular, blunt, mostly 3 - 5 mm, grey like the **L** margin, 1 - 1.5 cm apart; terminal **Sp** subulate, stout, rounded below, flat and broad at the base above, rarely channelled, sharp to blunt at the tip, 3 - 5 cm; **Inf** erect, 2 - 3 m, 'spicate', flowering from near the base, part-**Inf** with 2 - 3 **Fl**; **Ped** slender, 2 - 2.5 cm; **Fl** 35 - 40 mm; **Ov** slender, fusiform, 17 - 20 mm; **Tep** green in bud, opening yellow, tube short, spreading, 1.5 - 2 mm, lobes equal, 17 - 18 mm.

A distinctive species without close relatives. Its placement within Group *Marginatae* is for convenience only (Gentry 1982: 148).

A. inaequidens Koch (Wochenschr. Vereines Beförd. Gartenbaues Königl. Preuss. Staaten 3: 28, 1860). **T** [neo]: Mexico, México (*Gentry & al. 19612* [US, DES, MEXU, MICH]). – **D:** C Mexico.

A. inaequidens ssp. **barrancensis** Gentry (Agaves Cont. North Amer., 342-344, ills., 1982). **T:** Mexico, Durango (*Gentry & Arguelles 22282* [US, ME-XU, DES]). – **D:** Mexico (Durango); mountainous slopes of deep barrancas in the pine-oak forest region, 1800 - 2400 m.

[2c] Differs from ssp. *inaequidens*: **Ros** large, 1.5 - 2 × 3 - 3.5 m; **L** mostly longer and narrower, 100 - 170 (-200) × 10 - 16 cm, margins nearly straight to undulate; terminal **Sp** subulate-acicular and longer, 4 - 6 cm; **Inf** broader and shorter, part-**Inf** 20 - 30, broadly spreading.

A. inaequidens ssp. **inaequidens** – **D:** Mexico (Jalisco, Hidalgo, Michoacán, México, D.F., Morelos, Puebla); rocky slopes in pine-oak forest, mostly 1800 - 2400 m. **I:** Gentry (1982: 325, 341).

Incl. *Agave amoena* hort. *ex* Lemaire *ex* Jacobi (s.a.); **incl.** *Agave mescal* Koch (1865); **incl.** *Agave crenata* Jacobi (1866); **incl.** *Agave megalacantha* Hemsley (1880); **incl.** *Agave reginae* hort. *ex* A. Berger (1912); **incl.** *Agave heterodon* hort. *ex* A. Berger (1915); **incl.** *Agave bourgaei* Trelease (1920).

[2c] Stem short; **Ros** openly spreading, medium-sized to large, solitary; **L** variable, broadly or narrowly lanceolate or oblanceolate, ascending to outcurving, thickly fleshy, concave above, esp. towards the rounded base, mostly 75 - 150 × 11 - 21 cm, light green to yellow-green, rarely faintly glaucous, margins undulate to repand and crenate; marginal teeth dimorphic, straight or variously curved, the flattened bases longer than the height of the teeth, commonly 8 - 10 mm long, chestnut-brown to dark brown, 2.5 - 4 cm apart, with few smaller intermittent teeth, larger teeth on broad prominences; terminal **Sp** stout, broadly deeply channelled above, 2.5 - 5.5 cm, dark brown, protruding into the **L** tissue below, sharply decurrent to the uppermost marginal teeth; **Inf** 5 - 8 m, 'paniculate', narrow, scape short, part-**Inf** compact, 30 - 50 in the upper ½ of the **Inf**; **Fl** 60 - 90 mm; **Ov** 30 - 40 mm, neck short; **Tep** reddish-purple, opening yellow, lobes reddish, tube 5 - 12 (-15) mm, lobes unequal, 25 - 30 (-34) mm.

Best distinguished from the closely related *A. hookeri* from the same region by its bright yellowish-green leaves (Gentry 1982: 341).

A. inaguensis Trelease (Mem. Nation. Acad. Sci. 11: 47, t. 103-105, 1913). **T** [syn]: Bahamas, Little Inagua (*Nash & Taylor 342* etc. [MO]). – **Lit:** Correll & Correll (1982). **D:** Bahamas (Little Inagua, Caicos Islands); open sandy flats and rocky dwarf coastal coppices.

[2p] **Ros** caespitose, freely suckering; **L** oblong or lanceolate, rather abruptly acute and flat, sometimes conduplicate, 40 - 100 × 6 - 9 cm, typically white-glaucous; marginal teeth more recurved and less uniform than in *A. nashii*, very narrowly triangular, 1 - 2 mm, almost continuously joined by a narrow blackish border, 3 - 9 mm apart; terminal **Sp** 2 - 3 cm, dark brown; **Inf** 4 - 5 m, 'paniculate',

part-**Inf** on slender outcurved **Br** in the upper ¼ of the **Inf**; **Ped** 5 - 10 mm; **Fl** ± 50 mm; **Ov** subfusiform, 25 - 30 mm; **Tep** 15 - 17 × ± 5 mm, yellow, tube open, ± 5 mm; **Fr** oblong-ellipsoid, 3 - 4 cm, broadly and shortly stipitate and beaked.

According to the protologue with the habit of *A. nashii*.

A. indagatorum Trelease (Mem. Nation. Acad. Sci. 11: 42, t. 92, 1913). **T:** Bahamas, Watling Island (*Britton & Millspaugh* 6155 [NY]). − **Lit:** Correll & Correll (1982). **D:** Bahamas (Watling Island); rocky soil in coastal coppices.

[2n] Acaulescent; **Ros** solitary; **L** lanceolate, gradually acute, somewhat concave, 150 - 250 × 20 - 25 cm, somewhat greyish and at first very glaucous beneath, margins between the teeth straight, somewhat membranous, at first slightly pink; marginal teeth straight or slightly recurved, narrowly triangular, not lenticular at the base, ± 1 mm, 5 - 12 mm apart; terminal **Sp** conical, nearly straight, involutely grooved to the middle, smooth, ± 1.2 cm, chestnut-brown, rather glossy, decurrent for about its own length; **Inf** to ± 9 m, 'paniculate', reportedly bulbilliferous; **Ped** 15 - 20 mm; **Fl** ± 55 mm; **Ov** ellipsoid, 20 - 25 mm, extended as a neck into the tube; **Tep** ± 20 mm, yellow, tube rather open, ± 6 mm; **Fr** narrowly oblong, thickly stipitate, acuminately pointed, 3.5 - 6 × ± 2 cm.

A. intermixta Trelease (Mem. Nation. Acad. Sci. 11: 32, t. 64, 1913). **T:** Hispaniola, Haiti (*Parry* s.n. [Herb. Engelmann [MO?]]). − **D:** Haiti.

[2l] **Ros** and **L** unknown (!); **Ped** slender, 15 - 25 mm; **Fl** ± 65 mm, congested at the ends of the part-**Inf**; **Ov** elongated-fusiform, 35 mm, longer than the **Tep**; **Tep** yellow (?), tube narrowly conical, ± 8 mm, lobes 20 × 4 mm; **Fil** inserted almost at the throat, 30 mm; **Fr** rather broadly pear-shaped and oblong, stipitate and beaked, 4 × 2 cm; **Se** 8 × 5 mm.

See the note for *A. antillarum*.

A. involuta (McVaugh) Thiede & Eggli (KuaS 50(5): 110, 1999). **T:** Mexico, Jalisco (*Bauml & Voss* 1466 [RSA, MICH]). − **D:** Mexico (N Jalisco); grassy openings and hillsides in dry oak or pine-oak forest, 1500 - 2000 m, flowers March to April.

≡ *Manfreda involuta* McVaugh (1989).

[3a3] Plants 85 - 140 cm; **L** up to 10 or more, linear, tightly folded and often appearing tubular, 30 - 50 × 0.2 - 0.5 cm, margins (and veins of the lower face) papillose (sometimes obscurely so); **Inf** 'spicate', flowering part 10 - 20 cm, with 3 - 7 **Fl**; **Fl** sessile, ascending at maturity; **Ov** ellipsoid, 6 - 9 mm; **Tep** greenish with purple cast, drying glaucous greenish-purple, tube narrowly funnel-shaped, slenderly cylindrical towards the base but not constricted there, 7 - 13 mm, lobes spreading-ascending

or somewhat recurved, 8 - 10 mm, usually shorter than the tube; **Fil** apparently purplish, surpassing the tube by ± 20 mm; **Sty** longer than the **St**; **Fr** and **Se** not known.

According to the protologue (McVaugh 1989: 232-233) known from 2 collections only and of uncertain affinities due to its apparently unique combination of a very early flowering season, very narrow involute leaves, and slender tepal tubes longer than the lobes. With its papillate leaf margins, the ovary protruding into the tube and the cylindrical tube not narrowed above the ovary, the taxon fits well into the *A. guttata* Subgroup, however.

A. isthmensis García-Mendoza & Palma Cruz (Sida 15(4): 565-568, ills., 1993). **T:** Mexico, Chiapas (*García-Mendoza & al.* 4177 [MEXU, BRIT/SMU, DES, ENCB]). − **D:** Mexico (Oaxaca, Chiapas).

[2k] **Ros** compact, 17 - 32 × 25 - 36 cm, young plants with numerous rhizomatous offsets, adult plants with axillar offsets; **L** 84 - 132, ovate, narrowed towards the base, apex truncate, concave, scabrous below, 8 - 10 × 5 - 8 cm, margins deeply crenate; marginal teeth deltoid, (2-) 3 - 4 mm, brown-reddish, < 1 cm apart; terminal **Sp** sinuous, slightly applanate, 1.2 - 1.5 cm, brown-reddish, decurrent for 6 - 10 mm; **Inf** 1.7 - 2.2 m, 'paniculate', oblong, part-**Inf** (6-) 20 - 25; **Ped** 3 - 4 mm; **Fl** succulent, 38 - 46 mm; **Ov** 16 - 21 mm; **Tep** yellow, tube funnel-shaped, trisulcate, 4 - 5 mm, lobes 19 - 21 mm.

According to the protologue closest to *A. potatorum* and *A. seemanniana* (as *A. pygmaea*) and distinct by its compact small many-leaved rosettes with glaucous leaves with scabrous lower face and the rhizomatous as well as axillar offsetting.

A. jaiboli Gentry (US Dept. Agric. Handb. 399: 89-94, ills., 1972). **T:** Mexico, Sonora (*Gentry* 21177 [US]). − **D:** Mexico (Sonora); short-tree forest and oak woodland, 300 - 1000 m. **I:** Gentry (1982: 325, 345-346).

[2c] **Ros** usually open, 0.6 - 1 × 1.4 - 2 m, solitary; **L** linear to lanceolate, usually straightly ascending to spreading, sometimes incurved, widest at or above the middle, gradually narrowed below, long-acuminate, plane to conduplicate, 60 - 100 × 8 - 12 cm, green to yellowish-green, margins not or narrowly horny, decurrent from the **Sp** base for less than the **Sp** length; marginal teeth curved down- or upwards, 5 - 8 mm, reddish-brown, larger teeth ± 2 - 3 cm apart, on small regular prominences, smaller intermittent teeth 1 to several, 1 - 4 mm; terminal **Sp** subulate, terete, smooth, 3 - 4 cm, reddish-brown, shiny; **Inf** 6 - 8 m, 'paniculate', part-**Inf** small, diffuse, 12 - 15 in the upper ½ - ⅓ of the **Inf**, with 4 **Fl**; **Ped** short; **Fl** ± 60 mm; **Ov** 25 - 30 mm the 3 - 4 mm long neck, green; **Tep** yellow, flushed ferrugineous, tube funnel-shaped, 9 - 11 mm, lobes unequal, 22 - 23 mm.

Distinguished within Group *Crenatae* by its narrow almost ensiform leaves (Gentry 1982: 345).

A. jaliscana (Rose) A. Berger (Agaven, 38, 1915). **T** [lecto]: Mexico, Jalisco (*Pringle* 1850 [US, BM, BR, F, G, GH, LE, M, NY, P, VT]). – **D:** Mexico (Sonora, Jalisco, Michoacán); loose black loam or rocks, in pine-oak forests and on grassy slopes, flowers early November to early April.

≡ *Manfreda jaliscana* Rose (1903) ≡ *Polianthes jaliscana* (Rose) Shinners (1966).

[3a2] Plants medium-sized (for Subgen. *Manfreda*), solitary or caespitose; **R** half-fleshy; storage rhizome oblong, 1.3 - 2.2 cm ∅, spreading rhizomes cylindrical; **L** 5 - 10, spreading, linear, channelled, tip acute, with a short point, herbaceous, minutely scattered-papillate, (41-) 49 - 78 (-93) × 0.6 - 1.4 (-2.8) cm, green, unspotted (dried); margins with very narrow hyaline band, minutely papillate (at 12.5× magnification); remains of **L** bases separating into stiff fibres, 4 - 7 cm; **Inf** 1 - 1.55 m, 'spicate', flowering part elongate, semidense, 11 - 38 (-42) cm, with 11 - 40 sessile or occasionally shortly pedicellate **Fl**; mature **Fl** nearly erect; **Ov** narrowly ellipsoid, (6-) 8 - 13 (-16) mm, with a neck; **Tep** tube narrowly funnel-shaped, straight, slightly constricted above the **Ov**, 4 - 10 mm, lobes oblong, narrow, recurved, 9 - 17 mm; **Sty** exceeding the tube by (56-) 62 - 84 (-98) mm; **Sti** clavate, trigonous, deeply furrowed; **Fr** globose to oblong, 1.1 - 2.7 × 1.2 - 1.7 cm; **Se** 3 - 4 × 4 - 6 mm.

Easy to recognize in flower by its long-exserted stamens, the short tepal tube and long lobes. The very long narrow leaves with papillose margins are also characteristic (McVaugh 1989).

A. jarucoensis A. Álvarez (Revista Jard. Bot. Nac. Univ. Habana 1(1): 5-11, 1981). **T:** Cuba (*Álvarez* 41680 [HAJB]). – **Lit:** Álvarez de Zayas (1985: with ills.). **D:** W Cuba; limestone rocks and cliffs.

[2m] **Ros** solitary; **L** many, lanceolate, weak, flexuous, fleshy, coriaceous, slightly concave in the lower ⅔, 100 - 120 × 12 - 16 cm, green, slightly opaque, margins straight; marginal teeth triangular, straight or weakly curved towards the **L** base, 1 - 4 mm, castaneous-reddish, base reddish, 4 - 15 mm apart; terminal **Sp** acicular, straight, pungent, 1.7 - 2.2 cm, slightly lustrous, chestnut-brown to grey, not decurrent; **Inf** 2 - 4 m, 'paniculate', part-**Inf** tripartite, up to 45 cm; **Ped** 5 - 15 mm; **Fl** 48 - 60 mm; **Ov** subcylindrical or fusiform, 18 - 24 mm; **Tep** greenish-yellow, tube 5 - 7 mm, lobes 13 - 16 mm; **Fr** oblong to nearly rounded, shortly apiculate, walls thick, 22 - 34 × 15 - 16 mm.

According to the protologue similar to *A. papyrocarpa*, but differing in its more robust habit, larger flowers, and thick fruit walls.

A. karatto Miller (Gard. Dict., Ed. 8, no. 6, 1768). **T** [neo]: St. Kitts (*Britton & al.* s.n. [NY]). – **D:** Windward Islands (Anguilla, Antigua, Barbuda, La Désirade, Montserrat, Nevis, Saba, St. Eustatius, St. Kitts, St. Martin). **I:** KuaS 48: 98-99, 1997.

Incl. *Agave keratto* Haworth (1819) (*nom. inval.*, Art. 61.1); **incl.** *Agave salm-dyckii* Baker (1877); **incl.** *Furcraea gigantea* Boldingh (1909) (*nom. illeg.*, Art. 53.1); **incl.** *Agave nevidis* Trelease (1913); **incl.** *Agave obducta* Trelease (1913); **incl.** *Agave scheuermaniana* Trelease (1913); **incl.** *Agave trankeera* Trelease (1913); **incl.** *Agave vangrolae* Trelease (1913).

[2o] **Ros** solitary; **L** lanceolate, erect, arching, acute, concave above, 130 - 175 × ± 20 cm, (greyish-) green, rather glossy (or dull), margins nearly straight or shallowly concave, at first reddish; marginal teeth straight or variously curved or reflexed, triangular, sometimes with lenticular bases, 2 - 3 mm, brownish, 5 - 15 (-20) mm apart; terminal **Sp** grooved, smooth, polished at the tip, 1 - 1.5 cm, black, ultimate apex recurved-mucronate, 3 - 4 mm, decurrent, dorsally deeply immersed into the green **L** tissue; **Inf** 5 - 10 m, 'paniculate', oblong, part-**Inf** spreading, in the upper ⅓ of the **Inf**, freely bulbilliferous; **Fl** 60 - 65 mm, often aborting; **Ov** subfusiform, 30 - 35 mm; **Tep** golden-yellow, tube openly conical, 7 mm deep, lobes 22 mm; **Fr** broadly oblong, stipitate and short-beaked, to 4.5 × 2 cm, basal stalk 2 - 3 mm.

The broader circumscription of the species here applied follows Hummelinck (1938), Hummelinck (1987) and Hummelinck (1993) and includes several species separated from *A. karatto* by Trelease, some of which were still upheld by recent authors such as Howard & al. (1979).

A. karwinskii Zuccarini (Flora 15: 2(Beiblatt 2): 98, 1832). **T** [neo]: Mexico, Oaxaca (*Gentry* 12049 [US, DES, MEXU, MICH]). – **D:** Mexico (Puebla, Veracruz, Oaxaca); arid regions, 1550 - 1850 m. **I:** Gentry (1982: 555, 578).

Incl. *Agave laxa* Salm-Dyck (1834); **incl.** *Agave karwinskiana* Herbert (1837); **incl.** *Agave corderoyi* Baker (1877); **incl.** *Agave bakeri* Ross (1894).

[2f] Arborescent, stem 2 - 3 m, apparently forming clonal colonies with spreading rhizomes, **Ros** extending down the stem from the stem tip with **L** reflexing along the stems with age; **L** linear-lanceolate, ascending to radiately spreading, narrowed and thickened towards the base, acuminate, involute towards the base of the terminal **Sp**, convex below, guttered or concave above, 40 - 65 × 3 - 7 cm, green, margins straight; marginal teeth delicate, nearly straight to cuspidate and flexuous, pyramidal, 3 - 5 mm (middle of the lamina), dark brown, 2 - 4 cm apart; terminal **Sp** variable, subulate or conical with thickened base, 1.5 - 4 cm, dark brown to greyish and corroding at the base, decurrent or not; **Inf** ± 3 - 3.5 m, 'paniculate', openly diffuse, oval, part-**Inf** 10 - 15 in the upper ⅓ of the **Inf**; **Fl** small, 45 - 57 mm; **Ov** angularly cylindrical, slightly 6-

grooved, 20 - 30 mm, neck short; **Tep** greenish to pale yellow, with ferrugineous tinge, tube 10 - 11 mm, lobes unequal, 11 - 19 mm.

Easily recognized by its stem-forming tall habit with relatively small leaves and small flowers (Gentry 1982: 579).

A. kerchovei Lemaire (Ill. Hort. 11: 64, 1864). **T:** [lecto − icono]: US. − **D:** Mexico (Hidalgo, Puebla, Oaxaca), semi-arid highlands, 1400 - 1875 m. **I:** Gentry (1982: 126, 150). **Fig. IV.c**

≡ *Agave poselgeri* var. *kerchovei* (Lemaire) A. Terracciano (1885); **incl.** *Agave kerchovei* var. *brevifolia* hort. *ex* Besaucèle (s.a.); **incl.** *Agave kerchovei* var. *glauca* hort. *ex* Besaucèle (s.a.); **incl.** *Agave kerchovei* var. *inermis* Ortgies (s.a.); **incl.** *Agave kerchovei* var. *miniata* hort. *ex* Besaucèle (s.a.); **incl.** *Agave kerchovei* var. *variegata* hort. *ex* Besaucèle (s.a.); **incl.** *Agave beaucarnei* Lemaire (1864); **incl.** *Agave kerchovei* var. *diplacantha* Lemaire (1864); **incl.** *Agave kerchovei* var. *distans* Lemaire (1864); **incl.** *Agave kerchovei* var. *macrodonta* Lemaire (1864); **incl.** *Agave kerchovei* var. *pectinata* Baker (1877); **incl.** *Agave horrida* Hort. A. Berger (1898) (*nom. illeg.*, Art. 53.1); **incl.** *Agave expatriata* Rose (1900); **incl.** *Agave convallis* Trelease (1920); **incl.** *Agave dissimulans* Trelease (1920); **incl.** *Agave inopinabilis* Trelease (1920).

[1g] Stem short; **Ros** openly spreading, medium-sized, solitary or caespitose, branching commonly from the lower **L** axils; **L** 80 - 100 and more in mature **Ros**, generally lanceolate, straight to slightly curved, rigid, thick at the base, tip long-acuminate, convex below, plane to hollowed above, 40 - 100 (-125) × 5 - 12 cm, light yellowish-green to green, rarely pruinose, margins generally thick and horny, continuous and straight; marginal teeth variable, straight to variously curved, broadly flattened, larger teeth 8 - 15 mm, 2 - 5 (-7) cm apart, smaller teeth irregularly occuring, grey, margins rarely completely and **L** tip generally toothless; terminal **Sp** stout, deeply channelled above, 3 - 6 cm, brown to grey; **Inf** 2.5 - 5 m, 'spicate', densely flowered in the upper ½ - ⅔; **Fl** 38 - 46 mm; **Ov** fusiform, 18 - 21 mm, neck constricted; **Tep** greenish to purplish, tube openly spreading, 4 - 6 mm, lobes subequal, 15 - 20 mm. − *Cytology:* 2n = 120.

Typically easily distinguishable by the long lanceolate leaves prominently armed with large variable remotely spaced teeth, usually, but not always, lacking along the acuminate apex. The taxon is apparently closely related to *A. ghiesbreghtii* (Gentry 1982: 152).

A. lagunae Trelease (Trans. Acad. Sci. St. Louis 23(3): 143, pl. 21, 1915). **T:** Guatemala (*Trelease* 10 [ILL]). − **D:** Guatemala; mesophytic mountain forest, 1000 m, only known from the region of the type locality.

[2k] **Ros** openly spreading, medium-sized, solitary; **L** few, linear to lanceolate, acuminate, plane to concave, finely asperous, 40 - 70 × 8 - 12 cm, glaucous-green, margins nearly straight; marginal teeth mostly curved, slender, bases slightly elevated, larger teeth 4 - 6 mm (middle of the lamina), dark brown, 1 - 3 cm apart; terminal **Sp** subulate, groove openly channelled to narrow, 3 - 4 cm, dark brown; **Inf** 3 - 4 m, 'paniculate', rather open, part-**Inf** rather small, 15 - 20 in the upper ½ of the **Inf**; **Fl** slender, 60 - 70 mm; **Ov** slender, cylindrical, 39 - 45 mm, neck grooved, long, slender, constricted; **Tep** yellow, tube funnel-shaped, 7 - 10 mm, lobes 18 - 21 mm.

A poorly known species.

A. lechuguilla Torrey (in Emory, Rep. US Mex. Bound. 213, 1858). **T:** USA, Texas (*Wright* 682 [US 125459]). − **D:** USA (S New Mexico, Texas), Mexico (Chihuahua, Coahuila, Nuevo León, Tamaulipas, Durango, Zacatecas, San Luis Potosí, Querétaro, Hidalgo, México, D.F.); Chihuahuan Desert, (500-) 950 - 2300 m. **I:** Gentry (1982: 126, 155). **Fig. IV.d**

Incl. *Agave univittata* var. *recurvispinis* hort. *ex* Besaucèle (s.a.); **incl.** *Agave univittata* var. *viridis* hort. *ex* Besaucèle (s.a.); **incl.** *Agave univittata* var. *zonata* hort. *ex* Besaucèle (s.a.); **incl.** *Agave univittata* var. *foliis striatis* hort. *ex* Besaucèle (s.a.) (*nom. inval.*, Art. 24.2); **incl.** *Agave caerulescens* Salm-Dyck (1859) ≡ *Agave lophantha* var. *caerulescens* (Salm-Dyck) Jacobi (1864) ≡ *Agave univittata* var. *caerulescens* (Salm-Dyck) H. Jacobsen (1973); **incl.** *Agave poselgeri* Salm-Dyck (1859) ≡ *Agave lophantha* var. *poselgeri* (Salm-Dyck) A. Berger (1915); **incl.** *Agave lophantha* var. *gracilior* Jacobi (1864) ≡ *Agave univittata* var. *gracilior* (Jacobi) H. Jacobsen (1973); **incl.** *Agave lophantha* var. *subcanescens* Jacobi (1864) ≡ *Agave univittata* var. *subcanescens* (Jacobi) H. Jacobsen (1973); **incl.** *Agave lophantha* var. *brevifolia* Jacobi (1867) ≡ *Agave univittata* var. *brevifolia* (Jacobi) H. Jacobsen (1973); **incl.** *Agave multilineata* Baker (1888); **incl.** *Agave lophantha* var. *angustifolia* A. Berger (1915) ≡ *Agave univittata* var. *angustifolia* (A. Berger) H. Jacobsen (1973); **incl.** *Agave lophantha* var. *pallida* A. Berger (1915); **incl.** *Agave lophantha* var. *tamaulipasana* A. Berger (1915) ≡ *Agave univittata* var. *tamaulipasana* (A. Berger) H. Jacobsen (1973).

[1g] **Ros** rather open, small, mostly 30 - 50 × 40 - 60 cm, freely suckering; **L** few, linear-lanceolate, mostly ascending to erect, sometimes falcately spreading, thick, stiff, deeply convex below, concave above, 25 - 50 × 2.5 - 4 cm, light green to yellow-green, sometimes marked with green, margins straight and continuous, light brown to grey, easily separable from dry **L**; marginal teeth typically deflected, weak and friable, regular in size, 2 - 5 mm, brown or mostly light grey, mostly 1.5 - 3 cm apart, 8 - 20 on each margin; terminal **Sp** conical to subu-

late, strong, 1.5 - 4 cm, greyish; **Inf** 2.5 - 3.5 m, 'spicate', scape generally glaucous, part-**Inf** mainly with 2 - 3 **Fl**, rarely ascending with longer stalks (2 - 15 cm) and several- to many-flowered; **Fl** 30 - 45 mm; **Ov** fusiform, 15 - 22 mm, neck constricted; **Tep** yellow or frequently tinged with red or purple, tube open, 2.5 - 4 mm, lobes subequal, 13 - 20 mm. − *Cytology:* 2n = 120.

Usually easily recognizable by its widely suckering habit and narrow leaves with down-slanted teeth on straight margins. The taxon is a widespread characteristic component of varied Chihuahuan Desert communities and an important source of hard fibres (Gentry 1982: 154).

A. legrelliana Jacobi (Hamburg. Gart.- & Blumenzeit. 21: 567, 1865). − **D:** W Cuba. **I:** Berger (1915: 210).

Incl. *Agave americana* Sagra (1850) (*nom. illeg.*, Art. 53.1); **incl.** *Agave melanacantha* Lemaire *ex* Jacobi (1865); **incl.** *Agave laurentiana* Jacobi (1866); **incl.** *Agave coccinea* Lachaume (1876) (*nom. illeg.*, Art. 53.1); **incl.** *Agave legrelliana* var. *breviflora* Trelease (1913).

[21] **Ros** solitary; **L** variously lanceolate, subacuminate, concave, sometimes conduplicate, 100 - 200 × 20 - 30 cm, dark green, margins often concave; marginal teeth usually downcurved below, narrowly triangular, acuminately tapered, 2 - 6 mm, 1 - 1.5 (- 2) cm apart, or from abrupt green prominences whose tops harden lunately; terminal **Sp** conically subulate, a little curved, openly or flatly grooved below the middle or involute, smooth, 1.5 - 2 cm, brown, dull, scarcely decurrent; **Inf** 6 - 8 m, 'paniculate', amply ovoid throughout, part-**Inf** on ascending-recurved **Br**, not known to be bulbilliferous; **Ped** 20 - 30 mm; **Fl** (55-) 70 - 80 mm; **Ov** oblong, somewhat contracted at the base and at the top, ≤ 40 - 45 mm; **Tep** deep orange, tube ± 15 mm, lobes 30 mm; **Fr** rather narrowly oblong, stipitate and slightly beaked, 4 - 5 × 1.5 - 2 cm.

A. ×leopoldii Hort. *ex* G. Nicholson (Dict. Gard. Suppl., 82, 1900). − **Fig. III.a, III.g**

Garden hybrid *A. filifera* × *A. filifera* ssp. *schidigera* (Berger 1915: 76).

A. longibracteata (Verhoek-Williams) Thiede & Eggli (KuaS 50(5): 110, 1999). **T:** Mexico, Michoacán (*Verhoek-Williams & al.* 613 [US, BH, MEXU]). − **D:** Mexico (Michoacán); among rocks, ± 2070 m. **I:** Verhoek-Williams (1978).

≡ *Manfreda longibracteata* Verhoek-Williams (1978).

[3a2] Plants large (for Subgen. *Manfreda*), reproducing vegetatively by stoloniferous rhizomes from the base of the parent rhizome, rhizome globose, to 5 cm ∅; **R** wiry; **L** up to 14, erect-spreading, linear-lanceolate to lanceolate, broadly channelled,

sometimes undulate, brittle, herbaceous, tip obtuse, with a very short point, 21 - 39 × 1.5 - 3.5 (-4.3) cm, bright green, rarely with magenta spots below on the basal ⅓, margins entire, slightly revolute; remains of **L** bases coarsely fibrous, to 6 cm; **Inf** to 1.25 m, 'spicate', flowering part open, with 15 - 21 **Fl**; floral **Bra** large, narrowly triangular; **Fl** sessile, spreading outwards from the axis, scent citronella-like; **Ov** oblong, protruding into the tube, 12 - 15 mm; **Tep** greenish-grey on the outer face, sometimes flushed with brown or darker green above, golden-green within, tube narrowly funnel-shaped, slightly curved, 15 - 20 mm, lobes oblong, upper lobes erect-spreading, lower lobes reflexed, 12 - 15 mm; **Fil** flattened; **Sty** exserted for 22 - 25 (-32) mm; **Fr** ellipsoid, 2 - 2.6 × 1 cm; **Se** 3 - 4 × 3 - 4.5 mm.

According to the protologue (Verhoek-Williams 1978) vegetatively most noticeable for its crisp bright green leaves with a broad round channel; in bud or flower recognizable by the long floral bracts. It appears to be closest to *A. scabra* (as *Manfreda brachystachya*).

A. longiflora (Rose) G. D. Rowley (Repert. Pl. Succ. 26: 4, 1977). **T:** USA, Texas (*Runyon* 10 [US, NY]). − **Lit:** Cházaro Basáñez & Machuca Núñez (1995). **D:** USA (Texas), Mexico (Tamaulipas); clay slopes, dry gravelly hills and prairies in sandy loam overlying caliche, flowers September. **I:** Addisonia 7: pl. 244, 1922, as *Runyonia*.

≡ *Runyonia longiflora* Rose (1922) ≡ *Manfreda longiflora* (Rose) Verhoek-Williams (1975); **incl.** *Runyonia tenuiflora* Rose *in sched.* (s.a.) (*nom. inval.*, Art. 29.1); **incl.** *Runyonia tubiflora* Rose *in sched.* (s.a.) (*nom. inval.*, Art. 29.1); **incl.** *Polianthes runyonii* Shinners (1966).

[3a1] Plants medium-sized (for Subgen. *Manfreda*), rhizome to 6.5 × 2 cm; **R** fleshy; **L** 3 - 7 (-15), lanceolate, channelled, fleshy, tip acute, with a medium-sized point, to 26.5 × 1.4 (-2 in cultivation) cm, green with darker green or brown spots over the whole **L**, margins with coarse distantly-spaced cartilaginous (occasionally retrorse) teeth; membranous **L** bases covering the plant base, 2 - 5 cm; **Inf** to 50 (-96 in cultivation) cm, 'spicate', flowering part 8 - 20 (-35 in cultivation) cm, with 10 - 21 densely or laxly arranged **Fl**; **Fl** sessile, erect; **Ov** ellipsoid, 4 - 6 mm; **Tep** tube straight, narrowly funnel-shaped, 23 - 36 mm, lobes oblong, revolute, tip obtuse, with a small tuft of **Ha**, 8 - 14 (-19) mm; **St** very short, attached at the mouth of the tube; **Anth** 5 - 6 mm; **Sty** included; **Sti** 3-lobed, papillate; **Fr** depressed-globose, 0.9 - 1 × 1 - 1.3 cm; **Se** 3 × 4 mm.

This species is closely related to *A. maculosa*, from which it is distinguished by the longer narrow tepal tube, almost sessile anthers, and included styles, and smaller and more flattened fruits (Verhoek-Williams 1975: 224).

A. longipes Trelease (Mem. Nation. Acad. Sci. 11: 36, t. 63, 1913). **T:** Jamaica (*Maxon* 1624 [not indicated]). − **Lit:** Adams (1972). **D:** Jamaica (St. Andrews); local on well-drained slopes, 1000 - 1200 m.

[2l] **L** as in the broader-leaved forms of *A. sobolifera*, curved; marginal teeth narrowly triangular, often appressed-recurved; terminal **Sp** sometimes much compressed and conical, more strongly and persistently flattened on the upper face and less involute, often grey; **Inf** 'paniculate', not known to be bulbilliferous; **Ped** 20 mm; **Fl** larger, 60 - 70 mm; **Ov** oblong-fusiform, 30 - 40 mm; **Tep** yellow, tube openly conical, 6 - 8 mm, lobes 20 - 25 mm; **Fil** 50 - 60 mm; **Fr** unknown.

Similar to *A. sobolifera*, but with larger flowers and longer filaments (Adams 1972).

A. lophantha Schiede (Linnaea 4: 582, 1829). − **D:** USA (Texas), Mexico (Coahuila, Nuevo León, Tamaulipas, San Luis Potosí, Veracruz, Puebla); frequent on limestone, 30 - 1500 m. **I:** Gentry (1982: 126, 157, 160). **Fig. IV.e**

Incl. *Agave mezortillo* hort. (s.a.) (*nom. inval.*, Art. 29.1); **incl.** *Agave univittata* Haworth (1831); **incl.** *Agave heteracantha* Zuccarini (1832) ≡ *Agave univittata* var. *heteracantha* (Zuccarini) Breitung (1959).

[1g] Stem sometimes visible on old **Ros**; **Ros** small, 0.3 - 0.6 × 0.5 - 1 m, solitary or surculose; **L** numerous, linear to lanceolate, radiating, patulous, rather thin, pliable, somewhat thickened towards the base and rounded below, plane to concave above, generally 30 - 70 × 3 - 5 cm, light green to yellow-green, with or without pale mid-stripe, margins horny, undulate to crenate; marginal teeth straight or mildly curved, slender, single or occasionally double, mostly 4 - 8 mm, 1 - 2 cm apart, on broad low prominences; terminal **Sp** subulate, flattened above at the base, small, 1 - 2 cm, ferrugineous to grey; **Inf** slender, 'spicate', **Fl** in the upper ½, part-**Inf** with 1 - 2 **Fl** or also with 3 - 7 **Fl** on short stalks; **Ped** 5 - 10 mm; **Fl** 35 - 47 mm; **Ov** fusiform, 18 - 22 mm, neck short or long (5 - 7 mm) and constricted; **Tep** light grey-glaucous-green to yellow, tube short, open, 2 - 4 mm, lobes subequal, 14 - 20 mm.

This species has often been misinterpreted. It is distinguished from the closely related *A. lechuguilla* by its flatter leaves with sinuous to undulate firm borders, tubercles usually with at least one double set of teeth (frequently more), and with teeth that are more slender and closely set (Gentry 1982: 159).

A. lurida Aiton (Hort. Kew. 1: 472, 1789). **T** [neo]: Ex cult. (*Masters* s.n. [K]). − **D:** Mexico (Oaxaca); semi-arid tropical forest, 1850 m. **I:** Gentry (1982: 276, 292-293).

Incl. *Agave vera-cruz* Miller (1768); **incl.** *Agave mexicana* Lamarck (1783); **incl.** *Agave verae-crucis* Haworth (1812); **incl.** *Agave magni* Desfontaines (1815); **incl.** *Agave mangui* Desfontaines (1815) (*nom. inval.*, Art. 61.1?); **incl.** *Agave lepida* D. Dietrich (1840); **incl.** *Agave polyphylla* C. Koch (1860); **incl.** *Agave vernae* A. Berger (1915); **incl.** *Agave breviscapa* A. Berger *ex* Roster (1916).

[2a] Stem short; **Ros** radially symmetrical, 1.2 - 1.7 × 2.4 - 3.4 m, solitary or only rarely surculose; **L** linear-lanceolate, stiffly ascending to outcurving, concave to guttered and thinning beyond the slightly narrowed base, 110 - 150 × 12 - 18 cm, dull green to glaucous-grey, margins nearly straight; marginal teeth very regular, larger teeth 4 - 6 mm, mostly 1 - 2 cm apart, smaller and closer together towards the **L** base, with low black bases, on low protuberances, cusps usually deltoid-flattened, straight or curved, brown to greyish; terminal **Sp** conically subulate, 3 - 4.5 cm, greyish-brown, decurrent for several cm; **Inf** 6 - 7 m, 'paniculate', part-**Inf** ≥ 20, ascending, diffusely spreading, several times compound, open, in the upper ⅓ - ½ of the **Inf**; **Fl** 58 - 65 mm; **Ov** 28 - 34 mm, neck constricted, grooved; **Tep** greenish-yellow, tube funnel-shaped, 9 - 11 mm, lobes about equal, 18 - 24 mm.

Long only known from cultivation in Europe and without close relatives. *A. vera-cruz* and *A. mexicana* would have priority over *A. lurida* if their identity with this species can be ascertained. *A. lurida* was frequently mentioned in the older literature, but its identity remains doubtful. It was rediscovered after ± 200 years by Gentry on one spot in Oaxaca (Gentry 1982: 293), but Ullrich (1991c) regards the neotypification of Gentry and the correlation of the re-collected material with the name *A. lurida* as doubtful.

A. macroacantha Zuccarini (Flora 15: 2(Beiblatt 2): 97, 1832). **T** [neo]: Mexico, Puebla (*Gentry & al.* 20242 [US]). − **D:** Mexico (Puebla [esp. around Tehuacán], Oaxaca). **I:** Gentry (1982: 578). **Fig. IV.f**

≡ *Agave flavescens* var. *macroacantha* (Zuccarini) Jacobi (1864); **incl.** *Agave pugioniformis* Zuccarini (1832); **incl.** *Agave flavescens* Salm-Dyck (1834); **incl.** *Agave macracantha* Herbert (1837) (*nom. inval.*, Art. 61.1?) ≡ *Agave flavescens* var. *macracantha* (Herbert) Jacobi (1865); **incl.** *Agave bessereriana* Van Houtte (1868); **incl.** *Agave besseriana* Jacobi (1869); **incl.** *Agave subfalcata* Jacobi (1869); **incl.** *Agave macrantha* Jacobi (1869) (*nom. inval.*, Art. 61.1?); **incl.** *Agave besseriana* [?] *candida* Jacobi (1870); **incl.** *Agave besseriana* [?] *longifolia glauca* Jacobi (1870) (*nom. inval.*, Art. 24.2); **incl.** *Agave besseriana* [?] *longifolia viridis* Jacobi (1870) (*nom. inval.*, Art. 24.2); **incl.** *Agave besseriana* [?] *hystrix* hort. *ex* Hooker (1871); **incl.** *Agave linearis* Jacobi (1871); **incl.** *Agave oligophylla* Baker (1877); **incl.** *Agave sudburyensis* Baker (1877); **incl.** *Agave paucifolia* Baker (1878) (*nom. illeg.*, Art. 53.1); **incl.** *Agave integrifolia* Baker (1888); **incl.** *Agave macroacantha* var. *integrifolia* Trelease (1907); **incl.** *Agave macroacantha* var. *latifolia* Trelease (1907); **incl.** *Agave macroacantha* var. *planifolia* A. Berger (1915).

[2f] Stem short; **Ros** small, eventually 25 - 40 cm, commonly caespitose; **L** numerous, linear, rigid, radiately spreading, patulous, acuminate, 25 - 35 × 2.5 - 3 cm, bluish-grey-glaucous, margins straight or undulate; larger marginal teeth 3 - 4 mm, dark brown, irregularly spaced 1 - 3 cm apart, with slender cusps mostly curved from small low bases; terminal **Sp** subulate, straight to sinuous, rounded below, flat above, 3 - 3.5 cm, dark brown to grey, not decurrent; **Inf** ± 2 m, 'paniculate', slender, part-**Inf** shortly spreading, 10 - 14 in the upper ½ of the **Inf**, sometimes bulbilliferous; **Fl** 50 - 56 mm; **Ov** angular-fusiform, 25 - 30 mm, neck constricted, deeply grooved; **Tep** pruinose green with purple tinge, tube 14 mm, lobes ± equal, 13 - 16 mm, quickly wilting at anthesis.

Trelease (1907) gives a detailed review of the complicated taxonomic history throughout European literature and lists many synonyms and misapplications of Zuccarini's name. His treatment is followed here.

A. maculosa Hooker (CBM 85: t. 5122 + text, 1859). **T:** [lecto − icono]: l.c. t. 5122. − **D:** USA (S Texas), Mexico (Nuevo León, Hidalgo, Puebla, San Luis Potosí, Tamaulipas, C Veracruz); dry chaparral, on slopes or between rocks, and in moist oak woods, 10 - 1830 m, flowers mainly March to June. **I:** CBM 85: t. 5122, 1859; Berger (1915: 30); Addisonia 18: t. 601, 1933/34.

≡ *Manfreda maculosa* (Hooker) Rose (1903) ≡ *Polianthes maculosa* (Hooker) Shinners (1966); **incl.** *Agave maculata* Engelmann *ex* Torrey (1859) (*nom. illeg.*, Art. 53.1); **incl.** *Agave maculosa* var. *minor* Jacobi (1868); **incl.** *Agave maculosa* var. *brevituba* Engelmann (1875) ≡ *Agave maculata* var. *brevituba* (Engelmann) Mulford (1896) (*incorrect name*, Art. 11.4).

[3a1] Plants medium-sized; **L** erect-arching, linear-lanceolate, deeply channelled, 14 - 44 × 1.2 - 2.7 (-3.9 in cultivation) cm, dark green, unspotted or spotted with lighter green and brown or green, spots round to elliptic, sometimes glaucous, margins usually with small distantly spaced teeth; **Inf** 0.6 - 1.4 (-1.8 in cultivation) m, 'spicate', flowering part (7.5-) 14 - 22 (-29; to 48 in cultivation) cm, with 7 - 29 (-41) spreading **Fl**; **Ov** 9 - 16 (-19) mm; **Tep** tube 6 - 16 mm, lobes (6-) 9 - 13 (-16) mm, yellow-green or mahogany-brown inside; **Sty** longer than the lobes by up to 4 or shorter by up to 10 mm; **Fr** globose to oblong, 1.6 - 1.8 (-2.5) × 1.3 - 1.6 cm; **Se** 4 - 5 × 3 - 4 mm.

This species is, together with its putative nearest relative *A. longiflora*, unusual in the *Manfreda* Group by having a white to yellowish perianth that darkens to rose, therewith closely approaching the *Polianthes* Group (Verhoek-Williams 1975: 217).

A. mapisaga Trelease (CUSNH 23: 130, 1920). **T:** Mexico, D.F. (*Trelease* 147 [MO]). − **D:** Mexico; cultivated only.

A. mapisaga var. **lisa** Gentry (Agaves Cont. North Amer., 604, ill., 1982). **T:** Ex cult. (*Gentry* 21980 [US, DES]). − **D:** Mexico; cultivated only.

[2b] Differs from var. *mapisaga*: **Ros** gigantic, 2 - 2.5 m tall, sparsely suckering; **L** larger, 200 - 275 × 25 - 30 cm; **Fl** 80 - 90 mm; **Ov** 47 - 55 mm, tube 14 mm, lobes unequal, outer 19 - 21 mm.

This appears to represent a clonal variety with unknown provenance. According to Gentry (1982: 604) it is the largest-growing *Agave*. Only *A. franzosinii* and *A. atrovirens* may reach similar or even larger sizes.

A. mapisaga var. **mapisaga** − **D:** Mexico (Tamaulipas, Zacatecas, Hidalgo, México, Oaxaca); cultivated only. **I:** Gentry (1982: 597, 603).

[2b] Stem short, massive; **Ros** openly spreading, large, 2 - 2.4 × nearly 4 - 4.8 m, surculose; **L** linear, spreading to ascending, sometimes re- or inflexed, base very thickly fleshy, long-acuminate, upwards guttered, 185 - 250 × 19 - 25 cm, green, pale glaucous, or zonate, margins straight to repand; marginal teeth small, brown, mostly 4 - 6 cm apart, cusps 2 - 5 mm from low bases; terminal **Sp** conical-subulate, narrowly grooved above, 3 - 5 cm, dark to greyish-brown, long decurrent; **Inf** massive, 7 - 8 m and more, 'paniculate', part-**Inf** widely spreading, heavy and densely several times compound, 20 - 25 per **Inf**; **Fl** 80 - 100 mm; **Ov** 40 - 52 mm, green, neck short, not constricted; **Tep** in bud frequently reddish, opening yellow, tube funnel-shaped, 14 - 21 mm, lobes unequal, 22 - 27 mm.

This is a cultivar distinguished from the related *A. salmiana* by its longer linear leaves without the sigmoid apical bend characteristic for the latter, but is less often cultivated than *A. salmiana* (Gentry 1982: 603).

A. margaritae Brandegee (Proc. Calif. Acad. Sci., ser. 2, 2: 206, 1889). **T:** Mexico, Baja California (*Brandegee* s.n. [UC]). − **D:** Mexico (Baja California Sur: Santa Margarita and Magdalena Islands). **I:** KuaS 40(6): centre page pullout 1989/18.

Incl. *Agave connochaetodon* Trelease (1912).

[2h] **Ros** compact, small, caespitose; **L** 40 - 50, ovate to broadly lanceolate, thick, fleshy and rigid, narrowed above the base, shortly acuminate, concave above, 12 - 25 × 7 - 10 cm, glaucous-grey to yellowish-green, margins crenate; marginal teeth variously curved or flexed, weakly attached, 4 - 5 mm or to 8 - 15 mm (middle of the lamina), reddish-brown to greyish, 1 - 1.5 cm apart, on moderate

to prominent tubercles; terminal **Sp** subulate, 2 - 3 cm, shortly decurrent; **Inf** 2 - 3.5 m, 'paniculate', slender, part-**Inf** 6 - 12 in the upper ⅓ of the **Inf**; **Fl** 45 - 50 mm; **Ov** fusiform, 25 - 30 mm; **Tep** light yellow, tube ± 10 mm, lobes 15 mm.

Distinguished from all other taxa from Baja California esp. by its short broad leaves and the deep flower tube (Gentry 1982: 389).

A. marmorata Roezl (Belgique Hort. 33: 238, 1883). **T:** [neo − icono]: Curtis's Bot. Mag., 1912: t. 8442. − **D:** Mexico (Puebla [esp. around Tehuacán], Oaxaca). **I:** Gentry (1982: 508, 513-514); KuaS 43(3): centre page pullout 1992/10. **Fig. IV.a, IV.b**

Incl. *Agave todaroi* Baker (1888).

[2j] Stem short; **Ros** openly spreading, large, 1.2 - 1.3 × 2 m, solitary, rarely surculose; **L** 30 - 50, broadly lanceolate, frequently undulate, thick at the base, infolding along the middle of the lamina, involute at the spine base, convex below, flat above, generally roughly scabrous, mature **L** 100 - 135 × 20 - 30 cm, glaucous-grey to light green, sometimes zonate, margins crenate, with fleshy prominences; marginal teeth flattened, mostly 6 - 12 mm, chestnut-brown to dark brown, 2 - 5 cm apart, cusps from very broad bases, mostly straight, intermittent teeth few or none; terminal **Sp** usually shortly conical, 1.5 - 3 cm, rarely shortly decurrent; **Inf** 5 - 6.5 m, 'paniculate', part-**Inf** large, diffusely several times compound, 20 - 25 in the upper ½ of the **Inf**; **Fl** small, 40 - 48 mm; **Ov** cylindrical, 20 - 25 mm, light green, neck not constricted, scarcely grooved; **Tep** brilliant yellow, tube shallowly funnel-shaped, 5 - 6 mm, lobes equal, 14 - 16 mm.

Distinguished by the shortly conical terminal spines, the coarsely rough greyish leaves with strong prominences, and the small golden-yellow flowers in large diffuse 'panicles' (Gentry 1982: 512).

A. maximiliana Baker (Gard. Chron., ser. nov. 1877: 201, 1877). **T** [neo]: Ex cult. (*Anonymus* s.n. [K]). − **Lit:** McVaugh (1989). **D:** Mexico (Sinaloa, Durango, Zacatecas, Nayarit, Jalisco, Colima); dry rocky mountain slopes in the oak- and pine-forest zone, 930 - 2000 (-2700) m. **I:** Gentry (1982: 325, 348-349, 351).

Incl. *Agave gustaviana* hort. *ex* Baker (1877); **incl.** *Agave crenata* A. Berger (1911) (*nom. illeg.*, Art. 53.1); **incl.** *Agave conjuncta* A. Berger (1915); **incl.** *Agave katharinae* A. Berger (1915) ≡ *Agave maximiliana* var. *katharinae* (A. Berger) Gentry (1982).

[2d] **Ros** acaulescent or short-stemmed, of medium size, solitary; **L** usually broadly (ob-) lanceolate, curved, straight or slightly recurved, softly fleshy, generally 40 - 80 × 10 - 20 cm, mostly pale glaucous pruinose over yellow-green to green, or bluish-glaucous, margins variously repand to undulate or crenate, with strong prominences; marginal teeth heteromorphic, larger teeth variously curved, compressed, 6 - 10 mm (middle of the lamina), 1.5 - 3 cm apart, cusps slender, from elongate low (sometimes confluent) bases, intermittent teeth numerous, variable; terminal **Sp** slenderly conical, straight, smooth, 2.5 - 4 cm, brown or chestnut-brown to grey, shortly decurrent at the base; **Inf** 5 - 8 m, 'paniculate', narrow, part-**Inf** small, rather rounded, 15 - 25 (-30) in the upper ½ of the **Inf**; **Fl** slender, 52 - 65 mm; **Ov** 28 - 35 mm, neck short or long; **Tep** greenish-yellow, frequently flushed with rufous, tube openly funnel-shaped, 5 - 9 (-12) mm, lobes subequal, 15 - 22 mm; **Fr** shortly oblong, stipitate, tip rounded, 3.5 - 5 × 1.7 - 2 cm; **Se** with wavy testa, finely punctate, marginal wing abruptly raised, 5.5 - 6 × 4.5 - 5 mm.

Var. *katharinae* is included in the synonymy of the species according to McVaugh (l.c.).

A. mckelveyana Gentry (CSJA 42: 225-228, ills., 1970). **T:** USA, Arizona (*Gentry* 21979 [US]). − **Lit:** Turner & al. (1995). **D:** USA (W-C Arizona); rocky slopes in Chaparral and juniper associations, 8500 - 2200 m. **I:** Gentry (1982: 356). **Fig. IV.g**

[2h] **Ros** small, 20 - 40 cm, solitary or suckering; **L** rather few, linear or lanceolate, firmly spreading, broadest in the middle, 20 - 35 × 3 - 5 cm, light glaucous green or yellowish-green, margins nearly straight or undulate; marginal teeth curved downwards, rather friable, small to medium-sized, larger teeth in the middle of the lamina, 4 - 8 mm, greyish with reddish tips, mostly 1 - 3 cm apart, tubercles low; terminal **Sp** subulate, 1.5 - 4 cm, chestnut-brown to grey, shortly decurrent; **Inf** 2 - 3 m, 'paniculate', narrow, part-**Inf** small, compact, 10 - 19 in the upper ½ of the **Inf**; **Fl** small, 30 - 40 mm; **Ov** 16 - 22 mm incl. the constricted neck, light green; **Tep** openly spreading, yellow, tube open, 3 - 4.5 mm, lobes unequal, 12 - 23 mm.

Ecologically separated from its close relative *A. deserti* ssp. *simplex*, which is confined to lower elevations (Gentry 1982: 390). It hybridizes with *A. deserti* ssp. *simplex* and *A. utahensis* ssp. *utahensis* (Hodgson 1999).

A. michoacana (Cedano & al.) Thiede & Eggli (KuaS 50(5): 112, 1999). **T:** Mexico, Michoacán (*Escobedo* 1485 [IEB]). − **D:** Mexico (Michoacán); wet meadows, 2200 - 2700 m, flowers July to September. **I:** Cedano M. & al. (1993).

≡ *Polianthes michoacana* Cedano & al. (1993).

[3b1] Plants erect, scapose; **R** unknown; **L** 4 - 10, linear, long attenuate, 27.7 - 63.6 × 0.35 - 1 cm, margins and keel papillose, scape **L** triangular-linear, tip attenuate, all **L** yellowish-green, glabrous, basal **L** from an ovoid white bulb 2 × 1 - 2 cm, covered with the remains of **L** bases apically desintegrating into fibres; **Inf** 0.4 - 1.2 m, with up to 3 groups of sessile paired **Fl**; **Ov** cylindrical, 8.5 - 10

mm; **Tep** white, tube basally erect, upper part gradually flaring, 80 - 100 × 3 - 5 (middle of the tube) mm, lobes triangular, attenuate, 10 - 15 × 5 - 6 mm; **Anth** 6 - 11 mm; **Sty** columnar, slightly triangular; **Sti** lobes oblong, tip obtuse, reflexed, 1.5 - 2 mm; **Fr** ellipsoid, 2.3 - 2.5 × 1.4 - 1.5 cm; **Se** flattened, obovate-clavate, asymmetrical, 4.2 mm.

According to the protologue closely related to *A. dolichantha* (as *Polianthes longiflora*) (Cedano M. & al. 1993).

A. millspaughii Trelease (Mem. Nation. Acad. Sci. 11: 41, t. 87-88, 1913). **T**: Bahamas, Great Exuma (*Britton & Millspaugh* 3091 [NY]). – **D**: Bahamas; low coppices and scrublands.

[2n] Acaulescent; **Ros** solitary; **L** narrowly oblanceolate, concave, to 125 × 15 cm, green, somewhat glossy, margins between the teeth nearly straight; marginal teeth straight and spreading or occasionally reflexed, sometimes with upcurved tips, narrowly triangular, 3 - 5 mm, brown to nearly black, mostly 15 - 25 mm apart, scarcely lenticular at the base; terminal **Sp** triquetrous, conical, straight, round-grooved to about the middle or occasionally involute, smooth, rather dull, 1.5 - 2 cm, red-brown, decurrent; **Inf** to ± 10 m, 'paniculate', oblong, part-**Inf** with subascending **Br**, in the upper ⅔ of the **Inf**; **Ped** ± 10 mm; **Fl** ± 50 mm; **Ov** fusiform, 25 mm; **Tep** 15 - 20 × 4 mm, yellow, tube conical, ± 7 mm; **Fr** shortly oblong, shortly stipitate and beaked, 20 - 35 mm.

A. missionum Trelease (Mem. Nation. Acad. Sci. 11: 37-38, t. 72-75, 1913). **T**: St. Thomas (*Trelease* 15 [not indicated]). – **D**: Puerto Rico, Virgin Islands, St. Thomas. **I**: Succulenta 65: 162-163, 1986.

Incl. *Agave vivipara* Oldendorp (1777) (*nom. illeg.*, Art. 53.1); incl. *Agave morrisii* Eggers (1889) (*nom. illeg.*, Art. 53.1); incl. *Agave eggersiana* Trelease (1913).

[2l] **Ros** solitary; **L** broadly lanceolate, gradually acute, concave, occasionally conduplicate, 250 - 275 × 20 cm, dark green or very slightly greyish, rather glossy, margins nearly straight; marginal teeth straight, gently curved, or bent in either direction, conspicuously triangular, 3 - 5 mm, brown to nearly black, mostly 1 - 1.5 cm apart, smaller teeth from often confluent lenticular bases; terminal **Sp** somewhat triquetrously subulate, straight or a little upcurved, round-grooved to about the middle or occasionally involute, smooth, 1.5 - 2.5 cm, brown, or grey in age, somewhat glossy, decurrent; **Inf** 5 - 7 m, 'paniculate', oblong, part-**Inf** on somewhat ascending **Br**, in the upper ⅔ of the **Inf**; **Fl** 55 mm; **Ov** oblong-fusiform, 30 mm; **Tep** yellow, tube ± 7 mm, lobes 15 - 20 mm; **Fr** broadly oblong or somewhat turbinate, stipitate and beaked, 3 - 4 × 2 - 2.5 cm.

A. mitis Martius (Del. Sem. Hort. Bot. Monac. 1848: [], 1848). **T** [neo]: Mexico, Tamaulipas (*Gentry* 20077 [MEXU, DES, US]). – **D**: Mexico (Tamaulipas, San Luis Potosí, Hidalgo).

A. mitis var. **albidior** (Salm-Dyck) Ullrich (Succulentes 16(1): 32, 1993). **T** [neo]: Ex cult. (*Anonymus* Kew no. 109 [K]). – **D**: Mexico (Hidalgo); only known from the type locality. **I**: Gentry (1982: 217, 224, as *A. celsii* var. *albicans*).

≡ *Agave micrantha* var. *albidior* Salm-Dyck (1859); **incl.** *Agave albicans* Jacobi (1865) ≡ *Agave micrantha* var. *albicans* (Jacobi) A. Terracciano (1885) ≡ *Agave celsii* var. *albicans* (Jacobi) Gentry (1982); **incl.** *Agave ousselghemiana* Jacobi (1868); **incl.** *Agave concinna* Hort. Angl. *ex* Baker (1877) (*nom. illeg.*, Art. 53.1).

[1f] Differs from var. *mitis*: **L** pale glaucous; **Tep** lobes larger, 20 - 27 mm.

In cultivation since ± 1850. A small colony in the Barranca de Metztitlán appears assignable to this variety (Gentry 1982: 223-224).

A. mitis var. **mitis** – **D**: Mexico (Tamaulipas, San Luis Potosí, Hidalgo). **I**: Gentry (1982: 217, 221, as *A. celsii*). **Fig. V.c**

Incl. *Agave celsii* var. *celsii*; **incl.** *Agave micrantha* Salm-Dyck (1855); **incl.** *Agave celsii* Hooker (1856) ≡ *Agave bollii* var. *celsii* (Hooker) A. Terracciano (s.a.) (*incorrect name*, Art. 11.1); **incl.** *Agave rupicola* Regel (1858); **incl.** *Agave micracantha* Salm-Dyck (1859); **incl.** *Agave densiflora* Regel (1863); **incl.** *Agave bouchei* Jacobi (1865) ≡ *Agave rupicola* var. *bouchei* (Jacobi) A. Terracciano (1885); **incl.** *Agave celsiana* Jacobi (1865); **incl.** *Agave haseloffii* Jacobi (1866); **incl.** *Agave oblongata* Jacobi (1868); **incl.** *Agave botteri* Hemsley (1876).

[1f] **Ros** branching axillary, forming large long-lived dense clumps; **L** ovate, oblong, or spatulate, ascending to outcurving, thickly soft-fleshy, short-acuminate, convex below, guttered or concave above, 30 - 60 (-70) × 7 - 13 cm, green to light grey-glaucous, margins straight to undulate; marginal teeth sometimes with ciliate crests, small, 1 - 3 mm, whitish to reddish-brown, closely spaced; terminal **Sp** acicular, weak, 1 - 2 cm, brownish, decurrent along the **L** tip for 1 - 6 cm and more; **Inf** 1.5 - 2.5 m, 'spicate', densely bracteate and flowered, becoming lax at fruiting time, part-**Inf** with geminate **Fl**; **Fl** fleshy, 40 - 60 mm; **Ov** 13 - 20 mm, without neck; **Tep** green outside, yellow to reddish or lavender to purplish within, tube funnel-shaped, 10 - 17 mm, lobes dimorphic, 12 - 18 mm. – *Cytology:* 2n = 60.

Distinguished by its small compact rosettes, broad delicately denticulate leaves, and densely clavate inflorescences (Gentry 1982: 222). Gentry wrongly named this taxon *A. celsii*, and the nomenclature was finally clarified by Ullrich (1993a).

A. montana Villarreal (Sida 17(1): 191-195, ills., 1996). **T:** Mexico, Nuevo León (*Villarreal & al. 8120* [MEXU, ANSM, ENCB]). – **D:** Mexico (Nuevo León); limestone slopes, 3200 - 3400 m, only known from the type locality.

[2b] **Ros** semiglobose, compact, 0.9 - 1.35 × 1.4 - 1.65 m, solitary; **L** 84 - 112, regular in 12 - 16 rows, shortly elliptic, base broadened, apex acuminate, slightly concave to nearly flat, 30 - 40 × 15 - 17 cm, yellowish-green, margins straight, near the tip bordered brown-purple; marginal teeth antrorse and retrorse, greyish, 2.5 - 3.5 cm apart, 16 - 18 per margin; terminal **Sp** 3 - 5 cm; **Inf** 3.5 - 4.5 m, 'paniculate', ovate, with dentate **Bra**, part-**Inf** 20 - 30 in the upper ⅔ of the **Inf**; **Ped** 1 - 1.5 cm; **Fl** 60 - 70 mm; **Ov** 30 mm; **Tep** yellow, tube 20 mm, lobes 20 mm.

According to the protologue related to *A. parrasana* of Group *Parryanae* and esp. to *A. gentryi* (= *A. macroculmis sensu* Gentry) where it was formerly erroneously included despite being clearly distinct in many features. The descriptions of both Gentry (1982) and Ullrich (1990i) for the last-mentioned taxon combine characters of both species.

A. moranii Gentry (Occas. Pap. Calif. Acad. Sci. 130: 58, ills. (pp. 59-61), 1978). **T:** Mexico, Baja California (*Gentry & McGill 23287* [US, DES, SD]). – **Lit:** Turner & al. (1995). **D:** Mexico (Baja California: S San Pedro Martír); small desertic area, 450 - 1850 m. **I:** Gentry (1982: 356, 393).

[2h] Stem short; **Ros** large, 1 - 1.5 × 2 m, solitary; **L** triangularly long-lanceolate, straightly ascending to spreading, rigid, rounded beneath, deeply guttered, 70 - 120 × 8 - 12 cm, light to yellowish-green, sometimes glaucous, margins towards the **L** tip white-horny; marginal teeth sinuously curved, flattened, 6 - 12 mm (middle of the lamina and below), light grey, 2 - 4 cm apart, base broad or continuous with the **L** margin, teeth reduced and more remote towards the **L** tip; terminal **Sp** stout, broadly grooved above, 4 - 6 cm, nearly white, tip chestnut-brown, decurrent to the middle of the lamina; **Inf** 4 - 5 m, 'paniculate', part-**Inf** closely spaced, compact, large, 20 - 30 per **Inf**; **Ped** slender, 1 - 3 cm; **Fl** 50 - 70 mm; **Ov** fusiform, 25 - 40 mm, neck short, grooved, thick; **Tep** bright yellow, tube broadly funnel-shaped, 4 - 6 mm.

Distinguished from all other taxa in Group *Deserticolae* by its large solitary rosettes with large long rigid leaves with an apical horny margin, stout scapes and relatively congested panicles. It appears to be close to *A. avellanidens* and *A. gigantensis* (Gentry 1982).

A. murpheyi F. Gibson (Contr. Boyce Thompson Inst. Pl. Res. 7: 83, fig. 1, 1935). **T:** USA, Arizona (*Gibson s.n.* [Herb. Boyce Thompson Arboretum]). – **D:** USA (Arizona), Mexico (Sonora); "arborescent desert" within the Sonoran Desert, moun-tainous slopes or bajadas, 460 - 930 m. **I:** Gentry (1982: 421, 442-443).

[2g] **Ros** compact, 0.6 - 0.8 × ± 1 m, freely suckering; **L** linear, firm, straight, short-acuminate, 50 - 65 × 6 - 8 cm, light glaucous-green to yellowish-green, frequently lightly cross-zoned, with clearly visible impressions left by the central bud, margins undulate; marginal teeth regular, small, 3 - 4 mm, bases brown, cusps becoming grey, 1 - 2 cm apart; terminal **Sp** conical, very shortly grooved or flattened above, short, 1.2 - 2 cm, dark brown becoming greyish; **Inf** 3 - 4 m, 'paniculate', short, compact, part-**Inf** compact, short, 10 - 15 in the upper ¼ or ⅓ of the **Inf**, richly bulbilliferous, rarely producing **Fr**; **Fl** 65 - 75 mm; **Ov** thick, rounded, 32 - 40 mm, neck scarcely narrowed; **Tep** pale waxy green, tips purplish to brownish, tube deep, urceolate, 16 - 20 mm, lobes unequal, outer 15 - 19 mm. – *Cytology:* 2n = 60.

The species appears to be at least partially associated with old living sites of native Americans (Gentry 1982: 443), which is also true for *A. delamateri* and *A. decipiens*. It hybridizes with *A. chrysantha* (Hodgson 1999).

A. nanchititlensis (Matuda) Thiede & Eggli (KuaS 50(5): 110, 1999). **T:** Mexico, México (*Matuda 37640* [MEXU]). – **D:** Mexico (México); rocky sloping oak or oak-pine woods, flowers in January. **I:** Ullrich (1989).

≡ *Manfreda nanchititlensis* Matuda (1974).

[3a2] Plants small (for Subgen. *Manfreda*), daughter **Ros** arising in the **Ax** of the current year's **Ros**; **R** numerous, half-fleshy, spreading horizontally from the rhizome base; rhizome slender, 1.5 - 2 cm; **L** 4 - 9, upright-arching, linear, roundly channelled, tough-fibrous, 21 - 39 × 0.4 - 0.8 (-1) cm, densely spotted on both faces with irregular brown or dark green spots, veins on both faces with single rows of papillae, margins with a thin maroon line, entire or minutely papillate like the veins; remains of the **L** bases fibrous, 5.5 - 7 cm; **Inf** 35 - 94 cm, 'spicate', flowering part very open, 11 - 20 cm, with 7 - 10 flowering nodes; **Fl** sessile, horizontal when mature; **Ov** long-ellipsoid, 11 - 17 mm; **Tep** green with purple flush on the upper part, tube cylindrical, narrow, straight or slightly arched, 25 - 30 × 3 (middle of the tube) mm, lobes revolute, 6 - 10 mm, yellowish-green on both faces; **Sty** exserted, 45 - 56 mm; **Sti** clavate, deeply furrowed, broadly expanded at maturity; **Fr** ellipsoid, 1.5 - 1.9 × 1.1 - 1.2 cm; **Se** lunate, 3 mm.

This is the most slender species in the subgenus *Manfreda* and is therefore easily recognized by its nearly grass-like leaves, its lax inflorescence, and its long narrow floral tube (Verhoek-Williams 1975: 289).

A. nashii Trelease (Mem. Nation. Acad. Sci. 11: 45-46, t. 101-103, 1913). **T** [syn]: Bahamas, Inagua

(*Nash & Taylor* s.n. [NY, MO?]). – **Lit:** Correll & Correll (1982). **D:** Bahamas (Inagua); dwarf scrub and scrublands on sandy-rocky soils.

[2p] **Ros** solitary; **L** attenuate-oblong, concave, 30 - 50 × 4 - 5 cm, grey-green, sometimes purple-tinged, somewhat glaucous and with transverse bands, margins between the teeth nearly straight; marginal teeth straight or somewhat curved, acuminately triangular, sometimes nearly or quite confluent, scarcely 2 mm long, usually 3 - 5 mm apart; terminal **Sp** smooth, somewhat polished and recurved or upcurved towards the end, conically tapering, narrowly slit-grooved to beyond the middle, 0.3 - 1.5 cm, purplish-brown, decurrent; **Inf** 3.5 - 4 m, 'paniculate', part-**Inf** very lax, on slender outcurved **Br**, in the upper ⅓ of the **Inf** or more; **Ped** 5 - 10 mm; **Fl** ± 35 mm; **Ov** subfusiform to ovoid, 20 mm; **Tep** ± 13 × 3 mm, light yellow, tube openly conical, 3 mm; **Fr** oblong to oblong pear-shaped, slightly stipitate and beaked, 2 - 2.5 × 2 cm.

A. nayaritensis Gentry (Agaves Cont. North Amer., 515-516, ills., 1982). **T:** Mexico, Nayarit (*Gentry* 21167 [US, DES, MEXU]). – **D:** Mexico (Nayarit); volcanic cliff edges, 600 - 700 m, known only from the type locality.

[2j] **Ros** open, medium-sized, solitary, rarely surculose; **L** few, lanceolate, rather floppy, etiolated when growing in shady conditions (in habitat), narrowed towards the base, widest above the middle, long-acuminate, somewhat asperulous, 85 - 115 × 12 - 15 cm, light green, margins undulate to straight; marginal teeth small, 1 - 3 mm, chestnut-brown or darker, regularly spaced 1 - 1.5 cm apart, with scattered minute intermittent teeth; terminal **Sp** conical, with short narrow groove above, 0.9 - 1.5 cm, dark brown, not decurrent; **Inf** 3 - 4 m, 'paniculate', diffuse, ovate in outline, scape short, part-**Inf** widely spreading, several times compound, 14 - 15 in the upper ½ of the **Inf**; **Fl** small, 40 - 45 mm; **Ov** rounded-trigonous, 20 - 25 mm, neck short, furrowed; **Tep** bright yellow, tube broadly funnel-shaped, 4 mm, lobes subequal, 15 - 17 mm.

Without close relationship to other members of Group *Marmoratae* (Gentry 1982: 516).

A. neglecta Small (Fl. Southeast. US, 289, 1903). **T** [lecto]: USA, Florida (*Weber* s.n. [MO, NY]). – **D:** USA (Florida); sandy beaches. **I:** Gentry (1982: 627).

[2a] Stem short, 30 - 40 cm; **Ros** large, 1.3 - 1.7 m, suckering freely; **L** broadly lanceolate, ascending or arching, or reflexed in age, thickened and narrowed towards the base, acuminate, concave, 100 - 150 × 15 - 25 cm, pale green, glaucous, margins nearly straight; marginal teeth fine, small, closely set below, margin becoming toothless above; terminal **Sp** acicular, 2.5 cm, scarcely decurrent; **Inf** very tall, to 8 - 10 m, 'paniculate', open, broad, part-**Inf** diffusely compound, 18 - 20 in the upper ⅓

- ½ of the **Inf**, bulbilliferous; **Fl** 55 mm; **Tep** greenish-yellow, lobes ± 23 mm.

A hardly known species apparently close to *A. weberi* and *A. desmetiana* (Gentry 1982: 628). It is most probably of cultivated origin. Ullrich (1990d: 106) places it in Group *Agave*, which contrasts Gentry's placement in Group *Viviparae* (as *Sisalanae*).

A. neomexicana Wooton & Standley (CUSNH 16(4): 115, pl. 48, 1913). **T:** USA, New Mexico (*Standley* 541 [US 498333]). – **D:** USA (S New Mexico, SW Texas), Mexico (Coahuila?). **I:** Gentry (1982: 522, 537). **Fig. V.d**

≡ *Agave parryi* var. *neomexicana* (Wooton & Standley) McKechnie (1949) ≡ *Agave parryi* ssp. *neomexicana* (Wooton & Standley) B. Ullrich (1992).

[2i] **Ros** rather flat-topped, small to medium-sized, freely suckering; **L** few to many, lanceolate, rigid, usually broadest near the middle, mostly rather shortly acuminate, thickly rounded below, concave above, 20 - 45 × 5 - 12 cm; marginal teeth nearly straight or curved, slender, mostly 5 - 7 mm (above the middle of the lamina), dark brown to greyish, 1 - 3 cm apart; terminal **Sp** subulate to acicular, upper face flat, with a broad shallow groove, 2.5 - 4 cm, decurrent for 1 to several teeth; **Inf** 3 - 4 m, 'paniculate', part-**Inf** compact, mostly 10 - 17 in the upper ½ of the **Inf**; **Fl** 55 - 67 mm; **Ov** slender, fusiform, 32 - 38 mm, neck furrowed, constricted, 4 - 7 mm; **Tep** red to orange in bud, opening yellow, **OTep** reddish-tipped, tube funnel-shaped, 12 - 14 mm, lobes nearly equal, 15 - 20 mm. – *Cytology:* $2n = 120$.

Distinguished from its closest relative *A. parryi* by generally having the rosettes in smaller groups, more slender leaves, and smaller inflorescences with fewer part-inflorescences (Gentry 1982: 537). The occurence in Coahuila is based on a single doubtful specimen only and needs verification.

A. neonelsonii Thiede & Eggli (KuaS 50(5): 112, 1999). **T:** Mexico, Durango (*Nelson* 4630 [US?]). – **D:** Mexico (Durango).

Incl. *Polianthes nelsonii* Rose (1903).

[3b1] **L** several, linear, margins serrulate, from an oblong bulb, bulb tunics (= remains of dead **L**) thin, scape **L** much reduced; **Inf** erect, 'spicate', glabrous, ± 40 cm, with 2 - 5 paired sessile **Fl**; **Ov** 9 - 16 (-19) mm; **Tep** tube strongly curved downwards near the middle, very slender below, ± 50 mm, white, lobes short, tip rounded; **Fil** 3 mm (free part), attached near the mouth of the tube; **Anth** 2 mm; **Sti** exserted.

An insufficiently known species and apparently recorded from 2 collections only. The new name for *Polianthes nelsonii* was necessary to avoid homonymy with *A. nelsonii* Trelease 1912.

A. neopringlei Thiede & Eggli (KuaS 50(5): 112,

1999). **T**: Mexico, Jalisco (*Pringle* 5438 [GH, US]). – **Lit**: McVaugh (1989). **D**: Mexico (Durango, Nayarit, Aguascalientes, Guanajuato?, Jalisco, Guerrero, Morelos, San Luis Potosí); grasslands, grassy opening and hillsides in pine-oak forest regions, wet meadows and pastures, 800 - 2200 m, flowers August to October. **I**: Rose (1903: fig. 2, as *Polianthes durangensis*).

Incl. *Polianthes pringlei* Rose (1903); **incl.** *Polianthes durangensis* Rose (1903).

[3b1] Plants glabrous, subcaulescent; stem often 5 - 6 cm, from a narrowly ovoid bulb; **R** fleshy; **L** (1-) 5 - 10 in a basal **Ros** (or 1 - 6 additional **L** a few cm above the base of the scape), erect or nearly so, linear-attenuate, 10 - 25 (-45) × (0.1-) 0.3 - 0.7 (-1) cm, margins ± papillose, sometimes also papillose on the veins of the lower face; **Inf** 25 - 45 (-65) cm, 'spicate', flowering part 8 - 12 (-20) cm, with 3 - 7 widely spaced flowering nodes with paired **Fl**; **Fl** sessile, fragrant; **Ov** erect; **Tep** white, sometimes pale pink, or white inside and pink outside, tube ascending from the base, commonly smoothly curved outwards ± in the middle or below, slenderly cylindrical below, narrowly funnel-shaped above, 30 - 70 mm, 2.5 - 4 mm (distally) ∅, lobes subequal in 2 series, ascending-spreading or finally recurved, elliptic to oblong or oblanceolate, (5-) 10 - 15 mm; **St** included; **Sty** included, with 3 elliptic-oblong or ovate lobes ± 1.5 mm; **Fr** broadly ellipsoid, 1 - 1.3 × 0.8 - 1 cm; **Se** 3.5 - 4 mm.

The new name for *Polianthes pringlei* was necessary to avoid homonymy with *A. pringlei* Engelmann *ex* Orcutt 1883. The name of the heterotypic synonym *P. durangensis* could not be used because of *A. durangensis* Gentry.

A. nizandensis Cutak (CSJA 23(5): 143-145, ills., 1951). **T**: Mexico, Oaxaca (*Cutak* 19 [MO]). – **Lit**: Ullrich (1991j: with ills.). **D**: Mexico (Oaxaca). **Fig. V.e**

[1b] **Ros** open, small, surculose; **L** few, linear-lanceolate, patulous, sparsely fibrous, rather brittle or pliable, ± straight, convex below, plane above, 20 - 30 × 1.5 - 2.5 cm, green with pale midstripe, margins finely serrulate; terminal **Sp** conical, not pungent, small, 4 - 8 mm; **Inf** 1 - 2 m, 'spicate', sparsely flowered in the upper ¼, part-**Inf** with 2 - 4 **Fl**; **Ped** geminate, 6 - 10 mm; **Fl** 35 - 40 mm; **Ov** cylindrical, 12 - 15 mm, neck short, not constricted; **Tep** pale yellow; tube shortly funnel-shaped, 3 - 4 mm, lobes 15 - 16 mm.

A very distinct species without close relatives. It does not fit well into any section or group (Gentry 1982: 75). This prompted Ullrich (1991j) to erect a section (Sect. *Nizandensae*) of its own for the species, which is not followed here.

A. obscura Schiede (Linnaea 5: 464, 1830). **T** [neo]: Mexico, Veracruz (*Gentry & al.* 20417 [US, DES, MEXU]). – **Lit**: Cházaro Basáñez (1981);

Ullrich (1990j: with ills.). **D**: Mexico (Tamaulipas, San Luis Potosí, Veracruz, Puebla, Oaxaca). **I**: Gentry (1982: 217, 229, 231).

Incl. *Agave myriacantha* hort. *ex* Besaucèle (s.a.); **incl.** *Agave densiflora* var. *angustifolia* hort. *ex* Besaucèle (s.a.) (*nom. illeg.*, Art. 53.1); **incl.** *Agave hookeri* hort. *ex* Besaucèle (s.a.) (*nom. illeg.*, Art. 53.1); **incl.** *Agave micracantha* Baker (s.a.) (*nom. illeg.*, Art. 53.1); **incl.** *Agave densiflora* var. *foliis striatis aureis* hort. *ex* Besaucèle (s.a.) (*nom. inval.*, Art. 24.2); **incl.** *Agave densiflora* Hooker (1857) ≡ *Agave polyacantha* var. *densiflora* (Hooker) A. Terracciano (1885); **incl.** *Agave chloracantha* Salm-Dyck (1859); **incl.** *Agave uncinata* Jacobi (1865); **incl.** *Agave xalapensis* Roezl *ex* Jacobi (1865) ≡ *Agave polyacantha* var. *xalapensis* (Roezl *ex* Jacobi) Gentry (1982); **incl.** *Agave polyacantha* Jacobi (1865) (*nom. illeg.*, Art. 53.1); **incl.** *Agave lamprochlora* Jacobi (1868); **incl.** *Agave muilmannii* Jacobi (1870); **incl.** *Agave caribaea* Verschaffelt (1873); **incl.** *Agave attenuata* var. *subdenudata* hort. *ex* Trelease (1892); **incl.** *Agave engelmannii* Trelease (1892); **incl.** *Agave flaccifolia* A. Berger (1915).

[1f] **Ros** openly spreading, medium-sized, solitary or caespitose; **L** lanceolate-acuminate to oblong and short-acuminate, straightly ascending to upcurving, tissue firm, finely fibrous, lamina narrowed above the base, broadest in the middle, usually plane, 35 - 65 × 7 - 10 cm, green or yellow-green, passing into glaucous, margins generally straight, not horny except for the thinly decurrent terminal **Sp**; marginal teeth deltoid, 2 - 6 mm, reddish to dark brown, closely spaced or up to 5 - 12 mm apart; terminal **Sp** acicular, rounded below and above, small, 0.5 - 2.5 (-3.5) cm, dark brown; **Inf** 2 - 3 m, 'spicate', laxly or densely flowered in the upper ⅓ - ½, rarely bulbilliferous, part-**Inf** mostly with geminate **Fl**; **Ped** short, stout, 2 - 3 mm; **Fl** 46 - 51 mm; **Ov** cylindrical, 17 - 20 mm, green, neck short, not constricted; **Tep** reddish, tube funnel-shaped, 7 - 9 mm, lobes unequal, 19 - 23 mm. – *Cytology*: 2n = 60.

Recognized by its slender flowers with the rounded, ungrooved ovary, and well-developed teeth on the elongate-lanceolate leaves (Gentry 1982: 229, as *A. polyacantha*). Gentry misapplied the name *A. polyacantha* Haworth to these plants, but this name is of uncertain status. Ullrich (l.c.) consequently re-established the oldest available name, *A. obscura* Schiede. This name was in turn misapplied by both Trelease and Gentry to a taxon recently described by Ullrich (l.c.) as *A. horrida* ssp. *perotensis* (see there).

A. ocahui Gentry (US Dept. Agric. Handb. 399: 72-76, ills., 1972). **T**: Mexico, Sonora (*Gentry & Arguelles* 16637 [US, DES, MEXU]). – **D**: Mexico (Sonora).

Related to other Sonoran species of Group *Serru-*

latae (as *Amolae*) (*A. vilmoriniana*, *A. chrysoglossa*), based on the smooth narrow leaves with unarmed margins, the prolifically flowering inflorescences and the small slender yellow flowers with shallow tubes and tepals clasping the filaments (Gentry 1982: 78).

A. ocahui var. **longifolia** Gentry (Agaves Cont. North Amer., 78, ills. (p. 79-80), 1982). **T:** Mexico, Sonora (*Gentry* 11610 [US, DES]). − **D:** Mexico (E-C Sonora); scattered in the mountain region.

[1b] Differs from var. *ocahui*: Stem thick and round; **Ros** solitary; **L** linear-lanceolate, straightly ascending or recurving, sometimes falcate, mature **L** 60 - 80 (-90) × 2 - 3 (near the base) cm.

These larger, more robust and longer-leaved plants maintain these features even when cultivated together with the shorter-leaved var. *ocahui*. Varietal status is appropriate since they appear not to be geographically isolated from var. *ocahui* (Gentry 1982: 78-79).

A. ocahui var. **ocahui** − **D:** Mexico (NE Sonora); cliffs and outcrops of volcanic rocks, 500 - 1500 m. **I:** Gentry (1982: 67, 76-77). **Fig. V.f**

[1b] Stem short; **Ros** dense, green, *Yucca*-like, 0.3 - 0.5 × 0.5 - 1 m, solitary; **L** numerous, linear-lanceolate, erect to ascending, some older **L** declined or falcate, mostly stiff, widest at the base, plane above, surface smooth, minutely and densely punctate in fine lines, 25 - 50 × 1.5 - 2.5 cm, green, margins straight, lined with a narrow reddish-brown firm border detachable in dried **L**; marginal teeth none; terminal **Sp** weak, rather brittle, 1 - 2 cm, pruinose-grey over brown; **Inf** slender, ± 3 m, 'spicate', scape with numerous narrow chartaceous **Bra**, densely flowered from 1 - 1.5 m above the base, part-**Inf** with geminate **Fl**; **Fl** 30 - 38 mm; **Ov** 15 - 20 mm, neck constricted; **Tep** yellow, tube broadly funnel-shaped, 2 - 4 mm, lobes subequal, 14 - 16 mm.

A. ornithobroma Gentry (Agaves Cont. North Amer., 117-119, ills., 1982). **T:** Mexico, Sinaloa (*Gentry* 18358 [US, DES]). − **Lit:** McVaugh (1989). **D:** Mexico (Sinaloa, Nayarit); hot tropical lowland savanna.

[1d] Stem short; **Ros** asymmetrical, small, solitary to caespitose, suckering sparingly at maturity; **L** few, narrowly linear, straight-ascending to frequently curving to one side of the **Ros**, or falcate, short-acuminate, convex below from the base to the tip, convex above from the base to the middle of the lamina, smooth, 60 - 75 × 0.5 - 0.8 cm, light green to reddish, margins filiferous, reddish to white; terminal **Sp** subulate, weak, fraying, 0.6 - 1 cm; **Inf** 2.5 - 3 m, 'spicate', slender, laxly flowered in the upper ½ of the **Inf**, part-**Inf** with geminate **Fl**; **Ped** 5 - 8 mm; **Fl** slender, 30 - 48 mm; **Ov** small, 12 - 17 mm; **Tep** green with reddish or purplish flush, tube narrowly funnel-shaped, triquetrous, 9 - 13 mm, lobes about equal, 10 - 17 mm. − *Cytology:* 2n = 180.

Closely related to *A. geminiflora*, but separable by its caespitose habit, small few-leaved rosettes and slender inflorescences with small flowers (Gentry 1982: 118).

A. oroensis Gentry (Agaves Cont. North Amer., 294-296, ills., 1982). **T:** Mexico, Zacatecas (*Gentry & Enghard* 23592 [US, DES, MEXU]). − **D:** Mexico (N Zacatecas); cultivated only.

[2a] **Ros** low, openly spreading, solitary or suckering; **L** linear-lanceolate, straight to recurving, narrow and thickly convex below towards the base, long-acuminate, guttering upwards, slightly asperous, 80 - 100 × 8 - 10 cm, green, margins straight to repand; marginal teeth mostly straight, 3 - 6 mm (middle of the lamina), greyish, mostly 2 - 3 cm apart, smaller and more closely spaced towards the **L** base; terminal **Sp** acicular, narrowly grooved above for ½ of its length, 2.5 - 3 cm, greyish, finely decurrent to the uppermost teeth; **Inf** 5 - 6 m, 'paniculate', part-**Inf** laxly flowered, spreading, 12 - 16; **Fl** very slender, 70 - 75 mm; **Ov** fusiform, 34 - 37 mm, greenish, neck grooved; **Tep** pink in bud, opening yellow, tube 16 - 18 mm, lobes unequal, 20 - 21 mm.

A local cultivar well-characterized by its thick narrow green leaves and esp. the broad open pink-budded panicles with flowers with a tube constricted at the mouth (Gentry 1982: 294).

A. pachycentra Trelease (Trans. Acad. Sci. St. Louis 23(3): 135, 1915). **T:** Guatemala, Dept. Progreso (*Trelease* 2 [ILL]). − **D:** S Mexico (Oaxaca, Chiapas), Guatemala, El Salvador, Honduras; tropical deciduous forest, thorn forest, 300 - 1240 m. **I:** Gentry (1982: 486-487).

Incl. *Agave eichlamii* A. Berger (1915); **incl.** *Agave eichlamii* var. *interjecta* A. Berger (1915); **incl.** *Agave opacidens* Trelease (1915); **incl.** *Agave tenuispina* Trelease (1915); **incl.** *Agave weingartii* A. Berger (1915).

[2k] Stem short; **Ros** rather open, to 1 × 1.5 - 2 m, solitary, rarely surculose; **L** variable, broadly lanceolate, gradually narrowed towards the base, acuminate, plane to guttered, asperous above, rougher or scabrous below, mostly 60 - 100 × 12 - 18 cm, glaucous-white to yellowish or pale green, margins generally undulate, with sinuses between the teeth; marginal teeth variable, mostly 5 - 10 mm (middle of the lamina), brown, 1 - 3 cm apart, cusps straight or variously curved above low broad bases; terminal **Sp** finely subulate to nearly conical from a broad base, broadly to narrowly grooved above, scabrous, generally 4 - 6 cm, long-decurrent to the upper teeth; **Inf** 4 - 6 m, 'paniculate', open, rather irregular, scape usually crooked, young white-pruinose, part-**Inf** 20 - 30, small, on rather long stalks;

Fl 45 - 62 mm; **Ov** 25 - 35 mm, green, neck constricted, grooved; **Tep** yellow, **OTep** frequently reddish at the tips, tube 6 - 11 mm, lobes subequal, 13 - 20 mm. – *Cytology:* 2n = 120.

Highly variable in leaf characters (Gentry 1982: 487).

A. palmeri Engelmann (Trans. Acad. Sci. St. Louis 3: 319-320, 1875). – **D:** USA (Arizona, New Mexico), Mexico (N Sonora, Chihuahua); oak woodland and grama grassland, 930 - 1850 m. **I:** Gentry (1982: 423, 444). **Fig. V.g**

[2g] **Ros** rather open, 0.5 - 1.2 × 1 - 1.2 m, solitary, rarely suckering with age; **L** lanceolate, rather rigid, thick at the base, usually narrowed above the base, long-acuminate, convex below, somewhat guttered, mostly 35 - 75 × 7 - 10 cm, pale green to light glaucous-green or reddish-tinged, margins almost straight or undulate, with or without small tubercle-like bases to the teeth; marginal teeth variously curved, rather regular, slender, closely set, sometimes with smaller intermittent teeth; terminal **Sp** acicular, strong, shortly and openly grooved above the base, 3 - 6 cm, chestnut-brown or brown to aging grey; **Inf** 3 - 5 m, 'paniculate', broad, open, scape short, part-**Inf** horizontal, 8 - 12 in the upper ⅓ of the **Inf**; **Fl** narrow, 45 - 55 mm; **Ov** 25 - 30 mm, shiny green, neck short; **Tep** yellow to pink below, conspicuously red to brownish on the calloused tips, tube 12 - 14 mm, lobes dimorphic, outer 10 - 13 mm. – *Cytology:* 2n = 60.

Shows introgression with *A. chrysantha* (see there) and *A. shrevei* (Gentry 1982: 446).

A. palustris (Rose) Thiede & Eggli (KuaS 50(5): 112, 1999). **T:** Mexico, Nayarit (*Rose* 1943 [US]). – **D:** Mexico (Nayarit); swamps. **I:** Rose (1903: fig. 1).

≡ *Polianthes palustris* Rose (1903).

[3b1] **R** unknown; basal **L** 2 - 4, base attenuate, parallel veins prominent, 20 - 30 × 0.8 - 1.5 cm, stem **L** 3 or 4, becoming much reduced above; **Inf** ± 0.4 m, 'spicate' with erect scape and 3 - 5 pairs of **Fl**; lower **Fl** sessile or to 5 mm pedicellate, upper **Fl** almost sessile; **Fl** scented like the cultivated tuberose; **Tep** 30 - 60 mm, outcurved near the middle or just below, distally flaring, mouth of the tube very slightly oblique, lobes ovate, somewhat spreading, obtuse or obtusely pointed, 5 - 6 mm; **Anth** not exserted; **Fr** and **Se** unknown.

Known from the type collection only and never recollected (McVaugh 1989).

A. papyrocarpa Trelease (Mem. Nation. Acad. Sci. 11: 44, t. 95-97, 1913). **T:** Cuba, Isla de Pinos (*Curtiss* 335 [NY]). – **Lit:** Álvarez de Zayas (1985). **D:** Cuba (Isla de Pinos).

A. papyrocarpa ssp. **macrocarpa** A. Álvarez (Revista Jard. Bot. Nac. Univ. Habana 5(3): 7, ills.,

1985). **T:** Cuba, Isla de Pinos (*Álvarez* 43981A [HAJB]). – **D:** Cuba (Isla de Pinos).

[2m] Differs from ssp. *papyrocarpa:* **Inf** with less compact part-**Inf**; **Fr** larger.

Based on a single slightly differing population only. Its taxonomic separation appears doubtful.

A. papyrocarpa ssp. **papyrocarpa** – **Lit:** León (1946). **D:** Cuba (Isla de Pinos).

[2m] **Ros** solitary; **L** oblong to elongate-oblanceolate, gradually acute, somewhat concave, sometimes a little conduplicate above, 75 - 125 × 15 cm, at first slightly glaucous and rather dull, margins nearly straight or concave on young plants; marginal teeth straight or variously and unequally curved mostly downwards, triangular from scarcely or slightly dilated bases, 1 - 4 mm, 1 - 2.5 cm apart, occasionally with 1 or several minute intermittent teeth; terminal **Sp** usually a little curved and somewhat conically subulate, narrowly grooved below the middle, smooth or slightly granular below, somewhat polished towards the end, 0.8 - 1.5 cm, brown, not decurrent; **Inf** 4 m, 'paniculate', part-**Inf** few, very laxly arranged on slender outcurved **Br** in the upper ½ or more of the **Inf**; **Fl** ± 40 mm; **Ov** fusiform, 20 mm; **Tep** light yellow, tube conical, 4 mm, lobes 15 mm; **Fr** globose-oblong, not stipitate and little beaked, thin-walled, 2 - 2.5 × 1.5 - 2 cm.

A. parrasana A. Berger (Notizbl. Königl. Bot. Gart. Berlin 4: 250, 1906). **T:** Mexico, Coahuila (*Purpus* s.n. [US]). – **D:** Mexico (SE Coahuila); limestone mountains, 1400 - 2480 m. **I:** KuaS 43(5): centre page pullout 1992/15.

Incl. *Agave wislizeni* ssp. *parrasana* (A. Berger) Gentry (1975) (*incorrect name*, Art. 11.4).

[2i] **Ros** compact, small, 30 - 50 cm ⌀, solitary, with few or no suckers; **L** 40 - 60 per **Ros**, ovate, closely imbricate, thick, rigid, short-acuminate to merely acute, plane to concave above, generally 20 - 30 × 10 - 15 cm, frequently light grey to bluish-glaucous; marginal teeth straight to curved, slender from small bases, 5 - 10 (-15) mm, largest near the **L** tip, rapidly becoming smaller further down, greyish-brown, 1 - 2.5 cm apart; terminal **Sp** slender from a broad base, flat to openly grooved above, 2 - 3 (-4) cm, dark brown to greyish, sharply decurrent to the uppermost teeth; **Inf** 3 - 4 m, 'paniculate', ellipsoid, part-**Inf** compact, 10 - 15; **Fl** 50 - 60 mm; **Ov** 25 - 30 mm, neck short, not constricted; **Tep** flushed red or purple, opening pale yellow, tube cylindrical, 13 - 14 mm, lobes subequal, 13 - 15 mm.

Easily distinguished by its short, broad, abruptly short-acuminate leaves. It differs from all other taxa in Group *Parryanae* by its purplish-coloured large succulent bracts on the scape, which cover the budding part-inflorescences (Gentry 1982: 538).

A. parryi Engelmann (Trans. Acad. Sci. St. Louis

3(20): 311-313, 1875). **T**: USA, Arizona (*Rothrock* 274 [MO]). – **Lit**: Ullrich (1992f). **D**: SW USA, NW Mexico.

≡ *Agave applanata* var. *parryi* (Engelmann) Mulford (1896).

Distinguished by its compact, freely suckering, many-leaved, light green to greyish rosettes (Gentry 1982: 539). Berger (1915: 179) ascribed the name to Haage & Schmidt (Cat., 14, 1873). The name *A. scabra* Salm-Dyck 1858, though inappropriate as to its meaning, would have priority but is an illegal later homonym (Ullrich 1992f). The 3 varieties recognized by Gentry (l.c., 542 etc.) were synonymized by Ullrich (1992f), as they were regarded as mere ecotypes with size modified by more humid or arid conditions and merging into typical plants in cultivation, as also stated by Gentry. His decision is based on mere literature study, and the varieties were recently accepted by Hodgson (1999), and this is followed here.

A. parryi var. **couesii** (Engelmann *ex* Trelease) Kearney & Peebles (J. Washington Acad. Sci. 29(11): 474, 1939). – **D**: USA (C Arizona); open slopes in grassland and pine-oak woodland, 1100 - 2100 m. **I**: Gentry (1982: 542).

≡ *Agave couesii* Engelmann *ex* Trelease (1911); **incl.** *Agave parryi* fa. *integrifolia* Breitung (1963).

[2i] Differs from var. *parryi*: **Ros** 35 - 55 × 40 - 65 cm; **L** smaller, 25 - 42 (-47) × 6.5 - 11 cm; **Fl** smaller, 43 - 58 (-60) mm; **Ov** 20 - 34 mm; **Tep** tip more densely papillate, tube 6 - 9 mm. – *Cytology:* 2n = 120.

This taxon represents a variant of smaller growth from the NW border of the species' range. However, small-leaved forms occur at random elsewhere (Gentry 1982). *A. parryi* fa. *integrifolia* represents a toothless variant found in populations of this variety; such aberrations are widespread in the genus and do not merit formal taxonomic recognition. The taxon hybridizes with *A. chrysantha* (Hodgson 1999).

A. parryi var. **huachucensis** (Baker) Little *ex* Benson (Amer. J. Bot. 30(3): 235, 1943). **T**: USA, Arizona (*Pringle* s.n. [K, NY]). – **D**: USA (SE Arizona); Mexico (NE Sonora, W Chihuahua?); open slopes in oak woodland and pine forests, 1550 - 2150 m.

≡ *Agave huachucensis* Baker (1888) ≡ *Agave applanata* var. *huachucensis* (Baker) Mulford (1896).

[2i] Differs from var. *parryi*: **Ros** more robust, 45 - 75 × 75 - 85 cm; **L** larger, 32 - 65 × 10 - 20 cm; **Inf** broader; **Fl** larger, 62 - 81 mm; **Ov** 34 - 47 mm; **Tep** tube 8 - 9 mm.

An upland variant with larger growth.

A. parryi var. **parryi** – **D**: USA (C and SE Arizona, SW New Mexico), Mexico (W Chihuahua, W Durango); open rocky slopes in grama grasslands,

oak woodland, pine-oak-forest, and chaparral, 1200 - 2800 m. **I**: Gentry (1982: 522, 540); KuaS 42(11): centre page pullout 1991/32. **Fig. VI.d**

Incl. *Agave parryi* ssp. *parryi*; **incl.** *Agave scabra* ssp. *scabra*; **incl.** *Agave americana* var. *latifolia* Torrey (1859); **incl.** *Agave scabra* Salm-Dyck (1859) (*nom. illeg.*, Art. 53.1); **incl.** *Agave wislizeni* Engelmann (1875) (*nom. illeg.*, Art. 52.1); **incl.** *Agave marcusii* De Smet (1876); **incl.** *Agave noah* Nickels (1894); **incl.** *Agave chihuahuana* Trelease (1911); **incl.** *Agave patonii* Trelease (1911); **incl.** *Agave marcusea* hort. *ex* Trelease (1912); **incl.** *Agave marensii* hort. *ex* Trelease (1912); **incl.** *Agave parayi* hort. *ex* Trelease (1912) (*nom. inval.*, Art. 61.1); **incl.** *Agave parreyi* hort. *ex* Trelease (1912) (*nom. inval.*, Art. 61.1); **incl.** *Agave paryi* hort. *ex* Trelease (1912) (*nom. inval.*, Art. 61.1); **incl.** *Agave payrii* hort. *ex* Trelease (1912) (*nom. inval.*, Art. 61.1).

[2i] **Ros** compact, globose, (35-) 40 - 60 × 60 - 75 cm, freely suckering; **L** 100 - 160 per **Ros**, linear-ovate, closely imbricate, rigid, thick, short-acuminate, mostly (18-) 25 - 50 × (4.5-) 8 - 12 cm, glaucous-grey to light green; marginal teeth mostly rather straight on a nearly straight margin, small, largest above the middle of the lamina, 3 - 7 mm, dark brown to greyish, mostly 1 - 2 cm apart; terminal **Sp** nearly flat above, 1.5 - 3 cm, dark brown to grey with age, decurrent to the 1. or 2. teeth; **Inf** 4 - 6 m, 'paniculate', stout, part-**Inf** stout, 20 - 36 in the upper ½ of the **Inf**; **Fl** mostly 60 - 77 mm; **Ov** (27-) 30 - 47 mm, neck long, 6 - 9 mm, mildly constricted and grooved; **Tep** pink to red in bud, opening yellow, tube 9 - 12 mm, lobes subequal, 18 - 24 mm; **Fr** on stout **Ped**, 3.5 - 5 × 1.5 - 2 cm, shortly stipitate, beaked, strong-walled; **Se** 7 - 8 × 5 - 6 mm, semicircular in outline. – *Cytology:* 2n = 60, 120.

A. parryi var. **truncata** Gentry (Agaves Cont. North Amer., 543-545, ills., 1982). **T**: Mexico, Zacatecas-Durango (*Gentry & Gilly* 10566 [US, DES, MEXU, MICH]). – **D**: Mexico (Durango / Zacatecas border); only known from the region of the type locality, 2450 m.

[2i] Differs from var. *parryi*: **L** very small (sometimes only 7 - 15 cm long), broad, tip acute to truncate.

A diminutive variant at the SE border of the species's range.

A. parvidentata Trelease (J. Washington Acad. Sci. 15(17): 395, 1925). **T**: El Salvador (*Calderon* 2085 [US 1169884-5]). – **D**: El Salvador. **I**: Gentry (1982: 488).

Incl. *Agave compacta* Trelease (1927).

[2k] **Ros** subcaulescent, dense, to 1 × 1.7 m, solitary; **L** numerous, ascending to out- or incurving, ovate-lanceolate, contracted into the thick base, plane, acuminate, 80 - 100 × 15 - 25 cm, pale green

to light grey-glaucous, margins straight; marginal teeth deltoid from lenticular bases, nearly straight, 3 (-5) mm (middle of the lamina), 1 - 2 cm apart, reduced up- and downwards; terminal **Sp** acicular, involutely grooved to above the middle, smooth, ± 5 cm, dull light brown, decurrent for more than its length; **Inf** ± 2.5 m, 'paniculate', dense, oblong, scape short, part-**Inf** globose, > 30, bulbilliferous; **Fl** slender, 40 - 50 mm; **Ov** fusiform, 20 - 25 mm; **Tep** yellow, tube openly conical, ± 5 mm, lobes subequal, 15 - 20 mm.

Closest to *A. pachycentra* and *A. wercklei*. Characteristic are its very short-peduncled inflorescences branching from the level of the upper leaf tips (Gentry 1982: 488). Similar plants exhibiting this feature also occur in Chiapas (Mexico) (Ullrich 1992a). Ullrich (l.c.) proposes *A. calderonii* Trelease (1923) as oldest name for *A. parvidentata*. Lott & García-Mendoza (1994), however, treat *A. calderonii*, which is known from the type collection only, as a name of doubtful identity tentatively assignable to Group *Vivipara* (as *Rigidae*).

A. parviflora Torrey (in Emory, Rep. US Mex. Bound. 214, 1859). **T:** Mexico, Sonora (*Schott* s.n. [US, NY]). – **D:** USA (Arizona), Mexico (Sonora).

A. parviflora ssp. **flexiflora** Gentry (US Dept. Agric. Handb. 399: 56-57, 1972). **T:** Mexico, Sonora (*Gentry* 16638 [US, DES, MEXU]). – **D:** Mexico (Sonora); grama grasslands, 650 - 1500 m. **I:** Gentry (1982: 196, 202).

[1e] Differs from ssp. *parviflora*: **L** dimorphic, linear to lanceolate, 6 - 10 × 1 cm or 15 - 18 × 1.2 cm; terminal **Sp** whitish; **Inf** 1.5 - 2.5 m, part-**Inf** with 1 - 3 (mostly 2) **Fl**; **Fl** saccate; **Tep** with **Anth** and **Sty** bent downwards at anthesis; **Ov** 6 - 8 mm, tube 3 - 4 mm, lobes 3.5 - 5 mm.

A. parviflora ssp. **parviflora** – **Lit:** Ullrich (1990f: with ills.). **D:** USA (Arizona), Mexico (Sonora). **I:** Gentry (1982: 196, 202-203).

Incl. *Agave hartmanii* S. Watson (1891).

[1e] **Ros** very small, 10 - 15 × 15 - 20 cm, solitary or caespitose; **L** oblong-linear, widest at or above the middle, convex below, plane above, 6 - 10 × 0.8 - 1 cm, green, both faces with white impressions from the central bud, margins conspicuously white-filiferous; marginal teeth minute, near the **L** base only; terminal **Sp** weakly subulate, 5 - 8 mm, brown to greyish-white; **Inf** 1 - 1.8 m, 'spicate', laxly flowered through the upper ½, this part frequently reddish, part-**Inf** with 2 - 4 **Fl**; **Fl** 13 - 15 mm; **Ov** proper 4 - 5 mm, neck 2 mm; **Tep** pale yellow, tube urceolate, 5 mm, lobes slightly unequal, 2 - 3 mm.

This taxon has the smallest flowers in the genus. It is closely related to *A. polianthiflora* from which it is only separable with certainty by the distinctive flowers (Gentry 1982: 201).

A. ×peacockii Croucher (Gard. Chron. 1873: 1400, fig. 283, 1873). **T:** [icono]: Curtis's Bot. Mag. 1901: t. 7757. – **D:** Mexico (Hidalgo, Puebla, Oaxaca); calcareous hills. **I:** Gentry (1982: 126, 166).

≡ *Agave ghiesbreghtii* var. *peacockii* (Croucher) A. Terracciano (1885) ≡ *Agave roezliana* var. *peacockii* (Croucher) Trelease (1920).

A. ×peacockii is the putative natural hybrid between the sympatric *A. kerchovei* and *A. marmorata*. This is above all suggested by the inflorescences that are intermediate between Subgen. *Littaea* and Subgen. *Agave* (Gentry 1982: 165-166). A detailed morphometric analysis was provided by Valverde & al. (1996).

A. pedunculifera Trelease (CUSNH 23: 134, 1920). **T:** Mexico, Sinaloa (*Rose* 1713 [US]). – **Lit:** McVaugh (1989). **D:** Mexico (Sinaloa, Nayarit, Jalisco, Michoacán, Guerrero, Oaxaca); mountain slopes in tropical deciduous or oak forest, 300 - 2200 m. **I:** Gentry (1982: 67, 80-81).

[1b] **Ros** caulescent, solitary; **L** symmetrically ascending-horizontal, soft, thickened, narrowed and convex at the base, plane to concave, mostly ovate-acuminate and 50 - 70 × 15 - 18 cm, or lanceolate and 80 - 90 × 11 - 15 cm, pale green to glaucous-white, margins narrowly lined with brown or white, with close denticles 0.5 - 2 mm long, otherwise smooth; terminal **Sp** acicular, weak, ± 1 cm; **Inf** erect or recurving, 2 - 3 m, 'spicate', flowering from near the **L** tips, part-**Inf** with 2 or 4 **Fl**; **Ped** geminate; **Fl** slender, 37 - 52 mm; **Ov** cylindrical, slender, 20 - 27 mm, neck not constricted; **Tep** yellow, tube shallowly funnel-shaped, 2 - 6 mm, lobes equal, ± 22 mm.

Closely related to *A. attenuata* based on leaf and flower characters, but distinguished by being nearly stemless throughout all observed populations. *A. pedunculifera* exhibits considerable variability in leaf form and size and depth of the flower tube, but the different forms are linked by intermediates (Gentry 1982: 80).

A. pelona Gentry (US Dept. Agric. Handb. 399: 76-80, ills., 1972). **T:** Mexico, Sonora (*Gentry & Arguelles* 19898 [US, DES, MEXU]). – **Lit:** Turner & al. (1995). **D:** Mexico (Sonora); limestone rocks and cliffs. **I:** Gentry (1982: 170).

[1g] **Ros** subcaulescent, compact, 40 - 60 × 60 - 80 cm, solitary; **L** many, linear-lanceolate, erect to ascending, thick, stiff, sometimes slightly narrowed towards the base, long-acuminate, rounded below, plane above, epidermis smooth, minutely punctate, waxy, 35 - 50 × 3 - 5 cm, shiny dark green, turning reddish to purplish during drought or with age, margins with smooth white firm border; marginal teeth none; terminal **Sp** strong, sharp, sharply angled below, grooved or plane above, 4 - 7 cm, white to reddish, decurrent as a white border down the **L** margins; **Inf** 2 - 3 m, 'spicate', flowering through

the upper ½, part-**Inf** with geminate **Fl**; **Ped** 30 - 50 mm; **Fl** campanulate, 45 - 50 mm; **Ov** slender, 20 mm incl. neck, light green; **Tep** dark red, tube openly funnel-shaped, 8 - 9 mm, lobes 18 mm.

Without close relatives in Group *Marginatae* and geographically isolated from its remaining species. It may, however, be misplaced in this group and perhaps belongs to Group *Filiferae*, as indicated by its funnel-shaped flower tube with nectariferous inner lining and red tepals with recurved lobes (Gentry 1982: 169).

A. pendula Schnittspahn (Z. Gartenbau-Vereins Darmstadt 6: 7, 1857). **T** [neo]: Ex cult. (*Anonymus s.n.* [K]). − **D:** Mexico (Veracruz, Chiapas). **I:** Gentry (1982: 217, 227-228).

    Incl. *Agave aloina* Koch (1860); **incl.** *Agave sartorii* Koch (1860); **incl.** *Agave pulcherrima* hort. *ex* C. Koch (1865) (*nom. illeg.*, Art. 53.1); **incl.** *Agave rubrocincta* Jacobi (1868); **incl.** *Agave rufocincta* Jacobi (1868); **incl.** *Agave caespitosa* Todaro (1876) ≡ *Agave sartorii* var. *caespitosa* (Todaro) A. Terracciano (1885).

    [1f] Stem short; **Ros** open, spreading, branching axillary; **L** 20 - 30, slenderly lanceolate, softly fleshy, ascending to somewhat outcurving, rounded below, plane to concave above, 50 - 75 × 5 - 11 cm, green to yellow-green, frequently with pale yellow central stripe, margins not horny, denticulate with brown denticles ± 1 mm long; terminal **Sp** small, 5 - 8 mm, brown, not decurrent; **Inf** 1.3 - 1.8 m, 'spicate', slender, drooping, laxly flowered in the upper ⅓ - ½ of the **Inf**, part-**Inf** with solitary or geminate **Fl**; **Fl** 30 - 45 mm; **Ov** 10 - 15 mm, neck short, not constricted; **Tep** greenish or tinged with lavender, whitish inside, tube funnel-shaped, 6 - 13 mm, lobes about equal, 14 - 16 mm. − *Cytology:* 2n = 60.

Very distinctive with its overhanging inflorescence. In contrast to Gentry (l.c.), Berger (1915: 60) regards *A. sartorii* as an earlier valid name for this taxon, since he cites *A. pendula* as 'Schnittspahn *ex* Jacobi 1865'. This nomenclatural problem needs further study.

A. petiolata Trelease (Mem. Nation. Acad. Sci. 11: 20, t. 8, 1913). **T:** Curaçao (*Boldingh* A8 [MO?]). − **Lit:** Hummelinck (1993: with ills.). **D:** Leeward Islands (Curaçao).

    Incl. *Agave lurida* Hamelberg (1898) (*nom. illeg.*, Art. 53.1).

    [2q] **Ros** caulescent for < 1 m, suckering (?); **L** lanceolate, rather abruptly contracted into a long neck at the base, gradually acute, ± 110 × 17 cm, blue-glaucous, margins nearly straight between the teeth or prominences; marginal teeth straight or variously curved, narrowly triangular from half-round bases 5 - 10 mm wide and sometimes raised on abrupt green prominences, teeth 5 mm, purplish-chestnut-brown, 1.5 - 3 (-5) cm apart; terminal **Sp** acicular, ± flexuous, round-grooved to or

beyond the middle, granular-roughened below, smooth and polished towards the tip, 2.5 - 6 cm, chestnut-brown, shortly decurrent; **Inf** unknown; **Fl** 35 - 40 mm; **Ov** fusiform, 15 mm, tube open, 5 mm, lobes 15 mm; **Fr** unknown.

A curious plant, in leaf armature suggesting some of the Mexican species grown for Pulque (Trelease l.c.).

A. petrophila García-Mendoza & E. Martínez (Sida 18(2): 627, 1998). **T:** Mexico, Guerrero (*Martínez & al.* 2639 [MEXU, BRIT, ENCB, K, MO]). − **Lit:** García-Mendoza & Martínez Salas (1998: with ills.). **D:** Mexico (Guerrero, Oaxaca); rocky slopes on calcareous soil, 850 - 1300 m.

    Incl. *Agave gracilis* García-Mendoza & E. Martínez (1998) (*nom. illeg.*, Art. 53.1).

    [1a] Stems procumbent, to 1 m; **Ros** semiglobose, compact, 50 - 80 cm ⌀, caespitose; **L** > 100 per **Ros**, linear, plane, flexible, subcoriaceous, with longitudinal striae, 40 - 70 × 0.4 - 0.9 cm, glaucous to glaucous-green, margins yellowish, finely denticulate; terminal **Sp** weak, brownish-reddish; **Inf** 1.8 - 2 m, 'spicate', erect or slightly inclined, **Fl** in the upper ¼ - ½ of the **Inf**; **Ped** 1 mm; **Fl** campanulate, 20 - 22 (-25) mm; **Ov** cylindrical, 7 - 10 × 2 - 4 mm, slightly penetrating into the **Tep** tube; **Tep** oblong, 9 - 11 × 2.5 - 3.5 (-4.5) mm, green, tips dark reddish, tube 3 - 4 mm; **Fr** globose, 0.9 - 1 × 0.8 - 0.9 cm, dark brownish; **Se** 3 - 3.5 × 2 - 2.5 mm black.

This taxon shares morphological characteristics with *A. dasylirioides*, but differs in its small caespitose rosettes, smaller and narrower leaves, erect or slightly inclined inflorescences, much smaller flowers and smaller and globose fruits (García-Mendoza & Martínez Salas 1998). The species was first illegitimately named *A. gracilis* (l.c.). A plant from Oaxaca referred to *A. dasylirioides* by Ullrich (1990g) may belong here.

A. planifolia S. Watson (Proc. Amer. Acad. Arts 22: 479, 1887). **T:** Mexico, Chihuahua (*Pringle* 1141 [GH, VT]). − **D:** Mexico (Sonora, Chihuahua); sandy banks near streams and in oak wood regions, fall-flowering.

    ≡ *Manfreda planifolia* (S. Watson) Rose (1903) ≡ *Polianthes planifolia* (S. Watson) Shinners (1966).

    [3a3] Plants medium-sized (for Subgen. *Manfreda*); rhizome globose; **L** 4 - 5, spreading-arching, oblong to elliptic, narrowed towards the clasping basal pseudopetiole, tip acuminate, with a short point, channelled in the petiolar portion and at the tip, nearly flat in the middle, semisucculent, drying leathery, smooth, 21 - 30.5 × (2.5-) 3.1 - 6.1 cm, unspotted, margins with a narrow hyaline band, minutely regularly denticulate; remains of **L** bases coarsely fibrous, surrounding the rhizome, 3.8 - 5 cm; **Inf** 1.2 - 1.5 m, 'spicate', flowering part short, 11 cm in cultivation, with 6 - 14 sessile **Fl**; mature **Fl** spreading; **Ov** 15 mm; **Tep** tube cylindrical, at a

slight angle to the **Ov**, 5 - 7 mm, lobes oblong, reflexed, 12 - 19 mm; **Sty** exceeding the tube by ± 5 mm; **Sti** clavate, trigonous; **Fr** ovoid, 1.8 cm; **Se** 5 mm.

Probably most closely related to *A. guttata*, but distinctive because of its elliptic leaves with the appreciably narrowed and clasping base and acuminate tip, as well as the more N range (Verhoek-Williams 1975: 257).

A. platyphylla (Rose) Thiede & Eggli (KuaS 50(5): 112, 1999). **T:** Mexico, Jalisco (*Rose* 2598 [US]). – **Lit:** McVaugh (1989: with ill.). **D:** Mexico (S Durango, S Zacatecas, Jalisco); grasslands, rocky mesas among grasses, hillsides in pine-oak forests, 1500 - 2500 m, flowers August to November. **I:** Rose (1903).
≡ *Polianthes platyphylla* Rose (1903).

[3b1] Plants glabrous; **R** fleshy, tapering, to 3 - 7 cm, in dense clusters at the base of the bulbs; **L** 2 - 10 in a basal **Ros**, lying flat on the ground, lanceolate-elliptic to narrowly ovate, base narrowed to form a shortly subpetiolate part 5 mm wide or less, tip long-acute, 7 - 15 (excl. base) × 1.2 - 3 cm, margins narrow, smooth, hyaline, **L** bases imbricate, broadly expanded, rigid, chestnut-brown or yellowish, forming a narrowly ovoid bulb; **Inf** 40 - 70 cm, 'spicate', flowering part 10 - 20 cm, with few- to 10-flowered nodes; **Fl** essentially sessile, becoming horizontal or deflexed; **Ov** ellipsoid, erect at anthesis or nearly so; **Tep** white or cream-coloured, shaded with rose at the base, lobes pink, whole **Tep** pink with age, finally deep rose, tube strongly curved outwards just above the **Ov**, 1.5 mm ⌀ near the base, 2.5 - 3 mm ⌀ at **Fil** insertion, mouth oblique, 13 - 16 mm, lobes subequal, rounded or ovate, 2 - 2.5 (-3) mm; **St** included; **Sty** slightly exserted at maturity, with 3 flat obtuse lobes 1 - 1.3 mm long; **Fr** broadly ellipsoid to subglobose, ± 0.7 - 1 × 0.7 - 1 cm; **Se** 2.5 - 3 mm.

A. polianthes Thiede & Eggli (KuaS 52: [in press], 2001). – **Lit:** McVaugh (1989); Ullrich (1993b: with ills.). **D:** Cultivated only and not known from the wild.
Incl. *Polianthes tuberosa* Linné (1753) ≡ *Agave tuberosa* (Linné) Thiede & Eggli (1999) (*nom. illeg.*, Art. 53.1); **incl.** *Polianthes gracilis* Link (1821) ≡ *Polianthes tuberosa* var. *gracilis* (Link & Otto) Baker (1888); **incl.** *Polianthes tubulata* Sessé & Moçiño (1894); **incl.** *Polianthes tuberosa* fa. *plena* Moldenke (1948).

[3b1] Plants glabrous; **R** fleshy; **L** 6 - 10 in a basal **Ros** from a bulbous base, linear, soft, deeply channelled in the basal ½, to 30 - 60 × 1 - 1.5 cm, bright green, sometimes reddish near the base, sometimes with brown spots on the lower face; **Inf** 60 - 100 cm, 'spicate', flowering part 20 cm or more, laxly flowered, with up to 20 or more flowering nodes with paired **Fl**; **Fl** mostly sessile,

fragrant, 25 - 40 mm; **Tep** waxy white, base upright or strongly ascending, tube smoothly outcurved from below the middle, funnel-shaped above the curvature, expanding to the very slightly oblique mouth, there 7 - 8 mm ⌀, lobes subequal, elliptic-ovate, obtusely pointed, often 15 - 18 × 7 - 10 mm; **St** included; **Sty** included, with 3 oblong-ovate recurved lobes 2.5 mm long.

This is the "Tuberose" or "Nardo" grown in large quantities for the flower market, cultivated in Europe at least since 1601, when it was first illustrated by Clusius. Old European illustrations are discussed by Ullrich (1993b). Especially common is a form with double ('filled') flowers, *Polianthes tuberosa* fa. *plena* Moldenke 1948, which should better be treated as cultivar *Agave polianthes* 'Plena'. The species is at present not known from the wild and already Linné based his description on cultivated material from India. It is most probably of Mexican origin and possibly native to the region around Guadalajara in the state of Jalisco, where its putative ally *A. dolichantha* has recently been rediscovered in the wild (see note there).

A. polianthiflora Gentry (US Dept. Agric. Handb. 399: 51-54, ills., 1972). **T:** Mexico, Chihuahua (*Gentry* 8013 [US, DES, MEXU]). – **D:** Mexico (Sonora, Chihuahua); rock outcrops in pine-oak forest, 1250 - 2000 m. **I:** Gentry (1982: 196, 202-203); KuaS 40(12): centre page pullout 1989/36. **Fig. V.a, V.b**

[1e] **Ros** small, 10 - 20 × 20 - 30 cm, solitary or caespitose; **L** linear-lanceolate, widest in the middle, convex below, plane above, 10 - 20 × 1 - 1.3 cm, green, both faces with white impressions from the central bud, margins conspicuously white-filiferous; marginal teeth minute, near the **L** base only; terminal **Sp** weak, 0.7 - 1 cm, greyish; **Inf** 1.2 - 2 m, 'spicate', axis red, part-**Inf** in the upper ½ of the **Inf**, usually with 2 (or 1 or 3) **Fl**; **Ped** short; **Fl** 37 - 42 mm; **Ov** 9 - 12 mm, red; **Tep** pruinose, pink, tube long, very narrow and curved below, 22 - 32 mm, lobes subequal, 4 - 7 mm.

Differs from all other species of this subgenus by its long tubular flowers similar to those of the former genus *Polianthes* with very short lobes. Moreover, the flowers are not proterandrous, as in other Agaves.

A. portoricensis Trelease (Mem. Nation. Acad. Sci. 11: 38, t. 76-82, 1913). **T:** Puerto Rico (*Trelease* 7 p.p. [MO?]). – **D:** Puerto Rico, Culebra.

[2l] **Ros** solitary; **L** broadly lanceolate, subacuminate, somewhat conduplicate-concave, 100 - 150 × 15 - 20 cm, dark green, glossy, ± lightly glaucous when young, margins ± concave; marginal teeth straight or retrorse, conspicuously triangular from lenticular bases, 2 - 5 mm, mostly 1.5 - 3 cm apart; terminal **Sp** conically subulate, somewhat curved, sometimes compressed from the sides and basally

thickened, shallowly grooved or involute nearly to the end, smooth, 1 - 1.5 (-2) cm, chocolate- or chestnut-brown, glossy, decurrent for several times its length and dorsally immersed into the green **L** tissue; **Inf** 5 - 6 m, 'paniculate', narrowly oblong, part-**Inf** on nearly horizontal **Br**, in the upper ½ or more of the **Inf**, bulbilliferous; **Fl** ± 55 mm; **Ov** oblong-fusiform, 30 - 35 mm; **Tep** greenish-yellow, tube conical, ± 7 mm, lobes 15 mm; **Fr** subglobose, stipitate, ± beaked, 2.5 - 3 × 2 - 2.5 cm.

A. potatorum Zuccarini (Flora 15:2(Beiblatt 2): 96-97, 1832). − **D:** Mexico (Puebla, Oaxaca); semi-arid highlands with pine-oak forests, 1240 - 2300 m. **I:** Gentry (1982: 468, 491).

Incl. *Agave potatorum* var. *minor* hort. (s.a.) (*nom. inval.*, Art. 29.1); **incl.** *Agave scolymus* Karwinsky *ex* Salm-Dyck (1834); **incl.** *Agave elegans* hort. *ex* Salm-Dyck (1859); **incl.** *Agave latifolia* hort. *ex* Salm-Dyck (1859); **incl.** *Agave pulchra* hort. *ex* Salm-Dyck (1859); **incl.** *Agave quadrata* Lemaire (1864); **incl.** *Agave saundersii* Hooker *fil.* (1865); **incl.** *Agave verschaffeltii* Lemaire (1868) ≡ *Agave potatorum* var. *verschaffeltii* (Lemaire) A. Berger (1915); **incl.** *Agave auricantha* hort. *ex* Baker (1888).

[2k] **Ros** compact to openly spreading, small, solitary; **L** 50 - 80 (to >100) per **Ros**, ovate to shortly lanceolate, softly fleshy but rather rigid, thickened and narrowed towards the base, plane to somewhat hollowed above, mostly 25 - 40 × 9 - 18 cm, glaucous-white to green, margins undulate to deeply crenate with tubercle-like prominences, esp. above the middle of the lamina; marginal teeth on slender variously curved cusps from low broad bases, 5 - 10 mm and more, chestnut-brown to greyish-brown, mainly 1 - 3 cm apart; terminal **Sp** broad at the base, sharply pointed, sinuous, broadly grooved to flat above, 3 - 4.5 cm, chestnut-brown to greyish-brown, sharply decurrent as a ridge to the uppermost teeth; **Inf** 3 - 6 m, 'paniculate', **Bra** red to purplish, part-**Inf** small, compact, 15 - 30 in the upper ¼ - ½ of the **Inf**; **Fl** 55 - 80 mm; **Ov** 25 - 50 mm; **Tep** frequently tinged red or purplish in bud, light green to yellowish, tube cylindrical to funnel-shaped, 10 - 17 mm, lobes unequal, 13 - 24 mm.

A very polymorphic species widely distributed in horticulture.

A. potosina Robinson & Greenman (Proc. Amer. Acad. Arts 29: 393-394, 1894). **T** [lecto]: Mexico, San Luis Potosí (*Pringle* 3745 [GH, B, BM, BR, F, G, GH, K, M, MEXU, MO, NY, P, US, VT]). − **D:** Mexico (Coahuila, San Luis Potosí, Zacatecas); dry desert and limestone mesas, flowers in June. **I:** Piña Luján (1985: 29, 59-61).

≡ *Manfreda potosina* (Robinson & Greenman) Rose (1903) ≡ *Polianthes potosina* (Robinson & Greenman) Shinners (1966); **incl.** *Delpinoa gracillima* Ross (1897).

[3a1] Plants small; rhizome 2 - 3.4 × to 1.1 cm; **R** very fleshy; **L** 2 - 7, fleshy, recurved, lanceolate, channelled, to 16 × 1.4 cm, margins with irregular cartilaginous teeth, these coarse, broad and usually truncate, blunt, usually incised at the tip, occasionally retrorse, 2 - 5 (-14) mm apart; remains of **L** bases covering the plant base, membranous, 4 - 9.5 cm; **Inf** (15.5-) 24 - 54 (-75) cm, 'spicate', flowering part 9 - 29.5 cm, semidense to open above, with 7 - 31 nodes, **Fl** rarely paired, erect, green; **Ov** ellipsoid, 3 - 6 mm; **Tep** tube straight, constricted above the **Ov**, 6 - 14 mm, lobes erect, 2 - 5 mm; **Fil** varying in length but falling in 2 size classes attached at the base and the middle of the tube, exceeding the tube; **Sty** equalling the tube; **Sti** clavate, trigonous; **Fr** ± globose, (0.9-) 1 - 1.3 (-1.5) × 0.8- 1.2 cm; **Se** 2 - 3 × 3 - 4 mm.

The short style and several lengths of the filaments are characteristic features of this species (Verhoek-Williams 1975: 184).

A. potrerana Trelease (CUSNH 23: 138, 1920). **T:** Mexico, Chihuahua (*Pringle 802* [MO, B, NY, UC, US]). − **Lit:** McVaugh (1989). **D:** Mexico (Chihuahua, N Coahuila, Zacatecas); oak-pine grassland, 1500 - 2000 m. **I:** Gentry (1982: 126, 173).

[1g] **Ros** thick-stemmed, regularly spreading, 0.7 - 1 × 1.5 - 2 m, solitary; **L** numerous, lanceolate, straight, rigid, widest below the middle, convex below, roundly guttered above, mostly 40 - 80 × 6 - 7 cm, glaucous to light green, margins horny, continuous, straight, firm, brown towards the base, grey above; marginal teeth mostly straight, generally small, 2 - 4 mm, commonly 2 - 3 cm apart, lacking or reduced to serrations below the middle of the lamina; terminal **Sp** acicular, sharply angled below, flat to broadly canaliculate above, 2.5 - 4 cm, light brown to grey; **Inf** 4 - 7 m, 'spicate', stout, straight or arching, densely flowered through the upper ⅔, part-**Inf** with 2 - 4 **Fl**; **Ped** geminate, 4 - 15 mm; **Fl** 46 - 58 mm; **Ov** slender, 25 - 32 mm, neck smooth, constricted; **Tep** pink to red or yellow, tube 3 - 6 mm, lobes nearly equal, 17 - 24 mm.

Distinct and without close relatives in Group *Marginatae*. It differs by its solitary habit, tall inflorescences with large red flowers, and the long-acuminate leaves with reduced or lacking teeth on the lower ½ of the margin (Gentry 1982: 174).

A. pratensis A. Berger (Agaven, 37, 1915). **T:** Mexico, Nayarit (*Rose* 1994 [US, K, MEXU]). − **Lit:** McVaugh (1989). **D:** Mexico (Nayarit); small grassy openings along little streams, flowers in August; only known from the type collection.

Incl. *Manfreda rubescens* Rose (1903) ≡ *Polianthes rubescens* (Rose) Shinners (1966).

[3a3] Plants small (for Subgen. *Manfreda*); **R** fibrous; rhizome oblong, ≥ 1.5 × 1.1 - 1.2 cm; **L** 3 - 6, erect, linear-lanceolate, tip acute, fibrous, 17.3 - 28 × 0.65 - 1 cm, green, unspotted, margins with a nar-

row white cartilaginous band, continuously minutely papillate, but smooth to the touch; remains of **L** bases forming a dense mass of fine stiff light brown fibres, 4.2 - 8 cm, previous year's **L** bases intact; **Inf** 60 - 85 cm, 'spicate', flowering part dense, 5.4 - 6.2 cm, with 5 - 9 sessile **Fl**; **Ov** ellipsoid, 7 - 10 mm; **Tep** dark (purple, fide Verhoek-Williams (1975: 245)), tube very short, 3 - 5 mm, connected to the **Ov** without constriction, lobes erect, 2 - 5 mm; **Sty** equalling the tube; **Sti** clavate, trigonous; **Fr** ± globose, (0.9-) 1 - 1.3 (-1.5) × 0.8- 1.2 cm; **Se** 2 - 3 × 3 - 4 mm.

This species is perhaps nearest to *A. guttata*, from which it differs by its purple (vs. greenish-yellow) flowers and its different leaves (Verhoek-Williams 1975: 246).

A. producta Thiede & Eggli (KuaS 50(5): 112, 1999). **T:** Mexico, Guerrero (*Chisholm* s.n. [US 11260]). – **D:** Mexico (Guerrero).

Incl. *Polianthes elongata* Rose (1903).

[3b1] Plants 80 - 90 cm tall; stem bulb-like at the base, bulb 1.2 - 3.5 cm ∅; **L** elongate, oblanceolate, 30 × 1 - 1.2 (near the tip) cm, green, hardly if at all glaucous, flat above, trough-shaped below; scape **L** 6 - 7, reduced above, becoming **Bra**-like; **Inf** 'spicate', with ≥ 20 geminate **Fl**, scape reddish at the base, glaucous above, glabrous throughout; **Bra** ovate-linear, acuminate, 10 - 15 mm, as long as the **Ped**, reddish; **Tep** overall 2 cm long, red, tube slender, curved just above the base and almost at a right angle to the axis of the **Ov**, lobes somewhat spreading, short, rounded; **Anth** tips just exceeding the mouth of the **Pet** tube; **Fr** and **Se** unknown.

A hardly known species. When transferring *Polianthes elongata* to *Agave*, a new name was necessary to avoid homonymy with *A. elongata* Jacobi (1865).

A. promontorii Trelease (Annual Rep. Missouri Bot. Gard. 22: 50, 1912). **T:** Mexico, Baja California (*Nelson & Goldman* 7437 [US]). – **D:** Mexico (Baja California Sur); granitic mountains, 900 - 1800 m. **I:** Gentry (1982: 310, 320).

[2d] Stem thick; **Ros** open, large, 1 - 2 and more × 2 - 2.5 m, solitary; **L** lanceolate, straight to arching, fleshy-succulent, stiff, thick at the base, usually concave above, 100 - 150 × 11 - 17 cm, green to light glaucous-green, margins ± straight; marginal teeth straight to curved, regular, mostly 4 - 8 mm, reddish-brown, 5 - 10 mm apart; terminal **Sp** conically subulate, narrowly sulcate above, 3 - 5 cm, dark brown, shortly decurrent; **Inf** 5 - 9 m, 'paniculate', massive, part-**Inf** diffuse, 25 - 30 in the upper ½ of the **Inf**; **Fl** campanulate, 60 - 75 mm; **Ov** 36 - 42 mm, neck narrowed; **Tep** red to purplish in bud, tube 14 - 15 mm, lobes equal, 14 - 16 mm.

Clearly distinct from both *A. aurea* and *A. capensis* in the size of leaves and inflorescences (Gentry 1982: 321).

A. pubescens Regel & Ortgies (Gartenflora 23: 227, t. 804, 1874). **T:** Ex cult. BG St. Petersburg (*Anonymus* s.n. [LE]). – **D:** Mexico (Morelos, Oaxaca, Chiapas); rocky slopes in mountain regions, 365 - 1830 m, flowers in August.

≡ *Agave brachystachys* var. *pubescens* (Regel & Ortgies) A. Terracciano (1885) ≡ *Manfreda pubescens* (Regel & Ortgies) Verhoek-Williams (1975) (*nom. inval.*, Art. 29.1).

[3a2] Plants medium-sized (for Subgen. *Manfreda*); **R** half-fleshy; rhizome cylindrical, 1.5 × 1.7 cm; **L** 3 - 4 (up to 9 in cultivation), lanceolate, coriaceous, recurved-spreading, slightly channelled, slightly undulate, tip acute, with a short point, 18 - 28 × 2.1 - 3.2 cm, upper face green, lower face paler, both faces spotted with dark brown and densely pubescent, margins with a narrow hyaline band, entire, revolute; remains of **L** bases 4.5 - 9.5 cm; **Inf** 63 - 184 cm, 'spicate', flowering part elongate, with 10 - 19 (nearly) sessile spreading-horizontal **Fl**; **Ov** ellipsoid, 7 - 12 mm; **Tep** green, tube narrowly funnel-shaped, nearly straight, 13 - 22 × 4 (middle of the tube) mm, lobes much revolute, 9 - 13 mm; **Sty** first bent downwards, at maturity straight, exserted for 23 - 45 mm; **Sti** clavate, trigonous; **Fr** oblong, ± 2 × 1.2 cm; **Se** unknown.

With the exception of *A. maculata* the only species in Subgen. *Manfreda* with pubescent leaves, but distinguished by its generally larger size and much more exserted stamens and styles (Verhoek-Williams 1975: 299). Further collections might, however, bridge the gap between both species.

A. pumila De Smet *ex* Baker (Handb. Amaryll., 172, 1888). **T:** US, DES, HBG. – **D:** Known from cultivation only. **I:** Gentry (1982: 175-176); KuaS 42(5): centre page pullout 1991/14.

Incl. *Agave simonis* hort. *ex* A. Berger (1915).

[1g] Plants dimorphic, **juvenile form** persisting for 8 - 12 years; **Ros** small, 5 - 8 cm ∅, surculose; **L** ovate-orbicular, thickly succulent, broader than long, base broadly clasping, rounded below, deeply concave above, 2 - 4 × 3 - 4 cm, greyish-green, upper face striped, margins thin, friable, white; marginal teeth several, weak and small; terminal **Sp** conical, flexuous, small; **mature form** thick-stemmed; **Ros** open, short, 40 - 50 × 60 - 70 cm, not suckering; **L** deltoid-lanceolate, rigid, patulous, tickened at the base, upper face concave, lower face convex, 30 - 38 × 4 - 4.5 cm, greyish-green, without stripes below, margins narrowly horny, detaching, white; marginal teeth small, weak, 1 - 2 mm, 1 - 1.5 cm apart; terminal **Sp** conical, slender, 1.5 cm, decurrent along the **L** edges and along the keel in the middle of the lower face; **Inf**, **Fl** and **Fr** unknown.

Long known only from the stunted juvenile form, but developing into large 'normal' rosettes when given enough space. Such a dimorphism is unknown in other Agaves. The plant may represent a natural hybrid, probably *A. victoriae-reginae* × *A. lechuguilla* (Gentry 1982: 175).

A. revoluta Klotzsch (Allg. Gartenzeitung 8: 274, 1840). **T:** Ex cult. BG Berlin (*Anonymus* s.n. [B]). – **D:** Mexico (México); clay bluffs, flowers in July.

≡ *Manfreda revoluta* (Klotzsch) Rose (1903) ≡ *Polianthes revoluta* (Klotzsch) Shinners (1966).

[3a2] Plants small (for Subgen. *Manfreda*); rhizome 1.8 cm ⌀; **L** 5, linear-lanceolate, revolute, somewhat channelled, undulate, thin, tip broadly acute, with a short point, with a marked midrib region and closely set veins, 12.3 - 18.8 × 1.5 - 2 cm, margins with a narrow hyaline band, entire or finely papillate to erose-papillate; remains of **L** bases finely fibrous, 0.5 cm; **Inf** 0.8 - 1.2 m, 'spicate', flowering part elongate or crowded; **Fl** sessile, erect; **Ov** narrowly ellipsoid, 7 - 12 mm; **Tep** tube narrowly funnel-shaped, 12 - 14 mm, lobes oblong, thin; **Sty** exceeding the tube for 18 - 35 mm; **Sti** trigonous; **Fr** and **Se** unknown.

This species belongs to the *A. scabra* Group within Subgen. *Manfreda*. It differs from other members by its short revolute leaves. The anonymous specimen at B appears to represent the type material of Klotzsch (Verhoek-Williams 1975: 296).

A. rhodacantha Trelease (CUSNH 23: 117, 1920). **T** [neo]: Mexico, Nayarit (*Gentry & Gilly* 10704 [ARIZ]). – **Lit:** McVaugh (1989). **D:** Mexico (Sonora, Sinaloa, Nayarit, Jalisco, Puebla, Oaxaca); moister mountain slopes, 50 - 1000 m. **I:** Gentry (1982: 581).

[2f] Stem none or 50 - 90 cm; **Ros** truncate, large, 2 - 3 × 3 - 5 m, solitary or caespitose; **L** linear, hard-fibrous, rigid, straight, much thickened and scarcely narrowed at the base, smooth, 140 - 250 × 8 - 15 cm, green to faintly glaucous-green, margins straight to undulate; marginal teeth curved upwards, firm, slender, very sharp, regular, mostly 4 - 8 mm, dark brown, mostly 1 - 3 cm apart; terminal **Sp** conical but frequently with subulate tip, with short open groove above, 1 - 2.5 cm, dark brown; **Inf** 7 - 9 m, 'paniculate', broad, scape short, part-**Inf** large, remote, 35 - 45 per **Inf**; **Fl** 55 - 65 mm (dried and relaxed); **Ov** fusiform, 25 - 35 mm incl. the short neck; **Tep** green, yellowing at anthesis, tube urceolate, 8 - 10 mm, lobes subequal, 16 - 23 mm.

Distinguished from its close relative *A. vivipara* (as *A. angustifolia*) by its very long rigid leaves and large inflorescences with large long-stipitate fruits (Gentry 1982: 582), but the size differences may possibly be due to the moist habitat.

A. rosei Thiede & Eggli (KuaS 50(5): 112, 1999). **T:** Mexico, Nayarit (*Rose* 2178 [US]). – **D:** Mexico (Nayarit); in a deep canyon; only known from the type collection.

Incl. *Polianthes montana* Rose (1903).

[3b1] **L** not narrowed at the base, ± 30 × 0.7 cm,

margins obscurely papillose; **Inf** 1.1 m, 'spicate', flowering part ± 18 cm, with ± 10 flowering nodes; **Ped** (lowest, on faded **Fl**) 6 mm, otherwise 2.5 - 4 mm; **Tep** white, tube curved near the base, 14 - 15 mm, lobes ovate, 2.5 - 3 × 1.5 - 2 mm; **Anth** 4.3 - 4.7 mm; **Fil** (free parts) ± 7 mm.

McVaugh (1989) suggests the possibility that the specimen regarded as type, which much resembles *A. duplicata* (≡ *Polianthes geminiflora*), has erroneously been substituted for the original white-flowering type specimen of *A. rosei* (≡ *Polianthes montana*). The new name was necessary to avoid homonymy with *A. montana* Villarreal 1996.

A. rutteniae Hummelinck (Recueil Trav. Bot. Néerl. 33: 238, 1936). **T** [syn]: Aruba (*Hummelinck* 19a+b [U]). – **D:** Leeward Islands (Aruba); debris of igneous rocks.

[2q] **Ros** ± 0.9 - 1.5 m ⌀, suckering; **L** few, narrowly elliptic or lanceolate, straight or very slightly S-curved, acute, tip usually slightly curved upwards, lower face rounded to rather sharply conduplicate, 40 - 70 × 7.5 - 9 cm; marginal teeth slender-aciculate from small tubercles, usually somewhat recurved below the middle of the **L**, 4 - 5 mm (5 - 7 mm below the middle), 9 - 17 per 10 cm; terminal **Sp** acicular, straight, often somewhat flexuous, narrowly and usually shallowly grooved below or beyond the middle, involute or slightly involute towards the base, smooth, 2.2 - 2.8 cm, decurrent; **Inf** usually 2 - 3.5 m, 'paniculate', oblong or obovate, part-**Inf** few, on slightly S-curved ± ascending **Br**, in the upper ¼ - ⅓ of the **Inf**, forming **Fr** and at the same time freely bulbilliferous; **Fl** tube conical, 7.5 mm, lobes 14 - 16 mm; **Fr** shortly oblong, stipitate, not or nearly not beaked, 2.4 - 2.8 × 1.5 - 1.8 cm.

Hummelinck (1938) regarded this species as different from *A. vicina* (as *A. vivipara*) mainly on account of its flowers only.

A. salmiana Otto *ex* Salm-Dyck (Bonplandia 7: 88, 1859). – **D:** Mexico.

A. salmiana ssp. **crassispina** (Trelease) Gentry (Agaves Cont. North Amer., 609, ills. (pp. 597, 609), 1982). **T:** Mexico, San Luis Potosí (*Trelease* s.n. [MO ?]). – **D:** Mexico (Coahuila, Zacatecas, San Luis Potosí, Guanajuato, Hidalgo, Puebla). **I:** Gentry (1982: 597, 609).

≡ *Agave crassispina* Trelease (1920).

[2b] Differs from ssp. *salmiana*: **Ros** smaller, 0.8 - 1.2 m; **L** fewer and smaller, 60 - 90 × 16 - 25 cm (rarely larger), margins undulate to crenate; marginal teeth firm, with a broad base, mostly 7 - 12 mm, dark brown becoming grey with age, 1 - 3 cm apart.

This ssp. represents the extensive wild populations of *A. salmiana* (Gentry 1982: 610).

A. salmiana ssp. **salmiana** – **Lit:** McVaugh (1989).

D: Mexico (Coahuila, Durango, Zacatecas, San Luis Potosí, Colima, Hidalgo, Puebla); cultivated only. **I:** Gentry (1982: 597, 606, 610).

Incl. *Agave caratas* hort. *ex* Besaucèle (s.a.); **incl.** *Agave dyckii* hort. *ex* Besaucèle (s.a.); **incl.** *Agave salmiana* var. *contorta* hort. *ex* Besaucèle (s.a.); **incl.** *Agave latissima* auct. (s.a.) (*nom. illeg.*, Art. 53.1); **incl.** *Agave jacobiana* Salm-Dyck (1859); **incl.** *Agave tehuacanensis* Karwinsky *ex* Salm-Dyck (1859) (*nom. illeg.*, Art. 59.1); **incl.** *Agave potatorum* C. Koch (1860) (*nom. illeg.*, Art. 53.1); **incl.** *Agave montezumae* Hort. Belg. *ex* Jacobi (1864); **incl.** *Agave salmiana* var. *recurvata* Jacobi (1866); **incl.** *Agave atrovirens* W. Neubert (1867) (*nom. illeg.*, Art. 53.1); **incl.** *Agave coarctata* Jacobi (1868); **incl.** *Agave lehmannii* Jacobi (1868); **incl.** *Agave mitriformis* Jacobi (1868) ≡ *Agave salmiana* var. *mitriformis* (Jacobi) Cels (s.a.); **incl.** *Agave cochlearis* Jacobi (1870); **incl.** *Agave quiotifera* Trelease *ex* Ochoterena (1913); **incl.** *Agave compluviata* Trelease (1914); **incl.** *Agave atrovirens* var. *sigmatophylla* A. Berger (1915); **incl.** *Agave salmiana* var. *angustifolia* A. Berger (1915); **incl.** *Agave whitackeri* hort. *ex* A. Berger (1915); **incl.** *Agave potatorum* hort. *ex* A. Berger (1915) (*nom. illeg.*, Art. 53.1).

[2b] Stem short, thick; **Ros** massive, 1.5 - 2 × ± 3 - 4 m, surculose; **L** broadly linear-lanceolate, thickly fleshy, acuminate, tip sigmoidally curved, concave to guttered upwards, 100 - 200 × 20 - 35 cm, green to glaucous-greyish, margins of the upper ½ of the **L** often ± repand, sometimes with small prominences; marginal teeth mostly 5 - 10 mm (middle of the lamina), brown to greyish-brown, 3 - 5 cm apart, cusps straight to curved from low broad bases; terminal **Sp** subulate, stout, long, 5 - 10 cm, dark brown, grooved above for over ½ its length, long decurrent (sometimes to the middle of the lamina) as heavy horny margin; **Inf** 7 - 8 m, 'paniculate', broad, stout, scape **Bra** large, fleshy, imbricate, part-**Inf** large, several times compound, 15 - 20 in the upper ½ of the **Inf**; **Fl** 80 - 110 mm; **Ov** 50 - 60 mm, green, neck not constricted; **Tep** yellow, tube large, funnel-shaped, 21 - 24 mm, lobes unequal, 18 - 25 mm. − *Cytology:* 2n = 120.

Consisting of many forms cultivated in the pulque industry. It is generally recognizable by its broad, heavy, well-armed green leaves with long-acuminate sigmoid tips and large peduncular bracts subtending broad large pyramidal 'panicles' (Gentry 1982: 605).

A. salmiana var. **ferox** (Koch) Gentry (Agaves Cont. North Amer., 611, ill., 1982). **T** [neo]: Ex cult. La Mortola (*Anonymus* s.n. [K]). − **D:** Mexico (México, Puebla, Oaxaca); mainly cultivated but apparently also spontaneous.

≡ *Agave ferox* Koch (1860); **incl.** *Agave coelum* hort. *ex* Besaucèle (s.a.); **incl.** *Agave bonnetiana* Peacock *ex* Baker (1877).

[2b] Differs from ssp. *salmiana*: **Ros** 1 - 1.5 × ± 2 - 3 m; **L** broadly oblanceolate, outcurving, thick, 70 - 90 × 23 - 30 cm, light shiny green, margins crenate with strong prominences; marginal teeth 10 - 14 mm, on prominent tubercles; **Fl** more slender, 70 - 85 mm. − *Cytology:* 2n = 120?.

An easily recognizable variant of uncertain systematic status (Gentry 1982: 611).

A. scabra Ortega (Nov. Pl. Descr. Dec. 2: 13, 1797). **T:** [icono]: Cavanilles, Icones, t. 27, 1803. − **Lit:** McVaugh (1989: fig. 37); Ullrich (1992g). **D:** Mexico (widespread from Durango to Chiapas and Veracruz), Guatemala, Honduras, El Salvador, possibly Nicaragua; rocky slopes in pine-oak forests and ecotones with tropical deciduous forests and Matorral, 200 - 2800 m, flowers June to February but mainly August to September. **I:** Matuda (1961: 67-68, figs. 8-9, as *Manfreda pringlei* and *M. brachystachys*).

≡ *Manfreda scabra* (Ortega) McVaugh (1989); **incl.** *Agave brachystachys* Cavanilles (1802) ≡ *Manfreda brachystachys* (Cavanilles) Rose (1903) ≡ *Polianthes brachystachys* (Cavanilles) Shinners (1966); **incl.** *Agave spicata* De Candolle (1813) (*nom. illeg.*, Art. 53.1); **incl.** *Agave polyanthoides* Schiede *ex* Schlechtendal & Chamisso (1831); **incl.** *Agave saponaria* Lindley (1838); **incl.** *Agave humilis* M. Roemer (1847); **incl.** *Agave brachystachys* var. *strictior* Jacobi & C. D. Bouché (1865); **incl.** *Agave sessiliflora* Hemsley (1880) ≡ *Manfreda sessiliflora* (Hemsley) Matuda (1961); **incl.** *Agave langlassei* André (1901); **incl.** *Manfreda oliveriana* Rose (1903) ≡ *Agave oliveriana* (Rose) A. Berger (1915) ≡ *Polianthes oliveriana* (Rose) Shinners (1966).

[3a2] Plants large (for Subgen. *Manfreda*), reproducing vegetatively by buds from the rhizome; **R** fibrous, half-fleshy; rhizome large, oblong, to 7 cm ∅; **L** 4 - 9, erect-spreading, broadly or narrowly linear-lanceolate, coriaceous to herbaceous, usually deeply channelled in the lower part, often gently undulate, (25.5-) 37 - 77 (-91) × (1-) 1.6 - 4.8 (-6.5) cm, green, often glaucous, sometimes spotted with maroon, tip acute, with a short point, veins prominent on the lower face, each vein usually with a single row of papillate cells, margins with a narrow hyaline band, entire to papillate like the veins; remains of **L** bases separating into coarse fibres, 5 - 12 cm; **Inf** 1 - 2.5 m, 'spicate', flowering part elongate, lax, (10-) 23 - 47 (-82) cm, with 17 - 46 (-58) flowering nodes; **Fl** usually sessile (rarely lower or all **Fl** pedicellate), fairly succulent; **Ov** narrowly ellipsoid, (8-) 10 - 20 mm; **Tep** green, frequently with a brownish flush on the lower side, tube narrowly funnel-shaped, slightly curved, not markedly constricted above the **Ov**, (9-) 13 - 38 mm, lobes oblong, recurved, 9 - 20 (-23) mm, golden-green or brownish-maroon on the upper part, tips swollen and cucullate; **Sty** exceeding the tube by 24 - 37 (-

61

74) mm; **Sti** clavate, trigonous; **Fr** oblong, 1.8 - 2.9 × 1.1 - 1.6 cm; **Se** 2 - 4 × 4 - 5 mm.

McVaugh (1989: 234) has replaced the well-established name *Manfreda brachystachys* (based on *Agave brachystachys*, as '*brachystachya*') by the new combination *Manfreda scabra*, based on the earlier name *Agave scabra*. *A. scabra* is the most widely distributed species of Subgen. *Manfreda* and with the exception of the Guatemalan *A. fusca* the only one reaching Central America. It seems to be quite variable, but is characterized by leaves with prominent veins with a row of papillae and margins which are equally papillate, as well as the elongate open inflorescence, semihorizontal flower position, sinuous flower shape, and the tepal tube, which is longer than the ovary and the and lobes (McVaugh 1989).

The pollination biology of this taxon was dealt with by Eguiarte & Búrquez (1987) and Eguiarte (1988).

A. scaposa Gentry (Agaves Cont. North Amer., 303-304, ills., 1982). **T:** Mexico, Oaxaca (*Gentry* 22472 [US, DES]). – **D:** Mexico (Puebla, Oaxaca). **I:** Gentry (1982: 304).

[2a] Stem short; **Ros** large, 1.5 - 1.7 m, broad, solitary; **L** 60 - 70, broadly lanceolate, outcurving to spreading, coriaceous, heavily succulent, slightly narrowed above the thick base, upper face almost plane to concave, 100 - 115 × 20 - 25 cm, light green to yellowish-green, frequently glaucous, margins straight to crenate; marginal teeth numerous, dark brown, close-set, sometimes on small prominences, confluent or 1 - 2 cm apart, cusps 3 - 8 mm from broad flattened bases, with few smaller intermittent teeth placed at random; terminal **Sp** subulate, base conical, 2.5 - 6 cm, dark brown, decurrent to ¼ - ½ of the lamina; **Inf** 7 - 9 m, 'paniculate', scape 5 - 7 m, part-**Inf** 25 - 40 in the upper ¼ of the **Inf**; **Fl** and **Fr** unknown.

A. schottii Engelmann (Trans. Acad. Sci. St. Louis 3: 305-306, 1875). **T:** USA, Arizona (*Schott* s.n. [US, MO]). – **Lit:** Turner & al. (1995). **D:** USA (Arizona, New Mexico), Mexico (Sonora).

A. schottii var. **schottii** – **D:** USA (Arizona, New Mexico), Mexico (Sonora). **I:** Gentry (1982: 196); KuaS 43(12): centre page pullout 1992/35. **Fig. VI.e**

Incl. *Agave geminiflora* var. *sonorae* Torrey (1859) ≡ *Agave sonorae* (Torrey) Mearns (1907); **incl.** *Agave schottii* var. *serrulata* Mulford (1896); **incl.** *Agave mulfordiana* Trelease (1920); **incl.** *Agave schottii* var. *atricha* Trelease (1920).

[1e] **Ros** small, densely caespitose; **L** narrowly linear, straight, incurved, or falcate, pliable, widest at the base, deeply convex below, flat or somewhat convex above, smooth, 25 - 40 (-50) × 0.7 - 1.2 cm, yellowish-green to green, margins with a narrow border and sparse brittle threads; terminal **Sp** delicate, rather weak and brittle, 8 - 12 mm, greyish; **Inf** 1.8 - 2.5 m, 'spicate', slender, frequently crooked, flowering in the upper ¼ - ⅓, part-**Inf** with 1 - 3 **Fl**; **Ped** stout, 3 - 5 mm; **Fl** 30 - 40 mm; **Ov** 10 - 14 mm incl. the 4 - 6 mm long neck, greenish-yellow; **Tep** yellow, tube deeply funnel-shaped, 9 - 14 mm, lobes unequal, 10 - 16 mm. – *Cytology:* 2n = 60.

The flowers of *A. schottii* have a long tubular appearance due to the slender tube and the long narrow neck of the ovary. The taxon is easily confused with narrow-leaved forms of *A. felgeri*, but the latter has a short flower tube (Gentry 1982: 207).

A. schottii var. **treleasei** (Toumey) Kearney & Peebles (J. Washington Acad. Sci. 29: 474, 1939). **T:** USA, Arizona (*Toumey* s.n. [Herb. Toumey [not located]]). – **D:** USA (Arizona). **I:** Gentry (1982: 207).

≡ *Agave treleasei* Toumey (1901).

[1e] Differs from var. *schottii*: **L** larger, thicker and wider (1.5 - 2.5 cm), deep green.

A doubtful variant in need of better study (Gentry 1982: 207), which occurs sympatrically with var. *schottii* (Gentry 1972: 77). It appears to be based on scattered aberrant specimens. At the type locality, only few plants were found in 1940 (Benson & Darrow 1981: 68) and none later in the 80ies (Reichenbacher 1985: 103). Hodgson (1999), however, gives new distributional records and an altitudinal range of 600 - 1500 m in desert scrub.

A. sebastiana Greene (Bull. Calif. Acad. Sci. 1: 214, 1885). **T:** Mexico, Baja California (*Greene* s.n. [CAS]). – **Lit:** Turner & al. (1995). **D:** Mexico (Isla San Benito, Isla Cedros and Isla Natividad off the coast of Baja California). **I:** Gentry (1982: 645-646).

≡ *Agave shawii* var. *sebastiana* (Trelease) Gentry (1949); **incl.** *Agave disjuncta* Trelease (1912).

[2e] **Ros** elongate, medium-sized to rather large, 0.6 - 1.2 m ∅; **L** broadly linear to ovate, shortly acuminate, thick and rigid, sometimes slightly narrowed towards the base, rounded below, plane to slightly hollowed above, generally 25 - 45 × 8 - 24 cm, light yellowish- to greyish-green, with imprints left by the central bud, margins usually horny, dark brown; marginal teeth frequently down-flexed, slender, larger teeth (middle of the lamina) 5 - 10 mm, reddish-brown, 1 - 2 cm apart, or smaller and more numerous; terminal **Sp** stout, variously grooved above, 2 - 3 cm (rarely shorter), black to somewhat grey; **Inf** 2 - 3 m, 'paniculate', short, widely spreading, rounded to nearly flat, scape stout, with deltoid scarious appressed peduncular **Bra**, part-**Inf** large, 8 - 12 in the upper ¼ of the **Inf**; **Fl** 70 - 90 mm; **Ov** 35 - 55 mm; **Tep** green in bud, opening yellow, tube broadly funnel-shaped, 14 - 20 mm, lobes 16 - 25 mm.

Closely related to *A. shawii*, but differing significantly in the pale green somewhat glaucous leaves with more slender teeth, the smaller more remote and scarious peduncular bracts and the broader flatter inflorescences (Gentry 1982: 646).

A. seemanniana Jacobi (Abh. Schles. Ges. Vaterl. Cult., Abth. Naturwiss. 1868: 154, 1868). **T** [neo]: Honduras (*Gentry* 20684 [US, DES, MEXU]). – **Lit:** Ullrich (1992c). **D:** Mexico (Oaxaca, Chiapas), Guatemala, Honduras, N Nicaragua; dry rocky slopes, 400 - 2200 m. **I:** Gentry (1982: 496-499).
≡ *Agave scolymus* var. *seemanniana* (Jacobi) A. Terracciano (1885); **incl.** *Agave seemannii* hort. *ex* Besaucèle (s.a.) (*nom. inval.*, Art. 61.1); **incl.** *Agave caroli-schmidtii* A. Berger (1915); **incl.** *Agave guatemalensis* A. Berger (1915); **incl.** *Agave seemanniana* var. *perscabra* Trelease (1915); **incl.** *Agave tortispina* Trelease (1915); **incl.** *Agave pygmaea* Gentry (1982) ≡ *Agave seemanniana* ssp. *pygmaea* (Gentry) B. Ullrich (1992); **incl.** *Agave pygmae* Gentry (1982) (*nom. inval.*, Art. 61.1).
[2k] **Ros** compact, small to medium-sized, solitary; **L** ovate to broadly lanceolate or spatulate, thickly succulent, thickened and strongly narrowed at the base, plane to hollow-upcurved, generally 30 - 50 × 12 - 20 cm, light glaucous to yellowish-green, margins undulate to sharply crenate; marginal teeth mostly straight or some curved, deltoid, 5 - 10 mm, rarely much larger, dark to greyish-brown, 1 - 3 cm apart, usually on conspicuous marginal prominences; terminal **Sp** subulate, very broad at the base, broadly grooved above, 2 - 4 cm, dark brown to greyish, conspicuously decurrent as a sharp ridge to the upper marginal teeth; **Inf** 3 - 4 m, 'paniculate', ovate in outline, rather open, scape short, part-**Inf** spreading, 18 - 30 in the upper ½ of the **Inf**; **Fl** slender, 50 - 70 mm; **Ov** slender, fusiform to cylindrical, 25 - 38 mm, green, neck lightly furrowed; **Tep** yellow, tube broadly funnel-shaped, 7 - 11 mm, lobes slightly unequal, 13 - 24 mm.
This taxon exhibits considerable variation in leaf characters. Within its geographical range, it is recognizable by the small compact rosettes with broad plane leaves markedly narrowed at the base (Gentry 1982: 497-498).

A. shaferi Trelease (Mem. Nation. Acad. Sci. 11: 35, t. 57, 1913). **T:** Cuba (*Shafer* 3800 [MO?]). – **Lit:** León (1946). **D:** E Cuba.
[2l] **Ros** unknown; **L** elongate-lanceolate, rather gradually pointed, ± 75 × 10 cm, green, margins between the marginal teeth slightly concave; marginal teeth slightly curved upwards or downwards, triangular from lenticular bases, ± 1 mm, brown, 1 - 2 cm apart; terminal **Sp** conically subulate, unguiculately recurved, openly V-grooved to the middle, smooth, 1 cm, brown, dull, not decurrent; **Inf** 'paniculate', 6 - 7 m; **Fl** 50 mm; **Ov** fusiform, 25 - 30

mm; **Tep** bright yellow, tube conical, 5 - 6 mm, lobes 14 mm; **Fr** unknown.

A. shawii Engelmann (Trans. Acad. Sci. St. Louis 3: 314-316, 370, 1875). **T:** USA, California (*Hitchcock* s.n. [MO]). – **Lit:** Turner & al. (1995). **D:** USA (S California), Mexico (Baja California).

A. shawii ssp. **goldmaniana** (Trelease) Gentry (Occas. Pap. Calif. Acad. Sci. 130: 93, 1978). **T:** Mexico, Baja California (*Nelson & Goldman* 7151 [US]). – **D:** Mexico (C Baja California); 5 - 700 m. **I:** Gentry (1982: 636, 640); KuaS 41(9): centre page pullout 1990/26.
≡ *Agave goldmaniana* Trelease (1912).
[2e] Differs from ssp. *shawii*: **Ros** medium to large; **L** lanceolate rather than linear-ovate, more acuminate, longer, 40 - 70 × 10 - 18 cm. – *Cytology:* 2n = 60.
Representing the ecotype of the more arid interior habitats.

A. shawii ssp. **shawii** – **D:** USA (S California), Mexico (Baja California); coastal in sagebrush communities. **I:** Gentry (1982: 636, 640).
Incl. *Agave orcuttiana* Trelease (1912); **incl.** *Agave pachyacantha* Trelease (1912).
[2e] Stem short to long (2 m), erect to decumbent, frequently branching from **L** axils; **Ros** compact, small to medium-sized, solitary or caespitose; **L** ovate to linear-ovate, thick, fleshy, rigid, shortly acuminate, plane to slightly hollowed above, slightly asperous, 20 - 50 × 8 - 20 cm, glossy light to dark green; marginal teeth very variable in size and shape, straight or variously curved, 5 - 20 mm (middle of the lamina), decreasing in size below, reddish to dark brown or dark grey, usually 1 - 2 cm apart or rarely confluent; terminal **Sp** acicular, straight or sinuous, broad at the base, openly grooved above, 2 - 4 cm, dark reddish-brown to grey, decurrent as horny margin for 8 - 10 cm or along the entire **L**; **Inf** 2 - 4 m, 'paniculate', scape with closely imbricate large purple succulent **Bra** closely investing the part-**Inf**, these dense, horizontal to ascending, commonly 8 - 14 per **Inf**; **Fl** 75 - 100 mm; **Ov** 35 - 50 mm, greenish; **Tep** frequently purplish or red in bud, opening yellow or reddish, tube amply funnel-shaped, 12 - 16 mm, lobes unequal, 25 - 38 mm. – *Cytology:* 2n = 60.

A. shrevei Gentry (Publ. Carnegie Inst. Washington 527: 95, 1942). **T:** Mexico, Chihuahua (*Gentry* 2028 [CAS]). – **D:** NW Mexico.

A. shrevei ssp. **magna** Gentry (Agaves Cont. North Amer., 451-453, ills., 1982). **T:** Mexico, Chihuahua (*Gentry & Bye* 23360 [US, DES, MEXU]). – **D:** Mexico (Sonora, Chihuahua, Sinaloa).
[2g] Differs from ssp. *shrevei*: **Ros** 1.4 - 1.7 × up to 2.5 m, mostly solitary; **L** outcurving, guttered,

thickened and broadened towards the base, finely asperous, mature **L** mostly 120 - 150 × 15 - 25 cm, margins remotely crenate; marginal teeth along most of the lamina, 6 - 10 (-15) mm, mostly 3 - 5 cm apart, on pronounced prominences, frequently with small intermittent teeth; terminal **Sp** 3.5 - 6 cm; **Inf** 6 - 7 m, part-**Inf** 20 - 30 per **Inf**.

The main difference from ssp. *shrevei* is the larger size (Gentry 1982: 451).

A. shrevei ssp. **matapensis** Gentry (US Dept. Agric. Handb. 399: 115-117, ills., 1972). **T:** Mexico, Sonora (*Gentry* 11607 [US 2540344]). – **D:** Mexico (Sonora). **I:** Gentry (1982: 423, 455).

[2g] Differs from ssp. *shrevei*: **Ros** suckering late and sparingly; larger marginal teeth in the middle of the lamina down-flexed; **Fl** smaller; **Ov** 22 - 40 mm incl. the short unconstricted neck, tube 15 - 20 mm, outer lobes 11 - 16 mm.

A. shrevei ssp. **shrevei** – **D:** Mexico (Sonora, Chihuahua); open rocky limestone slopes in oak woodland and pine-oak forests, 930 - 1850 m. **I:** Gentry (1982: 423, 448-449). **Fig. VI.f**

[2g] **Ros** small to medium-sized, suckering with maturity; **L** ovate, short-acuminate, 20 - 35 × 8 - 10 cm, or lanceolate, acuminate and 50 - 60 × 12 - 18 cm, generally narrowed above the base, firm, thick, straight or outcurving near the tip, light grey, glaucous; marginal teeth variable, straight or flexed up- or downwards, larger teeth 5 - 10 mm (middle of the lamina), dark brown to grey, on small to pronounced prominences; terminal **Sp** acicular, stout, with a narrow or open groove from the base to above the middle, mostly 2.5 - 5 cm, brown; **Inf** 2.5 - 5 m, 'paniculate', part-**Inf** ascending, small, 8 - 16 in the upper ⅓ of the **Inf**; **Fl** persisting erect, slender, 60 - 70 mm; **Ov** 25 - 35 mm incl. the constricted neck; **Tep** light green to pale yellow, tips red to purplish, tube cylindrical or urceolate, 18 - 23 mm, lobes unequal, outer 10 - 12 mm.

Well distinguished by its broad light glaucous-grey leaves with margins bearing prominences with well-developed brown teeth, and the leathery perianth with a deep tube (Gentry 1982: 448).

A. sileri (Verhoek-Williams) Thiede & Eggli (KuaS 50(5): 111, 1999). **T:** USA, Texas (*Siler* s.n. [BH 69-518B]). – **Lit:** Verhoek-Williams (1978: with ill.). **D:** USA (S Texas), Mexico (Tamaulipas); open areas on clay soil, flowers April to July.

≡ *Manfreda sileri* Verhoek-Williams (1978); **incl.** *Manfreda variegata* var. *sileri* Verhoek-Williams (1975) (*nom. inval.*, Art. 29.1).

[3a1] Plants large (for Subgen. *Manfreda*), reproducing vegetatively by buds from the **Ax** of the **L** of the parent **Ros** or by buds from the rhizome, rhizome globose; **R** fleshy; **L** spreading, ovate-lanceolate, channelled and undulate or flat, long-attenuate

towards the tip, tip acute with a medium-sized point, succulent, brittle, (14-) 25 - 39 × 2.2 - 4.8 cm, light green, spotted, glaucous except over the spots, spots darker green or brown, large, round to elliptic, usually confluent, margins with a cartilaginous band, minutely denticulate, teeth of several sizes, irregularly spaced, often retrorse; remains of **L** bases membranous, not separating into fibres; **Inf** 2.4 - 2.6 m, 'spicate', flowering part dense, 28 - 39.5 cm, with 27 - 46 (-81) **Fl**; **Fl** sessile, nearly erect; **Ov** ellipsoid, (10-) 12 - 20 mm; **Tep** glaucous-green on the outer face, golden-green on the inner face, tube broadly campanulately funnel-shaped, (7-) 9 - 15 (-22) mm, lobes revolute, oblong, (7-) 10 - 21 mm; **Sty** straight, exserted for 44 - 66 (-95) mm; **Sti** clavate-capitate, trigonous; **Fr** oblong, 2.3 - 3.1 × 1.6 - 1.9 cm; **Se** 5 - 6 × 5 mm.

According to the protologue (Verhoek-Williams 1978) closely related to *A. variegata*, but different by its larger size, spreading and only shallowly channelled glaucous leaves spotted with large brown markings.

A. singuliflora (S. Watson) A. Berger (Agaven, 31, 1915). **T** [lecto]: Mexico, Chihuahua (*Pringle* 1142 [GH, US]). – **D:** Mexico (Chihuahua, Durango, Zacatecas); cool slopes in the pine-oak forest region, 1675 - 2590 m, flowers late June to early October.

≡ *Bravoa singuliflora* S. Watson (1887) ≡ *Manfreda singuliflora* (S. Watson) Rose (1903) ≡ *Polianthes singuliflora* (S. Watson) Shinners (1966).

[3a2?] Plants medium-sized (for Subgen. *Manfreda*); **R** fleshy; rhizome small, 1.7 × 1.2 cm; **L** 2 - 8 (-14), sprawling, linear-lanceolate, channelled, semisucculent, tip acute, with a medium-sized point, 17 - 34 × 0.4 - 1.3 (-1.5) cm, glaucous, occasionally red-speckled at the base, margins bordered by a narrow hyaline band; remains of the **L** bases 4 - 8 cm; **Inf** 45 - 116 cm, 'spicate', flowering part open, 5.2 - 28 (-46) cm, with 5 - 18 (-26) usually sessile horizontal **Fl** (lower or rarely all **Fl** pedicellate); **Ov** nearly erect, at a narrow angle to the **Inf** axis, ellipsoid, 4 - 10 mm; **Tep** green or green with a brown-maroon streak on the lower parts, tube narrowly funnel-shaped, arched so that the mouth faces downwards, 15 - 23 (-27) mm, lobes oblong, revolute, 7 - 12 (-18) mm; **Sty** exceeding the tube for 5 - 12 (-15) mm, white; **Sti** clavate, deeply fissured; **Fr** globose to oblong, 1.5 - 2.3 × 1.3 - 1.7 cm; **Se** 4 × 3 mm.

Differing from all other members in the *Manfreda* Group by the extreme curvature of the perianth (Verhoek-Williams 1975: 263).

A. sisalana Perrine (Trop. Pl., 8, 9, 16, 47, 60, 86, 1838). **T** [neo]: Mexico, Chiapas (*Gentry* 16434 [US, DES]). – **D:** Cultivated only; nearly worldwide in tropical regions. **I:** Gentry (1982: 621).

≡ *Agave rigida* var. *sisalana* (Perrine) Engelmann (1875); **incl.** *Agave houlettii* Jacobi (1866); **incl.**

Agave houlletiana Cels *ex* Jacobi (1866); **incl.** *Agave laevis* hort. *ex* Baker (1892); **incl.** *Agave sisalana* var. *armata* Trelease (1913) ≡ *Agave sisalana* fa. *armata* (Trelease) hort. (s.a.) (*nom. inval.*, Art. 29.1); **incl.** *Agave sisalana* fa. *marginata* Medina (1955); **incl.** *Agave sisalana* fa. *medio-picta* Medina (1955).

[2f] Stem 0.4 - 1 m; **Ros** 1.5 - 2 m, suckering with elongate rhizomes; **L** ensiform, fleshy, 90 - 130 × 9 - 12 cm, green, somewhat slightly zoned in youth; young **L** with few minute marginal teeth, mature **L** usually without marginal teeth; terminal **Sp** subulate, shortly shallowly grooved above, 2 - 2.5 cm, dark brown, somewhat lustrous, not decurrent; **Inf** 5 - 6 m, 'paniculate', ellipsoid, scape short, part-**Inf** 10 - 15 (-25) in the upper ½ of the **Inf**, bulbilliferous after flowering; **Fl** 55 - 65 mm, unpleasantly scented; **Ov** shortly fusiform, 20 - 25 mm, nearly neckless; **Tep** greenish-yellow, tube broadly urceolate, 15 - 18 mm, lobes equal, 17 - 18 mm. − *Cytology:* 2n = ± 138, 149, 150.

Easily recognizable by its green unarmed mature leaves with short dark brown conical to subulate non-decurrent terminal spine. The taxon appears to represent a sexually sterile clone that is widely cultivated in fibre plantations and could be of hybrid origin within the *A. vivipara*-complex (as *A. angustifolia*) (Gentry 1982: 628-629). Ullrich (1990d) consequently removed it from Gentry's Group *Sisalanae* and placed it in Group *Viviparae* (as *Rigidae*).

A. sobolifera Salm-Dyck (Hort. Dyck., 307, 1834). − **Lit:** Trelease (1913); Adams (1972); Proctor (1984: with ill.). **D:** Cayman Islands, Jamaica; dry rocky well-drained hillsides.

Incl. *Agave morrisii* Kent (s.a.) (*nom. illeg.*, Art. 53.1); **incl.** *Agave americana* Lamarck (1783) (*nom. illeg.*, Art. 53.1); **incl.** *Agave ornata* Jacobi (1865); **incl.** *Agave morrisii* Baker (1887); **incl.** *Agave laetevirens* hort. *ex* A. Berger (1915).

[2l] **Ros** solitary; **L** variously lanceolate, massive, curved, gradually acute or somewhat subacuminate, often deeply and conduplicately or undulately concave, ± 125 - 200 × 15 - 24 cm, 9 cm thick near the base, rather light green, somewhat glossy, margins ± concave; marginal teeth curved or reflexed-triangular (rarely straight), 1 - 4 mm, glossy dark brown, 5 - 15 mm apart, often hardened on the tops of green prominences of the margin; terminal **Sp** conical, nearly straight, slightly flattened, grooved or slightly involutely channelled below the middle when mature, smooth, somewhat glossy, 1.5 - 2.5 cm, reddish-brown, not decurrent; **Inf** 5 - 9 m, 'paniculate', oblong, part-**Inf** on rather short spreading **Br**, above the middle of the **Inf**, freely bulbilliferous; **Fl** ± 50 mm; **Ov** narrowly fusiform, 15 - 20 (-25) mm, from slightly shorter to longer than the **Tep**; **Tep** 12 - 19 mm, golden-yellow to light orange, tube open, 5 - 7 mm, lobes ± 20 mm; **Fr** narrowly oblong, turbinately narrowed at the base, shortly beaked at the tip, 4.5 - 5 × 1.3 - 2 cm.

See Trelease (1913: 33) on the difficult interpretation of this name.

A. sobria Brandegee (Proc. Calif. Acad. Sci., ser. 2, 2: 207, 1889). **T:** Mexico, Baja California (*Brandegee* 2 [UC, DS]). − **Lit:** Turner & al. (1995). **D:** Mexico (Baja California Sur).

A. sobria ssp. **frailensis** Gentry (Occas. Pap. Calif. Acad. Sci. 130: 54-56, ills., 1978). **T:** Mexico, Baja California (*Gentry & Cech* 11264 [US]). − **D:** Mexico (Baja California Sur: Cape region). **I:** Gentry (1982: 356, 401-402).

[2h] Differs from ssp. *sobria*: **Ros** compact, small, sparingly caespitose; **L** more numerous, broadly lanceolate, mostly 20 - 35 × 6 - 8 cm, glaucous-green to bluish-glaucous, margins with pronounced prominences; marginal teeth numerous, smaller, mostly 6 - 10 mm, chestnut-brown to greying, closely spaced; terminal **Sp** frequently sinuous or contorted, 3 - 4 cm; **Inf** with 10 - 15 part-**Inf**; **Fl** slender, 45 - 63 mm; **Ov** cylindrical, 25 - 40 mm, lobes 4 - 6 mm wide.

A. sobria ssp. **roseana** (Trelease) Gentry (Occas. Pap. Calif. Acad. Sci. 130: 54, 1978). **T:** Mexico, Baja California (*Rose* 16854 [US]). − **D:** Mexico (Baja California Sur: Espírito Santo Island and adjacent mainland). **I:** Gentry (1982: 356, 401-402); KuaS 45(9): centre page pullout 1994/26.

≡ *Agave roseana* Trelease (1912) ≡ *Agave sobria* var. *roseana* (Trelease) I. M. Johnston (1924).

[2h] Differs from ssp. *sobria*: **Ros** openly spreading; **L** broadly lanceolate, frequently twisted, acuminate, 35 - 50 × 7 - 10 cm, yellow-green, margins with prominent prominences, tubercles 1 - 1.5 mm; marginal teeth flexuous, few, large, larger teeth 10 - 25 mm, remote; terminal **Sp** sinuous to contorted, 5 - 7 cm; **Inf** with 8 - 12 part-**Inf**; **Fl** 45 - 65 mm, lobes 4 - 5 mm wide.

A. sobria ssp. **sobria** − **D:** Mexico (Baja California Sur); widely scattered but common in the Sierra de la Giganta, sea-level to 1070 m. **I:** Gentry (1982: 356, 397-398).

Incl. *Agave affinis* Trelease (1912); **incl.** *Agave carminis* Trelease (1912); **incl.** *Agave slevinii* I. M. Johnston (1924).

[2h] Stem short or none; **Ros** open, 0.5 - 1.5 m ∅, usually caespitose; **L** few, linear to lanceolate, straight to curved, long-acuminate, thick and convex below towards the base, plane to somewhat concave above, 45 - 80 × 5 - 10 cm, bright glaucous-grey, frequently cross-zoned, margins undulate to tuberculate; marginal teeth variously curved or straight, flattened, base broad, mostly 5 - 10 mm, base grey, reddish towards tips, mostly 3 - 4 cm apart; terminal **Sp** acicular, narrowly grooved above, mostly 3 - 6 cm; **Inf** 2.5 - 4 m, 'paniculate', slender,

part-**Inf** compact, nearly globose, 12 - 20 per **Inf**; **Fl** slender, 45 - 55 mm; **Ov** tapering at the base, 25 - 35 mm, neck short, scarcely constricted; **Tep** pale yellow, tube broadly funnel-shaped, 3 - 4 mm, lobes ± equal, 17 - 22 × 3 - 4 mm. − *Cytology:* 2n = 60.

Distinguished by the slender flowers with long narrow tepals as well as by the very light-glaucous long-lanceolate leaves with remote marginal teeth (Gentry 1982: 396).

A. spicata Cavanilles (Anales Ci. Nat. 5(15): 261, 1802). **T:** Ex cult. Madrid (*Anonymus* s.n. [MA]). − **Lit:** Ullrich (1995); Ullrich (1996). **D:** Not known from the wild; possibly Mexico (Hidalgo: Real del Monte?). **I:** Gentry (1982: 86, as *A. yuccaefolia*).

≡ *Agave yuccaefolia* var. *spicata* (Cavanilles) A. Terracciano (1885) (*incorrect name*, Art. 11); **incl.** *Agave yuccaefolia* var. *viridis* hort. *ex* Besaucèle (s.a.); **incl.** *Agave yuccaefolia* F. Delaroche (1811); **incl.** *Agave spicata* Gussone (1825) (*nom. illeg.*, Art. 53.1); **incl.** *Agave hookeri* C. Koch (1865); **incl.** *Agave cohniana* Jacobi (1866); **incl.** *Agave yuccaefolia* var. *caespitosa* A. Terracciano (1885).

[1b] Stem short or none; **Ros** open, small to medium-sized, suckering; **L** rather few, linear, recurving with maturity, soft, pliable, scarcely succulent, weakly and finely fibrous, convex below, concave above, 50 - 65 × 3 - 3.5 cm, mostly green with pale midstripe, sometimes reddish- or purple-spotted, margins finely serrulate with unequal denticles; terminal **Sp** conical to subulate, 3 - 8 mm, brown; **Inf** 2 - 3 m, 'spicate', slender, arching, part-**Inf** mostly with geminate **Fl**; **Ped** short; **Fl** 40 mm, unpleasantly scented; **Ov** 16 - 18 mm, neck short; **Tep** greenish-yellow, tube narrowly cylindrical, ± 8 mm, lobes 15 - 16 mm.

Ullrich (l.c.) replaced *A. yuccaefolia* (the name used by Gentry (1982: 85-86) for this plant) by the older name *A. spicata*. It is a very distinct species without close relatives within Group *Serrulatae* (as *Amolae*), which lead Ullrich (l.c.) to place it in a section of its own (Sect. *Yuccaefoliae* (A. Terracciano) Ullrich).

A. stictata Thiede & Eggli (KuaS 50(5): 111, 1999). **T:** [icono]: Martius, Ausw. merkw. Pfl., t. 13, 1831. − **D:** Mexico (México, Guerrero); rocky slopes and moist shady areas in oak woods, 1370 - 1830 m, flowers mid-July to mid-September. **I:** Piña Luján (1986: 17, as *Manfreda maculata*).

Incl. *Polianthes maculata* Martius (1831) ≡ *Manfreda maculata* (Martius) Rose (1903).

[3a2] Plants of small to medium size (for Subgen. *Manfreda*); **R** half-fleshy to fibrous, often extending horizontally from the base of the rhizome; rhizome 1.5 - 2.5 (-3.5) × 0.7 - 1.5 cm; **Ros** base surrounded by fibrous **L** bases forming an ovoid bulb-like underground portion of 3.5 - 5.5 × 0.8 - 1.3 cm; **L** 2 - 6, narrowly to broadly (ob-) lanceo-

late, narrowed towards the base, tip acute and short-pointed, only slightly channelled, coriaceous, undulate, 9 - 26 × (0.8-) 1 - 3.5 (-4.5) cm, lower face paler green, often with large elliptic dark green or brown spots scattered densely over both faces, densely pubescent on both faces with straight simple **Ha** 0.6 - 0.8 mm long, margins with a narrow hyaline band, entire; remains of **L** bases membranous, 1.8 - 4 (-4.5) cm; **Inf** 21 - 96 cm, 'spicate', flowering part lax, 7 - 25.5 cm, with 4 - 22 usually sessile (rarely shortly pedicellate) **Fl**; **Ov** ellipsoid, 5 - 12 mm; **Tep** tube straight, not constricted above the **Ov**, 10 - 19 × ± 4 (middle of the tube) mm, lobes oblong, revolute, 6 - 11 mm; **Sty** exceeding the tube for 9 - 17 (-19) mm; **Sti** clavate, trigonous; **Fr** subglobose to oblong, 1.2 - 1.6 × 0.8 - 1.1 cm; **Se** 3 - 4 × 2 - 3 mm.

Differentiated from most other species of Subgen. *Manfreda* by the bulbous portion formed by the leaf bases, and by the pubescent leaves narrowed into a petiolar portion; pubescent leaves are otherwise only found in *A. pubescens* (Verhoek-Williams 1975: 303-304). When transferring *Manfreda maculata* to the genus *Agave*, a new epithet was necessary because of the earlier name *A. maculata* Regel 1856.

A. striata Zuccarini (Flora 15: 2(Beiblatt 2): 98, 1832). **T:** K [neo]. − **D:** Mexico (Coahuila, Nuevo León, Tamaulipas, Durango, Zacatecas, San Luis Potosí, Querétaro, Hidalgo); limited to drier valleys and plains with annual rainfall < 500 mm.

A. striata ssp. **falcata** (Engelmann) Gentry (Agaves Cont. North Amer., 245, ills. (pp. 236, 246-247), 1982). **T** [lecto]: Mexico, Coahuila (*Wislizenus* 312 [MO]). − **D:** Mexico (Coahuila, Nuevo León, Durango, Zacatecas, San Luis Potosí); sandy coarse rocky soils on bajadas, slopes and plains in shrub and succulent deserts, 1000 - 2000 m.

≡ *Agave falcata* Engelmann (1875); **incl.** *Agave californica* Jacobi (1868) ≡ *Agave striata* var. *californica* (Jacobi) A. Terracciano (1885); **incl.** *Agave paucifolia* Todaro (1877); **incl.** *Agave californica* Baker (1877) (*nom. illeg.*, Art. 53.1); **incl.** *Agave falcata* var. *espadina* A. Berger (1915); **incl.** *Agave falcata* var. *microcarpa* A. Berger (1915).

[1a] Differs from ssp. *striata*: **L** fewer, straight to falcate, rigid, more xerophytic, broader, 30 - 60 × 0.8 - 1.8 cm, margins serrulate.

Intergrades gradually into ssp. *striata* (Gentry 1982: 245).

A. striata ssp. **striata** − **D:** Mexico (Coahuila, Nuevo León, Tamaulipas, Durango, Zacatecas, San Luis Potosí, Querétaro, Hidalgo, Puebla); drier valleys and plains. **I:** Gentry (1982: 236, 243-244). **Fig. VI.b, VI.c**

Incl. *Agave recurva* Zuccarini (1845) ≡ *Agave striata* var. *recurva* (Zuccarini) Baker (1877); **incl.**

Agave ensiformis hort. *ex* Baker (1877); **incl.** *Agave striata* var. *mesae* A. Berger (1915).

[1a] Stem short; **Ros** compact, 0.5 - 1 × 0.5 - 1.2 m, often forming large dense clusters 2 - 3 m broad by axillary branching; **L** many, linear, straight to arching, thick, rather turgid, convex above, smooth or scabrous along the keels and below, mostly 25 - 60 × 0.5 - 1 cm, pale green to red or purplish, brownish at the tip below the terminal **Sp**, striate, margins cartilaginous, ≤ 1 mm wide, pale yellow, scabrous or minutely serrulate; terminal **Sp** subulate, very pungent, 1 - 5 cm, reddish-brown to dark grey; **Inf** erect, 1.5 - 2.5 m, 'spicate', rather laxly flowered above the long scape, part-**Inf** with mostly geminate **Fl**; **Fl** tubular, 30 - 40 mm; **Ov** rounded-triangular, grooved, 12 - 15 mm, neckless; **Tep** greenish-yellow or red to purple, tube 14 - 20 mm, lobes about equal, 5 - 7 mm. − *Cytology:* 2n = 60.

The taxon is represented by extensive populations varying in growth habit, leaf forms, and to a lesser extent in flower structure. *A. echinoides* Jacobi (inflorescence unknown) is a doubtful synonym either of this species or of *A. stricta* (Gentry 1982: 245).

A. stricta Salm-Dyck (Bonplandia 7(7): 94-95, 1859). **T** [neo]: Mexico, Puebla (*Gentry & al.* 20226 [US, DES, MEXU]). − **Lit:** Ullrich (1990c). **D:** Mexico (Puebla [esp. Tehuacán valley], N Oaxaca). **I:** Gentry (1982: 236, 248). **Fig. VI.a**

≡ *Agave striata* var. *stricta* (Salm-Dyck) Baker (1877) ≡ *Agave striata* ssp. *stricta* (Salm-Dyck) B. Ullrich (1990); **incl.** *Agave hystrix* Hort. Belg. *ex* Jacobi (1870).

[1a] Stem elongate and branching, decumbent with age, 1 - 2 m; **Ros** often densely caespitose; **L** very numerous, long-lanceolate, linear, upcurved to straight, rigid, widest near the base, somewhat keeled above and below, 25 - 50 × 0.8 - 1 cm, green, striate, margins thin, pale yellow, cartilaginous, scabrous-serrulate; terminal **Sp** acicular, 1 - 2 cm, grey, decurrent along the margin, at the base bordered with the brownish **L** tip; **Inf** 1.5 - 2.5 m, 'spicate', straight or crooked, part-**Inf** with geminate **Fl**; **Fl** ascending to outcurved, 25 - 30 mm; **Ov** 8 - 11 mm, neckless; **Tep** red to purplish, tube funnel-shaped, 8 - 10 mm, lobes equal, 8 - 10 mm.

Distinguished from the vegetatively very similar *A. striata* by its short flower tube, equalled or exceeded in length by the tepals. Ullrich (l.c.) emphasizes vegetative similarities and suggests subspecific rank under *A. striata*, but he is not followed by Zamudio Ruiz & Sánchez Martinez (1995).

A. stringens Trelease (CUSNH 23: 114, 1920). **T:** Mexico, Jalisco (*Trelease* s.n. [MO]). − **Lit:** McVaugh (1989). **D:** Mexico (Jalisco: Río Blanco near Guadalajara).

[2f] **Ros** unknown; **L** concave, thin and recurving, ≥ 60 × 1 - 2 cm, very glaucous, margins carti-

laginous, nearly straight; marginal teeth curved, very sharp and slender, 1 - 2 mm, red or brown, scarcely 5 mm apart; terminal **Sp** conical, ± 8 × 2 mm, dark; **Inf** unknown.

Hardly known and in need of recollection.

A. subsimplex Trelease (Annual Rep. Missouri Bot. Gard. 22: 60, 1912). **T:** Mexico, Sonora (*Rose* 16811 [US]). − **Lit:** Turner & al. (1995). **D:** Mexico (Sonora); strictly coastal. **I:** Gentry (1982: 356, 405-406).

[2h] **Ros** low-spreading, small, 20 - 35 × 50 - 70 cm ∅, solitary or caespitose; **L** variable, lanceolate to ovate, thick, rigid, long- to short-acuminate, only a little narrowed towards the base, rounded below, hollowed above, 12 - 35 × 3 - 5 cm, grey-glaucous or light yellow-green, or sometimes purple-tinged, margins nearly straight or with strong prominences; marginal teeth variable, friable, straight or variously curved, rarely 2-tipped, larger teeth 3 - 15 mm, brown or more often yellowish-grey; terminal **Sp** subulate, frequently sinuous, shallowly grooved above, 2 - 4 cm, glaucous-grey, not or only a little decurrent; **Inf** 2 - 3.5 m, 'paniculate', slender, narrow, part-**Inf** short, 5 - 8 per **Inf**; **Fl** 40 - 45 mm; **Ov** ± 25 mm, with unconstricted long neck (5 mm); **Tep** yellow to pink, tube shallow, 3 - 4 mm, lobes 12 - 15 mm.

Closely related to *A. deserti* and *A. cerulata* (Gentry 1982: 405).

A. tecta Trelease (Trans. Acad. Sci. St. Louis 23: 145, pl. 26-27, 1915). **T:** Guatemala (*Trelease* 17 [ILL]). − **D:** Guatemala (region of Quezaltenango); cultivated only. **I:** Gentry (1982: 597, 613-614).

[2b] Stem very thick and broad; **Ros** semi-globose, open, 2 × 4 m, broad, freely suckering; **L** broadly lanceolate, straightly ascending, base deeply convex and thick, becoming thinner upwards, acuminate, concave to guttered, 100 - 160 × 30 - 40 cm, margins undulate; marginal teeth 8 - 10 mm (middle of the lamina), dull brown, 2 - 6 cm apart, cusps flattened, triangular, straight or curved from low bases; terminal **Sp** subulate, narrowly shortly grooved above, 5 - 7 cm, dull brown, decurrent or not; **Inf** 5 - 7 m, 'paniculate', massive; **Fl** 85 - 95 mm; **Ov** 38 - 43 mm, neck grooved, not constricted; **Tep** red-tinged in bud, greenish-yellow, tube funnel-shaped, 17 - 18 mm, lobes unequal, outer 32 - 33 mm.

Geographically isolated from other members of Group *Salmianae*; possibly a remnant of former use and now cultivated as a fence plant (Gentry 1982: 614).

A. tenuifolia Zamudio & E. Sánchez (Acta Bot. Mex. 37: 47-52, ills., 1995). **T:** Mexico, Querétaro (*Carranza* 1905 [IEB]). − **D:** Mexico (Tamaulipas, Querétaro); limestone slopes in pine-oak forest, 450 - 1500 m.

[1a] **Ros** lax, caespitose, forming dense groups by axillary or rhizomatous branching; **L** 30 - 50 (-90), linear, subcoriaceous, very thin and flexible, young straight, mature recurved, (29-) 50 - 100 (-130) × 0.25 - 0.5 (middle) - 1.3 (base) cm, green, striate, margins horny, < 1 mm wide, light green or hyaline, shortly serrulate; terminal **Sp** conical-subulate, 0.4 - 1 cm, coffee-reddish; **Inf** (0.9-) 1.5 - 1.75 (-2.3) m, 'spicate', straight, thin, lax, scape **Bra** triangular, long-cuspidate, (3-) 7 - 10 (-25) cm, decreasing in size towards the **Inf** tip, part-**Inf** (19-) 23 - 25 (-44) per **Inf**, with geminate **Fl** in the upper ⅓ of the **Inf**; **Fl** tubular, 23 - 30 mm; **Ov** cylindrical, sulcate, (6-) 9 - 12 mm; **Tep** yellow-greenish, tube 12 - 15 mm, lobes 4 - 7 mm.

According to the protologue close to *A. striata* ssp. *striata* but differing in its lax rosettes, fewer larger and pliable recurved leaves and its lax 'spikes' with geminate flowers. It differs from all other members of group *Striatae* in having fewer and longer leaves, lax inflorescences, and short stamens which are only shortly exserted.

A. tequilana F. A. C. Weber (Bull. Mus. Hist. Nat. (Paris) 8: 220, ills., 1902). **T:** [icono]: l.c. fig. 1. – **Lit:** McVaugh (1989). **D:** Mexico (Sonora, Sinaloa, Jalisco, Michoacán, Oaxaca); cultivated only. **I:** Gentry (1982: 555, 583).

Incl. *Agave palmaris* Trelease (1920); **incl.** *Agave pedrosana* Trelease (1920); **incl.** *Agave pesmulae* Trelease (1920); **incl.** *Agave pseudotequilana* Trelease (1920); **incl.** *Agave subtilis* Trelease (1920).

[2f] Stem short, thick, mature 30 - 50 cm; **Ros** radiately spreading, 1.2 - 1.8 m, surculose; **L** lanceolate, ascending to horizontal, firmly fibrous, mostly rigidly spreading, narrow and thickened towards the base, widest in the middle, acuminate, 90 - 120 × 8 - 12 cm, generally glaucous bluish- to grey-green, sometimes cross-zoned, margins straight to undulate or repand; marginal teeth generally regular in size or spacing or rarely irregular, mostly 3 - 6 mm (middle of the lamina), light to dark brown, 1 - 2 cm apart, with slender cusps curved from low pyramidal bases, rarely teeth remote and longer; terminal **Sp** flattened or openly grooved above, generally short, 1 - 2 cm, rarely longer, base broad, dark brown, decurrent or not; **Inf** 5 - 6 m, 'paniculate', densely branched, part-**Inf** large, dense, diffusely several times compound, 20 - 25 per **Inf**; **Fl** 68 - 75 mm; **Ov** 32 - 38 mm, neck short, not constricted; **Tep** green, tube funnel-shaped, 10 mm, lobes subequal, 25 - 28 mm. – *Cytology:* 2n = 60.

Distinguished from its close relative *A. vivipara* (as *A. angustifolia*) by larger leaves, thicker stems, and heavier more diffuse inflorescences with relatively large flowers with rather short tubes, albeit these differences are of degree rather than clear-cut (Gentry 1982: 583).

A. tequilana is important as source of the distilled liquor Tequila. For this purpose, it is cultivated in large plantations, esp. around the town of Tequila (Jalisco).

A. thomasae Trelease (Trans. Acad. Sci. St. Louis 23: 138, 1915). **T:** Guatemala (*Trelease* 19 [ILL]). – **D:** Guatemala; pine-forest zone, 2000 - 2800 m. **I:** Gentry (1982: 501).

[2k] **Ros** openly spreading, medium-sized, solitary or moderately suckering; **L** broadly lanceolate, softly succulent, pliable, narrowed and thickened towards the base, acuminate, plane to mildly guttered, smooth to slightly asperous below, 60 - 120 × 12 - 17 cm, pruinose or light glaucous to pale green; marginal teeth minute, 1 - 2 mm, 1 - 2 cm apart, reduced to denticles below; terminal **Sp** subulate to acicular, shallowly grooved above, 3 - 4.5 cm, dark brown; **Inf** 5 - 8 m, 'paniculate', narrow, scape short, part-**Inf** congested, roundish, 30 - 60 in the upper ½ - ⅚ of the **Inf**; **Fl** 60 - 70 mm; **Ov** strongly trigonous, 30 - 38 mm, tapering from the tube or with a short grooved neck; **Tep** purple to yellow, tube funnel-shaped, 6 - 11 mm, lobes unequal, 19 - 29 mm.

One of the few suckering Agaves in Guatemala. Distinguished by its soft grey-pruinose leaves with minute teeth and variously coloured flowers with strongly trigonous ovaries (Gentry 1982: 500).

A. titanota Gentry (Agaves Cont. North Amer., 176-180, ills., 1982). **T:** Mexico, Oaxaca (*Gentry & Tejeda* 22474 [US, DES, MEXU]). – **D:** Mexico (N Oaxaca); limestone canyon, 1070 - 1200 m, known only from the type locality. **Fig. VI.g**

[1g] **Ros** subcaulescent, openly spreading, medium-sized, solitary or sparingly surculose; **L** linear-ovate, broad, rigid, thick towards the base, short-acuminate, apex involute above, keeled and convex below, plane or concave above, finely granular, 35 - 55 × 12 - 14 cm, alabaster-white, margins horny, undulate to crenate, widest towards **L** tip (3 - 5 mm), continuous to the base or nearly so; marginal teeth variable, larger teeth 8 - 12 (-20) mm, variably spaced, **L** tip sometimes toothless for the uppermost 8 - 12 cm; terminal **Sp** broadly conical, keeled and protruding below, 3 - 4 cm, dark brown to grey; **Inf** erect, ± 3 m, 'spicate', flowering in the upper ½, part-**Inf** with geminate **Fl**; **Ped** geminate, 1 - 2 cm; **Fl** 45 - 50 mm; **Ov** 22 - 25 mm, pale greenish, neck constricted; **Tep** yellow, tube broadly funnel-shaped, 2 - 4 mm, lobes 21 - 24 mm.

Distinctive within Group *Marginatae* with its broad glaucous-white leaves (Gentry 1982: 179).

A. toumeyana Trelease (CUSNH 23: 140, 1920). **T:** USA, Arizona (*Toumey* 442 [US]). – **D:** USA (C Arizona).

A. toumeyana ssp. **bella** (Breitung) Gentry (Agaves Cont. North Amer., 211, ills. (pp. 212-213),

1982). **T:** USA, Arizona (*Breitung & Gibbons* 18153 [CAS]). – **D:** USA (C Arizona); open stony slopes and benches in juniper chaparral, 1250 - 1560 m. **Fig. VII.b**

≡ *Agave toumeyana* var. *bella* Breitung (1960).

[1e] Differs from ssp. *toumeyana*: **L** 100 or more, linear, more equal, smaller, 9 - 20 cm, margins replaced by denticles in the lower ½ of the **L**; **Inf** smaller. – *Cytology:* 2n = 60.

A. toumeyana ssp. **toumeyana** – **D:** USA (C Arizona); open limestone or volcanic rocky ledges, highland desert, chaparral, or lower pine forest, 625 - 1400 m. **I:** Gentry (1982: 196, 208). **Fig. VII.c**

Incl. *Agave toumeyana* var. *toumeyana*.

[1e] **Ros** small, densely caespitose; **L** 40 - 70, linear-lanceolate, straight or falcate or upcurving, rather rigid, thickly convex towards the base, plane above, of unequal length, smooth, 20 - 30 × 1.5 - 2 cm, light green or yellowish, both faces with impressions from the central bud, margins fine, brown with white threads, sometimes serrulate at the base; terminal **Sp** subulate, with a short narrow groove above, 1 - 2 cm, brown to greyish; **Inf** 1.5 - 2.5 m, 'spicate', densely or laxly flowered in the upper ⅓, part-**Inf** with geminate **Fl**; **Ped** basally united, short; **Fl** 18 - 25 mm; **Ov** 10 - 15 mm, neck slender, bent, 3 - 5 mm; **Tep** saccate, curved downwards, green, lobes whitish, tube broadly spreading, angled, 2 - 4 mm, lobes subequal, 7 - 9 mm.

The species suggests a large version of *A. parviflora*, but the leaves are always more acuminate, the flowers larger, the filaments are inserted higher up in the tube, and the lobes are more elongate (Gentry 1982: 210). Both ssp. hybridize with *A. chrysantha* (Hodgson 1999).

A. triangularis Jacobi (Nachtr. Versuch syst. Glied. Agaveen 2: 149, 1869). **T** [neo]: Mexico, Puebla (*Gentry* 23399 [DES, MEXU]). – **D:** Mexico (Puebla, Oaxaca); limestone soils, 1700 - 1900 m. **I:** Gentry (1982: 181-182). **Fig. VII.f**

≡ *Agave horrida* var. *triangularis* (Jacobi) Baker (1877); **incl.** *Agave regeliana* hort. *ex* Jacobi (1868) (*nom. illeg.*, Art. 53.1); **incl.** *Agave rigidissima* Jacobi (1869); **incl.** *Agave kerkhovei* hort. *ex* Jacobi (1870) (*nom. illeg.*, Art. 53.1); **incl.** *Agave hanburyi* Baker (1892); **incl.** *Agave triangularis* var. *subintegra* Trelease (1920).

[1g] Stem short; **Ros** rigid, slow-growing, widely surculose, forming open clusters; **L** deltoid-lanceolate, straight, rigid, thick at the base, long-acuminate, upper face concave, lower face convex, finely asperous, 30 - 60 × 5 - 7 (middle of the lamina) cm, olivaceous or light yellowish-green, finely flecked with brownish-red, margins horny, continuous, straight, 1 - 2 mm wide, greyish; marginal teeth present or lacking, straight or curved, few, small, 2 - 3 mm, remote, 3 - 5 cm apart, or large, 5 - 9 mm, and 1 - 2 cm apart, grey; terminal **Sp** conical to subu-

late, usually straight, 2.5 - 4 cm, greyish; **Inf** unknown.

Recognizable by its thick rigid deltoid olivaceous leaves. Toothless forms are common (Gentry 1982: 181).

A. tubulata Trelease (Mem. Nation. Acad. Sci. 11: 45, t. 99-100, 1913). **T:** Cuba (*Britton & al.* 9746 [MO]). – **Lit:** León (1946); Álvarez de Zayas (1985). **D:** W Cuba.

Incl. *Agave ekmannii* var. *microdonta* Trelease *in sched.* (s.a.) (*nom. inval.*, Art. 29.1); **incl.** *Agave ekmannii* Trelease (1926); **incl.** *Agave tubulata* ssp. *brevituba* A. Álvarez (1985).

[2m] **Ros** not described; **L** broadly lanceolate, gradually acute or subacuminate, sometimes conduplicate, 60 - 75 (-90) × 15 - 20 cm, rather glossy green, margins with teeth on green prominences or repand between the teeth; marginal teeth prevailingly upcurved towards the **L** tip and recurved towards the base, lunate rather than lenticular at the base, 1 - 3 mm, 1.5 - 2 cm apart, slender-cusped; terminal **Sp** acicularly conical, somewhat upcurved or flexuous, round-grooved or involute below the middle, smooth, 1.5 cm, brown, dull, decurrent; **Inf** 'paniculate', 2 - 5 m; **Fl** 30 - 35 mm; **Ov** 15 mm; **Tep** yellow, tube narrowly funnel-shaped, 6 - 8 mm, lobes 12 mm; **Fr** broadly oblong, shortly but distinctly stipitate and beaked, 2 - 3.5 × 1.2 - 1.5 cm.

The recently described ssp. *brevituba* differs but very slightly, and since material cited for the new taxon was also mentioned in the protologue by Trelease, it is here included in the synonymy.

A. underwoodii Trelease (Mem. Nation. Acad. Sci. 11: 37, t. 67-71, 1913). **T:** Cuba (*Trelease* 2+3 [MO?]). – **Lit:** León (1946). **D:** W Cuba.

Incl. *Agave morrisii* Worsley (1895) (*nom. illeg.*, Art. 53.1); **incl.** *Agave americana* Millspaugh (1900) (*nom. illeg.*, Art. 53.1).

[2l] **Ros** solitary; **L** ± narrowly lanceolate, gradually or in the broader forms acuminately pointed, concave, 100 - 200 × 20 - 25 cm, green, margins straight or somewhat concave; marginal teeth straight or somewhat curved or occasionally hooked (mostly downwards), rather strongly triangular from lenticular bases, 2 - 5 mm, brown or chestnut-brown, ± 1 (-2 or even 3) cm apart, exceptionally on green prominences; terminal **Sp** triquetrously conical or somewhat subulate, straight or slightly upcurved, openly grooved to or beyond the middle or involute, smooth or a little roughened, 1.5 - 2.5 cm, brown, rather dull, decurrent and somewhat dorsally intruded into the green **L** tissue; **Inf** 4 - 8 m, 'paniculate', broad, part-**Inf** on sharply recurved **Br**, in the upper ¾ or more of the **Inf**, not known to be bulbilliferous; **Fl** 50 - 55 mm; **Ov** 25 - 35 mm; **Tep** golden-yellow, tube conical, conduplicate, ± 8 mm, lobes 15 - 20 mm; **Fr** narrowly oblong, stipitate and beaked, 4 - 4.5 × 1.5 cm.

Rather variable (color of the terminal spine and the marginal teeth, shape of the groove of the terminal spine, length of the tepal lobes, exsertion of the filaments) according to the protologue.

A. utahensis Engelmann (in S. Watson, Bot. US Geol. Expl. 40. Parallel, 5: 497, 1871). **T** [lecto]: USA, Utah (*Palmer* s.n. [MO, US]). – **D:** USA (California, Utah, Nevada, Arizona).

≡ *Agave haynaldii* var. *utahensis* (Engelmann) A. Terracciano (1885).

A. utahensis ssp. **kaibabensis** (McKelvey) Gentry (Agaves Cont. North Amer., 259, 1982). **T:** USA, Arizona (*McKelvey* 4381 [A]). – **D:** USA (N Arizona). **I:** Gentry (1982: 252). **Fig. VII.d**

≡ *Agave kaibabensis* McKelvey (1949) ≡ *Agave utahensis* var. *kaibabensis* (McKelvey) Breitung (1960).

[1h] Differs from ssp. *utahensis*: Stem forming a short trunk with age; **Ros** larger, 40 - 60 cm ∅, usually solitary; **L** mostly 30 - 50 × 3 - 5 cm, light to bright green, younger **L** frequently pruinose-glaucous; **Inf** 3.5 - 5 m, stout, scape 4 - 6 cm ∅, part-**Inf** with 4 - 12 clustered **Fl**; **Fl** as in ssp. *utahensis* but larger.

A. utahensis ssp. **utahensis** – **D:** USA (California, Utah, Nevada, Arizona). **I:** Gentry (1982: 252, 258, 260-261).

Incl. *Agave utahensis* var. *utahensis*; **incl.** *Agave newberryi* Engelmann (1875); **incl.** *Agave scaphoidea* Greenman & Roush (1929) ≡ *Agave utahensis* var. *scaphoidea* (Greenman & Roush) M. E. Jones (1930); **incl.** *Agave utahensis* var. *nevadensis* Engelmann *ex* Greenman & Roush (1929) ≡ *Agave nevadensis* (Engelmann *ex* Greenman & Roush) Hester (1943); **incl.** *Agave utahensis* var. *discreta* M. E. Jones (1930); **incl.** *Agave eborispina* Hester (1943) ≡ *Agave utahensis* var. *eborispina* (Hester) Breitung (1960); **incl.** *Agave utahensis* fa. *nuda* hort. *ex* E. & B. Lamb (1978) (*nom. inval.*, Art. 32.1c?).

[1h] **Ros** rather compact, small, 18 - 30 × (15-) 25 - 40 cm ∅, caespitose; **L** 70 - 80, linear-lanceolate, stiff, straight or falcate or upcurving, convex below, plane to concave above, mostly 15 - 30 × 1.5 - 3 cm, light greyish- to yellow-green, in dwarf forms also bluish grey-glaucous; marginal teeth blunt, thick, detachable, larger teeth mostly 2 - 4 mm, brown-ringed around the bases, light grey, 1 - 2.5 cm apart; terminal **Sp** acicular, 2 - 4 cm, light grey or ivory-white, decurrent for 1 - 3 cm; **Inf** 2 - 4 m, 'spicate', lax or congested, scape 2 - 3 cm ∅, part-**Inf** with 2 - 8 clustered **Fl**; **Fl** sessile, urceolate, 25 - 31 mm; **Ov** 15 - 20 mm, neck long, 4 - 6 mm, constricted; **Tep** yellow, tube broadly funnel-shaped, very short, 2.5 - 4 mm, lobes nearly equal, 9 - 12 mm. – *Cytology:* 2n = 120.

A. utahensis var. *nevadensis* and var. *eborispina*, both accepted by Gentry (1982), are included in the synonymy of ssp. *utahensis* by Little (1981) and McKinney (1993), and this concept is adopted here. Gentry (l.c., 261) also mentions 'transitional forms'. Both varieties are best regarded as mere dwarf forms with large elongated spines from montane limestone outcrops. *A. utahensis* (ssp. *kaibabensis*) includes the N-most *Agave*-localities where temperatures may drop down to -18°C; plants from these populations may be hardy outdoors in C Europe with some protection from moisture.

A. variegata Jacobi (Hamburg. Gart.- & Blumenzeit. 21: 459-462, 1865). – **D:** USA (S Texas), Mexico (Nuevo León, Hidalgo, Puebla, San Luis Potosí, Tamaulipas, C Veracruz, naturalized in Yucatán); dry chaparral sites, on slopes, or between rocks, and in moist oak woods, 10 - 1830 m, flowers mainly March to June. **I:** Berger (1915: 35); Addisonia 17: pl. 569, 1932.

≡ *Manfreda variegata* (Jacobi) Rose (1903) ≡ *Polianthes variegata* (Jacobi) Shinners (1966); **incl.** *Manfreda tamazunchalensis* Matuda (1966); **incl.** *Manfreda xilitlensis* Matuda (1966).

[3a1] Plants medium-sized (for Subgen. *Manfreda*); **L** erect-arching, linear-lanceolate, deeply channelled, 14 - 44 × 1.2 - 2.7 (-3.9 in cultivation) cm, dark green, unspotted, or spotted with lighter green and brown or green, spots round to elliptic, sometimes glaucous, margins usually with small distantly spaced teeth; **Inf** 0.6 - 1.4 (-1.8 in cultivation) m, 'spicate', flowering part (7.5-) 14 - 22 (-29; to 48 in cultivation) cm, with 7 - 29 (-41) spreading **Fl**; **Ov** 9 - 16 (-19) mm; **Tep** tube 6 - 16 mm, lobes (6-) 9 - 13 (-16) mm, inside yellow-green or mahogany-brown; **Sty** exceeding the tube; **Fr** globose to oblong, 1.6 - 1.8 (-2.5) × 1.3 - 1.6 cm; **Se** 4 - 5 × 3 - 4 mm.

This species is notable within Subgen. *Manfreda* for its succulent minutely toothed and often spotted leaves, its tall inflorescence stalks, and flowers with short campanulate tubes and long-exserted stamens and styles (Verhoek-Williams 1975: 194).

A. vicina Trelease (Mem. Nation. Acad. Sci. 11: 19, t. 4, 10, 1913). **T** [syn]: Aruba (*Boldingh* 3&5 [MO]). – **Lit:** Hummelinck (1936). **D:** Leeward Islands (Aruba, Bonaire, Curaçao, Margarita). **I:** Hummelinck (1993).

Incl. *Agave vivipara* Crantz (1766) (*nom. illeg.*, Art. 53.1); **incl.** *Agave vivipara* var. *cabaiensis* Hummelinck (1936); **incl.** *Agave vivipara* var. *cuebensis* Hummelinck (1936).

[2q] **Ros** nearly acaulescent, ± 1 - 1.2 m ∅, suckering; **L** very broadly lanceolate, subacuminate, flatly concave, (17-) 40 - 60 (-75) × (5-) 12 - 20 cm, somewhat transiently glaucous, with age rather glossy green, margins a little concave; marginal teeth commonly upcurved in the upper ½ of the **L**

and recurved in the lower ½, slender from lunate bases, 3 - 4 mm, 1 - 1.5 cm apart, often on green prominences; terminal **Sp** triquetrously acicular, somewhat flexuous, narrowly round-grooved to the middle and involute below, smooth, polished towards the tip, 2.5 - 3 cm, red-brown, shortly decurrent; **Inf** scarcely 3 m, narrowly oblong, part-**Inf** on ascending **Br** in the upper ½ or more of the **Inf**, freely bulbilliferous; **Fl** 40 - 45 mm; **Ov** oblong, 20 - 25 mm; **Tep** yellow, tube open, ± 4 mm, lobes 15 mm; **Fr** broadly oblong, very shortly stipitate and beaked, 3 × 2.5 cm. − *Cytology:* 2n = 60.

This is the first-named Caribbean species in the genus, for long known under the misapplied name *A. vivipara* − a name erroneously used for many other Caribbean species. The nomenclatural confusion was clarified by Wijnands (1983), who showed that *A. vivipara* actually represents the oldest name for the widespread continental taxon previously referred to as *A. angustifolia*, which is consequently renamed here (see there). *A. vicina* is the oldest available binomial that clearly refers to the Caribbean plants in question. See also the note for *A. cocui*.

A. victoriae-reginae T. Moore (Gard. Chron., ser. nov. 4(94): 484-485, fig. 101, 1875). **T:** [lecto − icono] l.c. fig. 101. − **Lit:** Ullrich (1991d). **D:** Mexico (S Coahuila, C Nuevo León, NE Durango). **I:** Gentry (1982: 126, 183, 185). **Fig. VII.e**

Incl. *Agave victoriae-reginae* fa. *variegata* hort. (s.a.) (*nom. inval.*, Art. 29.1); **incl.** *Agave victoriae-reginae* var. *compacta* hort. (s.a.) (*nom. inval.*, Art. 29.1); **incl.** *Agave victoriae-reginae* var. *stolonifera* hort. (s.a.) (*nom. inval.*, Art. 29.1); **incl.** *Agave considerantii* Carruel (1875); **incl.** *Agave nickelsiae* Gosselin (1895) ≡ *Agave victoriae-reginae* fa. *nickelsiae* (Gosselin) Trelease (1920); **incl.** *Agave victoriae-reginae* var. *laxior* A. Berger (1912); **incl.** *Agave ferdinandi-regis* A. Berger (1915); **incl.** *Agave victoriae-reginae* fa. *dentata* Breitung (1960); **incl.** *Agave victoriae-reginae* fa. *latifolia* Breitung (1960); **incl.** *Agave victoriae-reginae* fa. *longifolia* Breitung (1960); **incl.** *Agave victoriae-reginae* fa. *longispina* Breitung (1960); **incl.** *Agave victoriae-reginae* fa. *viridis* Breitung (1960); **incl.** *Agave victoriae-reginae* fa. *ornata* Breitung (1960) (*nom. inval.*, Art. 37.1); **incl.** *Agave victoriae-reginae* fa. *stolonifera* H. Jacobsen (1960) (*nom. inval.*, Art. 37.1).

[1g] **Ros** acaulescent or (in cultivation) with a short stem, compact, very variable, small, solitary, surculose or caespitose; **L** linear to ovate, rigid, thick, short, generally closely imbricate, rounded at the tip, rounded to sharply keeled below, plane to concave above, 15 - 20 (-25) × 4 - 6 cm, green with conspicuous white markings, margins white-horny, continuous to the **L** base, 2 - 5 mm wide; marginal teeth usually none; terminal **Sp** 1 or 3, trigonous-conical, subulate, very broad at the base, 1.5 - 3 cm,

black; **Inf** erect, 3 - 5 m, 'spicate', densely flowered in the upper ½, part-**Inf** with 2 - 3 **Fl**; **Ped** forking, stout, 40 - 46 mm; **Fl** 40 - 46 mm; **Ov** thickly fusiform, 18 - 24 mm, neck short; **Tep** variously coloured, frequently tinged red or purple, tube shallow, funnel-shaped, 3 mm, lobes ± equal, 18 - 20 mm. − *Cytology:* 2n = 60, 120.

Typically easily identified by the toothless, thick and rigid leaves with white markings on both faces. The extreme variation is added to by apparent introgressive hybrid swarms with *A. asperrima* and *A. lechuguilla* (Gentry 1982: 185). The lectotype cited above was designated by Ullrich (1990e) and supersedes the neotype designation by Gentry (1982: 184).

A. vilmoriniana A. Berger (RSN 12: 503, 1913). **T** [lecto]: Mexico, Guadalajara (*Rose & Hough* 4833 [US]). − **Lit:** McVaugh (1989); Turner & al. (1995). **D:** Mexico (S Sonora, Chihuahua, Sinaloa, Durango, Zacatecas, Aguascalientes, Jalisco); on volcanic barranca cliffs, 600 - 1700 m. **I:** Gentry (1982: 67, 83-84).

Incl. *Agave edwardii* Trelease ms. (s.a.) (*nom. inval.*, Art. 29.1); **incl.** *Agave eduardii* Trelease (1920); **incl.** *Agave houghii* hort. *ex* Trelease (1920); **incl.** *Agave mayoensis* Gentry (1942).

[1b] **Stem** short; **Ros** ± 1 × 2 m, solitary; **L** linear-lanceolate, arching, pliable, deeply guttered, broadest at the base, heavily thickened towards the base, long acuminate, concave to conduplicate above, smooth, 90 - 180 × 7 - 10 cm, light to yellowish-green, margins with a fine brown continuous border ± 1 mm wide, scaly with age; marginal teeth none; terminal **Sp** acicular, 1 - 2 cm, brown to greyish-brown; **Inf** 3 - 5 m, 'spicate', densely flowered from 1 to 2 m above base, bulbilliferous or not, part-**Inf** with 2 - 4 (-8) **Fl**; **Ped** 1 - 2 times forked, 8 - 20 mm; **Fl** 35 - 40 mm; **Ov** 15 - 20 mm incl. a neck 3 - 4 mm long; **Tep** yellow, tube open, shallow, 4 mm, lobes equal, 14 - 17 mm.

A. virginica Linné (Spec. Pl. [ed. 1], 323, 1753). **T:** USA, Virginia (*Clayton* 498 [LINN, BM]). − **Lit:** Diggs & al. (1999: with ills.). **D:** S and E USA (from Missouri and Texas to the Atlantic).

≡ *Manfreda virginica* (Linné) Salisbury *ex* Rose (1899) ≡ *Polianthes virginica* (Linné) Shinners (1966).

The species is aberrant in being the only member of Subgen. *Manfreda* occurring in temperate regions. In addition to its distribution, the taxon is easily recognized by it slender green flowers with erect lobes and the style that is shorter than the stamens (Verhoek-Williams 1975: 170, 172).

A. virginica ssp. **lata** (Shinners) Thiede & Eggli (KuaS 52: [in press], 2001). **T:** USA, Texas (*Daly* 61 [BRIT / SMU]). − **D:** USA (S Oklahoma, N-C

Texas); mainly in the Backland Prairie, flowers mid-June to mid-July.

≡ *Agave lata* Shinners (1951) ≡ *Polianthes lata* (Shinners) Shinners (1966) ≡ *Manfreda virginica* ssp. *lata* (Linné) O'Kennon & al. (1999).

[3a4] Differs from ssp. *virginica*: **L** 4 - 10, 12 - 18 × (2-) 3- 8 (-10) cm, 3 - 6× as long as broad; **Inf** scape near the base 6 - 10 mm ∅; **Ov** 7 - 8 mm; **Tep** tube 19 - 27 mm, lobes 2.5 - 3 mm wide at the base; **Anth** 13 - 17 (-20) mm.

Verhoek-Williams (1975) included this taxon in the synonymy of the widespread and variable *Manfreda virginica*. Recently, Diggs & al. (1999) re-emphasized the clear, albeit overlapping differences between *A. lata* and *A. virginica*, and recognized the former at subspecies rank.

A. virginica ssp. **virginica** – **D:** S and E USA (from Missouri and Texas to the Atlantic); wooded areas, on rocky and sandy soils, flowers mid-July to mid-August.

Incl. *Agave pallida* Salisbury (1796); **incl.** *Agave virginica* var. *tigrina* Engelmann (1875) ≡ *Manfreda tigrina* (Engelmann) Small *ex* Rose (1903) ≡ *Manfreda virginica* var. *tigrina* (Engelmann) Rose (1903) ≡ *Agave virginica* fa. *tigrina* (Engelmann) Palmer & Steyermark (1935) ≡ *Agave tigrina* (Engelmann) Cory (1936) ≡ *Polianthes virginica* fa. *tigrina* (Engelmann) Shinners (1966); **incl.** *Allibertia intermedia* Marion (1882); **incl.** *Agave alibertii* Baker (1883).

[3a4] Plants medium-sized (for Subgen. *Manfreda*); rhizome 1 - 2.5 (-5) × 1 - 2.3 (-2.6) cm; **R** numerous, half-fleshy; **L** ± 10, semisucculent, spreading, oblanceolate to linear-lanceolate, 12 - 15 (-30) × 1 - 4.5 cm, 7 - 15× as long as broad, usually plain green with red spots and speckles near the base, upper face frequently sparsely or densely spotted, sometimes plain green, margins with a narrow cartilaginous band, entire or with regularly or irregularly spaced short or medium-sized prickles; remains of **L** bases membranous, 1.8 - 4 (-4.5) cm; **Inf** 0.75 - 1.9 (-2.12) m, scape 4 - 7 mm ∅ near the base; flowering part 14 - 68 cm, with 10 - 44 (-61) **Fl** (lower **Fl** rarely paired) in a dense 'spike'; **Fl** nearly erect, green; **Ov** (4-) 5 - 7 mm; **Tep** tube narrow, at a slight angle to the **Ov** and constricted above it, 9 - 16 mm, lobes erect, 1.5 mm wide at the base, sometimes tinged with purple; **Anth** 8 - 10 mm; **Sty** shorter than the **St**; **Sti** white, 3-lobed; **Fr** globose, 1 - 1.8 (-2.5) × 1 - 1.7 cm; **Se** 3 - 5 × 4 - 6 mm. – *Cytology:* n = 30.

A. vivipara Linné (Spec. Pl. [ed. 1], 323, 1753). **T:** [lecto – icono]: Commelin, Prael. Bot. 65: t. 15. – **D:** Mexico (widespread), C America to Panama; naturalized in RSA.

Incl. *Aloe vivipara* Crantz (1766).

A. vivipara 'Marginata' (Hort. *ex* Gentry) P. I.

Forster (Brittonia 44: 74, 1992). **Nom. inval.**, Art. 28 Note 4. – **I:** Gentry (1982: 565).

≡ *Agave angustifolia* var. *marginata* hort. *ex* Gentry (1982) (*nom. inval.*, Art. 36.1, 37.1).

[2f] Differs from var. *vivipara*: Stems 30 - 60 cm; **L** numerous, with narrow white or yellow margins.

Representing a horticultural selection only, now widely distributed around the world as an ornamental (Gentry 1982: 564).

A. vivipara 'Variegata' (Trelease) P. I. Forster (Brittonia 44: 75, 1992).

≡ *Agave angustifolia* var. *variegata* Trelease (1908).

[2f] Differs from var. *vivipara*: **L** with unusual broad marginal white stripe, remainder of the **L** silvery grey or milky.

Representing a horticultural selection only, which is reported to have arisen in the Botanical Garden of the College of Science at Poona, India, about 1895 (Gentry 1982: 567).

A. vivipara var. **deweyana** (Trelease) P. I. Forster (Brittonia 44: 74, 1992). **T:** Mexico, Tamaulipas (*Dewey* 649 [US]). – **D:** Cultivated only: Mexico (Tamaulipas, Veracruz).

≡ *Agave deweyana* Trelease (1909) ≡ *Agave angustifolia* var. *deweyana* (Trelease) Gentry (1982).

[2f] Differs from var. *vivipara*: **L** narrow in the type (5 - 6 cm), but wider in later collections, generally 110 - 115 × 7 - 10 cm; marginal teeth more remote in some collections.

This taxon is "not well marked", since various clones appear to be cultivated under this name (Gentry 1982: 564). This and the following varieties apparently represent mere cultivated selections and cultivars and should better be named as such.

A. vivipara var. **letonae** (Taylor *ex* Trelease) P. I. Forster (Brittonia 44: 74, 1992). **T:** El Salvador (*Milner* s.n. [MO]). – **D:** Cultivated only (for fibre): Guatemala, El Salvador. **I:** Gentry (1982: 565-566).

≡ *Agave letonae* Taylor *ex* Trelease (1925) ≡ *Agave angustifolia* var. *letonae* (Taylor *ex* Trelease) Gentry (1982).

[2f] Differs from var. *vivipara*: Stem broad; **Ros** robust; **L** nearly white.

A. vivipara var. **nivea** (Trelease) P. I. Forster (Brittonia 44: 74, 1992). **T:** Guatemala, Progreso Dept. (*Trelease* 11 [MO]). – **D:** Cultivated only (for fibre and fences): Guatemala. **I:** Trelease (1915b: pl. 22).

≡ *Agave nivea* Trelease (1915) ≡ *Agave angustifolia* var. *nivea* (Trelease) Gentry (1982).

[2f] Differs from var. *vivipara*: Stem short; **L** long, 130 - 140 × 9 - 10 cm, characteristically dull bluish-grey; **Inf** unknown.

A. vivipara var. **sargentii** (Trelease) P. I. Forster

(Brittonia 44: 75, 1992). **T:** [lecto – icono]: Annual Rep. Missouri Bot. Gard., 22: pl. 100-101, 1912. – **D:** Cultivated only: Mexico (México, Puebla?).

≡ *Agave angustifolia* var. *sargentii* Trelease (1912).

[2f] Differs from var. *vivipara*: Stem ± 25 cm; **Ros** dwarf; **L** numerous, narrowly oblong-lanceolate, minutely roughened, 25 - 30 × 2.5 - 3 cm, greyish-green; marginal teeth 1 - 2 mm, nearly black, glossy; **Inf** ± 1 m, part-**Inf** few, bulbilliferous; **Fl** 40 mm.

A. vivipara var. **vivipara** – **Lit:** León (1946); McVaugh (1989); Turner & al. (1995). **D:** Mexico (widespread from Sonora and Chihuahua to the S), Belize, Costa Rica, Honduras, Nicaragua, El Salvador, Panama; mainly tropical savannas, thorn forest, and tropical deciduous forests at low to middle elevations (to 1500 m, rarely more); naturalized in RSA. **I:** Gentry (1982: 555, 560); Hummelinck (1993). **Fig. VII.a**

Incl. *Agave angustifolia* var. *angustifolia*; **incl.** *Agave lurida* Jacquin (1790) (*nom. illeg.*, Art. 53.1); **incl.** *Agave angustifolia* Haworth (1812); **incl.** *Agave flaccida* Haworth (1812); **incl.** *Agave rigida* Spin (1812) (*nom. illeg.*, Art. 53.1); **incl.** *Agave jacquiniana* Schultes (1829) ≡ *Agave lurida* var. *jacquiniana* (Schultes) Salm-Dyck (1861) ≡ *Agave vera-cruz* var. *jacquiniana* (Schultes) Ascherson & Graebner (1906); **incl.** *Agave punctata* Salm-Dyck (1834); **incl.** *Agave rubescens* Salm-Dyck (1834) ≡ *Agave angustifolia* var. *rubescens* (Salm-Dyck) Gentry (1982) ≡ *Agave vivipara* var. *rubescens* (Salm-Dyck) P. I. Forster (1992); **incl.** *Agave flaccida* Salm-Dyck (1834) (*nom. illeg.*, Art. 53.1); **incl.** *Agave ixtli* Karwinsky *ex* Salm-Dyck (1837); **incl.** *Agave serrulata* Karwinsky *ex* Otto (1842) (*nom. illeg.*, Art. 53.1); **incl.** *Agave vivipara* Wight (1853) (*nom. illeg.*, Art. 53.1); **incl.** *Agave elongata* Jacobi (1865); **incl.** *Agave excelsa* Jacobi (1866); **incl.** *Agave flavovirens* Jacobi (1866); **incl.** *Agave erubescens* Ellemeet (1871); **incl.** *Agave ixtlioides* Hooker (1871) (*nom. illeg.*, Art. 53.1); **incl.** *Agave excelsa* Baker (1877) (*nom. illeg.*, Art. 53.1) ≡ *Agave ixtli* var. *excelsa* (Baker) A. Terracciano (1885); **incl.** *Agave spectabilis* Todaro (1878) (*nom. illeg.*, Art. 53.1); **incl.** *Agave sobolifera* var. *serrulata* A. Terracciano (1885); **incl.** *Agave rigida* var. *elongata* Anonymus (1893) (*nom. illeg.*, Art. 53.1); **incl.** *Agave rigida* A. Berger (1898) (*nom. illeg.*, Art. 53.1); **incl.** *Agave aboriginum* Trelease (1907); **incl.** *Agave endlichiana* Trelease (1907); **incl.** *Agave lespinassei* Trelease (1907); **incl.** *Agave wightii* Drummond & Prain (1907); **incl.** *Agave zapupe* Trelease (1907); **incl.** *Agave bergeri* Trelease *ex* A. Berger (1915); **incl.** *Agave donnell-smithii* Trelease (1915); **incl.** *Agave kirchneriana* A. Berger (1915); **incl.** *Agave prainiana* A. Berger (1915); **incl.** *Agave sicaefolia* Trelease (1915); **incl.** *Agave vivipara* var. *woodrowii* hort. *ex* A. Berger (1915); **incl.** *Agave*

pacifica Trelease (1920); **incl.** *Agave panamana* Trelease (1920); **incl.** *Agave yaquiana* Trelease (1920); **incl.** *Agave owenii* I. M. Johnston (1924); **incl.** *Agave prolifera* Schott *ex* Standley (1930); **incl.** *Agave costaricana* Gentry (1949); **incl.** *Agave breedlovei* Gentry (1982) (*nom. inval.*, Art. 34.1a).

[2f] Nearly acaulescent or stem to 20 - 60 (-90) cm; **Ros** radiately spreading, surculose; **L** linear to (very broadly) lanceolate, ascending to horizontal, mostly rigid, hard-fleshy, fibrous, narrowed and thickened towards the base, full-grown generally (40-) 60 - 120 × 3.5 - 10 (-20) cm, light green to glaucous-grey, margins straight to undulate, sometimes thinly cartilaginous; marginal teeth generally small, 2 - 5 mm, rarely longer, commonly reddish-brown or dark brown, evenly and closely spaced or remote, from low narrow bases, cusps slender, ± curved; terminal **Sp** variable, conical to subulate, flat to shallowly grooved above, 1.5 - 3.5 cm, dark brown, greying with age, not or thinly decurrent; **Inf** ± 3 - 5 m, 'paniculate', open, part-**Inf** horizontally spreading to ascending, 10 - 20, sometimes (freely) bulbilliferous; **Fl** (40-) 50 - 65 mm; **Ov** small, tapering at the base, 20 - 30 mm, neck short; **Tep** green to yellow, quickly wilting, drying reflexed along the tube, tube (4-) 8 - 16 mm, lobes unequal, (15-) 18 - 24 mm. – *Cytology:* 2n = 60, 120, 180.

The name *A. angustifolia* Haworth, under which the widespread Mexican / Central American taxon was since long known, represents a homotypic synonym of *A. vivipara* Linné and is also considered taxonomically identical (Wijnands 1983). The name *A. vivipara* was misapplied to several Caribbean species, but is now established here to replace *A. angustifolia*. For the Caribbean species previously named *A. vivipara*, the name *A. vicina* has to be used.

The plants from Panama separated as *A. panamana* (Cseh 1993) merely represent the S-most element in the complex (Lott & García-Mendoza 1994).

A. vivipara var. *rubescens* apparently represents a superfluous separate name for a common narrow-leaved wild form occuring irregularly within the species's range in Mexico (Gentry 1982: 567). Since the leaf measurements given by Gentry fall well into the range of var. *vivipara*, it is included in the synonymy here.

A. vivipara is by far the most wide-ranging species of the family and exhibits an extensive range of variation esp. in vegetative features. It is a sun-loving taxon (although also not rarely found with etiolated growth in light shade) occuring in nearly all vegetation types (albeit mainly in the 'tierra caliente') (Gentry 1982: 561).

A. vizcainoensis Gentry (Occas. Pap. Calif. Acad. Sci. 130: 67-69, ills., 1978). **T:** Mexico, Baja California (*Gentry* 7469 [US, ARIZ, DES, DS, MEXU]). – **D:** Mexico (Baja California Sur). **I:** Gentry (1982: 408).

[2h] **Ros** open, 30 - 50 × 50 - 90 cm, surculose or solitary; **L** few, lanceolate, thickly fleshy, rather rigid, narrowed above the base, broadest in the middle, 25 - 40 × 6 - 10 cm, glaucous-grey to green, sometimes reddish, margins undulate, horny above with decurrent terminal **Sp**; marginal teeth nearly straight or curved, slender or broadly flattened, 5 - 10 mm (middle of the lamina), dark brown to greyish, 1 - 3 cm apart; terminal **Sp** stoutly subulate, mostly rather straight, shallowly grooved above, 2.5 - 4 cm, brown to greyish, long-decurrent; **Inf** 2 - 3 m, 'paniculate', part-**Inf** spreading, 8 - 15 in the upper ½ of the **Inf**; **Fl** 65 - 75 mm; **Ov** 36 - 41 mm, green, neck 6 - 8 mm; **Tep** yellow, tube funnel-shaped, 8 - 12 mm, lobes 21 - 26 mm.

Appears to be related to *A. margaritae*, which differs in its small size and short broad long-toothed leaves (Gentry 1982: 407).

A. wallisii Jacobi (Nachtr. Versuch syst. Glied. Agaveen 2: 162, 1870). **T:** Colombia (*Wallis* s.n. [not indicated]). – **Lit:** Berger (1915). **D:** Colombia (mountains of the Río Cauca region).

[2o?] Acaulescent; **L** relatively few, broadly lanceolate, at first erect, somewhat recurved, older spreading to all sides, recurved towards the tip, base thickly fleshy, becoming thin and fibrous upwards, broadest somewhat above the middle, somewhat flatly keel-like thickened below, lower ½ flat or flatly guttered above, upper ½ flatly concave above, tip grooved, fresh green, dull, slightly glaucous, margins somewhat undulate and recurved; marginal teeth very small, triangular, curved upwards, at first pergamentaceous with brownish tip, later horny, chestnut-brown, on flat fleshy basis; **Inf** unknown.

Jacobi based his description on an undeveloped young plant (Berger l.c.). See also *A. cundinamarcensis*.

A. warelliana Baker (Gard. Chron., ser. nov. 1877: 264, fig. 53, 1877). **T:** Ex cult. La Mortola (*Anonymus* s.n. [K, US]). – **D:** Not known from the wild. **I:** Gentry (1982: 232).

[1f] **Ros** (sub-) acaulescent, rather robust, ± 1 × 1.7 m, sparsely surculose, branching axillary; **L** dense, lanceolate-spatulate, erect or spreading, thickly fleshy, slightly constricted above the base, acuminate, upper face flat, lower face convex, 70 - 75 × 13 - 14 (6 - 7 at the base) cm, light pale green or shiny glaucous, margins finely serrulate, brown, denticles 1 mm, 2 mm apart; terminal **Sp** straight, 1.8 - 2 cm, brown, long-decurrent; **Inf** ± 5 m, 'spicate', scape ± 2 m; **Fl** 90 - 95 mm; **Ov** smooth, 40 mm, light green; **Tep** yellow within, violet on the outside and brown-spotted, tube 14 - 15 mm, lobes 35 mm.

This is the largest species in Group *Polycephalae*, esp. notable by its large flowers. It seems close to *A. chiapensis*, but differs in its larger flowers with smaller bractlets and the finely serrulate brownish leaf margins (Gentry 1982: 231).

A. weberi Cels *ex* Poisson (Bull. Mus. Hist. Nat. (Paris) 17: 230-232, 1901). **T** [neo]: USA, Texas (*Gentry & al.* 20003 [US, DES, MEXU]). – **D:** Known from cultivation only. **I:** Gentry (1982: 621, 632).

Incl. *Agave franceschiana* Trelease *ex* A. Berger (1912).

[2a] **Ros** open, 1.2 - 1.4 × 2 - 3 m, freely suckering; **L** lanceolate, rather softly fleshy, pliable, straight to recurving, widest in the middle, narrowed below, concave or guttered above, 110 - 160 × 12 - 18 m, green or pruinose-greyish esp. in youth; marginal teeth usually absent along the upper ⅓ - ½ of the **L**, margins denticulate below, teeth 1 - 2 mm, ≤ 1 cm apart, rarely toothless throughout; terminal **Sp** subulate, openly shallowly grooved above in the upper ½, 3 - 4.5 cm, brown to greyish, decurrent for several cm; **Inf** 7 - 8 m, 'paniculate', tall, open, part-**Inf** several times compound, diffuse, sometimes bulbilliferous; **Fl** 70 - 80 mm; **Ov** 33 - 40 mm, pale green, neck short, grooved, not constricted; **Tep** bright yellow, tube rather urceolate, 18 - 20 mm, lobes subequal, 20 - 24 mm.

Placed in Group *Agave* by Ullrich (1990d: 106), in contrast to Gentry's placement in Group *Viviparae* (as *Sisalanae*).

A. wendtii Cházaro (Cact. Suc. Mex. 42(4): 95, 1997). **T:** Mexico, Veracruz (*Cházaro & Flores Macías* 6645 [XAL, WIS]). – **D:** Mexico (Veracruz); tropical evergreen forest, known only from the type locality.

[1f] **Ros** lax, to 45 cm ∅; **L** lanceolate-oblong, fleshy, brittle, 10 - 25 × 2.5 - 3.5 cm, glaucous when young, later turning light green; marginal teeth small, concolorous with the lamina; terminal **Sp** 1 cm, dark; **Inf** 1 m, 'spicate', dense, part-**Inf** with 2 - 3 **Fl**; **Fl** 20 - 29 mm.

According to the protologue closely related to *A. pendula*.

A. wercklei F. A. C. Weber *ex* Wercklé (Monatsschr. Kakt.-kunde 17(5): 71-72, 1907). **T:** Costa Rica (*Wercklé* s.n. [US]). – **Lit:** Horich (1973); Ullrich (1992b); both with ills. **D:** Costa Rica; Pacific slopes, hot regions in sparse grassland. **I:** Gentry (1982: 502).

[2k] Ros compact, 1 - 2 m ∅, solitary; **L** variable, ovate to lanceolate, ascending to outcurved, thick and robust, broad, thickly fleshy, narrowed at the base, short-acuminate, tip inrolled, plane to concave towards the tip, young with rough surface below the tip, 70 - 150 cm, green to white glaucous, with impressions from the central bud, margins smoothly rounded, straight to undulate; marginal teeth deltoid, small, 3 - 4 mm, black; terminal **Sp** conical at the base tapering into an acicular point, narrowly

grooved above, 2 - 3 cm, dark brown or black, finely shortly decurrent; **Inf** 8 m, 'paniculate', large, profuse, part-**Inf** short, branched, umbellate, dense, ± 45 per **Inf**, bulbilliferous; **Fl** 62 mm; **Ov** elongate, narrowed at both ends, 40 mm; **Tep** golden-yellow, tube openly funnel-shaped, 8 - 9 mm, lobes 17 mm.

The large thick leaves, the 'panicles' reaching deeply down the inflorescence axis, and the flowers point towards a relationship with *A. parvidentata* (Gentry 1982: 501). Ullrich (l.c.) clarified the correct author citation of the species.

A. willdingii Todaro (Hort. Bot. Panorm. 2: 36, t. 32, 1878). – **Lit:** Trelease (1913); Álvarez de Zayas (1985). **D:** Only known from cultivation; probably originating from Cuba.

[2m] **Ros** solitary; **L** rather few, broadly oblong-lanceolate, gradually acute, slightly concave and conduplicate, 60 - 80 × 15 cm, light green or slightly glaucous, margins slightly concave; marginal teeth variously curved, acuminately triangular or somewhat lenticular at the base and on green prominences, 1 - 3 mm, 1 - 1.5 cm apart; terminal **Sp** conical, slit-grooved below the middle, smooth, 1 - 1.5 cm, brown, dull, scarcely decurrent; **Inf** 4 - 5 m, 'paniculate', part-**Inf** few, very lax, on outcurved slender **Br** in the upper ⅓ or more of the **Inf**; **Fl** 30 mm; **Ov** nearly cylindrical, 15 mm; **Tep** orange, tube openly conical, 4 mm, lobes 10 - 12 mm; **Fr** unknown.

Not definitely known as wild plant (Trelease 1913: 43), but possibly originating from Cuba, as indicated by the small orange-yellow flowers. According to Trelease to be compared with *A. antillarum*. The species epithet is variously spelled 'wildingii', 'willdinghii' etc.

A. wocomahi Gentry (Publ. Carnegie Inst. Washington 527: 96, 1942). **T:** Mexico, Chihuahua (*Gentry* 1989 [CAS, ARIZ, DES]). – **Lit:** McVaugh (1989). **D:** Mexico (Sonora, Chihuahua, Sinaloa, Durango); rocky limestone mountain slopes in pine-oak forests, 1400 - 2500 m. **I:** Gentry (1982: 423, 457-458).

[2g] **Ros** 0.8 - 1.3 × 1.5 - 2 m, not suckering; **L** mostly lanceolate to linear-lanceolate, rarely ovate, ascending to depressed with age, rather rigid, somewhat narrowed towards the base, plane, 30 - 90 × 9 - 25 cm, dark to glaucous-green, margins straight to undulate; marginal teeth variously curved, below the middle of the lamina frequently down-curved, with smaller irregular intermittent teeth, larger teeth 1 - 2 cm, dark brown to glaucous-brown; terminal **Sp** stout, usually sinuous, flattened or with a broad groove, 3 - 6 cm, short- or long-decurrent; **Inf** 3 - 5 m, 'paniculate', open, part-**Inf** small, 8 - 15 in the upper ⅓ of the **Inf**; **Fl** erect, 65 - 85 mm; **Ov** cylindrical, 34 - 40 mm incl. a 2 - 5 mm long neck, light green; **Tep** yellow, tube deeply funnel-shaped, 18 - 22 mm, lobes dimorphic, 15 - 23 mm.

Distinguished from *A. shrevei* by its dark green leaf colour and more remote teeth. Easily confused with *A. bovicornuta* (Group *Crenatae*) with lighter yellowish-green leaves conspicuously narrowed just above the base and with different flowers (Gentry 1982: 456).

A. xylonacantha Salm-Dyck (Bonplandia 7: 92, 1859). **T:** [neo – icono]: Curtis's Bot. Mag. 1867, t. 5660. – **D:** Mexico (Tamaulipas, San Luis Potosí, Guanajuato, Querétaro, Hidalgo); dry limestone slopes and valleys, > 900 m. **I:** Gentry (1982: 126, 186-187).

Incl. *Agave cornuta* Hort. Belg. *ex* Besaucèle (s.a.); **incl.** *Agave xylonacantha* var. *mediopicta* Trelease (s.a.); **incl.** *Agave maximiliana* hort. *ex* Besaucèle (s.a.) (*nom. illeg.*, Art. 53.1); **incl.** *Agave xylonacantha* var. *latifolia* Jacobi (s.a.) (*nom. inval.*, Art. 29.1?); **incl.** *Agave xylonacantha* var. *macracantha* Jacobi (s.a.) (*nom. inval.*, Art. 29.1?); **incl.** *Agave xylonacantha* var. *torta* Jacobi (s.a.) (*nom. inval.*, Art. 29.1?); **incl.** *Agave xylonacantha* var. *vittata* Jacobi (s.a.) (*nom. inval.*, Art. 29.1?); **incl.** *Agave vittata* Regel (1858); **incl.** *Agave amurensis* Jacobi (1864); **incl.** *Agave kochii* Jacobi (1866); **incl.** *Agave splendens* Jacobi (1870) ≡ *Agave heteracantha* var. *splendens* (Jacobi) A. Terracciano (1885); **incl.** *Agave perbella* hort. *ex* Baker (1877); **incl.** *Agave vanderdonckii* hort. *ex* Baker (1877); **incl.** *Agave xylacantha* hort. (1877) (*nom. inval.*, Art. 61.1); **incl.** *Agave hybrida* hort. *ex* Baker (1887); **incl.** *Agave carchariodonta* Pampanini (1907) ≡ *Agave univittata* var. *carchariodonta* (Pampanini) Breitung (1959); **incl.** *Agave noli-tangere* A. Berger (1915); **incl.** *Agave xylonacantha* var. *horizontalis* hort. *ex* A. Berger (1915).

[1g] **Stem** short; **Ros** openly spreading, solitary or caespitose; **L** ensiform-lanceolate, rather rigid, broadest in the middle, long-acuminate, rounded below, plane to concave above, 45 - 90 × 5 - 10 cm, green to yellowish-green, sometimes glaucous, with or without a pale mid-stripe, margins horny, continuous, straight between the remote conspicuous prominences but looping over the prominences; marginal teeth broadly flattened, thickly terminating the broad prominences, frequently 3- to 5-tipped, 8 - 15 mm, light grey, commonly 2 - 5 cm apart; terminal **Sp** trigonous-subulate, stout, 2.5 - 5 cm, light grey; **Inf** erect, 3 - 6 m, 'spicate', long tapering, flowering in the upper ½ - ⅔, part-**Inf** with 2 - 3 **Fl**; **Fl** 40 - 50 mm; **Ov** fusiform, 20 - 27 mm; **Tep** greenish to pale yellow, tube 3 - 5 mm, lobes ± equal, 15 - 20 mm. – *Cytology:* 2n = 60.

Related to *A. lophantha*. The highly convoluted leaf margins with large flattened several-tipped teeth are like an exaggeration of the forms known from *A. lophantha* (Gentry 1982: 188).

A. zebra Gentry (US Dept. Agric. Handb. 399: 126-130, ills., 1972). **T:** Mexico, Sonora (*Gentry*

21984 [US]). – **Lit:** Turner & al. (1995). **D:** Mexico (Sonora: Sierra del Viejo, Cerro Quituni); limestone slopes, 700 - 1000 m. **I:** Gentry (1982: 508, 517-518). **Fig. VII.g**

[2j] **Ros** rather open, low-spreading, 0.4 - 0.6 × 1 - 1.6 m, mostly solitary; **L** lanceolate, arcuate, thick, rigid, narrowed above the base, broadest near the middle, deeply guttered, scabrous, 50 - 80 × 12 - 17 cm, light grey-glaucous, conspicuously cross-zoned, margins strongly undulate; marginal teeth variously curved, strong, flattened, large, mostly 1 - 2 cm (- middle of the lamina), grey with chestnut-brown tips, 1 - 3 cm apart, bases broad, low, scabrous, on conspicuous prominences; terminal **Sp** acicular, mostly very narrowly grooved above, 3.5 - 7.5 cm, yellowish-brown to light grey, scabrously decurrent for 5 - 10 cm to the uppermost teeth; **Inf** 6 - 8 m, 'paniculate', narrow, part-**Inf** small, 7 - 14 in the upper ⅓ - ¼ of the **Inf**; **Fl** small, 40 - 55 mm; **Ov** slender, 25 - 32 mm, neck slightly constricted, 6-sulcate; **Tep** yellow, tube funnel-shaped, 6 - 7 mm, lobes ± equal, 12 - 15 mm.

BESCHORNERIA

J. Thiede

Beschorneria Kunth (Enum. Pl. 5: 844, 1850). **T:** *Furcraea tubiflora* Kunth & Bouché. – **Lit:** García-Mendoza (1987). **D:** Mexico; dry rocky woodlands to cloud forests. **Etym:** For Friedrich W. C. Beschorner (1806 - 1873), German physician and botanist, director of the Institute of Public Assistance and the Lunatic Asylum at Owinsk, Poland.

Mostly acaulescent or rarely caulescent-arborescent (*B. albiflora*) rhizomatous **Ros** plants, caespitose with age; **L** ± linear-lanceolate, narrowed at the base, base inflated to form a **L** sheath, lamina tough, carinate-canaliculate, glabrous, midrib fleshy, tip a long soft point, margins entire or minutely denticulate; **Inf** racemes or few-branched panicles, straight or arching over, scape and **Bra** red, pink or yellow; **Bra** broad and long, conspicuous, coloured; **Ped** present; **Fl** pendulous, actinomorphic, 2 - 5 together in remote fascicles; **Tep** lanceolate, free but connivent to form a tube-like structure, greenish, yellowish or red, apical part slightly spreading; **ITep** carinate on the outside, papillose or puberulous on the inside; **Fil** filiform, slightly thickened at the base, papillose, ± as long as the **Tep**; **Anth** ± oblong; pollen released in monads or tetrads; **Ov** inferior, oblong, trigonous, 6-sulcate, 3-locular; **Sty** filiform, papillose, as long as or exceeding the **St**, basally swollen into 3 ridges; **Sti** obscurely 3-lobed, ciliate; **Fr** ± cylindrical loculicidal capsules; **Se** plano-convex, flat, shining black to blackish. – *Cytology:* x = 30.

With its long and merely soft leaves and the inflorescences with large colourful bracts, the genus is atypical for the family and might be mistaken as belonging to the *Bromeliaceae*, with some similarity to, e.g., *Billbergia*. García-Mendoza (1987) produced a full taxonomic revision of the genus, but this apparently remained unpublished.

In warmer regions, plants in good condition may flower each year from previous years' suckers, whereas flowering of greenhouse plants in northern regions may occur in long intervals only. The plants are just winterhardy in the warmest parts of the British Isles.

The following names are of unresolved application but are referred to this genus: *Beschorneria californica* hort. (s.a.) (*nom. inval.*, Art. 29.1); *Beschorneria dubia* Carrière (1877); *Beschorneria galeottii* Jacobi (1864); *Beschorneria glauca* hort. ex W. Watson (1889); *Beschorneria multiflora* hort. ex K. Koch (1859); *Beschorneria parmentieri* Jacobi (1864); *Beschorneria pumila* Jacobi (1864); *Beschorneria schlechtendalii* Jacobi (1864); *Beschorneria superba* Hort. Hamburg ex Baker (1888); *Beschorneria verlindeniana* Jacobi (1864).

B. albiflora Matuda (Anales Inst. Biol. UNAM, Ser. Bot. 43(1): 51-55, 1974). **T:** Mexico (*MacDougall 359-A* [MEXU]). – **Lit:** Lott & García-Mendoza (1994). **D:** Mexico (Oaxaca, Chiapas), Guatemala; evergreen forest or scrub.

Incl. *Beschorneria chiapensis* Matuda (1986) (*nom. inval.*, Art. 32.1c).

Caulescent-arborescent, stems 0.5 - 3 (-8) m, erect, sometimes prostrate; **L** erect, gradually narrowed towards the base, chartaceous, 60 - 90 (-125) × 5.5 - 7 (-10) cm, green-glaucous, tip not hardened, margins entire, rarely denticulate; **Inf** open, 1.5 - 2.5 m; scape red; **Bra** red; **Ped** 10 - 35 mm; **Fl** (50-) 60 - 85 mm, 2 - 4 grouped together; **Tep** linear-spatulate to linear-oblong, 25 - 35 (-45) × 2 - 5 mm, connivent, tips slightly spreading, red; **St** as long as the **Tep**; **Anth** 5 - 7 mm; **Ov** 25 - 40 × 3 - 6 mm, intensely red; **Fr** 50 - 70 × 20 - 30 mm. – *Cytology:* 2n = 60.

This is the only arborescent species in the genus. All others with the exception of the short-stemmed *B. wrightii* and *B. tubiflora* are acaulescent. – *B. chiapensis* appears to belong here.

B. calcicola García-Mendoza (Herbertia, ser. 3, 42: 28-30, ills., 1986). **T:** Mexico, Oaxaca (*García-Mendoza & Lorence 720* [MEXU, ENCB]). – **D:** Mexico (SE Puebla, NW Oaxaca, Veracruz); limestone rocks, 1900 - 2400 m, flowering May to August / September.

Acaulescent; **Ros** dense; **L** erect or sometimes ± recurved, linear, conduplicate, rigid, upper face scabrous and carinate, lower face somewhat rough, 30 - 50 × 0.3 - 0.6 cm, glaucous (drying greenish-yellow), **L** sheath triangular, 4 - 6 × 1.5 - 2 cm, yellowish, **L** margins denticulate; **Inf** racemose, 1.15 - 2.3 m, with 16 - 30 **Fl**, scape 0.8 - 1 m, pinkish or yellowish; scape **Bra** 7 - 13, pink; floral **Bra** 10 - 17, pinkish to scarious; **Ped** 0.7 - 2 cm, articulate at the tip; **Fl** pendulous, (35-) 40 - 50 mm, 1 - 2 (-3) grouped together; **Tep** linear to linear-spatulate, (20-) 25 - 33 × 2 - 5 mm, outside puberulent, inside papillose, with a conspicuous midrib, connivent to form a tubular structure with only the tips spreading, tips and inside white, outside pink or yellowish; **Fil** subulate, dilated at the base for 1 - 1.5 mm, papillose; **Anth** linear-oblong, 3 - 6 mm, pale green; **Ov** oblong to subglobose, 6-angled, puberulous, 13 - 20 × 2 - 4 mm, pinkish or yellowish; **Sty** exceeding the **St**, sometimes exceeding the **Tep**, papillose; **Fr** erect, subglobose, 20 - 28 × 15 - 18 mm; **Se** 5 - 7 × 4 - 5 mm, shiny black.

Most closely allied to *B. tubiflora* according to the protologue.

B. rigida Rose (CUSNH 12: 262, 1902). **T:** Mexico, San Luis Potosí (*Palmer* 593 [US 570098]). – **D:** Mexico (Guanajuato, Puebla, San Luis Potosí, Tamaulipas).

Acaulescent; **L** numerous, erect, rather rigid, roughened on both faces, 30 × ≤ 2 cm, narrowing into a long-acuminate tip; **Inf** ± 1 m; **Bra** large, 15 - 20 cm, purplish; **Fl** 45 mm, in groups of 2 - 4; **Tep** somewhat scabrous, dull, usually greenish-yellow; **St** shorter than the **Tep**; **Fr** oblong, 3 cm; **Se** black.

Insufficiently known. According to the protologue, this taxon was first regarded as belonging to *B. tubiflora*, but differs in its narrower erect leaves, which are rough on both faces, and more numerous dull flowers.

B. septentrionalis García-Mendoza (Cact. Suc. Mex. 33(1): 3-5, ills, (2): 52 [erratum], 1988). **T:** Mexico, Tamaulipas (*García-Mendoza & Ramos* 2903 [MEXU]). – **D:** Mexico (Tamaulipas); cloud forest, above 1400 m.

Acaulescent, rhizomatous; **Ros** caespitose; **L** 10 - 20, recurved, oblanceolate, basis dilated, glabrous on both faces, 70 - 90 (-105) × (5-) 6 - 9 (-13) cm, 1.8 - 2.5 (-3.3) cm broad at the basal constriction, brilliant green, tip shortly acuminate, margins finely denticulate with 1 - 3 (-4) denticles per mm; **Inf** cymose-paniculate, 1.5 - 2.5 m, with 4 - 7 part-**Inf** 9 - 25 (-50) cm long, overall with 90 - 130 **Fl**; scape carmine-red; scape **Bra** 4 - 5, oblanceolate, to 30 cm, carmine-red; floral **Bra** 12 - 30, lanceolate to deltoid, reddish to translucent; **Ped** (1-) 3.5 - 4.5 (-6) cm; **Fl** (50-) 55 - 60 (-65) mm, 2 - 4 grouped together; **Tep** linear-oblong to oblong-spatulate, inside papillose, outside glabrous, 25 - 30 × 2 - 8 mm, carmine-red, tip and margins yellowish; **Fil** subulate, papillose, 2 - 4 mm shorter than the **Tep**; **Anth** oblong-elliptic, 5 - 7 mm; **Ov** slightly 6-sulcate, 25 - 30 (-33) × 2 - 8 mm, carmine-red; **Fr** ovate, 35 - 50 (-65) × 25 - 35 mm, green; **Se** shining black.

Closest to *B. yuccoides* according to the protologue.

B. tubiflora (Kunth & Bouché) Kunth (Enum. Pl. 5: 844, 1850). – **D:** Mexico (San Luis Potosí, Hidalgo); pine-oak forest.

≡ *Furcraea tubiflora* Kunth & Bouché (1847); **incl.** *Beschorneria toneliana* Jacobi (s.a.); **incl.** *Beschorneria tonelii* Jacobi (1874).

Acaulescent or short-stemmed; **Ros** with 12 - 20 **L**; **L** tufted, ± recurved, linear, thickened and triangular at the base, ± contracted into a flat and thick pseudopetiole below the middle, long- or short-acuminate, scabrous on both faces, with fine longitudinal stripes, ± 30 - 60 × 2.5 - 5 (-6.25) cm, glaucous-green to very glaucous, margins denticulate; **Inf** ± 1 - 1.2 m, scape bright red-purple, terminating in an erect simple raceme, **Br** drooping (?); **Bra** lanceolate, to 12, purple-red; **Fl** pendulous, 2 - 4 together; **Tep** free, to 25 mm, longer than the **St**, reddish-green or brownish-green or purple and red, glabrous on the outer face; **Ov** to 13 mm; **Fr** not known. – *Cytology:* 2n = 60.

The report of *B. tonelii* for Chiapas (Breedlove 1986) most certainly represents a misidentification for *B. albiflora*.

B. wrightii Hooker *fil.* (CBM 127: t. 7779 + text, 1901). – **D:** Mexico (México).

Incl. *Beschorneria pubescens* A. Berger (1906).

Stems to ± 45 cm; **Ros** with ± 50 **L**; **L** large, densely crowded, spreading or recurved, ensiform, rather stiff and fleshy along the midrib, 60 - 150 × 5 cm, glaucous, nearly smooth below, rough only near the tip, base broadened, very thick and biconvex, tip narrowed into a long brown stiff point, margins very narrowly scarious, finely and deeply denticulate; **Inf** 1.2 - 2 m, pyramidal, rather slender, richly branched, scape bright red, tall; **Bra** ovate; **Fl** in fascicles; **Tep** greenish, fading to yellow, (weakly) pubescent.

Allied to *B. yuccoides* ssp. *dekosteriana* according to Oliver & Bailey (1927).

B. yuccoides K. Koch (Wochenschr. Gärtnerei Pflanzenk. 2: 337, 1859). – **D:** Mexico (Hidalgo, Puebla, Veracruz); 2700 - 3000 m.

B. yuccoides ssp. **dekosteriana** (K. Koch) García-Mendoza (Monocot. Mexic. Syn. Florist. 1(1): 30, 1993). – **D:** Mexico (Hidalgo, Puebla, Veracruz).

≡ *Beschorneria dekosteriana* K. Koch (1864); **incl.** *Beschorneria decosteriana* Baker (1883) (*nom. inval.*, Art. 61.1); **incl.** *Beschorneria argyrophylla* hort. *ex* W. Watson (1889) (*nom. inval.*, Art. 32.1c).

Differs from ssp. *yuccoides*: **Ros** with 15 - 20 **L** or more; **L** long acuminate, tapering very gradually from the middle towards both ends, ± 60 × 6 - 7.5 cm, light grey, glaucous; **Inf** scape light brown; **Fl** pendulous, ± 38 mm; **Fr** clavate.

The name was first used by García-Mendoza (1987) in his unpublished revision of the genus, but the combination was only validated in 1993.

B. yuccoides ssp. **yuccoides** – **D:** Mexico (Hidalgo); 2700 - 3000 m. **I:** Matuda (1967: as *B. hidalgorupicola*); Eggli (1994: 87, as *B. bracteata*). **Fig. VIII.a, VIII.b**

Incl. *Beschorneria bracteata* Jacobi *ex* Baker (1864); **incl.** *Beschorneria viridiflora* Hort. Hamburg *ex* Anonymus (1892); **incl.** *Beschorneria hidalgorupicola* Matuda (1967).

Acaulescent; **L** 20 - 35, erect, linear-lanceolate, base broadened, then narrowed to 1.25 cm above the base, attenuate, (sub-) coriaceous, upper face glabrous, lower face scabrous, 40 - 60 (-90) × 3.3 - 3.5 (-10) cm, grey-green to green, glaucous, tip acuminate, margins finely denticulate; **Inf** 1 - 1.8 (-3.2) m, scape dark red, tip overhanging at first, later erect, part-**Inf** in the upper ⅔, up to 20 and up to 30 cm long, drooping, red to reddish brown; **Bra** red; **Ped** short, 0.4 - 3.5 (-30) mm; **Fl** 40 - 50 mm, glabrous to glabrescent; **Tep** linear-oblong, oblong or spatulate, acute, 33 - 40 × 3.5 - 7 mm, free but narrowly connivent to form a tube-like structure, red or green-yellowish and with reddish tinge, tip green, upper ± 10 mm somewhat spreading and slightly pilose; **Fil** filiform, 35 mm, included; **Anth** 3.5 - 6 mm; **Ov** cylindrical, 20 × 5 mm, dark red, neck slightly constricted; **Sty** filiform; **Sti** slightly pilose; **Fr** oblong to subglobose, 30 - 40 × 15 - 25 mm; **Se** flat, 7 - 8 × 3.5 - 5 mm, black. – *Cytology:* 2n = 60.

Differs from *B. tubiflora* by the broader and shorter acuminate leaves, which narrow more prominently below the middle (Oliver & Bailey 1927). *B. hidalgorupicola* is included in the synonymy here following Galván Villanueva (1990). This is the only taxon of the genus more often seen in cultivation. The plant is hardy outdoors in the Mediterranean or even in S England (Ullrich 1991e).

FURCRAEA

J. Thiede

Furcraea Ventenat (Bull. Sci. Soc. Philom. Paris 1: 65, 1793). **T:** *Agave cubensis* Jacquin [Lectotype, designated by Britton, Fl. Bermuda, 80, 1918 (fide ING).]. – **Lit:** Drummond (1907); Trelease (1910); Trelease (1915a); Trelease (1915b); Trelease (1920); Lott & García-Mendoza (1994); Álvarez de Zayas (1996a). **D:** C and S Mexico, C America to Panama, Caribbean Region, Colombia, Venezuela, Peru, Bolivia, Brazil, Paraguay (the S American range apart from Colombia, Peru and Bolivia, as well as major parts of the Caribbean distribution might be exclusively anthropogenic). **Etym:** For Antoine F. de Fourcroy (1755 - 1809), French politician and chemist, 1784 director at the Jardin des Plantes in Paris.

Incl. *Fourcroea* Haworth (*nom. inval.*, Art. 61.1).

Incl. *Roezlia* hort. *non* Regel (*nom. inval.*, Art. 29.1).

Incl. *Funium* Willemet (1796). **T:** *Funium pitiferum* Willemet.

Incl. *Furcroea* De Candolle (1806) (*nom. inval.*, Art. 61.1).

Incl. *Furcroya* Rafinesque (1814) (*nom. inval.*, Art. 61.1).

Incl. *Fourcroya* Sprengel (1817) (*nom. inval.*, Art. 61.1).

Plants strictly monocarpic; stem none or a thick trunk to 6 m; **L** densely crowded, large, lanceolate, long and narrow, thin and flexible or rather thick and stiff, tip a short firm point, margins entire, denticulate or coarsely toothed; **Inf** tall lax terminal panicles to 13 m, part-**Inf** on long lateral **Br**, often bulbilliferous; **Fl** pendulous, bracteate, pedicellate, solitary or fasciculate in groups of 2 - 5, often all or in part replaced by bulbils; **Tep** principally equal but **ITep** often (generally?) ± larger, ovate-oblong, almost free to the base, white or greenish-white; **Fil** 3 + 3, shorter than the **Tep** and affixed to their bases, dilated below the middle, subulate distally, included; **Anth** linear-oblong, dorsifixed; **Ov** inferior, oblong, usually shortly rostrate at the tip; **Sty** columnar, swollen into 3 basal ridges; **Sti** small, capitate or shortly trilobate; **Fr** oblong or ovoid loculicidal 3-valvate capsules; **Se** flattened, black.

The genus is in urgent need of a critical revision and embraces many ill-known taxa of uncertain circumscription. Some species recognized may merely represent early anthropogenic selections, cultivars or hybrids. With a maximum length of 13 m (Verhoek-Williams 1998), the genus apparently has the largest inflorescences of any plant.

Ullrich (1991h) regards the publication of the generic name *Furcraea* as cited above as not effectively published under Art. 29.1 of the ICBN, since printed material was distributed only to correspondents of the "Société Philomatique" and not to the 'general public'. According to Ullrich, the name was only effectively published in the reprint of 1802, and so becomes antedated by *Funium* Willemeet. Ullrich's interpretation is, however, not followed here, since distribution to correspondents of a society complies with the provisions of Art. 29.1 (if not, numerous periodicals published by societies would have to be regarded as 'not effectively published').

Furcraea can (possibly artificially) be divided as follows:

[1] Sect. *Furcraea* (incl. Sect. *Spinosae* Drummond 1907, *nom. inval.*, Art. 22.2): Stems none or short and < 2 m; **L** firm-textured, green or a little glaucous, not striate, margins with conspicuous ± distant teeth (occasionally teeth few or absent; **Bra** much shorter than the **Ped**. – Possibly a paraphyletic hold-all.

[2] Sect. *Serrulatae* Drummond 1907 (incl. Subgen. *Roezlia* Baker 1888; incl. Ser. *Flexiles* Baker 1879): Stems conspicuous, plants sometimes arborescent; **L** flexible, glaucous, striately rou-

ghened, margins closely beset with minute denticles; **Inf** pubescent; **Bra** much longer than the **Ped**.

The following names are of unresolved application but are referred to this genus: *Agave aspera* Jacquin (1762) ≡ *Furcraea aspera* (Jacquin) M. Roemer (1847); *Agave noackii* Jacobi (1865) ≡ *Furcraea noackii* (Jacobi) hort. *ex* Baker (1877); *Agave stenophylla* Jacobi (1866); *Agave vivipara* Miller (1768) (*nom. illeg.*, Art. 53.1); *Agave vivipara* Willdenow (1799) (*nom. illeg.*, Art. 53.1); *Agave vivipara* Arruda da Cámara (1810) (*nom. illeg.*, Art. 53.1); *Furcraea agavephylla* Brotero *ex* Schultes (1829); *Furcraea aitonii* Jacobi (1869); *Furcraea albispina* Hort. Panorm. *ex* Baker (1893); *Furcraea altissima* Todaro *ex* Franceschi (1900); *Furcraea atroviridis* Jacobi & Goeppert (1866); *Furcraea cubensis* Haworth (1819) (*nom. illeg.*, Art. 53.1) ≡ *Agave cubensis* (Haworth) Sprengel (1825) (*nom. illeg.*, Art. 53.1); *Furcraea cubensis* var. *inermis* Baker (1881); *Furcraea demouliniana* Jacobi (1867); *Furcraea depauperata* Jacobi (1866); *Furcraea elegans* Todaro (1876); *Furcraea ghiesbreghtii* hort. *ex* Jacobi (1867); *Furcraea lipsiensis* Jacobi (1869); *Furcraea macra* Hort. Parmentier *ex* Jacobi (1866); *Furcraea pugioniformis* Hort. Verschaffelt *ex* Todaro (1876); *Furcraea rigida* Landry *ex* Jacobi (1867) (*nom. illeg.*, Art. 53.1); *Furcraea roezlii* Eichler (1881); *Furcraea roezlii* var. *atropurpurea* Hort. De Smet (1876); *Furcraea sobolifera* Hort. Cels *ex* Jacobi (1867); *Furcraea stricta* Jacobi (1867); *Furcraea valleculata* Jacobi (1867).

F. acaulis (Kunth) B. Ullrich (Quepo 6: 69, 1992). – **D:** Peru. **I:** Ullrich (1992h: as *F. humboldtiana*).
≡ *Yucca acaulis* Kunth (1816); **incl.** *Furcraea humboldtiana* Trelease (1910).
[1] **Ros** acaulescent or with a stem finally reaching 3 m; **L** spreading, lanceolate-ensiform, almost flat, smooth, ± 1.5 m × 12.5 - 15 cm, marginal teeth bifid, reflexed, 3 - 5 mm, usually in divergent pairs from the tops of green prominences, 25 - 62 mm apart, but toothless forms also known; **Inf** 7.2 - 9 (-12) m with a long scape; **Fl** pendent, ± 50 - 62 mm; **Tep** oblong, 32 - 38 mm, ± obtuse, light yellow, **ITep** slightly broader; **Ov** triquetrous, ± 20 - 25 mm; **Sty** triquetrous; **Sti** trifid.
Insufficiently known. See also under *F. tuberosa*.

F. andina Trelease (in L. H. Bailey, Stand. Cycl. Hort. 3: 1305, 1915). **T:** Peru (*Furlong* s.n. [MO]). – **Lit:** Ullrich (1992h: with ills.). **D:** Peru (Ancash, Cuzco, Huanuco, Junín, La Libertad, Lima); Andean grasslands, 1500 - 3500 m. **I:** Rauh (1958: 142, identification uncertain).
Incl. *Furcraea deledevantii* Rivière (1902); **incl.** *Furcraea delevantii* Rivière (1902) (*nom. inval.*, Art. 61.1); **incl.** *Furcraea altissima* hort. *ex* Trelease (1915) (*nom. illeg.*, Art. 53.1).

[1] **Ros** acaulescent; **L** large, oblong-lanceolate, marginal teeth prominent, curved, remote, normally reaching 6 mm or more, almost as large as distant; **Inf** short-stalked, bulbilliferous, bulbils conical-ovoid; **Fr** cuboid.
Insufficiently known. Trelease (1915a) includes the prioritable name *F. deledevantii* here, which is followed by Macbride (1936) and Brako & Zarucchi (1993).

F. antillana A. Álvarez (Anales Inst. Biol. UNAM, Ser. Bot. 67(2): 331-335, ills., 1996). **T:** Cuba, La Habana (*Álvarez* 63654 [HAJB]). – **D:** Greater Antilles: Cuba, Hispaniola, Puerto Rico; mainly semideciduous forests or dry coastal scrub, flowers July to September.
[1] Stems short, to 50 cm, not rhizomatous; **L** numerous, 90 - 110, straight, narrowly lanceolate, nearly plane to slightly canaliculate, slightly folded towards the tip, rigidly coriaceous, often asperous, (0.6-) 0.9 - 1.2 (-2) m × 5 - 10 cm, light green to somewhat yellowish, opaque, margins straight between the teeth, marginal teeth triangular, straight or normally somewhat reflexed, 2 - 5 (-7) mm, chestnut-brown to nearly black, on deltoid prominences, decurrent, 2 - 5 cm (0.4 - 2 cm at the base) apart, sometimes lacking in the upper ⅓ of the **L**, **L** tip acute, not or inconspicuously mucronate; **Inf** 4 - 6 (-8) m, narrowly fusiform, part-**Inf** (20-) 40 - 70 (-90) cm, ascending in the upper ⅔ of the **Inf**, bulbilliferous, bulbils narrowly fusiform; **Fl** 2 - 3 grouped together; **Ped** 4 - 10 mm; **Fl** pendent, campanulate, (25-) 32 - 40 mm; **Tep** elliptic, (12-) 14 - 19 (-27) × 5 - 8 mm, whitish, outside greenish; **Fil** 10 - 20 mm; **Ov** triquetrous, (13-) 18 - 20 mm; **Sty** 10 - 20 mm; **Fr** oblong, beaked, 2.5 - 5 × 1.6 - 3 cm.
Variable in leaf and flower characters according to the protologue, esp. as influenced by different edaphic conditions.

F. bedinghausii Koch (Wochenschr. Vereines Beförd. Gartenbaues Königl. Preuss. Staaten 6(30): 233-235, 1863). **T:** [lecto − icono]: Belg. Hort. 13(11): t. ad p. 327, 1863. – **Lit:** McVaugh (1989: with ill.); Ullrich (1991i). **D:** Mexico (Jalisco, Distrito Federal, Hidalgo, Michoacán, México); mountain slopes and summits, 2650 - 3500 m. **I:** Matuda (1961: 69-70); Benítez B. (1986: 62). **Fig. VIII.c**
≡ *Furcraea longaeva* ssp. *bedinghausii* (Koch) B. Ullrich (1991) (*nom. inval.*, Art. 33.2); **incl.** *Furcraea flaccida* Hort. Panorm. *ex* Hort. Kew (s.a.); **incl.** *Yucca parmentieri* Ortgies (1859); **incl.** *Agave argyrophylla* hort. *ex* Koch (1862); **incl.** *Agave toneliana* Hort. *ex* Morren (1863); **incl.** *Roezlia regia* Hort. *ex* Lemaire (1863); **incl.** *Yucca argyrophylla* Hort. *ex* Lemaire (1863); **incl.** *Yucca toneliana* hort. *ex* Koch (1863); **incl.** *Beschorneria multiflora* hort. *ex* C. Koch (1863) (*nom. illeg.*, Art. 53.1); **incl.** *Roezlia bulbifera* Roezl (1881); **incl.** *Furcraea*

roezlii André (1887) (*nom. illeg.*, Art. 53.1); **incl.** *Yucca pringlei* Greenman (1898).

[2] Stems erect, thick, to 5 - 8 × 0.3 - 0.4 m; **Ros** 2 - 3 m ∅; **L** first ascending-spreading, later spreading to pendent, forming a dry skirt below the **Ros**, lanceolate, stiff, ensiform, narrowed below the middle, long-attenuate to a subulate (but not spiny) tip, flat to concave or plicate, upper face striate and roughened by projections from the longitudinal veins, (35-) 70 - 120 × (4-) 6 - 10 (in the middle) cm, green, somewhat glaucous, marginal teeth minute, irregularly spaced, pale, deltoid, ± 2 teeth per mm; **Inf** erect, pyramidal, (2.5-) 4 - 5 (-8) × up to 2 m, pubescent, part-**Inf** 30 - 65 (-100) cm, freely bulbilliferous, bulbils elongate; scape 0.5 - 1.5 m; **Ped** 3 - 5 mm; **Fl** 40 mm, 2 - 4 grouped together; **Tep** elliptic or oblong-elliptic, 18 - 20 × 5 - 7 mm, greenish-white, outside pilose; **Fr** oblong-ovoid, apiculate, 4 - 7 × ± 3 cm; **Se** 10 - 12 × 6 - 8 mm.

Ullrich (1991i) suggested subspecific rank under *F. longaeva*, to which *F. bedinghausii* appears to be closely related.

F. boliviensis Ravenna (Pl. Life 34: 151-153, ill., 1978). **T:** Bolivia, Mizque (*Ravenna* 2305 [Herb. Ravenna]). – **D:** Bolivia (Mizque); rocky slopes, 2600 - 3500 m, infrequent.

[1] Stems short, stout, sometimes prostrate, 30 - 40 × 10 - 15 cm; **Ros** 0.9 - 1 (incl. stem) × 1 - 1.4 m; **L** often spreading, ensiform, thick, rigid, slightly narrowed near the base, moderately channelled, up to 45 - 55 (rarely more) × 8 - 10 cm, opaquely ashgreen, terminal **Sp** none but **L** tip an acute pungent point, marginal teeth small, uncinate, not exceeding 3 mm in length, rather close together; **Inf** unknown.

According to the protologue related to the Mexican *F. pubescens* (?= *F. undulata*; doubtful and no arguments given) and the Peruvian *F. andina*. *F. boliviensis* is the only native Bolivian species; other species reported for Bolivia (*F. occidentalis* aff., *F. foetida*) appear to represent garden escapes (Ravenna l.c.).

F. cabuya Trelease (Ann. Jard. Bot. Buitenzorg 3(Suppl. 2): 906, t. 36, 45, 1910). **T:** Costa Rica (*Worthen & Dewey* s.n. [ILL]). – **D:** SE Mexico, C America.

F. cabuya var. **cabuya** – **D:** Mexico (Yucatán), Honduras, Nicaragua, Costa Rica, Panama (cultivated only); thorn forests, savannas and pine forests, (50-) 300 - 1400 m, frequently (or exclusively?) cultivated. **I:** Berry (1995).

[1] **Ros** subcaulescent to caulescent, stems to 1 m, covered with old **L**; **L** lanceolate, abruptly narrowed above the base, openly concave, acute, semisucculent, coriaceous, 1.5 - 2 m × 14 - 22 cm, green, young glaucous, margins straight between the teeth, marginal teeth deltoid, strong, normally antrorse, 5 - 8 (-11) mm, yellowish to chestnut-brown, decur-

rent, 2.5 - 4.5 (-5) cm apart, along the whole margin, **L** tip with a conical **Sp** 1 - 3 (-5) × 1 - 1.3 mm, reddish or dark chestnut-brown; **Inf** 5 - 10 m, part-**Inf** to 1 m, finely puberulent to glabrous, sometimes bulbilliferous, bulbils elongated; scape long; **Bra** much shorter than the **Ped**; **Ped** glabrous, 3 - 6 (-12) mm; **Fl** (45-) 55 - 62 mm, 3 - 6 grouped together; **Tep** elliptic, overlapping parts papillose, (26-) 30 - 36 × (8-) 11 - 15 mm, light green to yellowish-green; **ITep** broader, 13 - 18 mm broad; **Fil** 11 - 16 mm; **Ov** 23 - 28 mm; **Sty** 16 - 22 mm; **Fr** unknown.

F. cabuya var. **integra** Trelease (Ann. Jard. Bot. Buitenzorg 3(Suppl. 2): 907, 1910). **T:** Costa Rica (*Worthen* s.n. [ILL?]). – **Lit:** Lott & García-Mendoza (1994). **D:** Costa Rica, Panama; 100 - 600 m, cultivated only.

[1] Differs from var. *cabuya*: **L** normally completely without marginal teeth, or rarely with some teeth near the **L** base, 2 - 3 (-6) mm, terminal **Sp** normally absent, or 2 × 0.7 mm; **Fl** (37-) 45 - 52 mm; **Tep** narrowly elliptic to elliptic, (22-) 25 - 28 × 7 - 14 mm; **Fil** 10 - 14 mm; **Ov** 20 - 25 mm; **Sty** 14 - 19 mm; **Fr** oblong to subquadrangular, 6 × 4.5 cm.

Appears to represent merely an unarmed selection with smaller flowers.

F. cahum Trelease (Ann. Jard. Bot. Buitenzorg 3(Suppl. 2): 908, 1910). **T** [lecto]: Mexico, Yucatán (*Schott* 809 [F]). – **D:** Mexico (Campeche, Quintana Róo, Yucatán); tropical semideciduous forests, to 100 m.

[1] **Ros** shortly caulescent, stem to 1 m; **L** ensiform, narrowed to 2.5 - 3.5 cm above the base, broadly acuminate, plane, 1.6 - 2.1 (-2.4) m × 6.5 - 9 cm, brilliant green, margins ± straight between the teeth, marginal teeth antrorse or straight, 2 - 4 mm, reddish to black, 2 - 4 (-5) cm apart, decurrent over the slightly deltoid base, terminal **Sp** 2 - 6 × 1.5 - 2.5 mm; **Inf** 4 - 5 m, **Br** and **Fl** minutely papillose-puberulent, richly bulbilliferous; scape long; **Bra** much shorter than the **Ped**; **Ped** 4 - 6 (-10) mm, puberulent; **Fl** 40 - 45 (-50) mm, 2 - 4 grouped together; **Tep** elliptic, 20 - 25 × 9 - 12 mm, yellowish-green; **Fil** 10 - 14 mm; **Ov** 20 - 27 mm; **Fr** 5 × 3 - 3.5 cm; **Se** 9 - 12 × 5 - 8 mm.

Cultivated for fibres, and possibly not distinct from *F. hexapetala* (Lott & García-Mendoza 1994).

F. foetida (Linné) Haworth (Synops. Pl. Succ., 73, 1812). **T:** [icono]: Plukenet, Almag. t. 258: fig. 2, 1700. – **Lit:** Lott & García-Mendoza (1994). **D:** C America?, Greater and Lesser Antilles, Trinidad, N South America (mainly or exclusively cultivated); widely cultivated in Africa and Asia.

≡ *Agave foetida* Linné (1753); **incl.** *Furcraea gigantea* var. *medio-picta* Trelease (s.a.); **incl.** *Agave foetida* Aublet (1775) (*nom. illeg.*, Art. 53.1); **incl.** *Agave foetida* Lamarck (1784) (*nom. illeg.*, Art.

53.1); **incl.** *Furcraea gigantea* Ventenat (1793) ≡ *Agave gigantea* (Ventenat) D. Dietrich (1840) (*nom. illeg.*, Art. 53.1); **incl.** *Funium piliferum* Willemet (1796); **incl.** *Furcraea madagascariensis* Haworth (1819) ≡ *Agave madagascariensis* (Haworth) Salm-Dyck (1822); **incl.** *Agave commelynii* Salm-Dyck (1834) ≡ *Furcraea commelynii* (Salm-Dyck) Kunth (1850); **incl.** *Furcraea gigantea* var. *willemetiana* M. Roemer (1847); **incl.** *Furcraea tuberosa* Hasskarl (1856) (*nom. illeg.*, Art. 53.1); **incl.** *Furcraea tuberosa* Hort. Belg. (1860) (*nom. illeg.*, Art. 53.1); **incl.** *Furcraea barrillettii* Jacobi (1869); **incl.** *Furcraea viridis* Hemsley (1885); **incl.** *Furcraea watsoniana* Hort. Sander (1898) ≡ *Furcraea gigantea* var. *watsoniana* (Hort. Sander) Drummond (1907); **incl.** *Furcraea variegata* hort. *ex* Trelease (1915).

[1] Stems none or short; **L** broad, obovate-lanceolate, ± flat, undulate, somewhat asperous below, 1.5 - 2.5 m × 18 - 25 cm, fresh bright green, margins entire, somewhat wavy, basally with a few trigonous hooked teeth, otherwise teeth absent; **Inf** to 8 - 10 m, rather narrow, scape long, richly branched, scarcely to freely bulbilliferous, bulbils short; **Fl** 40 - 50 mm; **Tep** equalling the **Ov**, 20 - 25 mm, greenish-white; **Ov** 20 - 25 mm.

At present not reliably known from Mesoamerica (Lott & García-Mendoza 1994).

F. guatemalensis Trelease (Trans. Acad. Sci. St. Louis 23(3): 149, t. 32, 1915). **T:** Guatemala (*Trelease* 23 [ILL]). – **Lit:** Lott & García-Mendoza (1994). **D:** Guatemala, Belize, Honduras, El Salvador; rocky slopes in pine-oak forest, 700 - 2300 m.

Incl. *Furcraea melanodonta* Trelease (1915).

[1] **Ros** subcaulescent; **L** (narrowly) lanceolate to typically almost ensiform, moderately concave, acute, smooth or slightly roughened, 1.3 - 2 (-2.25) m × 7 - 10 (-15) cm, opaque green to grey, tip with a robust **Sp**, subulate, grooved at the base, 2 × 1 mm, margins somewhat outcurved, straight between the prominences, marginal teeth (strongly) upcurved, decurrent on moderate fleshy elevations, 3 - 5 (-7) mm, 10 - 30 (-45) mm apart, red-brown, at first pale at the base, later becoming chestnut-brown; **Inf** glabrous, open, 2 - 5 m, bulbilliferous, bulbils ovoid-globose, without a leafy tuft; **Bra** much shorter than the **Ped**; **Fl** 40 - 45 mm; **Tep** oblong-elliptic, 20 × 6 - 11 mm, pale green or greenish-white; **Fil** 10 - 12 mm; **Ov** 15 - 20 mm; **Fr** globose-cuboidal, stipitate, beaked, 4 - 5 × 3.5 - 4 cm; **Se** 20 × 12 - 20 mm.

Plants from Mexico (Chiapas) placed here by Lott & García-Mendoza (1994) may represent an undescribed species at present under study (García-Mendoza 1999).

F. guerrerensis Matuda (Anales Inst. Biol. UNAM 36: 114, 1966). **T:** Mexico, Guerrero (*Kruse* 8 [MEXU]). – **Lit:** McVaugh (1989). **D:** Mexico

(Guerrero); oak forest, 500 m; only known from the type collection. **Fig. VIII.f**

[1] Stems none; **L** 25 - 35, narrowly ensiform, bases dilated, broadly acuminate, concave in the upper part, nearly plane in the lower part, to 1.5 - 1.75 m × 12 - 15 cm, dark green on both faces, marginal teeth incurved, deltoid, chestnut-brown, 1 - 2 cm apart, terminal **Sp** very small, hardly 1 - 1.5 mm, chestnut-brown; **Inf** erect, lax, 8 - 10 m; scape **Bra** distant; **Fl** pedicellate; **Tep** unequal, semirhomboid, greenish-yellow; **OTep** 30 - 35 × 12 mm; **ITep** 35 × 20 mm; **Fil** basally dilated, 15 mm; **Anth** oblong, 7 mm; **Sty** columnar; **Sti** hardly capitate.

Closely related to *F. guatemalensis*, but distinguished by its much shorter leaves, the columnar (instead of triquetrous) style and incurved instead of straight marginal teeth (Matuda l.c.). McVaugh (1989) depicts and describes a plant from S Nayarit, Jalisco, Colima and México possibly belonging here.

F. hexapetala (Jacquin) Urban (Symb. Antill., 4: 152, 1903). **T:** Cuba, La Habana (*Jacquin* s.n. [BM]). – **Lit:** Álvarez de Zayas (1996a: with ills.). **D:** Bahamas, Greater Antilles (W Cuba, Jamaica, Hispaniola); semideciduous forests and xeromorphic scrub, esp. abundant on anthropogenic sites, to 750 (-1250) m, flowers September to January.

≡ *Agave hexapetala* Jacquin (1760); **incl.** *Agave odorata* Persoon (s.a.); **incl.** *Agave cubensis* Jacquin (1763) ≡ *Furcraea cubensis* (Jacquin) Ventenat (1793) ≡ *Furcroya cubensis* (Jacquin) Ventenat (1796) (*incorrect name*, Art. 11.3); **incl.** *Furcraea macrophylla* Baker (1899).

[1] Stems thick, to 1 m tall, rhizomatous, sometimes with numerous basal offsets; **L** up to 80, straight, lanceolate, nearly plane in the centre, canaliculate towards the tip, slightly scabrous on the lower face, coriaceous, (1-) 1.15 - 1.75 (-2) m × 8 - 10 (-15) cm, bright green, **L** tip canaliculate, acute, inconspicuously mucronate, marginal teeth strongly upcurved, 6 - 11 mm, on deltoid bases, normally decurrent, 3 - 7 (-12) cm apart, yellowish to brownish; **Inf** to 8 (-10) m, broad, deltoid, part-**Inf** lax, in the upper ¾ of the **Inf**, to 1.6 m, pyramidally branched, bulbilliferous, bulbils ovoid, to 45 × 25 mm; **Ped** 4 - 10 mm; **Fl** solitary or clustered, pendulous, 2 - 4 grouped together, campanulate, (30-) 38 - 46 (-50) mm; **Tep** oblong, (17-) 21 - 25 (-30) × 6 - 10 mm, whitish; **Fil** 15 - 30 mm; **Ov** 17 - 21 mm; **Sty** 15 - 30 mm; **Fr** broadly oblong, base constricted and deeply sulcate, tip beaked, 3 - 5 × 2.5 - 4 cm; **Se** numerous, flat, 12 - 14 × 4 - 6 mm.

See note under *F. cahum*.

F. longaeva Karwinsky & Zuccarini (Flora 15: 2 (Beiblatt 2): 94-95, 1832). **T:** [icono]: Acta Acad. Leop.-Carol. Nat. Cur. 16(2): 666-668, t. 48, 1833. – **Lit:** Trelease (1915b: with ills.); Ullrich (1991i: with ills.). **D:** Mexico (Guerrero, Oaxaca, Puebla).

Incl. *Beschorneria floribunda* hort. *ex* Koch (1862); **incl.** *Furcraea longa* J. J. Smith (1897) (*nom. illeg.*, Art. 52.1?).

[2] **Ros** caulescent, stems tall, to 5 m or more, unbranched; **L** rigidly outcurved, (narrowly) lanceolate, subacuminate, concave, to 2 m × 8 - 15 cm, grey, margins with minute denticles; **Inf** 5 - 13 m, broadly conical, bulbils unknown; scape short; **Fl** 30 - 40 mm, pubescent; **Tep** rather shorter than the **Ov**; **Ov** 20 - 25 mm; **Fr** oblong, narrowed below; **Se** 4 × 6 mm.

The species appears to exhibit the largest inflorescences of any plant, and Verhoek-Williams (1998) mentions 13 m as maximum size. Plants may flower already after 25 (or perhaps even 7 or 8?) years, in contrast to earlier estimations of up to 400 years (Ullrich 1991i).

F. macdougallii Matuda (Cact. Suc. Mex. 1(2): 24-26, ills., 1955). **T:** Mexico (*MacDougall* 269 [MEXU]). – **Lit:** Lott & García-Mendoza (1994). **D:** Mexico (Puebla, Oaxaca); tropical deciduous and thorn forests on calcareous soils, 800 - 1000 m.

[1] Arborescent, stems 6 - 9 m, slender and unbranched; **L** numerous, young **L** erect to patent, old **L** reflexed and persistent, linear, gradually narrowed towards the base, concave, coriaceous, gradually acuminate, scabrous on both faces, 1.2 - 1.45 m × 6 - 7 cm, green, tip slightly hardened, rounded, reddish, margins strongly armed, teeth small, 1 - 3 mm; **Inf** 5 - 8 m, robust, much-branched, part-**Inf** puberulent to tomentose, 1 - 1.5 m, in the upper ½ of the **Inf**; **Bra** much shorter than the **Ped**; **Ped** 5 - 10 mm, puberulent; **Fl** 37 - 40 mm, 2 - 4 grouped together; **Tep** narrowly elliptic to elliptic, papillose, 15 - 22 × 3 - 6 mm, inside green, outside white; **Fil** 10 - 13 mm; **Ov** trigonous, cylindrical, ± 20 mm, with a neck 5 - 8 mm ∅; **Sty** abruptly dilated below but not strongly trigonous, 13 - 16 mm; **Fr** oblong-trigonous, coriaceous, 5 - 7 × 3 - 3.5 cm, inner face yellowish, outer face blackish.

A sterile collection from Chiapas first provisionally placed here by Lott & García-Mendoza (1994) was later described as a new species, *F. niquivilensis*.

F. niquivilensis Matuda *ex* García-Mendoza (Novon 9(1): 42-45, ills., 1999). **T:** Mexico, Chiapas (*García-Mendoza & al.* 6441 [MEXU, ENCB, K, MO]). – **D:** Mexico (Chiapas); at present only known from cultivation at settlements, 1800 - 2650 m.

[2] Arborescent, stems 1 - 3 × 0.3 - 0.4 m, unbranched, covered with old dry **L**; **Ros** 4 - 5 m ∅; **L** 80 - 150, erect, lanceolate, fibrous, coriaceous, scabrous or muricate on both faces, (1.7-) 1.9 - 2.1 m × 12 - 14 (base 7 - 8.5) cm, green, tip mucronate, 1 - 4 mm, dark chestnut-brown, margins straight, teeth antrorse or erect at the base, 5 - 6 (-8) × 3 - 4 (at the base) mm, chestnut-brown, base yellowish,

decurrent, on small prominences, (1-) 2 - 4 cm apart (middle of the **L**); **Inf** 6 - 9 m, pyramidal, puberulent, part-**Inf** to 2.3 m, bulbilliferous, bulbils (4-) 5.5 - 6.5 × (3-) 4.5 - 6 (-6.5) cm; **Ped** 5 - 10 mm, puberulent; **Fl** (7-) 7.5 - 8 cm, 1 - 3 together; **Tep** oblong, glabrous, but **ITep** papillose on the overlapping parts, (30-) 40 - 45 × 11 - 14 mm, greenish-white, outer faces tinged reddish; **Fil** 20 - 25 mm; **Ov** cylindrical, puberulent, 35 - 38 × 4 - 6 mm, green; **Sty** dilated below, trigonous, papillose, 25 - 28 mm; **Fr** and **Se** unknown.

Sterile collections of this plant were provisionally included under *F. macdougallii* (Lott & García-Mendoza 1994), but the clear differences in vegetative and esp. fertile features merit species status. According to the protologue, both species appear to be closely related and share leaves with both faces scabrous, as well as puberulent inflorescences, pedicels and ovaries.

F. occidentalis Trelease (BJS 50 (Beiblatt 111): 5, 1913). **T:** Peru (*Weberbauer* 1687 [B?]). – **Lit:** Ullrich (1992h). **D:** Peru (Ancash, Huanuco, Loreto, Lima); rocky slopes, 500 - 2500 m.

[1] Stems none or short (?); **L** narrowly oblong, ± 65 × 10 cm, margins minutely aculeate, teeth ± deltoid, straight or slightly retrorse, minute, 1 mm, yellow-brown or blackish, terminal **Sp** obtuse and semiglobose, minute, 0.5 × 1 mm; **Inf** 6 m, glabrous, freely bulbilliferous; **Fl** ± 50 mm; **Tep** ± 30 mm, greenish-white; **Ov** ± 20 mm.

Insufficiently known. Most of the references to *F. cubensis* by Weberbauer (1911) concern this species (Macbride 1936). It is a typical element of the W hill country of Peru (Macbride 1936).

F. quicheensis Trelease (Trans. Acad. Sci. St. Louis 23(3): 148, t. 29, 1915). **T:** Guatemala (*Cook* 421 [US 692146]). – **Lit:** Lott & García-Mendoza (1994). **D:** Mexico (Chiapas), Guatemala; oak forests, 2300 - 3300 m.

[1] **Ros** caulescent, stems 1 - 1.5 (-2) m, single or with few **Br**; **L** linear-lanceolate, gradually narrowed towards the base, broadly attenuate, applanate, subcoriaceous, asperous, 0.9 - 1.2 (-1.8) m × 6 - 11 cm, green-glaucous, tip narrowly rounded and obtuse, hardened, margins narrow, subcartilaginous, yellow, with straight minute yellowish denticles 1 - 2 mm apart; **Inf** narrow, 2 - 5 × ≤ 1 m, lower part-**Inf** much reduced, otherwise part-**Inf** < 60 cm; bulbils unknown; **Bra** 2 - 3× longer than the **Ped**; **Ped** glabrous, 20 - 35 mm, reddish; **Fl** (40-) 50 - 70 mm, 6 - 10 grouped together; **Tep** elliptic, 30 - 60 × 10 mm, green-yellowish, margins white, glabrous; **Fil** 10 - 13 mm; **Ov** glabrous, 20 - 35 mm; **Sty** ± 15 mm; **Fr** oblong, shortly rostrate, 5 - 8 × 2 - 3 cm; **Se** 10 × 6 mm.

Cultivated for its fibres. Ullrich (1991i) suggests subspecific rank under *F. longaeva*, to which *F. quicheensis* is apparently closely related.

F. samalana Trelease (Trans. Acad. Sci. St. Louis 23(3): 149, t. 30-31, 1915). **T:** Guatemala (*Trelease 20* [ILL]). − **Lit:** Lott & García-Mendoza (1994). **D:** Mexico (Chiapas), Guatemala, El Salvador; rocky slopes in scrub or pine forests, 200 - 2700 m.

[1] Stems none or up to 0.5 (-2) m; **L** lanceolate, upper part long-acute, channelled, almost smooth, 1 - 2 m × 10 - 15 cm, green or very slightly greyish, margins broad, flatly outcurved, concave and horny between the teeth, marginal teeth slender, mainly incurved, decurrent over low fleshy elevations, up to 7 mm, reddish chestnut-brown?, (1-) 2 - 5 (-6) cm apart, lacking in the upper ½ to ⅔ of the **L**, terminal **Sp** normally lacking, or 1 - 2 mm, reddish; **Inf** 3 - 5 m, oblong-paniculate, part-**Inf** in the upper ¾ of the **Inf**, sometimes with abundant large bulbils, these conical-ovoid, with a tuft of **L**; **Bra** much shorter than the **Ped**; **Ped** glabrous, 3 - 6 (-9) mm; **Fl** 50 - 55 mm, 1 - 3 grouped together; **Tep** elliptic to broadly elliptic, glabrous, (25-) 30 - 40 × 7 - 17 mm, greenish-yellow; **Fil** 12 - 14 mm; **Ov** 16 - 25 × 2 - 4 mm.

Cultivated for its fibres.

F. selloa K. Koch (Wochenschr. Vereines Beförd. Gartenbaues Königl. Preuss. Staaten 3: 22, 1860). − **D:** Not recorded. **I:** Jacobsen (1981: t. 85: 1-3).

Incl. *Furcraea selloa* var. *edentata* Trelease (s.a.) ≡ *Furcraea selloa* fa. *edentata* (Trelease) H. Jacobsen (1954) (*nom. inval.*, Art. 33.2); **incl.** *Furcraea selloa* var. *marginata* Trelease (s.a.); **incl.** *Agave cubensis* var. *striata* hort. (s.a.) (*nom. inval.*, Art. 29.1); **incl.** *Furcraea flavoviridis* Hooker (1860); **incl.** *Furcraea lindenii* Jacobi (1869); **incl.** *Furcraea lindenii* André (1874) (*nom. illeg.*, Art. 53.1) ≡ *Furcraea cubensis* var. *lindenii* (André) Hort. Kew (1897); **incl.** *Furcraea tuberosa* Franceschi (1900) (*nom. illeg.*, Art. 53.1).

[1] Stems finally to 0.9 - 1.5 m; **L** numerous, spreading, narrowly lanceolate, ensiform, much narrowed towards the base, concave and revolute or plicate, very asperous, ± 1 - 1.25 m × 7 - 10 cm, somewhat shining dark green, marginal teeth large, 5 - 6.5 mm, ± 3.3 - 4 cm apart, hooked, variously curved, brown; **Inf** to 6 m tall, glabrous, laxly branched, freely bulbilliferous; **Fl** 40 - 65 mm; **Tep** ± 25 mm; **Ov** ± 17 mm.

Described from cultivated material apparently originating from Quetzaltenango, Guatemala, but at present not certainly known from C America (Lott & García-Mendoza 1994). Several variegated or toothless horticultural variants have been described. Even if *F. flavoviridis* (publ. Feb. 1860) should be definitely conspecific, it is antedated by *F. selloa* (publ. Jan. 1860) (Drummond 1907).

F. stratiotes J. B. Petersen (Bot. Tidsskr. 37: 306, 1922). **T:** Nicaragua (*Oersted s.n.* [C]). − **D:** Nicaragua.

[1] Stems none or **Ros** subacaulescent; **L** ± 50,

linear-lanceolate, acuminate, narrowed to 1.5 - 2 cm above the base, 35 - 53 × 2.5 - 3.5 cm, glaucous, tip mucronate, margins straight between the teeth, marginal teeth geminate, 1.5 - 3 mm, 1.5 - 2.5 (-4.5) cm apart; **Inf** 2.8 m, panicle 80 cm, with 3 part-**Inf**, bulbilliferous, bulbils to 3.5 × 1.5 cm, strongly compressed (in pressed specimens only ?) with 3 - 5 **L**; **Bra** small, acuminate, entire, much shorter than the **Ped**; **Fl** 22 mm, solitary; **Tep** 14 × 6 mm, whitish, **ITep** somewhat broader; **Fil** 2.5 mm; **Ov** 8 × 2 mm; **Sty** 5 mm; **Fr** unknown.

Only known from the type material based on plants cultivated in Copenhague (Lott & García-Mendoza 1994).

F. tuberosa (Miller) W. T. Aiton (Hort. Kew., ed. 2 2: 303, 1811). **T:** [icono]: Plukenet, Almag., 19, 1700. − **Lit:** Drummond (1907: with ills.); Álvarez de Zayas (1996a: with ills.). **D:** Greater Antilles (Cuba, Jamaica, Hispaniola, Puerto Rico), Lesser Antilles (all islands); frequent near roads or settlements, flowering December to March.

≡ *Agave tuberosa* Miller (1768); **incl.** *Agave tuberosa* Lamarck (1784) (*nom. illeg.*, Art. 53.1); **incl.** *Agave tuberosa* Aiton (1789) (*nom. illeg.*, Art. 53.1); **incl.** *Furcraea spinosa* O. Targioni Tozzetti (1808) ≡ *Agave spinosa* (O. Targioni Tozzetti) Steudel (1840); **incl.** *Yucca superba* Roxburgh (1814); **incl.** *Agave gigantea* Tussac (1818); **incl.** *Agave vivipara* Maycock (1830) (*nom. illeg.*, Art. 53.1); **incl.** *Agave cubensis* Hasskarl (1856) (*nom. illeg.*, Art. 53.1); **incl.** *Furcraea geminispina* Jacobi (1866) ≡ *Furcraea tuberosa* var. *geminispina* (Jacobi) Trelease (1927); **incl.** *Furcraea interrupta* Hort. van Houtte *ex* Jacobi (1869); **incl.** *Furcraea tuberosa* Fenzl *ex* Baker (1879) (*nom. illeg.*, Art. 53.1); **incl.** *Agave gigantea* Baker (1888) (*nom. illeg.*, Art. 53.1); **incl.** *Agave campanulata* Sessé & Moçiño (1894).

[1] Stems none or short, hardly 30 cm; **Ros** semi-globose in outline; **L** up to 60, broadly oblong-lanceolate, nearly flat, moderately caniculate towards both ends, subcoriaceous, smooth, 1.1 - 1.5 (- 1.8) m × 10 - 15 (-17) cm, bright green, tip acute, slightly caniculate, mucro 1 - 2 mm, margins between the teeth straight, marginal teeth simple and recurved or geminate, 5 - 10 (-13) mm, decurrent, 2 - 6 (-12) cm apart, brown-reddish; **Inf** 5 - 8 m, fusiform, part-**Inf** lax, in the upper ⅔ of the **Inf**, to 80 cm, bulbilliferous, bulbils numerous, ovoid; **Ped** 6 - 9 mm; **Fl** (38-) 42 - 51 (-55) mm, 1 - 3 grouped together; **Tep** oblong, (18-) 21 - 27 (-30) × 6 - 9 mm, greenish-whitish; **ITep** slightly broader than the **OTep**; **Fil** 15 - 25 mm; **Ov** 20 - 25 mm; **Sty** 15 - 25 mm; **Fr** unknown.

Drummond (1907), Trelease (1915a), and Álvarez de Zayas (1996a) placed *F. geminispina* in the synonymy of *F. tuberosa*, in contrast to a placement in the synonymy of *F. acaulis* (Ullrich 1992h). Since *F. geminispina* was described by Jacobi from small cultivated plants of unknown origin and as it is not typified, its identity can possibly never be solved unambiguously.

F. undulata Jacobi (Abh. Schles. Ges. Vaterl. Cult., Abth. Naturwiss. 1869: 170, 1869). – **D:** Not certainly known from the wild.

Incl. *Furcraea pubescens* Todaro (1879); **incl.** *Furcraea pubescens* Baker (1892) (*nom. illeg.*, Art. 53.1).

[1] **Ros** (almost) acaulescent; **L** numerous, spreading, narrowly lanceolate, long-acuminate, base strongly thickened on both faces, keeled below, concave, smooth, to 1.3 - 1.5 m × 7 cm, fresh to dark olive-green, margins wavy, marginal teeth along the whole margin, triangular, upcurved, 5 mm, ± 1.7 - 3 cm apart, brown, terminal **Sp** obtuse, brown; **Inf** 4.5 - 7 m, richly branched, scape rather short, part-**Inf** finely pubescent, bulbilliferous, bulbils ovoid; **Fl** ± 55 mm, finely pubescent, fragrant; **Tep** greenish-yellow; **OTep** 35 × 12 - 14 mm; **ITep** 20 mm broad; **Ov** 25 mm. – *Cytology:* 2n = 120.

A species of uncertain status. Jacobi attributed it to Chiapas and Tabasco (Mexico), but no material from S Mexico matches the desription (Lott & García-Mendoza 1994). Distribution records for El Salvador, the Lesser Antilles or Puerto Rico are apparently all doubtful.

HESPERALOE

J. Thiede

Hesperaloe Engelmann *ex* S. Watson (in S. Watson, Bot. US Geol. Expl. 40. Parallel, 5: 497, 1871). – **Lit:** Gentry (1972); Starr (1995); Starr (1997). **D:** USA (Texas), N Mexico (Sonora, Coahuila, Nuevo León, San Luis Potosí). **Etym:** Gr. 'hespera', evening; for the occurrence in North America (i.e. in the West, where the sun disappears in the evening); and for the superficial similarity with *Aloe* (*Aloaceae*).

Acaulescent perennials; main **R** thick and fleshy, with many additional fibrous **R**; **Ros** monocarpic, caespitose with short to long rhizomes, forming grass-like clumps with bulbous fibrous bases; **L** few to many, linear-elongate, succulent, fibrous, either thin, narrow and arching to recurved, or thick, broad and stiffly erect, caniculate, either tightly packed or widely separated and forming large rings, **L** tip frayed or a hard **Sp**, **L** margins narrow, brown or white, filiferous, fibres thin and tightly curled to thick and nearly straight, white or grey; **Inf** terminal, from the centre of mature **Ros**, ascending, to 4 m, racemose to paniculate, with 3 - 8 part-**Inf** in the upper ½; **Ped** arising from indeterminate lateral spurs, either on the main stalk or on side **Br**; **Fl** stipitate, not opening in sequence, 6-merous, narrowly campanulate with ± connivent **Tep**, in indeterminate clusters on unequal **Ped**; **Per** appearing tubular to narrowly to broadly campanulate or rotate-cam-

panulate; **Tep** with fleshy keel, about equal, essentially free but united on a fleshy nectariferous **Rec**, coloured with combinations of green, white, and purplish-brown to red, pink, salmon, or coral-red to rarely yellow; **Fil** inserted on the **Rec** or adnate to the base of the **Tep**, included; **Anth** dorsifixed, sagittate, introrse, included to exserted; **Ov** superior, ovoid to oblong, trigonous, with 3 locules, each with numerous ovules in 2 ranks; **Sty** elongate but included in the **Per**; **Sti** distinctly capitate, fringed with papillae; **Fr** septicidal woody capsules, stipitate, beaked or not, transversely rugose, persistent; **Se** large, black, flat, thin. – *Cytology:* x = 30.

According to recent molecular and morphological phylogenies, *Hesperaloe* is closest to the monotypic genus *Hesperoyucca*, and both represent sister groups (see Bogler & Simpson (1995), Clary & Simpson (1995) and Bogler & Simpson (1996)). Both genera show the following putative shared derived features: Leaves with papillate epidermal cells arranged over the veins, stamens adnate to the tepal base or lower part of the tepals, styles slender, and stigmas fringed with papillae. *Hesperaloe* differs from *Hesperoyucca* esp. in its coarsely grasslike habit, the filiferous leaf margins, and the flower colour (never purely whitish) and ± connivent tepals. Differences from *Yucca* are as mentioned above for *Hesperoyucca*, plus the capitate stigma and the colourful ± connivent tepals.

The geographical range of *Hesperaloe* is remarkable: 2 species are found W of the Sierra Madre Occidental (which represents an important floristic continental divide), 3 species are mainly confined to the E Chihuahuan Desert region.

The flowers of *Hesperaloe* are pollinated by either hummingbirds as well as bees (*H. parviflora*) or by bats and hawkmoths (*H. funifera* and *H. nocturna*). *H. campanulata* combines both syndromes and is pollinated by bats and hawkmoths during the night and by hummingbirds and bees the following day, when the flowers close somewhat to form a tube (Starr 1997).

H. campanulata G. D. Starr (Madroño 44(3): 285-286, ills., 1997). **T:** Mexico, Nuevo León (*Starr 93-001* [ARIZ, MEXU, MO, TEX]). – **Lit:** Starr (1995: with ills.). **D:** Mexico (C Nuevo León); open Chihuahuan Desert scrub, limestone slopes and hillsides, 100 - 550 m, flowers March to October.

Ros moderately caespitose, forming clumps to 0.6 - 1 × 1 m; **L** stiff and erect to slightly spreading, linear-lanceolate, tapering towards the tip, slightly caniculate, 60 - 105 × 1.5 - 2.6 cm (widest point ⅓ from the base), medium green, margins finely filiferous; **Inf** to 3 m, unbranched racemes or panicles with 2 - 5 **Br** in the upper ⅓; **Fl** tubular-campanulate to broadly campanulate, 20 - 22 × 20 - 22 mm; **Ped** 8 - 13 mm; **OTep** linear to linear-lanceolate, 18 - 22 × 4 - 8 mm, inside white, outside pink with broad white margins; **St** included; **Fil** 14 - 15 mm, adnate to the **Tep** base for 3 mm; **Anth** 3 mm; **Ov** 6 × 4 mm; **Sty** 9 - 13 mm, included; **Fr** woody capsules, globose or oblong, 20 - 30 (excl. beak) × 20 - 25 mm, with a sharp 4 - 11 mm long beak; **Se** black, 6 - 9 × 5 - 6 mm.

Vegetatively looking like a small *H. funifera*, but easily separated by the flower colour. Distinguished from *H. parviflora* by more open flowers and lighter green less channelled leaves (Starr 1995).

H. funifera (Koch) Trelease (Annual Rep. Missouri Bot. Gard. 13: 36, pl. 3-4, 1902). **T** [neo]: Mexico, Coahuila (*Engard & Gentry* 23241 [ARIZ]). – **Lit:** Ullrich (1990a); Starr (1995). **D:** S USA, N Mexico.

≡ *Yucca funifera* Koch (1862) ≡ *Agave funifera* (Koch) Lemaire (1864).

H. funifera ssp. **chiangii** G. D. Starr (Madroño 44(3): 289-290, ills., 1997). **T:** Mexico, San Luis Potosí (*Garcia Moya* s.n. [DES]). – **D:** Mexico (San Luis Potosí, probably also S Nuevo León and SW Tamaulipas); locally common on flats and open slopes.

Differs from ssp. *funifera*: **Ros** long-rhizomatous, forming wide clumps or fairy rings to 2 m ∅; **L** stiff and erect, lanceolate, not arching, deeply canaliculate, 1 - 1.5 m × 5 - 6 cm (when flattened), medium to dark green, marginal fibres coarse, 2 - 3 mm ∅, white to grey near point of attachment, straight to slightly coiled.

This subspecies was already collected by C. G. Pringle in 1891. Ullrich (1990a) mentions localities for *H. funifera* in S Nuevo León and SW Tamaulipas, which may represent new records for *H. funifera* ssp. *chiangii*.

H. funifera ssp. **funifera** – **D:** USA (C-SW Texas), Mexico (E and C Coahuila, N Nuevo León); calcareous lowlands and foothills, 500 - 1000 m, flowers April to August.

Incl. *Hesperaloe davyi* Baker (1898).

Ros forming clumps to 1.5 m ∅; **L** stiff and erect, linear-lanceolate or lanceolate, not arching, caniculate, 1 - 2 m × 3 - 4 cm (when flattened), tapering from the middle towards the tip, light or yellowish-green, **L** tip with a **Sp**, **L** margins brown, medium to coarsely filiferous, fibres loosely coiled, 1 mm ∅, white or grey; **Inf** 2 - 4 m, with 3 - 8 part-**Inf** mostly in the upper ½; **Ped** 5 - 6 mm; **Fl** in indeterminate fascicles, rotate-campanulate, 25 mm, opening in the morning, closing in the evening of the same day; **Tep** white inside, 17 - 20 mm; **ITep** outside green and white with a narrow mid-stripe tinged brownish-purple, 8 - 9 mm wide; **OTep** outside green at the base, upper ⅔ reddish-purple, 6 - 7 mm wide; **St** included; **Ov** 10 - 12 × 4 - 5 mm; **Fr** woody capsules, globose or broadly oblong, 25 - 23

× 25 - 35 mm, sharply beaked, beak 2 - 4 mm; **Se** 8 - 9 × 5 - 7 mm, black. – *Cytology:* 2n = 60.

Mature plants are easily recognizable by their stiff long leaves. Young plants are difficult to distinguish from *H. parviflora*, but the leaves are greener and stiffer in *H. funifera* (Starr 1995).

H. nocturna Gentry (Madroño 19(3): 74-78, 1967). **T:** Mexico, Sonora (*Gentry & Felger* 19988 [US]). – **Lit:** Gentry (1972: with ills.). **D:** Mexico (N-C Sonora); 950 - 1150 m, flowers April to July.

Ros very dense, densely caespitose and forming clumps 1 - 2 m ∅; **L** upright and arching, linear, striate, tip long-attenuate, deeply canaliculate, flat towards the base, 1 - 1.5 m × 1 - 2 cm (at the base), tip an acicular and pungent **Sp**, fraying with age, **L** margins narrow, brown, white-filiferous, fibres irregularly wavy, white; **Inf** slender panicles, to 1.5 - 4 m, with 2 - 3 part-**Inf** in the upper ½; **Ped** 14 - 16 mm; **Fl** campanulate-rotate, 24 - 30 mm, nocturnal, in indeterminate fascicles; **Tep** reflexed at anthesis, 15 - 25 mm, buds pruinose pink to lavender, tube 2 - 3 mm; outside of the **ITep** at anthesis with broad reddish-purple mid-stripe, 8 - 9 mm wide; outside of the **OTep** reddish with greenish-brown mid-stripe, 6 - 7 mm wide; **St** included; **Fil** equalling the **Sty**, 8 - 9 mm, attached to the base of the **Tep** for 3 mm; **Anth** sagittate, versatile, 8 - 9 mm; **Ov** oblong, trigonous, 10 × 4 mm; **Sty** stout, 8 mm, included; **Sti** capitate, papillate; **Fr** woody capsules, depressed-ovoid or oblong, shortly beaked, rugose, 30 - 40 × 25 - 45 mm; **Se** black, 11 × 8 mm.

Easily identifiable by its long narrow leaves and nocturnal flowers (Starr 1995).

H. parviflora (Torrey) J. M. Coulter (CUSNH 2: 436, 1894). **T** [lecto]: USA, Texas (*Wright* 1908 [GH, NY]). – **D:** USA (C Texas), Mexico (NW Coahuila); in Creosote Bush desert, oak and chaparral zones, 600 - 2000 m, flowers March to September. **Fig. IX.a**

≡ *Yucca parviflora* Torrey (1859); **incl.** *Hesperaloe yuccoides* hort. (s.a.) (*nom. inval.*, Art. 29.1); **incl.** *Aloe yuccaefolia* A. Gray (1867) (*nom. illeg.*, Art. 52.1) ≡ *Hesperaloe yuccaefolia* (A. Gray) Engelmann (1871) (*nom. illeg.*, Art. 52.1); **incl.** *Hesperaloe engelmannii* Krauskopf *ex* S. Watson (1879) ≡ *Hesperaloe parviflora* var. *engelmannii* (Krauskopf *ex* S. Watson) Trelease (1902).

Ros densely caespitose, forming clumps to 1 m ∅; **L** arching, linear, narrowing towards the tip, 30 - 60 (-120) × 0.8 - 1.8 cm (at the base), dark green, margins finely filiferous, fibres tightly curled; **Inf** panicles to 1 - 2.5 m, part-**Inf** few, mainly in the upper ½; **Fl** tubular or oblong-campanulate, 25 - 35 mm, diurnal, in indeterminate fascicles; **Tep** pressed together at anthesis, 15 - 20 × 4 - 8 mm (**ITep** 17 mm long), salmon-coloured, coral-red, pink, or rosy-red (also yellow in a horticultural selection); **St** included; **Fil** elongate, 7 - 13 mm, attached to the base of the **Tep** for 1 mm; **Anth** 2 - 3 mm; **Ov** ovoid, small, 4 - 6 × 3 - 4 mm; **Sty** slender, elongate, included, 12 - 13 mm; **Fr** woody, ovoid or oblong-ovoid, 30 - 40 × 25 - 30 mm, rugose, beaked; **Se** 9 - 10 × 6 - 7 mm, black.

Distinguished by the combination of narrow, mainly salmon-coloured to reddish flowers and relatively short and dark green leaves (Starr 1995).

H. tenuifolia G. D. Starr (Madroño 44(3): 293-294, ills., 1997). **T:** Mexico, Sonora (*Meyer & Jenkins 9063* [ARIZ]). − **Lit:** Starr (1995: with ills.). **D:** Mexico (S Sonora: Cerro Agujudo); dry rhyolithic hilltops in pine-oak forest, 1500 m (only known from the type locality), flowers April to May.

Ros open, sparsely caespitose and forming small clumps to 50 cm ∅; **L** few, arching, narrowly linear, tapering towards the tip, 50 - 100 × 0.5 - 1 cm (at the base), margins thin, finely filiferous, fibres not tightly curled, white; **Inf** racemes or narrow panicles with 2 - 3 part-**Inf**, to 1.5 - 2 m; **Fl** rotate, 13 × 10 mm, nocturnal; **OTep** linear, 13 × 5 mm, outside dark pinkish-red, inside white with reddish margin; **ITep** ovate, 15 × 8 mm, outside dark pinkish-red with white margin, inside white; **St** included; **Fil** 9 mm, attached to the base of the **Tep** for 2 mm; **Anth** 3 mm; **Ov** 6 × 3 mm; **Sty** 4 mm; **Fr** woody, ovoid, 20 - 30 × 20 - 25 mm, beak none or 1 mm; **Se** black, 10 × 5 - 7 mm.

Very easily recognized by its long and very thin leaves with finely textured slightly curly marginal fibres. The very short open flowers cannot be confused with those of other species (Starr 1995).

HESPEROYUCCA

J. Thiede

Hesperoyucca (Engelmann) Baker (BMI 1892(5): 8, 1892). **T:** *Yucca whipplei* Torrey. − **D:** W USA, NW Mexico. **Etym:** Gr. 'hespera', evening; for the occurrence in W North America (i.e. in the West, where the sun disappears in the evening); and for the similarity to *Yucca* (*Agavaceae*).
≡ *Yucca* Sect. *Hesperoyucca* Engelmann (1873)

Ros sessile, sometimes stem rhizomatous, single or caespitose; **L** linear or rarely narrowly lanceolate, rigid and sword-like to flexible and frequently falcate, plano-convex or subtriquetrous, or keeled on both faces, 25 - 115 × 0.5 - 4 cm, ± grey-green, finely striate, base expanded, ± 4 - 7 × 4 - 7 cm, ± white to greenish, margin thin, horny, without fibres, teeth ± finely serrulate, end-**Sp** sharp; **Inf** terminal and **Ros** monocarpic, large, dense, cylindrical or somewhat slenderly ellipsoidal, 1.4 - 8 m with a bracteate scape 0.9 - 4.5 m long; **Fl** densely arranged, usually broadly expanding, pendent, ± globose, 3.5 - 5 cm, very fragrant; **Tep** broadly lanceolate, nearly equal, 30 - 65 × 8 - 25 mm, white, tips generally purple, tube none; **Fil** straight, linear below, tip

angled, club-like, attached to the lower part of the **Tep**, so that they are pulled away from the **Ov** as the **Fl** opens; pollen uniquely glutinous; **Ov** stout, 8 - 12 × 6 - 10 mm; **Sty** short, slender; **Sti** distinctly capitate, green towards center, fringed with elongated translucent papillae; **Fr** obovoid, strictly loculicidally dehiscent, 3 - 5 cm; **Se** flat, thin, smooth, without marginal wing, 6 - 7 × 8 mm, dull black. − *Cytology:* n = 30.

Hesperoyucca is re-established here as a monotypic genus based on recent phylogenetic studies and clear character differences from *Yucca*. With the exception of Trelease (1902), all authors included *Hesperoyucca* as a section or subgenus within *Yucca*. Recent molecular studies by Bogler & Simpson (1995), Bogler & Simpson (1996) and Clary & Simpson (1995) and structural phylogenies shown by Clary & Simpson (1995) clearly revealed a position independent of *Yucca* as sister group of *Hesperaloe* (in the structural phylogeny of Hernández Sandoval (1995), however, *Hesperoyucca* is associated with *Yucca* and not with *Hesperaloe*). *Hesperoyucca* and *Hesperaloe* again either represent the sister group of *Yucca* or of the remaining genera of *Agavaceae* (Bogler & Simpson l.c., Clary & Simpson l.c.).

Hesperoyucca differs clearly from *Yucca* (data in brackets) in forming a definite bulb in the seedling stage (Webber 1953: pl. 53) (absent, needs further study), its capitate stigma (vs. 6-lobed), its strictly loculicidally dehiscent fruits (vs. indehiscent or, if dehiscent, commonly septicidal, occasionally also septicidal and loculicidal), its filaments basally attached to the tepals and without apical thickenings (vs. filaments not attached to the tepals, but held close to the ovary and bent outwards near the swollen apex). The often very large inflorescences of *Hesperoyucca* by far exceed inflorescence size in *Yucca*, and unbranched plants ("ssp. *whipplei*") are monocarpic, whereas some branched plants ("ssp. *caespitosa*") develop new rosettes from the leaf axils of very young plants; both features are unknown in *Yucca*.

Haines (1941) recognized 5 varieties based largely on growth form (rosettes single or multiple by either branching or produced by rhizomes). Since wild populations often contain plants of different "varieties" (Keeley & Tufenkian 1983) and seeds from one capsule may even produce all possible growth forms (DeMason 1984), no infraspecific taxa are recognized here.

H. whipplei (Torrey) Baker (BMI 1892(5): 8, 1892). − **Lit:** Turner & al. (1995). **D:** USA (SW California, Arizona: W Grand Canyon), Mexico (N Baja California, N Baja California Sur, NW Sonora: Pinacate region). **I:** Bolliger (1998); Hochstätter (2000a). **Fig. VIII.d, VIII.e**
≡ *Yucca whipplei* Torrey (1859); **incl.** *Yucca californica* Groenland (1858); **incl.** *Yucca graminifolia*

Wood (1868) (*nom. illeg.*, Art. 53.1); **incl.** *Yucca whipplei* var. *caespitosa* M. E. Jones (1929) ≡ *Yucca whipplei* ssp. *caespitosa* (M. E. Jones) A. L. Haines (1942); **incl.** *Yucca whipplei* var. *parishii* M. E. Jones (1929) ≡ *Yucca whipplei* ssp. *parishii* (M. E. Jones) A. L. Haines (1941); **incl.** *Yucca whipplei* ssp. *typica* A. L. Haines (1941) (*nom. inval.*, Art. 24.3); **incl.** *Yucca whipplei* ssp. *intermedia* A. L. Haines (1942) ≡ *Yucca whipplei* var. *intermedia* (A. L. Haines) J. M. Webber (1953); **incl.** *Yucca whipplei* ssp. *percursa* A. L. Haines (1942) ≡ *Yucca whipplei* var. *percursa* (A. L. Haines) J. M. Webber (1953); **incl.** *Yucca newberryi* McKelvey (1947) ≡ *Yucca whipplei* ssp. *newberryi* (McKelvey) Hochstätter (2000); **incl.** *Yucca peninsularis* McKelvey (1947); **incl.** *Yucca whipplei* ssp. *eremica* Epling & A. L. Haines (1957).

Description as for the genus.

Y. californica is here listed as synonym with considerable doubt and would have priority if it is indeed conspecific.

H. whipplei is winterhardy in protected sites outdoors in Central Europe and may reach flowering size in as little as 13 years (Bolliger 1998).

YUCCA

J. Thiede

Yucca Linné (Spec. Pl. [ed. 1], 319, 1753). **T:** *Yucca aloifolia* Linné [Lectotype, designated by Britton & Shafer, North Amer. Trees, 151, 1908 (fide ING).]. – **Lit:** Trelease (1902); McKelvey (1938); McKelvey (1947); Webber (1953); Reveal (1977); Matuda & Piña Lujan (1980); Hochstätter (2000b). **D:** S Canada, N, C and S USA, Mexico, possibly Guatemala; cultivated worldwide. **Etym:** Name first used 1557 in a German travelogue and probably derived from a name used on Hispaniola through Span. 'yuca', which is, however, used for the edible root tubers of Cassava, and that was perhaps erroneously applied to *Yucca* for the edible flowers of some species.
Incl. *Iuka* Adanson (1763) (*nom. inval.*, Art. 61.1). **T:** *Yucca aloifolia* Linné.
Incl. *Clistoyucca* (Engelmann) Trelease (1902). **T:** *Clistoyucca arborescens* (Torrey) Trelease [*nom. illeg.*, ≡ *Yucca brevifolia* Engelmann].
Incl. *Samuela* Trelease (1902). **T:** not designated.

Woody perennials, terrestrial (very rarely epiphytic: *Y. lacandonica*); stems none, short, or thick and arborescent, then usually ± branched; **Ros** terminal; **L** mostly numerous, ± ensiform, nearly linear, thin and flexible or thicker and very rigid, margins entire, horny, often desintegrating into fibres, terminal **Sp** often present; **Inf** large panicles; **Fl** pedicellate, usually ± pendent, ± campanulate to globose, large, fleshy; **Tep** 3 + 3, all similar and subequal in size, mostly white or whitish (or greenish or slightly reddish), **Tep** tube none, short or up to ± ½ of the **Tep** length; **St** 3 + 3; **Fil** fleshy, clavate, or slightly swollen beneath the small versatile **Anth**, pubescent or at least papillose; **Ov** superior, 3-locular; **Sty** very short or none, with 3 short branches with a 2-lobed **Sti** each; **Fr** many-seeded loculicidal capsules with ± intruding dorsal false septa, more rarely septicidal, or baccate and indehiscent; **Se** flat and thin, black. – *Cytology:* x = 30.

The genus includes ± 45 mostly xerophytic species. They are more xerophytic than succulent and therefore fall mostly outside the scope of this Lexicon. However, many species of *Yucca* are horticulturally important or represent dominant elements of arid vegetations, and in order to provide a complete treatment of the family *Agavaceae*, the genus is dealt with in full here.

Yucca is easily recognizable by the typical filiferous leaf margin, which is otherwise only found in *Hesperaloe* and a couple of *Agave* species. The mostly whitish wax-like pendent flowers in usually compact inflorescences are another diagnostic feature of the genus. There is a closely knit symbiosis between *Yucca* species and its pollinator, the Yucca Moth (Powell 1984): Females of the various species of Yucca Moths emerge in time at the onset of flowering of the Yuccas. Upon visiting a flower, they deposit an egg in one of the 3 chambers of the ovary, and subsequently actively collect some pollen, which is placed between the stigmas when the next flower is visited. Recent phylogenetic studies indicated this symbiosis to have independently originated both in *Yucca* and *Hesperoyucca* (Bogler & al. 1995).

Recent preliminary morphological and molecular phylogenies (Clary & Simpson 1995) indicate the necessity of a complete reclassification of *Yucca*, since none of the traditionally recognized series within the genus is monophyletic in either data set. These data suggested that the traditional separation into berry-fruited taxa (= Sect. *Yucca / Sarcocarpa*) and capsule-fruited species (Sect. *Chaenocarpa*) is artificial (see also the comment for *Y. linearifolia*). Since no updated classification is available, the traditional infrageneric division based on McKelvey (1938) and McKelvey (1947) is repeated here in short, albeit all of her series names are invalidly (Art. 32.1c) published:

[1] Sect. *Yucca* (incl. Sect. *Euyucca* Engelmann 1873, *nom. inval.*; incl. Ser. *Sarcoyucca* Engelmann 1873 / Trelease 1902, *nom. illeg.*): **R** fibrous; adult plants mainly stem-forming, rarely rhizomatous (*Y. endlichiana*); **L** of young plants (< 6 years) few, broadened and generally reddish; **Fr** indehiscent, representing ± large fleshy berries; **Se** rough, unwinged.

    [1a] Ser. *Faxoniana* McKelvey 1938, *nom. inval.* (incl. *Samuela* Trelease 1902).

    [1b] Ser. *Baccatae* McKelvey 1938, *nom. inval.*

    [1c] Ser. *Yucca* (incl. Ser. *Treculiana* McKelvey 1938, *nom. inval. et illeg.*).

[1d] Ser. *Heteroyucca* Trelease 1902.

[2] Ser. *Clistoyucca* Engelmann 1873 (≡ *Clistoyucca* (Engelmann) Trelease 1902): **Fr** indehiscent, dry and spongy; **Se** smooth, unwinged. Only *Y. brevifolia*.

[3] Sect. *Chaenoyucca* (Engelmann) Trelease 1902 (≡ Ser. *Chaenoyucca* Engelmann 1873): **R** of young plants bulbous, adult plants rhizomatous; **L** of young plants many, thin and greenish-glaucous; **Fr** dehiscent dry capsules soon becoming erect at maturity; **Se** smooth, winged or unwinged.

[3a] Ser. *Rupicolae* McKelvey 1947, *nom. inval.*

[3b] Ser. *Elatae* McKelvey 1947, *nom. inval.*

[3c] Ser. *Constrictae* McKelvey 1947, *nom. inval.*

[3d] Ser. *Harrimaniae* McKelvey 1947, *nom. inval.*

[3e] Ser. *Arkansanae* McKelvey 1947, *nom. inval.*

[3f] Ser. *Glaucae* McKelvey 1947, *nom. inval.*

[3g] *Y. filamentosa* and related species.

The genus *Yucca* is "one of the most difficult" of the USA (Reveal 1977) due to its complex nomenclature including many older names of uncertain application and horticultural names, as well as the variability of many taxa, which apparently often includes hybridization and introgression. The following synopsis can only represent a first step towards a better understanding of the genus, without intending at all to solve the many remaining problems.

Hochstätter (2000b) indicates the following species to be hardy outdoors in C Europe: *Y. arkansana, Y. angustissima, Y. baccata, Y. baileyi, Y. elata, Y. filamentosa, Y. glauca, Y. gloriosa, Y. harrimaniae, Y. pallida* and *Y. recurvifolia*. The following species need additional protection from moisture: *Y. faxoniana, Y. rupicola, Y. rostrata* and *Y. thompsoniana* (Hochstätter 2000b). – Vernacular name: "Palm Lily".

House plants sold as 'Yuccas' by the horticultural trade are usually species of the genus *Cordyline* (variously classified as *Asteliaceae* or [as used in this Lexicon] *Dracaenaceae*).

The following names are of unresolved application but are referred to this genus: *Yucca acutifolia* Truffaut (1869); *Yucca ×andreana* Deleuil (s.a.); *Yucca atkinsii* Trelease (1894); *Yucca barrancasecca* hort. *ex* Pasquale (s.a.); *Yucca ×carrierei* Deleuil (s.a.); *Yucca ×carrierei* André (1895); *Yucca conspicua* hort. *ex* Regel (1871) (*nom. illeg.*, Art. 53.1); *Yucca contorta* hort. *ex* Carrière (1858); *Yucca crinifera* Lemaire (1846); *Yucca desmetiana* Baker (1870); *Yucca ehrenbergii* Baker (1875); *Yucca ×ensifera* Deleuil (s.a.); *Yucca ensifolia* Baker (1870); *Yucca fuauxiana* hort. (s.a.); *Yucca gigantea* Lemaire (1859); *Yucca gracilis* Link *ex* Sweet (1830); *Yucca hanburii* Baker (1892); *Yucca horri-*

da Humboldt *ex* Steudel (1840); *Yucca howardsmithii* Trelease (1937); *Yucca ×juncea* Deleuil (s.a.); *Yucca ×karlsruhensis* Graebner (1903); *Yucca ×massiliensis* Deleuil (s.a.); *Yucca mexicana* Sessé & Moçiño (1894) (*nom. illeg.*, Art. 53.1); *Yucca nitida* W. Watson (1906); *Yucca pitcairnifolia* Karwinsky *ex* G. Don (1839); *Yucca rubra* hort. *ex* Lavallée (1877); *Yucca spinosa* Kunth (1822); *Yucca stenophylla* Steudel (1840); *Yucca ×striatula* Deleuil (s.a.); *Yucca ×sulcata* Deleuil (s.a.); *Yucca toneliana* Lemaire (1865) (*nom. illeg.*, Art. 53.1); *Yucca ×treleasei* Sprenger (1901); *Yucca vomerensis* Sprenger (s.a.).

Y. aloifolia Linné (Spec. Pl. [ed. 1], 319, 1753). **T:** [lecto – icono]: Dillenius, Hort. Eltham., t. 323: fig. 416, 1732. – **Lit:** Matuda & Piña Lujan (1980: with ills.); Lott & García-Mendoza (1994). **D:** Mexico (probably only native in Veracruz and Yucatán); plains and slopes in tropical deciduous forests, to 1800 m.

Incl. *Yucca draconis* Linné (1756) ≡ *Yucca aloifolia* var. *draconis* (Linné) Engelmann (1873); **incl.** *Yucca haruckeriana* Crantz (1768); **incl.** *Yucca arcuata* Haworth (1819); **incl.** *Yucca conspicua* Haworth (1819) ≡ *Yucca aloifolia* var. *conspicua* (Haworth) Engelmann (1873); **incl.** *Yucca crenulata* Haworth (1819); **incl.** *Yucca serrulata* Haworth (1819); **incl.** *Yucca tenuifolia* Haworth (1819); **incl.** *Yucca armata* Steudel (1840); **incl.** *Yucca aloifolia* var. *stenophylla* Bommer (1859); **incl.** *Yucca parmentieri* hort. *ex* Carrière (1859); **incl.** *Yucca yucatana* Engelmann (1873) ≡ *Yucca aloifolia* var. *yucatana* (Engelmann) Trelease (1902); **incl.** *Yucca purpurea* hort. *ex* Baker (1880); **incl.** *Yucca quadricolor* hort. *ex* Baker (1880); **incl.** *Yucca tricolor* hort. *ex* Baker (1880).

[1c] Arborescent with stems to 8 m, slender, erect, simple or densely branched, sometimes with offsets; **L** rigid, patent, flattened or slightly concave, 25 - 60 × 2.5 - 6 cm, brilliant dark green, tip acute (pungent), margins rather horny, denticulate; **Inf** paniculate, pendent, tomentose; **Fl** globose, to 5 × 10 cm; **Tep** ovate, 30 - 40 × 15 - 22 mm, whitish with purple or green tinge towards the base; **Fil** slightly papillose, 8 - 10 mm; **Ov** oblong, basally constricted, 15 mm; **Fr** fleshy berries, ellipsoid, prismatic, 3.5 - 5 × 2 - 2.6 cm, blackish, pulpa purple; **Se** ovoid, thick, 5 - 6 × 6 - 7 mm.

Widely cultivated as foliage plant in (sub-) tropical gardens as well as indoors, esp. in the form of variegated cultivars.

Y. angustissima Engelmann *ex* Trelease (Annual Rep. Missouri Bot. Gard. 13: 58, 1902). **T:** USA, Arizona (*Bigelow* s.n. [MO 148375 + 148376]). – **Lit:** Hochstätter (2000b). **D:** USA (S Utah, N and C Arizona, W New Mexico).

For differences from *Y. elata* see there.

Y. angustissima var. **angustissima** – Lit: Reveal (1977); Hochstätter (2000b); both with ills. **D:** S USA (SW Utah, N Arizona, W New Mexico); desert flats or mesas, often in sandy places or near sandstone outcrops, 1050 - 2550 m.

[3b] Stems none to short and procumbent, 10 - 40 cm, or caulescent and erect, to 1 m; **Ros** compact, solitary or in small to large clumps to 3 m ∅; **L** rigidly spreading, flexible, linear, base broad, tip tapering, long-acuminate, flatly convex to flat and keeled, rarely canaliculate, 25 - 60 (-75) × 0.4 - 1.5 cm, pale yellow to blue-green, margins entire, cream to tan or reddish-brown, forming few fine slightly curled fibres, terminal **Sp** 3 - 7 mm; **Inf** erect, scape 0.2 - 2.5 m, glabrous or finely pubescent, racemose, simple, flowering part 0.2 - 1.5 (-2) m, well above the **L**, with few part-**Inf**; **Ped** slender, 1 - 2.5 (-4) cm; **Fl** pendent, campanulate to globose, 3 - 6.5 cm; **Tep** elliptic to ovate, **ITep** broader than the **OTep**, lanceolate, white to cream or greenish-white, often tinged with rose or rose-purple, tube 3 - 7 mm; **Ov** (0.7-) 1 - 2.5 cm; **Fr** dry capsules, commonly with a deep central constriction, oblong-cylindrical, 3.5 - 7.5 × 2 - 3 cm; **Se** thin, 5 - 7 mm, dull black. – *Cytology:* n = 30.

Y. angustissima var. **avia** Reveal (in Cronquist & al., Intermountain Fl. 6: 534, ill. (p. 535), 1977). **T:** USA, Utah (*Jones* 5639a [US]). – Lit: Reveal (1977). **D:** USA (C Utah); mainly on loamy-rocky soils.

≡ *Yucca angustissima* ssp. *avia* (Reveal) Hochstätter (1999).

[3b] Differs from var. *angustissima*: **L** 40 - 60 cm; **Fl** 3.5 - 4.5 cm; **Sty** 7 - 10 mm.

This taxon is distinguished by minor quantitative features only and is included in the synonymy of var. *angustissima* by USDA (2001) but may be recognized due to its geographical isolation.

Y. angustissima var. **kanabensis** (McKelvey) Reveal (in Cronquist & al., Intermountain Fl. 6: 534, ill. (p. 535), 1977). **T:** USA, Utah (*McKelvey* 4347A [A]). – Lit: Reveal (1977). **D:** USA (S Utah, N Arizona); sandy places.

≡ *Yucca kanabensis* McKelvey (1947) ≡ *Yucca angustissima* ssp. *kanabensis* (McKelvey) Hochstätter (1999).

[3b] Differs from var. *angustissima*: **L** 45 - 75 (-150) cm; **Inf** 2 - 4.5 m, scape 1 - 1.5 m, flowering part 1 - 2 m; **Fl** 5.5 - 6.5 cm; **Ov** 3 - 3.5 cm; **Sty** 5 - 8 mm; **Fr** moderatly constricted, larger, 4.5 - 7.5 cm.

Y. angustissima var. **toftiae** (S. L. Welsh) Reveal (in Cronquist & al., Intermountain Fl. 6: 534, ill. (p. 535), 1977). **T:** USA, Utah (*Welsh* 11935a [BRY, NY, US]). – Lit: Reveal (1977). **D:** USA (Utah); sandy alluvium and sandstone outcrops and mesas.

≡ *Yucca toftiae* S. L. Welsh (1975) ≡ *Yucca angustissima* ssp. *toftiae* (S. L. Welsh) Hochstätter (1999).

[3b] Differs from var. *angustissima*: **L** 25 - 60 (-70) cm; **Inf** 2 - 4.5 m, scape 1.2 - 2.5 m, flowering part 0.2 - 2 m; **Fl** 3 - 4.5 (-5.2) cm; **Ov** 1.5 - 2.5 cm; **Sty** 3 - 10 mm; **Fr** moderately constricted, larger, 4.5 - 7.5 cm.

Y. arkansana Trelease (Annual Rep. Missouri Bot. Gard. 13: 63, tt. 30, 31, 83, fig. 7, 92, 1902). **T:** USA, Arkansas (*Engelmann* 182 [MO]). – **D:** USA (S-C, N and E Texas, C and NE Oklahoma, W to SW Arkansas); prairie plains or flat stony hills, dry slopes, 850 - 2000 m, flowers late April to mid-May. **I:** Hochstätter (1999a).

Incl. *Yucca angustifolia* var. *mollis* Engelmann (1873) ≡ *Yucca glauca* var. *mollis* (Engelmann) Branner & Coville (1888); **incl.** *Yucca arkansana* var. *paniculata* McKelvey (1947) ≡ *Yucca louisianensis* var. *paniculata* (McKelvey) Shinners (1956).

[3e] Acaulescent or stems short, to 15 cm; **Ros** 1 or several in small and lax groups, asymmetrical; **L** ascending, or sometimes somewhat recurved, base ± stiff, major part flexible and weak, 20 - 60 (-100) × 1 - 2.5 (at the base 0.3 - 0.7) cm, broader in the middle, straight, upper face flat, somewhat concave at the tip, lower face convex, margins whitish at first, papery, with short curled fibres, with age almost fibreless, terminal **Sp** acute, straw-coloured; **Inf** 0.6 - 1 (-2) m, **Fl**-bearing part starting at the height of the **L**, lower part little-branched and with few **Fl**; **Fl** campanulate or nearly tubular, 3 - 6 cm, whitish-cream, somewhat tinged with greenish or reddish; **Tep** elliptic to oblong or lanceolate, 2 - 5 cm broad, margins irregular, sometimes roughly dentate, tomentose; **Ov** oblong to cylindrical, thickly robust; **Fr** dry capsules, oblong, cylindrical, constricted in the middle, walls thick, 4 - 6.5 (-7) × 2 cm ; **Se** black, shiny, 1 × 0.5 cm.

A variable taxon. *Y. arkansana* var. *paniculata* from the E range may be an E extension of the species with a taller paniculate infloresence (McKelvey 1947). It appears to approach *Y. louisianensis* and is also included in the synonymy of the latter by some authors such as Kartesz (1996) and USDA (2001).

Y. baccata Torrey (in Emory, Rep. US Mex. Bound. 2(1): 221, 1859). **T** [lecto]: USA, New Mexico (*Bigelow* s.n. [NY]). – Lit: Reveal (1977). **D:** S USA, NW Mexico.

Incl. *Yucca filamentosa* Wood (1868) (*nom. illeg.*, Art. 53.1); **incl.** *Yucca fragilifolia* Baker (1870); **incl.** *Yucca scabrifolia* Baker (1870); **incl.** *Yucca filifera* hort. *ex* Engelmann (1873).

For details of the typification see Reveal (1977). An ethnobotanical study of the species was presented by Potter-Bassano (1991).

Y. baccata var. **baccata** – **Lit:** Reveal (1977: with ills.). **D:** S USA (SE California, S Nevada, S Utah, N Arizona, Colorado, Texas), NW Mexico (N Chihuahua); dry slopes, 250 - 2000 m, flowers April to June. **I:** Matuda & Piña Lujan (1980).

Incl. *Yucca baccata* fa. *parviflora* McKelvey (1938).

[1b] Acaulescent or rarely with short stems; **Ros** asymmetrical and rather open, mostly simple and (50-) 60 - 75 × 130 - 150 cm, or clumped and 1 - 5 m ∅; **L** at the base spreading, central **L** more erect, straight, deeply canaliculate, rigid, (30-) 50 - 70 (-75) × 2.5 - 4 (flattened 3 - 6) cm, dark green, margins of the upper ½ separating, forming broad coarse recurved to curly fibres, **L** tip with a stout terminal **Sp**, stiff, 1.5 - 7 mm; **Inf** erect, short, as long as or longer than the **L**, to 1.3 m, scape and axis mostly green, with ± 15 part-**Inf**; **Ped** 0.7 - 4 cm; **Fl** pendent, campanulate, 6 - 13 cm; **Tep** lanceolate, 4 - 10 cm, dorsally red-brown, ventrally creamy-white, tube 7 - 12 mm; **Ov** (3-) 5 - 7 (-7.5) cm; **Fr** fleshy berries, ellipsoid, (10-) 15 - 17 × (3-) 5 - 6.5 cm, upper ⅓ constricted; **Se** 7 - 11 mm.

Y. baccata var. **vespertina** McKelvey (Yuccas Southwest US 1: 45, 1938). **T:** USA, Arizona (*McKelvey* 2167 [A]). – **Lit:** Reveal (1977: with ills.). **D:** S USA (SE California, S Nevada, S Utah, Arizona). **Fig. IX.b**

≡ *Yucca vespertina* (McKelvey) S. L. Welsh (1993).

[1b] Differs from var. *baccata*: **L** falcate, rather narrow, blue-green, glaucous, marginal fibres fine, wiry; **Inf** shorter to just slightly longer than the **L**, scape and axis mostly reddish-purple, with few part-**Inf**.

A poorly defined taxon according to several authors.

Y. baileyi Wooton & Standley (CUSNH 16: 114, 1913). **T:** USA, New Mexico (*Standley* 7638 [US 686602]). – **Lit:** Webber (1953); Reveal (1977: with ills.); Hochstätter (2000b). **D:** SW USA.

Based on its relatively small rosettes with narrow green strongly filiferous leaves, as well as on the distribution area, *Y. baileyi* appears to be related to *Y. elata* and *Y. angustissima* (Hochstätter 2000b).

Y. baileyi var. **baileyi** – **D:** SW USA (SE Utah, S Colorado, NE Arizona, E New Mexico); dry forest floors to grasslands, infrequent on exposed sandstone rims, 1200 - 2400 m, flowers April to June. **I:** Hochstätter (2000b).

Incl. *Yucca standleyi* McKelvey (1947).

[3b] Acaulescent; **Ros** solitary or 3 - 15 in clumps of 0.5 - 2 m ∅, branching from subterranean stems; **L** somewhat crowded, divergently spreading, somewhat rigid to flexible, linear, upper face flat, lower face convex, (20-) 25 - 60 (-100) × 0.3 - 0.8 cm, pale or yellow green, margins entire, white,

becoming separate and forming conspicuous fine curly fibres, tip gradually tapering towards a short terminal **Sp** 3 - 5 mm long; **Inf** racemose, simple, scape 1 - 10 cm, glabrous, **Fl**-bearing for up to 50 cm, included or just barely exceeding the **L**; **Ped** 1 - 2 cm; **Fl** pendent, campanulate to globose, 5 - 6.5 cm; **Tep** ovate to obovate, greenish-white, usually deeply tinged with purple esp. on the outer face, tube 3 - 7 mm; **Ov** 2 - 2.5 cm; **Fr** dry capsules, oblong-cylindrical, 4 - 7 × 2.5 - 5 cm, not or only slightly constricted; **Se** thin, with broad marginal wing, 6 - 10 mm.

Y. baileyi var. **intermedia** (McKelvey) Reveal (in Cronquist & al., Intermountain Fl. 6: 532, 1977). **T:** USA, New Mexico (*McKelvey* 4902 [A]). – **D:** SW USA (C New Mexico); 1500 - 2000 m.

≡ *Yucca intermedia* McKelvey (1947) ≡ *Yucca baileyi* ssp. *intermedia* (McKelvey) Hochstätter (1999); **incl.** *Yucca intermedia* var. *ramosa* McKelvey (1947).

[3b] Differs from var. *baileyi*: Stems short, erect.

Webber (1953) assumed a hybrid origin for *Y. intermedia*, which is rejected by Reveal (1977) because of the high degree of viable pollen set, the production of mature fruits, and the lack of any of the putative parents as suggested by Webber in its distribution area.

Y. baileyi var. **navajoa** (J. M. Webber) J. M. Webber (Yuccas Southwest, 51, 1953). **T:** USA, New Mexico (*Webber* s.n. [US 1872608]). – **D:** S USA (Arizona, New Mexico); chapparral and juniper woodlands on coarse gravelly soils or sandstone ledges, 1580 - 1980 m, flowers usually early June.

≡ *Yucca navajoa* J. M. Webber (1945).

[3b] Differs from var. *baileyi*: Subacaulescent; **Ros** forming dense clumps mainly through branching of the above-ground stems, smaller, more symmetrical; **L** shorter, 11 - 41 × 0.8 cm, broader in comparison to the length; **Se** without broad marginal wing.

Y. brevifolia Engelmann (in S. Watson, Bot. US Geol. Expl. 40. Parallel, 5: 496, 1871). **T** [lecto]: USA, California (*Bigelow* s.n. [NY, PANS, US]). – **Lit:** Reveal (1977); Benson & Darrow (1981); both with ills. **D:** SW USA.

≡ *Clistoyucca brevifolia* (Engelmann) Rydberg (1918); **incl.** *Yucca brevifolia* fa. *kernensis* Hochstätter (2000) (*nom. inval.*, Art. 32.1c).

Y. brevifolia var. **brevifolia** – **D:** SW USA (SE California, S Nevada, SW Utah, W Arizona); dry slopes and mesas, 850 - 2200 m. **Fig. IX.c, IX.e**

Incl. *Yucca draconis* var. *arborescens* Torrey (1857) ≡ *Yucca arborescens* (Torrey) Trelease *ex* Merriam (1893) (*nom. illeg.*, Art. 52.1) ≡ *Clistoyucca arborescens* (Torrey) Trelease (1902) (*nom. illeg.*, Art. 52.1); **incl.** *Yucca brevifolia* fa. *herbertii* J.

M. Webber (1953) ≡ *Yucca brevifolia* var. *herbertii* (J. M. Webber) Munz (1958).

[2] Arborescent, to (3-) 5 - 12 (-15) m, frequently with a single main trunk, **Br** usually from 1 - 3 m above the ground; **Ros** broad, flat- or round-topped, 0.3 - 1 (-1.5) × 0.3 - 0.5 m; **L** straight, upper face plane and lower face convex, or triquetrous, rigid, 15 - 35 × 0.7 - 1.5 cm, green, base whitish, tip with a stiff **Sp** 7 - 12 mm long, margins entire, thin, horny, minutely denticulate; **Inf** erect, short, (25-) 30 - 55 cm, broad, densely flowered, with numerous part-**Inf**; **Ped** 0.7 - 1.2 (-2.5) cm; **Fl** ellipsoid to globose, (3-) 4 - 7 cm; **Tep** lanceolate to oblong, greenish-white to cream-coloured, tube at least ½ of the length of the **Tep**; **Ov** 2.5 - 3 cm; **Fr** dry, indehiscent, ellipsoid, rather spongy, 6 - 8.5 (-10) × 3 - 5 (-6.5) cm; **Se** flat, thin, 8 - 11 mm, dull black.

For details of the typification see Reveal (1977). This taxon is one of the characteristic elements of the Mohave Desert of the SW USA, where it often dominates the landscape. Vernacular name: "Joshua Tree".

Y. brevifolia var. **jaegeriana** McKelvey (J. Arnold Arbor. 16: 269, 1935). **T:** USA, California (*McKelvey* 2732 [A]). – **D:** SW USA (SE California, S Nevada, SW Utah, W Arizona); hills and alluvial fans of the Upper Mojave Desert, 850 - 1500 m.

Incl. *Yucca brevifolia* var. *wolfei* Jones (1935) (*nom. inval.*, Art. 29.1).

[2] Differs from var. *brevifolia*: Stems smaller, mostly 1.8 - 3.6 (-4.5) m, trunks mostly < 37.5 cm ⌀; lowest **Br** usually within 0.9 m above the ground; **L** mostly 10 - 20 (-25) × 0.6 - 1 cm.

This taxon is a variant of smaller growth from the Upper Mojave Desert. It intergrades with var. *brevifolia* and is thus placed in the synonymy of the latter by Reveal (1977) and McKinney & Hickman (1993), but is otherwise kept separate by Kartesz (1996) and USDA (2001).

Y. campestris McKelvey (Yuccas Southwest US 173, t. 62-63, 1947). **T:** USA, Texas (*McKelvey* 2849 [A]). – **D:** USA (W Texas); sand dunes. **I:** Hochstätter (1999a); Hochstätter (2000b).

[3f] Acaulescent or stems short, 0.5 - 1 m; **Ros** in small to large and dense groups, lax; **L** upper face flat, lower side convex, to 65 × 0.6 (in the middle) cm, bluish-green, margins white to grey, finely fibrous, later glabrous; **Inf** 0.5 - 1.5 (-2) m, **Fl**-bearing part starting between (rarely above) the **L**, part-**Inf** many, thin, fragile, ascending; **Fl** globular, 10 - 12 cm; **Tep** 1.5 - 2.5 cm broad, upper margin irregular, toothed, slightly tomentose, fading greenish, sometimes somewhat tinged with rose; **Ov** oblong-ovoid, 1.2 - 2 cm; **Fr** dry capsules, symmetrical or constricted, 4 - 5 × 3 - 4.5 cm, reddish-brown, aged grey; **Se** black, shiny, large, 1 × 1 cm.

The species occurs in dense stands in a relatively small area (Hochstätter l.c.).

Y. capensis L. W. Lenz (CSJA 70(6): 289-293, ills., 1998). **T:** Mexico, Baja California Sur (*Lenz* 4501 [RSA]). – **D:** Mexico (Baja California Sur: Cape region); thorn scrub or tropical deciduous forest, 0 - 1000 m.

[1c] Stems 1 - 5.5 m, solitary, or plants becoming rhizomatous and group-forming with several un- or few-branched stems from the base, in age often decumbent; **L** narrowed above the expanded base, canaliculate in the middle, flat distally, rather thin, flexible, to 100 × 5 cm, margins dark grey, smooth to somewhat scabrous, without fibres, tip sharp but without a distinct terminal **Sp**; **Inf** broadly ellipsoid, many-flowered, scape short, not exceeding the **L**, densely tomentose to glabrous; **Fl** flat or saucer-shaped to more subglobose, to 10 cm ⌀; **Tep** elliptic, abruptly attenuate to the narrow tips, to 5 × 2.5 - 3 cm, cream-coloured; **Fr** fleshy berries, pendent, oblong-cylindrical, 5 × 11.5 cm; **Se** unknown.

According to the protologue, the species was formerly confused with the widespread coastal *Y. valida*, but differs clearly in its long wide leaves without fibrous margins, and long slender stems eventually falling down, as was first noted by Lenz (1992). It is regarded as being related with a group of mountain-dwelling mainland yuccas (*Y. schottii* auct., *Y. madrensis* and *Y. jaliscensis*).

Y. carnerosana (Trelease) McKelvey (Yuccas Southwest US 1: 24, t. 6-7, 1938). **T:** Mexico, Coahuila (*Pringle* 3912 [MO]). – **Lit:** Webber (1953); Matuda & Piña Lujan (1980: with ills.). **D:** SE USA (Texas: S-C Brewster County), NW Mexico (Chihuahua, Coahuila, Zacatecas, San Luis Potosí, Nuevo León, Tamaulipas); dry slopes in desert scrub or pine-oak forest, 850 - 2200 m, flowers March to April.

≡ *Samuela carnerosana* Trelease (1902).

[1a] Stems generally simple (very rarely 1- or 2-times branched in the upper part), sometimes forming groups of stems united at the base, 1.5 - 6 (to ≥ 10) m; **L** rigid, constricted near the base, 50 - 100 × 5 - 7.5 cm, bluish-green, margins richly filiferous; **Bra** persistent, white; **Inf** with large and strong scape, ellipsoid, exserted from the **L**, densely branching; **Fl** 45 - 90 mm, strongly scented; **OTep** 67 - 94 × 13 - 21 mm, **ITep** 65 - 93 × 20 - 28 mm, tube 17 - 30 mm; **Ov** 6 - 9 mm ⌀; **Fr** fleshy berries, oblong, 5 - 7.5 × 4 cm; **Se** 7 - 9 × 8 - 10 mm.

Y. coahuilensis Matuda & Piña Lujan (Pl. Mex. Gen. Yucca, 120-122, ills., 1980). **T:** Mexico, Coahuila (*Matuda* 38790 [UNAM]). – **D:** Mexico (Coahuila); grassland and small-leaved desert scrub, ± 360 m, flowers May to June.

[3a] Acaulescent; **L** many, canaliculate, 73 - 80 × 1 - 1.2 cm, margins white or greyish, hardly filiferous, terminal **Sp** very pungent; **Inf** 2.2 - 2.5 m; **Tep** lanceolate or ovate-lanceolate, to 40 × 12 - 16

mm, white; **Ov** cylindrical, 22 mm; **Fr** dry capsules, oblong-globose, to 7 × 3.5 cm.

Y. constricta Buckley (Proc. Acad. Nat. Sci. Philadelphia 1862: 8, 1863). **T:** USA, Texas (*Buckley* s.n. [PH?]). − **Lit:** Webber (1953); Hochstätter (2000b); both with ills. **D:** USA (Texas); 280 - 1230 m.

    Incl. *Yucca glauca* var. *constricta* Hort. Mesa Garden (s.a.) (*nom. inval.*, Art. 30.3); **incl.** *Yucca angustifolia* Carrière (1860) (*nom. illeg.*, Art. 53.1); **incl.** *Yucca albo-spica* hort. *ex* van Houtte (1867) (*nom. inval.*, Art. 32.1c?); **incl.** *Yucca polyphylla* Baker (1870).

    [3c] Acaulescent or stems sometimes to 1 (-1.5) m; **Ros** in small to larger lax groups, with 100 - 200 **L**; **L** grass-like, flexible, weak, sometimes appearing somewhat stiff, narrow at the base, broader in the middle, 20 - 50 × 1 - 2.5 cm, light to dark green, bluish-green, margins white, old grey to green, fibrous, curly but fibres soon eroding away, terminal **Sp** sharp; **Inf** 2.5 (-3) m, scape often longer than the **Fl**-bearing part, **Fl** high above the **L**, part-**Inf** from just at the base, with few to many ascending spreading **Br**; **Fl** tubular, 25 - 40 mm; **Tep** elliptic, thin, acute, pale greenish-white; **Ov** oblong-cylindrical, 2 cm; **Fr** dry capsules, oblong-cylindrical, 3 - 4.5 × 1.5 - 2 cm; **Se** black, 0.8 × 0.5 cm.

    Hochstätter (2000b) suggested that this species and *Y. campestris* (both dry-fruited) exhibit close affinities with the Yuccas of the E USA (*Y. filamentosa*, fruits dry; *Y. gloriosa*, fruits corky; *Y. recurvifolia*, fruits fleshy) based on their usually broader sharper grass-like leaves, and similar shape and colour of the style.

Y. decipiens Trelease (Annual Rep. Missouri Bot. Gard. 18: 228, 1907). **T:** Mexico, San Luis Potosí (*Anonymus* s.n. [MO?]). − **Lit:** Matuda & Piña Lujan (1980: with ills.). **D:** Mexico (Durango, Zacatecas, San Luis Potosí, Jalisco, Aguascalientes, Guanajuato); well-drained plains with deep soil, 1800 - 2400 m.

    [1c] Arborescent, stems to 15 m, **Br** numerous, to 90; **L** linear-ensiform, nearly plane, not very rigid, to 58 × 2.5 cm, shiny on both faces, margins with numerous curled greyish fibres; **Inf** scape overtopping the **L**, **Inf** ± conical, erect or somewhat curved, to 1 m; **Ped** to 2.5 cm; **Fl** many; **Tep** 40 - 55 × 11 - 18 mm; **Fr** fleshy berries, pendent, oblong, 5 - 8.8 × 2.5 - 3.2 cm, rostrate; **Se** 8 × 2 mm.

Y. declinata Laferrière (CSJA 67(6): 347-348, ills., 1995). **T:** Mexico, Sonora (*Gentry* 16615 [ARIZ 267477, ARIZ, US]). − **D:** Mexico (Sonora), open woodland, volcanic and limestone soils.

    [1b] Arborescent, stems thick, robust, 3 - 6 m, branching and forming a crown, suckering at the base when fully grown; **L** deflexed towards the stem, straight, canaliculate, 50 - 140 × 5 - 6 cm, yel-

lowish-green, margins smooth, with age becoming frayed into threads; **Inf** 1 - 1.3 m, glabrous, usually inclined; **Fl** small; **Tep** lanceolate, 4 - 5 × 0.8 - 1.2 cm, white; **Fil** pubescent, 1.1 - 1.8 cm; **Ov** elongate, 3.5 - 5 cm; **Fr** indehiscent, oblong, tapering at the base, 15 - 20 cm; **Se** flat, slightly ovoid, 1 - 1.5 cm ∅, black.

    The type collection was previously tentatively identified as *Y. grandiflora* by Gentry (1972: 162). Since Gentry's fieldnotes indicated that he considered the plant significantly different from its close relatives *Y. grandiflora* and *Y. arizonica*, Laferrière formally described the plant based on Gentry's specimens and notes. According to the protologue, *Y. declinata* is most distinctive in its horizontally oriented inflorescences. In addition, it differs from *Y. grandiflora* in its smaller flowers and glabrous rachis, from *Y. arizonica* (here treated as synonym of *Y. ×schottii*) in its taller habit, larger leaves, and more open inflorescences with flowers with shorter tepals and shorter stamens, and from *Y. schottii* auct. by its elongate ovary and glabrous inflorescence.

Y. elata Engelmann (Bot. Gaz. (Crawfordsville) 7: 17, 1882). **T** [lecto]: USA, Arizona (*Rothrock* 382 [US]). − **Lit:** Webber (1953); Reveal (1977); Matuda & Piña Lujan (1980); all with ills. **D:** SW USA, NW Mexico.

    Reveal (1977) placed under *Y. elata* all those plants from the SW USA with paniculate inflorescences with only the uppermost flowers arranged in racemes. These plants are almost always caulescent, in contrast *Y. angustissima*, which includes plants that are generally acaulescent and have inflorescences that are almost always strictly racemose. *Y. elata* is easily recognizable in its native range by its elegant crown of narrow flexible finely filiferous leaves with thin white margins on well-developed trunks (Gentry 1972).

    Webber (1953) assumes hybridization with *Y. glauca, Y. constricta* and *Y. baileyi* wherever they co-occur.

Y. elata var. **elata** − **D:** USA (S Arizona, S and C New Mexico, W Texas), N Mexico (N Chihuahua); desert places on sandy and gravelly soils, ± 500 - 2000 m, flowers April to June.

    Incl. *Yucca elata* var. *magdalenae* Hort. Mattern (s.a.) (*nom. inval.*, Art. 29.1); **incl.** *Yucca angustifolia* var. *radiosa* Engelmann (1871) ≡ *Yucca radiosa* (Engelmann) Trelease (1902); **incl.** *Yucca angustifolia* var. *elata* Engelmann (1873) (*nom. illeg.*, Art. 52.1).

    [3b] Arborescent with 1 to several stems 0.3 - 4.5 m tall, often branched above, solitary or in large clumps; **Ros** large; **L** numerous, divergent, finally reflexing and persisting as a dry skirt on the trunk, narrowly linear, lower face convex, upper face flat, striate, 30 - 90 × 0.5 - 0.7 (-1.3) cm, pale to yellow-

green, margins white to greenish-white, finely filiferous, tip long acuminate with a short **Sp**; **Inf** scape 0.5 - 1.5 (-2) m, glabrous, green to reddish or yellowish, relatively slender, extending well beyond the **L**, flowering part 0.5 - 1.5 (-3) m, ellipsoid, paniculate; **Ped** (0.7-) 1 - 2.5 cm; **Fl** pendent, campanulate to somewhat globose; **Tep** ovate to obovate or broadly elliptic, 35 - 50 × 15 - 25 mm, **ITep** broader, white to cream-coloured, or tinged with green or pink, tube 2 - 7 mm; **Ov** oblong-cylindrical, with deep **Ca** sutures, abruptly terminating in a **Sty** of 15 - 20 × ± 8 mm; **Fr** dry capsules, oblong-cylindrical, rather thin-walled, 5 - 8 × 3 - 6 cm, smooth, whitish; **Se** 7 - 10 × 9 - 14 mm. − *Cytology:* n = 30.

Y. elata var. **utahensis** (McKelvey) Reveal (in Cronquist & al., Intermountain Fl. 6: 533, 1977). **T**: USA, Utah (*McKelvey* 4167 [A]). − **D**: SW USA (SW Utah, N Arizona); 850 - 2200 m, flowers late April to early June.
≡ *Yucca utahensis* McKelvey (1947) ≡ *Yucca elata* ssp. *utahensis* (McKelvey) Hochstätter (1999).
[3b] Differs from var. *elata*: Acaulescent or more often caulescent with procumbent stems of 0.6 - 1.3 m; **L** 20 - 70 × 0.7 - 2 cm; **Ov** 2.5 - 3.5 cm; **Fr** 5 - 6 cm.
Webber (1953) interpreted this and the following variety as hybrid populations. This is regarded as being unjustified by Reveal (1977).

Y. elata var. **verdiensis** (McKelvey) Reveal (in Cronquist & al., Intermountain Fl. 6: 533-534, 1977). **T**: USA, Arizona (*McKelvey* 2752 [A]). − **D**: SW USA (C and S Arizona); 900 - 2000 m, flowers May to mid-June.
≡ *Yucca verdiensis* McKelvey (1947) ≡ *Yucca elata* ssp. *verdiensis* (McKelvey) Hochstätter (2000).
[3b] Differs from var. *elata*: Caulescent, stems distinct but short; **L** shorter, 25 - 45 × 0.4 - 0.7 (-1.3) cm; **Ov** 2 - 2.5 cm; **Fr** smaller, 4 - 4.5 cm.

Y. elephantipes Regel (Gartenflora 9: 35, 1859). − **Lit**: Lott & García-Mendoza (1994). **D**: Mexico (Chiapas), Guatemala (probably cultivated only).
Incl. *Yucca guatemalensis* Baker (1872); **incl.** *Yucca ghiesbreghtii* hort. *ex* Baker (1880); **incl.** *Yucca gigantea* Baker (1880); **incl.** *Yucca lenneana* Baker (1880); **incl.** *Yucca mooreana* Hort. Peacock *ex* Baker (1880); **incl.** *Yucca roezlii* hort. *ex* Baker (1880); **incl.** *Yucca mazelii* hort. *ex* W. Watson (1889).
[1c] Arborescent, stems 3 - 10 m, numerous from a thickened-inflated trunk-like base, slender and densely branched in the upper parts; **L** patent, narrowed to 1.5 - 2 cm above the base, flat or slightly canaliculate, (35-) 50 - 100 × 5 - 7 cm, brilliant dark green, tip acute, margins finely denticulate, with a yellow border; **Inf** paniculate, erect, dense,

surpassing the **L** only with the upper ¼; **Fl** pendent, globose; **Tep** narrowly ovate; **ITep** somewhat broader than the **OTep**, 30 - 50 × (10-) 15 - 20 mm, white to whitish; **Fil** 8 - 10 mm; **Ov** oblong, not constricted at the base, 10 - 15 mm; **Fr** fleshy berries, elllipsoid, 7 - 8 × 4.5 cm, pulpa greenish to whitish; **Se** 8 - 10 mm.
Widely cultivated as an ornamental in (sub-) tropical gardens. The flowers are reportedly edible. Lott & García-Mendoza (1994) and USDA (2001) apparently erroneously used the younger synonym *Y. guatemalensis* for this species.

Y. endlichiana Trelease (Annual Rep. Missouri Bot. Gard. 18: 229, t. 15-17, 1907). **T**: Mexico, Coahuila (*Endlich* s.n. [MO]). − **D**: Mexico (Coahuila); arid Chihuahuan Desert scrub, ± 1200 m. **I**: Matuda & Piña Lujan (1980). **Fig. IX.d**
[1b] Acaulescent; **Ros** rhizomatous, surculose; **L** few, erect, rigid, thick, semicircular in cross-section at the base, conduplicate further up, to 50 × 1.5 cm, bluish-green, brown-reddish at the base, margins finely fibrous, chestnut-brown, terminal **Sp** conical, short; **Inf** much shorter than the **L**, part-**Inf** with up to 6 **Fl**; **Ped** filiform, 2.5 cm; **Per** whitish, outside with brownish-red tinge, tube ovate, acute, 18 × 5 mm; **Fil** short, finely papillose; **Ov** oblong; **Fr** fleshy berries, pendent, subglobose to narrowly ellipsoid, 3 × 2 - 2.5 cm; **Se** 5 × 6 - 7 mm.
This is possibly the most succulent species of *Yucca*. It certainly is very desirable and suitable for collections due to its small size.

Y. faxoniana (Trelease) Sargent (Man. Trees [ed. 1], 121, fig. 106, 1905). **T**: USA, Texas (*Anonymus* s.n. [A?, MO?]). − **D**: SE USA (Texas), N Mexico (Chihuahua, Coahuila); dry slopes in desert scrub, 500 - 1500, flowers March to April. **Fig. IX.f, X.a**
≡ *Samuela faxoniana* Trelease (1902); **incl.** *Yucca australis* Trelease (1894) (*nom. illeg.*, Art. 53.1); **incl.** *Yucca macrocarpa* Sargent (1895) (*nom. illeg.*, Art. 53.1); **incl.** *Yucca australis* var. *valida* M. E. Jones (1929) (*nom. inval.*, Art. 43.1).
[1a] Stems simple, rarely branched 1 - 2× in the upper part, sometimes forming dense groups branched at the base, 2 - 6.5 m; **L** rigid, constricted near the base, 85 - 120 × 5 - 7.5 cm, bluish-green, margins richly filiferous; **Bra** persistent, white, sometimes rose-coloured; **Inf** with a short strong scape, exceeding the **L** for ½ - ¾, narrowly conical, openly branched; **Fl** 4 - 7 cm; **OTep** 55 - 87 × 15 - 17 mm, **ITep** 54 - 85 × 19 - 20 mm, tube 10 - 18 mm; **Ov** narrowly ovoid, 6 - 8 mm ∅; **Fr** fleshy berries, beaked, with adhering **Tep** remains, 3 - 9 × 2.5 - 3 cm; **Se** rough, 5 - 8 × 7 - 10 mm.

Y. filamentosa Linné (Spec. Pl. [ed. 1], 319, 1753). − **D**: E USA (New Jersey, Maryland, Virginia, West Virginia, North Carolina, Tennessee, South Carolina, Georgia, Florida, Alabama, Mississippi, Louisiana); sandy soils, flowers from mid-spring to early summer.

Incl. *Yucca glauca* Noisette *ex* Sims (1826) (*nom. illeg.*, Art. 53.1); **incl.** *Yucca filamentosa* var. *mexicana* S. Schauer (1847); **incl.** *Yucca filamentosa* var. *recurvifolia* Alph. Wood (1861); **incl.** *Yucca exigua* Baker (1871); **incl.** *Yucca filamentosa* var. *grandiflora* Baker (1872); **incl.** *Yucca filamentosa* var. *angustifolia* Engelmann (1873); **incl.** *Yucca filamentosa* var. *bracteata* Engelmann (1873); **incl.** *Yucca filamentosa* var. *laevigata* Engelmann (1873); **incl.** *Yucca filamentosa* var. *latifolia* Engelmann (1873); **incl.** *Yucca filamentosa* fa. *genuina* Engelmann (1873) (*nom. inval.*, Art. 24.3); **incl.** *Yucca antwerpensis* hort. *ex* Baker (1880).

[3g] Stems (almost) none, hidden by the **L** when present; **Ros** stoloniferous, clump-forming; **L** erect to spreading and recurved, oblanceolate, flexible, very clearly narrowed towards the base, rather abruptly tapering towards the tip, 50 - 75 × 2 - 4 cm, green or slightly glaucous, margins inrolled at the tip, otherwise splitting into stout curled fibres; **Inf** to 4.5 m, **Fl**-bearing part well above the **L**; **Fl** pendent, campanulate, 5 - 7 cm; **Tep** abruptly mucronate, 5 - 7 cm, white tinged with green, yellow or cream; **Fr** dry capsules, oblong, 3.8 - 5 × 2 cm; **Se** thin, flat, winged, 6 mm.

Very close to and possibly not distinct from *Y. flaccida* (see there for differences). Hardy outdoors in C Europe and therefore widely cultivated in many selections, including variegated forms.

Y. filifera Chabaud (Rev. Hort. 48: 432-434, 1876). **Nom. illeg.**, Art. 53.1. – **Lit:** Matuda & Piña Lujan (1980: with ills.); McVaugh (1989: with ills. as *Y. australis*). **D:** C Mexico (Chihuahua, Coahuila, Nuevo León, S Zacatecas, San Luis Potosí, Aguascalientes, NE Jalisco, Tamaulipas, Guanajuato, Querétaro, Hidalgo, Michoacán, México); plains with arid desert scrub, 500 - 2400 m, flowers January to March.

≡ *Yucca canaliculata* var. *filifera* (Chabaud) Fenzi (1889); **incl.** *Yucca baccata* var. *australis* Engelmann (1873) ≡ *Yucca australis* (Engelmann) Trelease (1892).

[1c] Arborescent, stems to 10 - 13 m tall, much-branched, old plants with up to 40 **Br**, trunk short, to 1.5 m ∅; **L** linear-oblanceolate, constricted near the base, rigid, generally asperous on both faces, 30 - 60 × 2 (-3.5) cm, margins with numerous spiralled white fibres, esp. on young **L**, terminal **Sp** stout, 1 - 3 cm, dark; **Inf** pendent, 0.6 - 1.5 m, ± cylindrical, obscurely puberulent to glabrous; **Ped** to 2.7 cm; **Per** white; **OTep** 30 - 52 × 7 - 25 mm, **ITep** somewhat broader; **Fil** 10 - 15 mm; **Ov** 18 - 20 × 4 - 5 mm; **Fr** fleshy berries, pendent, oblong, 5 - 8.8 × 2.7 - 3.3 cm, beak 0.2 - 0.7 cm, with the flavour and consistency of dates; **Se** somewhat rugose, 8 × 2 mm.

This name is unfortunately a later homonym of *Y.*

filifera hort. *ex* Engelmann 1873. McVaugh (1989) used the name *Y. australis* instead, but in order to avoid a name change, *Y. filifera* Chabaud should be proposed for conservation.

This is the most widely distributed species and it forms an important element of the tree stratum of the (Chihuahuan) desert scrub of C Mexico (Matuda & Piña Lujan l.c.).

Y. flaccida Haworth (Suppl. Pl. Succ., 34, 1819). – **D:** E USA (Texas, Oklahoma, Missouri, Arkansas, Louisiana, Tennessee, Mississippi, Alabama, Georgia, North Carolina, South Carolina, Florida); (semi-) open sites in pine scrub or woodland and coastal sands, flowers in spring.

≡ *Yucca filamentosa* var. *flaccida* (Haworth) Engelmann (1873) ≡ *Yucca filamentosa* fa. *flaccida* (Haworth) Voss (1895); **incl.** *Yucca concava* Haworth (1819) ≡ *Yucca filamentosa* var. *concava* (Haworth) Baker (s.a.) ≡ *Yucca filamentosa* fa. *concava* (Haworth) Voss (1895); **incl.** *Yucca glaucescens* Haworth (1819) ≡ *Yucca filamentosa* var. *glaucescens* (Haworth) Baker (s.a.) ≡ *Yucca flaccida* var. *glaucescens* (Haworth) Trelease (s.a.) ≡ *Yucca filamentosa* fa. *glaucescens* (Haworth) Voss (1895); **incl.** *Yucca puberula* Haworth (1828) ≡ *Yucca filamentosa* var. *puberula* (Haworth) Baker (s.a.) ≡ *Yucca filamentosa* fa. *puberula* (Haworth) Voss (1895); **incl.** *Yucca orchioides* Carrière (1861) ≡ *Yucca filamentosa* fa. *orchioides* (Carrière) Voss (1895); **incl.** *Yucca orchioides* var. *major* Baker (1877) ≡ *Yucca flaccida* var. *major* (Baker) M. L. Rehder (s.a.); **incl.** *Yucca smalliana* Fernald (1944) ≡ *Yucca filamentosa* var. *smalliana* (Fernald) Ahles (1964).

[3g] Acaulescent or stems short; **Ros** stoloniferous, clump-forming, dying slowly after flowering; **L** lanceolate, erect, gradually tapering towards the tip, flattened, glabrous, arching or curved in the middle with age, 40 - 80 × 1 - 4 (-5) cm, margins filiferous, terminal **Sp** pungent; **Inf** 0.9 - 4.25 m, scape 0.5 - 2.75 m, glabrous or pubescent; **Ped** 1.5 - 3 cm; **Fl** 4 - 5 cm; **Ov** 15 mm, pale green; **Tep** lanceolate to elliptic, tip obtuse, 3 - 5 × 1 - 3 cm, white, creamy-white or light greenish-white; **Fr** dehiscent capsules, oblong, inversely pear-shaped or conical, to 3.5 - 4 × 1.5 - 2 cm; **Se** 6 - 8 × 5 - 6 mm.

Very close to and probably better regarded as a variety of *Y. filamentosa*, from which it differs mainly in minor morphological features (thinner and narrower leaves and smaller, narrower flowers vs. thick and stiff leaves and larger flowers in the latter).

Y. glauca Nuttall (Cat. Pl. Upper Louisiana no. 89, 1813). **T** [neo]: USA, Montana (*Hochstätter* 1178.69 [SRP]). – **D:** S Canada, USA.

Hochstätter (1998: 74) provided a somewhat cryptic neotypification for this name. He later listed this neotype under *Y. glauca* [fa.] *montana* Hochstätter 1998 (*nom. nud.*).

Y. glauca var. **glauca** – **Lit:** Webber (1953); Hochstätter (2000b); both with ills. **D:** S Canada (S Alberta), USA (Montana, North Dakota, South Dakota, Minnesota, Wyoming, Nebraska, Iowa, Colorado, Kansas, Missouri, Texas, Oklahoma, Arkansas, Louisiana); common in Great Plains grasslands, badlands and mountains, 800 - 2600 (-2800) m, flowers May to July.

    Incl. *Yucca glauca* var. *arkansana* Hort. Mesa Garden (s.a.) (*nom. inval.*, Art. 29.1); **incl.** *Yucca glauca* var. *baileyi* Hort. Mesa Garden (s.a.) (*nom. inval.*, Art. 29.1); **incl.** *Yucca glauca* var. *elata* Hort. Mesa Garden (s.a.) (*nom. inval.*, Art. 29.1); **incl.** *Yucca glauca* var. *radiosa* Hort. Mesa Garden (s.a.) (*nom. inval.*, Art. 29.1); **incl.** *Yucca angustifolia* Pursh (1814); **incl.** *Yucca glauca* var. *rosea* D. M. Andrews (1934); **incl.** *Yucca glauca* fa. *montana* Hochstätter (1998) (*nom. inval.*, Art. 32.1c); **incl.** *Yucca glauca* ssp. *albertana* Hochstätter (2000).

    [3f] Acaulescent or stems short to 30 cm; **Ros** first single but soon clumped, groups dense, 0.8 - 2.5 m ∅; **L** divergently spreading, linear, upper face flat, lower face convex, occasionally triquetrous or nearly flat, flexible, striate, (20-) 50 - 70 × 0.5 - 1.1 cm, pale green or pallid, margins white or greenish-white, soon finely filiferous, terminal **Sp** short, acute, brownish; **Inf** (0.4-) 0.9 - 1.25 m, scape 24 - 53 cm, **Fl**-bearing part from between the **L**, part-**Inf** usually none or rarely few, abortive at the base; **Fl** globose or campanulate; **Tep** acute, 46 - 61 × 26 - 42 mm, greenish-white, commonly tinged with purple and shiny; **Ov** obovate, 20 × 9 - 13 mm, white or rarely greenish-white, abruptly terminating into the **Sty**; **Fr** dry capsules, oblong-cylindrical, 58 - 62 × 45 - 53 mm, beaked; **Se** 7 - 9 × 8 - 10 mm, with broad marginal wing, black.

    The measurements given by Hochstätter (1998) partly differ considerably from those of Webber (l.c.) reproduced here. The species represents a characteristic element of the prairie grassland plains in the mid-western USA.

    Y. glauca ssp. *albertana* differs in minor quantitative features only that fall well within the range given by Webber (l.c.). It might represent an artificial segregate for the smallest N-most form within the complex only, and is preliminarily included in the synonymy here.

Y. glauca var. **stricta** (Sims) Trelease (Annual Rep. Missouri Bot. Gard. 13: 61, tt. 25-27, 1902). **T** [neo]: USA, Kansas (*McKelvey* 2842 [A]). – **D:** USA (SE Colorado, SW Kansas, NE New Mexico, NW Oklahoma); 900 - 1500 m, flowers April to June. **I:** Hochstätter (1998).

    ≡ *Yucca stricta* Sims (1821) ≡ *Yucca angustifolia* var. *stricta* (Sims) Voss (1895) ≡ *Yucca glauca* ssp. *stricta* (Sims) Hochstätter (1999); **incl.** *Yucca glau-*

ca var. *gurneyi* McKelvey (1947) ≡ *Yucca glauca* ssp. *gurneyi* (McKelvey) Hochstätter (1999) (*nom. inval.*, Art. 29.1?).

    [3f] Differs from var. *glauca*: Plants generally more robust; **L** to 75 cm; **Inf** larger, > 1.8 m, tip long, racemose; **Se** smaller, ± 8 × 4 mm.

    Kartesz (1996) and USDA (2001) erroneously both use the younger synonym var. *gurneyi* for this taxon. Hochstätter (1998) selected a neotype.

Y. gloriosa Linné (Spec. Pl. [ed. 1], 319, 1753). – **D:** SE USA (North Carolina, South Carolina, Georgia); costal dunes, flowers in spring.

    Incl. *Yucca integerrima* Stokes (1812); **incl.** *Yucca obliqua* Haworth (1819) ≡ *Yucca gloriosa* fa. *obliqua* (Haworth) Voss (1895); **incl.** *Yucca rufocincta* Haworth (1819) ≡ *Yucca gloriosa* fa. *rubrocincta* (Haworth) Voss (1895); **incl.** *Yucca superba* Haworth (1819) (*nom. illeg.*, Art. 53.1); **incl.** *Yucca acuminata* Sweet (1828) ≡ *Yucca gloriosa* fa. *acuminata* (Sweet) Voss (1895); **incl.** *Yucca gloriosa* var. *marginata* Carrière (1859); **incl.** *Yucca gloriosa* var. *tristis* Carrière (1859); **incl.** *Yucca gloriosa* var. *variegata* Carrière & Hort. Belg. (1859); **incl.** *Yucca japonica* hort. *ex* Carrière (1859); **incl.** *Yucca pendula* Sieber *ex* Carrière (1859); **incl.** *Yucca ellacombei* hort. *ex* Baker (1870); **incl.** *Yucca patens* André (1870); **incl.** *Yucca pruinosa* Baker (1870) ≡ *Yucca gloriosa* fa. *pruinosa* (Baker) Voss (1895); **incl.** *Yucca tortulata* Baker (1870) ≡ *Yucca gloriosa* fa. *tortulata* (Baker) Voss (1895); **incl.** *Yucca gloriosa* fa. *planifolia* Engelmann (1873); **incl.** *Yucca gloriosa* fa. *plicata* Engelmann (1873); **incl.** *Yucca gloriosa* var. *planifolia* Engelmann (1873); **incl.** *Yucca gloriosa* var. *plicata* Engelmann (1873); **incl.** *Yucca plicata* hort. *ex* C. Koch (1873); **incl.** *Yucca plicatilis* hort. *ex* C. Koch (1873); **incl.** *Yucca gloriosa* var. *genuina* Engelmann (1873) (*nom. inval.*, Art. 24.3); **incl.** *Yucca gloriosa* var. *flexilis* Trelease (1902).

    [1d] Stems woody, to 5 m, simple or rarely ultimately branched; **L** lanceolate, mainly stiff, erect, ascending or recurved, somewhat narrowed towards the base, flexible, flat or pleated near the tip, smooth, 40 - 70 × 4 - 6 cm, glaucous when young, green or bluish-green when old, margins entire or with a few inconspicuous denticles, opaque, brown, often becoming frayed; **Inf** large, erect, 1.65 - 2.7 m, to 45 cm broad, **Fl** starting 40 - 50 cm above the **L** tips; **Ped** to 2 cm; **Fl** pendent, campanulate; **Tep** oblong-lanceolate, 4 - 5 × 2 - 2.5 cm, greenish-white, cream-coloured or reddish; **Fr** berry-like, not fleshy and indehiscent-leathery, pendent, 6-ribbed, 5.5 - 8 cm; **Se** unwinged, lustrous, 5 - 7 mm ∅, black.

    Y. gloriosa and its closest relative *Y. recurvifolia* (for differences see there) differ from all other species of Sect. *Yucca* in possessing indehiscent fruits, which are corky-leathery and not fleshy. In its growh habit, the species is similar to *Y. aloifolia*, but its habitat is more mound-like due to the terminal branching mode in contrast to a branching more from the middle with a trunk thus appearing more open. Commonly cultivated in the S USA and persisting at old homesites. – Probably the most widely grown *Yucca* species with many cultivars, hybrids and selections. *Y. ellacombei* hort. *ex* Baker (1870) represents another name for the horticultural selection *Y. gloriosa* 'Nobilis'.

Y. grandiflora Gentry (Madroño 14: 51-53, 1957). **T:** Mexico, Sonora (*Gentry* 11601 [US]). – **Lit:** Gentry (1972); Matuda & Piña Lujan (1980); both with ills. **D:** N Mexico (Sonora); grassy slopes in open woodland, on volcanic or limestone rocks, 600 - 1300 m, flowers February to April.

[1b] Stems branched at the base and above, 4 - 6 m tall; **L** ascending to descending, persisting dry and deflected when old and forming a skirt on the trunk, slightly narrowed above the base, smooth, 70 - 100 (-140) × 4 - 5 cm, dark green, margins narrow, brown, terminal **Sp** stout, broadly grooved, (chestnut-) brown; **Inf** erect or deflexed, open, 0.7 - 1 m, scape 10 - 30 cm, part-**Inf** densely white-tomentose, horizontal or slightly ascending; **Ped** short to nearly none; **Fl** erect or divergent, large; **Tep** spreading, ovate, thin, bluntly mucronate, connate at the base, 6 - 9 cm, creamy-white; **Ov** elongate, 4.5 - 6 cm; **Fr** fleshy berries, large.

Y. harrimaniae Trelease (Annual Rep. Missouri Bot. Gard. 13: 59, 1902). **T:** USA, Utah (*Trelease* s.n. [MO?]). – **Lit:** Reveal (1977); Hochstätter (2000b); both with ills. **D:** SW USA. **I:** Hochstätter (1999a).

Incl. *Yucca coloma* Andrews (1926).

Y. harrimaniae var. **harrimaniae** – **D:** SW USA (E-C Nevada, Utah, W and C Colorado, NE Arizona, N New Mexico); desert slopes and foothills, mostly on limestone, usually 1000 - 2500 m, flowers April to July. **I:** Hochstätter (1998).

Incl. *Yucca harrimaniae* var. *gilbertiana* Trelease (1907) ≡ *Yucca gilbertiana* (Trelease) Rydberg (1918) ≡ *Yucca harrimaniae* ssp. *gilbertiana* (Trelease) Hochstätter (2000); **incl.** *Yucca nana* Hochstätter (1998).

[3d] Acaulescent; **Ros** forming dense to open clumps of 3 - 20 **Ros**, 30 - 80 cm ∅; **L** linear to lanceolate or spatulate-lanceolate, canaliculate, lower face convex, base broad, rigid and stiff, striate, 10 - 50 × 0.7 - 4 cm, rather glaucous, grey- or blue- to deep green, margins at first papery, soon separating into long fine to coarse curly fibres, tip tapering gradually to a short stiff ivory-coloured terminal **Sp**; **Inf** erect, racemose or rarely with few short part-**Inf**, scape 10 - 40 cm, **Fl**-bearing part 35 - 70 cm; **Ped** 1 - 2 cm; **Fl** pendent, broadly campanulate, 4 - 6 (-6.5) cm; **Tep** ovate, fleshy, **ITep** broader than the **OTep**, white to pale green or yellowish to

greenish-yellow, commonly tinged with purple, tube 2 - 4 mm; **Ov** 1.5 - 2 cm; **Fr** dry capsules, cylindrical, (2.2-) 4 - 5 (-6) × 2 - 3 cm, deeply constricted towards the middle, opening mainly above the constriction only; **Se** 5 - 6 mm, dull black.

The recently described *Y. nana* clearly represents a redundant description of a small-growing local variant within this variable taxon.

Y. harrimaniae var. **neomexicana** (Wooton & Standley) Reveal (in Cronquist & al., Intermountain Fl. 6: 530, 1977). **T:** USA, New Mexico (*Standley* 6208 [US?]). – **Lit:** Reveal (1977); Hochstätter (2000b); both with ills. **D:** SW USA (SE Colorado, adjacent New Mexico).

≡ *Yucca neomexicana* Wooton & Standley (1913) ≡ *Yucca harrimaniae* ssp. *neomexicana* (Wooton & Standley) Hochstätter (1999).

[3d] Differs from var. *harrimaniae*: **L** linear, narrower, only 7 - 20 mm broad; **Tep** pure white.

The recent transfer to subspecies level of this and other taxa by Hochstätter (1999b) is – though it conforms with current trends in other genera and families – not accepted here, since many additional combinations would be necessary to give a comparable treatment of the groups.

Y. harrimaniae var. **sterilis** Neese & S. L. Welsh (Great Basin Naturalist 45(4): 789, 1986). **T:** (*Welsh* 18461 [BRY]). – **Lit:** Reveal (1977); Hochstätter (2000b); both with ills. **D:** SW USA (NE Utah: Uintah Basin); salt desert scrub communities.

≡ *Yucca harrimaniae* ssp. *sterilis* (Neese & S. L. Welsh) Hochstätter (1999).

[3d] Differs from var. *harrimaniae*: Strongly rhizomatous and **Ros** often widely spaced; **L** flaccid, often reclining on the ground, typically curved, margins sparingly or not filiferous; **Inf** only to 40 cm; **Fr** not known to be formed.

A geographically isolated variety.

Y. jaliscensis Trelease (CUSNH 23(1): 92, 1920). **T** [lecto]: Mexico, Jalisco (*Pringle* 4392 [US [status?]]). – **Lit:** Matuda & Piña Lujan (1980: with ills.); McVaugh (1989). **D:** Mexico (Jalisco, Guanajuato, Colima); plains with deeper soils or moderate slopes, 1000 - 1600 m, flowers September to May or almost throughout the year.

Incl. *Yucca schottii* var. *jaliscensis* Trelease (1902).

[1c] Arborescent, stems to 12 m, much branched, **Br** often 5 - 8, long and upright; **L** diverging, concave, flexible, (30-) 40 - 100 × (4-) 6 - 7.5 (-8) cm, glaucous-green, margins hardly fibrous, tip sharp-pointed but scarcely differentiated as **Sp**; **Inf** narrowly ellipsoid, 0.5 - 1 m, erect, or drooping at **Fr** time, largely enclosed within the **L**, densely tomentose to canescent with thick blunt **Ha**, scape short; **Fl** subglobose; **Tep** narrowly lanceolate, 22 - 36 × 8 - 16 mm; **Ov** 5 - 7 mm ∅; **Fr** fleshy berries, ellip-

soid, narrowed at the base, asymmetrical, 6 - 12 × 2.5 - 3.8 cm; **Se** rugose, 5 - 7 × 7 - 10 mm.

See Gentry (1972: 161) on the type collection of this species. According to McVaugh (l.c.), it is often cultivated and spreading around settlements, but seldom found in the wild.

Y. lacandonica Gómez-Pompa & Valdés (Bol. Soc. Bot. México 27: 43-44, 1962). **T:** Mexico, Chiapas (*Anonymus* s.n. [not located]). – **Lit:** Matuda & Piña Lujan (1980: with ills.). **D:** Mexico (Veracruz, Tabasco, Campeche, Chiapas); humid tropical evergreen forests, flowers in May.

[1c] Plants strictly epiphytic, stems 2.5 - 3 m, upcurving apically; **L** narrowed at the base (1 cm broad), weak, rigid, to 65 × 6 cm, margins denticulate, 0.5 cm broad, yellowish, tip very acute; **Inf** scape short, ± 40 cm; **Fl** campanulate; **Tep** oblong-linear, 45 × 8 mm; **Fr** fleshy berries, immature **Fr** conical, 4 × 2 cm; **Se** unwinged, 4 - 5 × 2 - 3 mm.

This species is remarkable in being the only strictly epiphytic taxon in the *Agavaceae* (a few species of *Agave* may rarely occur as facultative epiphytes under humid conditions).

Y. linearifolia Clary (Brittonia 47(4): 394-396, ills., 1995). **T:** Mexico, Nuevo León (*Clary* 364 [MEXU, ANSM, MO, TEX, US]). – **D:** Mexico (Coahuila, Nuevo León); desert scrub on shale in shaded canyons, 1100 - 1300 m.

[1/3a] Stems 2 - 3.5 m, mostly single; **Ros** with somewhat flattened top; **L** numerous, linear, distally twisting slightly outwards, persistent when old, reflexing and completely covering the trunk, 34 - 38 × 0.4 - 0.5 (in the middle) cm, greyish-green to glaucous (new growth), margins thin, horny, pale yellow, minutely denticulate, terminal **Sp** 3 - 8 mm, dark reddish-brown to black; **Inf** 60 - 80 cm, erect, scape short, moderately branched, lower (= largest) part-**Inf** with 4 - 6 **Fl**; **Fl** campanulate; **OTep** elliptic, 30 - 33 × 15 mm, **ITep** obovate, 30 × 20 mm, creamy-white, tube none; **Ov** oblong-cylindrical, 9 - 10 mm ∅; **Fr** fleshy berries, indehiscent, asymmetrical, narrowly ovoid, tip constricted, 5 - 7 × 2.3 - 2.5 cm; **Se** polymorphic, dull, black, to 5 - 7 × 4 - 6 mm.

According to the protologue apparently close to the dry-fruited *Y. queretaroensis* and *Y. rostrata* (both Sect. *Chaenocarpa*), but different in its fleshy indehiscent fruits. The species is the only fleshy-fruited *Yucca* with narrow denticulate leaves.

Y. louisianensis Trelease (Annual Rep. Missouri Bot. Gard. 13: 64, 1902). **T:** USA, Arkansas (*Ball* 558 [MO 148578 [lecto?]]). – **Lit:** Hochstätter (1998: with ills.). **D:** USA (C and E Texas, S Arkansas, N to C Louisiana); sandy soils in lax pine- and oak-forests.

≡ *Yucca arkansana* ssp. *louisianensis* (Trelease) Hochstätter (1999); **incl.** *Yucca freemanii* Shinners

(1951) ≡ *Yucca arkansana* ssp. *freemanii* (Shinners) Hochstätter (2000).

[3e] Acaulescent; **Ros** single or in small groups; **L** ascending or recurved, grass-like, weak, flexible, 20 - 40 × 2 - 3 cm, bluish-green, margins with curled fibres; **Inf** 1 - 2.5 m, scape 0.5 - 1.5 m, somewhat longer than the **Fl**-bearing part, part-**Inf** ascending; **Fl** campanulate or nearly tubular, 3 - 6 cm, whitish-cream, somewhat tinged with greenish or reddish; **Tep** elliptic to oblong or lanceolate, 25 - 35 × 13 - 15 (-20) mm, margins irregular, tomentose, sometimes roughly dentate; **Ov** oblong to cylindrical, thickly robust; **Fr** dry capsules, oblong-ovoid, sometimes deeply constricted in the middle, 4 - 5 cm, dark brown to black; **Se** black, shiny, 8 - 10 × 4 mm.

This taxon is regarded as a possible synonym of *Y. constricta* by McKelvey (1947), whereas Hochstätter (1998) treats it as a subspecies of *Y. arkansana.*

Y. madrensis Gentry (US Dept. Agric. Handb. 399: 159, ills., 1972). **T:** Mexico, Sonora / Chihuahua (*Gentry* 21209 [US 2557499]). – **Lit:** Matuda & Piña Lujan (1980); Benson & Darrow (1981); both with ills. **D:** S USA (Arizona), N Mexico (Sonora, Chihuahua); on rocky volcanic and limestone slopes in oak woodlands and pine-oak forests, 1200 - 1650 m, flowers in summer. **I:** Hochstätter (1999c: as *Y. schottii*).

[1c] Stems at first simple, eventually surculose, short or to 3 - 5 m, rarely branched; **Ros** deeply leaved in young plants, reduced in old plants; **L** numerous, linear-lanceolate, thin, pliable, mostly straightly ascending to declined on the trunk, rarely recurving, nearly flat to slightly conduplicate, mostly 50 - 80 (-100) × (2-) 3.5 - 5 cm, bluish- (to yellowish-) green, margins thin, brown, friable, exfoliating with age, not or sparsely filiferous, terminal **Sp** weak, 0.5 - 1.5 cm, reddish-brown to grey; **Inf** short, rarely exceeding the **L**, 50 - 80 cm, densely to sparsely tomentose; **Fl** small, globose, ± 3.5 cm, broader than long, basally truncate; **Ov** thick, ± 25 mm incl. the **Sty**; abruptly tapering into the **Sti**; **Tep** ovate or ovate-lanceolate, 2 - 4 × 1 - 2 cm, white, **OTep** brownish, mucronate, tube 7 - 12 mm; **Fr** fleshy berries, rounded at the base, tapering at the tip, frequently irregularly constricted, 60 - 125 × 25 - 38 mm; **Se** thick, rough, 5 - 9 × 7 - 10 mm.

Y. madrensis is characterized by its pliable, light bluish-green leaves, short inflorescences with small flowers, and a summer-flowering habit (Gentry 1972). Lenz & Hanson (2000) recently provided a broader concept for the species as to include the common fleshy-fruited *Yucca* with long flexible blue-green leaves from the mountains of S Arizona, previously named *Y. schottii*. These plants are regarded as conspecific with *Y. madrensis*, which was previously only known from the Sonora / Chihuahua border in M Mexico.

Y. mixtecana García-Mendoza (Acta Bot. Mex. 42: 1-4, ills., 1998). **T**: Mexico, Oaxaca (*García-Mendoza & al.* 6198 [MEXU, BM, ENCB, MO, OAX, TEX]). – **D**: Mexico (S Puebla, NW Oaxaca); xerophytic scrub, 1370 - 2200 m.

[1c] Arborescent, stems 2.5 - 5 m, ± conical, simple or sparsely branched in the upper part, forming rhizomatous colonies of 10 - 25 individuals; **L** linear-lanceolate or linear, erect, 40 - 65 (-75) × 1.5 - 3 cm, glaucous to greenish-yellowish, deciduous when dry, margins entire, with a dark border, filiferous with fine and soft threads, terminal **Sp** 0.5 - 1 cm, dark brown, canaliculate; **Inf** 50 - 80 cm, erect, moderately branched, scape 20 - 30 cm, pilose, part-**Inf** 10 - 20 cm, pilose, with 15 - 20 **Fl** each; **Ped** (1-) 1.5 - 2 cm, pilose; **Fl** (1.5-) 2 - 2.5 (-3) cm, campanulate, pendent; **Tep** elliptic, 2 - 2.5 (-3) × 0.4 - 1 (-l.3) cm, **OTep** broadest, whitish to yellowish; **Ov** cylindrical, 1.5 - 2 (-3) cm; **Fr** fleshy berries, cylindrical, pendent, (3-) 5 - 8 × 2 - 2.5 cm; **Se** drop-shaped, black.

According to the protologue closest to *Y. periculosa* and *Y. jaliscensis*, from both of which it differs by its shorter, conical, slender, sparsely branched stems, narrower caducous leaves and much smaller flowers and fruits.

Y. necopina Shinners (Spring Fl. Dallas-Fort Worth Area 408, 1958). **T**: USA, Texas (*Shinners* 20/102 [TEX?]). – **Lit**: Diggs & al. (1999: with ill.). **D**: USA (N-C Texas); river terraces in sandy soils; flowers May to June.

[3e] Similar to *Y. arkansana*, but stems 1 - 3 m tall; **L** 50 - 80 × 1.5 - 4 cm, margins white, with curly fibres; **Inf** large, much-branched, well above the **L**, completely glabrous; **Tep** greenish-white; **Ov** 2× as long as the **Sty** and **Sti**.

A local endemic known from few populations only. The status of the taxon is disputed: Field and molecular studies support species rank, whereas the forthcoming treatment for the 'Flora of North America' will treat it as a synonym of *Y. arkansana*. The species appears to be closest to *Y. louisianensis*, which is distinguished by its usually narrower leaves and pubescent inflorescences (Diggs & al. 1999).

Y. pallida McKelvey (Yuccas Southwest US 2: 57, 1947). **T**: USA, Texas (*McKelvey* 2862 [A]). – **Lit**: Webber (1953); Diggs & al. (1999: with ills.); Hochstätter (2000b: with ills.). **D**: USA (N-C Texas); limestone outcrops or rocky prairies, 100 - 400 m, flowers April to June.

≡ *Yucca rupicola* fa. *pallida* (McKelvey) hort. (s.a.) (*nom. inval.*, Art. 29.1); **incl.** *Yucca rupicola* Trelease (1902) (*nom. illeg.*, Art. 53.1); **incl.** *Yucca rupicola* var. *edentata* Trelease (1912) (*incorrect name*, Art. 11.4) ≡ *Yucca pallida* var. *edentata* (Trelease) Cory (1952).

[3a] Acaulescent; **Ros** growing in small to large rhizomatous colonies of usually 10 - 30 **Ros**, usually distinctly separate from one another; **L** few, thin, flexible, straight when mature, acuminate, flat except for 1.3 - 2.5 cm below the tip, 20 - 35 × up to 3 cm, blue- to grey-green, margins flat, bright yellow, finely serrate; **Inf** 1 - 3 m; **Fl** campanulate, 5 - 6.5 cm, pendent; **Tep** narrowly to broadly elliptic, ovate, 2 - 3.5 cm, pale green, margins white, somewhat serrate; **Ov** oblong-cylindrical, 3 - 4 cm, pale blue or yellow-green; **Fr** dry capsules, 4 - 6 × 1.3 - 2.5 cm, yellowish-brown, later dark brown to black; **Se** small, dull, rough, surface sculptured, 4 - 6 × 2 - 3 mm, black.

Webber (1953) reports difficulties to distinguish *Y. pallida* from *Y. rupicola* and indirectly includes the former as synonym of the latter. However, Kartesz (1996), USDA (2001) and Diggs & al. (1999) as well as Hochstätter (2000b) all keep the species separate from its close relative *Y. rupicola*. It differs from *Y. rupicola* in forming larger clumps of 10 - 30 rosettes with straight flat leaves with flat and bright yellow margins (Diggs & al. l.c.) and flowers with longer tepals, a more stocky ovary, thicker style, as well as the erect and scarcely spreading stigma lobes (Hochstätter l.c.).

Y. periculosa Baker (Gard. Chron. 1870: 1088, 1870). – **Lit**: Matuda & Piña Lujan (1980: with ills.). **D**: Mexico (Veracruz, Tlaxcala, Puebla, Oaxaca); desert scrub, on plains with deeper soil or on moderate slopes, 1300 - 1600 m.

≡ *Yucca baccata* var. *periculosa* (Baker) Baker (1880); **incl.** *Yucca circinata* Baker (1870) ≡ *Yucca baccata* var. *circinata* (Baker) Baker (1880).

[1c] Arborescent, stems to 15 m, much-branched in age, **Br** ascending; **L** oblong or linear-lanceolate, 35 - 50 × 2 - 3.5 cm, margins finely fibrous; **Inf** broadly ovoid, erect or somewhat inclined, compact, scape enclosed within the **L**; **Ped** 10 - 15 mm; **Fl** expanded; **Tep** generally pubescent, to 35 × 10 - 12 mm; **Fr** fleshy berries, pendent, oblong, 5 - 8 × 2.5 - 3.2 cm.

Y. potosina Rzedowski (Ciencia (Mexico) 55(4-5): 90-91, 1955). – **Lit**: Matuda & Piña Lujan (1980: with ills.). **D**: Mexico (San Luis Potosí); slopes with shallow soil, submontane scrub or oak scrub, ± 1700 m, flowers June to July.

[1c] Arborescent, stems 2 - 7 m, poorly branched; **L** plane or somewhat canaliculate, 30 - 100 × 3 - 6 cm, margins dark brown, outer part grey, with thin and curled grey fibres, terminal **Sp** 2 - 3 cm, grey; **Inf** pendent, much extending the **L**, with very dense part-**Inf**; **Ped** subverticillate, 1.5 - 2.5 cm; **Tep** elliptic-oblong to obovate, truncate at the base, tip acute, mainly glabrous, **ITep** apically with pubescent margins, 25 - 50 × 8 - 20 mm, white; **Ov** 15 - 20 mm; **Fr** fleshy berries, oblong, 4 - 8 × 2.5 - 3.5 cm; **Se** obovate, 6 - 8 × 5 - 6 mm.

Y. queretaroensis Piña Lujan (Cact. Suc. Mex. 34(3): 51-56, ills., 1989). **T:** Mexico, Querétaro (*Piña Lujan* s.n. [MEXU 472851, ENCB, IZTA]). – **D:** Mexico (Querétero); known only from the type locality, 1300 m, flowers April to June.

[3a] Caulescent, stems generally unbranched, 3 - 5 m, forming small rhizomatous colonies of 3 - 10 stems; **L** numerous, linear, rigid, both faces convex, on both faces with a slight keel and 2 furrows between keel and margins, 40 - 50 × 0.3 - 0.5 cm, green, persistent, in age reflexed and covering the stems, margins horny, yellowish, finely denticulate, terminal **Sp** 0.5 - 1.5 cm, coffee-brown; **Inf** ovoid, erect, much-branched, 0.6 - 0.8 m, part-**Inf** to 14 - 16 cm, finely tomentose; **Ped** 1 - 2 cm; **Fl** pendent, campanulate to globose; **Tep** lanceolate to narrowly elliptic, 23 - 26 × 12 mm, whitish-cream, tube none; **Fr** and **Se** unknown.

See l.c. 35(3): 61-62, 1990, for details of the typification. Closest to *Y. thompsoniana* according to the protologue, but different in its higher and unbranched stems, biconvex and more slender leaves and smaller flowers.

Y. recurvifolia Salisbury (Parad. Lond. t. 31 + text, 1806). – **Lit:** Cullen (1986). **D:** SE USA (Louisiana, Mississippi, Alabama, Florida, Georgia); sandy soils in Gulf Coast plains, flowers in autumn.

≡ *Yucca gloriosa* fa. *recurvifolia* (Salisbury) Engelmann (1873); **incl.** *Yucca recurva* Haworth (1819); **incl.** *Yucca angustifolia* Karwinsky *ex* G. Don (1839) (*nom. illeg.*, Art. 53.1); **incl.** *Yucca flexilis* Carrière (1859); **incl.** *Yucca acuminata* hort. *ex* Carrière (1859) (*nom. illeg.*, Art. 53.1); **incl.** *Yucca angustifolia* hort. *ex* Carrière (1859) (*nom. illeg.*, Art. 53.1); **incl.** *Yucca longifolia* hort. *ex* Carrière (1859) (*nom. illeg.*, Art. 53.1); **incl.** *Yucca stenophylla* hort. *ex* Carrière (1859) (*nom. illeg.*, Art. 53.1); **incl.** *Yucca boerhaavii* Baker (1870) ≡ *Yucca flexilis* var. *boerhaavii* (Baker) Trelease (1902); **incl.** *Yucca semicylindrica* Baker (1870); **incl.** *Yucca eylesii* Hort. Peacock *ex* Baker (1880); **incl.** *Yucca falcata* Hort. Peacock *ex* Baker (1880); **incl.** *Yucca mexicana* hort. *ex* Baker (1880); **incl.** *Yucca nobilis* Hort. Peacock *ex* Baker (1880); **incl.** *Yucca peacockii* Baker (1880) ≡ *Yucca flexilis* var. *peacockii* (Baker) Trelease (1902); **incl.** *Yucca laevigata* hort. *ex* Nicholson (1887).

[1d] Arborescent, stems simple or sometimes branched, to 2.5 m; **L** ensiform, usually recurving in the upper ½, tapering towards the tip, pliable, 50 - 100 × 3.5 - 5 cm, mainly green but glaucous, margins narrowly yellow or brown; **Inf** 1.65 - 2.1 m, scape 0.9 - 1.1 m, narrowly ellipsoid in outline, barely exceeding the **L**; **Fl** to 7.5 cm ∅; **Tep** white or slightly greenish-white; **Fr** berries, erect, indehiscent-leathery, not fleshy, oblong, 6-winged or 6-ribbed, 2.5 - 4.5 cm; **Se** thin, 5 - 8 mm ∅.

Poorly known. Very close to and doubtfully distinct from *Y. gloriosa* (see also there), which needs further study. Major differences are its more pliable lax leaves, its inflorescence barely exceeding the leaves, its smaller and erect fruits and its autumnal flowering season (vs. mainly stiff erect leaves, inflorescences distinctly held above the leaves, larger and pendent fruits, and a spring-flowering season in *Y. gloriosa*).

Y. reverchonii Trelease (Annual Rep. Missouri Bot. Gard. 22: 102, 1911). **T:** USA, Texas (*Reverchon* s.n. [MO 148679]). – **Lit:** Webber (1953); Reveal (1977); Matuda & Piña Lujan (1980); all with ills. **D:** USA (Texas), Mexico (Coahuila); usually on rocky limestone ledges and gravelly plains in dense bush, 300 - 900 m, flowers May to mid-June.

[3a] Acaulescent, **Ros** single but in age forming small dense clumps 0.3 - 1 m ∅ with 1 - 25 **Ros**; **L** few, linear to somewhat broader towards the middle, canaliculate, quite rigid, straight, 25 - 55 × 1 - 2 cm, light glaucous-green, margins hyaline yellow or occasionally red or brown, minutely denticulate; **Inf** narrowly ovoid to narrowly pyramidal, scape slender, 46 - 110 cm, glabrous to heavily floccose, **Fl**-bearing part starting 25 - 42 cm above the **L**, 36 - 100 cm, part-**Inf** and **Fl** few; **Fl** pendent, campanulate to somewhat globose, expanding but little at anthesis; **Tep** ovate, sharply acuminate, 38 - 60 × 15 - 29 mm, white or greenish-white; **Ov** 4 - 6 mm ∅, tapering or rarely abruptly narrowed into the **Sty**; **Fr** dry capsules, ellipsoid, rarely constricted, with attenuate beak, 38 - 59 × 18 - 31 mm; **Se** flat, thin, unwinged, 5 - 6 × 6 - 7 mm, dull black.

This species is similar to *Y. rupicola* in its tall upright inflorescences, but best distinguished by its straight, thinner and longer leaves (vs. twisted in *Y. rupicola*) (Hochstätter 2000b). Its distribution lies between that of *Y. rupicola* and *Y. thompsoniana* (McKelvey 1947), leading to apparent hybridization with both species (Webber 1953).

Y. rigida (Engelmann) Trelease (Annual Rep. Missouri Bot. Gard. 13: 65, 1902). – **Lit:** Matuda & Piña Lujan (1980: with ills.). **D:** Mexico (Chihuahua, Coahuila, Durango); stony ravines and slopes in desert scrub, 1200 - 1500 m. **Fig. X.b**

≡ *Yucca rupicola* var. *rigida* Engelman (1873).

[3a] Stems to 4.5 m, **Br** none or few; **L** linear, slightly broadened in the middle, slightly canaliculate, thin, 42 - 61 × 1.2 - 1.7 cm, yellowish-green, glaucous, tip very pungent; **Inf** ellipsoid to ovoid, scape 30 - 70 cm, slightly pubescent, **Fl**-bearing part to 1 m, dense, part-**Inf** 28 - 40; **Fl** globose to campanulate; **Tep** narrowly oblong, acuminate, 42 - 50 × 11 - 20 mm; **Ov** 2 - 6 mm ∅; **Fr** dry capsules, strongly beaked, 35 - 70 × 18 - 25 mm; **Se** 4 × 6 - 7 mm.

Y. rostrata Engelmann *ex* Trelease (Annual Rep. Missouri Bot. Gard. 13: 58, 1902). **T:** Mexico, Coahuila (*Palmer* s.n. [MO 148694]). – **Lit**: Matuda & Piña Lujan (1980); Hochstätter (2000b); both with ills. **D**: USA (Texas), Mexico (Chihuahua, Coahuila); mountain slopes, canyon bottoms, plains and moderate slopes with desert scrub, 300 - 800 m, flowers March to May.

[3a] Arborescent, stems 1.8 - 3.2 (-4.5) m, erect, simple or with few **Br**; **Ros** frequently asymmetrical, rather small; **L** linear, broadest above the middle, flat to canaliculate, smooth on both faces, often twisted, 25 - 60 × 1.2 - 1.7 cm, glaucous, margins yellow, finely denticulate, terminal **Sp** very pungent; **Inf** sparsely pubescent, 0.6 - 2 m, ellipsoid to ovoid in outline, densely many-flowered, scape 0.3 - 1 m, part-**Inf** 28 - 40, to 38 cm; **Fl** globose to campanulate; **Tep** narrowly ovate, sharply acuminate, 42 - 52 × 11 - 20 mm, white; **Ov** 2 - 6 mm ∅; **Fr** ovoid to ellipsoid, rarely constricted, tip with a strong and curved beak, 3.5 - 7 × 1.8 - 2.5 cm.

Closely related to *Y. thompsoniana*, which may represent a N dwarfer variant (Webber 1953). It is best distinguished by its ± larger habit with larger inflorescences and longer and broader leaves with the widest part considerably above the middle, and smooth on both faces (vs. smaller-growing with smaller inflorescences and shorter and narrower leaves widest at or above the middle and ± scabrous on both faces in *Y. thompsoniana*).

Y. rupicola Scheele (Linnaea 23: 143, 1850). **T** [neo]: USA, Texas (*Hochstätter* 1179.91 [SRP]). – **Lit**: Reveal (1977); Matuda & Piña Lujan (1980); Hochstätter (2000b); all with ills. **D**: USA (Texas), probably adjacent Mexico; limestone ledges and grassy plains, dense bush and open woodland, 450 - 880 m.

Incl. *Yucca lutescens* Carrière (1858); **incl.** *Yucca tortilis* hort. *ex* Carrière (1858); **incl.** *Yucca rupicola* var. *tortifolia* Engelman (1873); **incl.** *Yucca tortifolia* Lindheimer *ex* Engelmann (1873) (*nom. inval.*, Art. 32.1).

[3a] Acaulescent, soon developing open clumps with 6 - 15 large **Ros**; **L** few, very broad towards the middle, concave or flat but oblique or undulate or twisted, slightly striate, flaccid, 30 - 58 × 2 - 4 cm, dark green, margins hyaline reddish-brown or occasionally yellow, minutely denticulate, tip pungent; **Inf** narrowly ovoid to narrowly pyramidal, scape slender, 36 - 152 cm, glabrous to slightly floccose, **Fl**-bearing part starting 24 - 48 cm above the **L**, 31 - 100 cm, part-**Inf** 8 - 16, 1 - 13 cm; **Fl** few, pendent, mainly campanulate, expanding but little, rarely somewhat globose and open; **Tep** ovate, sharply acuminate, 3.8 - 6.9 × 1.5 - 3 cm, white or greenish-white; **Ov** tapering or somewhat abruptly terminating into the **Sty**, 4 - 6 mm ∅; **Fr** dry capsules, ellipsoid or somewhat cylindrical, beaked, rarely constricted, 3.8 - 5.4 × 2 - 3 cm; **Se** 6 - 8 × 7 - 8 mm.

The species is best characterized by its clearly twisted concave leaves (Hochstätter 2000b). For differences from *Y. pallida* see there. Since the original type material collected by Lindheimer is lost, Hochstätter (l.c.) selected a neotype.

Y. schidigera Roezl *ex* Ortgies (Gartenflora 20: 110, 1871). **T**: USA, California (*Nuttall* s.n. [GH]). – **Lit**: Matuda & Piña Lujan (1980); Turner & al. (1995); both with ills. **D**: S USA (California, Nevada, Arizona), probably adjacent Mexico; gravelly mountain and valley slopes, desert or chaparral vegetation, 300 - 1800 m, flowers April to mid-May. **I**: Hochstätter (1999a). **Fig. X.c**

Incl. *Yucca californica* Nuttall *ex* Baker (1880) (*nom. inval.*, Art. 34.1c); **incl.** *Yucca mohavensis* Sargent (1896).

[1c] Plants commonly fruticose, or clumped, clumps rather tall, broad and open, stems (1-) 4 - 7 (-23), erect or somewhat assurgent, rarely to 2.5 m; **L** numerous, broadest near the middle, greater part rather deeply canaliculate, thick, very rigid, 33 - 105 × 2.5 - 5 cm, yellow-green, margins thick, with coarse somewhat curled fibres; **Inf** ellipsoid or with flattened tip, scape 0 - 15 cm, **Fl**-bearing part entirely within the **L** or to ½ of its length above the **L**, 0.5 - 1.25 m, part-**Inf** many; **Fl** many, dense, globose; **Tep** (broadly) lanceolate, 24 - 45 × 6 - 10 mm, white or cream-coloured, commonly tinged with lavender or purple; **Ov** rather stout, 5 - 8 mm ∅; **Fr** fleshy berries, variable, long and cylindrical, mostly constricted in the middle, 9 - 11.5 × 3 - 3.8 cm, usually tapering from the swollen base to a rather blunt tip 6 - 8.5 cm long; **Se** unwinged, 6 - 9 × 8 - 11 mm, dull black.

Y. ×schottii Engelmann *pro sp.* (Trans. Acad. Sci. St. Louis 3: 46, 1873). **T** [lecto]: USA, Arizona (*Schott* s.n. [NY]). – **Lit**: Webber (1953); Gentry (1972: with ill.); Turner & al. (1995: with ill. as *Y. arizonica*). **D**: S USA (Arizona, SW New Mexico?), N Mexico (Sonora, Chihuahua); rocky slopes (volcanic and limestone) in oak woodland and pine-oak forests, (350-?) 1200 - 1500 m. **I**: Hochstätter (1999a). **Fig. X.g** ('*Y. thornberi*')

Incl. *Yucca puberula* Torrey (1858) (*nom. illeg.*, Art. 53.1); **incl.** *Yucca brevifolia* Schott *ex* Torrey (1859) (*nom. inval.*, Art. 34.1c); **incl.** *Yucca brevifolia* Schott *ex* Engelmann (1873) (*nom. inval.*, Art. 32.1); **incl.** *Yucca macrocarpa* Engelmann (1881); **incl.** *Yucca treleasei* Macbride (1918) (*nom. illeg.*, Art. 53.1); **incl.** *Yucca arizonica* McKelvey (1935); **incl.** *Yucca thornberi* McKelvey (1935); **incl.** *Yucca confinis* McKelvey (1938); **incl.** *Yucca baccata* var. *brevifolia* Benson & Darrow (1943).

Lenz & Hanson (2000) recently clarified the nomenclatural and taxonomic confusion surrounding *Y. schottii*. According to their interpretation, the name *Y. schottii* Engelmann represents the earliest applicable name for hybrids between *Y. baccata*, *Y. elata* and *Y. madrensis*, and it is thus recognized as name for a collective hybrid species. The common fleshy-fruited *Yucca* with long flexible blue-green leaves from the mountains of S Arizona, previously wrongly named *Y. schottii*, is regarded as pertaining to *Y. madrensis.*

Y. tenuistyla Trelease (Annual Rep. Missouri Bot. Gard. 13: 53, t. 17-19, 1902). **T:** USA, Texas (*Lindheimer* s.n. [MO]). – **D:** USA (Texas); scrubland at the coast, spring-flowering.

[3c] Acaulescent; **L** mostly recurving, lanceolate, soft, 40 - 70 × 1 - 2 cm, margins whitish, tip scarcely pungent; **Inf** not described; **Tep** narrowly acute, white (?); **Ov** white; **Sty** oblong, white or green, often deeply divided; **Fr** dry capsules, stout, cylindrical, symmetrical, 5 - 6.5 × 2.5 - 3 cm; **Se** glossy, 8 - 10 × 7 - 8 mm.

This taxon (as well as *Y. louisianensis*) is regarded as a possible synonym of *Y. constricta* by McKelvey (1947), from which both differ only tenuously.

Y. thompsoniana Trelease (Annual Rep. Missouri Bot. Gard. 22: 101, t. 104-107, 1911). **T:** Mexico, Chihuahua (*Bigelow* s.n. [MO 148777]). – **Lit:** Matuda & Piña Lujan (1980); Hochstätter (2000b); both with ills. **D:** USA (Texas), Mexico (Chihuahua, Coahuila, Nuevo León), usually on exposed rocky knolls and slopes, 275 - 1350 m, flowers April to May.

Incl. *Yucca rostrata* var. *linearis* Trelease (1907) ≡ *Yucca linearis* (Trelease) D. J. Ferguson (1996); **incl.** *Yucca rostrata* fa. *integra* Trelease (1911).

[3a] Arborescent, stems 1 - 3, 0.7 - 2.6 m, erect with comparatively long ascending or diffusive **Br**; **Ros** frequently asymmetrical, rather small; **L** few, linear or somewhat broader towards the middle, flat or canaliculate to flat above and keeled below, striate, thin and flexible, 18 - 30 × 0.7 - 1.2 cm, margins horny, yellow or brownish, minutely denticulate; **Inf** narrowly ellipsoid to somewhat ovoid, scape 38 - 68 cm, glabrous or evanescently pubescent, **Fl**-bearing part starting 11 - 19 cm above the **L**, 52 - 82 cm, part-**Inf** 20 - 34, 2 - 22 cm; **Fl** globose to campanulate, broadly spreading at anthesis; **Tep** narrowly oblong, sharply acuminate, conspicuously veined, 35 - 67 × 12 - 35 mm, white; **Ov** slender, usually tapering into the **Sty**, 4 - 6 mm ⌀; **Fr** dry capsules, ellipsoid or somewhat ovoid, rarely constricted, 3.5 - 7 × 2 - 2.5 cm, with a long beak; **Se** flat, thin, unwinged, 5 - 6 × 6 - 7 mm, dull black.

Closely related to *Y. rostrata* (see there for differences).

Y. torreyi Shafer (in Britton & Shafer, North Amer. Trees, 157, fig. 117, 1908). **T:** USA, Texas (*Bigelow* s.n. [not extant]). – **Lit:** Reveal (1977); Matuda & Piña Lujan (1980); Benson & Darrow (1981); all with ills. **D:** SE USA (Texas), Mexico (Chihuahua, Coahuila, Nuevo León, Tamaulipas, Durango); plains with deeper soil and flat slopes in desert scrub, 450 - 1500 m. **I:** Hochstätter (1999a).

Incl. *Yucca baccata* var. *macrocarpa* Torrey (1859) ≡ *Yucca macrocarpa* (Torrey) Merriam (1893) (*nom. illeg.*, Art. 53.1); **incl.** *Yucca crassifila* Engelmann (1873) (*nom. inval.*, Art. 32.1); **incl.** *Yucca torreyi* fa. *parviflora* McKelvey (1938).

[1c] Stems forming groups of variable height, to 4.5 m, simple or sparsely branched; **L** rigid, caniculate, sometimes plane, scabrous on both faces, 30 - 103 × 3 - 5 cm, yellowish-green, margins thick, with thick straight fibres, terminal **Sp** short; **Inf** with a scape 0 - 10 cm long, ellipsoid, tip truncate, exceeding the **L** for ¼ - ½ of its length, 36 - 70 cm, densely many-flowered; **Fl** subglobose to campanulate; **Tep** very variable in shape and size; **OTep** 34 - 75 × 8 - 18 mm, cream-coloured with dark purple markings towards the base; **Ov** 4 - 8 mm ⌀; **Fr** fleshy berries, cylindrical and ovoid, narrowed towards the tip, 7 - 11.2 × 2.5 - 3.8 cm; **Se** 5 - 8 × 6 - 9 mm.

Y. treculiana Carrière (Rev. Hort. 1858: 580, 1858). **T:** USA / Mexico (*Trécul* 1496 [P]). – **Lit:** Webber (1953); Matuda & Piña Lujan (1980: with ill.). **D:** SE USA (S-C Texas), N Mexico (Coahuila, Nuevo León, Tamaulipas, Durango); plains with deeper soils in desert scrub, 100 - 1600 m, flowers March to April. **I:** Hochstätter (1999a).

Incl. *Yucca treculiana* var. *treculiana*; **incl.** *Yucca agavoides* hort. *ex* Carrière (1858); **incl.** *Yucca aspera* Regel (1858); **incl.** *Yucca recurvata* hort. *ex* Carrière (1858); **incl.** *Yucca revoluta* hort. *ex* Carrière (1858); **incl.** *Yucca undulata* hort. *ex* Carrière (1858); **incl.** *Yucca canaliculata* Hooker (1860) ≡ *Yucca treculiana* var. *canaliculata* (Hooker) Trelease (1902); **incl.** *Yucca longiflora* Buckley (1863); **incl.** *Yucca argospatha* Verlot (1868); **incl.** *Yucca longifolia* Engelmann (1873) (*nom. inval.*, Art. 32.1); **incl.** *Yucca cornuta* hort. *ex* Baker (1880); **incl.** *Yucca concava* hort. *ex* Baker (1880) (*nom. illeg.*, Art. 53.1); **incl.** *Yucca treculiana* var. *succulenta* McKelvey (1938).

[1c] Stems forming groups of variable height, simple or sparsely branched; **L** rigid, concave, conduplicate, rather scabrous, 50 - 100 × 2.5 - 5 cm, yellowish-green, margins entire or with sparse thin and straight fibres, terminal **Sp** very acute; **Inf** with a scape to 30 cm, ellipsoid, exceeding the **L** for ½ - ¾ of its length, densely many-flowered; **Bra** large in the lower **Inf** parts; **Fl** globose to semiglobose; **OTep** 29 - 45 × 11 - 21 mm, white or with rose tinge; **Ov** 4 - 6 mm ⌀; **Fr** fleshy berries, cylindrical, terminal part conical, 6.5 - 10 × 1.7 - 2.4 cm; **Se** rough, 4 - 5 × 5 - 6 mm.

Similar to *Y. torreyi*, but distinguishable by its smaller and semiglobose flowers with stout ovaries and by its more symmetrical rosettes with relatively broader, shorter and predominantly non-filiferous leaves (Webber 1953).

The name is usually written 'treculeana', but the taxon is named for the French botanist A. A. L. Trécul.

Y. valida Brandegee (Proc. Calif. Acad. Sci., ser. 2, 2: 208, t. 11., 1889). − **Lit:** Matuda & Piña Lujan (1980: with ills.); Turner & al. (1995). **D:** Mexico (Baja California, Baja California Sur); Pacific coastal plains and gentle slopes, to 800 m.

[1c] Arborescent and stems 3 - 12 m tall, or shrub-forming and branched nearly from the base, surculose; **L** many, oblanceolate, thin, 15 - 35 × 1.5 - 3 cm, green-yellowish, old dead **L** forming a skirt around the stem, margins with thick curved fibres; **Inf** short, to 30 cm, hidden between the **L** for ¼ - ½, slightly pubescent; **Ped** ± 2.5 - 3 cm; **Fl** campanulate, scented; **Tep** narrowly lanceolate, 2.5 - 3 cm; **Fil** pubescent, 10 - 12 mm; **Ov** oblong, tip abruptly conical; **Sti** sessile; **Fr** fleshy berries, oblong, 2.5 - 4.5 cm, nearly black; **Se** 7 × 1.5 mm, with rugose margin.

An important and characteristic constituent of the fog-influenced Pacific coastal desert scrub formations of the Baja California Peninsula. Plants from the Cape region hitherto included here were recently described as a distinct separate species, *Y. capensis* Lenz.

Aloaceae

Small herbaceous shrubby or pachycaul arborescent **L**-succulent perennials, rarely with a bulbous base; **R** fibrous, usually terete, sometimes tuberously thickened or fusiform; stem stout or slender, simple or branched, erect, decumbent or pendulous, short to several m tall, sometimes so short that the plants are described as acaulescent; **L** simple, alternate, amplexicaul, linear, deltoid, falcate, lanceolate or triangular, crowded in dense **Ros** at tips of stems and **Br** or at ground-level, sometimes widely spaced along stem, persistent for several years, usually distinctly succulent and mottled with whitish spots or striations, often prickly along the margins and sometimes on both sides, tip usually ending in a weak to fairly strong **Sp**, surfaces smooth or rough; **L** tissue usually with coloured exudate when broken; **Inf** racemes, panicles or rarely spikes, axillary, bracteate, peduncle scape-like, massive or slender; **Fl** hermaphroditic, 3-merous throughout, red, orange, yellow or white, rarely green; **Tep** in 2 whorls of 3, petaloid and often fleshy, connivent or connate into a straight or curved sometimes ventricose tube, limb ± regular or sometimes bilabiate; **St** 6, free, **Anth** with 2 thecae, opening by longitudinal slits, included or exserted; **Ov** compound, 3-locular, superior, placentation axile, sometimes with septal **Nec**, each locule with numerous ovules; **Sty** terminal, with punctate or discoid **Sti**; **Fr** loculicidal capsules, rarely fleshy and dehiscent berries; **Se** usually flattened or winged. − *Cytology:* x = 7, with a distinctive basic karyotype of 4 long and 3 short chromosomes. Most species are diploid, but some polyploidy and aneuploidy occur, esp. in *Aloe* and *Haworthia*, as summarized by Riley & Majumdar (1979).

Distribution: Africa, Arabian Peninsula, Socotra, Mascarene Islands, Madagascar.

Literature: Berger (1908); Riley & Majumdar (1979).

The family *Aloaceae* was for long regarded as tribe *Aloeae* (often erroneously as "Aloineae") of the family *Liliaceae*. It is now treated as a distinct family (Cronquist 1981), as here, or as subfamily *Alooideae* in the *Asphodelaceae*, as advocated by Dahlgren & al. (1985) or Smith & Wyk (1991). It differs from *Asphodelaceae*, though not entirely consistently, by having succulent leaves, tubular flowers, a basic bimodal complement of 4 long and 3 short chromosomes, and the presence of a parenchymatous cap at the phloem pole.

Aloaceae numbers ± 500 species in 6 genera, of which 5 are restricted to southern Africa. Numerous taxa of *Aloe, Gasteria* and *Haworthia* are much cultivated by succulent plant enthusiasts, and *Aloe* species are also often planted as conspicuous garden plants in warmer climates. Apart from limited medicinal and other traditional uses of several taxa of *Haworthia* and *Gasteria*, *Aloe vera* and *A. ferox* are widely used medicinally as purgatives, as ingredients in cosmetic products, and to treat skin irritations and burns, and the industries based on these 2 taxa are worth millions of dollars per year alone in RSA.

Numerous bigeneric and some trigeneric hybrids have been created and formally named. In addition to the hybrid genera covered here, the illegitimate name ×*Rowleyara* D. M. Cumming 1999 refers to the combination *Astroloba* × *Gasteria* × *Haworthia*.

Key to the genera:

1 **Fr** apically acuminate capsules; underground plant parts bulbous; **Fl** usually < 15 mm long:
Chortolirion
− **Fr** apically rounded or obtuse capsules, or berries; underground plant parts rhizomatous (if rarely bulbous then **Fl** > 15 mm long): **2**
2 **Tep** orange, apically connivent and ± greenish:
Poellnitzia
− **Tep** of various colours incl. orange, apically spreading or recurved: **3**
3 **Fl** pendulous at anthesis, **Per** tube curved upwards, often basally inflated: **Gasteria**
− **Fl** erect, suberect, spreading or pendent at anthesis, **Per** tube straight or curved downwards: **4**
4 **Per** bilabiate, < 15 mm long, mouth not upturned, colours clear to dirty white, greenish or yellowish: **Haworthia**
− **Per** regular (if rarely weakly bilabiate then **Fl** > 15 mm long), mouth upturned, colours various, usually bright: **5**
5 **Fl** usually brightly coloured, fleshy; **St** as long as or longer than the **Per**: **Aloe**
− **Fl** white or dull-coloured, flimsy; **St** included:
Astroloba

[G. F. Smith & L. E. Newton]

×ALGASTOLOBA

U. Eggli

×**Algastoloba** D. M. Cumming (Haworthiad 13(1): 20, 1999). − **Lit:** Cumming (1999). **Etym:** Hybrid formula that combines the names of the involved genera *Aloe*, *Gasteria* and *Astroloba*.

= *Aloe* × *Astroloba* × *Gasteria*. The 2 known combinations have not been formally named.

ALOE

L. E. Newton

Aloe Linné (Spec. Pl. [ed. 1], 1: 319, 1753). **T:** *Aloe disticha* Linné [Typification according to P. V. Heath, Calyx 2(4): 142, 1992.]. − **Lit:** Reynolds

(1950); Reynolds (1958); Reynolds (1966); Wood (1983); Carter (1994); Lavranos (1995); Wyk & Smith (1996); Gilbert & Sebsebe (1997); Sebsebe & Gilbert (1997); Rowley (1997); Glen & Hardy (2000). **D:** Africa, esp. S and E; Madagascar, Mascarene Islands, SW Arabia; mainly drier areas, a few in dry forests. **Etym:** From the Greek ('aloe'), Arabian ('alloch') and Hebrew ('ahalim') names for the plants.

Incl. *Kumara* Medikus (1786). **T:** *Kumara disticha* Medikus.

Incl. *Lomatophyllum* Willdenow (1811). **T:** *Lomatophyllum borbonicum* Willdenow [= *Aloe purpurea* Lamarck].

Incl. *Rhipidodendrum* Sprengel (1811). **T:** not typified.

Incl. *Phylloma* Ker Gawler (1813). **T:** *Phylloma aloiflorum* Ker Gawler.

Incl. *Rhipidodendron* Willdenow (1817) (*nom. inval.*, Art. 61.1). **T:** not typified.

Incl. *Pachidendron* Haworth (1821). **T:** not typified.

Incl. *Bowiea* Haworth (1827). **T:** *Aloe bowiea* Schultes & Schultes *fil.* [Lectotype selected by G. F. Smith, South Afr. J. Bot. 56(3): 303-308, 1990].

Incl. *Pachydendron* Dumortier (1829) (*nom. inval.*, Art. 61.1). **T:** not typified.

Incl. *Agriodendron* Endlicher (1836). **T:** *Aloe ferox* Miller.

Incl. *Succosaria* Rafinesque (1840). **T:** *Aloe spicata* Linné *fil.*

Incl. *Ariodendron* Meisner (1842). **T:** *Aloe ferox* Miller.

Incl. *Busipho* Salisbury (1866). **T:** not typified.

Incl. *Ptyas* Salisbury (1866). **T:** *Kumara disticha* Medikus.

Incl. *Chamaealoe* A. Berger (1905). **T:** *Bowiea africana* Haworth 1824.

Incl. ×*Lomataloe* Guillaumin (1931).

Incl. *Leptaloe* Stapf (1933). **T:** *Leptaloe albida* Stapf [E. Phillips, Gen. South Afr. Pl., ed. 2, 186, 1951, fide ING.].

Incl. *Guillauminia* Bertrand (1956). **T:** *Aloe albiflora* Guillaumin.

Incl. *Lemeea* P. V. Heath (1993). **T:** *Aloe haworthioides* Baker.

Perennial **L** succulents; acaulescent, shrubby, or arborescent; simple or branching; **L** rosulate, distichous or scattered along the stem, usually ± triangular, lanceolate or falcate, sometimes linear, amplexicaul, the margins usually armed with soft or pungent deltoid teeth, the teeth usually more crowded near the **L** base, glabrous, sometimes with short prickles on the surface, uniformly coloured or with whitish or pale green spots, surface smooth or rough, usually with a bitter-tasting yellow or brown exudate when broken, rarely containing fibres; **Inf** lateral, usually ± erect, simple or branching, with cylindrical or capitate racemes, sometimes with **Fl** secund; **Fl** bracteate, pedicellate, **Ped** often lengthening in **Fr**, rarely **Fl** sessile; **Fl** base rounded, truncate or attenuate, zygomorphic, cylindrical at the base, usually becoming trigonous and slightly compressed laterally above, usually curved, proterandrous, usually red or yellow, rarely white, usually glabrous, rarely puberulent, usually erect in bud and pendulous at anthesis; **OTep** 3, usually marginally fused at base, tips free; **ITep** 3, usually with margins free but dorsally adnate to outer **Per** tube, tips free; **St** 6, usually with **Anth** exserted partly or just clear of **Per** (1 - 6 mm), each group of 3 elongating and exserted, and then retracted in turn during anthesis; **Sty** simple with small capitate **Sti**; **Fr** usually capsules, sometimes berries; **Se** angular or flattened, black or brown, usually with narrow membranous wings. – *Cytology:* The majority of the species investigated are diploid, with a somatic number of 2n = 14. One hexaploid and several tetraploids are also known. Known chromosome counts are summarized by Riley & Majumdar (1979).

The monotypic genus *Guillauminia* was proposed for *Aloe albiflora*, but this was not recognized by Reynolds (1966). Heath (1994) later recognized *Guillauminia* and added 5 other species, but without any explanatory text to justify his action. In the same paper Heath added 2 more species to his genus *Lemeea*, proposed in 1993 for *Aloe* Sect. *Aloinella* A. Berger, again without explanatory text.

Some intergeneric hybrids are known, esp. between *Aloe* and *Gasteria* (= ×*Gasteraloe*). Many early intergeneric hybrids were documented by Rowley (1982). Interspecific hybrids have occurred spontaneously in the wild, and many were recorded by Reynolds (1950, 1966) and Newton (1998a). There has also been some artificial hybridization with the aim of producing worthwhile cultivars, the greatest number of named ones being in southern Africa (described in Aloe, Journal of the South African Succulent Society, between 1971 and 1978), and in the USA (Riley 1993), with a few in England (Brandham 1973). The following names refer to such intrageneric hybrids: *A. ×andrea, A. ×antoninii, A. ×borziana, A. ×cyanea, A. ×delaetii, A. ×deleuilii, A. ×desmetiana A. ×gigantea, A. ×grusonii, A. ×hertrichii, A. ×hoyeri, A. ×insignis, A. ×jacobseniana, A. ×laetecoccinea, A. ×luteobrunnea A. ×pallancae, A. ×panormitana, A. ×paxii, A. ×principis, A. ×pseudopicta, A. ×riccobonii, A. ×robertii, A. ×schimperi, A. ×schoenlandii, A. ×speciosa, A. ×spinosissima, A. ×todari, A. ×tomlinsonii, A. ×varvarii, A. ×weingartii* and *A. ×winteri*.

In the descriptions below, width of leaf or bract is width at the widest point, at or near the base; details of marginal teeth refer to the middle of the leaf; and diameter of flowers is as seen from the side. Character states that are common to the majority of taxa are not stated, only the exceptions being given. For example, unless otherwise stated, leaves are erect or slightly spreading, flat or slightly canaliculate above and convex below, inflorescences are erect, floral bracts are scarious, whitish with darker veins, flowers are glabrous, with cylindrical base and laterally compressed with trigonous shape above, anthers are exserted from the perianth by 2 - 6 mm, and fruits are capsules.

No satisfactory infrageneric classification has been proposed since that of Berger (1908), since which time numerous new species have been described. The diversity of character combinations in the genus has, so far, obscured the phylogenetic relationships. For species of southern Africa Reynolds (1950) enlarged Berger's scheme somewhat, but when dealing with tropical species, he (Reynolds 1966) created new groups without formal names or ranks, some of them clearly artificial. The development of a new infrageneric classification for the genus is a long way off and so for the present a completely artificial synopsis is presented below, based on growth habit and inflorescence form. Some very variable species appear in more than one group:

A. Plants acaulescent or with stems shorter than the ∅ of the **Ros**; **L** broad (> 2 cm) unless otherwise stated:
[1] Plants with linear or very narrow **L** (usually < 2 cm wide), solitary or forming groups.
[2] Plants always solitary; **Inf** usually unbranched or with 1 **Br**.
[3] Plants always solitary; **Inf** usually with 2 or more **Br**.
[4] Plants solitary or forming small groups; **Inf** usually unbranched or with 1 **Br**.
[5] Plants solitary or forming small groups; **Inf** usually with 2 or more **Br**.
[6] Plants suckering to form groups; **Inf** unbranched or with 1 **Br**.
[7] Plants suckering to form groups; **Inf** with 2 or more **Br**.

B. Plants with obvious stems; **L** broad (> 2 cm) unless otherwise stated:
[8] **Ros** on unbranched stems to 2 m tall.
[9] Trees with usually unbranched trunks to > 2 m tall; in some taxa suckering or branching at base to produce 2 or more trunks.
[10] Plants with linear or very narrow **L**, solitary or branching, sometimes pendulous.
[11] Plants forming low clumps, often dense.
[12] Plants forming large clumps (at least 1 m tall).
[13] Plants with decumbent, sprawling or pendulous stems.
[14] Plants with stems scrambling in or supported by surrounding vegetation.
[15] Plants forming shrubs to 1 m tall.
[16] Plants forming shrubs > 1 m tall.
[17] Trees with basal trunks and branching above.

C. [18] Plants of unknown or uncertain growth habit.

The following provisional name is to be validated when the plant is better known: *A. vulcanica* Lavranos & Collenette *nom. nud.* (2000).

The following names are of unresolved application but are referred to this genus: *Aloe abyssinica* Lamarck (1783); *Aloe agavefolia* Todaro (1875); *Aloe albispina* Haworth (1804) ≡ *Aloe mitriformis* var. *albispina* (Haworth) A. Berger (1908); *Aloe arabica* Salm-Dyck (1817) (*nom. illeg.*, Art. 53.1); *Aloe arborea* Forsskål (1775); *Aloe arborescens* var. *viridifolia* A. Berger (1908); *Aloe baumii* Engler & Gilg (1903); *Aloe brownii* Baker (1889); *Aloe chloroleuca* Baker (1877); *Aloe cinnabarina* Diels & A. Berger (1905); *Aloe commelinii* Willdenow (1811) ≡ *Aloe mitriformis* var. *commelinii* (Willdenow) Baker (1880); *Aloe commutata* Todaro (1876); *Aloe commutata* var. *tricolor* (Baker) A. Berger (1908); *Aloe congolensis* De Wildeman & T. Durand (1899); *Aloe consobrina* Salm-Dyck (1863); *Aloe constricta* Baker (1880); *Aloe decaryi* Guillaumin (1941); *Aloe defalcata* Chiovenda (1932); *Aloe deflexidens* Pillans (1935); *Aloe dorsalis* Haworth (1821); *Aloe drepanophylla* Baker (1875); *Aloe elizae* A. Berger (1911); *Aloe flavispina* Haworth (1804) ≡ *Aloe mitriformis* var. *flavispina* (Haworth) Baker (1880); *Aloe gasterioides* Baker (1880); *Aloe grahamii* Schönland (1903); *Aloe heteracantha* Baker (1880); *Aloe hexapetala* Salm-Dyck (1817); *Aloe humilis* var. *macilenta* Baker (1880); *Aloe leucantha* A. Berger (1905); *Aloe longiflora* Baker (1888); *Aloe macracantha* Baker (1880); *Aloe mitis* A. Berger (1908); *Aloe mitriformis* var. *humilior* Willdenow (1811); *Aloe mitriformis* var. *pachyphylla* Baker (1880); *Aloe mitriformis* var. *spinosior* Haworth (1804); *Aloe monteiroi* Baker (1889); *Aloe nobilis* Haworth (1812); *Aloe nobilis* Baker (1880) (*nom. illeg.*, Art. 53.1); *Aloe nobilis* var. *densifolia* Baker (1880); *Aloe obscura* Miller (1768); *Aloe obscura* A. Berger *ex* Schönland (1905) (*nom. illeg.*, Art. 53.1); *Aloe pallescens* Haworth (1821); *Aloe perfoliata* Linné (1753); *Aloe perfoliata* var. *obscura* Aiton (1789); *Aloe perfoliata* var. *serrulata* Aiton (1789) ≡ *Aloe serrulata* (Aiton) Haworth (1804); *Aloe picta* Thunberg (1785); *Aloe picta* var. *major* Willdenow (1799); *Aloe platylepsis* Baker (1877); *Aloe pungens* A. Berger (1908); *Aloe rossii* Todaro (1894); *Aloe rubescens* De Candolle (1799); *Aloe runcinata* A. Berger (1908); *Aloe salmdyckiana* var. *fulgens* (Todaro) A. Berger (1908); *Aloe saponaria* var. *obscura* Haworth (1804); *Aloe sigmoidea* Baker (1880); *Aloe sororia* A. Berger (1908); *Aloe spinulosa* Salm-Dyck (1822) ≡ *Aloe mitriformis* var. *spinulosa* (Salm-Dyck) Baker (1880); *Aloe spuria* A. Berger (1908); *Aloe stans* A. Berger (1908); *Aloe straussii* A. Berger (1912); *Aloe ucriae* A. Terraciano (1897) ≡ *Aloe arborescens* var. *ucriae* (A. Terraciano) A. Berger (1908); *Aloe venenosa* Engler (1893); *Aloe virens* Haworth (1804); *Aloe xanthacantha* Willdenow (1811) ≡ *Aloe mitriformis* var. *xanthacantha* (Willdenow) Baker (1880).

A. aageodonta L. E. Newton (CSJA 65(3): 138-140, ills., 1993). **T:** Kenya, Eastern Prov. (*Newton 3643* [K, EA]). − **D:** Kenya; rocky hills, 960 - 1250 m. **Fig. X.h**

[13] Caulescent, branching at the base; stem erect for 1 m, then becoming decumbent and sprawling to 2 m, 3 cm ⌀; **L** 12 - 20, laxly rosulate and persistent for 20 - 30 cm below, triangular, 50 × 8 cm, dull green, spotted on young shoots, surface smooth; marginal teeth 4 mm, pungent, uncinate, tips brown; exudate yellow, drying brown; **Inf** 70 cm, erect, with 6 - 10 **Br**; racemes with 20 - 40 secund **Fl**, lax at base but most **Fl** crowded near tip; **Bra** 4 - 6 mm; **Ped** 10 - 13 mm; **Fl** yellow or dark red, 30 mm, base attenuate, 7 - 8 mm ⌀ across **Ov**, narrowed to 5 - 6 mm above, scarcely widening towards apex; **Tep** free for 5 - 7 mm, tips spreading to 10 - 12 mm.

Red- and yellow-flowered plants occur together in one population.

A. ×abhaica Lavranos & Collenette (CSJA 72(2): 87, ills. (p. 85), 2000). **T:** Saudi Arabia, Asir Prov. (*Collenette 7165* [K]). − **D:** Saudi Arabia. **I:** Collenette (1999: 18).

= *A. pseudorubroviolacea* × *A. edentata*.

A. abyssicola Lavranos & Bilaidi (CSJA 43(5): 204-208, ills., 1971). **T:** Yemen (*Lavranos & Bilaidi 7490* [FI]). − **D:** Yemen; cliff faces, 900 m.

[3] Acaulescent or with very short stems, simple, pendulous; **L** up to 50, rosulate, pointing downwards, 50 × 12 cm, grey-green, upper surface flat becoming caniculate towards tip; marginal teeth obtuse, hard, dark, 1 mm, 35 - 40 mm apart; **Inf** to 60 cm, growing downwards with only tips of racemes curved upwards, with 5 - 6 short spreading **Br**; peduncle yellow-green; racemes lax; **Bra** ovate-lanceolate, 7 - 8 × 3 mm; **Ped** 8 - 9 mm, green; **Fl** yellow-green, 25 mm, base rounded, 7 mm ⌀ across **Ov**, narrowed to 5 mm above, enlarging to 8 mm at mouth; **Tep** free for 7 mm.

A. aculeata Pole-Evans (Trans. Roy. Soc. South Afr. 5: 34, 1915). **T:** RSA, Northern Prov. (*?Pole-Evans in Govt. Herb. 55* [PRE]). − **D:** Zimbabwe, RSA (Northern Prov.); rocky or stony mountain slopes, in hot semi-arid areas. **I:** Reynolds (1950: 448-449).

[5] Acaulescent or with procumbent stem to 80 cm, usually simple; **L** ± 30, densely rosulate, lanceolate-attenuate, to 60 × 8 - 12 cm, arcuate-erectly incurved, dull green to glaucous with scattered reddish-brown prickles; marginal teeth 5 - 6 mm, pungent, reddish-brown, 10 - 20 mm apart; **Inf** 1 m, with 2 - 4 **Br**; peduncle deep brown; racemes cylindrical, 40 - 60 × 7 cm, dense; **Bra** deltoid-acuminate, reflexed, 10 × 7 mm; **Ped** 2 - 3 mm; **Fl** le-

mon-yellow with green-orange veins on lobes, 25 - 40 mm, base rounded; **Tep** free for 14 mm; **St** exserted 15 mm. – *Cytology:* 2n = 14 (Müller 1941).

Leaf marginal teeth and surface prickles sometimes arise from white tuberculate bases, and racemes vary in length. Natural hybrids with other species have been reported (Reynolds 1950).

A. acutissima H. Perrier (Mém. Soc. Linn. Normandie, Bot. 1(1): 17, 1926). **T** [syn]: Madagascar (*Perrier* 1107 [P?]). – **D:** Madagascar.

A. acutissima var. **acutissima** – **D:** Madagascar; thin soil on rocks, 240 - 1200 m. **I:** Reynolds (1966: 495-497). **Fig. X.d**
[15] Caulescent, much branched; stem to 1 m, 2 - 3 cm ∅, erect, divergent or decumbent; **L** ± 20, subdensely rosulate and persistent for 20 - 30 cm below, lanceolate, long-attenuate, 30 × 4 cm, greygreen tinged reddish; marginal teeth 3 mm, pungent, pale brown, 1 cm apart; sheath green-striate; **Inf** 50 cm, with 2 - 3 **Br**; racemes cylindrical-acuminate, 10 - 15 × 5 - 6 cm, sub-dense; **Bra** deltoid, clasping the **Ped**, 10 - 15 mm; **Ped** 15 mm, reddish-scarlet; **Fl** 30 mm, reddish-scarlet, base attenuate, 5.5 mm ∅ across **Ov**, slightly narrowed above and enlarging towards throat; **Tep** free for 10 mm. – *Cytology:* 2n = 14 (Brandham 1971).

Variable in size of stems, leaves and flowers. Plants at higher altitudes are more robust.

A. acutissima var. **antanimorensis** Reynolds (JSAB 22(1): 27-29, ills., 1956). **T:** Madagascar (*Reynolds* 7792 [TAN, K, P, PRE]). – **D:** Madagascar; flat rock surfaces among thorn bushes. **I:** Reynolds (1966: 498).
[15] Differs from var. *acutissima*: Stems shorter, thinner; **L** 10 - 15 × 1.5 - 2 cm; **Inf** mostly simple; racemes shorter; **Ped** 10 mm; **Fl** 20 - 25 mm. – *Cytology:* 2n = 14 (Brandham 1971).

A. adigratana Reynolds (JSAB 23(1): 1-3, ills., 1957). **T:** Ethiopia, Tigre Prov. (*Reynolds* 8076 [PRE, K]). – **D:** Ethiopia; rocky hills, 1800 - 2700 m. **I:** Reynolds (1966: 215-216).
Incl. *Aloe abyssinica* Hooker *fil.* (1900) (*nom. illeg.*, Art. 53.1); **incl.** *Aloe eru* var. *hookeri* A. Berger (1908).
[15] Caulescent, branching; stem erect to 1 m, 12 cm ∅, decumbent to 2 m; **L** 16 - 20, rosulate, spreading and slightly recurved, ensiform, 60 - 80 × 15 cm, dull green with pale green spots; marginal teeth 10 mm, 25 - 35 mm apart, pungent; exudate drying deep brown; **Inf** 90 cm with 3 - 5 **Br**; racemes cylindrical-conical, 15 - 20 × 8 - 9 cm, dense; **Bra** deltoid, 8 × 3 mm; **Ped** 18 mm; **Fl** orange or yellow, 28 - 33 mm, clavate, base attenuate, 6 mm ∅ across **Ov**; **Tep** free for 14 - 16 mm.
On the 2 possible synonyms, Reynolds (1966) states "The following almost certainly belong here".

A. affinis A. Berger (in Engler, A. (ed.), Pflanzenr. IV.38 (Heft 33): 206, 1908). **T:** RSA, Mpumalanga (*Wilms* 1490 [not located]). – **D:** RSA (Mpumalanga); on sandstone or quartzite in mountainous areas, 1220 - 1520 m. **I:** Reynolds (1950: 243-244).
[3] Acaulescent, simple; **L** ± 20, densely rosulate, arcuate-erect, 30 - 45 × 9 - 11 cm, green with longitudinal dark lines, with prominent horny, reddish-brown margin; marginal teeth 5 - 8 mm, pungent, 10 - 15 mm apart; exudate dries pale yellow; **Inf** to 1 m with 5 - 10 **Br**; racemes cylindrical, to 25 cm, rather dense; **Bra** narrowly deltoid, 15 mm; **Ped** 15 mm; **Fl** dull brick-red, 45 mm, base rounded, 9 - 10 mm ∅ across **Ov**, abruptly narrowed to 5 - 6 mm above, then slightly decurved and enlarging towards mouth; **Tep** free for 10 mm; **St** exserted 0 - 1 mm.

Reynolds (1950) reports some variants with leaves obscurely spotted on the upper surface. Natural hybrids with other species have been reported by the same author.

A. africana Miller (Gard. Dict., Ed. 8, [no. 4], 1768). **T:** not typified. – **D:** RSA (Eastern Cape); mostly bush and scrub country. **I:** Reynolds (1950: 457-458). **Fig. X.e**
≡ *Pachidendron africanum* (Miller) Haworth (1821); **incl.** *Aloe perfoliata* var. β Linné (1753); **incl.** *Aloe perfoliata* var. *africana* Aiton (1789); **incl.** *Aloe pseudoafricana* Salm-Dyck (1817); **incl.** *Aloe africana* var. *angustior* Haworth (1819); **incl.** *Aloe africana* var. *latifolia* Haworth (1819); **incl.** *Aloe angustifolia* Haworth (1819) ≡ *Pachidendron angustifolium* (Haworth) Haworth (1821); **incl.** *Pachidendron africanum* var. *angustum* Haworth (1821); **incl.** *Pachidendron africanum* var. *latum* Haworth (1821); **incl.** *Aloe bolusii* Baker (1880).
[9] Caulescent, usually simple; stem erect, to 4 m, dead **L** persistent; **L** ± 30, densely rosulate, spreading to recurved, 65 × 12 cm, dull green to glaucous, upper face glabrous or with few reddish prickles near tip, lower face with median reddish prickles near tip; marginal teeth 4 - 5 mm, pungent, 15 mm apart; **Inf** 60 - 80 cm with 2 - 4 **Br**; racemes cylindrical-acuminate, 40 - 60 × 10 - 12 cm, dense; **Bra** ovate-lanceolate, 11 × 7 - 8 mm; **Ped** 5 - 6 mm; **Fl** yellow to yellow-orange, 55 mm, base rounded, 5 - 6 mm ∅ across **Ov**, enlarging to 8 mm at mouth, upper ½ markedly upcurved; **Tep** free for 19 mm; **St** exserted 15 mm.
Natural hybrids with other species have been reported (Reynolds 1950).

A. ahmarensis Favell & al. (CSJA 71(5): 257-259, ills., 1999). **T:** Yemen (*al-Gifri* 3776 [Herb. Univ. Aden]). – **D:** Yemen; valley bottoms and basalt lava flows in sparsely vegetated sandy plains, ± 500 m.
[7] Caulescent or usually acaulescent, solitary or usually suckering to up to 25 **Ros**; stem procumb-

ent, to 50 cm; **L** ± 12, densely rosulate, deltoid-arcuate, ascending, ± 30 × 9 cm, upper face pale pinkish-grey with waxy bloom, lower face greenish-grey with a bloom; marginal teeth 2 - 3 mm, dark brown, 18 - 30 mm apart; exudate honey-coloured; **Inf** 65 cm, ascending, with many **Br**, lower **Br** rebranched; racemes cylindrical, subdense, **Bra** ovate-deltoid, 6 × 2 mm; **Ped** 7 - 8 mm, pink; **Fl** bright coral-pink, yellowish at the tip, 33 mm, base rounded, 8 mm ∅ across **Ov**, narrowing above to 6 mm, widening to 7 mm at the tip; **Tep** free for 10 mm; **St** exserted 1.5 mm.

A. albida (Stapf) Reynolds (JSAB 13(2): 101, 103, 1947). **T:** RSA, Mpumalanga (*Galpin* 873 [PRE]). – **D:** RSA (Mpumalanga); stony ground or on rocks, 1450 - 1520 m. **I:** Reynolds (1950: 112-113, t. 2); FPA 51: t. 2010, 1990. **Fig. X.f**
≡ *Leptaloe albida* Stapf (1933); **incl.** *Aloe kraussii* var. *minor* Baker (1896) ≡ *Aloe myriacantha* var. *minor* (Baker) A. Berger (1908); **incl.** *Aloe kraussii* Schönland (1903) (*nom. illeg.*, Art. 53.1).
[1] Acaulescent, usually simple; **R** fusiform; **L** 6 - 12, rosulate, linear, 10 - 15 × 0.4 - 0.5 cm, dull green; marginal teeth 0.5 mm, 1 mm apart; **Inf** 9 - 15 cm, simple; raceme capitate, 2 - 5 × 5 cm with 8 - 16 **Fl**; **Bra** ovate-acuminate, 10 - 15 mm; **Ped** 10 - 15 mm; **Fl** dull creamy-white, green-tipped, 18 mm, base attenuate, narrowed slightly above **Ov** to bilabiate mouth; **Tep** free to base; **St** exserted 0 - 1 mm. – *Cytology:* 2n = 14 (Müller 1945: as *Leptaloe*).

A. albiflora Guillaumin (Bull. Mus. Nation. Hist. Nat., Sér. 2, 12: 353, 1940). **T:** Madagascar, Toliara (*Boiteau* 227 p.p. [P]). – **D:** Madagascar. **I:** Reynolds (1966: 406-407).
≡ *Guillauminia albiflora* (Guillaumin) A. Bertrand (1956).
[1] Acaulescent, suckering to form small clumps; **R** fusiform; **L** 10, rosulate, linear-attenuate, 15 × 1.5 cm, grey-green with many small dull-white spots, with narrow dull-white cartilaginous margin; marginal teeth 0.5 - 1 mm, soft to firm; **Inf** 30 - 36 cm, simple; raceme 9 cm, lax, with ± 18 **Fl**; **Bra** ovate-long acuminate, 5 - 6 × 2 mm; **Ped** 8 mm; **Fl** white, 10 mm, base rounded, campanulate, 14 mm ∅ across mouth; **Tep** free to base; **St** exserted 8 mm.

A. albovestita S. Carter & Brandham (Bradleya 1: 20-21, ills., 1983). **T:** Somalia, Burao Region (*Bailes* 214 [K]). – **D:** Somalia; rocky ground and rock crevices, often in deep shade, ± 1450 m.
[7] Acaulescent, suckering to form small clumps; **L** rosulate, lanceolate, 20 - 30 × 12 cm, glaucous, distinctly longitudinally striate, upper face with few paler spots, with narrow cartilaginous margin; marginal teeth 1.5 mm, red-brown, 5 - 10 mm apart; **Inf** 75 cm, with 2 - 3 **Br**; racemes cylindrical to subcapitate, 10 - 20 × 5 cm; **Bra** lanceolate, 15 × 3 mm;

Ped 18 mm; **Fl** dull pink with greyish tip, finely white-flecked and densely covered with white bloom; 25 - 33 mm, base rounded, 7 mm ∅ across **Ov**, narrowed to 4.5 mm above; **Tep** free for 7 mm. – *Cytology:* 2n = 14.

A. aldabrensis (Marais) L. E. Newton & G. D. Rowley (Excelsa 17: 59, 1997). **T:** Aldabra Archipelago, West Island (*Stoddart* 920 [K]). – **D:** Aldabra Archipelago (Astove Atoll [Seychelles]; limestone in mixed scrub or thickets, almost sea-level.
≡ *Lomatophyllum aldabrense* Marais (1975).
[8] Acaulescent or caulescent; stem erect or decumbent, to 2 m; **L** densely rosulate, linear-lanceolate to ensiform, long-attenuate, 60 - 100 × 4 - 10 cm, green, often tinged orange or red, with horny margin; marginal teeth small; **Inf** 15 - 35 cm, with 3 - 5 **Br**, rarely rebranched; racemes cylindrical, subdense; **Bra** ovate to deltoid, 2.5 - 4.5 mm; **Ped** 12 - 20 mm; **Fl** bright orange-red, 18 - 25 mm, base shortly attenuate; **Tep** free for slightly more than ½; **St** slightly exserted; **Fr** berries.

A. alfredii Rauh (CSJA 62(5): 232-233, ills., 1990). **T:** Madagascar (*Rauh* 68690 [HEID]). – **D:** C Madagascar; quartz fields, 1400 m.
[11] Caulescent, branching at base; stem 25 × 1.5 cm; **L** scattered along stem, linear, 30 × 1.5 cm, dark dull green; marginal teeth 1 mm, white; sheath 15 mm, red-brown with darker veins; **Inf** 60 cm, simple; raceme cylindrical, 8 - 10 × 5 - 6 cm, many-flowered; **Bra** triangular, 5 mm; **Ped** 5 mm; **Fl** lemon-yellow with green mid-stripe, 20 mm, base rounded.

A. alooides (Bolus) van Druten (BT 6(3): 544-545, 1956). **T:** RSA, Mpumalanga (*MacLea* s.n. in *BOL* 3011 [BOL, SAM]). – **D:** RSA (Mpumalanga); dolomite outcrops in mountains, 1700 - 2000 m. **I:** Reynolds (1950: 437-438, t. 58, as *A. recurvifolia*). **Fig. XI.a**
≡ *Urginea alooides* Bolus (1881) ≡ *Notosceptrum alooides* (Bolus) Bentham (1883); **incl.** *Aloe recurvifolia* Groenewald (1935).
[8] Caulescent; stem usually simple, rarely branched low down, erect, stout, 2 m, covered with remains of dead **L**; **L** densely rosulate, lanceolate-ensiform, long-attenuate, deeply canaliculate, arcuate-recurved, 130 × 18 cm, green, sometimes slightly reddish, usually with distinct reddish margin; marginal teeth 2 - 3 mm, usually curved towards **L** tip, 10 mm apart, brownish-tipped; **Inf** 1.3 m, erect, simple; raceme narrowly cylindrical, slightly acuminate, 80 × 4.5 cm, dense; **Bra** ovate-acute, 5 - 7 × 4 - 5 mm; **Ped** none; **Fl** sessile, lemon-yellow, 9 mm, base rounded, campanulate, 8 mm ∅ across mouth; **Tep** free to base; **St** exserted 7 - 8 mm.

A. ambigens Chiovenda (Pl. Nov. Min. Not. Ethiop. 1: 6, 1928). **T:** Somalia (*Puccioni & Stefanini* 447 [FI]). – **D:** Somalia; bushland on steep limestone rock faces, ± 250 m.

[15] Caulescent, branching; stem to 40 cm; **L** 5 - 15, laxly rosulate, to 20 × 2 - 3 cm, glaucous green, sometimes with a few paler spots; marginal teeth to 1 mm, white, 10 - 20 mm apart; **Inf** to 1 m, with up to 8 **Br**; racemes cylindrical, 4 - 12 cm, lax; **Bra** 5 × 3 mm; **Ped** 4 mm; **Fl** red or yellow, 20 - 25 mm, 7 mm ∅ across **Ov**; **Tep** free for 6 mm; **St** not exserted.

A. amicorum L. E. Newton (CSJA 63(2): 80-81, ills., 1991). **T:** Kenya, Eastern Prov. (*Newton* 3217 [K, EA]). – **D:** Kenya; ledges on steep rock face, 1450 m.

[13] Caulescent, branching sparsely near base; stem pendulous, 112 × 2.5 cm ∅, covered with dead sheathing **L** bases; **L** laxly rosulate but curving to give almost distichous appearance, falcate, 46 × 5.5 cm, bluish-green, tinged purplish-red in sun, sometimes with few scattered whitish spots on upper face, surface slightly rough, with narrow white cartilaginous margin; marginal teeth 1 mm, 8 - 10 mm apart, red-tipped; apex obtuse with 1 - 2 minute teeth; exudate yellow, drying brownish-yellow; **Inf** 76 cm, slightly ascending, almost horizontal, with 6 **Br**, lightly covered with white bloom; racemes with secund **Fl**, 8 - 27 cm, lax; **Bra** triangular, 3.5 × 2.5 mm; **Ped** 8 mm; **Fl** crimson with a light cover of white bloom, margins of lobes creamy-white, 28 mm, base truncate, 11 - 12 mm ∅ across **Ov**, narrowed to 7 mm above, lobes spreading to 10 mm; **Tep** free for 10 - 11 mm.

A. amudatensis Reynolds (JSAB 22(3): 136-137, ills., 1956). **T:** Uganda, Karamoja (*Reynolds* 7996 [PRE, EA, K]). – **D:** Kenya, Uganda; dry bushland, 1340 - 1800 m. **I:** Reynolds (1966: 77).

[6] Acaulescent, suckering to form small to large dense clumps; **L** ± 12, densely rosulate, lanceolate-attenuate, 22 × 5 cm, dull green with oval white spots in irregular transverse bands, with white-cartilaginous margin; marginal teeth 2 mm; **Inf** 50 - 65 cm, mostly simple, sometimes with 1 - 2 **Br**; racemes cylindrical, slightly conical, 8 × 6 - 7 cm, sub-lax; **Bra** ovate-deltoid, 10 × 3 mm; **Ped** 17 mm; **Fl** rose-pink to coral-red, 23 mm, base truncate, 9 mm ∅ across **Ov**, abruptly narrowed to 6 mm above, then enlarging towards mouth; **Tep** free for 7 mm; **St** exserted 0 - 1 mm. – *Cytology:* 2n = 14 (Brandham 1971).

A natural hybrid with *A. tweediae* has been reported (Reynolds 1966).

A. andongensis Baker (Trans. Linn. Soc. London, Bot. 1: 263, 1878). **T:** Angola, Cuanza Norte (*Welwitsch* 3729 [BM, K, LISU]). – **D:** Angola.

A. andongensis var. **andongensis** – **D:** Angola; exposed rocks, 1050 - 1525 m. **I:** Reynolds (1966: 347).

[15] Caulescent, branching; stem ascending, sometimes becoming decumbent, 30 - 60 cm; **L** ± 14, rosulate, persistent for 30 cm below, lanceolate-attenuate, 20 - 25 × 6 - 7 cm, dull grey-green, upper face mostly without spots, sometimes sparingly spotted, lower face usually with many crowded white spots near base, with slight cartilaginous margin; marginal teeth 2 - 3 mm, 5 - 7 mm apart; **Inf** 30 - 40 cm, with 2 - 3 **Br**; racemes subcapitate to cylindrical-acuminate, 6 - 12 × 6 - 8 cm; **Bra** lanceolate-acute, 5 - 8 × 3 mm; **Ped** 14 - 18 mm; **Fl** pale orange-scarlet, paler at mouth, 25 mm, base shortly attenuate, 5 - 6 mm ∅ across **Ov**, narrowed to 4 - 5 mm above, enlarging to 6 - 7 mm at mouth; **Tep** free for 8 mm. – *Cytology:* 2n = 14 (Brandham 1971).

A natural hybrid with *A. gossweileri* has been reported (Reynolds 1966).

A. andongensis var. **repens** L. C. Leach (JSAB 40(2): 115-116, ills., 1974). **T:** Angola, Cuanza Sul (*Leach & Cannell* 13950 [LISC, BM, BR, K, PRE, SRGH]). – **D:** Angola; slopes of rounded granite hills.

[13] Differs from var. *andongensis*: Stem prostrate, forming large spreading clumps; **L** smaller, proportionately narrower, more copiously white-spotted, the spots tending to be in wavy transverse bands; marginal teeth more crowded.

A. andringitrensis H. Perrier (Mém. Soc. Linn. Normandie, Bot. 1(1): 41, 1926). **T:** Madagascar, Fianarantsoa (*Perrier* 13637 [P]). – **D:** Madagascar; gravelly places, 2000 - 2600 m. **I:** Reynolds (1966: 452-453).

[3] Acaulescent, simple; **R** fusiform; **L** 15 - 20, rosulate, triangular, 50 × 7 cm, grey-green, surface very slightly asperulous; marginal teeth 1 mm, pinkish, 5 - 10 mm apart; **Inf** 80 - 90 cm, with 6 - 8 **Br**; racemes subcapitate, 6 - 10 cm, dense; **Bra** ovate-acute, 8 × 3 mm; **Ped** 25 mm; **Fl** dull orange to yellowish, 22 mm, base rounded, 4 mm ∅ across **Ov**, enlarging upwards to 9 mm at mouth; **Tep** free for 17 mm; **St** not exserted. – *Cytology:* 2n = 14 (Resende 1937).

A. angelica Pole-Evans (FPSA 14: t. 554 + text, 1934). **T:** RSA, Northern Prov. (*Pole-Evans* s.n. [PRE 13040]). – **D:** RSA (Northern Prov.); rocky slopes. **I:** Reynolds (1950: 471-472).

[9] Caulescent, simple or branched; stem 3 - 4 m, upper ½ covered with dead **L** remains; **L** densely rosulate, ensiform, youngest spreading, oldest much recurved, 80 × 10 - 12 cm, green, with brownish-red margin; marginal teeth 2 - 3 mm, brownish-red, pungent, 10 mm apart; **Inf** with up to 20 **Br**; racemes densely capitate, 8 - 10 × 8 - 10 cm; **Bra**

ovte-acute, 8 - 10 × 8 - 10 mm; **Ped** 25 mm; **Fl** greenish-yellow to yellow, 25 mm, ventricose with mouth slightly upturned, base rounded; **Tep** free for 18 mm; **St** exserted 15 mm. – *Cytology:* 2n = 14 (Riley 1959: 241).

Natural hybrids with other species have been reported (Reynolds 1950).

A. angolensis Baker (Trans. Linn. Soc. London, Bot. 1: 236, 1878). **T:** Angola, Loanda Prov. (*Welwitsch* 3728 [BM, LISU]). – **D:** Angola; wooded hills. **I:** Reynolds (1966: 311).

[18] Very shortly caulescent; **L** densely rosulate, lanceolate-ensiform, very fleshy, 60 × 4 - 5 cm, glaucous; marginal teeth 2 mm, 15 - 20 mm apart; **Inf** 90 cm, simple or with 1 - 3 arcuate-erect **Br**; racemes cylindrical, slightly acuminate, ± 10 cm, dense; **Bra** ovate-acute, 10 mm; **Ped** 3 - 6 mm; **Fl** sulphur-yellow, 20 - 24 mm, base rounded; **Tep** free for < 10 mm.

A little-known species, whose identity is in doubt. Reynolds (1966) suggested that it might belong to the *A. littoralis*-complex.

A. anivoranoensis (Rauh & Hebding) L. E. Newton & G. D. Rowley (Bradleya 16: 114, 1998). **T:** Madagascar (*Rauh* 22864 [HEID]). – **D:** NE Madagascar; limestone rocks in deciduous forest. **I:** Rauh (1998: 99, as *Lomatophyllum*).

≡ *Lomatophyllum anivoranoense* Rauh & Hebding (1998).

[11] Caulescent; stems 1 or branching at the base, to 30 cm, thin at base, thicker towards tip, up to 2 cm ∅; **L** 8 - 10, scattered along stem, erect when young, curved downwards when older, linear, up to 30 × 2 cm, canaliculate, tip long attenuate and often spirally coiled, blue-green; marginal teeth deltoid, 2 mm, 1 - 1.5 mm apart; sheath 1 - 2 cm, white tinged reddish; **Inf** simple, erect or ascending; raceme up to 10 cm, lax, with 15 - 30 **Fl**; **Bra** triangular, long-attenuate, up to 15 mm; **Fl** bright cinnabar-red, whitish with green midstripe at tip, 25 - 30 mm, scarcely constricted above **Ov**, slightly curved; **Tep** free almost to base; **St** exserted 10 mm; **Fr** berries, 15 × 12 mm.

The description of the protologue states that the outer tepals are united for 3 - 5 mm, but in the accompanying Fig. 17 they seem to be united almost to the tip.

A. ankaranensis Rauh & Mangelsdorff (KuaS 50(10): 273-275, ills., 2000). **T:** Madagascar (*Mangelsdorff* s.n. in *BG Heidelberg* 74805 [HEID]). – **D:** NW Madagascar; shady undergrowth on eroded limestone.

[2] Acaulescent; **Ros** mostly solitary, 35 - 40 cm ∅; **L** to 15, densely rosulate, spreading, curved upwards in drought, broadly lanceolate, 3.5 - 5 cm wide, dark green to brown-green with whitish-grey flecks and streaks; marginal teeth 2 mm, white, 5 - 8

mm apart; exudate yellow; **Inf** to 17 cm, erect, simple, sometimes with bulbils in **Ax** of sterile **Bra**; racemes cylindrical, 8 × 6 cm, lax, with ± 20 **Fl**; **Bra** triangular, brownish-red, > 3 mm; **Ped** ± 3 mm; **Fl** cinnabar-red, lobes paler with wide green midstripe, 25 - 27 mm, obscurely 3-angled, 6 mm ∅; **Tep** free for ± 5 mm; **St** scarcely exserted; **Fr** berries, globose, 10 - 15 mm ∅.

A. ankoberensis M. G. Gilbert & Sebsebe (KB 52(1): 146-147, 1997). **T:** Ethiopia, Shewa Region (*Ash* 2353 [K]). – **D:** Ethiopia; steep rocky slopes and cliffs, 3000 - 3500 m.

[13] Caulescent, usually simple; stem pendulous, to 6 m; **L** densely rosulate, 20 - 30 × 7 - 17.5 cm, dull greyish- to bluish-green; marginal teeth 2 - 3 mm, pale, usually with minute dark reddish-brown tip, 0.7 - 1.9 mm apart; **Inf** descending at base and curving upwards again with U-bend, with 1 - 6 **Br**; racemes cylindrical, 6 - 8 cm, dense; **Bra** ovate-lanceolate, acute, 14 - 25 × 5 - 6.5 mm; **Ped** (6-) 10 - 25 mm; **Fl** bright orange-red, 35 - 40 mm, 6 - 10 mm ∅ when pressed; **Tep** free for 12 - 22 mm.

A. antandroi (Decary) H. Perrier (Mém. Soc. Linn. Normandie, Bot. 1(1): 19, 1926). **T:** Madagascar (*Decary* s.n. [P]). – **D:** Madagascar; dry limestone rocks or rubble, 165 m. **I:** Reynolds (1966: 491-492).

≡ *Gasteria antandroi* Decary (1921); **incl.** *Aloe leptocaulon* Bojer (1837) (*nom. inval.*, Art. 32.1c).

[14] Caulescent, branching at or near base; **R** woody; stem usually partly supported by surrounding vegetation, 60 - 100 × 0.5 - 0.7 cm; **L** 12 - 20, laxly rosulate, triangular, 10 - 15 cm × 6 - 10 mm, grey-green, upper face sometimes with a few scattered small white spots, lower face with many dull white spots, apex obtusely rounded with ± 3 small soft white teeth; marginal teeth 0.5 - 1 mm, white to very pale brown, 7 - 10 mm apart; sheaths green-striate, 10 mm apart; **Inf** 16 cm, simple or with 1 **Br**; racemes subcapitate, 3 × 5 cm, lax, ± 10 **Fl**; **Bra** ovate-acute, 4 × 3 mm; **Ped** 8 mm; **Fl** scarlet, 22 mm, base truncate, 6 mm ∅ across **Ov**, slightly narrowed above and enlarging towards mouth; **Tep** free for 15 - 16 mm. – *Cytology:* 2n = 14 (Brandham 1971).

A. antsingyensis (Léandri) L. E. Newton & G. D. Rowley (Excelsa 17: 59, 1997). **T** [syn]: Madagascar (*Decary* 7936 [K]). – **D:** Madagascar; limestone, in shade. **I:** Rauh (1995a: 79, 329).

≡ *Lomatophyllum antsingyense* Léandri (1935).

[13] Caulescent; stem decumbent, to 1 m, 1 cm ∅; **L** scattered along stem, linear, 20 - 50 × 2 cm, dull green; marginal teeth to 1 mm, 10 mm apart; sheath 2 - 3 cm, striate, with white spots; **Inf** 10 - 12 cm, simple; racemes cylindrical, lax, with 15 - 30 **Fl**; **Bra** acute, ± 8 - 10 mm; **Ped** 8 - 10 mm; **Fl** red, green-tipped, ± 20 mm, base rounded, ± 8 mm ∅

across **Ov**, slightly narrowed above, then enlarging slightly to mouth; **Tep** free for 3 - 4 mm; **St** exserted to 1 mm; **Fr** berries.

A. arborescens Miller (Gard. Dict., Ed. 8, [no. 3], 1768). **T:** not typified. – **D:** Malawi, Moçambique, Zimbabwe, RSA; rocky slopes, sometimes in dense bush, to 2150 m. **I:** Reynolds (1950: 408-411); Reynolds (1966: 383).

Incl. *Aloe perfoliata* var. η Linné (1753); **incl.** *Aloe fruticosa* Lamarck (1783); **incl.** *Aloe arborea* Medikus (1783) (*nom. illeg.*, Art. 53.1); **incl.** *Aloe perfoliata* var. *arborescens* Aiton (1789); **incl.** *Catevala arborescens* Medikus (1789); **incl.** *Aloe frutescens* Salm-Dyck (1817) ≡ *Aloe arborescens* var. *frutescens* (Salm-Dyck) Link (1821); **incl.** *Aloe natalensis* J. M. Wood & M. S. Evans (1901) ≡ *Aloe arborescens* var. *natalensis* (J. M. Wood & M. S. Evans) A. Berger (1908); **incl.** *Aloe arborescens* var. *milleri* A. Berger (1908); **incl.** *Aloe arborescens* var. *pachythyrsa* A. Berger (1908).

[16] Caulescent, much branched; stem to 2 - 3 m, 30 cm ∅ at base, with dead **L** persistent for 30 - 60 cm; **L** densely rosulate, triangular, 50 - 60 × 5 - 7 cm, dull green to grey-green; marginal teeth 3 - 5 mm, usually curved towards apex, firm, 5 - 20 mm apart; **Inf** 60 - 80 cm, usually simple, sometimes with 1 **Br**; racemes conical to elongate-conical, 20 - 30 × 10 - 12 cm, dense; **Bra** ovate-acute to obtuse, 15 - 20 × 10 - 12 mm; **Ped** 35 - 40 mm; **Fl** scarlet, 40 mm, base rounded, 7 mm ∅ across **Ov**, slightly narrowed above, then enlarging towards mouth; **Tep** free to the base. – *Cytology:* 2n = 14 (Taylor 1925).

In spite of the specific epithet, this species is usually a large shrub, rarely a tree with a single trunk. It is the most widespread *Aloe* in South Africa, and very variable. Natural hybrids with other species have been reported (Reynolds 1950).

A. archeri Lavranos (CSJA 49(2): 74-75, 1977). **T** [neo]: Kenya, Rift Valley Prov. (*Powys & Archer* s.n. in *Newton* 3118 [K, EA]). – **Lit:** Newton (1992). **D:** Kenya; dry *Acacia* scrub, shade of trees, 1250 - 1800 m.

[13] Caulescent, branching from base; stem erect to 70 × 3.5 cm, then becoming decumbent to 4 m; **L** laxly rosulate and persisting for ± 60 cm below, triangular, 40 × 10 cm, dark green (on young shoots and seedlings, with scattered whitish spots), surface rough; marginal teeth 5 mm, firm, uncinate, 6 - 15 mm apart; exudate yellow, drying brownish-yellow; **Inf** 70 - 140 cm, with 6 - 12 **Br**, lower **Br** sometimes rebranched; racemes cylindrical, 10 - 22 cm, sub-dense; **Bra** ovate-lanceolate, 12 - 15 × 4 mm, closely imbricate in bud stage; **Ped** 10 mm; **Fl** red, with yellow margin on lobes, 22 - 25 mm, base shortly attenuate, 5 mm ∅ across **Ov**, narrowed slightly to 4.5 mm above, then enlarging to 7 mm at mouth; **Tep** free for 12 - 15 mm.

The original description of *A. archeri* included characters from 2 different taxa, the type (holotype) specimen illustrated in the protologue, and plants from another locality that were later named *A. murina* L. E. Newton. The holotype was deposited at EA but later found to be missing.

A. arenicola Reynolds (JSAB 4(1): 21-24, ills., 1938). **T:** RSA, Northern Cape (*Reynolds* 2574 [PRE, BOL]). – **D:** RSA (Northern Cape, Western Cape); coastal sandy plains. **I:** Reynolds (1950: 380-381).

[12] Caulescent, simple or branching to form low clumps; stem 1 m, 3 - 4 cm ∅, decumbent, apical 20 - 30 cm ascending and leafy; **L** ± 20, sub-densely rosulate, lanceolate-attenuate, tip usually a whitish **Sp**, 18 × 5.5 cm, bluish-green with many irregularly scattered oblong white spots, with whitish horny margin; marginal teeth 0.5 mm, 5 - 8 mm apart; **Inf** 50 cm, simple or with 1 - 2 **Br**; racemes densely capitate, 6 × 9 cm; **Bra** 10 × 3 - 4 mm; **Ped** 35 mm; **Fl** peach-red, paler at mouth, 40 mm, base rounded; **Tep** free for 20 mm.

A natural hybrid with *A. krapohliana* has been reported (Reynolds 1950).

A. argenticauda Merxmüller & Giess (Mitt. Bot. Staatssamml. München 11: 437-444, 1974). **T:** Namibia (*Merxmüller & Giess* 28216 [M, PRE, WIND]). – **D:** Namibia.

[6] Acaulescent or with short stems, suckering to form dense clumps of up to 50 **Ros**; **L** densely rosulate, 30 × 3 - 5 cm, grey-green, surface rough; marginal teeth 2 mm, dark brown, 8 mm apart; **Inf** 120 cm, simple or rarely with 1 **Br**; peduncle mostly covered with large silvery **Bra**; racemes 25 - 30 cm, dense; **Bra** lanceolate-acuminate, 50 - 70 × 8 - 12 mm, silvery; **Ped** 5 - 7 mm; **Fl** pinkish to dark red, greenish towards mouth, 30 - 35 mm.

A. aristata Haworth (Philos. Mag. J. 66: 280, 1825). **T** [neo]: RSA, Eastern Cape (*Reynolds* 1024 [PRE]). – **D:** Lesotho, RSA (Northern Cape, Western Cape, Eastern Cape, Free State, KwaZulu-Natal); sandy flats to grassy mountain slopes. **I:** Reynolds (1950: 169-171); CSJGB 43: 3-6, 1981.

Incl. *Aloe longiaristata* Schultes & Schultes *fil.* (1829); **incl.** *Aloe aristata* var. *leiophylla* Baker (1880); **incl.** *Aloe aristata* var. *parvifolia* Baker (1896); **incl.** *Aloe ellenbergeri* Guillaumin (1934).

[6,7] Acaulescent, rarely simple, usually in dense clumps of up to 12 **Ros**; **L** 100 - 150, lanceolate, tapering to a long dry awn-like **Bri**, 8 - 10 × 1 - 1.5 cm, green with scattered small white spots, more numerous and sometimes in ± transverse bands on lower face, lower face with several soft white prickles near tip, lower face with several soft white prickles in 1 or 2 median rows near tip; marginal teeth 1 - 2 mm, soft, white, 1 - 2 mm apart; **Inf** 50 cm, occasionally simple, usually with 2 - 6 **Br**; ra-

cemes laxly subcapitate, 15 - 20 × 12 - 15 cm, with 20 - 30 **Fl**; **Bra** 11 - 12 mm; **Ped** 35 mm; **Fl** jasperred on the upper face, paler beneath, 40 mm, base rounded, 7 mm ∅ across **Ov**, slightly narrowed to 6 mm above, enlarging towards mouth; **Tep** free for 7 mm. − *Cytology:* 2n = 14 (Resende 1937).

A. armatissima Lavranos & Collenette (CSJA 72(1): 22-23, ills. (incl. p. 21), 2000). **T:** Saudi Arabia, Hijaz Prov. (*Collenette* 3738 [K]). − **D:** Saudi Arabia; in forest on granite. **I:** Collenette (1999: 19).

[3] Acaulescent or with a short prostrate stem, solitary; **L** 10 - 14, densely rosulate, lanceolate, 30 - 50 × 9 - 17 cm, glaucous green, often with many paler spots; marginal teeth ± 5 mm, brown-tipped, 8 - 17 mm apart; **Inf** to 170 cm, erect, with 4 - 8 (-27) **Br**; racemes cylindrical, 25 - 55 cm, sub-dense; **Bra** lanceolate, 10 - 12 × 5 - 7 mm; **Ped** 6 - 8 mm; **Fl** yellow or rarely red, ± 38 mm, base shortly attenuate, 6 - 7 mm ∅ across **Ov**, scarcely constricted above; **Tep** free for 22 mm; **St** exserted 4 mm.

A. asperifolia A. Berger (BJS 36: 63, 1905). **T:** Namibia (*Stapf* 7 [Z]). − **D:** Namibia; very arid places, often dependent on night fog for moisture. **I:** Reynolds (1950: 312-313). **Fig. XI.g**

[7] Acaulescent, dense clumps of 20 - 40 **Ros**, or with short creeping rooting stems, simple or branched; **L** lanceolate-acuminate, ± falcate, 15 - 25 × 4 - 7 cm, glaucous to sometimes almost white, surface very rough; marginal teeth 2 - 3 mm, horny, brownish, 5 - 15 mm apart; **Inf** 50 - 75 cm, oblique, with 2 - 3 **Br**; racemes with subsecund **Fl**, 20 - 25 cm, sub-lax; **Bra** deltoid-ovate to deltoid-acuminate, 10 - 15 × 3 - 4 mm; **Ped** 6 - 8 mm; **Fl** scarlet, 28 mm, base shortly attenuate, 6 - 7 mm ∅ across **Ov**, slightly swollen above, narrowing towards mouth, mouth distinctly upturned; **Tep** free for 10 mm; **St** exserted 8 - 10 mm.

A. babatiensis Christian & I. Verdoorn (BT 6(2): 440-442, ills., 1954). **T:** Tanzania, Northern Prov. (*Pole-Evans & Erens* 872 [PRE, SRGH]). − **D:** Tanzania; exposed rocks, 1700 - 2100 m. **I:** Reynolds (1966: 358-359).

[16] Caulescent, branching to form dense shrubs; stem erect or divergent, sometimes decumbent, to > 1 m, 5 cm ∅; **L** ± 24, densely rosulate, persistent for 30 cm, lanceolate-attenuate, 25 × 8 - 9 cm, olive-green tinged reddish, surface slightly glossy; marginal teeth 4 - 5 mm, reddish-brown, ± 10 mm apart; **Inf** 65 cm, with 2 - 4 **Br**; racemes cylindrical-acuminate, 20 - 25 × 10 cm, sub-dense; **Bra** ovate-acute, 30 × 15 mm, imbricate in bud stage; **Ped** 20 - 25 mm; **Fl** salmon-pink, 38 - 40 mm, base truncate, 7 - 8 mm ∅ across **Ov**, slightly narrowed above, then enlarging to mouth; **Tep** free for 8 - 10 mm. − *Cytology:* 2n = 14 (Cutler & al. 1980).

In spite of the specific epithet, this species does not occur near Babati; the nearest population is 54 km west of Babati, at a higher altitude.

A. bakeri Scott-Elliot (JLSB 29: 60, 1891). **T:** Madagascar, Toliara (*Scott-Elliot* 2957 [K]). − **D:** Madagascar; shallow soil and crevices on rocky hills, 40 m. **I:** Reynolds (1966: 414-416). **Fig. XI.h**

≡ *Guillauminia bakeri* (Scott-Elliot) P. V. Heath (1994).

[11] Caulescent, branching and suckering from the base to form dense clumps of up to 100 or more shoots; stem 10 - 20 × 0.5 - 0.7 cm; **L** ± 12, scattered along stem for 5 - 8 cm, triangular, attenuate-acute, 7 cm × 8 mm, green tinged reddish, sometimes with pale green spots (few on upper face, many on lower face); marginal teeth 1 mm, firm, white, 1 - 2 mm apart; sheath 0.5 - 1 cm; **Inf** 25 - 30 cm, simple; raceme subcapitate, 3 - 4 cm, lax, 8- to 10-flowered; **Bra** ovate-acuminate, 3 × 1.5 mm; **Ped** 10 - 12 mm; **Fl** apricot-scarlet at base, becoming orange then yellow towards mouth, the lobes greenish-tipped, 23 mm, base shortly attenuate, narrowed slightly above **Ov**, then enlarging towards mouth; **Tep** free to base; **St** exserted 0 - 1 mm. − *Cytology:* 2n = 14 (Brandham 1971).

A. ballii Reynolds (JSAB 30(3): 123-125, ills., 1964). **T:** Zimbabwe, Melsetter Distr. (*Bullock* 37/1 [SRGH]). − **D:** Moçambique, Zimbabwe.

A. ballii var. **ballii** − **D:** Moçambique, Zimbabwe; in crevices on rock faces, 380 - 400 m. **I:** Reynolds (1966: 11-12). **Fig. XI.b**

[10] Caulescent, branching freely with up to 50 **Ros**; stem pendulous on steep rocks, spirally coiled and 1 - 1.5 m, 0.9 cm ∅; **L** 7 - 12, distichous, triangular-acute, 20 - 30 × 1 cm, green with white or pale green elongated spots near base; marginal teeth minute, white, 2 - 4 mm apart, absent towards apex; **Inf** 50 - 60 cm, oblique to horizontal or arching downwards, simple; raceme cylindrical-acuminate, up to 44 × 4 cm, sub-lax, with 40 - 50 **Fl**; **Bra** ovate-acute, 3 × 2 mm; **Ped** 14 - 20 mm; **Fl** flame-scarlet to pale reddish-orange, 12 - 16 mm, slightly campanulate, base attenuate, 4 mm ∅ across **Ov**, enlarging to the wide mouth; **Tep** free to base; **St** not exserted. − *Cytology:* 2n = 14 (Brandham 1971).

A. ballii var. **makurupiniensis** Ellert (CSJA 70(3): 130-131, ills., 1998). **T:** Zimbabwe/Moçambique (*Ellert* 525 [SRGH]). − **D:** Moçambique, Zimbabwe; open grassy woodland on quartzite ridges and slopes, 400 - 900 m.

[10] Differs from var. *ballii*: Erect, almost acaulescent, suckering from the base only, with up to 12 **Ros**; **L** up to 48.5 cm; **Inf** up to 73 cm, erect.

A. ballyi Reynolds (JSAB 19(1): 2, 1953). **T:** Kenya, Coast Prov. (*Reynolds* 6378 [PRE, EA, K]).

– **D:** Kenya, Tanzania; dense bush and riverine thickets, 900 - 1500 m. **I:** Reynolds (1966: 325-326). **Fig. XI.c**

[9] Caulescent, simple; stem erect, 5 - 6 m, 10 - 15 cm ∅, without persistent dead **L**; **L** ± 25, densely rosulate, spreading when young, becoming greatly recurved, lanceolate-long attenuate, 90 × 14 cm, grey-green, surface smooth; marginal teeth 4 - 5 mm, pungent, sometimes uncinate, white, 10 - 15 mm apart; exudate colourless; **Inf** 60 cm, sub-oblique, with ± 20 **Br**; racemes terminal, cylindrical, lateral with ± 20 subsecund to secund **Fl**, to 14 cm, lax; **Bra** ovate-acute, 5 × 5 mm; **Ped** 10 mm; **Fl** carmine to reddish-orange, greyish-tipped, 33 mm, base rounded, 9 mm ∅ across **Ov**, very slightly narrowed above; **Tep** free for 22 mm. – *Cytology:* 2n = 14 (Brandham 1971).

Broken leaves give off a strong odour reminiscent of rats or mice, and the sap is poisonous.

Tanzanian populations west of Arusha mentioned by Reynolds (1966) are distinct, and they were later described as *A. elata*.

A. barberae Dyer (Gard. Chron., ser. nov. 1: 568, 1874). **T:** RSA, KwaZulu-Natal (*Anonymus* s.n. [K]). – **D:** Moçambique, RSA (KwaZulu-Natal), Swaziland; dense bush and low forest. **I:** Reynolds (1950: 499-501, t. 58-59, as *A. bainesii*).

≡ *Aloe bainesii* var. *barberae* (Dyer) Baker (1896); **incl.** *Aloe bainesii* Dyer (1874).

[17] Caulescent, copiously branching dichotomously; stem erect, to 18 m, 1 - 3 m ∅ at base; **L** ± 20, densely rosulate, ensiform, 60 - 90 × 7 - 9 cm, dull green; marginal teeth 2 - 3 mm, dull white, brownish-tipped, firm horny, 10 - 25 mm apart; **Inf** 40 - 60 cm, usually with 2 **Br**; racemes cylindrical, slightly acuminate, 20 - 30 × 8 - 10 cm, dense; **Bra** linear-acuminate, 10 × 1 mm; **Ped** 10 mm; **Fl** rose to rose-pink, 33 - 37 mm, base rounded; **Tep** free for 28 - 32 mm; **St** exserted 15 mm. – *Cytology:* 2n = 14 (Resende 1937: as *A. bainesii*).

Wyk & Smith (1996) report a small variant (± 2 m tall at maturity) in Moçambique. In most literature this species appears as *A. bainesii* Dyer, a name published in the same paper as *A. barberae*. A few months later Dyer published a note in which the 2 taxa were united, and he chose *A. barberae* as the name for the united species. This later note was overlooked until attention was drawn to it by Smith & al. (1994). In accordance with ICBN Art. 11.5, Dyer's choice of name must be used.

A. bargalensis Lavranos (CSJA 45(3): 116-117, ills., 1973). **T:** Somalia, Bosaso Region (*Lavranos & Bavazzano* 8459 [FI]). – **D:** Somalia; shallow soil on limestone, ± 300 m.

[6] Acaulescent or shortly caulescent, suckering at base; **L** rosulate, lanceolate, somewhat falcate, 30 - 40 × 6 cm, green with irregular white spots and prominently furrowed, with dark green striations;

marginal teeth 1 - 2 mm, 12 - 25 mm apart; **Inf** 70 - 120 cm, simple or rarely with 1 - 2 **Br**; racemes narrowly cylindrical-acuminate, 30 - 40 cm, sub-dense; **Bra** ovate-deltoid, long-acute, to 15 × 5 mm; **Ped** 5 - 7 mm; **Fl** reddish, yellow towards mouth, 30 mm, base attenuate, 5 mm ∅ across **Ov**, narrowed to 4 mm ∅ above, then enlarging to 7 mm at mouth; **Tep** free for 15 - 16 mm.

A. belavenokensis (Rauh & Gerold) L. E. Newton & G. D. Rowley (Excelsa 17: 59, 1997). **T:** Madagascar, Taolanaro (*Rauh 73987* [HEID]). – **D:** Madagascar; coastal forest on sand. **Fig. XI.d**

≡ *Lomatophyllum belavenokense* Rauh & Gerold (1994).

[1] Very shortly caulescent, stoloniferous and forming groups to 1 m ∅; stem subterranean, to 1.5 × 1 - 1.5 cm; **L** ± 10, densely rosulate, linear-attenuate, 20 - 30 × 1 cm, green, red-brown to violet at base; marginal teeth 1 mm, white to pale red; exudate white, becoming yellow in air; **Inf** 25 - 40 cm, simple; racemes cylindrical, 10 - 20 cm, lax, with 10 - 15 (- 20) **Fl**; **Bra** acute, 3 - 5 mm; **Ped** 10 - 15 mm; **Fl** bright cinnabar-red, whitish with green veins in upper ⅓, 20 - 25 mm, base rounded, 6 - 7 mm ∅ across **Ov**, slightly narrowed above, then enlarging to mouth; **Tep** free for 15 - 20 mm; **St** not exserted; **Fr** berries.

A. bella G. D. Rowley (Repert. Pl. Succ. 23: 12, 1974). **T:** Somalia, Bosaso Region (*Lavranos & Bavazzano* 8458 [FI]). – **D:** Somalia; stony coastal plain and limestone plateau, ± 30 m. **I:** CSJA 45: 118-119, 1973, as *A. pulchra*.

Incl. *Aloe pulchra* Lavranos (1973) (*nom. illeg.*, Art. 53.1).

[7] Acaulescent or shortly caulescent, suckering to form dense clumps; stem decumbent; **L** rosulate, deltoid, acute, rather falcate, to 50 × 11 cm, pinkish-brown with red cartilaginous margin; marginal teeth small, red, ± 20 mm apart; **Inf** to 1 m, with up to 5 **Br**; racemes conical to subcapitate, 6 - 11 cm, dense; **Bra** deltoid-acute, 8 - 10 × 4 mm; **Ped** 6 - 7 mm; **Fl** bright red, becoming green towards mouth, 27 mm, base rounded, 8 - 9 mm ∅ across **Ov**, slightly narrowed above towards mouth; **Tep** free for 7 - 9 mm.

A. bellatula Reynolds (JSAB 22(3): 132-134, ills., 1956). **T:** Madagascar, Fianarantsoa (*Millot* s.n. in *Reynolds* 6591 [PRE, P, TAN]). – **D:** Madagascar; mountain slopes, ± 1500 m. **I:** Reynolds (1966: 402-404). **Fig. XI.e**

≡ *Guillauminia bellatula* (Reynolds) P. V. Heath (1994).

[1] Acaulescent, suckering to form dense clumps; **L** ± 16, densely rosulate, linear-attenuate, 10 - 13 × 0.9 - 1 cm, dark green with many pale green spots, surface rough; marginal teeth 1 mm, soft, cartilaginous, smaller or absent towards apex; **Inf** 60 cm,

simple or with 1 **Br**; racemes cylindrical-acuminate, 12 - 16 × 4 cm, lax with 35 **Fl**, denser above; **Bra** deltoid-acuminate, 4 - 6 × 2 mm; **Ped** 12 mm; **Fl** light coral-red, 13 mm, campanulate, base attenuate, 6 mm ∅ across **Ov**; **Tep** free for 7 mm; **St** exserted 0 - 1 mm. – *Cytology:* 2n = 14 (Brandham 1971).

A. berevoana Lavranos (KuaS 49(7): 161-162, ills., 1998). **T:** Madagascar (*Röösli & Hoffmann* s.n. [P, MO]). – **D:** W Madagascar, sandstone cliffs near sea level.
[11] Caulescent, branching at base to form very large groups; stems ascending, to 60 cm; **L** 8 - 10, laxly arranged, lanceolate, tip acute, 30 × 3 cm, grass-green, with longitudinal striations; marginal teeth 2 mm, dark green with whitish tip, 8 - 12 mm apart; **Inf** to 60 cm, with 3 **Br**; raceme 10 - 12 cm, lax, few-flowered; **Bra** 5 mm, triangular; **Ped** 10 mm; **Fl** dark red, 17 mm, 4 mm ∅; **Tep** free for 5 mm; **Fr** not known.

A. bernadettae Castillon (Adansonia, sér. 3, 22(1): 136-138, ills., 2000). **T:** Madagascar (*Castillon 2* [P]). – **D:** Madagascar; on gneissic rocks.
[13] Caulescent, suckering; stem procumbent, to 1.5 m and 6 cm ∅; **L** 15 - 20, rosulate, 70 × 3 - 7 cm, green; marginal teeth 2 mm, curved, 7 - 10 mm apart; **Inf** 80 cm, erect, simple; racemes cylindrical, 15 - 20 × 5 cm, dense, with 80 - 100 sessile **Fl** and a tuft of sterile **Bra** at the tip; **Bra** 10 × 4 mm; **Fl** lemon-yellow, 16 mm, 6 mm ∅ across **Ov**, campanulate; **Tep** free for 11 - 14 mm; **St** exserted ± 5 mm.

A. bertemariae Sebsebe & Dioli (KB 55(3): 679-681, ills., 2000). **T:** Ethiopia, Harerge Region (*Dioli 4* [ETH, K]). – **D:** Ethiopia; sandy clay soil in *Acacia* woodland, 300 - 400 m.
[4] Acaulescent, suckering to form groups of up to 4 **Ros**; **L** 13 - 15, rosulate, erect or slightly recurved, triangular to lanceolate, strongly canaliculate, becoming tubular in dry condition, 50 - 65 × 8 - 9 cm, both faces densely spotted with 2 × 0.5 - 0.9 mm whitish-green blotches, longitudinally striped, lower face of most **L** with 3 - 5 brown prickles 1 - 2 mm long; marginal teeth 1 - 2 mm, brown, 15 - 25 mm apart; exudate dark yellow, drying brown; **Inf** 1 - 2 m, simple; racemes cylindrical-conical, 50 - 80 cm, subdense; **Bra** triangular-acuminate, deflexed, 9 - 12 × 3 - 4 mm, with many brown veins; **Ped** 4 - 7 mm; **Fl** dark coral-red, whitish at mouth, cylindrical-trigonous, 26 - 28 × 5 mm, slightly constricted above **Ov**, minutely pubescent; **Tep** free for 8.7 - 9.3 mm; **St** not exserted; **Fr** ovoid, 14 - 17 × 5 - 6 mm, shortly pubescent, transversely ridged.

A. betsileensis H. Perrier (Mém. Soc. Linn. Normandie, Bot. 1(1): 48, 1926). **T:** Madagascar, Toliara (*Perrier 13676* [P]). – **D:** Madagascar; rocky grassland, 800 - 1400 m. **I:** Reynolds (1966: 482-483).

[3] Acaulescent, simple; **L** 20 - 30 (-50), densely rosulate, triangular with apex slightly twisted, 30 - 40 × 7 - 9 cm, dull green tinged reddish, with reddish margin; marginal teeth 2 - 3 mm, reddish, pungent, 8 - 12 mm apart; exudate drying yellow; **Inf** 60 cm and simple in young plants, 70 cm to > 1 m with 3 - 5 **Br** in old plants; racemes cylindrical, 30 - 35 × 4 - 5 cm, very dense, **Fl** sessile, arranged in 13 spirally twisted rows; **Bra** ovate-obtuse, 8 - 10 × 6 - 8 mm, fleshy, reddish; **Fl** yellow with orange tips, 15 mm, slightly campanulate, base rounded, 7 mm ∅ across **Ov**, enlarging to 9 mm at mouth; **Tep** free to base.
Flowering starts on the sunny side of the raceme, where the flowers are more orange in colour.

A. bicomitum L. C. Leach (Kirkia 10: 385-386, 1977). **T:** Tanzania, Western Prov. (*Reynolds 8948* [PRE, EA, K]). – **D:** Tanzania, Zambia; rock outcrops, 1400 m. **I:** Reynolds (1966: 174-175, as *A. venusta*).
Incl. *Aloe venusta* Reynolds (1959) (*nom. illeg.*, Art. 53.1).
[5] Acaulescent, simple or in small groups; **L** ± 20, densely rosulate, triangular, 50 × 9 cm, dull grey-green with many elliptic pale green spots, more crowded towards base, smaller and more crowded on the lower face, with continuous pinkish margin; marginal teeth 3 mm, uncinate, pinkish, 8 - 10 mm apart; **Inf** 1 - 1.2 m, with ± 10 **Br**; racemes cylindrical-conical, 15 - 20 cm, dense; **Bra** ovate-cuspidate, 11 × 10 mm, somewhat fleshy in the middle, imbricate in bud stage; **Ped** 13 mm; **Fl** pale scarlet, minutely pubescent, 32 mm, base very shortly attenuate, 7 mm ∅ across **Ov**, slightly narrowed above, then enlarging towards mouth; **Tep** free for 12 mm.

A. boiteaui Guillaumin (Bull. Mus. Nation. Hist. Nat., Sér. 2, 14: 349, 1942). **T:** Madagascar, Toliara (*Boiteau* s.n. [P]). – **D:** S Madagascar (Toliara: near Fort Dauphin). **I:** Reynolds (1966: 420).
≡ *Lemeea boiteaui* (Guillaumin) P. V. Heath (1994).
[1] Acaulescent or with short stem, suckering; **L** ± 10, rosulate, triangular, 15 - 20 × 1.4 cm, olive-green, with narrow pinkish cartilaginous margin; marginal teeth 0.5 - 1 mm, pinkish, 2 mm apart, absent towards apex; **Inf** 10 - 15 cm, simple; raceme 5 cm, lax, with ± 10 **Fl**; **Bra** deltoid-acute, 5 × 3 mm; **Ped** 10 mm; **Fl** bright scarlet, 25 mm, base shortly attenuate, slightly narrowed above **Ov**, then enlarging towards mouth; **Tep** free for 18 mm.

A. boscawenii Christian (JSAB 8(2): 165-167, ills., 1942). **T:** Tanzania, Tanga Distr. (*Boscawen 7* in *Christian 902* [SRGH, EA, K, PRE]). – **D:** Tanzania; scrub on sandy soil along coast, to 60 m. **I:** Reynolds (1966: 365).
[14] Caulescent, branching near base; **Br** 1 - 2 m

× 5 - 7 cm, erect for 20 - 30 cm, when longer sprawling or supported by surrounding vegetation; **L** scattered along stem for 20 - 30 cm, ovate-lanceolate, 44 - 50 × 8 cm, light green, with narrow cartilaginous margin, surface smooth, exudate yellow; marginal teeth 2 - 3 mm, brown-tipped, pungent, 7 - 18 mm apart; **Inf** ± 90 cm, with 3 - 9 **Br**, lower **Br** sometimes rebranched; racemes cylindrical, 10 - 12 × 7 cm, lax below, more dense towards apex; **Bra** long-acuminate, 7 × 3 mm; **Ped** 18 mm; **Fl** yellow, becoming brownish towards apex, 30 mm, base attenuate, 9 mm ∅ across **Ov**, very slightly narrowed above; **Tep** free for 18 mm.

A. bosseri Castillon (Adansonia, sér. 3, 22(1): 135-137, ills., 2000). **T:** Madagascar (*Castillon* 1 [P]). − **D:** Madagascar; limestone cliffs.

[6, 7] Acaulescent or caulescent to 40 cm, suckering at the base; **L** 10 - 15, rosulate, tip attenuate, 30 - 70 × 4 cm, blue-green to yellowish-green, upper face finely lineate, margins a 1 mm wide whitish or rose border; marginal teeth absent or 0.1 - 0.5 mm, 0.5 - 1 mm apart; **L** sheath ± 1 cm; **Inf** 70 cm, erect or ascending, simple or with 2 - 6 **Br**; racemes cylindrical, ± 10 cm, lax, with 15 - 20 **Fl**; **Bra** ovate-acute, 2.5 - 3.5 mm; **Ped** 12 - 15 mm; **Fl** red, with green tips, inner lobes whitish, 20 - 25 mm, 8 mm ∅ across **Ov**; **Tep** free for 16 - 21 mm; **St** not or only slightly exserted.

A. bowiea Schultes & Schultes *fil.* (Syst. Veg. 7(1): 704, 1829). **T:** [neo − icono]: K [unpubl. drawing]. − **Lit:** Smith (1990). **D:** RSA (Eastern Cape); Valley Bushveld, dense thickets. **I:** Aloe 28: 9, 1991.

Incl. *Bowiea africana* Haworth (1824) ≡ *Chamaealoe africana* (Haworth) A. Berger (1905).

[1] Acaulescent, suckering to form dense groups; **R** fusiform; **L** 18 - 25, rosulate-multifarious, linear, 10 - 15 × 1.25 cm, pale glaucous green with scattered white spots, more numerous on lower face; marginal teeth minute, soft, white; **Inf** ± 45 cm, simple; racemes cylindrical, ± 15 cm, lax; **Bra** deltoid; **Ped** 1 - 2 mm; **Fl** greenish-white, 8 - 15 mm, clavate, base shortly attenuate, enlarging above **Ov** to mouth; **Tep** free to base; **St** exserted 6 - 8 mm.

A. boylei Baker (BMI 1892: 84, 1892). **T:** RSA, KwaZulu-Natal (*Allison* s.n. in *Boyle* s.n. [K]). − **D:** RSA.

Incl. *Aloe micracantha* Pole-Evans (1923) (*nom. illeg.*, Art. 53.1); **incl.** *Aloe agrophila* Reynolds (1936).

A. boylei ssp. **boylei** − **D:** RSA (Eastern Cape, KwaZulu-Natal, Mpumalanga, Northern Prov.). **I:** Wyk & Smith (1996: 253).

[4] Caulescent, simple or branching; stem short or up to 20 × 6 cm with age, simple or branching at base to form groups of 6 - 12 shoots; **L** ± 10 - 14,

rosulate, lanceolate to ensiform, 50 - 60 × 1.5 - 3 cm, dark green, upper face sometimes lineate and with few scattered elliptic spots near base, lower face with many white spots from base to middle, with soft white margin ± 2 mm; marginal teeth 1 - 3 mm, soft, white, 2 - 5 mm apart, smaller and more crowded towards **L** tip; **Inf** 40 - 60 cm, simple; raceme capitate, 10 - 12 × 10 - 12 cm, with ± 40 **Fl**; **Bra** ovate-acuminate, ± 20 - 23 mm; **Ped** 40 - 45 mm; **Fl** salmon-pink, greenish-tipped, 30 - 35 mm, base attenuate, 11 - 12 mm ∅ across **Ov**, narrowing to 8 - 9 mm at mouth; **Tep** free almost to base; **St** exserted 0 - 1 mm. − *Cytology:* 2n = 14 (Müller 1945).

A. boylei ssp. **major** Hilliard & B. L. Burtt (Notes Roy. Bot. Gard. Edinburgh 42(2): 252, 1985). **T:** RSA, KwaZulu-Natal (*Hilliard & Burtt* 8438 [E, NU]). − **D:** RSA (KwaZulu-Natal). **I:** Reynolds (1950: 154-155).

[4] Differs from ssp. *boylei*: Plants more robust; **L** 7 cm broad; **Fl** 40 mm.

Plants regarded by Reynolds (1950) as a "weak form" of this species (also described as *A. agrophila*) match Baker's description, and plants regarded by Reynolds as typical are now ssp. *major.*

A. brachystachys Baker (CBM 121: t. 7399 + text, 1895). **T:** "Zanzibar" (*Kirk* s.n. [K]). − **D:** Tanzania; rock outcrops in montane bushland and at edge of mist forest, ± 2000 m. **I:** CSJA 48: 279-280, 1986, as *A. schliebenii*.

Incl. *Aloe lastii* Baker (1901); **incl.** *Aloe schliebenii* Lavranos (1970).

[4] Acaulescent, or with short procumbent stem with age, simple or suckering to form small groups; **L** 20 - 30, rosulate, lanceolate-attenuate, 30 - 60 × 4 - 8 cm, yellowish-green, red-brown in sun, longitudinally lineate; marginal teeth 1 - 3 mm, brown-tipped, curved, 10 mm apart; **Inf** 40 - 100 cm, simple or occasionally with 1 short **Br**; raceme cylindrical, 15 - 20 cm, dense; **Bra** ovate, 12 - 14 × 5 - 10 mm, fleshy, red, becoming scarious with age, imbricate in bud stage; **Ped** 16 - 22 mm; **Fl** pale orange-red, yellow-green at mouth, 25 - 32 mm, clavate, base attenuate, 4 - 5 mm ∅ across **Ov**, not narrowed above, enlarging to 7 - 8 mm at mouth; **Tep** free for 8 - 10 mm.

One of many plants sent to Kew from Zanzibar by the governor, Sir John Kirk, without locality information. Most of these plants had been collected on the African mainland, in parts of Kenya and Tanzania.

A. branddraaiensis Groenewald (FPSA 20: t. 761 + text, 1940). **T:** RSA, Mpumalanga (*van der Merwe* s.n. [PRE 24208]). − **D:** RSA (Mpumalanga); Bushveld in frost free areas. **I:** Reynolds (1950: 219-220).

[5] Acaulescent, usually simple; **L** 20 - 25, rosu-

late, sometimes subdistichous or somewhat spirally twisted, lanceolate-attenuate, ± 35 × 8 - 10 cm, green with many longitudinal dull whitish striations and many irregular, somewhat H-shaped, whitish spots; marginal teeth 2 - 3 mm, pale brown, deflexed, 10 - 15 mm apart; **Inf** 1 - 1.5 m, with many **Br**, lower **Br** rebranched; racemes capitate, 3 - 6 × 7 cm, with ± 15 **Fl**; **Bra** deltoid-acuminate, 8 mm; **Ped** 20 mm, shorter on lateral racemes; **Fl** dull scarlet at base, paler at mouth, 27 mm, base rounded, 5.5 mm ⌀ across **Ov**, narrowed to 3.5 mm above, then enlarging to 6 mm at mouth; **Tep** free for 7 mm; **St** exserted 1 mm. – *Cytology:* 2n = 14 (Brandham 1971).

A natural hybrid with *A. burgersfortensis* has been reported (Reynolds 1950).

A. brandhamii S. Carter (Fl. Trop. East Afr., Aloaceae, 32-33, ills., 1994). **T:** Tanzania, Iringa Distr. (*Carter & al.* 2600 [K, DAR, EA, NHT]). – **D:** Tanzania; light shade on rocky slopes, 750 - 1200 m.

[7] Acaulescent or shortly caulescent, suckering to form small groups; stem to 1 m on old plants, ascending; **L** densely rosulate, lanceolate, 50 - 80 × 10 - 20 cm, dull dark green, often bronzed, surface smooth; marginal teeth 2 - 3 mm, pungent, brown-tipped, 10 - 20 mm apart; exudate drying yellow; **Inf** 1.5 - 2 m, with up to 25 **Br**, lower **Br** sometimes rebranched; racemes with **Fl** secund, 15 - 30 cm, lax; **Bra** ovate, 12 - 15 × 5 - 7 mm; **Ped** 5 - 9 mm; **Fl** coral-pink, paler at mouth, minutely white-flecked, 30 - 40 mm, base rounded, 7 - 8 mm ⌀ across **Ov**, scarcely narrowed above; **Tep** free for ± 10 - 13 mm.

A. brevifolia Miller (Gard. Dict. Abr. ed. 6, [no. 8], 1771). **T:** not preserved. – **D:** RSA.

Incl. *Aloe perfoliata* var. δ Linné (1753).

A. brevifolia var. **brevifolia** – **D:** RSA (Western Cape); open bushland. **I:** Reynolds (1950: 184-186). **Fig. XII.a**

Incl. *Aloe prolifera* Haworth (1804).

[6] Acaulescent, suckering to form dense clumps; **L** 30 - 40, densely rosulate, lanceolate-deltoid, tip a firm **Sp**, 6 × 2 cm, glaucous, lower face with a few soft prickles in median line or irregular in upper ⅓; marginal teeth 2 - 3 mm, whitish, 10 mm apart; **Inf** 40 cm, simple; raceme conical, 15 × 7 cm, subdense; **Bra** ovate-lanceolate, 15 mm; **Ped** 15 mm; **Fl** pale scarlet, 38 mm, base truncate, very slightly narrowed above **Ov**, slightly enlarging to mouth; **Tep** free to base. – *Cytology:* 2n = 14 (Resende 1937).

A natural hybrid with *A. mitriformis* has been reported (Reynolds 1950).

A. brevifolia var. **depressa** (Haworth) Baker (JLSB 18: 160, 1880). **T:** not typified. – **D:** RSA (Western Cape); shale cliff. **I:** Reynolds (1950: 189).

≡ *Aloe depressa* Haworth (1804); **incl.** *Aloe perfoliata* var. ζ Linné (1753); **incl.** *Aloe serra* De Candolle (1799) ≡ *Aloe brevifolia* var. *serra* (De Candolle) A. Berger (1908).

[6] Differs from var. *brevifolia*: **L** ± 60, 12 - 15 × 6 cm, surface smooth or with white subtuberculate spots in upper ½; marginal teeth 2 - 4 mm; **Inf** 60 cm; **Bra** ± 15 mm; **Ped** to 20 mm; **Fl** 40 mm, flame-scarlet. – *Cytology:* 2n = 14 (Resende 1937).

The largest of the varieties.

A. brevifolia var. **postgenita** (Schultes & Schultes *fil.*) Baker (JLSB 18: 160, 1880). **T:** not typified. – **D:** RSA (Western Cape). **I:** Reynolds (1950: 187).

≡ *Aloe postgenita* Schultes & Schultes *fil.* (1830); **incl.** *Aloe prolifera* var. *major* Salm-Dyck (1817).

[6] Differs from var. *brevifolia*: **L** 10 - 13 × 4 cm; intermediate in size between var. *brevifolia* and var. *depressa*. – *Cytology:* 2n = 14 (Resende 1937).

A natural hybrid with *A. mitriformis* has been reported (Reynolds 1950).

A. breviscapa Reynolds & P. R. O. Bally (JSAB 24(4): 176-177, t. 19, 1958). **T:** Somalia, Northern Region (*Reynolds* 8542 [PRE, EA, K]). – **D:** Somalia; arid gypsum plains, ± 1400 m. **I:** Reynolds (1966: 267-268).

[7] Caulescent, suckering to form small to large dense clumps, with almost horizontal **Ros**; stem short or to 50 cm, decumbent, to 1 m with age; **L** ± 24, densely rosulate, lanceolate-attenuate, 30 - 35 × 8 - 10 cm, bluish-grey tinged reddish; marginal teeth absent or few in basal ¼, 1 - 2 mm, blunt, 10 mm apart; exudate drying yellow; **Inf** ± 50 cm, arcuate-ascending, with 4 - 8 **Br**; racemes cylindrical, 20 - 25 × 6 cm, lax; **Bra** ovate-acute, 6 × 3 mm; **Ped** 10 - 14 mm; **Fl** scarlet with a bloom, greenish at mouth, 26 - 30 mm, base shortly attenuate, 8 mm ⌀ across **Ov**, slightly narrowed above; **Tep** free for 10 mm.

A. broomii Schönland (Rec. Albany Mus. 2: 137, 1907). **T:** RSA, Cape Prov. (*Broom* s.n. [not located]). – **D:** RSA, Lesotho.

A. broomii var. **broomii** – **D:** Lesotho, RSA (Eastern Cape, Western Cape, Northern Cape, Free State); rocky slopes, 1400 - 1900 m, flowers in September. **I:** Reynolds (1950: 163-165).

[4] Caulescent, usually simple; stem short or up to 1 m, procumbent, usually simple, sometimes dividing into 2 or 3, covered with dried **L**; **L** densely rosulate, ovate-lanceolate, acuminate with pungent terminal **Sp**, 30 × 10 cm, green, obscurely lineate, with reddish-brown horny margin; marginal teeth 1 - 2 mm, reddish-brown with paler tips, 10 - 15 mm apart; **Inf** 1 - 1.5 m, mostly simple, rarely with 1 **Br**; racemes cylindrical, slightly acuminate, to 100 × 6 -

8 cm, very dense; **Bra** lanceolate-acute, 30 × 15 mm, rather fleshy, white to pale lemon with brownish tips; **Ped** 1 - 2 mm; **Fl** pale lemon, 20 - 25 mm, base rounded, enlarging above **Ov** and narrowing to mouth, completely hidden by **Bra**; **Tep** free to base; **St** exserted 12 mm.

Natural hybrids with other species have been reported (Reynolds 1950).

A. broomii var. **tarkaensis** Reynolds (JSAB 2(2): 72-73, ills., 1936). **T:** RSA, Eastern Cape (*Reynolds* 1777 [PRE]). — **D:** RSA (Eastern Cape); rocky slopes, flowers February - March. **I:** Reynolds (1950: 166).

[4] Differs from var. *broomii*: more luxuriant, **L** 2 - 3× broader at base; **Bra** dry and much shorter; **Ped** 4 mm; **Fl** 30 mm, not hidden by **Bra**.

A. brunneodentata Lavranos & Collenette (CSJA 72(2): 86, ill. (p. 84), 2000). **T:** Saudi Arabia (*Collenette* 5826 [K]). — **D:** Saudi Arabia; on granite, ± 1800 m. **I:** Collenette (1999: 20).

[2] Acaulescent, solitary; **L** densely rosulate, 28 - 35 × 7 cm, pale bluish-green; marginal teeth short, brown; **Inf** ± 60 cm, simple or with 1 **Br**; racemes cylindrical, lax; **Bra** 12 - 15 × 4 - 6 mm; **Fl** reddish, downy, 24 - 26 (-35) mm, base rounded, not constricted above **Ov**; **Tep** free for ± 11 mm; **St** not exserted.

Known from only one population.

A. brunneostriata Lavranos & S. Carter (CSJA 64(4): 206-208, ills., 1992). **T:** Somalia, Bari Region (Migurtein) (*Lavranos & Horwood* 10187 [K]). — **D:** Somalia; sandy plains, 640 m.

[7] Shortly caulescent, suckering from base to form small groups; **Br** erect or ascending, 40 cm; **L** up to 10, laxly rosulate, lanceolate, acute, 30 × 7 cm, creamy-yellow with many longitudinal reddish-brown lines, surface smooth; marginal teeth absent or < 0.5 mm, blunt, yellow; **Inf** 50 - 60 cm, with 6 - 7 (-12) **Br**; racemes with subsecund **Fl**, lax; **Bra** ovate-acute, 5 - 6 × 5 - 6 mm; **Ped** 5 - 6 mm; **Fl** yellow with greenish veins at lobe tips, 16 - 20 mm, base shortly attenuate, 5 - 6 mm ∅ across **Ov**, slightly narrowed above and enlarging to mouth; **Tep** free for ± 4 - 6 mm.

A. buchananii Baker (BMI 1895: 119, 1895). **T:** Malawi, Southern Prov. (*Buchanan* s.n. [K]). — **D:** Malawi; woodland, 1150 - 1600 m. **I:** Reynolds (1966: 29-31).

[4] Acaulescent or shortly caulescent, usually simple or 2 - 3 (rarely to 9) in clumps; **R** fusiform; stem to 20 cm; **L** distichous or becoming rosulate, triangular, 60 × 4 - 6 cm, green with few scattered elongated dull white spots towards base, lower face more copiously spotted, with narrow translucent cartilaginous margin; marginal teeth 0.5 mm, 8 - 15 mm apart, absent towards **L** tip; **Inf** 60 - 80 cm,

simple; raceme cylindrical-acuminate, 15 - 20 × 7 cm, dense; **Bra** ovate-acute, apiculate, 25 - 30 × 10 - 12 mm, fleshy, pale pink, imbricate in early bud stage; **Ped** 35 - 40 mm; **Fl** salmon-pink or light coral-red, greenish at mouth, 30 mm, base shortly attenuate, 10 - 11 mm ∅ across **Ov**, narrowing towards mouth; **Tep** free to base or almost to base; **St** exserted 0 - 1 mm.

A. buchlohii Rauh (KuaS 17(1): 2-4, ills., 1966). **T:** Madagascar, Toliara (*Rauh* M1381 [HEID]). — **D:** Madagascar; bare gneissic rocks, 100 m. **I:** Reynolds (1966: 431); FPA 52: t. 2047, 1992.

[4] Acaulescent, simple or in small groups; **L** 10 - 20, densely rosulate, lanceolate-attenuate with small apical **Sp**, 40 - 50 × 3 cm, green, sometimes tinged reddish, few spots near base; marginal teeth 3 mm, pungent, reddish-brown-tipped, 8 mm apart; **Inf** 60 cm, simple (or with 3 **Br** in cultivation); raceme subcapitate, 10 × 7 cm, sub-dense; **Bra** ovate-attenuate, 7 × 2.5 mm; **Ped** 15 - 20 mm; **Fl** pale yellow, or pale rose at base and yellowish upwards, 25 mm, base shortly attenuate, 7 mm ∅ across **Ov**, scarcely narrowed above; **Tep** free to base.

A. buettneri A. Berger (BJS 36: 60, 1905). **T:** Togo (*Büttner* s.n. [B]). — **D:** Benin, Ghana, Mali, Nigeria, Togo; grassland and savanna woodland, 250 - 900 m. **I:** Reynolds (1966: 45).

Incl. *Aloe barteri* Baker (1880); **incl.** *Aloe barteri* var. *dahomensis* A. Chevalier (1952) (*nom. inval.*, Art. 36.1); **incl.** *Aloe barteri* var. *sudanica* A. Chevalier (1952) (*nom. inval.*, Art. 36.1); **incl.** *Aloe paludicola* A. Chevalier (1952) (*nom. inval.*, Art. 36.1).

[5] Acaulescent, simple or rarely suckering at base forming small groups, the base enlarging to form a bulb towards the end of the growing season; **L** ± 16, rosulate, usually dying back in dry season, triangular, 35 - 55 cm, green, obscurely lineate, sometimes with few scattered whitish spots, with very narrow white to pale pink cartilaginous margin, surface smooth; marginal teeth 3 - 4 mm, firm, 10 - 15 mm apart; **Inf** 70 - 90 cm, with 3 - 5 **Br**; racemes cylindrical-conical to subcapitate, 15 × 7 cm, sub-dense; **Bra** deltoid-acute or lanceolate-acuminate, 10 - 15 × 6 - 8 mm; **Ped** 20 - 25 mm; **Fl** greenish-yellow to dull red, 38 mm, base rounded, 9 - 11 mm ∅ across **Ov**, narrowed to 6 - 8 mm above, enlarging to 9 - 11 mm and narrowing again to mouth; **Tep** free for 12 - 13 mm; **St** exserted 0 - 1 mm. — *Cytology:* 2n = 14 (Resende 1937).

Reynolds (1966) included *A. bulbicaulis* and *A. paedogona* in this taxon, but Carter (1994) treated them as 3 geographically distinct species. Further studies are required to determine their relationships.

Until 1963, only 1 species was reported from West Africa, *A. barteri*. Keay (1963) showed that the type of this name was a mixture of 2 taxa, *A. buettneri* and *A. schweinfurthii*.

A natural hybrid with *A. schweinfurthii* is known as *A. ×keayi* Reynolds (*pro. sp.*) (Newton 1976).

A. buhrii Lavranos (JSAB 37(1): 37-40, ills., 1971). **T:** RSA, Northern Cape (*Buhr s.n. in Lavranos* 8163 [PRE]). – **D:** RSA (Northern Cape); karroid veld on shale, isolated hilltops, ± 650 m. **I:** Wyk & Smith (1996: 127).

[7] Acaulescent, suckering to form dense clumps; **L** ± 16, rosulate, lanceolate-deltoid, 40 × 9 cm, glaucous with reddish tinge, distinctly striate and with irregular white elongate or H-shaped spots, with pale red cartilaginous 1.5 - 2 mm margin; marginal teeth < 1 mm, 3.5 mm apart or laterally confluent; **Inf** to 60 cm, with 7 - 15 **Br**, lower **Br** rebranched; racemes subcapitate, lax; **Bra** deltoid-acute, yellowish, 5 - 10 mm; **Ped** 20 - 25 mm; **Fl** orange-red, 25 - 27 mm, base truncate, 6 mm ∅ across **Ov**, abruptly narrowed to 4 mm above, then enlarging towards mouth; **Tep** free for 6 - 7 mm.

A. bukobana Reynolds (JSAB 20(4): 169-171, ills., 1955). **T:** Tanzania, North-Western Div. (*Reynolds* 7507 [PRE, EA, K]). – **D:** Tanzania; sandstone hills and rock outcrops, 1180 - 1460 m. **I:** Reynolds (1966: 108-109).

[7] Acaulescent, suckering freely to form small dense groups; **L** ± 16, densely rosulate, lanceolate-attenuate with the tip a small **Sp**, 30 × 8 cm, dull green with slight bloom above, grey-green below; marginal teeth 4 mm, firm, brownish-tipped, 10 mm apart; exudate drying yellow; **Inf** 70 - 90 cm, with up to 10 **Br**, lower **Br** sometimes rebranched; racemes narrowly conical-cylindrical, 30 - 40 cm, very lax; **Bra** ovate-acute, 4 × 3 mm; **Ped** 12 mm; **Fl** dull scarlet, paler at mouth, 30 - 35 mm, base rounded or shortly attenuate, 7 - 8 mm ∅ across **Ov**, enlarging slightly above; **Tep** free for 7 mm. – *Cytology:* 2n = 14 (Brandham 1971).

Natural hybrids with *A. macrosiphon* have been reported (Reynolds 1966).

A. bulbicaulis Christian (FPSA 16: t. 630 + text, 1936). **T:** Zambia (*Porter s.n.* [PRE 20587]). – **D:** Angola, Malawi, Tanzania, Zaïre, Zambia; grassland in open woodland, ± 1600 m. **I:** Reynolds (1966: 44).

Incl. *Aloe trothae* A. Berger (1905).

[3] Acaulescent, simple, the base enlarging to form a bulb 8 - 10 cm ∅ towards the end of the growing season; **L** ± 16, rosulate, usually dying back in dry season, ovate-lanceolate, to ± 50 × 15 cm, bright green, longitudinally striate, with whitish cartilaginous 1 - 2 mm margin, surface smooth; marginal teeth 1 mm, densely crowded; **Inf** to ± 60 cm, with 3 - 4 (-7) **Br**; racemes cylindrical, 10 - 20 × 7 cm, sub-dense; **Bra** ovate-acuminate, ± 12 × 8 mm; **Ped** ± 20 mm; **Fl** pale yellow to pinkish- or brownish-yellow, to 40 mm, base rounded, 9 - 11 mm ∅ across **Ov**, narrowed to 6 - 8 mm above, en-

larging to 9 - 11 mm and narrowing again to mouth; **Tep** free for 12 - 13 mm; **St** exserted 0 - 1 mm.

Reynolds (1966) included this in *A. buettneri*, but Carter (1994) treated them as geographically distinct species. Further studies are required to determine their relationships. Carter (1994) suggested that the poorly-known *A. trothae* possibly belongs here, in which case the name would have priority over *A. bulbicaulis*.

A. bulbillifera H. Perrier (Mém. Soc. Linn. Normandie, Bot. 1(1): 22, 1926). **T:** Madagascar (*Perrier* 11017 [P]). – **D:** Madagascar.

A. bulbillifera var. **bulbillifera** – **D:** Madagascar; montane forest, 300 - 800 m. **I:** Reynolds (1966: 455-456).

[5] Acaulescent, simple, or with short stem and basal **Br** in shaded situations; **L** 24 - 30, densely rosulate, acute, 40 - 60 × 8 - 10 cm, green; marginal teeth 15 mm, smaller towards tip, firm, 10 - 20 mm apart; exudate drying deep orange to purple; **Inf** 2 - 2.5 m, usually curved over sideways and often falling to sprawl on the ground, with up to 30 **Br**, the lowest to 1 m and with up to 12 secondary **Br**, bulbils developing in **Ax** of sterile **Bra** on **Br** below racemes; racemes cylindrical-acuminate, 20 - 25 cm, lax; **Bra** deltoid, 3 × 2 mm; **Ped** 8 - 10 mm; **Fl** scarlet, 25 mm, base truncate, 5 mm ∅ across **Ov**, slightly narrowed above, then enlarging to mouth; **Tep** free to base. – *Cytology:* 2n = 14 (Resende 1937).

Bulbil formation is rare in the genus (cf. *A. patersonii*).

A. bulbillifera var. **paulianae** Reynolds (JSAB 22(1): 26-27, ills., 1956). **T:** Madagascar, Mahajanga (*Paulian s.n. in Reynolds* 7656 [TAN, K, P, PRE]). – **D:** Madagascar; rocky mountain slopes, 270 m. **I:** Reynolds (1966: 457-458).

[5] Differs from var. *bulbillifera*: **L** ± 20; marginal teeth 3 mm, dull white; **Inf** 2 m, with 8 - 12 compact **Br** in upper ¼, the lowest to 30 cm and with 8 - 12 secondary **Br**, bulbils developing on main peduncle below lowest **Br**; **Bra** ovate-acute, 4 × 3 mm.

A. bullockii Reynolds (JSAB 27(2): 73-75, ills., 1961). **T:** Tanzania, Kahama Distr. (*Bullock* 3076 [K]). – **D:** Tanzania; woodland, 1220 m. **I:** Reynolds (1966: 39-40).

[2] Acaulescent, simple, geophytic, base enlarging to form a bulb, 3 - 4 cm ∅; **R** thick, fusiform; **L** 8 - 10, rosulate, linear-lanceolate, 10 - 20 × 2 - 3 cm, green, lineate, with narrow cartilaginous margin; marginal teeth 0.5 - 1 mm, soft, pale pink, 0.5 - 1 mm apart; **Inf** 35 - 50 cm, simple or with 1 **Br**; racemes cylindrical, 7 - 10 × 5 cm, semi-dense; **Bra** ovate-acute, deflexed, 8 - 10 × 5 - 6 mm; **Ped** 4 - 5 mm; **Fl** pale scarlet to coral-red, 30 mm, base trunc-

ate, 6 mm ∅ across **Ov**, narrowed to 4 mm above, then enlarging and narrowing again towards mouth; **Tep** free for 8 mm.

One of a few species with a bulbous base.

A. burgersfortensis Reynolds (JSAB 2(1): 31-34, ills., 1936). **T**: RSA, Mpumalanga (*Reynolds* 1465 [PRE, BOL]). – **D**: RSA (Mpumalanga); semi-arid Bushveld, sandy soil, 760 m. **I**: Reynolds (1950: 274-275).

[7] Acaulescent, simple or suckering to form small to large groups; **L** 10 - 20, densely rosulate, triangular-acuminate, up to 35 - 40 cm, upper face brownish-green with oblong scattered white spots ± in wavy transverse bands, lower face paler glaucous green, unspotted and somewhat striate; marginal teeth 3 - 5 mm, pungent, brown, 10 - 14 mm apart; **Inf** 1 - 1.3 m, with 4 - 9 arcuate-erect **Br**; racemes cylindrical-acuminate, 20 - 35 (-40) cm, lax or subdense; **Bra** deltoid-acuminate, slightly longer than **Ped**; **Ped** 10 - 15 mm; **Fl** dull reddish with a bloom, somewhat white-striped in upper ½, sometimes shading to orange at mouth, 30 mm, base rounded, 7 mm ∅ across **Ov**, narrowed above to 5 mm, enlarging towards mouth; **Tep** free for 7 mm; **St** exserted 0 - 1 mm. – *Cytology:* 2n = 14 (Brandham 1971).

Natural hybrids with other species have been reported (Reynolds 1950).

A. bussei A. Berger (in Engler, A. (ed.), Pflanzenr. IV.38 (Heft 33): 273, 1908). **T**: Tanzania, Mpwapwa Distr. (*Busse* 294 [B, BM, K]). – **Lit**: Carter (1994). **D**: Tanzania; rock outcrops and cliffs, 580 - 1500 m. **I**: Reynolds (1966: 72-74, as *A. morogoroensis*). **Fig. XII.b**

Incl. *Aloe morogoroensis* Christian (1940).

[6,7] Acaulescent, suckering to form dense groups; **L** ± 20; rosulate, ovate-lanceolate, attenuate, 20 - 30 × 5 - 6 cm, glossy green, usually flushed coppery-red, sometimes with few whitish spots on lower face, with narrow white cartilaginous margin; marginal teeth 2 - 5 mm, inclining towards **L** tip, white or pale yellowish, 7 - 15 mm apart; **Inf** 40 - 60 (-75) cm, simple or usually with 1 - 4 **Br**; racemes conical-cylindrical, 15 - 25 cm, ± dense; **Bra** ovate-acute, 4 - 6 × 3 mm; **Ped** 8 - 10 mm; **Fl** coral-red, yellowish at mouth, 28 - 35 mm, base attenuate, 6 mm ∅ across **Ov**, slightly narrowed above, enlarging to 8 mm and narrowing again at mouth; **Tep** free for 15 mm; **St** exserted 0 - 1 mm.

A. calcairophila Reynolds (JSAB 27(1): 5-6, ills., 1961). **T**: Madagascar, Fianarantsoa (*Descoings* 2114 [TAN, K]). – **D**: Madagascar; limestone hills, 1400 m. **I**: Reynolds (1966: 408-410). **Fig. XIII.g**

≡ *Guillauminia calcairophila* (Reynolds) P. V. Heath (1994); **incl.** *Aloe calcairophylla* hort. (s.a.) (*nom. inval.*, Art. 61.1).

[6] Acaulescent, suckering to form small groups; **L** ± 10, distichous, triangular-acuminate, 5 - 6 × 1.4

cm, dull grey-green; marginal teeth 2 - 3 mm, soft cartilaginous, 2 - 3 mm apart; **Inf** 20 - 25 cm, simple; raceme 3 - 4 cm, lax, with 8 - 10 **Fl**; **Bra** ovate-acute, 3 × 2 mm; **Ped** 5 - 6 mm; **Fl** white, 10 mm, base shortly attenuate, ventricose, enlarging to 4 mm ∅ above **Ov**, slightly constricted at mouth; **Tep** free for 5 mm; **St** not exserted.

A. calidophila Reynolds (JSAB 20(1): 26-28, ills., 1954). **T**: Ethiopia, Sidamo Prov. (*Reynolds* 7029 [PRE, K]). – **D**: Ethiopia, Kenya; hot arid plains, 1280 - 1460 m. **I**: Reynolds (1966: 217-218).

[12] Caulescent, branching to form small to large dense groups; stem short or to 2 m, decumbent with ascending tip; **L** ± 20, densely aggregated at stem tip, triangular, deeply canaliculate, strongly recurved, 80 × 16 cm, dull olive-green; marginal teeth 4 - 5 mm, dull white with reddish-brown tip, 20 - 25 mm apart; exudate drying deep brown; **Inf** 1 - 1.3 m, with ± 12 **Br**; racemes slightly conical-cylindrical, 10 - 13 × 5 cm, dense; **Bra** ovate-acuminate, 3 - 4 × 2 mm; **Ped** 10 mm; **Fl** scarlet, becoming orange towards mouth, 22 mm, clavate, base shortly attenuate, 6 mm ∅ across **Ov**, enlarging above; **Tep** free for 11 mm. – *Cytology:* 2n = 14 (Brandham 1971).

A. cameronii Hemsley (CBM 1903: t. 7915 + text, 1903). **T**: Malawi (*Cameron* s.n. [K]). – **D**: Zimbabwe, Malawi, Moçambique, Zambia.

A. cameronii var. **bondana** Reynolds (Aloes Trop. Afr. & Madag., 353, ills., 1966). **T**: Zimbabwe, Inyanga Distr. (*Reynolds* 8585 [SRGH, PRE]). – **D**: Zimbabwe; granite hills. **I**: FPA 51: t. 2011, 1990.

[8] Differs from var. *cameronii*: Stems mostly simple, 60 - 180 cm; **Fl** mostly yellowish to orange, 38 - 40 mm, more fleshy and somewhat clavate.

A. cameronii var. **cameronii** – **D**: Malawi, Moçambique, Zambia, Zimbabwe; shallow soil pockets on granite rocks; 1280 - 2070 m. **I**: Reynolds (1966: 349-351).

[16] Caulescent, branching at base; stem erect, to 150 × 3 - 4 cm, usually with dried **L** remains; **L** rosulate and persistent on apical 30 - 50 cm of stem, triangular, tip attenuate-acute, 40 - 50 × 5 - 7 cm, green, usually turning copper-red in winter; marginal teeth 2 - 3 mm, pungent, pale brown, 10 - 15 mm apart; **Inf** 60 - 90 cm, with 2 - 3 **Br**; racemes cylindrical, slightly acuminate, 10 - 15 × 7 - 8 cm, sub-dense; **Bra** ovate-acute, 2 × 3 mm; **Ped** 3 - 5 mm; **Fl** bright scarlet, 45 mm, sometimes slightly clavate, base rounded, 5 - 7 mm ∅ across **Ov**, slightly enlarging above, slightly narrowed towards mouth; **Tep** free for 12 - 15 mm. – *Cytology:* 2n = 14 (Ferguson 1926).

Natural hybrids with other species have been reported (Reynolds 1966).

A. cameronii var. **dedzana** Reynolds (JSAB 31(2):

167-168, ills., 1965). **T:** Malawi, Central Prov. (*Christian* 459 [SRGH, PRE]). – **D:** Malawi, Moçambique; rocky hills. **I:** Reynolds (1966: 353, flowers only).

[16] Differs from var. *cameronii*: Dense shrubs; **Br** 50 - 80 cm; **Inf** with racemes 20 - 25 cm.

A. camperi Schweinfurth (Bull. Herb. Boissier 2(app. 2): 67, 1894). **T:** Eritrea (*Schweinfurth* 514a [not located]). – **D:** Eritrea, Ethiopia; open areas in valleys, 1400 - 2530 m. **I:** Reynolds (1966: 212-213).

Incl. *Aloe abyssinica* Salm-Dyck (1817) (*nom. illeg.*, Art. 53.1); **incl.** *Aloe spicata* Baker (1896) (*nom. illeg.*, Art. 53.1); **incl.** *Aloe albopicta* hort. *ex* A. Berger (1908); **incl.** *Aloe eru* A. Berger (1908); **incl.** *Aloe eru* fa. *erecta* hort. *ex* A. Berger (1908); **incl.** *Aloe eru* fa. *glauca* hort. *ex* A. Berger (1908); **incl.** *Aloe eru* fa. *maculata* hort. *ex* A. Berger (1908); **incl.** *Aloe eru* fa. *parvipunctata* hort. *ex* A. Berger (1908); **incl.** *Aloe eru* var. *cornuta* A. Berger (1908).

[12] Caulescent, branching at base, sometimes forming groups 1 - 2 m $\emptyset$; stem erect, divergent or decumbent, to 100×6 - 10 cm; **L** 12 - 16, rosulate, persistent for 10 - 20 cm below, triangular, recurved, $50 - 60 \times 8$ - 12 cm, dark green, sometimes with few dull white lenticular spots, with reddish margin; marginal teeth 3 - 5 mm, pungent, brownish-red, 10 - 20 mm apart; **Inf** 70 - 100 cm, with 6 - 8 **Br**, lower **Br** sometimes rebranched; racemes cylindrical, $6 - 9 \times 6$ - 7 cm, dense; **Bra** deltoid, 2×2 mm; **Ped** 12 - 18 mm; **Fl** orange to yellow, 20 - 22 mm, cylindrical-clavate, base shortly attenuate, 5 mm $\emptyset$ across **Ov**, enlarging to 11 mm above; **Tep** free for ± 7 mm. – *Cytology:* $2n = 14$ (Resende 1937: as *A. eru*).

A. canarina S. Carter (Fl. Trop. East Afr., Aloaceae, 41-42, ills., 1994). **T:** Uganda, Karamoja Distr. (*Reynolds* 7951 [K, EA, PRE]). – **D:** Sudan, Uganda; open deciduous bushland, 1345 - 1570 m. **I:** Reynolds (1966: 304-305, as *A. marsabitensis*).

[7] Caulescent, suckering to form small groups; stem decumbent, to 80 cm, with dead **L** persistent; **L** densely rosulate, lanceolate-attenuate, $50 - 80 \times 10$ - 15 cm, dull greyish-green tinged reddish; marginal teeth 2 - 3 mm, pungent, brown-tipped, 10 - 18 mm apart; **Inf** to ± 1 m, with 15 - 20 **Br**, lower **Br** rebranched; racemes cylindrical, 10 - 15 cm, lax; **Bra** ovate-acuminate, $2 - 3 \times 2.5$ mm; **Ped** 6 - 7 mm; **Fl** yellow, 25 - 30 mm, base truncate, 10 - 12 mm $\emptyset$ across **Ov**, narrowed to 8 mm towards mouth; **Tep** free for ± 8 - 10 mm.

Included by Reynolds (1966) in *A. marsabitensis* (which is now sunk under *A. secundiflora*). The material in Sudan might represent a distinct species (Carter 1994).

A. cannellii L. C. Leach (JSAB 37(1): 41-46, ills.,

1971). **T:** Moçambique, Manica e Sofala Distr. (*Cannell* 33 [SRGH, LISC, PRE]). – **D:** Moçambique; grass tufts on almost vertical cliffs, ± 1500 m.

[6] Acaulescent, suckering to form dense tufts; **R** fleshy, sub-fusiform, with partially exposed sub-tuberous stock; **L** usually 4 - 5, rosulate, linear-acute, to 26×0.4 - 0.8 cm, green or with brownish tinge, with narrow hyaline margin, upper face sometimes with a few white spots, lower face more copiously white spotted; marginal teeth ± 0.25 mm, whitish, ± 1 mm apart; **Inf** 20 - 30 cm, simple; raceme cylindrical, ± 10 - 12.5 cm, lax, with 10 - 20 **Fl**; **Bra** ovate-acute, $4.5 - 6.5 \times 3$ - 3.5 mm; **Ped** 10 - 15 mm; **Fl** orange-scarlet becoming greenish towards the mouth, 20 - 22 (- 25) mm, base shortly attenuate, 3.5 - 4 mm $\emptyset$ across **Ov**, enlarging to ± 5 mm above, narrowed to 4 mm at mouth; **Tep** free to base; **St** not exserted.

A. capitata Baker (JLSB 20: 272, 1883). **T:** Madagascar, Antananarivo (*Baron* 897 [K]). – **D:** Madagascar.

A. capitata var. **capitata** – **D:** Madagascar; rock crevices and soil pockets on granite and gneiss mountains, 1550 - 1600 m. **I:** Reynolds (1966: 466-467); FPA 50: t. 1973, 1988. **Fig. XII.d**

Incl. *Aloe cernua* Todaro (1890).

[5] Acaulescent, or caulescent in shady places, usually simple; stem to 60 cm; **L** 20 - 30, densely rosulate, triangular-attenuate, tip slightly twisted, obtuse and shortly dentate, 50×6 cm, green tinged reddish, with brownish-red horny margin; marginal teeth 2 mm, pungent, reddish, 8 - 12 mm apart; exudate drying yellow; **Inf** ± 80 cm, with 3 - 4 **Br**; racemes densely capitate, $3 - 4 \times 9$ - 10 cm; **Bra** ovate-acute, 6×4 mm; **Ped** lowest 10 mm, becoming 25 - 30 mm above; **Fl** orange-yellow, 25 mm, narrowly campanulate, base rounded, 6 mm $\emptyset$ across **Ov**, enlarging to 10 mm at mouth; **Tep** free to base but cohering for 24 - 25 mm; **St** exserted 8 - 10 mm. – *Cytology:* $2n = 14$ (Satô 1937: as *A. capitata* var. *typica*).

A natural hybrid with *A. macroclada* has been reported (Bosser 1968: 510).

A. capitata var. **cipolinicola** H. Perrier (Mém. Soc. Linn. Normandie, Bot. 1(1): 39, 1926). **T:** Madagascar (*Perrier* 13225 [P]). – **D:** Madagascar; marble outcrops, 1250 - 1400 m. **I:** Reynolds (1966: 474-475).

[9] Differs from var. *capitata*: Caulescent; stem 2 - 3 m, simple; **L** 60 - 100, densely rosulate, tip acute with 2 - 3 small teeth, 60×5 - 6.5 cm, glossy green, sometimes tinged reddish; marginal teeth 3 - 4 mm; **Inf** ± 1 m, with 6 - 10 **Br**; **Ped** lowest almost nil. – *Cytology:* $2n = 14$ (Brandham 1971).

A. capitata var. **gneissicola** H. Perrier (Mém. Soc. Linn. Normandie, Bot. 1(1): 37, 1926). **T:** P. – **D:**

Madagascar; gneissic rocks, 600 - 1440 m. **I:** Reynolds (1966: 468-469).

[5] Differs from var. *capitata*: **L** ± 20, lanceolate, tip rounded with 3 - 5 very short teeth, 40 - 50 × 3.5 - 4 cm, to 5 cm (± 10 cm from base), grey-green to glaucous; **Inf** racemes with fewer **Fl**; **Bra** 10 × 5 mm, brittle; **Ped** lowest 5 mm, uppermost 35 mm; **Fl** 35 mm.

A. capitata var. **quartziticola** H. Perrier (Mém. Soc. Linn. Normandie, Bot. 1(1): 38, ills., 1926). **T:** Madagascar, Fianarantsoa (*Perrier* 11001 [P]). – **D:** Madagascar; quartzite (mostly) or basalt outcrops, 700 - 1560 m. **I:** Reynolds (1966: 470-473).

[5] Differs from var. *capitata*: **L** thick and fleshy, 30 - 40 × 9 - 12 cm, glaucous to bluish-grey; marginal teeth 3 - 4 mm, 10 - 20 mm apart; **Inf** 1 m; racemes 10 - 12 cm, sub-dense with more **Fl**; **Bra** narrowly deltoid, 7 - 10 × 3 - 4 mm; **Ped** lower 15 - 20 mm, uppermost 40 - 50 mm.

A natural hybrid with *A. macroclada* has been reported (Reynolds 1966).

A. capitata var. **silvicola** H. Perrier (Mém. Soc. Linn. Normandie, Bot. 1(1): 39, 1926). **T:** Madagascar, Mahajanga (*Perrier* 11012 [P]). – **D:** Madagascar; forest, epiphyte on tree trunks, or on very shaded rocks, 1000 - 1200 m.

[5] Differs from var. *capitata*: Stem short, lying on tree trunks or rocks; **L** 50 - 60 × 3 - 4 cm; marginal teeth very small, sometimes absent; **Bra** 7 × 6 mm, tip rounded; **Ped** lower 8 mm, uppermost to 40 mm; **Fl** 20 mm.

A. capmanambatoensis Rauh & Gerold (KuaS 51(11): 293-294, ills., 2000). **T:** Madagascar, Iharana (*Rauh & Gerold* 73669a [HEID]). – **D:** NE Madagascar; on granite.

[6] Acaulescent or shortly caulescent; **Ros** to 40 cm Ø, 10 cm tall, suckering to form groups; **L** ± 10, densely rosulate, long-attenuate, 20 - 25 × 4 cm, upper face dark reddish-green with longitudinal whitish flecks, lower face grey-green with less distinct flecks; marginal teeth small, broadly triangular, continuous; **Inf** to 1.2 m, simple or with short **Br**; racemes cylindrical, to 40 cm, lax, with ± 20 **Fl**; **Bra** narrowly lanceolate-attenuate, shorter than **Ped**; **Ped** to 45 mm; **Fl** cinnabar-red, lobes white with wide green veins, 25 - 40 mm, 7 mm Ø, not narrowed above **Ov**; **Tep** free for 15 - 30 mm; **St** not exserted.

A. carnea S. Carter (KB 51(4): 784-785, 1996). **T:** Zimbabwe, Eastern Prov. (*Leach & Wild* 11135 [K, LISC, PRE, SRGH]). – **D:** Zimbabwe; usually rocky ground, in grass and light *Brachystegia* woodland, 900 - 1375 m.

[5] Acaulescent or caulescent, simple or in small groups; stem to ± 20 cm, with dead **L** remains; **L** densely rosulate, ovate-lanceolate, to ± 30 × 6 - 10 cm, apical 10 cm soon drying, upper face glossy dark greyish-green with conspicuous elongated whitish spots, often in transverse bands towards base, lower face pale milky green without spots, ± lineate esp. near margin, with horny, often reddish-brown margin, surface smooth; marginal teeth ± 6 mm, pungent, dark red-brown, 8 - 12 mm apart; exudate drying yellow; **Inf** 75 - 200 cm, with 6 - 12 (-15) **Br**, the lowest sometimes rebranched; racemes capitate, 3 - 5 × ± 7 cm, dense, more lax at base; **Bra** linear-lanceolate, 8 - 12 × 2.5 mm; **Ped** 18 - 28 mm; **Fl** dull pale coral-pink (flesh-coloured) with whitish segment margins to base, 25 - 30 mm, 6 - 7 mm Ø across **Ov**, abruptly narrowed above **Ov**, then enlarging to mouth; **Tep** free for up to 10 mm; **St** exserted ± 2 mm.

A. castanea Schönland (Rec. Albany Mus. 2: 138, 1907). **T:** RSA, Mpumalanga (*Burtt-Davy* 2856 [GRA]). – **D:** RSA (Gauteng, Mpumalanga, Northern Prov.); woodland and bushland slopes, often in open flat country. **I:** Reynolds (1950: 439, t. 44). **Fig. XI.f**

[16] Caulescent, branching near base, sometimes rebranched with 10 - 20 crowns; stem 3 - 4 m; **L** densely rosulate, attenuate, 100 × 10 cm, glaucous, with thin pale brownish-red margin; marginal teeth 1.5 mm, uncinate, 8 - 10 mm apart; **Inf** 1.5 - 2 m, usually oblique, simple; raceme narrowly cylindrical, slightly acuminate, 70 - 100 cm, dense; **Bra** ovate-acute, 12 × 8 mm; **Ped** 3 mm; **Fl** reddish-brown, campanulate, 18 mm, base rounded, 6 mm Ø across **Ov**, enlarging to 15 mm at tip; **Tep** free to base or almost so; **St** exserted 12 - 15 mm. – *Cytology:* 2n = 14 (Müller 1945).

Natural hybrids with other species have been reported (Reynolds 1950).

A. castellorum J. R. I. Wood (KB 38(1): 25-26, t. 2, 1983). **T:** Yemen (*Wood* 2504 [K]). – **D:** Saudi Arabia, Yemen; exposed W-facing slopes with high incidence of mist, 1400 - 2400 m. **I:** Collenette (1999: 20).

[2] Acaulescent, simple; **L** rosulate, ensiform, acuminate, 30 × 3.5 cm, yellow-green; marginal teeth small, pale brown, ± 6 mm apart; **Inf** 1 - 1.5 m; racemes cylindrical, sub-dense; **Bra** ovate-cuspidate, ± 10 × 6 mm; **Ped** ± 5 mm; **Fl** yellow with green veins, ± 25 mm, not narrowed above **Ov**; **St** scarcely exserted. – *Cytology:* 2n = 14 (from the protologue).

In the protologue the inflorescence is described as 'branched', but the photograph shows a simple inflorescence.

A. catengiana Reynolds (Kirkia 1: 160, 1961). **T:** Angola, Benguela (*Reynolds* 9307 [PRE, K]). – **D:** Angola; hot arid bush country, 518 m. **I:** Reynolds (1966: 373-374).

[16] Caulescent, branching to form thickets 1 - 2

m or more ⌀; stem ascending, divergent or sprawling, 1.5 - 2 m, to 3 m when supported by bushes; **L** 16 - 20, scattered along stem for 30 cm, lanceolate-attenuate, 30 × 3 - 5 cm, pale yellowish grey-green, with very pale green lenticular spots, more numerous in lower ½; marginal teeth 3 mm, firm, reddish-brown-tipped, 8 - 10 mm apart; sheath 15 - 20 mm, lineate; **Inf** 40 cm, with ± 6 **Br**; racemes cylindrical-acuminate, laterals with sub-secund **Fl**, to 16 × 4 cm, lax; **Bra** ovate-acute, 5 × 3 mm; **Ped** 10 mm; **Fl** dull scarlet, 28 mm, base rounded, 7 mm ⌀ across **Ov**, very slightly narrowed above, then enlarging to mouth; **Tep** free for 10 mm. − *Cytology:* 2n = 14 (Brandham 1971).

A. cephalophora Lavranos & Collenette (CSJA 72(1): 20-21, ills., 2000). **T:** Saudi Arabia, Hijaz Prov. (*Collenette* 4981 [K, ZSS]). − **D:** Saudi Arabia; steep rocky slopes, ± 1400 m. **I:** Collenette (1999: 21).

[6,7] Acaulescent, suckering at base; **L** 8 - 10, densely rosulate, arcuate-ascending, triangular-attenuate, 34 - 36 × 4 - 4.5 cm, olive-green, surface rough; marginal teeth few, small, white; **Inf** ± 25 cm, erect, simple or with up to 2 **Br**; racemes capitate or subcapitate, dense; **Bra** ovate, 10 × 8 mm, whitish with 3 - 5 brown nerves; **Ped** 12 - 14 mm; **Fl** coral-red at base, becoming cream-yellow with green nerves towards apex, ± 35 mm, base shortly attenuate, funnel-shaped, 5 mm ⌀ across **Ov**, widening to 10 mm at mouth; **Tep** free for ± 20 mm; **St** exserted 4 mm.

A. chabaudii Schönland (Gard. Chron., ser. 3, 38: 102, 1905). **T:** Zimbabwe (*Brown* s.n. [GRA]). − **D:** Tanzania, Zaïre, Zambia, Zimbabwe, Malawi, Moçambique, Swaziland, RSA.

A. chabaudii var. **chabaudii** − **D:** Tanzania, Zaïre, Zambia, Zimbabwe, Malawi, Moçambique, Swaziland, RSA; on or at the base of granite outcrops, 380 - 1610 m. **I:** Reynolds (1950: 340-341, t. 27); Reynolds (1966: 103-104). **Fig. XII.e**

[7] Acaulescent or very shortly caulescent, suckering to form small to large dense groups; **L** ± 20, densely rosulate, ovate-lanceolate, acuminate, to 50 × 10 cm, dull grey-green to glaucous-green, obscurely lineate, sometimes with few scattered H-shaped spots, with narrow greyish cartilaginous margin; marginal teeth 1 - 2 mm, near base pale brownish, 5 - 10 mm apart, more brownish, more uncinate, and further apart towards **L** tip; **Inf** 60 - 80 (-100) cm, with 6 - 12 **Br**, the lowest 1 - 3 rebranched; racemes broadly cylindrical, slightly acuminate, 8 - 15 × 10 cm, sub-lax with 30 - 40 **Fl**; **Bra** ovate-acute or deltoid-acuminate, 3 - 5 mm; **Ped** lowest 20 - 25 mm, shorter upwards, oblique to almost horizontal; **Fl** pale brick-red, paler at mouth, 35 - 40 mm, base shortly attenuate, 7 mm ⌀ across **Ov**, narrowed to 5

mm above, enlarging towards mouth; **Tep** free for 8 mm. − *Cytology:* 2n = 14 (Müller 1945).

A very variable species. Natural hybrids with other species have been reported (Reynolds 1950, 1966).

A. chabaudii var. **mlanjeana** Christian (FPSA 18: t. 698 + text, 1938). **T:** Malawi, Southern Prov. (*Everett* s.n. in *Christian* 274 [PRE 23025]). − **D:** Malawi; rocky slopes, 1524 m. **I:** Reynolds (1966: 106).

[7] Differs from var. *chabaudii*: **L** 30 - 40 × 9 cm, green turning brick-red in the upper part, with white horny margin; marginal teeth 3 mm, 10 - 15 mm apart; **Inf** 40 - 50 cm, oblique; **Ped** 18 - 20 mm; **Fl** coral-red, 30 - 32 mm, 9 mm ⌀ across **Ov**; **Tep** free for 12 mm. − *Cytology:* 2n = 14 (Brandham 1971).

A. chabaudii var. **verekeri** Christian (FPSA 18: t. 699 + text, 1938). **T:** Zimbabwe (*Vereker* s.n. [PRE 23027]). − **D:** Moçambique, Zimbabwe; fire-protected habitats on rocky ground. **I:** Reynolds (1966: 107).

[7] Differs from var. *chabaudii*: **L** olive-green turning reddish in upper ½ in dry season, with white margin; marginal teeth 4 mm, uncinate; **Inf** subcapitate racemes, ± 8 cm; **Ped** 15 - 17 mm; **Fl** various shades of red and yellow, 32 mm.

A. cheranganiensis S. Carter & Brandham (Cact. Succ. J. Gr. Brit. 41(1): 4-6, ills., 1979). **T:** Kenya, Rift Valley Prov. (*Brandham* 1727 [K, EA]). − **D:** Kenya, Uganda; open deciduous woodland on rocky sandy plains and rocky slopes, 1220 - 1980 m.

[16] Caulescent, branching from base; stem to 2 m × 4 cm; **L** laxly crowded at stem tip, ovate-attenuate, to 40 × 5 cm, glaucous-green, with scattered whitish spots on young shoots; marginal teeth 3 mm, green with brown tips, 8 - 13 mm apart; exudate yellow; **Inf** 60 cm, usually with 2 **Br**; racemes cylindrical-acuminate, ± 20 × 7 cm, dense; **Bra** ovate-acuminate, 5 - 8 × 3 - 4 mm; **Ped** to 10 - 20 mm; **Fl** bright orange, yellow at mouth, 27 - 30 mm, base shortly attenuate, 7 mm ⌀ across **Ov**, not narrowed above, **OTep** reflexed; **Tep** free for 22 - 25 mm. − *Cytology:* 2n = 28 (Cutler & al. 1980).

A. chlorantha Lavranos (JSAB 39(1): 85-90, ills., 1973). **T:** RSA, Northern Cape (*Lavranos* 10024 [PRE]). − **D:** RSA (Northern Cape); dolerite outcrops, ± 1400 m. **I:** Wyk & Smith (1996: 131).

[4] Acaulescent or caulescent, simple or up to 10 **Ros** in groups; stem to 1.5 m, procumbent; **L** rosulate, deltoid, acute, slightly falcate, to 40 × 8 cm, bright green shading into purplish, ± striate, lower face often with white spots, with dark brown-red cartilaginous margin; marginal teeth 2 mm, dark brown-red, 10 - 30 mm apart; **Inf** to 1.6 m, simple or rarely with 1 **Br**; racemes cylindrical, to 35 cm,

sub-dense; **Bra** ovate-deltoid, acute, fleshy, yellow-green, 12 - 20 × 5 - 8 mm; **Ped** 15 - 22 mm; **Fl** yellow-green, 10 mm, base attenuate, enlarged to 4 mm ∅ above **Ov**, narrowing to mouth; **Tep** free to base.

A. chortolirioides A. Berger (in Engler, A. (ed.), Pflanzenr. IV.38 (Heft 33): 171, 1908). **T:** RSA, Mpumalanga (*Galpin* 490 [K]). – **D:** RSA, Swaziland.

Reynolds (1950) cites Thorncroft s.n. (BOL) as type, and this was followed by Glen & Hardy (1987), but only Galpin 490 is cited in the protologue and so this is automatically the type (ICBN Art. 7.3). – *Cytology:* 2n = 14 (Müller 1945).

A. chortolirioides var. **chortolirioides** – **D:** RSA (Mpumalanga), Swaziland; rocky ridges, 1300 - 1600 m. **I:** Reynolds (1950: 124-125).

Incl. *Aloe boastii* Letty (1934) ≡ *Aloe chortolirioides* var. *boastii* (Letty) Reynolds (1950).

[10] Caulescent, branching to form dense tufts of up to 50 **Br**; **R** fusiform; stem 10 - 20 × 2 - 3 cm, usually branching about the middle; **L** 15 - 20, multifarious, linear, 10 - 25 cm × 3 - 5 mm, dull green, lower face sometimes with few small white spots near base; marginal teeth 0.5 mm, white, cartilaginous, 2 - 3 mm apart; **Inf** 25 cm, simple; raceme capitate, ± 5 × 5 - 7 cm, with 18 - 20 **Fl**; **Bra** ovate-acuminate, 10 - 13 mm; **Ped** 20 - 25 mm; **Fl** scarlet to yellow, 30 - 35 mm, base shortly attenuate, enlarging above **Ov**; **Tep** free almost to base.

Flowering is induced by burning of surrounding grass (Reynolds 1950). A natural hybrid with *A. arborescens* has been reported (Reynolds 1950).

A. chortolirioides var. **woolliana** (Pole-Evans) Glen & D. S. Hardy (SAJB 53(6): 489-490, 1987). **T:** RSA, Mpumalanga (*Pole-Evans* s.n. [PRE 8320]). – **D:** RSA (Mpumalanga, Northern Prov.), Swaziland; 1670 m. **I:** Reynolds (1950: 128, as *A. woolliana*).

≡ *Aloe woolliana* Pole-Evans (1934).

[10] Differs from var. *chortolirioides*: Stem 15 - 25 cm; **L** 37 cm × 5 - 8 mm; **Inf** to 45 cm; **Fl** jasper-red or rose-pink, sometimes yellowish at the mouth, 30 - 40 mm.

A. christianii Reynolds (JSAB 2(4): 171-173, ills., 1936). **T:** Zimbabwe (*Reynolds* 1885 [PRE, K, SRGH]). – **D:** Angola, Malawi, Moçambique, Tanzania, Zaïre, Zambia, Zimbabwe; in partial shade in woodland or tall grass, 700 - 2000 m. **I:** Reynolds (1950: 309-301); Reynolds (1966: 187-188).

[8] Caulescent, simple or in small groups; stem erect or procumbent, to 150 × 10 - 15 cm, with dead **L** remains; **L** 30 - 40, densely rosulate, lanceolate-attenuate, 30 - 75 × 10 - 15 cm, upper face dull green, obscurely lineate, lower face dull bluish-green, very obscurely lineate; marginal teeth 3 - 5

mm, pungent, sometimes slightly uncinate, pinkish to pale brown, 10 - 20 mm apart; **Inf** 2 - 3 m, with 5 - 10 **Br**, lower **Br** sometimes rebranched; racemes cylindrical-acuminate, 25 - 30 cm, sub-dense with 40 - 50 **Fl**; **Bra** ovate-acute, 5 - 6 × 3 mm; **Ped** 8 - 12 mm; **Fl** coral-red with a bloom, 40 - 45 mm, base shortly attenuate, 9 - 10 mm ∅ across **Ov**, not narrowed above; **Tep** free for 15 mm. – *Cytology:* 2n = 14 (Brandham 1971).

A. chrysostachys Lavranos & L. E. Newton (CSJA 48 (6): 278-279, ills., 1976). **T:** Kenya, Eastern Prov. (*Classen* 14 [EA]). – **Lit:** Newton (1996). **D:** Kenya; soil pockets on gneissic outcrops, 900 - 1200 m.

Incl. *Aloe meruana* Lavranos (1980).

[7] Acaulescent or with very short stem, suckering sparsely from base forming small groups; **L** 14 - 18, densely rosulate, lanceolate-attenuate, ± 30 × 8 - 10 cm, bluish-green, usually purplish in exposed situations, with reddish cartilaginous margin, surface smooth; marginal teeth 4 mm, firm, uncinate, 8 - 12 mm apart; exudate drying yellow; **Inf** 30 - 40 cm, with 5 - 10 **Br**; racemes 6 - 10 cm, lax, with secund **Fl**; **Bra** deltoid-acute, 5 - 7 × 1.5 mm; **Ped** 10 - 12 mm; **Fl** dull red or yellow, 35 mm, base attenuate, 8 mm ∅ across **Ov**, narrowed to 7 mm above, enlarging to 8 mm at mouth; **Tep** free for 11 - 13 mm. – *Cytology:* 2n = 14 (from protologue).

Plants with red and yellow flowers occur in the same population, with yellow predominating.

A. ciliaris Haworth (Philos. Mag. J. 66: 281, 1825). **T:** RSA, Cape Prov. (*Bowie* s.n. [K [drawing]]). – **D:** RSA.

The species includes plants representing 3 levels of ploidy, recognized as varieties, the origin of which was discussed by Brandham & Carter (1990).

A. ciliaris var. **ciliaris** – **D:** RSA (Eastern Cape); amongst and supported by bushes and trees. **I:** Reynolds (1950: 353-354, t. 29).

Incl. *Aloe ciliaris* var. *flanaganii* Schönland (1903) ≡ *Aloe ciliaris* fa. *flanaganii* (Schönland) Resende (1943); **incl.** *Aloe ciliaris* fa. *gigas* Resende (1938).

[14] Caulescent, branching, scrambling; stem to 5 m or more, 1 - 1.5 cm ∅; **L** scattered along stem for 30 - 60 cm, linear-lanceolate, long acuminate, 10 - 15 × 1.5 - 2.5 cm, green; marginal teeth 1 mm, shorter towards **L** tip, firm, white, 3 mm apart; sheath 5 - 15 mm, obscurely green-lineate, upper margin opposite lamina with fringe of 2 - 4 mm cilia; **Inf** 20 - 30 cm, usually simple, sometimes with 1 **Br**; racemes cylindrical, 8 - 15 × 4 - 5 cm, sub-dense with 24 - 30 **Fl**; **Bra** ovate-acuminate; **Ped** ± 5 mm; **Fl** scarlet, yellowish-green at mouth, 28 - 35 mm, slightly clavate, base very shortly attenuate, gradually enlarging above **Ov**; **Tep** free for 6 mm. – *Cytology:* 2n = 42 (Müller 1945).

There is no evidence supporting reports such as Reynolds (1950) that *A. ciliaris* occurs wild in Kenya.

A. ciliaris var. **redacta** S. Carter (KB 45(4): 643, 1990). **T:** RSA, Eastern Cape (*Wisura* 2640 [K]). – **D:** RSA (Eastern Cape); sand dunes.

[14] Differs from var. *ciliaris*: **L** sheath cilia 1 - 2 mm; **Bra** 4 - 5 mm; **Ped** 12 - 14 mm; **Fl** 21 - 25 mm. – *Cytology:* 2n = 28 (tetraploid).

Morphologically and cytologically intermediate between the other 2 varieties, but apparently not of hybrid origin.

A. ciliaris var. **tidmarshii** Schönland (Rec. Albany Mus. 1: 41, 1903). **T:** RSA, Eastern Cape (*Schönland* 1587 [GRA, BOL]). – **D:** RSA (Eastern Cape); temperate mountain slopes. **I:** Reynolds (1950: 355-356, as *A. tidmarshii*).

≡ *Aloe ciliaris* fa. *tidmarshii* (Schönland) Resende (1943) ≡ *Aloe tidmarshii* (Schönland) F. S. Müller *ex* R. A. Dyer (1943).

[14] Differs from var. *ciliaris*: **L** 7 - 10 × 1.5 - 2 cm, sheath cilia < 1 mm; **Inf** racemes sub-lax; **Fl** 16 - 25 mm. – *Cytology:* 2n = 14 (Müller 1945).

A. citrea (Guillaumin) L. E. Newton & G. D. Rowley (Excelsa 17: 61, 1997). **T:** Madagascar (*Boiteau* 227 [P ?]). – **D:** Madagascar. **I:** Rauh (1995a: 329).

≡ *Lomatophyllum citreum* Guillaumin (1944).

[2] Acaulescent; **L** ± 16, densely rosulate, lanceolate-attenuate, often contorted towards tips, to 28 × 2.5 cm, dark green; marginal teeth 3 mm, green, 1 - 2 mm apart; **Inf** 15 cm, simple; racemes cylindrical, 6 cm, dense, with ± 20 **Fl**; **Bra** triangular, < 5 mm; **Ped** 6 - 8 mm; **Fl** lemon-yellow with green midrib, 25 mm, narrowed slightly above **Ov**, then enlarging slightly to mouth; **Tep** free for ± 5 mm; **Fr** berries.

A. citrina S. Carter & Brandham (Bradleya 1: 21-23, ills., 1983). **T:** Somalia, Ben Adir Region (*Bally & Melville* 15287 [K]). – **D:** Ethiopia, Kenya, Somalia; sandy plains with thorn bush, 90 - 200 (-990) m.

[5] Acaulescent, simple or with a few suckers; **L** rosulate, lanceolate-attenuate, to 60 × 15 cm, glaucous with many elongated white spots; marginal teeth 1.5 mm, green, brown-tipped, 15 - 35 mm apart; exudate drying purple; **Inf** to 2 m, with 2 - 6 **Br**; racemes cylindrical-acuminate, 25 - 50 × 5 cm, ± dense; **Bra** ovate-acute, to 12 × 4 mm, pubescent; **Ped** to 10 mm, pubescent; **Fl** pale lemon-yellow, shortly tomentose, 28 - 35 mm, base rounded, 6 mm ∅ across **Ov**, slightly narrowed above, enlarging to 8 mm at mouth; **Tep** free for 18 mm; **St** exserted 1 mm. – *Cytology:* 2n = 14 (from protologue).

A. classenii Reynolds (JSAB 31(4): 271-273, ills.,

1965). **T:** Kenya, Coast Prov. (*Classen* 128 in *Reynolds* 10117 [PRE, EA, K]). – **D:** Kenya; soil pockets on gneissic outcrops, ± 600 m. **I:** Reynolds (1966: 255-256).

[7] Acaulescent or with short stem to 50 cm, suckering from base forming dense groups; **L** ± 24, densely rosulate, lanceolate, tip with a **Sp**, 35 - 40 × 7 - 8 cm, dark olive-green, becoming reddish in exposed situations, lower face sometimes with few elongated pale spots near base, surface smooth; marginal teeth 5 mm, pungent, pale brownish, 10 - 15 mm apart; exudate drying yellow; **Inf** ± 60 cm, with ± 10 **Br**, lower **Br** rebranched; racemes cylindrical-acuminate, ± 7 × 4 cm, lax; **Bra** ovate-acute, 3 mm; **Ped** 8 - 10 mm; **Fl** dark wine-red with powdery bloom, 20 - 25 mm, base rounded, 7 mm ∅ across **Ov**, not narrowed above; **Tep** free for 10 mm.

A. claviflora Burchell (Trav. South. Afr. 1: 272, 1822). **T:** RSA, Northern Cape (*Burchell* s.n. [not located]). – **D:** Namibia, RSA (Northern Cape, Western Cape, Eastern Cape, Free State); well-drained flat stony ground or rocky slopes. **I:** Reynolds (1950: 320-321). **Fig. XII.f**

Incl. *Aloe schlechteri* Schönland (1903); **incl.** *Aloe decora* Schönland (1905).

[6,7] Acaulescent or with decumbent stems, usually forming hollow circular groups 1 - 2 m ∅; stem 10 - 20 cm; **Ros** decumbent; **L** 30 - 40, densely rosulate, ovate-lanceolate, ± 20 × 6 - 8 cm, glaucous, lower face 1- to 2-carinate in upper ⅓, the keel with 4 - 6 brownish prickles, 2 - 4 mm; marginal teeth 2 - 4 mm, pungent, brownish, ± 10 mm apart; **Inf** to 50 cm, oblique, almost horizontal, simple or with 1 - 4 **Br**; racemes cylindrical-acuminate, 20 - 30 cm, dense; **Bra** ovate-acute, reflexed, ± 15 × 6 - 8 mm; **Ped** 7 - 10 mm; **Fl** red with a bloom, green-tipped, paler and turning lemon-yellow to ivory after pollination, 30 - 40 mm, base long-attenuate, enlarging above **Ov** to 10 mm ∅; **Tep** free for ¼ - ½ of their length; **St** exserted 10 - 15 mm. – *Cytology:* 2n = 14 (Riley 1959: 241).

Natural hybrids with other species have been reported (Reynolds 1950).

A. collenetteae Lavranos (CSJA 67(1): 32-33, ills., 1995). **T:** Oman, Dhofar (*Collenette* 8945 [E]). – **D:** Oman (Dhofar Prov.); limestone cliffs and rocky grassy hillsides, ± 800 m.

[11] Caulescent, suckering from base; stem to 20 cm; **L** 5 - 8, rosulate, triangular-acute, 30 - 32 × 3 cm, bright green, sometimes with a few pale spots; marginal teeth 0.5 mm, cartilaginous, white, 10 mm apart, obsolete in upper ⅓; **Inf** ± 20 cm, usually with 2 **Br**; racemes to 70 mm, lax, with up to 22 **Fl**; **Bra** deltoid-acute, 2 - 4 mm; **Ped** 5 - 6 mm; **Fl** bright orange-red with green veins, ± 20 mm, base rounded, 4 mm ∅ across **Ov**, not narrowed above, narrowing slightly at mouth; **Tep** free for ± 7 mm.

A. collina S. Carter (KB 51(4): 781-782, 1996). **T:** Zimbabwe, Eastern Prov. (*Leach* 269 [K, B, LISC, SRGH]). – **D:** Zimbabwe; rocky hillsides, often in shelter of bushes, 1980 - 2200 m. **I:** Reynolds (1966: 88-89, as *A. saponaria*).

[3] Acaulescent, simple; **L** densely rosulate, ovate-lanceolate, to ± 25 × 7 - 11 cm plus apical portion soon drying, upper face dark green with conspicuous elongated whitish spots in transverse bands, lower face pale green usually without spots; marginal teeth ± 6 mm, pungent, reddish-brown, 1 - 1.5 mm apart; exudate drying purplish-brown; **Inf** to 75 cm, with 3 - 7 **Br**, lowest **Br** sometimes rebranched; racemes capitate-corymbose, 3 - 4 × ± 10 - 12 cm, very dense; **Bra** lanceolate-acuminate, 15 - 20 × 4 - 6 mm; **Ped** 30 - 40 mm; **Fl** bright orange-red, 35 - 40 mm, ± 9 mm ⌀ across **Ov**, abruptly narrowed above **Ov**, then enlarging slightly to mouth; **Tep** free for 27 - 30 mm; **St** exserted ± 2 mm.

This is included in *A. saponaria* in the sense of Reynolds (1966).

A. commixta A. Berger (in Engler, A. (ed.), Pflanzenr. IV.38 (Heft 33): 260-261, ills., 1908). **T:** RSA, Western Cape (*Wright* s.n. [K]). – **D:** RSA (Western Cape); between rocks and low shrubs, 1500 m. **I:** Reynolds (1950: 360-361).
　Incl. *Aloe perfoliata* var. α Linné (1753); **incl.** *Aloe gracilis* Baker (1880) (*nom. illeg.*, Art. 53.1).

[15] Caulescent, branching to form dense shrubs; stem suberect to erect, ± 100 × 2 - 2.5 cm; **L** scattered along stem for ± 30 cm, lanceolate-acuminate, to 20 × 3 cm, dull green; marginal teeth 1 - 2 mm, firm, white, 2 - 4 mm apart; sheath green-striatulate; **Inf** 30 - 35 cm, simple; raceme subcapitate, 5 - 7 cm, dense; **Bra** ovate-deltoid, acuminate; **Ped** ± 6 mm; **Fl** yellowish to orange, to 40 mm, base shortly attenuate, not narrowed above **Ov**; **Tep** free for 20 mm to almost the base. – *Cytology:* 2n = 14 (Snoad 1951).

A. comosa Marloth & A. Berger (BJS 38: 86, 1905). **T:** RSA, Western Cape (*Marloth* 3787 [BOL, GRA, PRE]). – **D:** RSA (Western Cape); slopes of hills and valleys. **I:** Reynolds (1950: 387-388).

[8] Caulescent, simple; stem to 2 m, with dead **L** remains persistent; **L** densely rosulate, lanceolate-ensiform, to 65 × 12 cm, upper face glaucous to somewhat brownish-pinkish, obscurely lineate, lower face bluish-green, with pinkish margin; marginal teeth 1 - 2 mm, brownish-red, 5 - 10 mm apart; **Inf** to 2.5 m or more, usually simple, sometimes with 1 **Br**; raceme narrowly cylindrical, 100 × 8 cm, dense; **Bra** lanceolate, long-acuminate, 40 mm, imbricate in bud stage and tufted at tip of raceme; **Ped** 20 mm, almost erect, at anthesis with base remaining appressed to axis and upper ⅓ markedly decurved; **Fl** usually rosy-cream to pinkish-ivory, sometimes deep pink with a bloom, 35 mm, ventri-

cose, base very shortly attenuate, enlarging above **Ov** to 12 mm ⌀ in middle, narrowing to mouth; **Tep** free for ± 23 mm; **St** exserted 10 mm. – *Cytology:* 2n = 14 (Riley 1959).

A. compressa H. Perrier (Mém. Soc. Linn. Normandie, Bot. 1(1): 33, 1926). **T:** Madagascar, Antananarivo (*Perrier* 12556 [P]). – **D:** Madagascar.

A. compressa var. **compressa** – **D:** Madagascar; quartzite rocks, 1000 - 1500 m. **I:** Reynolds (1966: 425-425).

[2] Acaulescent or shortly caulescent, simple; **L** 15 - 20, distichous, triangular, tip rounded, 12 - 15 × 5 cm, glaucous, surface smooth; marginal teeth 2 mm, flattened, green- or red-tipped, 4 - 5 mm apart; **Inf** 60 - 70 cm, simple, rarely with 1 **Br**; raceme almost ovate, 7 - 8 × 6 cm, dense, with 40 - 60 **Fl**; lower **Bra** long-acute, upper obtuse and emarginate, 20 - 24 × 10 - 17 mm, white with 3 brown veins; **Ped** 1 - 2 mm; **Fl** white, with red midstripe in upper ½, 25 - 33 mm; **Tep** free to base, connivent in lower ½, **OTep** recurved in upper ½.

A. compressa var. **paucituberculata** Lavranos (KuaS 49(7): 158-159, ills., 1998). **T:** Madagascar (*Röösli & Hoffmann* 7/95 [P, MO, TAN]). – **D:** C Madagascar; quartzite mountain, 1725 - 1950 m.

[2] Differs from var. *compressa*: **L** 9 × 4 cm, sparsely tuberculate.

A. compressa var. **rugosquamosa** H. Perrier (Mém. Soc. Linn. Normandie, Bot. 1(1): 34, 1926). **T:** Madagascar, Antananarivo (*Perrier* 10993 [P]). – **D:** Madagascar; quartzite rocks, ± 1350 m. **I:** Reynolds (1966: 426).

[2] Differs from var. *compressa*: More robust; **L** to 23 × 3 - 3.5 cm, greyish, upper face rough, with tiny obtuse protuberances; marginal teeth smaller, more crowded; **Inf** stouter, often branched; **Bra** ± 55 × 15 mm; **Fl** 55 mm.

A. compressa var. **schistophila** H. Perrier (Mém. Soc. Linn. Normandie, Bot. 1(1): 34, 1926). **T:** Madagascar, Fianarantsoa (*Perrier* 11005 [P]). – **D:** Madagascar; schistose rocks, ± 1400 m. **I:** Reynolds (1966: 427-428).

[2] Differs from var. *compressa*: **L** more numerous, more compressed, 12 × 2 - 2.5 cm; marginal teeth almost confluent; **Inf** with racemes 4 cm; **Fl** reddish, 22 mm.

A. comptonii Reynolds (Aloes South Afr. [ed. 1], 382-385, ills., 1950). **T:** RSA, Western Cape (*Reynolds* 5725 [PRE]). – **D:** RSA (Western Cape, Eastern Cape); flat stony country, gentle slopes, quartzite ridges and rock faces.

[7] Acaulescent or caulescent, usually in dense groups; stem short, rarely to 1 m; **Ros** to 60 cm ⌀, 40 - 50 cm tall; **L** ± 20, rosulate, lanceolate-attenu-

ate, to 30 × 9 cm, glaucous-green, sometimes tinged reddish, lower face obscurely carinate in upper ½, the keel with up to 6 prickles, larger near **L** tip; marginal teeth 2 - 3 mm, pale brown from white base, 10 - 15 mm apart; exudate drying deep orange; **Inf** 80 - 100 cm, with 3 - 5 (-8) **Br**; racemes subcapitate with rounded tip, sometimes broadly conical, to 15 × 9 - 10 cm, dense; **Bra** ovate-lanceolate, acuminate, 7 × 3 mm; lowest **Ped** 30 - 35 mm, slightly shorter above; **Fl** dull scarlet, 35 - 40 mm, base shortly attenuate, narrowed slightly above **Ov**, enlarging towards mouth; **Tep** free to base or almost so. – *Cytology:* 2n = 14 (Riley 1959).

Wyk & Smith (1996) suggest that this could be conspecific with *A. mitriformis.*

A. confusa Engler (Pfl.-welt Ost-Afr., Teil C, 141, 1895). **T:** Tanzania, Moshi Distr. (*Volkens* 410 [B?, BM]). – **D:** Kenya, Tanzania; rocky slopes and cliff faces, 885 - 1000 m. **I:** Reynolds (1966: 166-167).

[13] Caulescent, branching mostly at base to form dense tangled masses to 4 m ∅; stem decumbent or pendulous, to 1 m, 1.8 cm ∅; **L** scattered along stem for 20 - 30 cm, lanceolate-attenuate, mostly falcate, to 30 × 3 - 4 cm, grey-green, lower face usually with a few small white spots near base, surface smooth, with narrow white cartilaginous margin; marginal teeth 1 - 2 mm, firm, whitish, ± 10 - 15 mm apart; sheath ± 15 mm, usually green-striate, sometimes spotted; exudate yellow, rapidly turning deep purple in air; **Inf** ± 45 cm, oblique with racemes erect, simple or with 2 - 3 **Br**; racemes cylindrical-acuminate, 10 - 20 × 7 cm, sub-dense, with ± 25 **Fl**; **Bra** deltoid-acuminate, 4 - 7 × 2 - 3 mm; **Ped** 12 - 15 mm; **Fl** yellow, sometimes red, ± 25 - 30 mm, base shortly attenuate, ± 6 mm ∅ across **Ov**, slightly narrowed above, then enlarging to mouth; **Tep** free for 11 mm. – *Cytology:* 2n = 14 (Müller 1945).

A. congdonii S. Carter (Fl. Trop. East Afr., Aloaceae, 18, 20, ill. (p. 10), 1994). **T:** Tanzania, Iringa Distr. (*Congdon* 281 [K, EA, NHT, SRGH]). – **D:** Tanzania; rock outcrops, 1500 - 2375 m.

[6] Acaulescent, suckering to form large groups; **L** ± 8 - 10, densely rosulate, ovate-lanceolate, 15 - 25 × 3 - 5 cm, greyish-green tinged red, with many elongated whitish spots; marginal teeth 1 - 2 mm, pale, minutely brown-tipped, 5 - 8 mm apart; **Inf** 20 - 30 - 50 cm, simple or occasionally with 1 - 2 **Br**; racemes subcapitate to cylindrical, 5 - 10 cm, ± dense; **Bra** ovate-acuminate, 6 - 8 × 3 - 3.5 mm; **Ped** 18 - 22 mm; **Fl** orange-pink, 32 - 38 mm, base rounded, 8 mm ∅ across **Ov**, slightly narrowed above, then enlarging slightly to mouth; **Tep** free for ± 10 - 13 mm; **St** slightly exserted.

A. conifera H. Perrier (Mém. Soc. Linn. Normandie, Bot. 1(1): 47, 1926). **T:** Madagascar, Fianarantsoa (*Perrier* 13123 [P]). – **D:** Madagascar; soil

pockets on eroded granite; 1300 - 1500 m. **I:** Reynolds (1966: 478-479).

[2] Acaulescent or with short stem to 10 cm, simple; **L** 20 - 24, densely rosulate, lanceolate-attenuate, tip rounded with few short teeth, 16 × 4 - 4.5 cm, bluish-grey tinged reddish; marginal teeth 2 - 3 mm, pungent, reddish, 5 - 10 mm apart; **Inf** 50 cm, usually simple, rarely with 1 - 2 **Br**; racemes cylindrical, 10 - 15 (-20) × 3.5 cm, very dense; young raceme at first conical with densely imbricate **Bra**, resembling a narrow pine-cone; **Bra** obovate-cuspidate, 12 mm; **Ped** none; **Fl** lower ½ lemon, yellow near mouth, 14 mm, slightly campanulate-clavate, base rounded, 4 mm ∅ across **Ov**, enlarging above; **Tep** free to base.

A. cooperi Baker (Gard. Chron., ser. nov. 1: 628, 1874). **T:** RSA, KwaZulu-Natal (*Cooper* 1193 / 3263 [K]). – **D:** RSA, Swaziland.

A. cooperi ssp. **cooperi** – **D:** RSA (KwaZulu-Natal, Mpumalanga), Swaziland; grassland, sometimes marshy, and rocky slopes, to 1980 m. **I:** Reynolds (1950: 151-152).

   Incl. *Aloe schmidtiana* Regel (1879).

[4] Acaulescent or shortly caulescent, usually simple, sometimes suckering to form small groups; stem to 15 cm; **L** 16 - 20, distichous, sometimes spirally twisted to rosulate with age, triangular, plicate-carinate and V-shaped in cross section, 60 - 80 × 5 - 6 cm, upper face green, sometimes with few scattered spots near base, obscurely lineate, lower face with many white spots near base, with narrow white cartilaginous margin; marginal teeth 1 - 2 mm, firm, white, 1 - 2 mm apart; **Inf** to 1 m or more, simple; raceme broadly conical, to 20 × 10 - 14 cm, sub-dense, with ± 40 **Fl**; **Bra** ovate-acuminate, 20 - 32 mm; **Ped** 40 - 45 mm; **Fl** salmon-pink with green tips, 38 - 40 mm, base attenuate, 12 mm ∅ across **Ov**, narrowing to ± 9 mm near mouth; **Tep** free almost to base; **St** exserted 0 - 1 mm. – *Cytology:* 2n = 14 (Kondo & Megata 1943).

A. cooperi ssp. **pulchra** Glen & D. S. Hardy (FPA 49(3-4): t. 1944 + text, 1987). **T:** RSA, KwaZulu-Natal (*Harrison* 980 [PRE]). – **D:** RSA (KwaZulu-Natal), Swaziland; rocky hillsides.

[4] Differs from ssp. *cooperi*: **L** 40 - 80 × 2.5 - 6 cm, thicker, more sharply keeled, lower face with many white **Tu** near base, each with a **Ha**-like process; marginal teeth in basal ¼ only; **Inf** raceme conical to subcapitate; **Bra** clasping the **Ped**; **Fl** 35 - 45 mm.

A. corallina I. Verdoorn (FPA 45: t. 1788 + text, 1979). **T:** Namibia, Kaokoland (*Leistner & al.* 179 [PRE]). – **D:** Namibia; cliff faces.

[5] Caulescent, usually simple; stem 0.5 cm ∅; **L** densely rosulate, lanceolate-acuminate, ± 50 × 11 cm, grey-green with light bloom, upper face obscu-

rely lineate; marginal teeth on young **L** minute, red-brown, on mature **L** absent; **Inf** with 2 - 3 **Br**; racemes cylindrical, 17 - 28 cm, sub-dense; **Bra** oblong-acute, 10 - 15 mm; **Ped** 10 - 17 mm; **Fl** coral-red, ± 30 mm, base shortly attenuate, enlarged above **Ov** to middle, then narrowing to mouth; **Tep** free for 20 mm.

In the protologue the length of the stem is not given, but the habit drawing suggests that it is quite short.

A. crassipes Baker (JLSB 18: 162, 1880). **T:** Sudan, Equatoria Prov. (*Schweinfurth* 3765 [K]). — **D:** Sudan, Zambia; thick grass in woodland, ± 1280 m. **I:** Reynolds (1966: 185).

[3] Acaulescent or very shortly caulescent; **L** ± 20 - 24, densely rosulate, triangular, 40 × 7 cm, dull glaucous-green; marginal teeth 5 mm, firm, brown-tipped, 15 mm apart; **Inf** 50 - 60 cm, with ± 5 **Br**; racemes cylindrical-conical, 17 - 25 × 8 cm, dense; **Bra** deltoid-acute, 10 × 4 mm; **Ped** 14 mm; **Fl** dull yellowish-green, 38 mm, base truncate, 10 mm ∅ across **Ov**, very slightly narrowed above, very slightly enlarging to mouth; **Tep** free for 13 mm; **St** exserted 0 - 1 mm.

A. cremersii Lavranos (Adansonia, n.s., 14(1): 99-101, ills., 1974). **T:** Madagascar, Fianarantsoa (*Cremers* 2048 [P]). — **D:** Madagascar; quartzite summits.

[3] Caulescent, simple; stem erect, to 50 cm; **L** to 30, rosulate, to 60 × 5.5 cm, green tinged reddish, lineate; marginal teeth 2 mm, whitish or pale brown, 5 - 12 mm apart; **Inf** 1.5 m, with 2 - 3 **Br**; racemes cylindrical, sub-dense; **Bra** lanceolate, 7 × 2 mm; **Ped** 7 mm; **Fl** red, 18 - 20 mm, clavate, base attenuate, 3 mm ∅ across **Ov**, enlarging to 5 mm at mouth; **Tep** free almost to base; **St** exserted 0 - 1 mm.

A. cremnophila Reynolds & P. R. O. Bally (JSAB 27(2): 77-79, t. 13-14, 1961). **T:** Somalia (*Reynolds* 8450B [PRE]). — **Lit:** Brandham & al. (1994). **D:** Somalia; limestone cliff faces, 1980 - 2130 m.

[13] Caulescent, branching at base; stem pendulous, 10 - 20 × 0.8 - 1 cm; **L** 6 - 8, rosulate, lanceolate-attenuate, 10 × 2 cm, grey-green; marginal teeth 2 mm, pungent, pale brown, 3 - 5 mm apart; **Inf** 25 - 30 cm, base pointing downwards, then arcuate-ascending, simple; raceme cylindrical-conical, 10 - 12 × 5.5 cm, lax; **Bra** ovate-acute, 10 × 5 mm; **Ped** 10 - 12 mm; **Fl** scarlet, yellowish-green at mouth, 25 mm, very slightly clavate, base rounded, 5 mm ∅ across **Ov**, very slightly narrowed and then enlarging above; **Tep** free for 5 mm; **St** exserted 0 - 1 mm. — *Cytology:* 2n = 28 (Brandham & al. 1994).

A. cryptoflora Reynolds (JSAB 31(4): 281-284, ills., 1965). **T:** Madagascar, Fianarantsoa (*Fievet*

s.n. in *Reynolds* 11619 [PRE, K]). — **D:** Madagascar; granite. **I:** Reynolds (1966: 480-481).

[3] Acaulescent or shortly caulescent, simple; **L** 15 - 20, densely rosulate with slight spiral twist, lanceolate-attenuate, 20 - 25 × 6.5 cm, deep green slightly tinged reddish, with brownish-red margin; marginal teeth 2 - 3 mm, brownish-red, 5 - 10 mm apart; **Inf** 40 - 60 cm, usually with 1 **Br**; racemes cylindrical, very slightly conical, 14 × 3 cm, very dense; **Bra** ovate-orbicular-cuspidate, rounded and somewhat cupped, fleshy, pale green, 11 × 12 mm, closely imbricate in bud stage; **Ped** none; **Fl** greenish-yellow, becoming orange-yellow at the mouth, 10 mm, cylindrical-campanulate, base rounded, 3.5 mm ∅ across **Ov**; **Tep** free to base.

A. cryptopoda Baker (J. Bot. 1884: 52, 1884). **T:** Moçambique (*Kirk* 96 [K]). — **D:** Botswana, Malawi, Moçambique, Zambia, Zimbabwe, RSA (Northern Prov., North-West Prov., Mpumalanga), Swaziland; rocky slopes and inselbergs with little or no grass. **I:** Reynolds (1950: 331-333). **Fig. XIII.h**

Incl. *Aloe wickensii* Pole-Evans (1915); **incl.** *Aloe wickensii* var. *wickensii*; **incl.** *Aloe pienaarii* Pole-Evans (1915); **incl.** *Aloe wickensii* var. *lutea* Reynolds (1935).

[5] Caulescent or very shortly caulescent, usually simple, sometimes suckering to form small groups; **L** ± 40 - 50, densely rosulate, lanceolate-ensiform, or gradually attenuate from base, to 90 × 12 - 15 cm, deep green, reddish-green to bluish-green; marginal teeth 2 mm, pungent, reddish-brown, 5 - 7 mm apart; **Inf** 1.25 - 1.75 m, with 5 - 8 **Br**; racemes cylindrical-conical to cylindrical-acuminate, 25 - 35 cm, dense; **Bra** ovate-acuminate, ± 20 × 10 - 12 mm; **Ped** 15 - 20 mm; **Fl** scarlet, greenish-tipped, upper ½ becoming yellowish at anthesis, ± 35 - 40 (-45) mm, base truncate, not narrowed above **Ov**; **Tep** free to base. — *Cytology:* 2n = 14 (Müller 1945).

Natural hybrids with other species have been reported (Reynolds 1950).

A. cyrtophylla Lavranos (KuaS 49(7): 159-161, ills., 1998). **T:** Madagascar (*Röösli & Hoffmann* 36/95 [P, MO, TAN]). — **D:** S Madagascar; limestone, ± 1400 m.

[11] Caulescent, much branched from the base; stems erect, up to 30 cm; **L** in lax **Ros**, lanceolate, tip acute, the distal ½ strongly rolled back, 12 - 18 × 2 cm, dark green; marginal teeth 1.5 mm, whitish, 5 - 6 mm apart; **Inf** 30 - 50 cm, simple; raceme 22 cm, lax; **Bra** acute, 10 × 7 mm, whitish with 5 brown veins; **Ped** 7 mm; **Fl** coral-red, becoming yellowish and then green above, ± 28 mm, 4 mm ∅ across the **Ov**, slightly constricted above the **Ov**, then enlarging to 7 mm ∅, base shortly attenuate; **Tep** free for 9 mm; **Fr** unknown.

A. dabenorisana van Jaarsveld (JSAB 48(3): 419-424, ills., 1982). **T:** RSA, Northern Cape (*van*

Jaarsveld & Kritzinger 6426 [NBG]). – **D:** RSA (Northern Cape); vertical S- and SW-facing quartz cliffs, 900 - 1000 m. **I:** Wyk & Smith (1996: 111).

[5] Acaulescent or shortly caulescent, simple or in dense groups; stem to 30 × 2.7 cm; **Ros** hanging down or ascending; **L** in 4 - 5 vertical ranks when young, later spirally arranged, lanceolate-acuminate, to 24 × 5 cm, green, tinged red, obscurely striate; marginal teeth 2 mm, white, ± 10 mm apart; exudate drying orange-yellow; **Inf** 25 - 30 cm, arcuate and ascending, rarely simple, usually with 2 - 4 **Br**; racemes conical, 12 × 0.7 - 0.9 cm, lax; **Bra** lanceolate-acuminate, 7 × 2 mm; **Ped** 20 - 28 mm; **Fl** yellowish with red keel, 25 mm, 5 mm ∅ across **Ov**, enlarging to 8 mm near mouth; **Tep** free to base.

A. dawei A. Berger (Notizbl. Königl. Bot. Gart. Berlin 4: 246, 1906). **T** [neo]: Uganda, Mengo Distr. (*Reynolds* 7511 [PRE, EA, K]). – **D:** Kenya, Rwanda, Uganda, Zaïre; grassland and thickets, 800 - 1525 m. **I:** Reynolds (1966: 369-370).

Incl. *Aloe beniensis* De Wildeman (1921); **incl.** *Aloe pole-evansii* Christian (1940).

[16] Caulescent, branching; stem erect or decumbent, to 2 m × 6 - 8 cm, with dead **L** remains; **L** 16 - 20, laxly rosulate and persisting 30 cm below, lanceolate-attenuate, 40 - 60 × 6 - 9 cm, olive-green to dark green, sometimes tinged reddish, with dull white spots on young shoots; marginal teeth 2 - 4 mm, pungent, reddish-brown, 10 - 15 mm apart; exudate drying yellow; **Inf** 60 - 90 cm, with 3 - 8 **Br**; racemes broadly cylindrical-conical, 8 - 20 × 8 cm, sub-dense; **Bra** ovate-acute, 3 - 4 × 3 - 5 mm; **Ped** 10 - 15 mm; **Fl** scarlet, paler at mouth, 33 - 35 mm, base shortly attenuate, 8 mm ∅ across **Ov**, very slightly narrowed above; **Tep** free for 11 - 12 mm. – *Cytology:* 2n = 14 (Brandham 1971).

A. debrana Christian (FPA 26: t. 1016 + text, 1947). **T:** Ethiopia, Shewa Region (*McLoughlin* 812A [PRE 27173]). – **Lit:** Gilbert & Sebsebe (1997). **D:** Ethiopia; grassland and low scrub on volcanic rocks, usually gentle slopes, 2400 - 2700 m. **I:** Reynolds (1966: 279-280, as *A. berhana*). **Fig. XII.c**

Incl. *Aloe berhana* Reynolds (1957).

[5] Acaulescent or with very short stem, simple or suckering to form small groups; **L** 24 or more, densely rosulate, lanceolate-attenuate, 25 - 60 × 7.5 - 15 cm, dull green, with reddish-brown horny margin; marginal teeth 2 - 4 mm, reddish-brown, pungent, 0.8 - 15 mm apart; exudate drying brownish; **Inf** 1 m, with 8 - 12 **Br**, lower **Br** sometimes re-branched; racemes subcapitate, 5 - 10 × 6 - 7 cm, dense; **Bra** ovate-triangular, 3 - 5 × 1.5 - 2 mm; **Ped** 10 - 15 mm; **Fl** scarlet, paler at mouth, 17 - 20 (-35) mm, base very shortly attenuate, 8 mm ∅ across **Ov**, slightly narrowed above, then enlarging to the mouth; **Tep** free for 5 - 9 mm.

A. decorsei H. Perrier (Mém. Soc. Linn. Normandie, Bot. 1(1): 43, 1926). **T:** Madagascar, Fianarantsoa (*Perrier* 11002 [P]). – **D:** Madagascar; gneiss slopes, 1600 - 2200 m. **I:** Reynolds (1966: 451).

[2] Acaulescent, simple; **L** 12 - 15, rosulate, slightly falcate, 60 - 70 × 5 - 6 cm, green; marginal teeth 1 - 1.5 mm; **Inf** 80 - 120 cm, simple; raceme 18 - 30 cm, lax; **Bra** oblanceolate, shortly acute, 15 × 6 mm; **Ped** 3 - 4 cm; **Fl** yellow becoming green with white margins at tips, 20 - 22 mm, campanulate, base shortly attenuate; **Tep** free for 18 - 20 mm; **St** not exserted.

Little known; type locality not recorded precisely.

A. decurva Reynolds (JSAB 23(1): 15-17, ills., 1957). **T:** Moçambique (*Reynolds* 8200 [PRE, K, SRGH]). – **D:** Moçambique; soil pockets on granite slopes, 1060 m. **I:** Reynolds (1966: 247-248).

[4] Acaulescent or very shortly caulescent, simple, sometimes dividing into 2 **Ros**; **L** 20 - 24, densely rosulate, ensiform-attenuate, to 55 × 9 cm, dull green tinged reddish; marginal teeth 3 mm, pungent, 8 - 15 mm apart; exudate drying yellow; **Inf** to 90 cm, simple, rarely with 1 **Br**; raceme broadly cylindrical to slightly acuminate, 15 - 20 × 10 - 12 cm, very dense, pointing downwards on decurved peduncle; **Bra** ovate-obtuse, 2 × 3 mm; **Ped** 1 mm; **Fl** bright red or orange, 38 mm, ventricose, base narrowly rounded, 5 mm ∅ across **Ov**, enlarging above to 11 mm in middle, narrowing to mouth; **Tep** free for 33 mm; **St** exserted 10 - 12 mm.

A. delphinensis Rauh (CSJA 62(5): 230-232, ills., 1990). **T:** Madagascar (*Rauh* 68629a [HEID]). – **D:** Madagascar; granite rocks, 100 m.

[11] Caulescent, branching at base forming lax groups; stem erect or ascending, to 30 × 1 cm; **L** scattered along stem, linear-acute, 15 - 25 × 1.5 cm, green to red-green; marginal teeth small, reddish, crowded; sheath ± 1 cm, red-brown with darker veins; **Inf** 18 - 28 cm, simple; racemes cylindrical, ± 8 × 6 cm, lax, with ± 15 **Fl**; **Bra** triangular, 2 mm, red-brown; **Ped** 15 - 18 mm, almost horizontal in bud stage; **Fl** red becoming whitish with green midrib towards tip, 25 mm, narrowed very slightly above **Ov**.

A. deltoideodonta Baker (JLSB 20: 271, 1883). **T:** Madagascar (*Baron* 946 [K]). – **D:** Madagascar.

A. deltoideodonta var. **brevifolia** H. Perrier (Mém. Soc. Linn. Normandie, Bot. 1(1): 24, 1926). **T:** Madagascar, Toliara (*Perrier* 12740 [P]). – **D:** Madagascar; denuded sandstone rocks, ± 1000 m. **I:** Reynolds (1966: 437).

[18] Differs from var. *deltoideodonta*: **L** 10 × 5 cm; **Inf** 30 cm; racemes 6 - 10 cm, denser.

Little-known. The type specimen has narrower leaves (2.5 cm) than given in the protologue.

A. deltoideodonta var. **candicans** H. Perrier (Mém. Soc. Linn. Normandie, Bot. 1(1): 25, 1926). **T:** Madagascar, Fianarantsoa (*Perrier* s.n. [P]). – **D:** Madagascar; semi-denuded rocky slopes and pavements, 660 - 800 m. **I:** Reynolds (1966: 435-436).

[18] Differs from var. *deltoideodonta*: **L** 15 - 20 × 5 - 6 cm; **Inf** racemes cylindrical-conical, 10 - 20 cm; **Bra** ovate-acute, ± 15 × 7 - 8 mm, very white, usually deflexed near base. – *Cytology:* 2n = 14 (Brandham 1971).

A. deltoideodonta var. **deltoideodonta** – **D:** Madagascar. **I:** Reynolds (1966: 434).

Incl. *Aloe deltoideodonta* var. *typica* H. Perrier (1926) (*nom. inval.*, Art. 24.3).

[18] Acaulescent or shortly caulescent, probably suckering; **L** 12 - 16, densely rosulate, lanceolate-deltoid, 10 - 13 × 2.5 - 3 cm, green, probably without spots, with narrow straw-coloured cartilaginous margin; marginal teeth 2 mm, straw-coloured, 3 - 5 mm apart; **Inf** 40 - 60 cm, simple or with 1 - 2 **Br**; racemes narrowly cylindrical-acuminate and narrowing almost to a point, 15 - 20 cm, sub-dense; **Bra** lanceolate-deltoid, ± 10 mm, white; **Ped** 10 - 12 mm; **Fl** probably scarlet, ± 25 mm, base attenuate; **Tep** free for 10 mm; **St** exserted 0 - 1 mm.

Little-known; type locality unknown.

A. descoingsii Reynolds (JSAB 24(2): 103-105, ills., 1958). **T:** Madagascar, Toliara (*Descoings* 2440 [TAN, PRE]). – **D:** Madagascar.

≡ *Guillauminia descoingsii* (Reynolds) P. V. Heath (1994).

A. descoingsii ssp. **augustina** Lavranos (CSJA 67(3): 158-161, ills., 1995). **T:** Madagascar, Toliara (*Lavranos & al.* 29194 [P]). – **D:** Madagascar; limestone, in shade of bushes, ± 100 m.

[11] Differs from ssp. *descoingsii*: **L** narrower, greyish-green, with many white tubercles each bearing a **Bri**-like prickle, esp. on a lower surface keel towards tip; **Fl** almost cylindrical, with wide mouth.

A. descoingsii ssp. **descoingsii** – **D:** Madagascar; shallow soil at top of limestone cliffs, ± 350 m. **I:** Reynolds (1966: 411-412).

[11] Acaulescent or very shortly caulescent, suckering freely forming dense groups; **L** ± 8 - 10, densely rosulate, ovate-attenuate, to 3 × 1.5 cm, dull green with many dull white tubercles, surface rough; marginal teeth 1 mm, cartilaginous, white, 1 - 1.5 mm apart, smaller and becoming obsolete towards tip; **Inf** 12 - 15 cm, simple; raceme capitate, 12 × 25 mm, with ± 10 **Fl**; **Bra** ovate-acute, 2 × 1 mm; **Ped** 5 mm; **Fl** scarlet, paler to slightly orange at mouth, 7 - 8 mm, urceolate, base flat and shortly attenuate, 4 mm ∅ across **Ov**, narrowed to 3 mm at mouth; **Tep** free for 2 mm; **St** not exserted. – *Cytology:* 2n = 14 (Brandham 1971).

A. deserti A. Berger (BJS 36: 61, 1905). **T:** Tanzania, Pare Distr. (*Volkens* 2378 [B]). – **D:** Kenya, Tanzania; sandy stony soil in grass or at edge of thickets, 550 - 1825 m. **I:** Reynolds (1966: 338-340).

[15] Caulescent, branching sparsely; stem erect to 75 × 4 - 5 cm or decumbent to 1 m; **L** ± 16 - 20, laxly rosulate and persistent for 20 - 30 cm, lanceolate-attenuate, 40 - 45 × 7 - 8 cm, dull green tinged brownish, with or without elliptic whitish spots, surface slightly rough; marginal teeth 2 - 3 mm, pungent, pale brown, 10 - 15 mm apart; **Inf** 1.2 - 1.5 m, with 3 - 8 **Br**; racemes cylindrical to slightly acuminate, to 25 × 7 cm, sub-lax, limp and drooping in bud stage, becoming stiff and erect as **Fl** open; **Bra** ovate-acute, 12 - 15 × 6 - 8 mm, conspicuously white, closely imbricate in bud stage, becoming deflexed later; **Ped** 7 - 10 mm; **Fl** dull rose-pink with a bloom, paler at mouth, 28 - 35 mm, base shortly attenuate, 8 mm ∅ across **Ov**, narrowed slightly above and enlarging to mouth; **Tep** free for 14 - 18 mm.

A. dewetii Reynolds (JSAB 3(3): 139-141, ills., 1937). **T:** RSA, KwaZulu-Natal (*Reynolds* 2319 [PRE, BOL]). – **D:** RSA (KwaZulu-Natal, Mpumalanga), Swaziland; grassland. **I:** Reynolds (1950: 267-268).

[3] Acaulescent, simple; **L** ± 20, densely rosulate, lanceolate-attenuate, to 48 × 13 cm, dull green, upper face with many dull white elongated spots in irregular transverse bands or scattered irregularly, lower face without spots and obscurely lineate, with prominent horny brown margin, whole surface with dull glossy appearance; marginal teeth to 10 mm, pungent, brown, 10 - 15 mm apart; **Inf** to 2 m or more, with 10 **Br**, lower **Br** rebranched; racemes cylindrical-acuminate, to 40 × 7 cm; **Bra** deltoid, 20 × 3 mm; **Ped** to 15 mm; **Fl** dull scarlet with a bloom, 35 - 42 mm, base truncate, 14 mm ∅ across **Ov**, abruptly narrowed above to 6 - 7 mm, enlarging towards mouth; **Tep** free for 6 mm; **St** exserted to 1 mm. – *Cytology:* 2n = 14 (Müller 1945).

A. dewinteri Giess (BT 11: 120-122, ills., 1973). **T:** Namibia, Kaokoveld (*Buhr* s.n. in *Giess* 10990 [WIND, M, PRE]). – **D:** Namibia; vertical cliff faces.

[2,3] Acaulescent or shortly caulescent, simple; stem to 10 cm; **L** 14 - 22, rosulate, lanceolate-attenuate, to 50 × 15 cm, grey-green with light powdery bloom, with yellowish-brown horny margin, surface smooth; marginal teeth 1 - 2 mm, brown, 10 - 20 mm apart; **Inf** to 85 cm, simple or with 2 - 3 **Br**; racemes cylindrical-acuminate, 25 - 40 × 5 cm, dense; **Bra** oblong-obovate, obtuse to apiculate; **Ped** to 4 mm; **Fl** coral-pink, becoming yellowish to white when open, 30 - 33 mm, base attenuate, enlarging above **Ov** to 8 mm ∅ in middle, narrowing to mouth; **Tep** free for 20 - 25 mm.

A. dhufarensis Lavranos (CSJA 39(5): 167-171, ills., 1967). **T:** Oman (*Lavranos* 4337 [PRE]). – **D:** Oman; dry water courses and gravel banks, ± 250 m.

[3] Acaulescent, simple; **L** 14 - 20, rosulate, deltoid-acuminate, 45 × 14 cm, bluish-grey; marginal teeth absent except for few white teeth in very young plants; **Inf** to 90 cm, with 1 - 2 **Br**; racemes cylindrical-acuminate, ± 20 cm, lax; **Bra** deltoid-acute, to 12 mm; **Ped** 12 - 15 mm; **Fl** coral-red with a bloom, 28 - 30 mm, base attenuate, 9 mm ⌀ across **Ov**, not narrowed above; **Tep** free for 22 mm.

A. dichotoma Masson (Philos. Trans. 66: 310, 1776). **T:** not located. – **D:** Namibia, RSA (Northern Cape); rocky slopes. **I:** Reynolds (1950: 489-494).

≡ *Rhipidodendrum dichotomum* (Masson) Willdenow (1811); **incl.** *Aloe ramosa* Haworth (1804); **incl.** *Aloe montana* Schinz (1896) ≡ *Aloe dichotoma* var. *montana* (Schinz) A. Berger (1908).

[17] Caulescent, copiously branching dichotomously and forming dense rounded crowns; stem erect, to 9 m, to 1 m ⌀ at base; **L** ± 20, densely rosulate, lanceolate-linear, 25 - 35 × 5 cm, glaucous green, with very narrow brownish-yellow margin; marginal teeth 1 mm, smaller and becoming obsolete towards apex, brownish-yellow; **Inf** 30 cm, with 3 - 5 **Br**; racemes broadly cylindrical, slightly acuminate, to 15 × 9 cm; **Bra** acuminate, 5 - 7 × 3 mm; **Ped** 5 - 10 mm; **Fl** bright canary-yellow, 33 mm, ventricose, base shortly attenuate, enlarging to 14 mm ⌀ in middle; **Tep** free for 25 mm; **St** exserted 12 - 15 mm. – *Cytology:* 2n = 14 (Riley 1959: 241).

A. dinteri A. Berger (in Dinter, Neue Pfl. Deutsch-SWA, 14, 1914). **T:** Namibia (*Dinter* 2791a [SAM]). – **D:** Namibia; limestone rocks in low bush. **I:** Reynolds (1950: 211-212).

[3] Acaulescent, simple, 26 × 26 cm; **L** ± 12, trifarious, lanceolate-acuminate, plicate-carinate, towards tip V-shaped in cross section, 20 - 30 × 5 - 8 cm, chocolate-brown or deep brownish-green, with many elongated white spots in ± broken transverse bands, with narrow white cartilaginous margin, keel on lower face with 1 mm white cartilaginous margin; marginal teeth 0.5 mm, firm, white, 1 - 2 mm apart, becoming smaller and more crowded upwards, with similar teeth on keel; **Inf** 50 - 85 cm, with 3 - 8 **Br**; racemes cylindrical-acuminate, 15 - 20 × 7 cm, lax, with 30 - 40 **Fl**; **Bra** lanceolate-deltoid, very acuminate, a little shorter than **Ped**; **Fl** pale rose-pink with bluish bloom, paler to almost white at mouth, 28 - 30 mm, base rounded, 6.5 mm ⌀ across **Ov**, narrowed to 3.5 mm above, enlarging towards mouth; **Tep** free for 5 - 10 mm; **St** exserted 0 - 1 mm.

A. diolii L. E. Newton (CSJA 67(5): 277-279, ills.,

SEM-ills., 1995). **T:** Sudan, Equatoria Prov. (*Powys & Dioli* 824 [K, EA]). – **D:** Sudan; montane forest; 1760 m. **Fig. XII.g**

[11] Caulescent, branching at base to form loose clumps; stem sprawling, rooting on contact with soil, to 35 cm; **L** 8 - 10, laxly rosulate, triangular to slightly falcate, 25 × 2.5 cm, dull green with few scattered elongated white spots in young growth, few spots near base or none on older shoots, surface slightly rough; marginal teeth 1 mm, firm, uncinate, red-tipped, 2 - 8 mm apart, to 13 mm apart towards **L** tip; exudate yellow, drying brown; **Inf** 36 cm, simple or usually with 2 - 6 **Br**; racemes cylindrical, 9 cm, lax; **Bra** triangular, attenuate, 4 - 5 × 2 mm; **Ped** 10 mm; **Fl** light scarlet with light bloom at base, becoming pale yellow with red mid-stripe above, 20 - 25 mm, base truncate, 8 mm ⌀ across **Ov**, narrowed gradually to 6 mm above, lobes spreading to 8 mm; **Tep** free for 5 - 8 mm. – *Cytology:* 2n = 14 (from protologue).

A. distans Haworth (Synops. Pl. Succ., 78, 1812). **T:** not typified. – **D:** RSA (Western Cape); shallow soil on limestone rocks near coast. **I:** Reynolds (1950: 377-379).

Incl. *Aloe mitriformis* var. *angustior* Lamarck (1784); **incl.** *Aloe perfoliata* var. *brevifolia* Aiton (1789); **incl.** *Aloe brevifolia* Haworth (1804) (*nom. illeg.*, Art. 53.1); **incl.** *Aloe mitriformis* var. *brevifolia* Aiton (1810).

[13] Caulescent, branching at base; stem 2 - 3 m, 3 - 4 cm ⌀, sprawling and rooting, sometimes with more **Br** forming dense groups; **L** laxly rosulate and persistent for 50 cm, lanceolate, to 15 × 7 cm, glaucous-green, the tip a **Sp**, upper face sometimes with a few scattered whitish subtuberculate spots, lower face usually with several subtuberculate whitish spots mostly in lower ½; marginal teeth 3 - 4 mm, golden-yellow, 5 - 8 mm apart, with 2 - 4 similar prickles along a keel near tip of lower face; **Inf** 40 - 60 cm, with 3 - 4 **Br**; racemes capitate, to 8 × 10 cm, dense; **Bra** ± 8 × 5 mm; **Ped** 30 - 40 mm; **Fl** dull scarlet, ± 40 mm, base rounded, slightly narrowed above **Ov** and enlarging towards mouth; **Tep** free to base; **St** exserted to 1 mm. – *Cytology:* 2n = 14 (Brandham 1971).

Wyk & Smith (1996) suggest that this might be just a coastal variant of *A. mitriformis*.

A. divaricata A. Berger (BJS 36: 65, 1905). **T:** Madagascar (*Hildebrandt* 3047 [P, BM, K]). – **D:** Madagascar.

A. divaricata var. **divaricata** – **D:** Madagascar; arid bush and coastal thickets, to 800 m. **I:** Reynolds (1966: 504-506). **Fig. XII.h**

Incl. *Aloe sahundra* Bojer (1837) (*nom. inval.*, Art. 32.1c); **incl.** *Aloe vahontsohy* Decorse *ex* Poisson (1912); **incl.** *Aloe vaotsohy* Decorse & Poisson

(1912); **incl.** *Aloe vahontsohy* H. Perrier (1938) (*nom. inval.*).

[16] Caulescent, simple or sparsely branching at base or low down; stem to 3 m or more, with dead **L** remains; **L** 30 or more, rosulate and persistent for 50 - 100 cm below, ensiform, obtuse-attenuate, 60 - 65 × 7 cm, dull grey-green tinged reddish; marginal teeth 5 - 6 mm, pungent, reddish-brown, 15 - 20 mm apart; exudate drying yellow; **Inf** ± 1 m, with many spreading **Br**, the lowest with 8 - 10 secondary **Br**, producing a total of 60 - 80 racemes; racemes cylindrical-acuminate, 15 - 20 cm, lax, with ± 20 **Fl**; **Bra** deltoid, 4 × 2 mm; **Ped** 6 mm; **Fl** scarlet, 28 mm, base slightly rounded, 7 mm ∅ across **Ov**, narrowed to 6 mm above, enlarging to the mouth; **Tep** free to base. – *Cytology:* 2n = 14 (Brandham 1971).

A. divaricata var. **rosea** (Decary) Reynolds (Aloes Madag. Revis., 133, 1958). **T:** not typified. – **D:** Madagascar.

≡ *Aloe vaotsohy* var. *rosea* Decary (1921).

[16] Differs from var. *divaricata*: **Fl** light pink.

A. doei Lavranos (JSAB 31(2): 163-166, ills., 1965). **T:** Yemen (*Lavranos & Doe* 2264 [PRE]). – **D:** Yemen.

A. doei var. **doei** – **D:** Yemen (Audhali escarpment); ± 2250 m. **I:** Reynolds (1966: 130-131).

[5] Acaulescent, usually simple, occasionally suckering sparsely; **L** 12 - 14, rosulate, lanceolate, 30 - 35 × 8 - 10 cm, dull green, upper face sometimes with a few scattered white spots, with purplish-brown cartilaginous margin; marginal teeth to 3 mm, pungent, red-brown, 8 - 12 mm apart; **Inf** 70 cm, with 2 - 3 **Br**; racemes conical, 15 - 20 × 5 cm, lax; **Bra** deltoid, 3 - 6 × 2 - 3; **Ped** 5 - 7 mm; **Fl** yellow, sparsely covered with 8 - 9 mm white **Ha**, 25 - 26 mm, base rounded, 7 mm ∅ across **Ov**, scarcely narrowed above; **Tep** free for 10 - 14 mm.

A. doei var. **lavranosii** Marnier-Lapostolle (CSJA 42(6): 262, ills., 1970). **T:** Yemen (*Lavranos* 2842 [Herb. Les Cèdres]). – **D:** Yemen.

[5] Differs from var. *doei*: **Fl** yellowish-white, **Ha** longer and thicker.

A. dominella Reynolds (JSAB 4(4): 101-103, ills., 1938). **T:** RSA, KwaZulu-Natal (*Reynolds* 2094 [PRE]). – **D:** RSA (KwaZulu-Natal); rocky slopes, 1220 m. **I:** Reynolds (1950: 129-130).

[10] Caulescent, suckering freely forming tufts of up to 50 stems; stem to 15 cm, covered with dead **L** remains; **L** ± 20, multifarious, linear-lanceolate, ± 35 × 1 cm, dull green, lower face with many small white spots near base, with very narrow white cartilaginous margin; marginal teeth 0.5 - 1 mm, firm, white, 2 - 5 mm apart; **Inf** ± 35 cm, simple; raceme capitate, ± 4 × 8 cm, with ± 20 **Fl**; **Bra** ovate-

acuminate, to 15 × 3 mm; **Ped** to 20 mm; **Fl** yellow, 18 mm, slightly clavate, base shortly attenuate; **Tep** free to base.

A. dorotheae A. Berger (in Engler, A. (ed.), Pflanzenr. IV.38 (Heft 33): 263-264, 1908). **T:** Tanzania, Pangani Distr. (*Strauss* 435 [B]). – **D:** Tanzania; soil pockets on rock slabs, 600 - 685 m. **I:** Reynolds (1966: 69-70).

Incl. *Aloe harmsii* A. Berger (1908).

[6,7] Shortly caulescent, suckering to form dense groups; stem to 25 cm; **L** ± 20, rosulate, lanceolate-attenuate, to 30 × 5 - 6 cm, bright glossy green becoming reddish in drought, with scattered white spots, with narrow white cartilaginous margin, surface smooth; marginal teeth 3 - 5 mm, uncinate, white-tipped, 10 - 15 mm apart; exudate drying yellow; **Inf** to 40 - 60 cm, simple or rarely with 1 - 3 **Br**; racemes conical-cylindrical, 10 - 25 cm, subdense; **Bra** ovate-acute, 3 - 6 × 2 - 3 mm; **Ped** 4 - 10 mm; **Fl** coral-red becoming greenish-yellow at tip, or entirely yellow, to 27 - 35 mm, base shortly attenuate, 6 mm ∅ across **Ov**, slightly narrowed above, then enlarging to 8 mm and narrowed slightly to mouth; **Tep** free for 13 - 17 mm; **St** exserted 0 - 1 mm. – *Cytology:* 2n = 14 (Brandham 1971).

Probably only cultivated at the reported type locality, but matching plants are known in the wild elsewhere in Tanzania. *A. harmsii* is placed here with a question mark according to Carter (1994).

A. duckeri Christian (JSAB 6(4): 179-180, ills., 1940). **T:** Malawi, Northern Prov. (*Ducker* s.n. in *Burtt* 5862 [SRGH, BM, K]). – **D:** Malawi, Tanzania, Zambia; grassland and open woodland, 1370 - 2440 m. **I:** Reynolds (1966: 87).

[5] Acaulescent or shortly caulescent, simple, rarely suckering sparsely; stem to 30 cm; **L** ± 20, rosulate, lanceolate-attenuate, to 50 × 10 - 12 cm, dull green, obscurely lineate, usually with scattered white spots, with cartilaginous margin; marginal teeth 4 - 5 mm, pungent, red-brown, 10 - 20 mm apart; **Inf** 1 - 2 m, with 3 - 10 **Br**; racemes capitate, 5 - 6 × 8 - 10 cm, dense; **Bra** deltoid-attenuate, 10 - 12 × 3 mm; **Ped** 30 mm; **Fl** scarlet, orange or pink, 35 - 40 mm, base truncate, 10 mm ∅ across **Ov**, abruptly narrowed to 5 - 6 mm above, then enlarging to mouth; **Tep** free for 12 - 13 mm.

A. dyeri Schönland (Rec. Albany Mus. 1: 289, 1905). **T:** not typified. – **D:** RSA (Mpumalanga); in long grass or amongst trees and bushes on shady rocky slopes, 915 - 1370 m. **I:** Reynolds (1950: 249-250).

[5] Acaulescent or shortly caulescent, usually simple; **L** ± 20, densely rosulate, lanceolate-attenuate, ± 60 - 70 × 10 - 15 cm, upper face green to yellowish-green, usually with small elongated whitish spots scattered or in irregular transverse bands, lower face paler and usually more copiously spotted;

marginal teeth 5 - 6 mm, pungent, light brown, 10 - 15 mm apart; **Inf** 1.5 - 2 m, with up to 15 **Br**, lower **Br** rebranched; racemes cylindrical-acuminate, 12 - 25 × 9 cm, sub-dense near tip, lax below; **Bra** deltoid-acuminate, 10 - 13 mm; **Ped** 10 - 13 mm; **Fl** dull brick-red, 35 mm, base truncate, 9 mm ∅ across **Ov**, abruptly narrowed to 4.5 mm above, enlarging to 8 mm at mouth; **Tep** free for 8 mm. – *Cytology:* 2n = 14 (Müller 1945).

A. ecklonis Salm-Dyck (Monogr. Gen. Aloes & Mesembr. 5: t. 5 + text, 1849). **T:** RSA (*Ecklon s.n.* [not located]). – **D:** Lesotho, Swaziland, RSA (Eastern Cape, KwaZulu-Natal, Free State, Mpumalanga); grassland, rarely on rocky slopes, to 2070 m. **I:** Reynolds (1950: 145-147).

[4] Acaulescent or very shortly caulescent, simple or suckering to form groups of 12 or more shoots; **L** 14 - 20, multifarious, lanceolate, to 40 × 9 cm, dull green, somewhat veined, lower face sometimes with a few small white spots near base, with white cartilaginous margin; marginal teeth 1 - 3 mm, firm, white; **Inf** to 50 cm, simple; racemes densely capitate, ± 5 × 10 - 12 cm, with ± 40 **Fl**; **Bra** ovate-acuminate, ± 10 - 13 mm; **Ped** 30 - 40 mm; **Fl** yellow to red, 20 - 24 mm, ventricose, base attenuate, 7 mm ∅ in the middle; **Tep** free almost to base. – *Cytology:* 2n = 14 (Vosa 1982: 413).

A. edentata Lavranos & Collenette (CSJA 72(2): 86-87, ills. (p. 85), 2000). **T:** Saudi Arabia, Asir Prov. (*Collenette* 5160 [K]). – **D:** Saudi Arabia. **I:** Collenette (1999: 21).

[3] Acaulescent or with very short stem, solitary; **L** densely rosulate, spreading and recurved, lanceolate-attenuate, ± 30 × 6 - 7 cm, dark green, sometimes with white chalky streaks, surface rough; margin entire; **Inf** to 1 m, with ± 18 **Br**; racemes subcapitate, lax; **Bra** 6 × 2 - 3 mm; **Ped** 12 - 14 mm; **Fl** rose-pink, 20 - 22 mm, base rounded, 5 - 6 mm ∅ across **Ov**, scarcely constricted above; **Tep** free for ± 9 mm; **St** exserted ± 3 mm.

A. elata S. Carter & L. E. Newton (in S. Carter, Fl. Trop. East Afr., Aloaceae, 56-57, t. 4: lower right, 1994). **T:** Tanzania, Mbulu Distr. (*Greenway & Kanuri* 11776 [K, EA]). – **D:** Kenya, Tanzania; dense bush and dry forest on rocky slopes, 1050 - 1500 m. **I:** Reynolds (1966: 325, fig. 321, as *A. ballyi*). **Fig. XIII.a**

[9] Caulescent, usually simple, sometimes suckering at base; stem unbranched, to 6 (-10) m, to 12 cm ∅, without persistent dead **L**; **L** densely rosulate, becoming strongly recurved, lanceolate, to 100 × 10 cm, dark greyish-green, surface smooth; marginal teeth 4 - 5 mm, uncinate, white-tipped, 10 - 26 mm apart; exudate colourless; **Inf** to 65 cm, with up to 9 **Br**, lower **Br** rebranched; racemes cylindrical, 6 - 14 cm, lax; **Bra** ovate, 4 - 7 × 3 - 5 mm; **Ped** 8 - 10 mm; **Fl** scarlet in bud, becoming yellow at an-

thesis, 30 - 32 mm, base rounded, 10 mm ∅ across **Ov**; **Tep** free for 24 - 26 mm.

Broken leaves give off a slight odour reminiscent of rats. Material of this species was included in *A. ballyi* by Reynolds (1966).

A. elegans Todaro (Hort. Bot. Panorm. 2: 25, t. 29, 1882). **T:** Ethiopia, Axum Distr. (*Schimper* 927 [[lecto – icono]: l.c. t. 29]). – **Lit:** Gilbert & Sebsebe (1997). **D:** Eritrea, Ethiopia; open bush on stony slopes, 1615 - 2500 m. **I:** Reynolds (1966: 204-206).

Incl. *Aloe abyssinica* var. *peacockii* Baker (1880); **incl.** *Aloe vera* var. *aethiopica* Schweinfurth (1894) ≡ *Aloe aethiopica* (Schweinfurth) A. Berger (1905); **incl.** *Aloe schweinfurthii* Hooker *fil.* (1899) (*nom. illeg.*, Art. 53.1); **incl.** *Aloe peacockii* A. Berger (1905); **incl.** *Aloe percrassa* var. *saganeitiana* A. Berger (1908); **incl.** *Aloe abyssinica* A. Berger (1908) (*nom. illeg.*, Art. 53.1).

[5] Acaulescent or shortly caulescent with age, usually simple, rarely dividing into 2 or 3 **Ros**; stem (when present) to 30 cm, decumbent; **L** 16 - 20, densely rosulate, lanceolate-attenuate, tip with short teeth, ± 60 × 15 - 18 cm, grey-green, with reddish margin; marginal teeth 3 - 4 mm, pungent, brownish-red, 15 - 25 mm apart; exudate drying reddish-brown; **Inf** 1 m, with 8 **Br**, lower **Br** sometimes rebranched; racemes broadly cylindrical, subcapitate, to 8 × 7 cm, dense, buds spreading horizontally to slightly nutant; **Bra** ovate-acuminate, ± 8 × 2 - 3 mm; **Ped** 15 mm; **Fl** yellow, orange or scarlet, 25 - 30 mm, slightly clavate, base shortly attenuate, 6 mm ∅ across **Ov**, slightly narrowed above, then enlarging to mouth; **Tep** free for 13 - 15 mm.

A. elgonica Bullock (BMI 1932: 503, 1932). **T:** Kenya, Trans-Nzoia Distr. (*Lugard* 299 [K]). – **D:** Kenya; rocky slopes, often in grass, 1980 - 2380 m. **I:** Reynolds (1966: 360-361).

[12] Caulescent, branching to form dense clumps sometimes to 2 m ∅; stem erect or decumbent, to 1 m or more; **L** 20 - 24, densely rosulate, triangular-attenuate, to 40 × 9 cm, dark green, often reddish tinged, surface smooth; marginal teeth 8 - 9 mm, pungent, 10 - 15 mm apart; exudate drying yellow; **Inf** 50 - 70 cm, simple or with 3 - 4 **Br**; racemes cylindrical-conical, 18 × 8 - 9 cm, dense; **Bra** ovate-acute, 5 × 4 mm; **Ped** 20 - 25 mm; **Fl** orange-scarlet, yellowish at mouth, 40 mm, base shortly attenuate, 7 - 8 mm ∅ across **Ov**, narrowed slightly above, then enlarging to mouth; **Tep** free for 15 mm. – *Cytology:* 2n = 28 (Brandham 1971).

Mount Elgon straddles the Kenya-Uganda border, but *A. elgonica* is known only from the Kenyan side of the mountain.

A. ellenbeckii A. Berger (BJS 36: 59, 1905). **T:** Somalia (*Ellenbeck* 2340 [B]). – **D:** Ethiopia,

Kenya, Somalia; sandy soils in deciduous bushland, 700 - 1400 m. **I:** FPA 51: 5. 2012, 1990, as *A. dumetorum*.

Incl. *Aloe dumetorum* B. Mathew & Brandham (1977).

[6] Acaulescent, suckering to form clumps; **L** 6 - 12, rosulate, linear-lanceolate, 15 - 27 × 1 - 2.5 (-3) cm, yellowish-green with many white spots sometimes in irregular transverse bands, with cartilaginous margin, surface smooth; marginal teeth 0.5 - 1 mm, 2 - 6 mm apart; exudate pale yellow; **Inf** 50 - 75 cm, simple; racemes conical-cylindrical, 5 - 8 × 6 cm, lax; **Bra** linear-lanceolate, 5 - 10 × 2 mm; **Ped** 7 - 10 mm; **Fl** orange-red, becoming yellow at mouth, 24 - 30 mm, base truncate, 7 - 10 mm ∅ across **Ov**, abruptly narrowed to ± 5 mm above, then enlarging to mouth; **Tep** free for 8 - 10 mm.

Lavranos (1995) treats this as a "species of uncertain position", preferring to reserve judgement on its relationship with *A. dumetorum* until more material is found at or near the type locality. It should also be noted that Kenyan material currently in cultivation as *A. dumetorum* does not agree closely with the protologue of that name and could prove to be an as yet undescribed taxon.

A. eminens Reynolds & P. R. O. Bally (JSAB 24: 187-189, t. 30-32, 1958). **T:** Somalia, Erigavo Distr. (*Reynolds* 8435 [PRE, EA, K]). – **D:** Somalia; forested slopes, 1550 - 1830 m. **I:** Reynolds (1966: 386-388).

[17] Caulescent, irregularly branched; stem erect, to 15 m × 15 cm ∅ at base, more slender above; **L** 16 - 20, rosulate, triangular-obtuse, deeply canaliculate and recurved, 40 - 45 × 5 cm, dull green, with narrow white cartilaginous margin, surface smooth; marginal teeth 2 - 3 mm, blunt, white, 3 - 5 mm apart, becoming smaller to obsolescent above except for a few short crowded teeth in apical 4 - 5 cm; exudate yellow; **Inf** 50 - 60 cm, with 3 - 5 **Br**; racemes cylindrical-acuminate, 16 × 8 - 9 cm, subdense; **Bra** deltoid, 6 × 3 mm; **Ped** 10 mm; **Fl** red, 40 mm, base rounded, 12 mm ∅ across **Ov**, slightly narrowing towards mouth; **Tep** free for 32 mm.

A. enotata L. C. Leach (JSAB 38(3): 187-193, ills., 1972). **T:** Zambia, Abercorn Distr. (*Richards* s.n. in *Leach* 14796 [SRGH, K]). – **D:** Zambia; vertical cliffs.

[13] Caulescent, simple or suckering at the base; stem pendulous, to 60 × 3 cm; **L** ± 12, rosulate, strongly falcate, pointing downwards, 45 - 60 × 5 - 6 cm, pale green suffused with pink, with narrow whitish margin; marginal teeth small, pungent, usually uncinate, whitish, 20 - 40 mm apart; **Inf** ± 85 cm, pendulous, with 6 **Br**; racemes cylindrical-acuminate, 14 - 20 × ± 7.5 cm, lax; **Bra** ovate-acute, ± 4 × 3 mm; **Ped** to 17 mm; **Fl** dull red, obscurely purplish-striped, with a faint bloom, 25 - 28 mm, base shortly attenuate, 8 mm ∅ across **Ov**, nar-

rowed to ± 5.5 mm above, then enlarging to ± 6.5 mm at mouth; **Tep** free for 12 - 16 mm; **St** exserted 0 - 1 mm.

A. eremophila Lavranos (JSAB 31(1): 71-74, ills., 1965). **T:** Yemen, Hadhramaut (*Rauh & Lavranos* 13348 [HEID, K, PRE]). – **D:** Yemen (Hadhramaut); limestone plateau, 1375 - 1950 m. **I:** Reynolds (1966: 137-138).

[5] Acaulescent or with short procumbent stem, usually simple, sometimes forming groups of up to 6 **Ros**; **L** 10 - 22, rosulate, deltoid-acute, 35 × 10 cm, grey-green tinged brownish; marginal teeth 4 mm, pungent, dark brown, 10 - 15 mm apart, smaller and more crowded near base; **Inf** 40 - 75 cm, with 3 - 5 **Br**; racemes conical-cylindrical, to 22 × 5 cm, dense; **Bra** deltoid, 12 × 5 mm; **Ped** 7 - 8 mm; **Fl** scarlet, 30 mm, base shortly attenuate, 8 mm ∅ across **Ov**, slightly narrowed above, not enlarging to mouth; **Tep** free for 14 mm.

A. erensii Christian (FPSA 20: t. 797 + text, 1940). **T:** Kenya, Rift Valley Prov. (*Pole Evans & Erens* 1587 [PRE]). – **D:** Kenya, Sudan; rocky slopes and cliffs, 670 - 800 m. **I:** Reynolds (1966: 59-60).

[5] Acaulescent, simple or with a few suckers; **L** ± 16 - 19, densely rosulate, ovate-acute, 21 × 8 - 9 cm, grey-green with many elongated white spots in transverse bands, with narrow whitish translucent margin; marginal teeth 1 - 1.5 mm, whitish, 4 - 6 mm apart; **Inf** to 50 cm, with 2 - 7 **Br**; racemes cylindrical, laterals with subsecund **Fl**, to 21 cm, lax; **Bra** ovate-acuminate, 5 - 6 × 3 mm; **Ped** 8 - 9 mm; **Fl** pink with a bloom, 29 mm, base rounded, 10 mm ∅ across **Ov**, narrowed to 8 mm above, then enlarging to 9 mm at mouth; **Tep** free for 13 mm. – *Cytology:* 2n = 14 (Brandham 1971).

A. ericetorum Bosser (Adansonia, n.s., 8(4): 508-509, ills., 1968). **T:** Madagascar (*Bosser* 17143 [P]). – **D:** C Madagascar; sandy heath and moorland.

[3] Acaulescent, simple; **L** 15 - 20, rosulate, lanceolate, 18 - 19 × 3.5 - 4.5 cm, glaucous, with narrow cartilaginous margin; marginal teeth 1.5 - 2.5 mm, pale yellow, 0.7 - 1 mm apart; **Inf** 50 - 70 cm, with 1 - 2 **Br**; racemes cylindrical and lax to densely capitate, to 15 cm; **Bra** ovate-acute, 8 - 10 mm; **Ped** 15 - 30 mm; **Fl** yellow, 35 - 37 mm, base rounded, very slightly narrowed above **Ov** and slightly enlarging to mouth; **Tep** free for ± 18 - 24 mm.

A. erythrophylla Bosser (Adansonia, n.s., 8(4): 508-511, ill., 1968). **T:** Madagascar (*Bosser* 19583 [P]). – **D:** C Madagascar; montane gneiss and quartzite.

[2] Acaulescent or shortly caulescent, simple; **L** 6 - 8, rosulate, lanceolate-linear, acuminate, 10 - 17 × 2 - 4 cm, dark brownish-red or in the centre green tinged red; marginal teeth 1 - 1.5 mm, red, 4 - 6 mm

apart; **Inf** 15 - 40 cm, simple or with 1 **Br**; raceme pyramidal, 6 - 20 cm, lax, with 20 **Fl**; **Bra** ovate-subacute, 6 - 9 mm; **Ped** 7 - 12 mm; **Fl** red, 20 - 25 mm, base rounded, slightly narrowed above **Ov**; **Tep** free for 7 - 8 mm; **St** not exserted.

A. esculenta L. C. Leach (JSAB 37(4): 249-259, ills., 1971). **T:** Angola, Huila (*Leach & Cannell* 13818 [SRGH, BM, LISC, PRE]). − **D:** Angola, Botswana, Namibia, Zambia; sandy flats, ± 1000 m.

[7] Acaulescent or shortly caulescent, dividing and sometimes suckering at base, forming dense clumps; stem often decumbent, to 40 cm; **L** ± 20, rosulate, triangular-lanceolate, acute, 40 - 60 × 7 - 10.5 cm, grey or grey-green to pinkish-brown, with many white spots in irregular ± transverse bands, lower face usually with a row of sharp blackish-brown prickles with white bases on median line for ½ to ⅔ of **L** length, often sharply keeled; marginal teeth 3 - 5 mm, shiny brown, 10 - 20 mm apart; **Inf** 1.5 (-2.2) m, with 3 - 5 **Br**, lower **Br** sometimes rebranched; racemes cylindrical-acuminate, 30 - 40 × ± 6 cm, sub-dense; **Bra** ovate-acute, 20 - 27 × 10 - 11 mm; **Ped** 5 - 6 mm; **Fl** deep pink with pale yellow border on lobes, 28 - 30 mm, subclavate, base rounded, ± 6 mm ∅ across **Ov**, enlarging to ± 8 mm above middle; **Tep** free for 15 - 18 mm.

According to Leach (1971) there is a strong possibility that this could be *A. baumii* Engler & Gilg, at present treated as a taxon of uncertain status.

A. eumassawana S. Carter & al. (KB 51(4): 776, 1996). **T:** Eritrea (*Ash* 1816 [K]). − **D:** Eritrea; coastal bushland with *Acacia* and *Euphorbia cactus* on sandy soil, near sea level. **I:** Reynolds (1966: 154, as *A. massawana*).

[7] Acaulescent, or nearly so, suckering to form large groups; **L** ± 16, rosulate, lanceolate, 45 - 50 × 7 - 18 cm, dull grey-green, sometimes with few pale spots, surface smooth; marginal teeth 3 mm, tip reddish-brown, 1.5 - 2 mm apart; **Inf** 1.2 - 1.5 m, with 1 - 2 **Br**, all parts minutely papillose-puberulent when young, later glabrescent; racemes cylindrical-conical, (15-) 20 - 25 cm, lax; **Bra** ovate-triangular, tip acute to acuminate, 6.5 - 7 × 2.5 - 4 mm; **Ped** 3 - 4.5 (-7) mm; **Fl** pale scarlet or orange, lobes with paler yellowish margins, very minutely papillose, (18-) 20 - 21 mm, base rounded, uniformly ± 7 mm ∅ (when pressed); **Tep** free for ± 12 mm.

Reynolds included this taxon in his *A. massawana*, forming the epithet from the old locality name Massawa (now Mits'iwa) in Eritrea, but selecting a specimen from Tanzania as type. According to Carter & al. the Eritrean and Tanzanian plants are not conspecific. As the name *A. massawana* must be used for the Tanzanian material, the name of the newly described Eritrean species indicates that it is the true Massawa plant.

A. excelsa A. Berger (Notizbl. Königl. Bot. Gart.

Berlin 4: 247, 1906). **T** [lecto]: Zimbabwe, Bulawayo Distr. (*Eyles* 1240 [PRE]). − **Lit:** Leach (1977: 386-389). **D:** Moçambique, Malawi, Zambia, Zimbabwe, RSA.

A. excelsa var. **breviflora** L. C. Leach (Kirkia 10 (2): 387-389, ill., 1977). **T:** Moçambique, Zambesia Distr. (*Royle* s.n. in Leach 14111 [SRGH, PRE]). − **D:** Malawi, Moçambique; fire-protected rocky outcrops, 230 - 615 m.

[9] Differs from var. *excelsa*: **L** narrower, both faces without prickles; **Inf** with racemes narrower, usually less dense; **Fl** ± 20 mm.

A. excelsa var. **excelsa** − **D:** Malawi, Moçambique, Zambia, Zimbabwe, RSA (Northern Prov.); rocky hills and plains in open bush, 450 - 1525 m. **I:** Reynolds (1966: 315-316).

[9] Caulescent, simple; stem erect, to 4 m, with dried **L** remains; **L** ± 30, densely rosulate, oldest becoming recurved, triangular-attenuate, to 80 × 15 cm, dull green, lower face usually tuberculate-aculeate; marginal teeth 5 - 6 mm, pungent, reddish-brown, 15 - 20 mm apart; **Inf** 80 - 100 cm, with up to 14 **Br**; racemes cylindrical, 15 - 25 cm, very dense; **Bra** reflexed, 4 - 6 × 4 - 6 mm; **Ped** ± 1 mm; **Fl** red or orange, 30 mm, slightly ventricose, base rounded, 5 mm ∅ across **Ov**, enlarging to 7 mm in middle, then narrowing to mouth; **Tep** free for 22 - 23 mm; **St** exserted 8 - 10 mm. − *Cytology:* 2n = 14 (Brandham 1971).

A. falcata Baker (JLSB 18: 181, 1880). **T:** RSA, Northern Cape (*Zeyher* 1678 [K]). − **D:** RSA (Northern Cape, Western Cape); arid plains, 600 - 915 m. **I:** Reynolds (1950: 316-317).

[5] Acaulescent or shortly caulescent, usually in groups, with **Ros** lying almost on their sides; **L** ± 20, densely rosulate, lanceolate-acuminate, usually falcate, ± 30 × 7 cm, tip usually a **Sp**, grey-green to glaucous, surface rough, lower face with slight keel near tip with ± 6 prickles; marginal teeth 5 mm, horny, reddish-brown, 10 mm apart; **Inf** to 60 cm, with up to 10 **Br**, lower **Br** sometimes rebranched; racemes usually cylindrical-acuminate, to 30 (-40) × 7 cm, shorter and denser to longer and more lax; **Bra** deltoid-acuminate, 18 mm or slightly longer; **Ped** ± 18 mm; **Fl** dull reddish to pale scarlet, rarely yellow, 40 mm, base shortly attenuate, 7 mm ∅ across **Ov**, scarcely narrowed above; **Tep** free for 10 mm; **St** exserted 8 mm.

A. ferox Miller (Gard. Dict., Ed. 8, no. 22, 1768). **T:** not typified. − **Lit:** Wyk & Smith (1996); Viljoen & al. (1996). **D:** Lesotho, RSA (Western Cape, Eastern Cape, Free State, KwaZulu-Natal); arid bushland. **I:** Reynolds (1950: 460-465). **Fig. XIV.a**

≡ *Pachidendron ferox* (Miller) Haworth (1821); **incl.** *Aloe perfoliata* var. ε Linné (1753); **incl.** *Aloe* *perfoliata* var. γ Linné (1753); **incl.** *Aloe socotorina* Masson (1773); **incl.** *Aloe perfoliata* Thunberg (1785); **incl.** *Aloe perfoliata* var. *ferox* Aiton (1789); **incl.** *Aloe perfoliata* var. ζ Willdenow (1799); **incl.** *Aloe muricata* Haworth (1804); **incl.** *Aloe supralaevis* Haworth (1804) ≡ *Pachidendron supralaeve* (Haworth) Haworth (1821); **incl.** *Aloe pseudoferox* Salm-Dyck (1817) ≡ *Pachidendron pseudoferox* (Salm-Dyck) Haworth (1821); **incl.** *Aloe subferox* Sprengel (1826) ≡ *Aloe ferox* var. *subferox* (Sprengel) Baker (1880); **incl.** *Aloe ferox* var. *incurva* Baker (1880); **incl.** *Aloe ferox* var. *hanburyi* Baker (1896); **incl.** *Aloe galpinii* Baker (1901) ≡ *Aloe ferox* var. *galpinii* (Baker) Reynolds (1937); **incl.** *Aloe candelabrum* A. Berger (1906); **incl.** *Aloe ferox* var. *erythrocarpa* A. Berger (1908).

[9] Caulescent, simple; stem to 3 (-5) m, with dead **L** remains; **L** 50 - 60, densely rosulate, lanceolate-ensiform, to 100 × 15 cm, dull green, sometimes reddish tinged, surface glabrous or with few to many irregular prickles; marginal teeth ± 6 mm, reddish to reddish-brown, 10 - 20 mm apart; **Inf** with 5 - 8 **Br**; racemes cylindrical, slightly acuminate, 50 - 80 × 9 - 12 cm, narrowing to ± 6 cm at tip, very dense, buds horizontal; **Bra** ovate-acute, 8 - 10 × 3 - 5 mm; **Ped** 4 - 5 mm; **Fl** scarlet, sometimes orange, 33 mm, slightly ventricose-clavate, base rounded, enlarging above **Ov**, slightly narrowed at mouth; **Tep** free for 22 mm; **St** exserted 20 - 25 mm. – *Cytology:* 2n = 14 (Fernandes 1930), Vosa (1982: as *A. candelabrum*).

A very variable species, but Reynolds (1950) did not accept the varieties that have been described. According to Reynolds (1950) *A. horrida* Haworth possibly belongs here. Natural hybrids with other species have been reported by the same author.

A. fibrosa Lavranos & L. E. Newton (CSJA 48(6): 273-275, ills., 1976). **T:** Kenya, Eastern Prov. (*Archer* 410 [EA, K]). – **D:** Kenya, Tanzania; rocky slopes with dense scrub, 1500 - 2000 m.

[16] Caulescent, branching at base; stem erect to 2 m × 3 cm, when longer supported by surrounding vegetation or decumbent to 2.5 m, with dead **L** remains; **L** scattered along stem, lanceolate-acute, 30 - 35 × 3 - 6 cm, bright green, tinged brownish in exposed situations, with few spots on young shoots; marginal teeth 3 - 4 mm, firm, brown-tipped, 15 - 17 mm apart; **L** with internal fibres, esp. at base and in sheath; **Inf** to 1 m, simple or with 1 - 2 **Br**; racemes conical-cylindrical, 10 - 20 × 10 cm, dense; **Bra** ovate, 12 - 18 × 7 mm, whitish, imbricate in bud stage; **Ped** 20 - 25 mm; **Fl** orange-red, yellowish at mouth, 30 - 35 mm, base truncate, 8 - 9 mm ∅ across **Ov**, narrowed to 5 mm above, enlarging to 9 - 10 mm at mouth; **Tep** free for ± 10 - 11 mm; **St** exserted ± 1 mm. – *Cytology:* 2n = 14 (Cutler & al. 1980).

A. fievetii Reynolds (JSAB 31(4): 279-281, ills.,

1965). **T:** Madagascar, Fianarantsoa (*Rauh* 10332 [PRE, K, P, TAN]). – **D:** Madagascar; grass soil pockets on granite rocks, 1200 m. **I:** Reynolds (1966: 464-465).

[4] Acaulescent or shortly caulescent, usually simple; **L** 12 - 16, densely rosulate, lanceolate-attenuate, tip slightly twisted and with 3 - 4 small teeth, 35 × 5 - 6 cm, green with slight reddish tinge, with pinkish-red margin; marginal teeth 2 - 3 mm, pinkish-red, 7 - 10 mm apart; **Inf** to 50 cm, simple or with 1 **Br**; racemes corymbose-capitate, ± 9 × 4 cm, dense; **Bra** ovate-acute, 5 × 5 mm, reddish-brown with membranous edge; **Ped** 30 mm; **Fl** orange, 27 - 30 mm, slightly clavate, base rounded, 5 - 6 mm ∅ across **Ov**, enlarging above; **Tep** free for 14 mm; **St** exserted 0 - 1 mm.

A. fimbrialis S. Carter (KB 51(4): 779-781, ills., 1996). **T:** Zambia, Barotseland (*Fanshawe* 8938 [K, SRGH]). – **Lit:** Newton (1998b). **D:** Zambia, Tanzania; Miombo woodland, often on ant hills, 900 - 1250 m.

[2] Acaulescent, dividing dichotomously and suckering at base; **R** fibrous plus 7 - 10 thickened **R** to 1 cm ∅; **L** 6 - 10, rosulate, spreading, bases expanded to form a bulb to 5 cm ∅ and 3.5 cm tall, linear-lanceolate with tips turned to one side, to 40 × 2 - 3 cm, glossy mid-green, lower face slightly paler green, both surfaces with numerous minute white prickles in longitudinal rows overlying larger vascular bundles; marginal teeth 1 - 1.5 mm, soft, white, densely crowded, 0.5 - 1 mm apart; exudate bright yellow; **Inf** 45 - 90 cm, simple; raceme cylindrical, 20 - 30 × 6 - 8 cm, lax to sub-dense; **Bra** linear-lanceolate, attenuate, 16 - 25 × 3 - 6 mm, imbricate in bud stage; **Ped** 8 - 10 mm; **Fl** coral pink, lobes with yellow margins, 30 - 40 mm, base flat, 6 mm ∅ across **Ov**, narrowed to 5 mm above, then enlarging to mouth; **Tep** free for 10 - 13 mm; **St** exserted 2 mm.

A. fleurentiniorum Lavranos & L. E. Newton (CSJA 49(3): 113-114, ills., 1977). **T:** Yemen, Sana'a Prov. (*Lavranos* 11386 [E]). – **D:** Saudi Arabia, Yemen; rocky slopes, 1500 - 2350 m. **I:** Collenette (1999: 22).

[3] Acaulescent, simple; **L** 8 - 12, rosulate, lanceolate, 20 - 30 × 6 - 7 cm, dark green, tinged brownish in exposed situations, surface rough; marginal teeth 1 - 1.5 mm, firm, whitish; **Inf** 35 - 40 cm, with 3 - 6 **Br**; racemes cylindrical, lax, with 10 - 20 **Fl**; **Bra** acute, 6 - 8 mm; **Ped** 11 mm; **Fl** bright red, yellow at mouth, 31 - 33 mm, base very shortly attenuate, 8 mm ∅ across **Ov**, scarcely narrowed above; **Tep** free for 9 - 10 mm. – *Cytology:* 2n = 14 (from protologue).

A. fleuretteana Rauh & Gerold (KuaS 51(5): 121-123, ills., 2000). **T:** Madagascar (*Gerold* s.n. in *BG*

Heidelberg 75706 [HEID]). – **D**: Madagascar; bushland on granite.

[14] Caulescent, suckering at base forming dense groups; stems to 5 cm; **L** ± 10, loosely rosulate, lanceolate-attenuate, 13 × 8 cm, green with fine white striations and white flecks; marginal teeth pungent, deltoid, 2 mm; **Inf** to 35 cm, erect, simple; racemes cylindrical with 10 - 12 **Fl**, lax; **Bra** 10 - 15 mm; **Fl** shiny cinnabar-red, whitish with green mid-stripe above, 28 mm, base flat, scarcely constricted above **Ov**, then widening to 7 mm ∅; **Tep** free for ± 5 mm; **St** only slightly exserted.

Similar to *A. antandroi* according to the protologue.

A. flexilifolia Christian (JSAB 8(2): 167-169, ills., 1942). **T**: Tanzania, Lushoto Distr. (*Boscawen* s.n. in *Christian* 897 [SRGH, EA, K, PRE]). – **D**: Tanzania; rocky slopes and cliff faces, 1000 - 1220 m. **I**: Reynolds (1966: 363-364).

[13,15] Caulescent, much branched at base, forming groups to 2 m ∅; stem in deep soil, to 1 m × 6 - 7 cm, erect or ascending; in shallow soil and overhanging rock faces to 2 m × 5 cm, sprawling and pendulous; **L** rosulate and persistent for 30 cm below, ensiform-attenuate, becoming recurved or falcately decurved when stems pendulous, 50 × 6 - 7 cm, glaucous-green tinged bluish, with narrow pale cartilaginous margin; marginal teeth 1 - 2 mm, brownish, 10 - 20 mm apart; exudate drying brownish; **Inf** 50 - 65 cm, usually suboblique or basally curving downwards with racemes erect, with 6 - 8 **Br**; racemes cylindrical, to 12 × 7 - 8 cm, subdense; **Bra** ovate-deltoid, 5 - 6 × 3 mm; **Ped** 12 - 14 mm; **Fl** scarlet or brownish-red, paler at mouth, 33 - 35 mm, base rounded, 9 mm ∅ across **Ov**, slightly narrowed above, slightly enlarging towards mouth; **Tep** free for ± 11 - 12 mm.

The specific epithet was chosen because the type plant developed transverse folds in the leaves when they became deflexed, but this is not a constant character.

A. forbesii Balfour *fil.* (in Forbes & Ogilvie-Grant, Nat. Hist. Socotra, 511-512, t. 26B, 1903). **T**: [icono]: l.c. t. 26B. – **D**: Socotra. **I**: Reynolds (1966: 193-195).

[5] Acaulescent or shortly caulescent, simple or in small groups; **L** 16 - 20, densely rosulate, lanceolate-attenuate, 25 × 6 - 7 cm, dull green; marginal teeth 0.5 - 1 mm, pale, 4 - 8 mm apart; **Inf** 60 - 80 cm, with 5 or more **Br**; racemes cylindrical, to 10 - 25 × 5 cm, lax; **Bra** 3 - 4 × 1.5 mm; **Ped** 10 - 12 mm; **Fl** pale scarlet in lower ⅓, becoming yellowish above, 22 - 24 mm, base shortly attenuate, narrowed above **Ov**, enlarging towards mouth; **Tep** free for 6 - 7 mm. – *Cytology:* 2n = 14 (Brandham 1971).

A. fosteri Pillans (South Afr. Gard. 23: 140, 1933). **T**: RSA, Mpumalanga (*Pillans* s.n. [BOL 20447]).

– **D**: RSA (Mpumalanga); subtropical bushland, 610 - 910 m. **I**: Reynolds (1950: 253-254).

[5] Acaulescent or with stem to 20 cm, usually suckering to form small groups, sometimes simple; **L** ± 20, rosulate, lanceolate-attenuate, ± 40 × 8 - 10 cm, upper face pale dull green, obscurely lineate and with obscure dull white oval spots in ± transverse bands, lower face usually unspotted; marginal teeth 3 - 4 mm, pungent, slightly deflexed, 10 - 15 mm apart; exudate drying purplish; **Inf** to 2 m, with ± 8 **Br**, lower **Br** rebranched; racemes narrowly cylindrical-acuminate, to 40 × 6 - 7 cm, lax; **Bra** lanceolate-acuminate, usually a little longer than **Ped**; **Ped** 6 - 9 mm; **Fl** pale dull brick-red, somewhat pruinose, 30 mm, base truncate, 9 mm ∅ across **Ov**, abruptly narrowed to 5.5 mm above, enlarging towards mouth; **Tep** free for 8 mm. – *Cytology:* 2n = 14 (Müller 1945).

A. fouriei D. S. Hardy & Glen (FPA 49(3-4): t. 1941 + text, 1987). **T**: RSA, Mpumalanga (*Fourie* 3070 [PRE]). – **D**: RSA (Mpumalanga); steep grassy slopes on dolomite. **I**: Wyk & Smith (1996: 263).

[10] Caulescent, simple or forming small groups; stem ± 15 cm, with dead **L** bases on apical 1 cm; **L** distichous, triangular, 27.5 - 35 × 1 - 2.5 cm, grass-green; marginal teeth ± 0.4 mm, 2.5 mm apart; **Inf** ± 40 cm, simple; raceme capitate, sub-dense, with ± 20 **Fl**; **Bra** ± 16 × 8 mm; **Ped** 23 - 45 mm; **Fl** orange, becoming green towards tip, 35 - 40 mm, base rounded, 11 - 13 mm ∅ across **Ov**, narrowing above to 6 - 8 mm at mouth; **Tep** free to base; **St** not exserted.

A. fragilis Lavranos & Röösli (CSJA 66(1): 4-5, ills., 1994). **T**: Madagascar, Diego Suarez (*Lavranos* 28737 [ZSS, P]). – **D**: N Madagascar; crystaline basement rocks by sea-shore. **Fig. XIII.b, XIII.c**

[11] Acaulescent or shortly caulescent, caespitose forming large groups; **Br** procumbent; **L** rosulate, deltoid-acute, 3 - 5 (- 10) × 1.5 - 2 cm, glossy, dark glaucous-green, with many whitish-green spots, with narrow greenish-white margin; marginal teeth 1 - 1.5 mm, firm, greenish-white, often coalescent; **Inf** 20 - 60 cm, usually simple, rarely with 1 **Br**; racemes cylindrical, 10 - 15 cm, lax, with up to 25 **Fl**; **Bra** lanceolate, ± 2 × 1 mm; **Ped** 8 - 9 mm; **Fl** with carmine-red base, becoming creamy-white with green midrib towards mouth, 20 - 25 mm, base shortly attenuate, 4 mm ∅ across **Ov**, very slightly narrowed above, then enlarging slightly above and narrowing at mouth; **St** exserted 0 - 1 mm.

A. framesii L. Bolus (South Afr. Gard. 1933: 140 (June), 1933). **T**: RSA, Northern Cape (*Frames* s.n. [BOL 19186]). – **D**: RSA (Northern Cape, Western Cape); coastal sand flats. **I**: Reynolds (1950: 403-404). **Fig. XIV.b**

Incl. *Aloe amoena* Pillans (1933) ≡ *Aloe microstigma* ssp. *framesii* (L. Bolus) Glen & D. S. Hardy (2000).

[12] Caulescent, branching freely forming dense groups of up to 20 **Ros** and to 3 m ∅; stem decumbent; **L** densely rosulate, lanceolate-attenuate, 30 - 35 × 7 - 8 cm, dull grey-green to slightly bluish-green, with or without scattered white spots; marginal teeth ± 3 mm, pungent, reddish-brown, 10 mm apart; **Inf** to 70 cm, sometimes simple, mostly with 2 - 3 **Br**; racemes conical to cylindrical-acuminate, to 25 × 10 cm, sub-dense; **Bra** ovate-acute, 20 × 9 mm, reddish; **Ped** 25 - 30 mm; **Fl** dull scarlet, sometimes greenish-tipped, 35 mm, base rounded, not narrowed above **Ov**; **Tep** free to base. − *Cytology:* 2n = 14 (Riley 1959).

A. francombei L. E. Newton (Brit. Cact. Succ. J. 12(2): 54-55, ills., 1994). **T:** Kenya, Rift Valley Prov. (*Newton & al. 4130* [K, EA]). − **D:** Kenya; shade of shrubs on rocky slopes, 1520 - 1650 m. **Fig. XIII.d**

[15] Caulescent, branching at base; stem erect or ascending, to 40 cm; **L** 15 - 20, rosulate, triangular, to 31 × 7 cm, dull green with few or many white elliptic spots, mostly on lower face, becoming red-tinged in exposed situations, surface rough; marginal teeth to 4 mm, firm, brown-tipped, 7 - 9 mm apart; exudate yellow; **Inf** to 60 cm, with up to 8 **Br**; racemes cylindrical, 6 - 20 cm, sub-dense; **Bra** triangular, 10 - 12 × 3 mm, imbricate in bud stage; **Ped** 8 - 13 mm; **Fl** pale pink with minutely pustulate surface, lobes with white margins, 25 mm, base rounded, 5.5 mm ∅ across **Ov**, enlarging to 7 mm above, then narrowing to 6 mm at mouth, lobes spreading to 8 - 9 mm; **Tep** free for 12 - 14 mm.

A. friisii Sebsebe & M. G. Gilbert (KB 55(3): 683-686, ills., 2000). **T:** Ethiopia, Gamo Gofa Region (*Friis & al. 8931* [ETH, C, K]). − **D:** Ethiopia; under thickets on a rocky slope in deciduous woodland, 600 - 1600 m.

[13] Caulescent; stem erect or sprawling, to 20 cm, 2 - 4 cm ∅, simple or with up to 2 **Br**; **L** lax, narrowly elliptic, 25 - 35 × 3.5 - 5 cm, pale green with sparse whitish, sometimes obscure, spots; marginal teeth 1 - 2 mm, whitish, sometimes with brownish tips, 2 - 8 mm apart; **Inf** 50 - 75 cm, ascending, with 8 - 13 **Br**, lower **Br** rebranched; racemes cylindrical, 3 - 14 cm, lax; **Bra** ovate-acuminate, 2 - 5 × 1 - 3 mm, scarious with usually 3 reddish-brown veins; **Ped** 8 - 12 mm; **Fl** yellow with darker longitudinal veins, 22 - 25 × 7 - 10 mm, base rounded, slightly narrowed above **Ov**; **Tep** free for 7.3 - 8.3 mm; **St** not exserted.

A. fulleri Lavranos (CSJA 39(4): 125-127, ills., 1967). **T:** Yemen (*Fuller s.n.* in *Lavranos 4206* [PRE]). − **D:** Yemen; *Acacia* scrub on sandy soil, ± 900 m.

[3] Caulescent, simple; stem short; **L** ± 12, rosulate, ensiform, 45 × 8 cm, glaucous green, at times tinged yellowish, with reddish-brown cartilaginous margin; marginal teeth 1 - 2 mm, pungent, brown, 12 - 25 mm apart; **Inf** 40 - 70 cm, with 1 - 2 **Br**; racemes cylindrical-acuminate, 30 - 35 cm, lax; **Bra** 12 - 15 mm, pink at base; **Ped** 5 - 6 mm; **Fl** coral-red, tips of lobes pinkish-cream, outside densely but minutely papillate, esp. on lobes, 35 mm, base attenuate, 7 mm ∅ across **Ov**, narrowed to 5 mm above, then enlarging to mouth; **Tep** free for 15 mm.

Yellow-flowered plants have also been seen at the type locality.

A. gariepensis Pillans (South Afr. Gard. 23: 213, 1933). **T:** Namibia, Warmbad Div. (*Pillans 6557* [PRE]). − **D:** Namibia, N RSA (Northern Cape); rocky slopes. **I:** Reynolds (1950: 401-402). **Fig. XIII.e**

Incl. *Aloe gariusiana* Dinter (1928) (*nom. inval.*, Art. 32.1c).

[4] Acaulescent or shortly caulescent, usually simple, sometimes branching to form small groups; stem erect or procumbent, to 1 m when simple, with dead **L** remains; **L** densely rosulate, lanceolate-attenuate, 30 - 40 × 5 - 8 cm, dull green to reddish-brown, somewhat lineate, with or without white spots, with horny margin; marginal teeth 2 - 3 mm, pungent, reddish-brown, ± 10 mm apart; **Inf** 80 - 120 cm, simple; racemes narrowly cylindrical-acuminate, 35 - 50 × 7 cm, dense; **Bra** lanceolate, ± 25 × 8 mm, imbricate in bud stage; **Ped** 15 - 20 mm; **Fl** mostly yellow or greenish-yellow, sometimes reddish in bud, 23 - 27 mm, base rounded, enlarging towards mouth; **Tep** free to base.

A very variable species. Natural hybrids with other species have been reported (Reynolds 1950).

A. gerstneri Reynolds (JSAB 3: 133, 1937). **T:** RSA, KwaZulu-Natal (*Reynolds 2320* [PRE, BOL]). − **D:** RSA (KwaZulu-Natal); rocky slopes. **I:** Reynolds (1950: 455-456, t. 51).

[4,5] Acaulescent or shortly caulescent, usually simple; **L** 20 - 30, densely rosulate, lanceolate-ensiform, to 60 × 9 cm, dull grey-green, surface smooth, lower face sometimes with a few prickles in median line near tip; marginal teeth 4 mm, pungent, pale brown with white base, 10 - 15 mm apart; **Inf** to 1.3 m, simple in young plants, later with 1 - 2 **Br**; racemes cylindrical, slightly acuminate, to 36 × 6 - 7 cm, dense; **Bra** lanceolate, 18 × 5 mm; **Ped** 5 mm; **Fl** reddish-orange, 30 mm, slightly ventricose, base very shortly attenuate; **Tep** free for 15 - 17 mm; **St** exserted 13 mm. − *Cytology:* 2n = 14 (Müller 1945).

A. gilbertii T. Reynolds *ex* Sebsebe & Brandham (KB 47(3): 509, 512, ills., 1992). **T:** Ethiopia, Sidamo Region (*Gilbert & al. 9307* [K, ETH, UPS]). − **Lit:** Gilbert & Sebsebe (1997). **D:** Ethiopia.

A. gilbertii ssp. **gilbertii** – **D:** Ethiopia; woodland, hedgerows and sandy mountain slopes, 1300 - 1800 (-1900) m. **Fig. XIV.c**

[7] Acaulescent or usually shortly caulescent, usually suckering to form dense sprawling clumps, rarely simple; stem ascending, to 1.5 m; **L** densely rosulate, ascending with tips gently recurved, 40 - 60 × 9 - 11 cm, dark green or grey-green, often tinged brown or mauve, surface smooth; marginal teeth 3 - 5 mm, brown-tipped, 0.7 - 1.4 mm apart; **Inf** to 1.2 m, with 5 - 20 **Br**, lower **Br** sometimes rebranched; racemes cylindrical, 6 - 15 cm, lax; **Bra** ovate-acute, 4 - 6 × 2 - 3 mm; **Ped** 9 - 10 mm; **Fl** orange to red, 23 - 27 mm, cylindrical to somewhat clavate, 4.5 - 8 mm ⌀ when pressed; **Tep** free for 8 - 11 mm. – *Cytology:* 2n = 14 (from the protologue).

A. gilbertii ssp. **megalacanthoides** M. G. Gilbert & Sebsebe (KB 52(1): 151, 1997). **T:** Ethiopia, Gamo Gofa Region (*Gilbert & Phillips* 9135 [K, ETH, UPS]). – **D:** Ethiopia; 1200 - 1350 m.

[7] Differs from ssp. *gilbertii*: **L** spreading and deeply canaliculate, strongly recurved; **Fl** 27 - 28 mm.

A. gillettii S. Carter (KB 49(3): 417, 1994). **T:** Somalia, Bari Region (*Gillett* 23457 [K, EA, MOG]). – **D:** Somalia; bushland on rocky limestone slopes, 1340 - 1650 m.

[13] Caulescent, sparsely branched; stem decumbent, 0.5 - 1 cm ⌀; **L** laxly rosulate, linear-lanceolate, to ± 15 × 1.5 cm, dark grey-green, with many whitish spots; marginal teeth 1 mm, cartilaginous, 3 - 5 mm apart; exudate drying purplish; **Inf** to 25 cm, simple; raceme 3 - 5 cm, lax, with 8 - 12 **Fl**; **Bra** ovate-acuminate, 4 - 5 × 2 mm, white; **Ped** to ± 10 mm; **Fl** coral-red, green-tipped, 20 - 26 mm, 7 mm ⌀ across **Ov**; **Tep** free for 6 - 8 mm. – *Cytology:* 2n = 14 (Brandham & al. 1994).

A. glabrescens (Reynolds & P. R. O. Bally) S. Carter & Brandham (Bradleya 1: 23-24, ills., 1983). **T:** Somalia (*Reynolds* 8390 [PRE, EA, K]). – **D:** Somalia; arid plains and gypsum hills, 820 m. **I:** Reynolds (1966: 126, as *A. rigens* var. *glabrescens*).

≡ *Aloe rigens* var. *glabrescens* Reynolds & P. R. O. Bally (1958).

[7] Acaulescent or caulescent, suckering to form small to large groups; stem decumbent, to 50 cm; **L** 18 - 24, densely rosulate, triangular, 40 - 50 × 10 - 12 cm, grey-green tinged reddish; marginal teeth 3 mm, pungent, reddish-brown, 10 - 20 mm apart; **Inf** 75 - 100 cm, with 3 - 4 **Br**; racemes cylindrical-acuminate, 25 × 5 - 6 cm; **Bra** ovate-acute, 9 × 4 mm; **Ped** 5 mm; **Fl** strawberry-pink, paler to greenish at mouth, glabrous or minutely pubescent under a lens, 32 mm, base rounded, very slightly narrowed above **Ov**; **Tep** free for 9 mm. – *Cytology:* 2n = 14 (Carter & Brandham 1983).

A. glauca Miller (Gard. Dict., Ed. 8, no. 16, 1768). **T** [neo]: RSA, Western Cape (*Reynolds* 1967 [PRE]). – **D:** RSA.

A. glauca var. **glauca** – **D:** RSA (Northern Cape, Western Cape); dry scrub. **I:** Reynolds (1950: 197-202).

Incl. *Aloe perfoliata* var. κ Linné (1753); **incl.** *Aloe perfoliata* var. *glauca* Aiton (1789); **incl.** *Aloe rhodacantha* De Candolle (1799); **incl.** *Aloe glauca* var. *major* Haworth (1812); **incl.** *Aloe glauca* var. *minor* Haworth (1812); **incl.** *Aloe glauca* var. *elatior* Salm-Dyck (1817); **incl.** *Aloe glauca* var. *humilior* Salm-Dyck (1817).

[2] Shortly caulescent; **L** 30 - 40, densely rosulate, lanceolate, to 30 - 40 × 10 - 15 cm, very glaucous, obscurely lineate, lower face sparingly tuberculate-aculeate near tip; marginal teeth 4 - 5 mm, pungent, reddish-brown; **Inf** ± 60 - 80 cm, simple; raceme cylindrical-acuminate, 15 - 20 × 8 - 9 cm, dense; **Bra** ovate-deltoid, ± 30 × 10 mm, white, basally sub-amplexicaul; **Ped** 30 - 35 mm; **Fl** pink, greenish-tipped, ± 40 mm, base rounded, slightly narrowed above **Ov**, then enlarging and narrowing again to mouth; **Tep** free to base. – *Cytology:* 2n = 14 (Resende 1937).

Leaf characters are variable, and distinct variants are found in different localities. Natural hybrids with *A. arborescens* have been reported (Reynolds 1950).

A. glauca var. **spinosior** Haworth (Revis. Pl. Succ., 40, 1821). – **D:** RSA (Northern Cape, Western Cape); quartzite slopes.

Incl. *Aloe muricata* Schultes (1809) ≡ *Aloe glauca* var. *muricata* (Schultes) Baker (1880).

[2] Differs from var. *glauca*: **L** more spreading, less glaucous, lower face more tuberculate-aculeate; marginal teeth larger.

A. globuligemma Pole-Evans (Trans. Roy. Soc. South Afr. 5: 30, 1915). **T:** RSA, Northern Prov. (*Pole Evans* s.n. in *Govt. Herb.* 20 [PRE]). – **D:** Botswana, Zimbabwe, RSA (Northern Prov., Mpumalanga); dry bushland, 600 - 1300 m. **I:** Reynolds (1950: 444-447, t. 45); Reynolds (1966: 223-224). **Fig. XIV.d**

[7] Shortly caulescent, suckering and branching forming large dense groups; stem decumbent, to 50 cm; **L** ± 20, densely rosulate, lanceolate-attenuate, 45 - 50 × 8 - 9 cm, glaucous, with narrow dull white to pale pink cartilaginous margin; marginal teeth 2 mm, dull white, pale brown-tipped, mostly curved towards tip, to 10 mm apart; **Inf** to 1 m, with 8 - 18 **Br**, lower **Br** sometimes rebranched; racemes oblique with secund **Fl**, 30 - 40 cm, sub-dense; **Bra** ovate-acute, 6 mm; **Ped** 3 - 4 mm; **Fl** yellow to ivory, reddish-tinged near base, with a bloom, 26 mm, clavate, base rounded, 5 mm ⌀ across **Ov**, enlarging to 10 mm above, narrowing slightly at mouth;

Tep free for ± 18 mm; **St** exserted 12 mm. – *Cytology:* 2n = 14 (Müller 1941).

Natural hybrids with other species have been reported (Reynolds 1950).

A. gossweileri Reynolds (JSAB 28(3): 205-207, ills., 1962). **T:** Angola, Cuanza Sul Distr. (*Reynolds* 9760 [PRE, K, LISC]). – **D:** Angola; rocky hills, 850 - 1100 m. **I:** Reynolds (1966: 372).

[16] Caulescent, branching at base, forming thickets to several m ∅; stem ascending or divergent, 1 - 1.5 m × 3 - 4 cm ∅; **L** ± 16, rosulate, persistent for 10 - 20 cm below, lanceolate-attenuate, 30 × 5 cm, green, sometimes with spots; marginal teeth 3 - 4 mm, 15 mm apart; **Inf** 40 - 50 cm, with 6 - 8 **Br**; racemes with secund **Fl**, 10 - 15 × 5 - 6 cm, subdense; **Bra** ovate-acute, 3 × 2 mm; **Ped** 10 mm; **Fl** scarlet, paler at mouth, 30 mm, base rounded, 6 mm ∅ across **Ov**, slightly narrowed above, then enlarging slightly to mouth; **Tep** free for 10 - 12 mm.

A. gracilicaulis Reynolds & P. R. O. Bally (JSAB 24(4): 184-186, t. 27-28, 1958). **T:** Somalia (*Reynolds* 8428 [PRE, EA, K]). – **D:** Somalia; dry bushland, 1090 - 1160 m. **I:** Reynolds (1966: 309-310).

[9] Caulescent, simple or branching at base; stem to 4 m, 8 - 10 cm ∅ at base, ± 6 cm ∅ above, with dead **L** remains for 30 - 50 cm below tip; **L** ± 20, rosulate, ensiform-acute, 50 - 60 × 8 cm, greygreen, with 1 mm white cartilaginous margin; marginal teeth 1 mm, blunt, white, 2 - 10 mm apart; **Inf** 60 cm, with ± 10 **Br**, lower **Br** rebranched; racemes cylindrical, 5 - 6 × 5 cm, sub-dense; **Bra** ovate-acute, 3 × 2 mm; **Ped** 5 - 6 mm; **Fl** yellow, 18 mm, base rounded, 5 mm ∅ across **Ov**, enlarging slightly to mouth; **Tep** free for 12 mm.

A. gracilis Haworth (Philos. Mag. J. 66: 279, 1825). **T:** not typified. – **D:** RSA.

A. gracilis var. **decumbens** Reynolds (Aloes South Afr. [ed. 1], 358-359, 1950). **T:** RSA, Western Cape (*Muir* 5383 [PRE]). – **D:** RSA (Western Cape); rocks and cliffs, 180 - 365 m. **I:** Aloe 31: 46-47, 1994.

[13] Differs from var. *gracilis*: **Br** decumbent, to 75 × 1 cm; **L** ± 15 × 1.5 cm; sheath ± 10 mm; **Ped** 4 - 5 mm; **Fl** 28 - 33 mm.

A. gracilis var. **gracilis** – **D:** RSA (Eastern Cape); thickets on rocky slopes. **I:** Reynolds (1950: 357-359).

Incl. *Aloe laxiflora* N. E. Brown (1906).

[16] Caulescent, branching at base; stem erect, to 2 m × ± 2 cm ∅; **L** scattered along stem for 30 - 60 cm, lanceolate, 25 × 2.5 cm, dull green; marginal teeth 1 mm, firm, white, 2 - 5 mm apart; sheath 10 - 15 mm, pale green, striate; **Inf** 20 - 30 cm, simple or with 1 - 2 **Br**; racemes cylindrical to slightly conical, ± 10 cm, sub-dense, with 20 - 30 **Fl**; **Bra** del-

toid-acuminate, 5 × 2 - 3 mm; **Ped** 8 mm; **Fl** bright red to scarlet, yellowish at mouth, 40 - 45 mm, base rounded, scarcely narrowed above **Ov**; **Tep** free for ± 10 - 12 mm; **St** exserted 0 - 1 mm. – *Cytology:* 2n = 14 (Müller 1941: as *A. laxiflora*).

A. grandidentata Salm-Dyck (Observ. Bot. Hort. Dyck., 3, 1822). **T:** not typified. – **D:** Botswana, RSA (Northern Cape, North-West Prov., Free State); stony plains and ironstone slopes, frequently in the shade of low bushes. **I:** Reynolds (1950: 285-287). **Fig. XIV.e, XIV.g**

[7] Acaulescent or shortly caulescent, suckering to form dense groups; **L** 10 - 20, densely rosulate, lanceolate, ± 15 - 20 × 6 - 7 cm, apical 5 cm soon drying, brownish-green with many dull white oblong spots, usually in transverse bands, usually more prominent on lower face, with horny reddish-brown margin; marginal teeth 3 - 5 mm, pungent, reddish-brown, 8 - 10 mm apart; **Inf** ± 90 cm, with 4 - 7 **Br**; racemes cylindrical, slightly acuminate, to 20 cm, lax, with 20 - 30 **Fl**, sometimes shorter and denser; **Bra** deltoid-acuminate, 10 - 15 mm, white; **Ped** 10 - 15 mm; **Fl** dull reddish, rarely pink, lobes with 1.5 mm white border, 28 - 30 mm, clavate, base rounded, slightly narrowed above **Ov**, then enlarging to mouth; **Tep** free for ± 8 - 10 mm. – *Cytology:* 2n = 14 (Resende 1937).

Natural hybrids with other species have been reported (Reynolds 1950).

A. grata Reynolds (JSAB 26(2): 87-89, pl. 8-9, 1960). **T:** Angola, Bié Distr. (*Reynolds* 9246 [PRE, K, LUA]). – **D:** Angola; hillsides, mostly on rocks, 1770 m. **I:** Reynolds (1966: 117-118).

[7] Acaulescent or shortly caulescent, suckering to form dense groups; **L** 16 - 20, densely rosulate, lanceolate-attenuate, ± 20 - 25 × 7 - 8 cm, upper face green tinged reddish-brown, lower face paler glaucous green with many crowded pale green 1 mm ∅ spots in lower ¼; marginal teeth 2 - 3 mm, 5 - 8 mm apart; exudate drying pale yellow; **Inf** 70 - 90 cm, with 2 - 3 **Br**; racemes capitate or sub-capitate, 8 - 10 × 8 cm; **Bra** ovate-acute, 2 × 1.5 mm; **Ped** 20 mm; **Fl** scarlet, 25 - 28 mm, base shortly attenuate, 6 mm ∅ across **Ov**, narrowed to 5 mm above, then enlarging, narrowing slightly at mouth; **Tep** free for 7 mm.

A. greatheadii Schönland (Rec. Albany Mus. 1: 121, 1904). **T:** Botswana (*Schönland* 1616 [GRA, PRE]). – **D:** Botswana, Malawi, Moçambique, Zaïre, Zimbabe, RSA, Swaziland.

A. greatheadii var. **davyana** (Schönland) Glen & D. S. Hardy (SAJB 53(6): 490-491, 1987). **T:** RSA, Transvaal (*Burtt Davy* 1855 [GRA]). – **D:** RSA (Free State, Gauteng, KwaZulu-Natal, Northern Prov., North-West Prov.), Swaziland; grassland on rocky slopes. **I:** Reynolds (1950: 234-236, t. 13, as *A. davyana*).

≡ *Aloe davyana* Schönland (1904); **incl.** *Aloe longibracteata* Pole-Evans (1915); **incl.** *Aloe barbertoniae* Pole-Evans (1917); **incl.** *Aloe comosibracteata* Reynolds (1936); **incl.** *Aloe graciliflora* Groenewald (1936); **incl.** *Aloe labiaflava* Groenewald (1936); **incl.** *Aloe mutans* Reynolds (1936); **incl.** *Aloe verdoorniae* Reynolds (1936); **incl.** *Aloe davyana* var. *subolifera* Groenewald (1939).

[4] Differs from var. *greatheadii*: Sometimes suckering to form dense groups; **L** to 20 cm; **Inf** 60 - 100 cm, with 3 - 5 **Br**; racemes conical; **Bra** 20 - 25 mm; **Ped** 20 - 25 mm; **Fl** dull brick-red to pink, lobes with white border, 27 - 35 mm. − *Cytology:* 2n = 14 (Kondo & Megata 1943).

Wyk & Smith (1996) have "provisionally" recognized *A. longibracteata* as a separate species. Natural hybrids with other species have been reported (Reynolds 1950).

A. greatheadii var. **greatheadii** − **D:** Botswana, Malawi, Moçambique, Zaïre, Zimbabwe, RSA (Northern Prov.); grassland. **I:** Reynolds (1950: 232-233).

Incl. *Aloe pallidiflora* A. Berger (1905); **incl.** *Aloe termetophila* De Wildeman (1921).

[3] Acaulescent or shortly caulescent, simple; **L** densely rosulate, attenuate, ± 47 × 12 cm, upper face dark glossy green with many elongated whitish spots, near base usually confluent and forming transverse bands, lower face pale green, lineate, without spots, usually with brown cartilaginous margin; marginal teeth ± 5 mm, pungent, brown, to 20 mm apart; **Inf** to 1.75 m, with ± 4 - 6 **Br**; racemes cylindrical, 9 - 30 cm, lax when long, usually short and dense; **Bra** acuminate, ± 32 × 5 mm; **Ped** 12 mm; **Fl** white with broad red midrib, 32 mm, base rounded, 7 mm ⌀ across **Ov**, narrowed to 5 mm above, then enlarged to 8 mm; **St** exserted 0 - 1 mm.

A very variable taxon.

A. greenii Baker (JLSB 18: 165, 1880). **T:** not typified. − **Lit:** Smith & Crouch (1995). **D:** RSA (KwaZulu-Natal); below shrubs in hot valleys. **I:** Wyk & Smith (1996: 199).

[7] Acaulescent, stoloniferous and forming dense groups; **L** 12 - 16, densely rosulate, linear-lanceolate, attenuate, ± 40 - 45 × 7 - 8 cm, bright green, obscurely lineate, with many confluent oblong whitish spots in irregular transverse bands, more pronounced on lower face, with pale brownish margin; marginal teeth 3 - 4 mm, pale brown to pink, 8 - 10 mm apart; **Inf** 1 - 1.3 m, with 5 - 7 **Br**; racemes oblong-cylindrical, 15 - 25 cm, sub-dense; **Bra** lanceolate-deltoid acuminate, 10 mm; **Ped** 10 mm; **Fl** light to dark pink with a bloom, lobes with white border, 28 - 30 mm, base truncate, 7 mm ⌀ across **Ov**, abruptly narrowed to 4 mm above, enlarging to mouth; **Tep** free for 10 mm. − *Cytology:* 2n = 14 (Müller 1945).

A. grisea S. Carter & Brandham (Bradleya 1: 19-20, ills., 1983). **T:** Somalia, Hargeisa Region (*Bally* 9662 [K]). − **D:** Somalia; stony ground or rocks on mountain slopes, 1340 - 1590 m.

[5] Acaulescent, simple or suckering sparsely; **L** rosulate, triangular, to 25 × 15 cm, very glaucous with many white spots in irregular transverse bands, with 1 mm horny brown margin; marginal teeth 1 - 3 mm, 2 - 8 mm apart; **Inf** to 60 cm, with 2 - 3 **Br**, lower **Br** sometimes rebranched; racemes conical to subcapitate, ± 10 × 6 cm; **Bra** ovate-aristate, to 20 × 5 mm; **Ped** to 30 mm; **Fl** bright orange-scarlet, yellow at mouth, 23 mm, base rounded, 6 mm ⌀ across **Ov**, abruptly narrowed to 3.5 mm above, enlarging to mouth; **Tep** free for 5 mm; **St** exserted 0 - 1 mm. − *Cytology:* 2n = 14 (from protologue).

A. guerrae Reynolds (JSAB 26(2): 85-87, pl. 6-7, 1960). **T:** Angola, Bié Distr. (*Reynolds* 9218 [PRE, K, LUA]). − **D:** Angola; grassland with scattered bushes, 1220 - 1670 m. **I:** Reynolds (1966: 228-229).

[3] Acaulescent or shortly caulescent, simple, **Ros** usually tilted slightly to one side; **L** ± 24, densely rosulate, lanceolate-attenuate, 40 × 6 - 7 cm, dull green, obscurely lineate; marginal teeth 4 - 5 mm, pungent, pale brown or reddish-brown, 10 - 15 mm apart; exudate drying yellow; **Inf** 90 - 100 cm, with 8 - 10 **Br**, lower **Br** sometimes rebranched; racemes with secund **Fl**, oblique to almost horizontal, 20 cm, sub-dense; **Bra** ovate-acute, 6 - 8 × 4 mm; **Ped** 5 mm; **Fl** scarlet with a bloom, 40 mm, base truncate, 8 mm ⌀ across **Ov**, scarcely narrowed above; **Tep** free for 10 - 12 mm. − *Cytology:* 2n = 14 (Brandham 1971).

A. guillaumetii Cremers (Adansonia, n.s., 15(4): 498-501, ills., 1976). **T:** Madagascar (*Cremers* 2670 [P]). − **D:** Madagascar; eroded sandstone outcrops. **I:** FPA 52(1): t. 2046, 1992. **Fig. XIV.f**

[6] Acaulescent, suckering to form large mats; **L** 6 - 12, rosulate, triangular, 38 - 40 × 2 - 5 cm, bright green tinged red, with scattered small white spots; marginal teeth 1 mm, whitish, 2 - 5 mm apart; **Inf** 80 - 110 cm, simple; racemes cylindrical, 25 - 30 cm, lax, **Bra** lanceolate-acute, 4 × 2 mm; **Ped** 15 - 20 mm; **Fl** red at base becoming pink then green above, 25 mm, base shortly attenuate, 5 mm ⌀ across **Ov**, narrowed slightly above, then enlarged to 6 mm at mouth; **Tep** free for 9 mm.

A. haemanthifolia A. Berger & Marloth (BJS 38: 85, 1905). **T:** RSA, Western Cape (*Marloth* 3786 [BOL, PRE]). − **D:** RSA (Western Cape); in grass on rocky slopes in damp climate, 1200 - 1525 m. **I:** Reynolds (1950: 156-157).

[6] Acaulescent, usually suckering to form dense groups, rarely simple; **L** 10 - 16, distichous, lorate, tip obtuse to rounded and minutely crenate, to 18 × 8 cm, dull glaucous green with reddish margin, sur-

face smooth; marginal teeth absent; exudate colourless; **Inf** to 45 cm, simple; racemes capitate, 4 - 5 cm, with ± 30 **Fl**; **Bra** lanceolate-acuminate, to 25 × 6 - 7 mm, fleshy; **Ped** 25 - 35 mm; **Fl** scarlet, 38 mm, slightly clavate, base attenuate, above **Ov** gradually enlarging to mouth; **Tep** free almost to base; **St** not exserted.

A. hardyi Glen (FPA 49(3-4): pl. 1942 + 3 pp. of text, 1987). **T:** RSA, Mpumalanga (*Fourie* 3252 [PRE]). – **D:** RSA (Mpumalanga); rocky cliffs. **I:** Wyk & Smith (1996: 83).

[13] Caulescent, branching; stem pendulous; **L** 12 - 20, rosulate, pendulous, lanceolate, 40 - 70 × 5 - 8 cm, glaucous blue-green; marginal teeth small, scattered; **Inf** 50 - 70 cm, simple; racemes conical to subcapitate, sub-dense; **Bra** obovate-acute, 14 - 17 × 10 - 15 mm; **Ped** 15 - 30 mm; **Fl** pink to red, 25 - 35 mm, base rounded, slightly narrowed above **Ov**, then enlarging to mouth.

A. harlana Reynolds (JSAB 23(1): 9-10, t. 9, 1957). **T:** Ethiopia, Harar Prov. (*Reynolds* 8158 [PRE, K]). – **D:** Ethiopia; in low bush on stony slopes, 1650 - 1830 m. **I:** Reynolds (1966: 276).

[5] Acaulescent or shortly caulescent, simple or dividing into 2 - 4 **Ros**; **L** ± 24, densely rosulate, lanceolate-attenuate, 50 × 12 - 15 cm, pale to dark olive-green, sometimes glossy, lower face sometimes with a few dull elongated pale green blotches near base, usually with horny reddish-brown margin; marginal teeth 3 - 4 mm, pungent, reddish-brown, 10 - 15 mm apart; exudate drying deep brown; **Inf** ± 70 - 90 cm, with 6 - 8 **Br**; racemes conical-capitate or cylindrical-acuminate, to 20 cm; **Bra** ovate-acute, 10 × 5 - 7 mm, densely imbricate in bud stage; **Ped** 15 mm; **Fl** deep red, sometimes yellow, 33 mm, base shortly attenuate, 11 mm ⌀ across **Ov**, not narrowed above; **Tep** free for 10 mm. – *Cytology:* 2n = 14 (Brandham 1971).

A. haworthioides Baker (JLSB 22: 259, 1886). **T:** Madagascar (*Baron* 3424 [K]). – **D:** Madagascar.

≡ *Aloinella haworthioides* (Baker) Lemée (1939) (*nom. inval.*, Art. 43.1) ≡ *Lemeea haworthioides* (Baker) P. V. Heath (1993).

A. haworthioides var. **aurantiaca** H. Perrier (Mém. Soc. Linn. Normandie, Bot. 1(1): 50, 1926). **T:** Madagascar, Fianarantsoa (*Perrier* 14582 [P]). – **D:** Madagascar; on granite, 2000 m. **I:** Reynolds (1966: 397-398).

[4] Differs from var. *haworthioides*: **Inf** racemes bright orange-red, including axis, **Bra** and **Fl**.

A. haworthioides var. **haworthioides** – **D:** Madagascar; gneiss and quartzite rocks, 1200 - 1800 m. **I:** Reynolds (1966: 395-396).

[4] Acaulescent, simple or suckering to form dense groups; **R** fusiform; **L** ± 30, densely rosulate,

narrowly lanceolate-deltoid, tip a short pellucid point, 3 - 4 cm, dark grey-green, with many white pustules sometimes tipped with a short white **Ha**; marginal teeth 1 - 2 mm, narrowly deltoid, soft to firm, white, crowded; **Inf** 20 - 30 cm, simple; raceme cylindrical, slightly acuminate, 4 - 6 × 1.2 cm, sub-dense, with 20 - 30 **Fl**; **Bra** suborbiculate, obtuse, shortly mucronate, 5 mm; **Ped** none or negligible; **Fl** white to pale pink, 6 - 8 mm, slightly campanulate; **Tep** free to base.

A. hazeliana Reynolds (JSAB 25(4): 279-281, pl. 25-26, 1959). **T:** Zimbabwe, Melsetter Distr. (*Munch* s.n. in *Reynolds* 9031 [PRE, K]). – **D:** Moçambique, Zimbabwe; soil pockets and rock fissures, 1520 - 2135 m. **I:** Reynolds (1966: 25-26).

[10] Caulescent, branching sparsely at base; stem to 50 × 1.5 cm; **L** ± 12, distichous, scattered along stem for 15 cm, linear, tip obtuse, cuspidate and minutely dentate, to 20 × 1 - 1.5 cm, green, sometimes with a few small scattered pale green elliptic spots, lower face with many crowded pale green elliptic spots near base, with narrow hyaline margin; sheath 10 - 15 mm, with many spots; marginal teeth 0.5 mm, firm, white, 5 mm apart; **Inf** 30 - 40 cm, simple; raceme cylindrical, 8 - 10 × 4 cm, lax, with ± 18 **Fl**; **Bra** ovate-acute, 4 × 3 mm, slightly fleshy; **Ped** to 13 mm; **Fl** scarlet, green-tipped, 25 mm, base shortly attenuate, 6 mm ⌀ across **Ov**, not enlarging to mouth; **Tep** free to base.

A. helenae Danguy (Bull. Mus. Nation. Hist. Nat., Sér. 2, 1: 433, 1929). **T:** Madagascar, Toliara (*Decary* 3325 [P]). – **D:** S Madagascar; dense bush on littoral sand and limestone. **I:** Reynolds (1966: 513-514). **Fig. XIII.f**

[9] Caulescent, simple; stem to 4 m, to 20 cm ⌀, with dead **L** remains; **L** ± 40, densely rosulate, ensiform, greatly recurved, to 140 × 12 - 15 cm, green; marginal teeth 2 - 3 mm, pungent, pale green, 15 mm apart; **Inf** 40 - 60 cm, simple; racemes cylindrical-claviform, 15 × 9 cm, dense, with 300 - 400 **Fl**; **Bra** lanceolate-deltoid, acute, 12 × 6 mm, red-tipped, thick and fleshy; **Ped** 2 - 3 mm; **Fl** yellowish, reddish at mouth, 24 - 27 mm, cylindrical base becoming widely campanulate above, base obtuse, 4 mm ⌀ across **Ov**, slightly narrowed above, then enlarging to mouth; **Tep** free for 12 - 15 mm; **St** exserted 6 - 8 mm.

A. heliderana Lavranos (CSJA 45(3): 114-115, ills., 1973). **T:** Somalia, Bosaso Region (*Lavranos & Bavazzano* 8456 [FI]). – **D:** Somalia; limestone hills, ± 500 m.

[15] Caulescent, branching sparingly at base; stem erect, to 1 m × 4 cm ⌀, covered by dried **L** bases; **L** ± 20, rosulate, deltoid-acute, ± 20 × 6 cm, grey-green with few or many white spots, with white cartilaginous margin; marginal teeth to 2 mm, white, 5 - 10 mm apart; **Inf** to 60 cm, with ± 8 **Br**,

lower **Br** rebranched; racemes capitate or shortly cylindrical, **Fl** sometimes slightly secund, lax; **Bra** deltoid-acute, brownish, brittle, 3 - 5 mm; **Ped** 7 mm; **Fl** red or yellow, 20 - 22 mm, base rounded, 5 - 6 mm ∅ across **Ov**, slightly narrowed above, then enlarging to 5 - 6 mm at mouth; **Tep** free for 5 - 7 mm.

A. hemmingii Reynolds & P. R. O. Bally (JSAB 30(4): 221-222, ills., 1964). **T:** Somalia (*Bally 7146* [EA]). – **Lit:** Carter & al. (1984). **D:** Somalia. **I:** Reynolds (1966: 51-52).

[4] Acaulescent or very shortly caulescent, simple or in small groups; **L** ± 10, densely rosulate, ovate or lanceolate-attenuate, 10 - 12 × 3 - 3.5 cm, brownish-green with many dull white elongated streaks, lower face with white streaks smaller and more numerous; marginal teeth 2 mm, pungent, whitish and brown-tipped, 4 - 6 mm apart; **Inf** 30 - 35 cm, simple; racemes cylindrical, 10 - 15 × 5 cm, lax, with ± 18 **Fl**; **Bra** ovate-acuminate, 8 × 3 mm; **Ped** 6 - 8 mm; **Fl** flamingo-pink or pale rose-red, minutely spotted, 24 mm, base truncate, 8 mm ∅ across **Ov**, scarcely narrowed above; **Tep** free for 7 mm. – *Cytology:* 2n = 14 (Carter & al. 1984).

Carter & al. (1984) concluded that this is conspecific with *A. somaliensis*, but it is still recognized as distinct by Lavranos (1995).

A. hendrickxii Reynolds (JSAB 21(2): 51-53, ills., 1955). **T:** Zaïre, Oriental Prov. (*Reynolds 6300* [PRE, BR]). – **D:** Zaïre; granite outcrops, 2450 m. **I:** Reynolds (1966: 336-337).

[7] Acaulescent or shortly caulescent, branching freely to form dense groups; **L** ± 16, rosulate, lanceolate-attenuate, 22 × 4 cm, upper face dull green with greyish bloom, lower face grey-green sometimes with a few obscure dull greyish elongated spots; marginal teeth 3 mm, firm, greenish-white in lower ½ of **L**, brownish-tipped above, 5 - 8 mm apart; exudate drying yellow; **Inf** 40 - 50 cm, with 4 - 5 **Br**; racemes cylindrical-conical, 12 - 15 cm, laterals shorter, sub-dense; **Bra** deltoid, 14 × 3 - 4 mm; **Ped** 13 mm; **Fl** dull scarlet, 30 mm, base rounded, 4.5 mm ∅ across **Ov**, slightly narrowed above **Ov**, then enlarging to mouth; **Tep** free for 10 mm.

A. hereroensis Engler (BJS 10: 2, 1888). **T:** Namibia (*Marloth 1438* [PRE]). – **D:** Angola, Namibia, RSA (Northern Cape); arid stony deserts, ± 300 m. **I:** Reynolds (1950: 324-325); Reynolds (1966: 101).

Incl. *Aloe hereroensis* var. *hereroensis*; **incl.** *Aloe orpeniae* Schönland (1905) ≡ *Aloe hereroensis* var. *orpeniae* (Schönland) A. Berger (1908); **incl.** *Aloe hereroensis* var. *lutea* A. Berger (1908).

[5] Acaulescent or usually very shortly caulescent, sometimes in age with decumbent stem to 1 m, simple or in small groups; **L** ± 30, densely rosulate, lanceolate-deltoid, 30 - 50 × 6 - 9 cm, very glauc-

ous, obscurely lineate to sulcate, lower face sometimes with few or many irregularly scattered whitish spots; marginal teeth 3 - 4 mm, pungent, sometimes bifid, reddish-brown to brownish, 8 - 15 mm apart; **Inf** ± 1 m, with 4 - 8 **Br**, lower **Br** sometimes rebranched; racemes capitate, corymbose, 6 - 8 × 8 - 10 cm, dense; **Bra** lanceolate-deltoid, attenuate, 15 - 30 mm; **Ped** 30 - 50 mm; **Fl** usually orange, but also various shades of red and yellow, 25 - 35 mm, base rounded, enlarged slightly above **Ov**, narrowing at mouth; **Tep** free for 14 - 16 mm. – *Cytology:* 2n = 14 (Müller 1941).

Natural hybrids with other species have been reported (Reynolds 1950).

A. heybensis Lavranos (CSJA 71(3): 159-160, ills., 1999). **T:** Somalia, South S1 (*Lavranos & Bauer 27847* [UPS]). – **D:** S Somalia; granite inselberg, ± 700 m.

[11] Caulescent, branching at the base; stem to 45 cm, weak and becoming decumbent; **L** ± 12, rosulate, triangular, to 35 × 7 cm, glaucous to brownish-green, often with pale spots, surface glossy, smooth; marginal teeth deltoid, 4 mm, brown-tipped, 5 - 12 mm apart; exudate pale yellow, drying pale yellow; **Inf** 45 cm, erect, with up to 5 **Br**; racemes cylindrical, lax below, subcapitate towards tip; **Bra** triangular-acuminate, 10 × 2 mm, reddish-brown; **Ped** 14 mm; **Fl** dull red, glossy, becoming dull green towards tip, 25 mm, base very shortly attenuate, 9 mm ∅ across **Ov**, not narrowing above; **Tep** free for 10 - 12 mm; **St** exserted 2 - 3 mm.

A. hijazensis Lavranos & Collenette (CSJA 72(1): 23-24, ills., 2000). **T:** Saudi Arabia, Hijaz Prov. (*Collenette 2842* [K]). – **D:** Saudi Arabia; 2000 - 2500 m. **I:** Collenette (1999: 22).

[3] Acaulescent, solitary; **L** ± 15, densely rosulate, lanceolate, 25 - 35 × 8 - 11 cm, bright green; marginal teeth 1 - 2 mm, dark brown, 17 mm apart; **Inf** to 150 cm, erect, with up to 3 **Br**; racemes cylindrical, 30 - 50 cm, sub-dense; **Bra** lanceolate, 15 - 18 × 6 - 7 mm; **Fl** yellow, ± 30 mm, base rounded, 6 - 7 mm ∅ across **Ov**, not constricted above; **Tep** free for ± 19 mm; **St** exserted.

A. hildebrandtii Baker (CBM 1888: t. 6981 + text, 1888). **T:** Somalia (*Hildebrandt s.n.* [K]). – **Lit:** Brandham & al. (1994). **D:** Somalia; rocky slopes and dry bush, 1095 - 1615 m. **I:** Reynolds (1966: 341-343).

Incl. *Aloe gloveri* Reynolds & P. R. O. Bally (1958).

[13] Caulescent, branching at base; stem decumbent, to 1 m × 3 - 4 cm ∅; **L** scattered along stem for 30 cm below tip, lanceolate-attenuate, 20 - 30 × 4 - 6 cm, dull green, sometimes with a few white lenticular spots; marginal teeth 2 - 3 mm, pungent, reddish-brown, 8 - 10 mm apart; exudate drying orange-brown; **Inf** ± 50 cm, usually oblique, with 8 -

12 **Br**, lower **Br** sometimes rebranched; racemes cylindrical, slightly conical, **Fl** subsecund in oblique racemes, 10 - 18 × 5 cm, lax; **Bra** ovate-acute, 3 × 2 mm; **Ped** 10 - 15 mm; **Fl** yellow, orange or dull scarlet with a bloom, 26 - 30 mm, base rounded, 8 mm ∅ across **Ov**, above narrowing slightly to mouth; **Tep** free for 12 mm. − *Cytology:* 2n = 14 (Brandham & al. 1994).

Natural hybrids with *A. megalacantha* have been reported (Reynolds 1966).

A. hlangapies Groenewald (Tydskr. Wetensk. Kuns 14: 60-63, 1936). **T** [lecto]: RSA, Mpumalanga (*van der Merwe* 102 [PRE]). − **D:** RSA (KwaZulu-Natal, Mpumalanga); grassland. **I:** Reynolds (1950: 138-139).

Incl. *Aloe hlangapitis* Groenewald (1936) (*nom. inval.*, Art. 61.1); **incl.** *Aloe hlangapensis* Groenewald (1937) (*nom. inval.*, Art. 61.1).

[4] Acaulescent or shortly caulescent, simple or branching at base to form small groups; stem to 15 cm; **L** 10 - 14, distichous, lorate-acuminate, 35 - 45 × 5 - 6 cm, dull green, sometimes with few white spots, lower face usually with many white spots near base; marginal teeth ± 0.5 mm, soft, white, 5 - 15 mm apart; **Inf** 50 cm, simple; racemes capitate, to 7 × 9 - 10 cm, dense, with ± 50 **Fl**; **Bra** ovate-acuminate, 15 × 7 mm; **Ped** 25 mm; **Fl** apricot-yellow, greenish-tipped, 28 - 30 mm, base rounded, enlarged slightly above **Ov**, narrowing at mouth; **Tep** free for 23 - 25 mm.

Groenewald later changed the name to *A. hlangapitis* and then to *A. hlangapensis*, but these changes are illegitimate under the ICBN.

A. howmanii Reynolds (Kirkia 1: 156-157, t. 15, 1961). **T:** Zimbabwe, Melsetter Distr. (*Ball* 646 [PRE, SRGH]). − **D:** Zimbabwe; cliff faces, 1675 - 2375 m. **I:** Reynolds (1966: 17-18).

[10] Caulescent, simple or branching from base; stem pendulous, to 30 × 1.2 cm; **L** 6 - 12, distichous, linear, mostly falcately decurved, tip obtuse and shortly cuspidate, 15 - 20 × 1.2 - 1.5 cm, green, with 1.5 mm translucent hyaline margin, lower face sometimes with a few small white spots near base; marginal teeth absent or minute, 2 - 4 mm apart; **Inf** 20 - 25 cm, descending, with raceme upturned, simple; raceme subcapitate, slightly conical, 4 - 5 cm, sub-dense, with ± 12 - 18 **Fl**; **Bra** ovate-acute, 3 - 4 × 2 mm; **Ped** 10 - 15 mm; **Fl** scarlet, green-tipped, 24 mm, slightly ventricose, base shortly attenuate, 5 mm ∅ across **Ov**, enlarged slightly above, narrowing at mouth; **Tep** free to base.

A. humbertii H. Perrier (Bull. Mus. Nation. Hist. Nat., Sér. 2, 3: 692, 1931). **T:** Madagascar, Toliara (*Humbert* 6211 [P]). − **D:** S Madagascar; silicious rocks, 1800 - 1980 m.

[2] Acaulescent or shortly caulescent, simple; **L** 7

- 12, rosulate, triangular with tip rounded and dentate, 25 - 30 × 5 - 6 cm, green with horny margin; marginal teeth yellow, 3 - 6 mm apart; **Inf** 35 - 40 (-80) cm, simple or with 1 - 2 **Br**; racemes 8 cm, lax or dense; **Bra** lanceolate, shortly cuspidate, 11 × 5 mm; **Ped** absent; **Fl** red or yellow tinged-red, ± 20 mm; **Tep** free for 10 mm; **St** scarcely exserted.

A. humilis (Linné) Miller (Gard. Dict. Abr. ed. 6, no. 10, 1771). **T:** RSA, Cape Prov. (*Oldenland* s.n. [not extant]). − **D:** RSA (Western Cape, Eastern Cape); karroid vegetation, in shade or in the open. **I:** Reynolds (1950: 174-175). **Fig. XIV.h**

≡ *Aloe perfoliata* var. *humilis* Linné (1753) ≡ *Catevala humilis* (Linné) Medikus (1786); **incl.** *Aloe humilis* var. *humilis*; **incl.** *Aloe humilis* var. *incurva* Haworth (1804) ≡ *Aloe incurva* (Haworth) Haworth (1812); **incl.** *Aloe suberecta* Haworth (1804) ≡ *Aloe humilis* var. *suberecta* (Haworth) Baker (1896); **incl.** *Aloe suberecta* var. *acuminata* Haworth (1804) ≡ *Aloe acuminata* (Haworth) Haworth (1812) ≡ *Aloe humilis* var. *acuminata* (Haworth) Baker (1880); **incl.** *Aloe tuberculata* Haworth (1804); **incl.** *Aloe humilis* Ker Gawler (1804) (*nom. illeg.*, Art. 53.1); **incl.** *Aloe echinata* Willdenow (1809) ≡ *Aloe humilis* var. *echinata* (Willdenow) Baker (1896); **incl.** *Aloe acuminata* var. *major* Salm-Dyck (1817); **incl.** *Aloe humilis* subvar. *semiguttata* Haworth (1821); **incl.** *Aloe humilis* var. *semiguttata* Haworth (1821); **incl.** *Aloe subtuberculata* Haworth (1825) ≡ *Aloe humilis* var. *subtuberculata* (Haworth) Baker (1896); **incl.** *Aloe humilis* subvar. *minor* Salm-Dyck (1837); **incl.** *Aloe humilis* var. *candollei* Baker (1880).

[6] Acaulescent, suckering to form dense groups; **L** 20 - 30, rosulate, ovate-lanceolate, very acuminate, ± 10 × 1.2 - 1.8 cm, glaucous-green with a dewy bloom, obscurely lineate, upper side with few tubercles, lower face more copiously tuberculate and with a few irregular soft white prickles; marginal teeth 2 - 3 mm, soft, white; **Inf** 25 - 35 cm, simple; raceme cylindrical, to 10 cm, sub-dense, with ± 20 **Fl**; **Bra** lanceolate-acuminate, 25 - 35 mm; **Ped** 25 - 35 mm; **Fl** scarlet, sometimes orange, 35 - 42 mm, ventricose, base rounded, enlarged above **Ov**, narrowing at mouth; **Tep** free for 23 - 28 mm; **St** exserted to 1 mm. − *Cytology:* 2n = 14 (Satô 1937).

A very variable species. The published varieties were based on material cultivated in Europe. Reynolds (1950) suggested that there are only 2 distinct varieties but he did not change the classification. The varieties were not documented by Wyk & Smith (1996) and it is unlikely that they will be upheld in any future revision.

Natural hybrids with other species have been reported (Reynolds 1950).

A. ibitiensis H. Perrier (Mém. Soc. Linn. Normandie, Bot. 1(1): 30, 1926). **T:** Madagascar, Antananarivo (*Perrier* 13980 [P]). − **D:** C Madagascar;

quartzite cliffs and ledges, 1450 - 2000 m. **I:** Reynolds (1966: 443-444).

Incl. *Aloe ibityensis* hort. (s.a.) (*nom. inval.*, Art. 61.1).

[5] Acaulescent, usually simple; **L** 12 - 16, densely rosulate, lanceolate-acute, 25 - 30 × 7 cm, yellowish- to olive-green, prominently lineate-striate; marginal teeth 1 - 2 mm, firm, pale yellowish, 3 - 5 mm apart; **Inf** ± 80 cm, with 2 - 4 **Br**; racemes cylindrical-acuminate, 25 × 5 cm, sub-dense; **Bra** ovate-acute, 4 - 7 × 2 - 3 mm; **Ped** 14 mm; **Fl** scarlet, 26 mm, base attenuate, 4 - 5 mm ∅ across **Ov**, enlarging above; **Tep** free for 9 mm. – *Cytology:* 2n = 14 (Brandham 1971).

A. imalotensis Reynolds (JSAB 23: 68, 1957). **T:** Madagascar, Fianarantsoa (*Perrier* 11022 [P]). – **D:** Madagascar; rock outcrops of sandstone and shale, 270 - 770 m. **I:** Reynolds (1966: 445-447).

Incl. *Aloe deltoideodonta* var. *contigua* H. Perrier (1926) ≡ *Aloe contigua* (H. Perrier) Reynolds (1958); **incl.** *Aloe deltoideodonta* fa. *latifolia* H. Perrier (1938) (*nom. inval.*, Art. 36.1); **incl.** *Aloe deltoideodonta* fa. *longifolia* H. Perrier (1938) (*nom. inval.*, Art. 36.1); **incl.** *Aloe deltoideodonta* subfa. *variegata* Boiteau *ex* H. Jacobsen (1954) (*nom. inval.*, Art. 36.1).

[5] Acaulescent or shortly caulescent, simple or branching to form small groups; stem decumbent, to 20 × 3 cm; **L** 20 - 24, densely rosulate, ovate-acute, to 30 × 12 - 15 cm, very fleshy, dull bluish-green tinged reddish, obscurely lineate, with 1 mm pink to reddish cartilaginous margin; marginal teeth 1 - 1.5 mm, deltoid or obtuse, pink, 1 - 4 mm apart, sometimes contiguous; exudate drying yellow; **Inf** 50 - 60 cm, with 2 - 4 **Br**; racemes cylindrical, slightly acuminate, 10 - 20 × 7 cm, sub-dense; **Bra** ovate-acute, 7 - 10 × 3 - 4 mm; **Ped** 15 - 18 mm; **Fl** coral-red, 30 - 34 mm, base shortly attenuate, 6 mm ∅ across **Ov**, very slightly narrowed above, then enlarging slightly to mouth; **Tep** free almost to base.

A. ×imerinensis Bosser (Adansonia, n.s., 8(4): 510-512, ill., 1968). **T:** Madagascar, Centre (*Bosser* 17043 [P]). – **D:** Madagascar.

= *Aloe capitata* × *A. macroclada*.

A. immaculata Pillans (South Afr. Gard. 24: 25, 1934). **T:** RSA, Northern Prov. (*Stellenbosch Univ. Gard.* 6774 [BOL]). – **D:** RSA (Northern Prov.). **I:** Reynolds (1950: 239-240).

[3] Acaulescent or shortly caulescent, simple; stem to 10 cm; **L** 16 - 20, densely rosulate, lanceolate-attenuate, to 40 × 6 - 8 cm, apical 5 - 10 cm soon drying, upper face dull green to brownish-green, with distinct darker longitudinal striations, sometimes with a few scattered white spots, lower face greyish-green, sometimes obscurely finely striate, with horny brownish to reddish-brown margin; marginal teeth 4 - 5 mm, pungent, brownish to red-

dish-brown, 10 - 15 mm apart; exudate drying purple; **Inf** to 1 m, with 6 - 10 **Br**, lower **Br** sometimes rebranched; racemes subcapitate, 10 - 20 × 8 - 9 cm, dense; **Bra** lanceolate-deltoid, 10 - 15 mm; **Ped** 12 - 15 mm; **Fl** coral-red, 30 - 33 mm, base truncate, 7 mm ∅ across **Ov**, abruptly narrowed to 4 mm above, then enlarging to mouth; **Tep** free for 10 mm. – *Cytology:* 2n = 14 (Brandham 1971).

A. inamara L. C. Leach (JSAB 37(4): 259-266, ills., 1971). **T:** Angola, Cuanza-Sul Distr. (*Leach & Cannell* 14608 [LISC, PRE]). – **D:** Angola; almost vertical cliff faces.

[13] Caulescent, branching at base and more sparingly above, sometimes forming dense mats to 3 m; stem pendulous, to 2 m × 2 cm ∅, covered by dried **L** bases; **L** ± 9, rosulate, falcate, pointing donwards, 45 - 60 (-90) × 4 - 5 cm, pale yellowish-green, brown when exposed to sun, obscurely lineate with few to many small ± H-shaped whitish spots, lower face with more spots in ± transverse bands, with whitish or faint pink margin; marginal teeth 0.3 - 1 mm, whitish, often brown-tipped, 4 - 20 mm apart; exudate yellow, not bitter; **Inf** 40 - 55 cm, descending and upturned towards tip, with 4 - 6 **Br**; racemes cylindrical-conical to almost capitate, to 7.5 × 7.5 cm, sub-lax; **Bra** triangular-attenuate, 7.5 - 9 × ± 3 mm; **Ped** 22 - 27 mm; **Fl** dull red with greenish tip, 26 - 29 mm, base truncate, 8 mm ∅ across **Ov**, abruptly narrowed to ± 7 mm above, then enlarging and narrowing again to ± 6 mm at mouth; **Tep** free for 6.5 - 8 mm; **St** exserted 0 - 1 mm.

A. inconspicua Plowes (Aloe 23(2): 32-33, ills., 1986). **T:** RSA, KwaZulu-Natal (*Plowes* 7079 [PRE]). – **D:** RSA (KwaZulu-Natal); grassy glades among acacias on shale and sandstone. **I:** Wyk & Smith (1996: 267).

[1] Acaulescent, simple, with persistent **L** bases forming an ovoid bulb to 50 × 25 mm; **R** fleshy; **L** rosulate, linear-acuminate, 3 × 0.3 - 0.4 cm, green, lower face with elongated white spots on basal ½, with translucent margin; marginal teeth 0.5 mm, soft, translucent, 2 - 4 mm apart; **Inf** ± 1.5 m, simple; raceme cylindrical, ± 7 × 2 cm, with ± 30 **Fl**; **Bra** 13 mm; **Ped** absent; **Fl** green with semi-translucent whitish **Tep** margins, 15 mm, narrowing slightly at mouth; **Tep** free to base.

A. inermis Forsskål (Fl. Aegypt.-Arab., 74, 1775). **T:** Yemen (*Forsskål* s.n. [C †]). – **Lit:** Wood (1983); Lavranos (1992). **D:** Yemen; rocky slopes, 760 m. **I:** Reynolds (1966: 221).

[13] Caulescent, suckering and forming small to large groups; stem decumbent or ascending to erect, to 50 cm; **L** ± 12 - 16, rosulate and persistent for 20 cm, lanceolate or ensiform, attenuate, spreading and becoming decurved, 25 - 30 (- 45) × 5 - 7 cm, grey-green or dull pale olive-green, sometimes with few or many scattered small dull white lenticular spots

towards base, with whitish cartilaginous margin, surface rough; marginal teeth absent; **Inf** ± 70 cm, oblique, with 6 - 9 **Br**, lower **Br** sometimes re-branched; racemes with **Fl** secund on more oblique **Br**, to 15 cm, sub-dense; **Bra** ovate-acute, 4 × 2 - 3 mm, white; **Ped** 5 - 9 mm; **Fl** dull scarlet or yellow, 28 - 30 mm, base rounded, 7 - 8 mm ∅ across **Ov**, slightly narrowed above **Ov**, then slightly enlarging to mouth; **Tep** free for 7 mm. – *Cytology:* 2n = 14 (Wood 1983).

A natural hybrid with *A. vacillans* has been reported (Wood 1983).

A. integra Reynolds (FPSA 16: t. 607 + text, 1936). **T:** RSA, Mpumalanga (*Reynolds* 1650 [PRE]). – **D:** Swaziland, RSA (Mpumalanga); in grass on rocky slopes, 1520 m. **I:** Reynolds (1950: 141-142).

[4] Caulescent, usually simple, sometimes in groups of up to 6 **Ros**; **R** fusiform; stem to 20 × 4 - 6 cm ∅; **L** 15 - 30, rosulate, triangular-attenuate, 10 - 12 × 4 - 5 cm, apical 5 cm soon drying, glossy deep green, obscurely lineate; marginal teeth usually absent, sometimes very minute and 2 - 5 mm apart; **Inf** to 50 cm, simple; raceme pyramidal, 8 - 12 × 7 - 9 cm, sub-dense, terminating in dense tuft of reflexed purple **Bra**; **Bra** deltoid-acuminate, 15 - 20 mm, purple; **Ped** to 30 mm; **Fl** lemon-yellow to rich canary-yellow, 15 - 18 mm, base attenuate, enlarged above **Ov**, narrowing at mouth; **Tep** free to base. – *Cytology:* 2n = 14 (Müller 1945).

A. inyangensis Christian (FPSA 16: t. 640 + text, 1936). **T:** Zimbabwe (*Piers* s.n. in *Christian* 518 [PRE 28428]). – **D:** Zimbabwe.

A. inyangensis var. **inyangensis** – **D:** Zimbabwe; on rocks, 1525 - 2285 m. **I:** Reynolds (1966: 23-24).

[10] Caulescent, branching near base to form dense clumps to 2 m ∅; stem ± 20 × 1.5 - 2 cm, **L** ± 8 - 10, laxly distichous, 15 - 20 × 1.5 cm, dark green, with or without a few scattered spots near base, lower face usually with many very pale green lenticular spots near base, spots sometimes tuberculate, with white cartilaginous margin; marginal teeth 0.5 mm, 1 - 2 mm apart; exudate yellow; **Inf** 30 - 35 cm, simple; racemes cylindrical-acuminate, 6 - 8 cm, sub-lax, with ± 16 **Fl**; **Bra** lanceolate-attenuate, 25 × 9 mm, imbricate in bud stage; **Ped** 30 mm; **Fl** bright scarlet, greenish at mouth, 35 - 40 mm, base rounded, 7 mm ∅ across **Ov**, not narrowed above; **Tep** free to base; **St** not exserted.

A. inyangensis var. **kimberleyana** S. Carter (KB 51(4): 777-779, ill., 1996). **T:** Zimbabwe, Eastern Prov. (*Carter & al.* 2702 [K, SRGH]). – **D:** Zimbabwe; damp places on rocks and pendulous on cliffs, 1220 - 1900 m.

[10] Differs from var. *inyangensis*: Stem to 1 m; **Br** more robust; **L** to 40 × 2.5 cm, thick and fleshy.

A. isaloensis H. Perrier (Bull. Acad. Malgache, n.s. 10: 20, 1927). **T:** Madagascar, Fianarantsoa (*Perrier* 17232 [P, K]). – **D:** Madagascar; bush on sandstone slopes, 600 - 1200 m. **I:** Reynolds (1966: 493-495). **Fig. XVI.a, XVI.b**

[10] Caulescent, branching at base; stem erect, divergent or decumbent, to 30 (-50) × 0.8 - 1 (-1.2) cm; **L** 10 - 14, sub-lax and persistent for up to 10 cm, linear-attenuate, 13 - 20 × 1 - 1.3 cm, grey-green; marginal teeth 1 - 1.5 mm, firm, greenish to pale brown, 5 - 10 mm apart; sheath 0.5 cm, obscurely striate; exudate yellow-orange, very abundant; **Inf** 30 - 50 cm, simple or with up to 5 **Br**; racemes cylindrical, slightly acuminate, 10 - 14 × 4 - 5 cm, lax, with 20 - 30 **Fl**; **Bra** deltoid, 3 × 1.5 mm; **Ped** 6 - 7 mm; **Fl** scarlet, 22 mm, base shortly attenuate, 6 mm ∅ across **Ov**, narrowed to 5 mm above, then enlarging slightly to mouth; **Tep** free for 11 mm; **St** exserted 0 - 1 mm. – *Cytology:* 2n = 14 (Brandham 1971).

A. itremensis Reynolds (JSAB 22(1): 29-30, ills., 1955). **T:** Madagascar, Fianarantsoa (*Reynolds* 7706 [TAN, K, P, PRE]). – **D:** Madagascar; steep sandstone slopes, sometimes in shade, 1700 m. **I:** Reynolds (1966: 499-500).

[3] Acaulescent or shortly caulescent, simple; stem to 20 × 3 cm; **L** 12 - 16, rosulate, triangular-acute, to 30 × 4.5 cm, dull green tinged reddish; marginal teeth 1 - 1.5 mm, brownish, paler-tipped, 5 - 8 mm apart; **Inf** 1 - 1.2 m, with 2 - 3 **Br**; racemes cylindrical-acuminate, 15 - 20 cm, lax or sub-dense; **Bra** deltoid, 6 × 2 mm; **Ped** 8 - 9 mm; **Fl** scarlet, 25 mm, base attenuate, 5 mm ∅ across **Ov**, enlarging above and narrowed at mouth; **Tep** free to base.

A. jacksonii Reynolds (JSAB 21(2): 59-61, t. 5, 1955). **T:** Ethiopia, Ogaden (*Jackson* s.n. in *Reynolds* 6224 [PRE]). – **Lit:** Brandham & al. (1994). **D:** Ethiopia (Ogaden); rock outcrops. **I:** Reynolds (1966: 54-55).

[11] Caulescent, suckering to form groups to 50 cm ∅; stem erect or sprawling, to 20 × 0.8 - 1 cm; **L** 5 - 7, laxly rosulate, subulate-attenuate, 10 - 15 × 1 - 1.4 cm, upper face dull green with several scattered dull pale green lenticular spots towards base, lower face greyish-green with many dull pale green lenticular spots, scattered or sometimes in broken transverse bands, surface very slightly rough; marginal teeth 1 mm, firm, white, reddish-tipped, 3 - 6 mm apart; exudate drying yellow; **Inf** 30 cm, simple; raceme cylindrical, 6 - 8 cm, lax, with 16 - 20 **Fl**; **Bra** ovate-acuminate, 4 × 2.5 mm; **Ped** 5 - 7 mm; **Fl** bright scarlet, paler to almost white towards mouth, 27 mm, base truncate, 8 - 9 mm ∅ across **Ov**, narrowing slightly towards mouth; **Tep** free for 6 - 7 mm. – *Cytology:* 2n = 28 (Brandham 1971).

A. jucunda Reynolds (JSAB 19(1): 21-23, t. 11, 1953). **T:** Somalia (*Bally* 7157 in *Reynolds* 6223

[PRE]). – **D**: Somalia; on limestone, usually in shade, 1060 - 1680 m. **I**: Reynolds (1966: 49-50).

[11] Acaulescent or shortly caulescent, suckering to form dense groups to 50 cm ∅; **L** ± 12, densely rosulate, ovate-acuminate, 4 × 2 - 5 cm, dark green with many pale green to dull white spots, more numerous on lower face, surface smooth; marginal teeth 2 mm, pungent, reddish-brown, 3 - 4 mm apart; exudate yellow; **Inf** 33 cm, simple; raceme cylindrical, 13 × 5 cm, sub-dense, with ± 20 **Fl**; **Bra** ovate-acute, 5 × 3 mm; **Ped** 7 mm; **Fl** pale rose-pink, 20 mm, base shortly attenuate, 7 mm ∅ across **Ov**, slightly narrowed above, enlarging slightly to mouth; **Tep** free for 7 mm. – *Cytology:* 2n = 14, 21 (Brandham 1971).

A. juvenna Brandham & S. Carter (Cact. Succ. J. Gr. Brit. 41(2): 27-29, ills., 1979). **T**: Kenya (*Carter & Stannard* 5 [K]). – **D**: Kenya; in grass on rocky ridge, 2300 m.

[11] Caulescent, freely branching and suckering at base; stem erect to 25 × 0.7 cm, or decumbent to 45 × 0.12 cm, densely leafy; **L** persistent below stem tip, to 35 per 10 cm of stem, deltoid, to 4 × 2 cm, bright green with many elongated pale green spots, many raised into small prickles; marginal teeth 2 - 4 mm, cartilaginous, 4 - 6 mm apart; **Inf** 25 cm, simple or with 1 **Br**; raceme conical, to 8 × 6 cm, sub-dense; **Bra** ovate, ± 5 × 4 mm; **Ped** 13 - 18 mm; **Fl** coral-pink, greenish-yellow tipped, 27 mm, base attenuate, 8 mm ∅ across **Ov**, very slightly narrowed above, enlarging slightly to mouth; **Tep** free for 9 mm; **St** exserted 0 - 1 mm. – *Cytology:* 2n = 28 (Brandham 1971).

Common in cultivation, but known from only one locality in the wild.

A. ×keayi Reynolds (*pro sp.*) (JSAB 29(2): 43-44, t. 6-7, 1963). **T**: Ghana (*Keay & Adams* FHI-37757A [K, GC]). – **D**: Ghana.

= *A. buettneri* × *A. schweinfurthii* (Newton 1976).

A. kedongensis Reynolds (JSAB 19(1): 4-6, t. 3-4, 1953). **T**: Kenya, Rift Valley Prov. (*Reynolds* 6546 [PRE, EA, K]). – **D**: Kenya; dense bush, usually on rocky ground, 1825 - 2300 m. **I**: Reynolds (1966: 375-376).

≡ *Aloe nyeriensis* ssp. *kedongensis* (Reynolds) S. Carter (1980).

[16] Caulescent, branching at or near base, forming dense thickets; **Br** erect or sprawling, to 4 m, 3 - 7 cm ∅; **L** rosulate and persistent for 30 - 60 cm, lanceolate, 30 × 4 cm, grey-green to yellowish-green, surface smooth; marginal teeth 2 - 3 mm, reddish-brown-tipped, 10 - 15 mm apart; exudate drying yellowish; **Inf** to 75 cm, with 2 - 4 **Br**; racemes cylindrical, 10 - 20 × 8 cm, dense; **Bra** ovate-acute, 5 × 5 mm; **Ped** 20 - 25 mm; **Fl** scarlet, 35 mm, base shortly attenuate, 7 mm ∅ across **Ov**,

slightly narrowed above, then enlarging slightly to mouth; **Tep** free for 14 mm. – *Cytology:* 2n = 28 (Cutler & al. 1980).

A. kefaensis M. G. Gilbert & Sebsebe (KB 52(1): 140-141, 1997). **T**: Ethiopia, Keffa Region (*Lissanework* s.n. in *Sebsebe* 2411 [ETH, K, UPS]). – **D**: Ethiopia; wooded grassland, ± 1800 m.

[5?] Acaulescent; **L** rosulate, 35 - 45 × 8 - 11 cm, green, obscurely lineate, usually with sparse pale spots; marginal teeth 3 - 4.5 mm, pale, sometimes with minute dark tip, 1.2 - 1.9 mm apart; **Inf** ± 1.5 m, with few **Br**; racemes cylindrical, 30 - 35 cm, lax; **Bra** 11 - 14 × 4.5 - 6 mm; **Ped** 16 - 22 mm; **Fl** scarlet, 28 - 32 mm, 6 mm ∅ across **Ov**, narrowed to 4.5 - 5 mm above; **Tep** free for 6 - 7 mm; **St** exserted 0 - 1 mm.

A. keithii Reynolds (JSAB 3(1): 47-49, t. 5, 1937). **T**: Swaziland (*Reynolds* 1983 [PRE]). – **D**: Swaziland; deep red soil in mist belt of mountains. **I**: Reynolds (1950: 279).

[5] Caulescent, simple or in small groups; stem to 30 cm; **L** ± 20, densely rosulate, lanceolate-attenuate, 60 × 9 - 11 cm, green, indistinctly lineate, with many whitish spots in irregular transverse bands, lower face paler green, usually without spots; marginal teeth 6 - 8 mm, pungent, brownish, 15 - 25 mm apart; **Inf** to 1.75 m, with 5 - 8 **Br**, lower **Br** sometimes rebranched; racemes cylindrical-acuminate, 20 - 25 × 7 - 9 cm, sub-dense; **Bra** deltoid-acuminate, 12 - 15 mm; **Ped** 12 - 15 mm; **Fl** coral-red, 36 mm, base truncate, 11 - 12 mm ∅ across **Ov**, abruptly narrowed to 6 mm above, then enlarging to mouth; **Tep** free for 12 mm.

A. ketabrowniorum L. E. Newton (Brit. Cact. Succ. J. 12(2): 50-51, ills., 1994). **T**: Kenya, Eastern Prov. (*Brown* 136 [K, EA]). – **D**: Kenya; rocky slopes, 760 m. **Fig. XV.a**

[13] Caulescent, branching at base; stem decumbent, to 30 cm; **L** 6 - 10, rosulate, triangular to slightly falcate, to 38 × 4.5 cm, mid-green, often reddish-tinged, with a few scattered elliptic whitish spots, surface smooth, with white hyaline margin; marginal teeth 1 mm, firm, white or red-tipped, sometimes pointing backwards, 5 - 10 mm apart; exudate pale yellow, drying brown; **Inf** to 70 cm, with 7 - 8 **Br**; racemes with **Fl** subsecund, 3 - 14 cm, lax; **Bra** triangular, 4 × 2 mm; **Ped** 5 mm; **Fl** pale red, fading above to white with pale red midstripe on lobes, 25 - 28 mm, base rounded, 8 - 10 mm ∅ across **Ov**, narrowed to 6 - 7 mm above, enlarging to mouth; **Tep** free for 9 - 12 mm.

A. khamiesensis Pillans (South Afr. Gard. 1934: 24, 1934). **T**: RSA, Northern Cape (*Pillans* 6665 [BOL]). – **D**: RSA (Northern Cape); rocky slopes in mountains. **I**: Reynolds (1950: 405-406).

[8] Caulescent, simple or branched about middle;

stem to 1.5 m × 10 - 15 cm ∅, with persistent dead L bases; L densely rosulate, lanceolate-attenuate, ± 40 × 8 cm, dull green, obscurely lineate, usually with a few scattered elliptic white spots, more numerous on lower face; marginal teeth 2 - 4 mm, pungent, reddish-brown, 5 - 10 mm apart; **Inf** to 90 cm, with 4 - 8 **Br**; racemes long-conical, 25 - 30 × 9 cm, dense; **Bra** ovate-acute, 18 × 8 mm; **Ped** 25 mm; **Fl** orange-red, greenish-tipped, 30 - 35 mm, base rounded, very slightly narrowed above **Ov**; **Tep** free to base. – *Cytology:* 2n = 14 (Brandham 1971).

A. kilifiensis Christian (JSAB 8(2): 169-170, t. 3, 1942). **T:** Kenya, Coast Prov. (*Moggridge* s.n. in *Burtt* 5554 [SRGH, BM, K]). – **D:** Kenya; coral cliffs and dry bushland, 3 - 380 m. **I:** Reynolds (1966: 81). **Fig. XV.b**

[5] Acaulescent or shortly caulescent, simple or suckering to form small groups; stem to 30 cm; **L** ± 15, rosulate, lanceolate-attenuate, 27 × 7 cm, dull green, usually with many scattered elliptic or H-shaped whitish spots, surface smooth; marginal teeth 3 mm, horny, brown, 4 mm apart; exudate yellow; **Inf** to 57 cm, with 4 - 6 **Br**; racemes capitate, 8 × 8 cm, sub-dense, with ± 20 **Fl**; **Bra** triangular, 14 × 6 mm; **Ped** 16 mm; **Fl** deep wine-red, 30 mm, base truncate, 10 mm ∅ across **Ov**, abruptly narrowed to 6 mm above, then enlarged to 9 mm and narrowed again to mouth; **Tep** free for 11 mm.

A. kniphofioides Baker (HIP 1890: t. 1939, 1890). **T:** RSA, KwaZulu-Natal (*Tyson* 2829 [SAM, GRA, BOL]). – **D:** RSA (KwaZulu-Natal, Mpumalanga); grassy slopes, 1220 - 1525 m. **I:** Reynolds (1950: 122-123).

Incl. *Aloe marshallii* J. M. Wood & M. S. Evans (1897).

[1] Acaulescent, simple, with ovoid bulb-like underground swelling 6 - 8 × 5 - 6 cm; **R** fusiform; **L** ± 20, multifarious, linear, 20 - 30 × 0.6 - 0.7 cm, green; marginal teeth absent or minute, white; **Inf** to 55 cm, simple; raceme cylindrical, 10 - 15 cm, lax, with ± 12 - 16 **Fl**; **Bra** ovate-acuminate, to 15 mm; **Ped** to 15 mm; **Fl** scarlet, green-tipped, 35 - 50 mm, base rounded, not narrowed above **Ov**; **Tep** free for 6 - 8 mm; **St** exserted 0 - 1 mm. – *Cytology:* 2n = 14 (Müller 1945: as *A. marshallii*).

A. krapohliana Marloth (Trans. Roy. Soc. South Afr. 1: 408, 1908). **T:** RSA, Northern Cape (*Marloth* 4673 [PRE]). – **D:** RSA.

A. krapohliana var. **dumoulinii** Lavranos (JSAB 39(1): 41-43, ills., 1973). **T:** RSA, Northern Cape (*Lavranos & Butler* 8777 [PRE]). – **D:** RSA (Northern Cape); exposed quartzite hills.

[6] Differs from var. *krapohliana*: Dividing and suckering to form groups of up to 15 **Ros**; **L** deltoid and strongly incurved, shorter; **Inf** 15 cm; **Ped** 12 -

15 mm; **Fl** 25 - 27 mm, ventricose, 6 - 7 mm ∅ across **Ov**, enlarged to 8 - 9 mm above, then narrowing to 4 mm at mouth.

A. krapohliana var. **krapohliana** – **D:** RSA (Northern Cape, Western Cape); arid rocky slopes. **I:** Reynolds (1950: 179-181). **Fig. XVI.c**

[4] Acaulescent or shortly caulescent, usually simple, sometimes 2 - 3 **Ros**; stem to 20 cm; **L** 20 - 30, rosulate, lanceolate-acuminate, to 20 × 4 cm, glaucous, surface smooth; marginal teeth minute, white, 3 - 5 mm apart; **Inf** to 40 cm, simple or with 1 - 2 **Br**; racemes cylindrical, 14 × 6 cm, sub-dense; **Bra** lanceolate-acuminate, to 20 × 5 mm; **Ped** to 20 mm; **Fl** scarlet, greenish at mouth, 35 mm, base rounded, 7 mm ∅ across **Ov**, very slightly enlarged above; **Tep** free to base. – *Cytology:* 2n = 14 (Resende 1937).

A. kraussii Baker (JLSB 18: 159, 1880). **T:** RSA, KwaZulu-Natal (*Krauss* 275 [BM]). – **D:** RSA (KwaZulu-Natal); in grass on stony slopes. **I:** Reynolds (1950: 143-144).

[4] Acaulescent or very shortly caulescent, simple or in small groups; **R** fusiform; **L** 8 - 10, distichous, becoming rosulate with age, linear-acuminate, 30 - 40 (-60) × 3.5 (-5) cm, dull green, lower face sometimes with a few white spots near base, with narrow white cartilaginous margin; marginal teeth minute, obsolescent towards tip; **Inf** to 40 cm, simple; racemes capitate, dense, with 30 - 40 **Fl**; **Bra** lanceolate-deltoid, ± 15 mm; **Ped** 35 mm; **Fl** lemon-yellow to yellow, 16 - 18 mm, base attenuate, narrowed above **Ov** to mouth; **Tep** free almost to base.

A. kulalensis L. E. Newton & Beentje (CSJA 62(5): 251-252, ills., 1990). **T:** Kenya, Eastern Prov. (*Newton* 3219 [K, EA]). – **D:** N Kenya; steep slopes and rock faces, 1585 - 1890 m. **Fig. XV.d**

[13] Caulescent, branching at base; stem sprawling or pendulous, to 2 m × 2.15 cm ∅; **L** 20 - 25, lax, ± 1 cm apart, triangular, 18 - 25 × 1.8 - 3 cm, mid-green, surface smooth; marginal teeth 1 mm, colourless, 2 - 13 mm apart; exudate drying yellow; **Inf** to 32 cm, ascending, with 2 - 4 **Br**; racemes cylindrical, 2.5 - 15 cm, lax; **Bra** triangular, 5 - 6 × 1 mm; **Ped** 8 - 10 mm; **Fl** scarlet with yellow tip, 25 mm, base rounded, 7 mm ∅ across **Ov**, narrowed gradually to 5 mm above, enlarging to 6 mm at mouth; **Tep** free for 10 - 13 mm.

A. labworana (Reynolds) S. Carter (Fl. Trop. East Afr., Aloaceae, 28, 1994). **T:** Uganda, Karamoja Distr. (*Jackson* s.n. in *Reynolds* 7980 [PRE, EA, K]). – **D:** Sudan, Uganda; rocky outcrops, 1300 - 1500 m. **I:** Reynolds (1966: 292-293, as *A. schweinfurthii* var.).

≡ *Aloe schweinfurthii* var. *labworana* Reynolds (1956).

[7] Acaulescent, suckering to form groups; **L** 12 - 16, rosulate, lanceolate-attenuate, 60 - 80 × 7 - 8 cm, bluish-green with grey bloom, with many pale greenish lenticular spots; marginal teeth 5 - 6 mm, pungent, reddish-tipped, 12 - 20 mm apart; exudate drying yellowish; **Inf** to 90 cm, with 10 - 12 **Br**; racemes cylindrical, slightly conical, 8 - 9 × 5 cm, sub-dense; **Bra** ovate-acute, ± 2 mm; **Ped** 10 mm; **Fl** yellow, 28 mm, base rounded, 6 - 7 mm ∅ across **Ov**, very slightly narrowed above, then slightly enlarging to mouth; **Tep** free for 9 mm. – *Cytology:* 2n = 14 (Newton 1970: as *A. schweinfurthii* var.).

A. laeta A. Berger (in Engler, A. (ed.), Pflanzenr. IV.38 (Heft 33): 256-257, 1908). **T:** Madagascar, Antananarivo (*Catat* 1115 [P]). – **D:** Madagascar.

A. laeta var. **laeta** – **D:** Madagascar; on rocks, 1600 - 2200 m. **I:** Reynolds (1966: 438-439).

[2] Acaulescent or very shortly caulescent, simple; stem to 5 × 3.5 cm; **L** ± 24, densely rosulate, lanceolate-attenuate, to 20 × 7 - 8 cm, bluish-grey, obscurely lineate, with conspicuous narrow pink cartilaginous margin, surface rough; marginal teeth 2 mm, narrowly deltoid, firm, pink, confluent at base; **Inf** 40 - 60 cm, usually simple, sometimes with 2 - 3 **Br**; raceme capitate or subcapitate, 5 - 7 × 6 cm, dense; **Bra** deltoid-acute, 5 × 2.5 mm; **Ped** 20 - 25 mm; **Fl** crimson, 15 mm, base attenuate, enlarged above **Ov** to 7 mm, then narrowing to mouth; **Tep** free almost to base; **St** exserted 0 - 1 mm.

A. laeta var. **maniaensis** H. Perrier (Mém. Soc. Linn. Normandie, Bot. 1(1): 30, 1926). **T:** Madagascar, Fianarantsoa (*Perrier* 11003 [P]). – **D:** Madagascar; quartzite, ± 1400 m.

[2] Differs from var. *laeta*: **L** 8 × 2 cm, tip rounded; **Fl** straight, infundibuliform.

A. lateritia Engler (Pfl.-welt Ost-Afr., Teil C, 140, 1895). **T:** Tanzania, Moshi Distr. (*Volkens* 404 [B, BM]). – **D:** Ethiopia, Kenya, Tanzania.

A. lateritia var. **graminicola** (Reynolds) S. Carter (Fl. Trop. East Afr., Aloaceae, 17, 1994). **T:** Kenya, Central Prov. (*Reynolds* 6576 [PRE, EA, K]). – **D:** Ethiopia, Kenya; grassland and open bushland, 1675 - 2530 m. **I:** Reynolds (1966: 79-80, as *A. graminicola*). **Fig. XV.e**

≡ *Aloe graminicola* Reynolds (1953); **incl.** *Aloe solaiana* Christian (1940).

[7] Differs from var. *lateritia*: Acaulescent or very shortly caulescent, usually suckering to form dense clumps; stem to 50 cm; **L** with marginal teeth more pungent; **Inf** with racemes always capitate. – *Cytology:* 2n = 14 (Brandham 1971: as *A. graminicola*).

Natural hybrids with *A. nyeriensis* have been reported by Carter (1994).

A. lateritia var. **lateritia** – **D:** Kenya, Tanzania; grassland and open bushland, often on rocky slopes, 250 - 2125 m. **I:** Reynolds (1966: 96, fig. 100).

Incl. *Aloe boehmii* Engler (1895); **incl.** *Aloe campylosiphon* A. Berger (1904); **incl.** *Aloe amanensis* A. Berger (1905).

[5] Acaulescent, usually simple, sometimes suckering to form small groups; **L** 16 - 20, densely rosulate, lanceolate-attenuate, 25 - 50 × 5 - 10 cm, bright green, usually with elongated white spots in irregular transverse bands, surface smooth; marginal teeth 3 - 4 mm, pungent, brown, 10 - 15 mm apart; **Inf** to 1.25 m, with 3 - 8 **Br**; racemes capitate to subcapitate, 4 - 12 × 8 cm, dense, sometimes to 20 cm, lax; **Bra** linear-lanceolate, 10 - 20 (-25) × 4 mm; **Ped** 20 - 30 mm; **Fl** orange-red, sometimes yellow, usually glossy, 30 - 38 mm, base truncate, 8 - 10 mm ∅ across **Ov**, abruptly narrowed to ± 5 mm above, enlarging to mouth; **Tep** free for 10 - 13 mm. – *Cytology:* 2n = 14 (Resende 1937).

A. lavranosii Reynolds (JSAB 30(4): 225-227, t. 31-32, 1964). **T:** Yemen (*Lavranos* 1890 [PRE, K]). – **D:** Yemen; basalt hills, 1370 m. **I:** Reynolds (1966: 251-252).

[5] Acaulescent or shortly caulescent, usually in groups of up to 8 **Ros**, rarely simple; stem decumbent, to 15 cm; **L** 10 - 14, densely rosulate, deltoid-acute, somewhat falcate, tip with 2 - 4 small sharp teeth, 28 × 7.5 cm, olive-green tinged brownish, with pinkish-brown horny margin; marginal teeth 3 mm, pungent, brownish, 6 - 12 mm apart; **Inf** 60 cm, ascending to suberect, with 4 - 9 **Br**; racemes cylindrical-acuminate, 16 - 20 × 5 cm, sub-dense; **Bra** ovate-lanceolate, acute, 8 × 3 mm; **Ped** 8 mm; **Fl** bright yellow, 30 mm, base rounded, 7 mm ∅ across **Ov**, slightly narrowed above, enlarging slightly to mouth; **Tep** free for 10 mm.

A. leachii Reynolds (JSAB 31: 275, 1965). **T:** Tanzania (*Leach & Brunton* 10178 [PRE, K]). – **D:** Tanzania; open wooded grassland, 275 - 500 m. **I:** Reynolds (1966: 227).

[5] Acaulescent or shortly caulescent, simple or with a few suckers forming small groups; **L** ± 20, rosulate, lanceolate-attenuate, 35 × 6 cm, dark green tinged red; marginal teeth ± 5 mm, pungent, reddish-brown, 10 - 20 mm apart; **Inf** to 1 m, with 7 - 10 **Br**, lower **Br** sometimes rebranched; terminal racemes cylindrical, laterals sometimes with ± secund **Fl**, 15 - 20 × 7 cm, lax; **Bra** ovate-acute, 5 × 4 mm; **Ped** 6 - 8 mm; **Fl** scarlet, paler at mouth, 30 mm, base rounded, 7 mm ∅ across **Ov**, scarcely narrowed above; **Tep** free for 15 mm.

A. leandrii Bosser (Adansonia, n.s., 8(4): 505-507, ills., 1968). **T:** Madagascar, Centre-Est (*Bosser* 18550 [P]). – **D:** Madagascar; quartz sand at edge of forests, 900 - 1000 m.

[4] Acaulescent or shortly caulescent, simple or branching at base; **L** 8 - 15, rosulate, narrowly lan-

ceolate or linear, 14 - 15 × 1 - 2 cm, green; marginal teeth 1 - 1.5 mm, 3.5 - 8 mm apart; **Inf** 70 - 85 cm, simple; racemes cylindrical, 5 - 10 cm, lax; **Bra** ovate or oblong, 5 - 7 mm; **Ped** 15 - 25 mm; **Fl** yellow, 25 - 27 mm, base rounded, slightly narrowed above **Ov**; **Tep** free for 8 - 9 mm; **St** not exserted.

A. leedalii S. Carter (Fl. Trop. East Afr., Aloaceae, 9-11, ills., 1994). **T:** Tanzania, Njombe Distr. (*Leedal* 5523 [K]). − **D:** Tanzania; among rocks in crevices and on ledges, 2130 - 2950 m.

[10] Caulescent, suckering to form dense groups; stem ascending, to 5 × 0.5 - 1 cm, dead **L** sheath persistent; **L** rosulate, linear, to 30 × ± 1 cm, green; marginal teeth minute, cartilaginous, white, densely crowded; **Inf** to 30 cm, simple; raceme cylindrical, 5 - 8 cm, ± dense; **Bra** ovate-acute, 15 - 20 × 5 - 7 mm, orange-brown; **Ped** ± 25 mm; **Fl** bright orange, green-tipped, 25 - 30 mm, base obconical, 5 - 6 mm ∅ across **Ov**, very slightly narrowed above, slightly enlarging to mouth; **Tep** free almost to base; **St** scarcely exserted.

A. lensayuensis Lavranos & L. E. Newton (CSJA 48(6): 276-278, ills., 1976). **T:** Kenya, Eastern Prov. (*Classen* 92 [EA]). − **D:** Kenya; gneiss inselbergs, 700 - 2000 m.

[13] Caulescent, suckering sparsely at base; stem decumbent, to 1 m × 2.5 - 3 cm ∅; **L** ± 10, laxly rosulate, lanceolate-acute, 30 × 5 cm, bluish-grey, sometimes tinged pinkish, sometimes with a few scattered elongated whitish spots, with white or reddish cartilaginous margin, surface slightly rough; marginal teeth 1 mm, soft, white, 2 - 10 mm apart; exudate yellow, drying brownish; **Inf** ± 75 cm, with 8 - 12 **Br**; racemes cylindrical or with secund **Fl** on oblique **Br**, lax, with 5 - 12 **Fl**; **Bra** ovate-acute, 1 - 3 × 1.5 mm; **Ped** ± 8 mm; **Fl** pinkish-red with pronounced bloom and pinkish-cream margins on the lobes, 20 - 24 mm, base shortly attenuate, 6 - 10 mm ∅ across **Ov**, narrowed slightly above; **Tep** free for 6 - 7 mm. − *Cytology:* 2n = 14 (from protologue).

A. lepida L. C. Leach (JSAB 40(2): 102-106, ills., 1974). **T:** Angola, Huambo Distr. (*Baptista de Sousa* s.n. in *Leach* 14538A [LISC, SRGH]). − **D:** Angola; rock slopes in shade of trees.

[15] Caulescent, branching at base; stem erect, to 30 cm; **L** rosulate, ovate-attenuate, 20 - 28 × 7.5 - 9 cm, bright to dark yellowish deep green, with white spots in wavy transverse bands, smaller and more numerous on lower face; marginal teeth 3 - 7 mm, pungent, often hooked, brown-tipped, 6 - 12 mm apart; **Inf** 30 - 50 cm, with 1 - 2 **Br**; racemes cylindrical-acuminate, 20 × 7 - 8 cm, lax; **Bra** ovate-acuminate, 6 - 7 × 3 - 3.5 mm; **Ped** 15 - 20 mm; **Fl** pale orange-scarlet, somewhat yellowish-striped, 25 - 29 mm, base truncate, ± 5.5 mm ∅ across **Ov**, narrowed to ± 4 mm above, then enlarging to ± 6 mm at mouth; **Tep** free for 5 - 6 mm; **St** not exserted.

A. leptosiphon A. Berger (BJS 36: 66, 1905). **T:** Tanzania, Lushoto Distr. (*Engler* 1073a [B]). − **D:** Tanzania; steep rocky slopes, often in grass, 1200 - 1675 m. **I:** Reynolds (1966: 74, as *A. greenwayi*). **Fig. XVI.d**

Incl. *Aloe greenwayi* Reynolds (1964).

[4,5] Acaulescent or shortly caulescent, simple or suckering to form small groups; stem to 50 cm; **L** ± 16, densely rosulate, ovate-lanceolate, 20 - 35 × 5 - 7 cm, glossy green, upper face sometimes with scattered elongated whitish spots, lower face with many spots; marginal teeth 2 mm, pale green, 8 - 10 mm apart; exudate yellow, drying brownish; **Inf** 40 - 60 cm, simple or usually with 1 - 2 **Br**; racemes conical-cylindrical, 10 - 20 × 5 - 7 cm, dense; **Bra** ovate-acute, 10 - 11 × 4 - 5 mm; **Ped** 8 - 10 mm; **Fl** bright red to orange-scarlet with greenish-yellow mouth, or entirely yellow, 25 - 30 mm, base shortly attenuate, 5 - 6 mm ∅ across **Ov**, slightly narrowed above; **Tep** free for ± 8 - 10 mm. − *Cytology:* 2n = 14 (Brandham 1971: as *A. greenwayi*).

A. lettyae Reynolds (JSAB 3: 137, 1937). **T:** RSA, Northern Prov. (*Reynolds* 2339 [PRE]). − **D:** RSA (Northern Prov.); in long grass or among bushes, on slopes, 945 m. **I:** Reynolds (1950: 259-260).

[3] Acaulescent, simple; **L** ± 20, densely rosulate, lanceolate-attenuate, to 45 × 9 cm, dull green with many elongated dull white spots, lower face with larger more obscure spots in transverse bands; marginal teeth 3 - 4 mm, brownish, 10 - 15 mm apart; **Inf** to 1.75 m, with 8 - 12 **Br**, lower **Br** rebranched; racemes cylindrical, slightly acuminate, 20 - 25 × 8 - 9 cm; **Bra** deltoid-acuminate, 12 - 15 mm; **Ped** 12 - 15 mm; **Fl** rose-red, 38 - 42 mm, base rounded, 10 - 11 mm ∅ across **Ov**, abruptly narrowed to 6 mm above, then enlarging towards mouth; **Tep** free for 10 mm. − *Cytology:* 2n = 14 (Müller 1941).

A. lindenii Lavranos (CSJA 69(3): 149-151, ills., 1997). **T:** Somalia, Nugaal Region (*Lavranos & al.* 23377 [UPS]). − **D:** Somalia; gypsum and limestone hills, ± 150 m.

[2] Acaulescent, solitary; **L** to ± 20, densely rosulate, lanceolate-attenuate, 40 - 45 × 7 - 8 cm, bluish-grey, distinctly striate, sometimes with scattered pale spots, margins cartilaginous; marginal teeth 1 - 2 mm, 15 - 22 (-60) mm apart; **Inf** to 70 cm, usually simple, sometimes with 1 - 2 **Br**; racemes elongate-conical, 18 - 22 cm, dense; **Bra** deltoid-attenuate, 15 × 4 mm; **Ped** 3 mm; **Fl** pale red, 30 mm, base very shortly attenuate, 5 mm ∅ across **Ov**, scarcely narrowed above, then widening towards mouth; **Tep** free for 6 - 7 mm; **St** exserted 2 mm.

A. linearifolia A. Berger (BJS 57: 640, 1922). **T:** RSA, KwaZulu-Natal (*Rudatis* 1643 [B]). − **D:** RSA (KwaZulu-Natal); in sparse grass on stony slopes, 700 m. **I:** Reynolds (1950: 139-140).

[1] Acaulescent, simple or with 1 - 2 **Br** at base;

R fusiform, 6 mm ∅; **L** 6 - 8, usually distichous, linear, ± 25 × 0.5 - 0.8 cm, green, lower face with many white and brown spots near base; marginal teeth near base, minute, sometimes absent; **Inf** 20 - 25 cm, simple; raceme capitate, 2 cm, with 16 - 24 **Fl**; **Bra** pale brownish, to 10 × 5 mm; **Ped** 12 - 15 mm; **Fl** greenish-yellow to yellow, 12 mm, base attenuate, not narrowed above **Ov**; **Tep** free almost to base; **St** exserted 0 - 1 mm.

A. lineata (Aiton) Haworth (Trans. Linn. Soc. London 7: 18, 1804). **T:** not typified. – **D:** RSA.
≡ *Aloe perfoliata* var. *lineata* Aiton (1789).

A. lineata var. **lineata** – **D:** RSA (Western Cape, Eastern Cape); in dry scrub and grassland. **I:** Reynolds (1950: 203-205, t. 11).

Incl. *Aloe lineata* var. *glaucescens* Haworth (1821); **incl.** *Aloe lineata* var. *viridis* Haworth (1821).

[8] Caulescent, simple or branched, usually covered with dead **L**; stem erect, to 1.5 m; **L** 30 - 40, densely rosulate, lanceolate-acuminate, 30 - 40 × 7 - 9 cm, dull to bright green, distinctly lineate, with reddish-brown horny margin; marginal teeth 4 - 6 mm, pungent, reddish-brown, 5 - 15 mm apart; **Inf** 75 - 100 cm, simple; raceme conical, 20 - 30 cm, sub-dense; **Bra** deltoid-acute, ± 20 mm; **Ped** 40 mm; **Fl** salmon-pink, 45 - 50 mm, base truncate, 9 mm ∅ across **Ov**, narrowed to 8 mm above, then enlarged to 11 mm and narrowing slightly to mouth; **Tep** free to base or almost to base. – *Cytology:* 2n = 14 (Müller 1941).

Natural hybrids with *A. humilis* have been reported by Reynolds (1950).

A. lineata var. **muirii** (Marloth) Reynolds (Aloes South Afr. [ed. 1], 205, 1950). **T:** RSA, Western Cape (*Muir* 3627 [PRE]). – **D:** RSA (Western Cape); quartzite slopes. **I:** Reynolds (1950: 204).
≡ *Aloe muirii* Marloth (1929).

[8] Differs from var. *lineata*: **L** bright yellowish-green to somewhat orange-green, more distinctly lineate; marginal teeth larger. – *Cytology:* 2n = 14 (Resende 1937: as *A. muirii*).

Natural hybrids with other species have been reported (Reynolds 1950).

A. littoralis Baker (Trans. Linn. Soc. London, Bot. 1: 263, 1878). **T:** Angola, Luanda (*Welwitsch* 3727 [BM]). – **D:** Angola, Botswana, Moçambique, Namibia, Zambia, Zimbabwe, RSA; dry hills. **I:** Reynolds (1966: 317-319).

Incl. *Aloe rubrolutea* Schinz (1896); **incl.** *Aloe schinzii* Baker (1898).

[9] Caulescent, simple; stem erect, to 4 m, covered with dead **L**; **L** ± 30 - 40, densely rosulate, lanceolate-ensiform, acute, 60 × 10 - 13 cm, greygreen, sometimes with white spots, lower face sometimes with a few small brown prickles in me-

dian line; marginal teeth 3 - 4 mm, pungent, brown, 10 - 20 mm apart; **Inf** 1.5 m, with 8 - 10 **Br**, lower **Br** sometimes rebranched; racemes cylindrical-acuminate, 30 × 6 cm, dense; **Bra** lanceolate, 12 - 18 × 5 - 6 mm, white, usually deflexed; **Ped** 6 - 7 mm; **Fl** rose-pink to deep pink-scarlet, paler at mouth, 30 - 34 mm, base rounded, 6 mm ∅ across **Ov**, enlarging slightly above middle; **Tep** free for 15 - 17 mm. – *Cytology:* 2n = 14 (Koshy 1937).

Natural hybrids with *A. zebrina* have been reported (Reynolds 1966).

A. lolwensis L. E. Newton (CSJA 73: in press, 2001). **T:** Kenya (*Hartmann & Newton* 28585 [K, EA]). – **D:** Kenya; grassland, 1220 - 1550 m.

[7] Acaulescent, suckering to form dense clumps; **L** ± 30, densely rosulate, erect to slightly spreading, lanceolate, to 50 × 11 cm, glossy mid-green, surface smooth; marginal teeth 3 mm, pungent with tips pointing forwards, green with red-brown tips, 12 - 15 mm apart; exudate brownish-yellow, drying pale brown; **Inf** to 1.3 m, erect, with 7 - 9 **Br**, lower **Br** rebranched; racemes cylindrical, terminal raceme to 20 cm, 6 cm ∅, lateral racemes 6 - 15 cm, subdense; **Bra** ovate-acute, 3 × 2 mm; **Ped** 6 - 7 mm; **Fl** bright coral-red, lobes with pale margins, 28 - 33 mm, base rounded and very shortly attenuate, 6 mm ∅ across **Ov**, constricted to 5 mm above **Ov**, enlarging to 7 mm (side view) just below mouth; **Tep** free for 13 - 14 mm; **St** exserted 5 - 10 mm.

A. lomatophylloides Balfour *fil.* (JLSB 16: 22, 1877). **T:** Rodrigues Island (*Balfour* 1306 [K, E]). – **D:** Rodrigues Island.
≡ *Lomatophyllum lomatophylloides* (Balfour *fil.*) Marais (1975).

[3] Caulescent; stem decumbent, short, 3 - 4 cm ∅; **L** densely rosulate, lanceolate-attenuate to ensiform, 50 - 75 × 8 cm, dark green; **Inf** 10 - 12 cm, with 2 - 3 **Br**; racemes cylindrical, sub-dense; **Bra** lanceolate, 1.5 - 4 mm; **Ped** 20 - 23 mm; **Fl** orange-red, 14 - 21 mm, base attenuate, slightly narrowed above **Ov**, then enlarging to mouth; **Tep** free for 5 - 10 mm; **St** not exserted; **Fr** berries.

A. longistyla Baker (JLSB 18: 158, 1880). **T** [lecto]: RSA, Eastern Cape (*Bolus* 689 [K, PRE [photo]]). – **D:** RSA (Western Cape, Eastern Cape); sandy or stony ground, usually on flats. **I:** Reynolds (1950: 158-159, t. 6).

[4] Acaulescent, simple or branching at base forming groups of up to 10 **Ros**; **L** 20 - 30, densely rosulate, lanceolate, 12 - 15 × 3 cm, glaucous with a bloom, surface with soft to firm white prickles with tubercular base; marginal teeth 3 - 4 mm, firm, white, 5 mm apart; **Inf** 15 - 20 cm, simple; raceme conical, 11 × 11 cm, dense, with 40 - 50 **Fl**; **Bra** lanceolate-deltoid, acute, sometimes slightly fleshy, 25 - 30 × 12 - 15 mm; **Ped** 6 - 8 mm; **Fl** pale salmon-pink to rose-red, 55 mm, upper ⅓ curving up-

wards, base rounded, 9 mm ∅ across **Ov**, narrowed slightly at mouth; **Tep** free for 13 mm; **St** exserted 25 mm. – *Cytology:* 2n = 14 (Resende 1937).

A. luapulana L. C. Leach (JSAB 38(3): 185-188, ills., 1972). **T:** Zambia, Luapula Distr. (*Weeks* 1 in *Leach* 14754 [SRGH, PRE]). – **D:** Zambia; termite mounds and rocks near waterfall, ± 1230 m.

[3] Acaulescent, simple; **L** ± 16, rosulate, ovate-attenuate, ± 30 × 6 - 7.5 cm, pale green, obscurely striate, lower face greyish-green, more clearly striate; marginal teeth 1 - 3.5 mm, pungent, whitish at base, brownish-orange at tip, 2 - 5 mm apart near base, larger and 15 - 18 mm apart above; exudate drying yellow; **Inf** 1 - 1.2 m, with 6 **Br**; racemes cylindrical-acuminate, 16 - 26 cm, lax; **Bra** ovate-acute, 4 - 5 × 3.5 mm; **Ped** ± 7.5 mm; **Fl** coral-red with slight bloom, becoming yellowish at mouth, 29 - 33 mm, base truncate, 8 - 9 mm ∅ across **Ov**, narrowed to ± 5 mm above, then enlarging very slightly to mouth; **Tep** free for 12 - 16 mm.

A. lucile-allorgeae Rauh (Bradleya 16: 97-98, 100, ills., 1998). **T:** Madagascar, Tolanaro (*Humbert* 13738 [P]). – **D:** Madagascar; W slopes of gneiss hills, 800 m.

[6] Caulescent; stems up to 35 cm, with thin creeping part and thicker erect fertile part; **L** scattered along stem, gradually tapering to the tip, 5 - 8 × 3 cm; marginal teeth deltoid, 1 mm; sheath long; **Inf** up to 20 cm, simple; raceme with few **Fl**; **Bra** ovate, acute, 2 - 3 mm; **Ped** 7 mm; **Fl** red becoming paler above, with green tip, 20 mm, slightly constricted above **Ov**, base rounded; **Tep** free for < 6.5 mm; **St** not exserted; **Fr** not known.

A. luntii Baker (BMI 1894: 342, 1894). **T:** Yemen (*Lunt* 225 [K]). – **Lit:** Lavranos (1992). **D:** Oman, Yemen, Somalia; stony hills, 1000 m. **I:** Reynolds (1966: 222, as *A. inermis*).

[7] Caulescent, suckering to form small groups; stem to 30 cm; **L** 7 - 8, distichous at first, becoming densely rosulate, ensiform-acute, ± 30 × 5 cm, grey to brownish with a bloom, surface smooth; marginal teeth absent; **Inf** with 4 - 5 **Br**; racemes with secund **Fl**, lax; **Bra** 3 - 4 mm; **Ped** 4 - 6 mm; **Fl** dull red to coral-pink with conspicuous bloom, 26 - 28 mm, base shortly attenuate, slightly narrowed above **Ov**, enlarging again to mouth; **Tep** free for < ½ their length; **St** exserted 4 - 5 mm.

Reynolds (1966) included this name as a synonym of *A. inermis*, but Lavranos (1992) argued for its recognition as a distinct species.

A. lutescens Groenewald (FPSA 18: t. 707 + text, 1938). **T:** RSA, Northern Prov. (*van der Merwe* 1377 [PRE 23301 (= 23005)]). – **D:** RSA (Northern Prov.); between trees or shrubs on dry rocky or stony slopes. **I:** Reynolds (1950: 338).

[7] Caulescent, usually dividing to form dense

groups; stem erect or decumbent, to 80 cm; **L** ± 30, densely rosulate, lanceolate-attenuate, 50 - 60 × 8 - 9 cm, dull to semi-glossy yellowish-green; marginal teeth ± 2 mm, pinkish-brown, 3 - 5 mm apart; **Inf** 1 - 1.5 m, with 3 **Br** in one vertical plane; racemes cylindrical-acuminate, 35 - 40 × 7 cm; **Bra** ovate-acuminate, 15 × 15 mm; **Ped** to 15 mm; **Fl** yellow, sometimes with slightly reddish base, opening from scarlet and green-tipped buds, 30 - 35 mm, base rounded, not narrowed above **Ov**; **Tep** free to base. – *Cytology:* 2n = 14 (Brandham 1971).

A. macleayi Reynolds (JSAB 21(2): 55-57, t. 3-4, 1955). **T:** Sudan, Equatoria Prov. (*MacLeay* s.n. in *Reynolds* 6773 [PRE, EA, K]). – **D:** Sudan; amongst rocks in grassland, 1525 - 2440 m. **I:** Reynolds (1966: 299-300).

[3] Acaulescent, simple; **L** ± 24, densely rosulate, lanceolate-attenuate, to 50 × 12 cm, deep green to olive-green, obscurely lineate, with yellowish-white margins; marginal teeth 4 mm, firm, white, uncinate, 8 - 15 mm apart; exudate drying yellow; **Inf** 90 cm, slightly oblique but with racemes erect, with ± 9 **Br**; racemes cylindrical-conical, 22 × 5.5 cm, lax; **Bra** ovate-acute, pale brown, 3 × 4 mm; **Ped** 10 mm; **Fl** scarlet at base, becoming orange to yellowish towards mouth, 36 mm, base rounded, 7 mm ∅ across **Ov**, very slightly narrowed above; **Tep** free to base.

A. macra Haworth (Suppl. Pl. Succ., 45, 105, 1819). **T:** not typified. – **Lit:** Marais (1975). **D:** Réunion; dry hills.

≡ *Phylloma macrum* (Haworth) Sweet (1827) ≡ *Lomatophyllum macrum* (Haworth) Salm-Dyck *ex* Schultes & Schultes *fil.* (1829).

[2] Caulescent; stem to 30 cm; **L** 10 - 12, laxly rosulate, ensiform-attenuate, 30 - 35 × 3 cm, green with red margin; marginal teeth minute, red; **Inf** 30 cm, simple or with 1 **Br**; racemes cylindrical, 10 - 15 cm, dense; **Bra** lanceolate, ± 5 mm; **Ped** ± 10 mm; **Fl** reddish-orange, becoming yellow towards tip, 13 - 14 mm; **St** not exserted; **Fr** berries.

A. macrocarpa Todaro (Hort. Bot. Panorm. 1: 36, t. 9, 1875). **T:** Ethiopia, Tigrey Region (?) (*Schimper* s.n. [[icono]: l.c. t. 9]). – **Lit:** Gilbert & Sebsebe (1997). **D:** Benin, Cameroon, Eritrea, Ethiopia, Ghana, Mali, Nigeria, Sudan, Djibouti; rocky grassland and on rock ledges, 400 - 2000 m. **I:** Reynolds (1966: 92-95).

Incl. *Aloe commutata* Engler (1892); **incl.** *Aloe macrocarpa* var. *major* A. Berger (1908); **incl.** *Aloe edulis* A. Chevalier (1920); **incl.** *Aloe barteri* Schnell (1953) (*nom. illeg.*, Art. 53.1).

[5] Acaulescent or shortly caulescent, simple or suckering to form small groups; stems to 30 cm; **L** 16 - 20, densely rosulate, lanceolate to lanceolate-attenuate, 20 - 40 × 6 - 7 cm, green with many dull

white or pale greenish oval spots arranged irregularly or in transverse bands, surface smooth; marginal teeth 3 mm, pale brown, 8 - 10 mm apart; **Inf** 80 - 100 cm, with 3 - 5 **Br**; racemes cylindrical, 15 - 20 × 6 cm, lax; **Bra** 8 × 3 mm; **Ped** 12 - 15 mm; **Fl** scarlet, 25 - 35 mm, base truncate, 8 mm ∅ across **Ov**, abruptly narrowed to 5 mm above, then enlarging to mouth; **Tep** free for 6 - 7 mm; **St** exserted 0 - 1 mm. – *Cytology:* 2n = 14 (Satô 1937).

Some populations, including those in W Africa, have been named as *A. macrocarpa* var. *major*, but there is no discontinuity in characters between this and the typical variety. A natural hybrid with *A. schweinfurthii* has been seen in Nigeria (Newton, unpublished).

A. macroclada Baker (JLSB 20: 273, 1883). **T** [syn]: Madagascar (*Baron* 1178 [P?]). – **D:** Madagascar; grassland on dry mountain slopes, 700 - 1500 m. **I:** Reynolds (1966: 484-486).

[2] Acaulescent, simple; **L** ± 36, densely rosulate, ensiform-attenuate, 75 × 15 cm, green; marginal teeth 3 mm, pungent, orange-brown, paler towards base, ± 10 mm apart; **Inf** 1.75 (-2.4) m, simple or sometimes with 1 **Br**; raceme cylindrical, 60 - 75 (-100) × 7 cm, very dense; **Bra** ovate-acute, shortly cuspidate, 10 × 7 mm, reflexed at base; **Ped** 4 - 5 mm; **Fl** pale scarlet, greenish inside mouth, 20 - 25 mm, campanulate, base rounded, 6 mm ∅ across **Ov**, enlarged above to 20 mm across mouth; **Tep** free to base; **St** exserted 8 - 10 mm.

Natural hybrids with varieties of *A. capitata* have been reported by Bosser (1968) and Reynolds (1966).

A. macrosiphon Baker (FTA 7: 459, 1898). **T:** Tanzania, Bukoba Distr. (*Scott-Elliot* 8176 [K, BM]). – **D:** Kenya, Rwanda, Tanzania, Uganda; usually in shade of thickets or among rocks, 1125 - 1585 m. **I:** Reynolds (1966: 177-179).

Incl. *Aloe mwanzana* Christian (1940); **incl.** *Aloe compacta* Reynolds (1961).

[7] Acaulescent, suckering to form dense clumps; **L** ± 16, densely rosulate, lanceolate-attenuate, 50 - 70 × 5 - 10 cm, glossy dark green, tinged reddish-brown in the open, with many dull white to pale green elongated spots, spots smaller on lower face, surface smooth; marginal teeth 2 - 5 mm, pungent, brown-tipped, 8 - 15 mm apart; exudate yellow drying brownish; **Inf** 1 - 1.5 m, with 8 - 10 **Br**; racemes cylindrical-acuminate, to 20 × 6 cm, sub-dense; **Bra** ovate-acute, 10 - 15 × 5 - 8 mm, white; **Ped** 10 mm; **Fl** bright rose-pink, pale yellowish at mouth, 27 - 33 mm, base shortly attenuate, 7 mm ∅ across **Ov**, slightly narrowed above, then enlarging to mouth; **Tep** free for 9 - 11 mm. – *Cytology:* 2n = 14 (Brandham 1971).

Natural hybrids with other species have been reported (Reynolds 1966).

A. maculata Allioni (Auct. Syn., 13, 1773). **T:** [icono]: Commelin, Hort. Med. Amstel. 2: fig. 5, 1701. – **Lit:** Wyk & Smith (1996). **D:** Lesotho, RSA (Eastern Cape, Free State, KwaZulu-Natal, Mpumalanga, Western Cape), Swaziland; grassland to rocky slopes, to 1830 m. **I:** Reynolds (1950: 225-228, t. 12, as *A. saponaria*). **Fig. XVI.e**

Incl. *Aloe saponaria* var. *saponaria*; **incl.** *Aloe perfoliata* var. θ Linné (1753); **incl.** *Aloe perfoliata* var. λ Linné (1753); **incl.** *Aloe disticha* Miller (1768) (*nom. illeg.*, Art. 53.1); **incl.** *Aloe maculosa* Lamarck (1783); **incl.** *Aloe maculata* Medikus (1786) (*nom. illeg.*, Art. 53.1); **incl.** *Aloe perfoliata* var. *saponaria* Aiton (1789) ≡ *Aloe saponaria* (Aiton) Haworth (1804); **incl.** *Aloe umbellata* De Candolle (1799); **incl.** *Aloe saponaria* var. *latifolia* Haworth (1804); **incl.** *Aloe latifolia* Haworth (1812); **incl.** *Aloe leptophylla* N. E. Brown *ex* Baker (1880); **incl.** *Aloe saponaria* var. *brachyphylla* Baker (1880); **incl.** *Aloe leptophylla* var. *stenophylla* Baker (1896); **incl.** *Aloe saponaria* var. *ficksburgensis* Reynolds (1937).

[5] Acaulescent or shortly caulescent, simple or suckering to form dense groups; stem to 50 cm; **L** 12 - 20, densely rosulate, lanceolate, 25 - 30 × 8 - 12 cm, green with many oblong dull white spots arranged in wavy transverse bands, lower face lineate, usually without spots; marginal teeth 3 - 5 mm, pungent, brown, ± 10 mm apart; **Inf** 40 - 100 cm, with 3 - 7 **Br**; racemes capitate-corymbose, sometimes conical or round-topped, 10 - 12 × (8-) 12 - 16 cm, dense; **Bra** deltoid-acuminate, ± ⅓ to ½ as long as **Ped**; **Ped** (25-) 35 - 45 mm; **Fl** usually salmon-pink to orange, sometimes yellow or red, 35 - 45 mm, base truncate, 10 mm ∅ across **Ov**, abruptly narrowed to 6 mm above, then enlarging to the mouth; **Tep** free for 10 - 15 mm. – *Cytology:* 2n = 14 (Taylor 1925: as *A. saponaria*).

A very variable species. Gilbert & Sebsebe (1997) have suggested that the better known and long-used name *A. saponaria* should be conserved, esp. as it is the type of the well-defined Sect. *Saponariae* A. Berger. Natural hybrids with other species have been reported (Reynolds 1950: as *A. saponaria*).

A. madecassa H. Perrier (Mém. Soc. Linn. Normandie, Bot. 1(1): 23, 1926). **T:** Madagascar, Antananarivo (*Perrier* 13062 [P]). – **D:** Madagascar.

A. madecassa var. **lutea** Guillaumin (Bull. Mus. Nation. Hist. Nat., Sér. 2, 27: 86, 1955). **T:** Madagascar (*Perrier fil.* 221 [P]). – **D:** Madagascar.

[3] Differs from var. *madecassa*: **L** with marginal teeth rose; **Bra** almost colourless; **Ped** 20 mm; **Fl** lemon-yellow with green veins.

Reynolds (1966) states that the type plant had been cultivated in Paris for 26 years before being described, and the variety is doubtfully distinct.

A. madecassa var. **madecassa** – **D:** Madagascar. **I:** Reynolds (1966: 448-449).

[3] Acaulescent or almost so, simple; **L** ± 20, densely rosulate, lanceolate-acute, ± 25 × 7 - 9 cm, green, obscurely lineate, with narrow pink cartilaginous margin; marginal teeth 2 mm, pungent, pale pink, 5 - 8 mm apart; exudate drying white, like a milky froth; **Inf** 1 m or more, with 6 - 10 **Br**; racemes cylindrical-acuminate, to ± 20 × 5 - 6 cm, lax; **Bra** lanceolate, ± 9 × 3 mm, white; **Ped** 14 mm; **Fl** scarlet, ± 25 mm, slightly clavate, base shortly attenuate, ± 5 mm ∅ across **Ov**, enlarging above to near mouth; **Tep** free for 12 mm. – *Cytology:* 2n = 14 (Resende 1937).

The white frothy leaf exudate appears to be unique in the genus.

A. marlothii A. Berger (BJS 38: 87, 1905). **T** [syn]: Botswana (*Marloth* 3788 [B]). – **Lit:** Wyk & Smith (1996). **D:** Moçambique, Botswana, RSA, Swaziland.
　　Incl. *Aloe ferox* var. *xanthostachys* A. Berger (1908) (*incorrect name*, Art. 11.4); **incl.** *Aloe ferox* A. Berger (1908) (*nom. illeg.*, Art. 53.1); **incl.** *Aloe marlothii* J. M. Wood (1912) (*nom. illeg.*, Art. 53.1).

A. marlothii ssp. **orientalis** Glen & D. S. Hardy (FPA 49(3-4): pl. 1943 + text, 1987). **T:** RSA, KwaZulu-Natal (*Plowes* 2260 [PRE, LISC, SRGH]). – **D:** Moçambique, RSA (KwaZulu-Natal); sandy soil, lower altitudes than var. *marlothii*.

[8,12] Differs from var. *marlothii*: Often suckering to form clumps; stem erect or procumbent to oblique, to 1.75 m; **L** surface with few or no prickles; **Inf** with oblique racemes.

A. marlothii var. **bicolor** Reynolds (JSAB 2(1): 34, 1936). **T:** RSA, Mpumalanga (*Reynolds* 1440 [PRE]). – **D:** RSA; rocky hills.

[9] Differs from var. *marlothii*: **L** with fewer prickles; marginal teeth red; **Fl** greenish-white, opening from red buds.

Probably the hybrid *A. marlothii* × *A. rupestris.*

A. marlothii var. **marlothii** – **D:** Botswana, Moçambique, RSA (KwaZulu-Natal, Gauteng, Mpumalanga, Northern Prov., Northwest Prov.), Swaziland; rocky hills. **I:** Reynolds (1950: 480-482, t. 56). **Fig. XVI.f**
　　Incl. *Aloe supralaevis* var. *hanburyi* Baker (1896); **incl.** *Aloe spectabilis* Reynolds (1927).

[9] Caulescent, simple; stem erect, to 4 (-6) m, covered with dead **L**; **L** 40 - 50, densely rosulate, lanceolate-attenuate, 1 - 1.5 m × 20 - 25 cm, dull grey-green to glaucous, with few to many scattered reddish-brown 3 - 4 mm prickles, more numerous on lower face; marginal teeth 3 - 4 mm, pungent, reddish-brown, 10 - 15 mm apart; **Inf** ± 80 cm, with many **Br**, lower **Br** rebranched, with a total of 20 - 30 racemes; racemes horizontal to suboblique with

secund **Fl**, 30 - 50 × 5 - 6 cm, dense; **Bra** ovate to lanceolate-acute, ± 8 - 9 × 5 mm, brownish; **Ped** 5 mm; **Fl** orange to yellowish-orange, 30 - 35 mm, clavate to ventricose, base rounded, ± 7 mm ∅ across **Ov**, enlarging above and narrowing at mouth; **Tep** free for 20 - 23 mm; **St** exserted 15 mm. – *Cytology:* 2n = 14 (Resende 1937).

Natural hybrids with other species have been reported (Reynolds 1950).

A. massawana Reynolds (JSAB 25: 207-209, pl. 18-19, 1959). **T:** Tanzania, Dar es Salaam Distr. (*Reynolds* 8733 [PRE, EA, K]). – **D:** Tanzania; sandy soils near coast. **I:** Reynolds (1966: 154). **Fig. XV.f**
　　Incl. *Aloe kirkii* Baker (1894).

[7] Acaulescent or shortly caulescent, suckering at base to form dense clumps to 3 m ∅; **L** ± 16, densely rosulate, lanceolate-attenuate, ± 50 × 10 cm, dull grey-green, sometimes with a few white spots towards base; marginal teeth 2 - 3 mm, soft to firm, white with reddish-brown tips, 15 - 25 mm apart, smaller and closer towards base; exudate drying yellow; **Inf** 1.2 - 1.5 m, with 2 - 3 **Br**; racemes cylindrical-acuminate, 15 - 20 × 5 cm, sub-dense; **Bra** ovate-deltoid, 7 × 3 mm; **Ped** 6 mm; **Fl** pale scarlet, 30 - 32 mm, base rounded, 7 mm ∅ across **Ov**, slightly narrowed above, then enlarging slightly to mouth; **Tep** free for 12 mm.

Material from Massawa in Eritrea included here by Reynolds (1966) has recently been described as *A. eumassawana.*

A. kirkii is insufficiently known and placed here with a question mark by Carter (1994).

A. mawii Christian (JSAB 6(4): 186-188, t. 23, 1940). **T:** Malawi (*Christian* 942 [SRGH]). – **D:** Malawi, Moçambique, Tanzania; rocky slopes, 550 - 1830 m. **I:** Reynolds (1966: 237-240). **Fig. XVII.a**

[4,8] Acaulescent or usually caulescent, simple or with a few **Br**; stem erect, to 2 m × 10 - 12 cm; **L** 20 or more, densely rosulate, lanceolate-ensiform, to 60 × 10 cm, greyish green or green with bluish tinge, somewhat striate, with narrow reddish margin, surface smooth; marginal teeth 3 - 4 mm, pinkish-tipped, 7 - 15 mm apart; exudate pale yellow; **Inf** to 1 m, oblique, simple or with 1 **Br**; raceme oblique or horizontal, with **Fl** secund, 30 cm, dense; **Bra** triangular-acute, shortly cuspidate, 1 × 3 mm; **Ped** 1 - 2 mm; **Fl** red or orange, 35 - 40 (-48) mm, ventricose, base rounded, 7 mm ∅ across **Ov**, enlarging to 10 mm above, then narrowing to mouth; **Tep** free for 22 mm; **St** exserted ± 12 mm. – *Cytology:* 2n = 14 (Brandham 1971).

Acaulescent variants are at lower altitudes.

A. mayottensis A. Berger (in Engler, A. (ed.), Pflanzenr. IV.38 (Heft 33): 246, 1908). **T:** Comoros, Mayotte Island (*Boivin* 3071 [P]). – **D:** Comoros; in shade on hill slopes. **I:** Reynolds (1966: 507).

[15] Caulescent, branching at base and above; stem 50 × 3 - 4 cm; **L** ± 20, lanceolate-attenuate, 40 - 50 × 6 - 7 cm, green; marginal teeth 3 mm, pale yellow to pale brown, 10 - 15 mm apart; **Inf** ± 1 m, with 3 - 4 **Br**; racemes cylindrical-acuminate, to 20 × 6 cm, sub-dense; **Bra** ovate-acute, 6 × 4 mm; **Ped** ± 6 mm; **Fl** 28 mm, base shortly attenuate, 8 mm ∅ across **Ov**, very slightly narrowed above; **Tep** free for 14 mm.

A. mcloughlinii Christian (FPA 28: t. 1112 + text, 1951). **T:** Ethiopia, Harar Prov. (*McLoughlin* 826 [PRE]). − **D:** Ethiopia, Djibouti; open areas in scrub, 1160 m. **I:** Reynolds (1966: 64-65).

    Incl. *Aloe maclaughlinii* hort. (s.a.) (*nom. inval.*, Art. 61.1).

[5] Acaulescent or shortly caulescent, simple or usually in groups of up to 6 **Ros**; **L** 16 - 20, densely rosulate, lanceolate-attenuate, ± 40 - 50 × 7 cm, glossy green with many pale green elongated spots, the spots more numerous and larger on the lower face, surface smooth; marginal teeth ± 3 - 5 mm, firm, reddish-brown tipped, 10 - 15 mm apart; **Inf** 1 - 1.2 m, with 6 - 9 **Br**; racemes cylindrical-acuminate, to 15 × 6 cm, lax; **Bra** deltoid-acute, 5 × 2 mm; **Ped** 10 mm; **Fl** strawberry-red with slight bloom, 20 - 24 mm, base rounded, 9 - 10 mm ∅ across **Ov**, slightly narrowed above, then enlarging slightly to mouth; **Tep** free for 16 mm. − *Cytology:* 2n = 14 (Brandham 1971).

A. medishiana Reynolds & P. R. O. Bally (JSAB 24(4): 186-187, t. 29, 1958). **T:** Somalia (*Reynolds* 8441 [PRE, EA, K]). − **D:** Somalia; exposed rocky slopes, 1460 - 1525 m. **I:** Reynolds (1966: 307-308).

[8] Caulescent, simple or branching at base; stem erect, to 2 m × 3 - 3.5 cm; **L** ± 24, crowded along apical 20 cm of stem, ensiform, to 30 × 5.5 cm, grey-green, with narrow white cartilaginous margin; marginal teeth ± 1 mm, firm, white, 5 - 10 mm apart; **Inf** ± 50 cm, with 6 - 8 **Br**; racemes cylindrical, 8 - 10 × 5 cm, sub-dense; **Bra** ovate-acute, 2 - 3 × 2 mm, white; **Ped** 9 mm; **Fl** dull scarlet, 19 mm, base shortly attenuate, 5 mm ∅ across **Ov**, scarcely narrowed above; **Tep** free for 5 - 6 mm.

A. megalacantha Baker (FTA 7: 469, 1898). **T:** Ethiopia, Harerge Region (*Riva* 905 [FT]). − **Lit:** Gilbert & Sebsebe (1997). **D:** Ethiopia, Somalia.

A. megalacantha ssp. **alticola** M. G. Gilbert & Sebsebe (KB 52(1): 150, 1997). **T:** Ethiopia, Harerge Region (*Gilbert* 4080 [K, ETH]). − **D:** Ethiopia; edges of thickets on limestone slopes, 2100 - 2150 m.

[12] Differs from ssp. *megalacantha*: Often forming more compact clumps; stem erect; **L** marginal teeth ± 4 mm; **Bra** 11 - 12 × 2.5 - 4 mm; **Fl** 28 - 30 mm, 4 - 6 mm ∅ when pressed; **Tep** free for 10 - 12 mm.

A. megalacantha ssp. **megalacantha** − **D:** Ethiopia, Somalia; dry open bushland on rocky slopes and sandy plains, 1100 - 1850 m. **I:** Reynolds (1966: 295-297).

    Incl. *Aloe magnidentata* I. Verdoorn & Christian (1947).

[12] Caulescent, branching at base; stem sprawling, to 2 m, covered with dead **L**; **L** 24 or more, rosulate and persisting for 50 cm, lanceolate-attenuate, deeply canaliculate, recurved with tips pointing down, 60 - 80 × 13 - 15 cm, dull light green to bluish-green, with pinkish horny margin, surface rough; marginal teeth 4 - 6 mm, blunt, pinkish, reddish-brown tipped, 0.5 - 0.7 mm apart; **Inf** 0.5 - 1 m, with 6 - 13 **Br**, lower **Br** rebranched; racemes cylindrical-conical, 5 - 14 × 7 cm, sub-dense; **Bra** ovate-deltoid, 4 - 7 × 2 - 4 mm; **Ped** 8 - 15 mm; **Fl** yellow, orange or scarlet, 23 - 28 mm, base rounded, 5 - 7 mm ∅ when pressed, very slightly narrowed above, then enlarging slightly to mouth; **Tep** free for 12 - 14 mm. − *Cytology:* 2n = 14 (Brandham 1971).

Natural hybrids with other species have been reported (Reynolds 1966).

A. megalocarpa Lavranos (KuaS 49(7): 162-163, ills., 1998). **T:** Madagascar (*Lavranos & al.* 28728 [P]). − **D:** N Madagascar; dense dry forest, 30 m.

[2] Acaulescent, solitary; **L** ± 25, in spreading **Ros**, up to 55 cm, to 2.5 cm wide at base, widening to 3.5 cm in the middle, dark green, glossy; marginal teeth triangular, 2 mm, whitish, brown-tipped, firm, 5 - 15 mm apart; **Inf** up to 30 cm, simple, somewhat lax, with ± 15 **Fl**; **Bra** 2 - 4 mm; **Ped** 5 - 8 mm; **Fl** bright red, dark green at tip, 25 mm, 5 mm ∅, slightly curved; **Tep** free for 4 mm; **Fr** berries, ± 2 cm.

A. melanacantha A. Berger (BJS 36: 63, 1905). **T:** RSA, Northern Cape (*Drège* 2697 [W]). − **D:** Namibia, RSA.

A. melanacantha var. **erinacea** (D. S. Hardy) G. D. Rowley (Excelsa 9: 71, 80, 1980). **T:** Namibia, Lüderitz Distr. (*Hardy* 2619 [PRE]). − **D:** Namibia; rocky hills. **I:** FPA 48: t. 1885, 1984.

    ≡ *Aloe erinacea* D. S. Hardy (1971).

[4] Differs from var. *melanacantha*: **Ros** more compact; **L** biconvex, 8 - 16 × 3 - 4 cm; **Fl** ± 28 mm; **Tep** free to base.

A. melanacantha var. **melanacantha** − **D:** Namibia, RSA (Northern Cape); rocky slopes, 610 - 1100 m. **I:** Reynolds (1950: 182-183). **Fig. XVI.g**

[4] Caulescent, simple or usually in groups; stem erect and short or, with age, decumbent to 50 cm or

more, covered with dead **L**; **L** densely rosulate, deltoid-lanceolate, to 20 × 4 cm, tip a black **Sp**, dull deep green to brownish-green, lower face carinate in upper ½, the keel with ± 6 black prickles to 10 mm; marginal teeth 10 mm, pungent, black, 10 - 15 mm apart; **Inf** to 1 m, simple, sometimes with 1 **Br**; raceme cylindrical-attenuate, 20 - 25 × 8 cm, dense; **Bra** 25 × 7 mm; **Ped** 15 mm; **Fl** bright scarlet turning yellowish, ± 45 mm, base rounded, enlarged slightly above **Ov**, narrowing slightly at mouth; **Tep** free almost to base. − *Cytology:* 2n = 14 (Müller 1945).

A. menachensis (Schweinfurth) Blatter (Fl. Arab., 8: 463, 1936). **T:** Yemen (*Schweinfurth* 1685 [K]). − **D:** Yemen; exposed rocky slopes, 2200 - 2300 m. **I:** Reynolds (1966: 135).

≡ *Aloe percrassa* var. *menachensis* Schweinfurth (1894) ≡ *Aloe trichosantha* var. *menachensis* (Schweinfurth) A. Berger (1908).

[18] Caulescent; stem erect, to 50 cm; **L** densely rosulate, triangular-lanceolate, attenuate, 40 × 16 cm, greenish-purple, with purplish margins, lower face carinate near tip, keel sometimes with 1 - 3 prickles; marginal teeth 2 mm, blunt, brown-tipped, 15 - 20 mm apart; **Inf** "tall", with many **Br**; racemes cylindrical, 15 - 20 × 5 - 6 cm, dense; **Bra** 10 - 15 mm, reflexed; **Ped** 5 - 7 mm; **Fl** pale or reddish-scarlet, shortly white-tomentose, 30 mm; **Tep** free for 12 mm.

A little-known species, believed by Favell & al. (1999) to be a hybrid between *A. vacillans* and *A. tomentosa*.

A. mendesii Reynolds (JSAB 30(1): 31-32, t. 10, 1964). **T:** Angola (*Santos & Henriques* 1131 [LISC, LUAI, PRE]). − **D:** Angola, Namibia; vertical rock faces, 2220 m. **I:** Reynolds (1966: 170-171).

[13] Caulescent; stem pendulous, to 1 m × 4 cm ∅; **L** ± 10, falcate, hanging downwards, 50 × 7 - 8 cm, green, obscurely lineate; marginal teeth 1 - 2 mm, blunt, cartilaginous, 10 - 15 mm apart; **Inf** to 60 cm, pendulous with racemes arcuate-ascending, with 3 - 4 **Br**; racemes cylindrical-acuminate, 10 × 6 cm, sub-dense; **Bra** ovate-acute, 12 × 5 mm, deep pink, imbricate in bud stage; **Ped** 18 - 20 mm; **Fl** scarlet, 25 mm, slightly ventricose, base rounded, 4 mm ∅ across **Ov**, enlarged above towards mouth, narrowing just below mouth; **Tep** free for 20 mm.

A. menyharthii Baker (FTA 7: 459, 1898). **T:** Moçambique (*Menyharth* 1248 [K]). − **D:** Malawi, Moçambique.

A. menyharthii ssp. **ensifolia** S. Carter (KB 51(4): 783-784, 1996). **T:** Moçambique, Niassa Prov. (*Leach & Rutherford-Smith* 10876 [K, SRGH]). − **D:** Moçambique; grassland on sandy soils in open *Brachystegia* woodland, 570 - 1500 m.

[5] Differs from var. *menyharthii*: **L** lanceolate-attenuate (ensiform), 40 - 70 × 3 - 5.5 cm, tip not drying early; marginal teeth 2 - 4 mm; **Inf** with racemes 4 - 12 cm.

A. menyharthii ssp. **menyharthii** − **Lit:** Carter (1996). **D:** Malawi, Moçambique; grassland, 300 - 610 m. **I:** Reynolds (1966: 85, fig. 85, as *A. swynnertonii*).

[5] Acaulescent or very shortly caulescent, solitary or in small groups of 3 - 4 **Ros**; **L** ± 20, densely rosulate, lanceolate-attenuate, 25 - 40 × 6 - 9 cm, apical 5 - 10 cm soon drying, upper face dark green with oblong to confluent H-shaped spots in irregular transverse bands, lower face paler green, obscurely lineate, usually without spots; marginal teeth 4 mm, pungent, reddish-brown, < 1 mm apart; **Inf** 1.5 - 1.75 m, with 8 - 12 **Br**, lower **Br** sometimes re-branched; racemes capitate-corymbose, 6 - 8 × 7 - 8 cm, dense; **Bra** lanceolate, ± 10 - 12 mm; **Ped** 20 - 25 mm; **Fl** coral-red to pinkish scarlet, 25 - 30 mm, ± 9 mm ∅ across **Ov**, abruptly narrowed above, then gradually widening to mouth; **Tep** free for ± 8 mm; **St** exserted ± 2 mm.

This taxon was included in *A. swynnertonii* by Reynolds (1966).

A. metallica Engler & Gilg (in Warburg, Kunene-Sambesi Exped., 191, 1903). **T:** Angola, Bié Distr. (*Baum* 891 [B]). − **D:** Angola; sandstone rocks; 1300 - 1430 m. **I:** Reynolds (1966: 152).

[2,3] Acaulescent or shortly caulescent, simple; **L** ± 15, densely rosulate, lanceolate-attenuate, 25 - 40 × 7 - 9 cm, bluish-grey with metallic sheen (lost in cultivation), with slightly reddish-brown horny margin; marginal teeth 2 - 3 mm, pungent, reddish-brown, 10 - 20 mm apart; **Inf** to 1.2 m, simple or with few **Br**; racemes cylindrical-acuminate, to 35 × 6 cm, sub-dense; **Bra** lanceolate-acute, 18 - 20 × 8 mm, white, imbricate in bud stage; **Ped** 8 mm; **Fl** reddish-pink, 32 mm, base rounded, 7 mm ∅ across **Ov**, enlarging slightly towards mouth; **Tep** free for 13 mm.

A. meyeri van Jaarsveld (JSAB 47(3): 567-571, ills., 1981). **T:** RSA, Northern Cape (*van Jaarsveld* 6137 [NBG]). − **Lit:** Vorster (1983). **D:** Namibia, RSA (Northern Cape); S-facing cliffs. **I:** FPA 52: t. 2065, 1993.

Incl. *Aloe richtersveldensis* Venter & Beukes (1982).

[13] Caulescent, simple or branching at base, sometimes branching along stem; stem pendulous, to 1 m, with dead **L** for a short distance below **Ros**; **L** rosulate and persistent for short distance below, lanceolate-acuminate, 20 × 3.5 cm, glaucous with distinct powdery bloom, slightly striate; marginal teeth 2 mm, white, 5 - 8 mm apart; **Inf** 1.5 - 2.5 m, pendulous-recurved, simple or rarely branched; raceme capitate, ± 7 × 8 cm; **Bra** deltoid-acuminate, 5

× 3 mm; **Ped** 20 mm; **Fl** orange-red, green-tipped, to 20 mm, subclavate, base rounded, ± 3.5 mm ∅ across **Ov**, enlarging above to 4 - 5 mm at mouth; **Tep** free to base.

A. micracantha Haworth (Suppl. Pl. Succ., 105, 1819). **T:** RSA, Eastern Cape (*Burchell* 4482 [K]). – **Lit:** Smith & Mössmer (1995). **D:** RSA (Eastern Cape); well-drained sandy to stony soils; ± 500 m. **I:** Reynolds (1950: 148-150).
    Incl. *Aloe micracantha* Link & Otto (1825) (*nom. illeg.*, Art. 53.1).
    [4] Acaulescent or very shortly caulescent, simple or sometimes with 1 - 2 **Br**; **R** thick, fusiform; **L** 12 - 18, multifarious, deltoid-acuminate, to 50 × 0.3 cm, deep green to yellowish-green, with many white subtuberculate and subspinulescent spots; marginal teeth to 2 mm, firm, white, 1 - 3 mm apart; **Inf** 40 - 50 cm, simple; raceme capitate, 8 × 9 cm, dense, with ± 24 **Fl**; **Bra** ovate-acuminate, 35 mm, somewhat fleshy near base; **Ped** 35 mm; **Fl** salmon-pink, 38 mm, base shortly attenuate, slightly narrowed above **Ov**; **Tep** free to base. – *Cytology:* 2n = 14 (Müller 1941).

A. microdonta Chiovenda (Pl. Nov. Min. Not. Ethiop. 1: 7, 1928). **T:** Somalia (*Puccioni & Stefanini* 49 [FI]). – **D:** Somalia, Kenya; light sandy soils, 15 - 400 m. **I:** Reynolds (1966: 301-302).
    [12] Caulescent, branching, sometimes forming large dense groups; stem decumbent with tip ascending, to 1 m; **L** ± 16, densely crowded at stem tips, triangular, 50 - 70 × 9 - 11 cm, dull green to olive-grey, sometimes with reddish tinge, sometimes with a few dull white spots near base; marginal teeth 1 (-2) mm, whitish with pale brown tips, 5 - 14 mm apart; exudate drying yellow; **Inf** 1.3 m, with 8 - 12 **Br**, lower **Br** sometimes rebranched; racemes oblique or ascending, with ± secund **Fl**, 10 - 15 cm, lax; **Bra** ovate-acute, 2 - 4 × 2 mm; **Ped** 5 - 6 mm; **Fl** scarlet, sometimes paler at mouth, 23 mm, base rounded or shortly attenuate, 7 mm ∅ across **Ov**, slightly narrowed above **Ov**, then enlarging to mouth; **Tep** free for 10 mm.
    Natural hybrids with *A. rabaiensis* have been reported (Reynolds 1966).

A. microstigma Salm-Dyck (Monogr. Gen. Aloes & Mesembr. Sect. 26: fig. 4 (fasc. 6: fig. 11), 1854). **T:** not typified. – **D:** RSA (Western Cape, Eastern Cape); hot dry flat scrub country, sometimes on steep slopes. **I:** Reynolds (1950: 396-398).
    Incl. *Aloe juttae* Dinter (1923); **incl.** *Aloe brunnthaleri* A. Berger *ex* Cammerloher (1933).
    [4] Caulescent, simple or in small groups; stem usually decumbent, to 50 × 10 cm, covered with dead **L**; **L** densely rosulate, lanceolate-deltoid, 30 - 50 × 6.5 - 8 cm, green, sometimes with reddish tinge, usually obscurely lineate, frequently with several scattered white spots, lower face usually with

many spots, with reddish-brown cartilaginous margin; marginal teeth 2 - 4 mm, pungent, reddish-brown, 5 - 10 mm apart; **Inf** 60 - 80 cm, simple; raceme conical, 20 cm, to cylindrical-acuminate, 40 cm; **Bra** lanceolate-acute, ± ½ as long as **Ped**, deep brown; **Ped** 25 - 30 mm; **Fl** orange, becoming greenish-yellow, rarely red, 25 - 30 mm, slightly ventricose, base rounded, slightly enlarged above **Ov** towards mouth; **Tep** free to base. – *Cytology:* 2n = 14 (Resende 1937).
    Natural hybrids with other species have been reported (Reynolds 1950). Reynolds (l.c.) reports that forms of this species occur near Aus, Namibia.

A. millotii Reynolds (JSAB 22(1): 23-26, ills., 1956). **T:** Madagascar, Toliara (*Reynolds* 7840 [TAN, K, P, PRE]). – **D:** Madagascar (Toliara); xerophytic bush on limestone, 100 - 150 m. **I:** Reynolds (1966: 488-490).
    [11] Caulescent, with up to 20 **Br** at base; stem decumbent, divergent or ascending, to 25 × 0.7 - 0.9 cm; **Int** 5 - 10 mm; **L** 8 - 10, distichous in young shoots, becoming spiral to rosulate, persistent to 7 cm below stem tip, triangular, tip rounded with ± 5 soft white cartilaginous prickles, 8 - 10 × 7 - 9 cm, dull grey-green with reddish tinge, upper face sometimes with a few small dull white spots towards base, lower face with many scattered 2 × 1 - 1.5 mm white spots; marginal teeth 1 mm, white, 5 - 10 mm apart; **Inf** 12 - 15 cm, simple; raceme cylindrical, 3 - 5 × 4 - 5 cm, lax, with 6 - 8 **Fl**; **Bra** ovate-acute, 7 × 4 mm, reflexed; **Ped** 6 - 8 mm; **Fl** scarlet, paler at mouth, 22 mm, base truncate, 7 mm ∅ across **Ov**, narrowed to 6 mm above, then enlarging to mouth; **Tep** free to base; **St** exserted 1 mm. – *Cytology:* 2n = 14 (Brandham 1971).

A. milne-redheadii Christian (JSAB 6(4): 177-179, t. 18, 1940). **T:** Angola, Moxico Distr. (*Milne-Redhead* 4253 [SRGH, K]). – **D:** Angola, Zambia; ridge of hills, 1340 - 1370 m. **I:** Reynolds (1966: 110-111).
    [7] Acaulescent or shortly caulescent, suckering to form small to large groups; **L** ± 16 - 20, densely rosulate, ovate-lanceolate, tip acute, to 30 × 7 cm, brownish-green, obscurely lineate, usually with many pale spots, lower face usually with more spots, in wavy transverse bands; marginal teeth 3 mm, brownish, 10 - 15 mm apart; **Inf** 50 - 90 cm, with 1 - 7 **Br**; racemes cylindrical-acuminate, 20 - 25 × 8 cm, sub-dense; **Bra** ovate-acute, 6 × 3 mm; **Ped** 18 mm; **Fl** scarlet, 30 - 35 mm, base truncate, 8 mm ∅ across **Ov**, narrowed above, then enlarging to mouth; **Tep** free for 10 mm.

A. minima Baker (HIP 25: t. 2423 + text, 1895). **T:** RSA, KwaZulu-Natal (*Evans* 409 [NH]). – **D:** RSA, Swaziland.
    ≡ *Leptaloe minima* (Baker) Stapf (1933).

A. minima var. **blyderivierensis** (Groenewald) Reynolds (JSAB 13(2): 101, t. 15: fig. 2, 1947). **T:** RSA, Mpumalanga (*van der Merwe* 38 [PRE 21361]). – **D:** RSA (Mpumalanga); grassland. **I:** Reynolds (1950: 118-119).

≡ *Leptaloe blyderivierensis* Groenewald (1938).

[2] Differs from var. *minima*: **L** 4 - 6, distichous, slightly broader; **Inf** to 60 cm; **Fl** 12 mm.

A. minima var. **minima** – **D:** RSA (KwaZulu-Natal), Swaziland; grassland. **I:** Reynolds (1950: 118-119).

Incl. *Aloe parviflora* Baker (1901).

[2] Acaulescent, simple; **R** fusiform; **L** 6 - 10, rosulate, linear, 25 - 35 × 4 - 6 cm, green, lower face with many slightly tuberculate spots near base; marginal teeth minute, whitish, in lower ½ only; **Inf** 30 - 50 cm, simple; raceme capitate, ± 3 × 4 cm, dense, with ± 15 **Fl**; **Bra** ovate-acuminate, ± ½ as long as **Ped**; **Ped** 10 - 20 mm; **Fl** dull pinkish, 10 - 11 mm, base attenuate, narrowing above **Ov** towards mouth; **Tep** free to base; **St** exserted 0 - 1 mm.

A. mitriformis Miller (Gard. Dict., Ed. 8, no. 1, 1768). **T:** not typified. – **D:** RSA (Western Cape); usually flat rocky areas. **I:** Reynolds (1950: 372-376, t. 32). **Fig. XV.g**

Incl. *Aloe mitriformis* var. *humilior* Haworth (s.a.); **incl.** *Aloe perfoliata* var. ν Linné (1753); **incl.** *Aloe perfoliata* var. *mitriformis* Aiton (1789); **incl.** *Aloe perfoliata* var. ξ Willdenow (1799); **incl.** *Aloe mitriformis* var. *elatior* Haworth (1804); **incl.** *Aloe xanthacantha* Salm-Dyck (1854); **incl.** *Aloe parvispina* Schönland (1905).

[13] Caulescent, branching at base and above, forming dense sprawling groups; stem decumbent, to 2 m × 6 cm; **L** densely rosulate, ovate-lanceolate, to 20 × 10 - 15 cm, tip a single or bifid **Sp**, glaucous-green to green, rarely with spots, lower face usually carinate towards tip, the keel with ± 4 - 6 prickles; marginal teeth 4 - 6 mm, whitish to yellowish, 10 - 15 mm apart; exudate drying yellow; **Inf** 40 - 60 cm, with 2 - 4 **Br**; racemes capitate, 10 × 12 cm, dense; **Bra** lanceolate-acuminate, 10 × 5 - 6 mm; **Ped** 40 - 45 mm; **Fl** dull scarlet, 40 - 45 mm, base attenuate, slightly narrowed above **Ov**, then enlarging to mouth; **Tep** free to base. – *Cytology:* 2n = 14 (Riley 1959).

A very variable species, for which several varietal names have been published. Reynolds (1950) did not accept the varieties as distinct.

The name *A. parvispina* Schönland was once used for some populations of this taxon, and also for populations of a different taxon subsequently described as *A. comptonii*.

A. modesta Reynolds (JSAB 22(2): 85-86, ills., 1956). **T:** RSA, Mpumalanga (*Reynolds* 7626 [PRE]). – **D:** RSA (Mpumalanga); grassland on stony ground, 2100 m.

[2] Acaulescent, simple, with underground ovoid bulb-like swelling 2.5 × 2 cm; **R** fleshy; **L** 4 - 6, rosulate, linear-acute, 15 - 20 × 8 - 9 cm, dull deep green, lower face with many dull green lenticular spots near base; marginal teeth absent; **Inf** 25 - 30 cm, simple; raceme subcapitate, slightly conical, 3.5 - 4 × 3 - 3.5 cm, dense; **Bra** ovate-acuminate, 10 × 6 mm; **Ped** 1 mm; **Fl** yellowish-green, 13 mm, base rounded, 4 mm ∅ across **Ov**, not narrowed above; **Tep** free to base.

Apparently the only African species with fragrant flowers.

A. molederana Lavranos & Glen (FPA 50(2): t. 1982 + 6 pp. of text, diags., ill., 1989). Somalia (*Lavranos & Horwood* 10379 [PRE]). – **D:** Somalia; gypsum hills.

[5] Acaulescent or shortly caulescent, simple or in small groups; stem to 50 cm; **L** 12 - 16, rosulate, falcate, 25 - 40 × 5.5 - 9.5 cm, glaucous, surface smooth; marginal teeth absent or minute and sparse; exudate golden-yellow; **Inf** 40 - 70 cm, with up to 4 **Br**; racemes subcylindrical, to 16 cm, lax; **Bra** ovate or deltoid, acute, 7 - 8 × 3 - 4 mm; **Ped** 6 - 9 mm; **Fl** pink, tomentose, 24 - 28 mm, base rounded, slightly narrowed above **Ov**; **Tep** free for 9 mm.

A. monotropa I. Verdoorn (FPA 34: t. 1342 + text, 1961). **T:** RSA, Northern Prov. (*Smuts* 1560 [PRE]). – **D:** RSA (Northern Prov.); in shade on wooded dolomite slopes. **I:** Wyk & Smith (1996: 209).

[5] Caulescent, simple or sometimes suckering; stem ± decumbent, to 30 cm; **L** ± 20, laxly rosulate, persisting below stem tip, ovate-lanceolate, long-attenuate, ± 33 × 6 cm, green, lineate and with white oblong or H-shaped spots, lower face with spots sometimes in irregular transverse bands; marginal teeth ± 2 mm, brown-tipped; **Inf** ± 80 cm, with ± 6 **Br**, lower **Br** rebranched; terminal raceme cylindrical, ± 10 - 20 cm, laterals with **Fl** secund, 6 - 18 cm, lax; **Bra** deltoid-acuminate, 6 × 2.5 mm; **Ped** 7 - 11 mm; **Fl** old rose with light bloom, ± 30 mm, base truncate, 5 mm ∅ across **Ov**, abruptly narrowed above, then enlarging to mouth; **Tep** free for 7 mm.

Yellow-flowered plants are also reported (Dyer & Hardy 1969).

A. monticola Reynolds (JSAB 23(1): 7-9, t. 7-8, 1957). **T:** Ethiopia, Tigre Prov. (*Reynolds* 8118 [PRE, EA, K]). – **D:** Ethiopia (Tigre); volcanic mountain slopes, 2250 - 2370 m. **I:** Reynolds (1966: 281-282).

[5] Acaulescent or very shortly caulescent, usually simple; **L** 24 or more, densely rosulate, lanceolate-attenuate, 60 - 70 × 14 - 16 cm, glossy olive-green, with prominent brown horny margin; marginal teeth 6 mm, pungent, pale brown, 10 - 15 mm apart; exudate drying brownish; **Inf** 1 m, with ±

8 **Br**; racemes subcapitate, 6 - 8 × 8 cm; **Bra** lanceolate-attenuate, 15 - 20 × 6 - 7 mm, imbricate in bud stage; **Ped** 15 - 20 mm; **Fl** usually yellow, sometimes scarlet, 38 mm, base rounded, 8 mm ∅ across **Ov**, slightly narrowed above; **Tep** free for 14 mm.

A. morijensis S. Carter & Brandham (Cact. Succ. J. Gr. Brit. 41(1): 3-4, ills., 1979). **T:** Kenya, Rift Valley Prov. (*Bally* 17021 [K]). − **D:** Kenya, Tanzania; dry bush, and soil pockets on rocky slopes, 2400 - 2440 m.

[14] Caulescent, branching mostly at base; stem suberect and spreading, to 1 m × 1.5 cm; **L** scattered along stem, ovate-attenuate, to 17 × 3 cm, bright green with a few elongated pale spots, lower face darker green with more spots, surface smooth; marginal teeth 2 - 5 mm, green to brownish, 5 - 15 mm apart; sheath to 2 cm, striate, fibrous; exudate absent; **Inf** to 50 cm, simple or sometimes with 1 - 2 **Br**; racemes conical-acuminate, to 20 × 6 cm; **Bra** ovate-deltoid, aristate, to 15 × 8 mm; **Ped** 20 mm; **Fl** orange-scarlet becoming yellow towards tips, 28 mm, base attenuate, 6 mm ∅ across **Ov**, slightly narrowed above, then enlarging to 7 mm at mouth; **Tep** free for 7 mm; **St** exserted 0 - 1 mm. − *Cytology:* 2n = 14 (Cutler & al. 1980).

A. mubendiensis Christian (JSAB 8(2): 172-173, t. 5, 1942). **T:** Uganda, Toro Distr. (*Pole Evans & Erens* 1685 [SRGH, K, PRE]). − **D:** Uganda; granite outcrops, 1220 - 1370 m. **I:** Reynolds (1966: 260-261).

[7] Acaulescent or shortly caulescent, suckering to form large groups; **L** ± 16, densely rosulate, lanceolate, 30 - 35 × 6.5 cm, dull grey-green, obscurely lineate, sometimes with a few elongated lenticular whitish spots, rarely on lower face, with slight pinkish cartilaginous margin; marginal teeth 3 - 4 mm, reddish-brown, paler tipped, 10 - 15 mm apart; exudate drying yellowish; **Inf** 70 - 90 cm, with 8 **Br**, lower **Br** sometimes rebranched; racemes cylindrical, laterals with secund **Fl**, ± 10 × 6 - 7 cm, subdense; **Bra** ovate-acuminate, 3 - 4 × 2 - 3 mm; **Ped** 10 mm; **Fl** dark brick-red, paler at mouth, 30 mm, base rounded, 7 mm ∅ across **Ov**, slightly narrowed above, then enlarging slightly to mouth; **Tep** free for 8 - 9 mm.

A. mudenensis Reynolds (JSAB 3(1): 39-42, t. 1, 1937). **T:** RSA, KwaZulu-Natal (*Reynolds* 2029 [PRE, BOL]). − **D:** RSA (KwaZulu-Natal); dry scrub in warm valleys. **I:** Reynolds (1950: 245-246, t. 15).

[5] Caulescent, simple or in small groups; stem erect or decumbent, to 80 × 10 cm; **L** ± 20, densely rosulate, lanceolate-attenuate, 25 - 30 × 8 - 9 cm, bluish-green with many scattered white oblong spots, sometimes lineate, lower face paler, sometimes with oblong dull white spots irregular or ± in

transverse bands, lineate; marginal teeth to 7 mm, pungent, brown, 10 - 20 mm apart; exudate drying reddish-purple; **Inf** to 1 m, with 4 - 8 **Br**; racemes subcapitate, ± 12 × 8 - 9 cm, sub-dense; **Bra** deltoid-acuminate, usually slightly > ½ as long as **Ped**; **Ped** 20 - 25 mm; **Fl** salmon-orange, sometimes reddish, 35 mm, base truncate, 8 mm ∅ across **Ov**, abruptly narrowed to 5 mm above, then enlarging to mouth; **Tep** free for 9 mm. − *Cytology:* 2n = 14 (Riley 1959).

Natural hybrids with *A. marlothii* (as *A. spectabilis*) have been reported (Reynolds 1950).

A. multicolor L. E. Newton (Brit. Cact. Succ. J. 12(2): 51-52, ills., 1994). **T:** Kenya, Eastern Prov. (*Newton & al.* 4133 [K, EA]). − **D:** Kenya; open dry bush and among rocks, 960 - 1250 m.

[12] Caulescent, branching at base; stem erect to 1 m, then becoming decumbent, to 2 m; **L** rosulate and persistent for ± 20 cm below, triangular, 36 - 70 × 6.5 - 10 cm, mid-green, often with red tinge when exposed, lower face with white spots near base, mostly in longitudinal rows near midline, surface smooth; marginal teeth to 4 mm, firm, brown-tipped, 8 - 10 mm apart; exudate yellow; **Inf** 60 - 75 cm, with 5 - 8 **Br**, lower **Br** sometimes rebranched; racemes cylindrical, 5 - 11 cm, sub-dense; **Bra** ovate-acute, 10 - 11 × 6 - 7 mm, whitish; **Ped** 11 - 12 mm; **Fl** crimson at base, outer lobes crimson with orange-red margins, inner lobes yellow with white margins, 22 - 25 mm, clavate, base shortly attenuate, 4 - 5 mm ∅ across **Ov**, narrowed to 3.5 - 4 mm above, then enlarging to 7 mm and narrowing again to 5 mm at mouth; **Tep** free for ± 14 mm.

A. munchii Christian (FPA 28: t. 1091 + text, 1951). **T:** Zimbabwe, Melsetter Distr. (*Munch* 2 [SRGH]). − **D:** Moçambique, Zimbabwe; rock outcrops, 1525 - 2135 m. **I:** Reynolds (1966: 320-322).

[9] Caulescent, simple or sparingly branched, usually without persistent dead **L**; stem erect, to 5 m; **L** ± 24 - 30, densely rosulate, ensiform, 50 × 6 - 8 cm, dull grey-green with reddish tinge, with pinkish cartilaginous margin; marginal teeth 1 - 1.5 mm, pinkish, 10 - 15 mm apart; **Inf** 60 cm, with 2 - 3 **Br**; racemes conical-capitate, 10 - 12 × 14 cm, dense; **Bra** ovate-cuspidate, 14 × 10 - 12 mm; **Ped** 35 - 40 mm; **Fl** scarlet or orange, 45 mm, base very shortly attenuate, 7 - 8 mm ∅ across **Ov**, enlarging slightly to mouth; **Tep** free to base.

A. murina L. E. Newton (Taxon 41(1): 31-33, ills., 1992). **T:** Kenya, Rift Valley Prov. (*Newton* 3497 [K, EA]). − **D:** Kenya; soil pockets on rock slopes, 1500 m.

[3] Acaulescent or shortly caulescent, simple; stem prostrate, to 15 cm; **L** 15 - 20, rosulate, lanceolate, 30 - 40 × 5 - 10 cm, dark green, tinged reddish when exposed, surface rough; marginal teeth 2 - 3 mm, pungent, brown, 7 - 12 mm apart; exudate

very pale yellow; **Inf** to 1 m, with up to 13 **Br**; racemes with secund **Fl**, 5 - 20 cm, lax; **Bra** triangular, 6 × 2 mm; **Ped** 11 mm; **Fl** dull brownish-red with heavy white bloom, appearing reddish-grey, 25 mm, base shortly attenuate, 8 mm ∅ across **Ov**, narrowed to 7 mm above; **Tep** free for 8 - 11 mm.

Rare, known from a single locality only in the wild.

A. musapana Reynolds (JSAB 30(3): 125-126, t. 22, 1964). **T:** Zimbabwe, Melsetter Distr. (*Bullock* 36/1 [SRGH, K, PRE]). – **D:** Zimbabwe; steep rock faces, 1905 - 2060 m. **I:** Reynolds (1966: 21-22).

[10] Caulescent, branching at base or higher, forming dense groups; stem mostly pendulous, to 20 × 1 cm; **L** ± 10, distichous, linear-acute, 30 - 40 × 1.5 cm, dark green, sometimes with a few white spots near base, lower face with many dirty-white lenticular spots near base; marginal teeth minute, cartilaginous, white, near base only; **Inf** 30 - 40 cm, simple; raceme cylindrical-acuminate, 15 × 6 cm, sub-dense; **Bra** ovate-acute, 10 × 6 mm; **Ped** 20 mm; **Fl** scarlet, sometimes bright orange, pale green at mouth, 28 - 30 mm, base rounded, 6 mm ∅ across **Ov**, enlarging above to mouth; **Tep** free to base; **St** exserted 1 - 2 mm.

A. mutabilis Pillans (South Afr. Gard. 23: 168, 1933). **T:** RSA, Northern Prov. (*van Balen* s.n. [BOL 20477]). – **D:** RSA (Gauteng, Mpumalanga, Northern Prov., North-West Prov.); steep to vertical rock faces. **I:** Reynolds (1950: 418-419).

[13] Caulescent, simple or branching; stem procumbent or pendulous, to 1 m × 10 - 15 cm; **Ros** upturned; **L** densely rosulate, lanceolate, sometimes falcate, 60 - 70 × 8 - 9 cm, glaucous green to dull green, obscurely lineate, with narrow pale brownish-yellow margin; marginal teeth ± 2 mm, firm, pale yellow to orange-yellow, 15 - 25 mm apart; **Inf** 60 - 90 cm, arcuate-erect, usually simple or with 1 - 2 **Br**; racemes conical, 25 - 30 cm, dense; **Bra** oblong-obtuse, ± 13 × 5 mm; **Ped** 20 - 25 mm; **Fl** scarlet in bud, becoming greenish-yellow to yellow at anthesis, 30 - 35 mm, base shortly attenuate, enlarging above **Ov** towards mouth; **Tep** free to base. – *Cytology:* 2n = 14 (Müller 1941).

A. myriacantha (Haworth) Schultes & Schultes *fil.* (Syst. Veg. 7(1): 704, 1829). **T:** [neo – icono]: K [unpubl. drawing]. – **D:** Burundi, Rwanda, Kenya, Tanzania, Uganda, Zaïre, Zimbabwe, Malawi, RSA (Eastern Cape, KwaZulu-Natal); rocky grassland, 600 - 2630 m. **I:** Reynolds (1950: 116-117).

≡ *Bowiea myriacantha* Haworth (1827) ≡ *Leptaloe myriacantha* (Haworth) Stapf (1933); **incl.** *Aloe johnstonii* Baker (1887); **incl.** *Aloe caricina* A. Berger (1905); **incl.** *Aloe graminifolia* A. Berger (1905).

[1] Acaulescent, simple or suckering to form small groups; **R** fusiform, from a corm-like base; **L**

8 - 12, rosulate, linear, 25 × 0.8 - 0.1 cm, dull green with a few white spots near base, lower face with more spots near base, the spots tuberculate, almost spinulescent; marginal teeth minute, white, more distant or obsolescent towards **L** tip; **Inf** 20 - 25 cm, simple; raceme capitate, ± 4.5 × 6 cm, dense, with 20 - 30 **Fl**; **Bra** ovate-acuminate, 15 mm; **Ped** 15 mm; **Fl** mostly dull reddish-pink, rarely greenish-white, 20 mm, base attenuate, not or slightly narrowed above **Ov**, mouth bilabiate; **Tep** free to base; **St** exserted 0 - 1 mm. – *Cytology:* 2n = 14 (Brandham 1971).

A. mzimbana Christian (FPSA 21: t. 838 + text, 1941). **T:** Malawi, Northern Prov. (*Pole Evans & Erens* 643 [PRE]). – **D:** Tanzania, Zaïre, Zambia, Malawi; rock outcrops in woodland, 840 - 1860 m. **I:** Reynolds (1966: 112-113). **Fig. XVII.b**

[7] Acaulescent or shortly caulescent, suckering to form dense groups; **L** ± 20, densely rosulate, deltoid-ovate-lanceolate, 20 - 45 × 7 - 8 cm, greyish-green, obscurely striate, sometimes with a few scattered spots, with reddish-pink margin; marginal teeth 2 - 4 mm, reddish-pink, 8 - 10 mm apart; exudate drying yellow; **Inf** 30 - 80 cm, simple or with 2 - 8 **Br**; racemes cylindrical, 8 - 15 cm, dense; **Bra** ovate-lanceolate, 6 - 10 × 3 - 4 mm, fleshy; **Ped** 15 - 20 mm; **Fl** coral-red to scarlet, 35 mm, base attenuate, 8 mm ∅ across **Ov**, narrowed to 6 mm above, then enlarging to mouth; **Tep** free for 12 mm. – *Cytology:* 2n = 14 (Brandham 1971).

A. namibensis Giess (Mitt. Bot. Staatssamml. München 8: 123-126, 1970). **T:** Namibia, Swakopmund Distr. (*Giess* 9212 [WIND, M]). – **D:** Namibia; arid desert with night fogs.

[3] Acaulescent, simple; **L** densely rosulate, lanceolate, to 50 × 7 cm, glaucous, surface slightly rough; marginal teeth to 2 mm, pale, 1 mm apart; **Inf** to 95 cm, with 2 - 4 **Br**; racemes cylindrical-acuminate, to 45 × 6 cm, dense; **Bra** ovate-lanceolate, to 35 × 14 mm; **Ped** to 3 mm; **Fl** coral-pink with green tips, to 30 mm; **Tep** free to base; **St** exserted 10 - 12 mm.

A. namorokaensis (Rauh) L. E. Newton & G. D. Rowley (Bradleya 16: 114, 1998). **T:** Madagascar (*Rauh* 72140 [HEID]). – **D:** W Madagascar; humid cracks in limestone rocks. **I:** Rauh (1998: 99, as *Lomatophyllum*).

≡ *Lomatophyllum namorokaense* Rauh (1998).

[2] Acaulescent or shortly caulescent; **L** ± 9, laxly rosulate, lanceolate, 25 - 35 × 3 cm, dark green with narrow reddish margin; marginal teeth deltoid, 2 mm, 6 - 10 mm apart; sheath 3 cm; **Inf** 30 cm, simple; raceme 10 cm, sub-dense; **Bra** < 10 mm, reddish; **Ped** 10 mm; **Fl** bright cinnabar-red, 25 - 30 mm, slightly curved, constricted above the **Ov**, then enlarging again; **Tep** free to base; **Fr** berries, 15 mm.

A. ngongensis Christian (JSAB 8(2): 170-172, t. 4, 1942). **T:** Kenya, Rift Valley Prov. (*Pole Evans & Erens* 1129 [SRGH, EA, PRE]). — **D:** Kenya, Tanzania; rocky ground in open woodland or at the edge of thickets, 1370 - 1900 m. **I:** Reynolds (1966: 366-367, as *A. rabaiensis*). **Fig. XVI.h**

[16] Caulescent, branching at base; stem erect, to 1.5 m; **L** laxly rosulate and persistent below stem tip, lanceolate-attenuate, 30 - 60 × 5 - 10 cm, bluish-green, sometimes tinged purplish, surface smooth; marginal teeth 3 - 4 mm, brown-tipped, 5 - 10 mm apart; exudate brown; **Inf** to 60 cm, with 6 - 8 **Br**, lower **Br** sometimes rebranched; racemes subcapitate, ± 6 × 8 cm, dense; **Bra** lanceolate, 7 - 10 × 3 mm, pale brown; **Ped** 10 - 20 mm; **Fl** bright glossy scarlet, yellowish at mouth, 25 - 30 mm, base shortly attenuate, 8 mm ∅ across **Ov**, slightly narrowed above, then enlarging to mouth; **Tep** free for less than ⅓. — *Cytology:* 2n = 14 (Cutler & al. 1980: as *A. rabaiensis*).

Reynolds (1966) included this taxon in *A. rabaiensis*.

A. niebuhriana Lavranos (JSAB 31(1): 68-71, t. 13, 1965). **T:** Yemen (*Rauh & Lavranos* 3159 [PRE, K]). — **D:** Saudi Arabia, Yemen; rocky hills and sand banks, 250 - 500 m. **I:** Reynolds (1966: 121-123); Collenette (1999: 23).

[4,6] Acaulescent or shortly caulescent, simple on rocky ground, or suckering freely forming dense groups on sandy soil; stem decumbent; **L** 15 - 25, rosulate, lanceolate-attenuate, to 45 × 10 cm, greygreen tinged purplish; marginal teeth 1.5 - 2 mm, dark brown, 12 - 15 mm apart; **Inf** 50 - 100 cm, usually simple, sometimes with 1 - 2 **Br**; racemes conical, 12 - 30 × 5 - 6 cm, dense; **Bra** deltoid-acute, 8 × 3 - 4 mm; **Ped** 4 - 6 mm; **Fl** scarlet tipped greenish-yellow, rarely all greenish-yellow, usually shortly pubescent, 28 - 31 mm, base shortly attenuate, 6 - 7 mm ∅ across **Ov**, not narrowed above; **Tep** free for 19 - 21 mm.

A. nubigena Groenewald (Tydskr. Wetensk. Kuns 14: 135-137, 1936). **T:** RSA, Mpumalanga (*van der Merwe* 133 [PRE]). — **D:** RSA (Mpumalanga); shaded rock faces, often in spray zones of waterfalls, ± 1430 m. **I:** Reynolds (1950: 132).

[10] Caulescent, suckering and forming dense groups; stem to 25 × 2 cm; **L** usually distichous, sometimes rosulate, linear-acuminate, to 30 × 1.5 cm, green, obscurely lineate, sometimes with a few scattered white spots near base, lower face with white spots, more numerous towards base; marginal teeth minute, white, 1 - 2 mm apart, frequently absent; **Inf** to 30 cm, simple; raceme capitate, ± 5 - 6 × 5 - 6 cm, with 10 - 15 **Fl**; **Bra** ovate-acute, to 14 × 8 mm; **Ped** to 25 mm; **Fl** scarlet with green tips, 25 mm, slightly ventricose, base shortly attenuate, ± 6.5 mm ∅ in middle; **Tep** free to base; **St** exserted to 1 mm. — *Cytology:* 2n = 14 (Müller 1945).

A. nuttii Baker (HIP 1897: t. 2513 + text, 1897). **T** [syn]: Zambia, Northern Prov. (*Nutt* 1896 [K]). — **D:** Angola, Tanzania, Zaïre, Zambia, Malawi; grassland, often on rocky slopes, 1620 - 2650 m. **I:** Reynolds (1966: 33-35). **Fig. XVII.c**

Incl. *Aloe brunneo-punctata* Engler & Gilg (1903); **incl.** *Aloe corbisieri* De Wildeman (1921); **incl.** *Aloe mketiensis* Christian (1940).

[10] Caulescent, simple or suckering to form groups of up to 12 **Ros**; stem to 20 × 5 cm, often very short; **L** 16 - 20, rosulate, linear-acute, 40 - 50 × 1.5 - 2 cm, green, sometimes obscurely lineate, usually with a few dull pale greenish lenticular spots near base, esp. on lower face, with 0.5 - 1 mm white cartilaginous margin; marginal teeth ± 1 mm, soft, white, obsolescent towards tip; **Inf** 60 - 80 cm, simple; racemes cylindrical-acuminate, 15 - 25 × 8 - 9 cm, sub-dense; **Bra** ovate-acute, 15 - 20 × 10 mm, imbricate in bud stage; **Ped** 30 - 35 mm; **Fl** peach-red, strawberry-pink or salmon-pink, 38 - 42 mm, base shortly attenuate, 9 mm ∅ across **Ov**, not narrowed above; **Tep** free almost to the base or united in lower ¼; **St** exserted 0 - 1 mm.

A. nyeriensis Christian *ex* I. Verdoorn (FPA 29: t. 1126 + text, 1952). **T:** Kenya, Central Prov. (*Pole Evans & Erens* 1198 [PRE]). — **D:** Kenya; dry scrub, 1760 - 2120 m. **I:** Reynolds (1966: 380-381).

Incl. *Aloe ngobitensis* Reynolds (1953).

[16] Caulescent, branching at or near base and forming dense groups; stem erect, to 3 m × 7 cm, with dead **L** persistent; **L** ± 20, laxly rosulate and persisting to 50 cm below stem tip, lanceolate-attenuate, 50 - 60 × 7 cm, greyish-green, with white spots on young growth; marginal teeth 3 mm, pungent, 10 mm apart; sheath 2 - 4 cm; exudate yellow; **Inf** 60 cm, with 5 - 8 **Br**, lower **Br** sometimes rebranched; racemes cylindrical-conical, to 15 cm, sub-dense; **Bra** ovate-acute, 5 - 7 × 3 - 4 mm; **Ped** 15 - 20 mm; **Fl** glossy coral red to scarlet, 40 mm, base shortly attenuate, 8 - 9 mm ∅ across **Ov**, slightly narrowed above, then enlarging to mouth; **Tep** free for 15 mm. — *Cytology:* 2n = 14 (Brandham 1971: as *A. ngobitensis*), 2n = 28 (Cutler & al. 1980).

Natural hybrids with *A. lateritia* var. *graminicola* have been reported (Carter 1994).

A. occidentalis (H. Perrier) L. E. Newton & G. D. Rowley (Excelsa 17: 61, 1997). **T** [syn]: Madagascar (*Perrier* 1137 [P?]). — **D:** W Madagascar; on sand, limestone or basalt rocks in dry deciduous forests, mostly in shade. **I:** Rauh (1995a: 330).

≡ *Lomatophyllum occidentale* H. Perrier (1926).

[2,3,8] Acaulescent or caulescent, simple; stem erect, to 1 m × 6 - 10 cm; **L** 15 - 20, densely rosulate, lanceolate-attenuate, 80 - 100 × 10 - 12 cm, greenish-yellow; marginal teeth 4 mm, greenishwhite, 6 - 25 mm apart; **Inf** shorter than **L**, simple or with 3 - 5 **Br**; racemes cylindrical-conical, 12 -

26 cm, sub-dense, with 50 - 80 **Fl**; **Bra** triangular-acute, ½ as long or longer than **Ped**; **Ped** 5 - 10 mm; **Fl** purple to scarlet, green tipped, 26 - 30 mm, base rounded, very slightly narrowed above **Ov** and slightly enlarging to mouth; **Tep** free for 5 - 6 mm; **Fr** berries.

A. officinalis Forsskål (Fl. Aegypt.-Arab., 73, 1775). **T:** Yemen (*Forsskål* s.n. [C †]). – **D:** Saudi Arabia, Yemen; stony hillsides and sandy plains, 200 - 700 m. **I:** Collenette (1999: 23).

≡ *Aloe vera* var. *officinalis* (Forsskål) Baker (1880); **incl.** *Aloe maculata* Forsskål (1775) (*nom. illeg.*, Art. 53.1); **incl.** *Aloe vera* var. *angustifolia* Schweinfurth (1894) ≡ *Aloe officinalis* var. *angustifolia* (Schweinfurth) Lavranos (1965).

[6] Caulescent, usually suckering to form dense clumps; stem decumbent, short; **L** 10 - 12, densely rosulate, ensiform-attenuate, ± 60 - 70 × 6 - 12 cm, yellow-green, often with white spots, lower face sometimes with up to 9 median prickles, surface smooth; marginal teeth stout, crowded; **Inf** to 1 m, simple or with 1 - 3 **Br**; racemes cylindrical, 15 - 20 cm, lax; **Bra** ovate-acute or lanceolate, 10 mm; **Ped** 6 - 8 mm; **Fl** red or yellow to orange, 28 - 30 mm, slightly clavate; **Tep** free for ± 15 mm; **St** scarcely exserted. – *Cytology:* 2n = 14 (Wood 1983).

Wood (1983) treated this as a variety of *A. vera.*

A. oligophylla Baker (JLSB 20: 272, 1883). **T:** Madagascar (*Baron* 1207 [K]). – **D:** Madagascar.

≡ *Lomatophyllum oligophyllum* (Baker) H. Perrier (1926).

[1] Caulescent; **L** 2 - 4, rosulate, linear, 40 - 50 × 1.2 - 1.5 cm, green; marginal teeth 1 - 3 mm, green, 12 - 25 mm apart; **Inf** 20 - 28 cm, simple; raceme 5 - 8 cm, sub-dense; **Bra** deltoid, 1 mm; **Ped** 12 - 15 mm; **Fl** with base attenuate, otherwise unknown; **Fr** berries.

The type is in poor condition, and the plant has not been found again.

A. orientalis (H. Perrier) L. E. Newton & G. D. Rowley (Excelsa 17: 61, 1997). **T:** Madagascar (*Perrier* 5367 [P]). – **D:** Madagascar; sand dunes at sea level. **I:** Rauh (1995a: 330).

≡ *Lomatophyllum orientale* H. Perrier (1926).

[7] Acaulescent, suckering to form groups; **L** 20 - 25, densely rosulate, lanceolate-attenuate; marginal teeth greenish-white; **Inf** ½ as long as **L**, ascending, with 2 - 3 **Br**; racemes cylindrical-conical, sub-dense; **Bra** acute, 3 mm; **Ped** 22 - 24 mm; **Fl** pale red, 22 - 24 mm, slightly narrowed above **Ov**, then enlarging to mouth; **Tep** free for 11 - 12 mm; **Fr** berries.

A. ortholopha Christian & Milne-Redhead (BMI 1933: 478, 1933). **T:** Zimbabwe (*Eyles* 5448 [SRGH]). – **D:** Zimbabwe; rocky slopes on serpentine

hills, 1525 m. **I:** Reynolds (1966: 234-235). **Fig. XVII.h**

[3] Acaulescent, simple; **L** 30 or more, densely rosulate, lanceolate, to 50 × 12 - 14 cm, dull grey-green tinged pink, with pinkish to reddish-brown margin; marginal teeth to 4 mm, pungent, 4 (near base) - 20 (towards tip) mm apart; **Inf** 80 - 90 cm, with 2 - 3 (-5) **Br**; racemes almost horizontal, with secund **Fl**, 30 cm, very dense; **Bra** lanceolate-attenuate, 10 - 15 × 5 mm; **Ped** ± 8 mm; **Fl** orange-scarlet to blood-red, 40 mm, ventricose, base rounded, 6 - 7 mm ∅ across **Ov**, enlarged to 10 - 11 mm in middle, then narrowing to 6 - 7 mm at mouth; **Tep** free for 30 mm; **St** exserted 10 - 15 mm.

A. otallensis Baker (FTA 7: 458, 1898). **T:** Ethiopia, Sidamo Region (*Ruspoli & Riva* 1711 [B† ?, FI, K [drawings, fragment]]). – **D:** Ethiopia. **I:** Gilbert & Sebsebe (1992).

Incl. *Aloe boranensis* Cufodontis (1939).

[5] Acaulescent, simple or in small groups; **L** ± 24, rosulate, triangular-lanceolate, attenuate, 35 - 50 (-80) × 4 - 6.5 (-9.5) cm, grey-green, sometimes with round-oblong white spots in longitudinal rows, spots more numerous on lower face; marginal teeth 3 - 4.5 mm, reddish-brown, 8 - 10 mm apart; **Inf** with up to 12 **Br**; racemes cylindrical, 5 - 8 cm, dense; **Bra** ovate-obtuse, shortly cuspidate, 11 - 17 × 4 - 6.5 mm, white; **Ped** 7 - 12 mm; **Fl** pale pink with grey or yellow tip, 19 - 23 (-27) mm, base attenuate, slightly narrowed above **Ov**; **Tep** free for ± 10 mm; **St** exserted 0 - 1 mm.

A. pachygaster Dinter (RSN 19: 179, 1924). **T:** Namibia (*Dinter* 4736). – **D:** Namibia; arid slopes, 1525 m. **I:** Reynolds (1950: 314-315).

[6] Acaulescent or almost so, usually in dense groups of up to 20 **Ros**; **L** rosulate, lanceolate, 12 - 16 × 2.5 cm, bright grey-green, surface rough; marginal teeth 2 mm, yellow when young, becoming nearly black, 5.5 mm apart; **Inf** 90 cm, simple; racemes cylindrical, 35 - 45 cm, dense; **Bra** ovate-acute, 30 × 17 - 18 mm; **Ped** 5 - 6 mm; **Fl** coral-red, green-tipped, 32 - 34 mm, ventricose, base rounded, enlarged above **Ov** to 12 mm in middle, then narrowing to mouth; **Tep** free for 14 - 16 mm.

According to Reynolds (1950) a little-known species needing further investigation. His photographs show erect inflorescences, whereas Jankowitz (1975) shows plants with inflorescences almost horizontal.

A. paedogona A. Berger (J. Bot. 44: 57, 1906). **T:** Angola, Malange Distr. (*Gossweiler* 946 [BM, K]). – **D:** Angola, Namibia; grassland, ± 1800 m.

[5] Acaulescent, simple or rarely suckering to form small groups; **L** ± 16, rosulate, usually dying back in dry season, the bases enlarging to form a bulb towards the end of the growing season, lamina ensiform, 45 cm, green, obscurely lineate, some-

times with few scattered whitish spots, surface smooth; marginal teeth 3 mm, firm, 10 - 40 mm apart; **Inf** to 2 m, with 3 - 5 **Br**; racemes cylindrical-conical to subcapitate, 7 cm, sub-dense; **Bra** lanceolate-acuminate, long-cuspidate, 15 - 25 × 4 mm; **Ped** 25 - 30 mm; **Fl** yellow-green, ± 35 mm, base attenuate, 9 - 11 mm ∅ across **Ov**, narrowing to 6 - 8 mm above, enlarging to 9 - 11 mm and narrowing again to mouth; **Tep** free only at tips; **St** not exserted.

Reynolds (1966) included this in *A. buettneri*, but Carter (1994) treated them as geographically distinct species. Further studies are required to determine their relationships.

A. palmiformis Baker (Trans. Linn. Soc. London, Bot. 1: 263, 1878). **T:** Angola, Huila (*Welwitsch* 3726 [BM, K, LISU]). − **D:** Angola; sandstone rocks in woodland, 1250 - 1500 m. **I:** Reynolds (1966: 354-355).

[16] Caulescent, branching mostly at base; stem erect, to 1.5 m × 4 cm ∅; **L** ± 14, laxly rosulate and persisting to 30 cm below stem tip, lanceolate-attenuate, to 30 × 5 cm, dull green tinged reddish, lower face usually with many small pale green spots towards base; marginal teeth 4 - 5 mm, pungent, pale brown, 10 mm apart; sheath 10 mm, lineate; **Inf** 40 - 50 cm, with up to 4 **Br**; racemes cylindrical, slightly acuminate, 10 - 20 × 7 cm, sub-dense; **Bra** 2 - 3 × 2 mm; **Ped** 13 - 15 mm; **Fl** rose-scarlet, 30 mm, base very shortly attenuate, 5.5 mm ∅ across **Ov**, then slightly enlarging to mouth; **Tep** free for 10 mm; **St** exserted 1 mm. − *Cytology:* 2n = 14 (Brandham 1971).

A. parallelifolia H. Perrier (Mém. Soc. Linn. Normandie, Bot. 1(1): 31, 1926). **T:** Madagascar, Antananarivo (*Perrier* 13981 [P, K]). − **D:** Madagascar; quartzite mountains, 1800 - 2000 m. **I:** Reynolds (1966: 418-419).

[10] Caulescent, sparingly branched at base to form compact groups; stem erect, to 4 cm; **L** 4 - 7, laxly rosulate, linear, 15 × 1 cm, tip rounded with 4 - 5 pale brown 1 mm teeth, dull olive-green; marginal teeth to 2 mm, pale brown, 4 - 8 mm apart; **Inf** 30 - 40 cm, simple; raceme cylindrical and lax to subcapitate, with 12 - 18 **Fl**; **Bra** lanceolate-acute, 10 × 4 mm, fleshy at base; **Ped** 30 mm; **Fl** rose-pink to rose-red, almost white at tips, 30 mm, slightly campanulate, base shortly attenuate, 5 mm ∅ across **Ov**, enlarging above to 8 mm at mouth; **Tep** free to base; **St** exserted 0 - 1 mm.

A. parvibracteata Schönland (Rec. Albany Mus. 2: 139, 1907). **T:** Moçambique (*Burtt-Davy* 2853 [GRA]). − **Lit:** Wyk & Smith (1996). **D:** Moçambique, RSA (KwaZulu-Natal, Mpumalanga), Swaziland; mostly rock outcrops in flat grassland, 150 - 230 m. **I:** Reynolds (1950: 276-277).

Incl. *Aloe komatiensis* Reynolds (1936); **incl.**

Aloe pongolensis Reynolds (1936); **incl.** *Aloe decurvidens* Groenewald (1937); **incl.** *Aloe lusitanica* Groenewald (1937); **incl.** *Aloe pongolensis* var. *zuluensis* Reynolds (1937) ≡ *Aloe parvibracteata* var. *zuluensis* (Reynolds) Reynolds (1950).

[5,7] Acaulescent or shortly caulescent, solitary or in small groups or, usually, suckering to form large dense groups; stem sometimes to 40 cm; **L** 10 - 20, densely rosulate, usually spreading-decurved giving **Ros** a flattened appearance, lanceolate-attenuate, 30 - 40 × 6 - 10 cm, green to brownish-green, upper face only with many oblong dull white spots usually in ± transverse bands, sometimes obscurely lineate; marginal teeth 3 - 5 mm, pungent, brown, 10 - 15 mm apart; exudate drying deep purple to violet; **Inf** 1 - 2 m, with 4 - 8 **Br**, lower **Br** sometimes rebranched; racemes cylindrical, slightly acuminate, 15 - 20 (-40) × 6 - 7 cm, lax, with 40 - 50 **Fl**; **Bra** deltoid or lanceolate, acuminate, ± as long as **Ped**; **Ped** 6 - 15 mm; **Fl** coral-red or dull to glossy red, sometimes somewhat pruinose, 30 - 35 (-40) mm, base truncate, 7 - 9 mm ∅ across **Ov**, abruptly narrowed to 5 - 6 mm above, then enlarging to mouth; **Tep** free for 8 - 10 mm; **St** exserted 1 - 2 mm. − *Cytology:* 2n = 14 (Riley 1959).

Natural hybrids with *A. umfoloziensis* have been reported (Reynolds 1950).

A. parvicapsula Lavranos & Collenette (CSJA 72(2): 84, 86, ills., 2000). **T:** Saudi Arabia, Asir Prov. (*Collenette* 7526 [K]). − **D:** Saudi Arabia; montane thickets, ± 1350 m. **I:** Collenette (1999: 24).

[3] Acaulescent, solitary; **L** densely rosulate, ascending, rigid, lanceolate-attenuate, 60 × 13 cm, greyish-green; marginal teeth 2 - 3 mm, whitish, ± 14 mm apart; **Inf** to 75 cm, erect, usually with 2 - 3 **Br**; racemes cylindrical; **Bra** lanceolate, 12 - 15 × 5 - 6 mm; **Ped** 6 - 7 mm; **Fl** reddish, densely white-tomentose, to 35 mm, 6 - 7 mm ∅ across **Ov**, narrowing above towards mouth; **Tep** free for 17 - 20 mm; **St** scarcely exserted.

A. parvicoma Lavranos & Collenette (CSJA 72(1): 21-22, ills., 2000). **T:** Saudi Arabia, Asir Prov. (*Collenette* 8565 [K]). − **D:** Saudi Arabia; cliff top, surrounded by thick vegetation. **I:** Collenette (1999: 24).

[8] Caulescent, suckering at base; stem 10 (-30) × 2.5 cm, erect; **L** ± 15, crowded at the stem tip, narrowly deltoid, 35 - 46 × 3.5 - 5.5 cm, green; marginal teeth 1 - 2 mm, horny, ± 15 mm apart; **Inf** to 60 cm, erect, simple or with 1 - 2 **Br**; racemes cylindrical, sub-dense; **Bra** 8 - 10 × 4 - 7 mm; **Ped** 13 - 17 mm; **Fl** salmon-pink shading to yellowish at apex, 25 - 28 mm, base shortly attenuate, scarcely constricted above **Ov**; **Tep** free for 7 - 10 mm; **St** exserted 3 mm.

Close to, and possibly conspecific with, *A. rivierei*.

A. parvidens M. G. Gilbert & Sebsebe (KB 47(4): 650, 1992). **T:** Ethiopia, Sidamo Region (*Gilbert & al.* 7714A [ETH, K]). – **D:** Ethiopia, Kenya, Tanzania, Somalia; usually in shade on sandy or stony soil, 560 - 900 m. **I:** Reynolds (1966: 67-68, as *A. pirottae*).

[5] Acaulescent or shortly caulescent, simple or in small groups; **L** rosulate, lanceolate-attenuate, 25 - 42 × 4.5 - 6.5 (-9) cm, dark green to brownish with many elliptic pale spots, surface smooth; marginal teeth 1 - 2.5 mm, brown-tipped, 8 - 13 mm apart; **Inf** 1 - 1.4 m, with 2 - 10 **Br**; racemes cylindrical, sometimes with some **Fl** secund, 9 - 20 cm, lax; **Bra** ovate-acute, 5 - 6 × 3 - 4.5 mm; **Ped** 5.5 - 12 mm; **Fl** pale pink with yellowish lobes, 26 - 30 mm, base rounded, 6 - 8 mm ∅ across **Ov**, slightly narrowed above; **Tep** free for 6 - 10 mm; **St** exserted ± 1 mm. – *Cytology:* 2n = 14 (Brandham 1971: as *A. pirottae*).

A. pirottae sensu Reynolds (non A. Berger) belongs here.

A. parvula A. Berger (in Engler, A. (ed.), Pflanzenr. IV.38 (Heft 33): 172-173, 1908). **T:** Madagascar, Fianarantsoa (*Grandidier* s.n. [P]). – **D:** Madagascar; ± 2000 m. **I:** Reynolds (1966: 399-401).

≡ *Lemeea parvula* (A. Berger) P. V. Heath (1994); **incl.** *Aloe sempervivoides* H. Perrier (1926).

[4] Acaulescent, simple or sometimes forming small groups; **L** ± 24, densely rosulate, triangular-acute, 10 × 1.2 cm, pale bluish-grey with many firm white 0.5 - 1 mm prickles; marginal teeth 1 - 2 mm, soft to firm, white, 1 - 2 mm apart; **Inf** ± 35 cm, simple; racemes cylindrical, to 10 × 5 cm, lax, with ± 12 - 15 **Fl**; **Bra** ovate-deltoid, ± ½ as long as **Ped**; **Ped** 12 - 15 mm; **Fl** light coral-red, 26 mm, subventricose, base attenuate, enlarged to middle, then narrowing to mouth; **Tep** free for 7 mm; **St** exserted 0 - 1 mm.

A. patersonii B. Mathew (KB 32(2): 321-322, 1978). **T:** Zaïre (*Paterson* 515 [K]). – **D:** Zaïre; on rocks in full sun.

[2] Acaulescent; **L** 17, densely rosulate, lanceolate-attenuate, 30 × 5 cm, pale green with paler oblong blotches forming irregular bands; marginal teeth ± 1.5 mm, white or pale brownish, 9 - 10 mm apart; **Inf** 60 cm, with 1 **Br**; racemes cylindrical, 25 cm; **Bra** lanceolate, 10 - 20 mm, green, those at base of **Inf** with bulbils in **Ax**; **Ped** 14 mm; **Fl** salmon with greenish tinge at base, 30 mm, 6 mm ∅ across **Ov**, abruptly narrowed to 4 mm above, then enlarging to 8 mm at mouth; **Tep** free for 13 mm; **St** exserted to 1 mm. – *Cytology:* 2n = 14 (from the protologue).

Bulbil formation is rare in the genus (cf. *A. bulbillifera*).

A. pearsonii Schönland (Rec. Albany Mus. 2: 229, 1911). **T:** RSA, Northern Cape (*Pearson* 6091

[GRA]). – **D:** Namibia, RSA (Northern Cape); hot arid rocky slopes. **I:** Reynolds (1950: 366-368). **Fig. XVII.g**

[16] Caulescent, much branched at base or higher, forming dense shrubs to 2 m ∅; stem erect, to 2 m × 1.5 cm; **L** scattered along stem, ovate-acute to ovate-lanceolate, recurved to deflexed, 7 - 9 × 3 - 4 cm, green, obscurely lineate, reddish in drought; marginal teeth 1 - 2 mm, pungent, whitish to reddish, to 5 mm apart; **Inf** ± 40 cm, with 2 - 3 **Br**; racemes cylindrical, 9 - 15 × 6 - 7 cm, lax; **Bra** lanceolate-acuminate, 6 - 8 × 3 mm; **Ped** ± 20 mm; **Fl** yellow or brick-red, 25 mm, base attenuate, enlarging above **Ov** to mouth; **Tep** free for 12 - 13 mm. – *Cytology:* 2n = 14 (Resende 1937).

A. peckii P. R. O. Bally & I. Verdoorn (FPA 31: t. 1214 + text, 1956). **T:** Somalia (*Peck* s.n. in *Bally* 4283 [EA, PRE]). – **D:** Somalia; on gypsum soil, mostly in shade, 1525 - 1555 m. **I:** Reynolds (1966: 62-63).

[5] Acaulescent, simple or in small dense groups; **L** 14 - 16, densely rosulate, lanceolate, 16 × 6 cm, olive-green, usually with many elongated whitish-green spots; marginal teeth ± 3 - 4 mm, pungent, brownish, 6 - 10 mm apart; exudate drying deep brown; **Inf** 60 - 80 cm, with 6 - 8 **Br**; racemes cylindrical, 15 - 20 × 5 cm, sub-dense; **Bra** ovate-attenuate, 10 - 12 × 4 mm, pale brown; **Ped** 10 mm; **Fl** straw-coloured to greenish-yellow, **Tep** with broad pale border, giving **Fl** striped appearance, 25 - 30 mm, base shortly attenuate, 7 mm ∅ across **Ov**, slightly narrowed above, then enlarging slightly to mouth; **Tep** free for ± 14 mm. – *Cytology:* 2n = 14 (Brandham 1971).

A. peglerae Schönland (Rec. Albany Mus. 1: 120, 1904). **T:** RSA, Gauteng (*Pegler* 921 [BOL, GRA, PRE]). – **D:** RSA (Gauteng); rocky slopes. **I:** Reynolds (1950: 160-161, t. 7).

[4] Acaulescent or very shortly caulescent, simple or sometimes in small groups; stem decumbent; **L** ± 30, densely rosulate, lanceolate-acuminate, incurved at tip, the tip a pungent **Sp**, ± 25 × 7 cm, glaucous, becoming reddish at tip, usually with a few reddish to brown prickles with whitish tuberculate base in mid-line, sometimes 2 rows on lower face; marginal teeth to 6 mm, pungent, whitish, tipped reddish to brown, 15 mm apart; **Inf** 40 cm, simple; racemes cylindrical, slightly acuminate, to 25 × 7 - 8 cm, very dense; **Bra** ovate-acuminate, 16 × 7 mm; **Ped** 2 - 4 mm; **Fl** mostly greenish-cream tinged slightly reddish, 26 - 30 mm, ventricose, base rounded, enlarged above **Ov** to middle, then narrowing to mouth; **Tep** free almost to base; **St** exserted 15 - 20 mm, deep purple with orange **Anth**.

Natural hybrids with other species have been reported (Reynolds 1950).

A. pembana L. E. Newton (CSJA 70(1): 27-31,

ills., 1998). **T:** Tanzania (*Brown* 206 [K, EA]). – **D:** Tanzania (Pemba); light shade in coastal forest, on sandy soil, ± 3 m. **I:** Carter (1994: t. 3, as *Aloe* unidentified).

≡ *Lomatophyllum pembanum* (L. E. Newton) Rauh (1998).

[8] Caulescent, suckering at base to form dense clumps; stem erect, to 2 m × 8 cm; **L** up to 20, densely rosulate, lanceolate-attenuate, to 110 × 9.6 cm, bright glossy green, sometimes with white hyaline margin, surface smooth; marginal teeth 1.5 mm, white, sometimes brown-tipped, 6 - 10 mm apart; exudate pale yellow; **Inf** to 60 cm, erect, with 2 - 3 (-5) **Br** usually arising close together; racemes cylindrical, 5 - 18 × 7 - 8 cm, dense; **Bra** ovate, attenuate, 3 - 5 × 2 mm; **Ped** 16 - 20 mm; **Fl** buds red or yellow with green tips, fading rapidly to cream or white at anthesis; **Fl** creamy-white with green midstripe, 20 - 22 mm, base shortly attenuate, 6 mm ∅ across **Ov**, slightly narrowed to 5 mm above, then enlarging to 6 mm at mouth; **Tep** free for 12 - 16 mm; **St** exserted 5 mm; **Fr** berries.

A. pendens Forsskål (Fl. Aegypt.-Arab., 74, 1775). **T** [lecto]: Yemen (*Forsskål* 16 [C]). – **Lit:** Wood (1983). **D:** Yemen; steep rock faces, 1500 - 2300 m. **I:** Reynolds (1966: 164).

Incl. *Aloe variegata* Forsskål (1775) (*nom. illeg.*, Art. 53.1); **incl.** *Aloe arabica* Lamarck (1783); **incl.** *Aloe dependens* Steudel (1840).

[13] Caulescent, suckering at base; stem decumbent or pendulous, to 20 × 1 - 1.5 cm; **L** scattered along stem, subdistichous when young, later spiral to rosulate, ensiform-attenuate, deflexed and recurved, ± 20 × 1.5 cm, pale green or bronze with narrow horny reddish margin, variegated on young shoots; marginal teeth 1 mm, reddish, 5 - 7 mm apart; sheath 1 - 2 cm, white-striped and spotted; **Inf** ± 15 cm, ascending, simple; racemes cylindrical, slightly acuminate, 15 - 30 cm, sub-dense; **Bra** ovate-acute, 10 × 5 mm; **Ped** 6 mm; **Fl** usually yellow, rarely orange-red, ± 18 mm, not narrowed above **Ov**; **Tep** free for 14 mm. – *Cytology:* 2n = 14 (Wood 1983).

The description is based on Wood (1983), who collected material at the type locality. Schweinfurth, referred to by Reynolds (1966), did not visit the type locality.

A. penduliflora Baker (Gard. Chron., ser. 3, 4: 178, 1888). **T:** Kenya (*Kirk* s.n. [K]). – **Lit:** Carter & Reynolds (1990); Newton (1991). **D:** Kenya; rock slopes, sometimes in semi-shade at edge of bushes, 645 - 900 m. **I:** Ashingtonia 2: 16, 1975.

[16] Caulescent, branching at or near base forming clumps to 2 m ∅; stem erect (in deep soil) or sprawling (in shallow soil), to 130 × 3 cm, with dead **L** bases remaining; **L** laxly rosulate and scattered along stem for 30 - 40 cm below tip, lanceolate to somewhat ensiform, to 41 × 7.5 cm,

mid-green or with bluish bloom near base, with a few scattered whitish spots on young shoots, surface smooth; marginal teeth 2 mm, uncinate, white-tipped, 16 - 25 mm apart; exudate pale yellow drying brown; **Inf** 26 - 40 (-47) cm, descending at base and curving upwards again with U-bend, with 2 - 5 **Br**, lowest **Br** sometimes rebranched; racemes subcapitate, 5 - 18 × 8 cm, dense; **Bra** lanceolate-acuminate, to 10 × 3 mm; **Ped** 15 - 22 mm; **Fl** yellow, rarely red, 30 - 33 mm, base shortly attenuate, 9.5 - 10 mm ∅ across **Ov**, narrowed to 8 mm above; **Tep** free for 7 - 10 mm.

One of many plants sent to Kew from Zanzibar by the governor, Sir John Kirk, without locality information. Most of these plants had been collected on the mainland, in parts of Kenya and Tanzania. Glen & Hardy (1988) suggested that *A. penduliflora* was identical to *A. confusa*, but this was refuted by Carter & Reynolds (1990) and by Newton (1991).

A. percrassa Todaro (Hort. Bot. Panorm. 1: 81, t. 21, 1875). **T:** [lecto – icono]: l.c. t. 21. – **Lit:** Gilbert & Sebsebe (1997). **D:** Eritrea, Ethiopia; rocky slopes with sparse vegetation, 2100 - 2700 m. **I:** Reynolds (1966: 272-275).

≡ *Aloe abyssinica* var. *percrassa* (Todaro) Baker (1880); **incl.** *Aloe schimperi* Schweinfurth (1894) (*nom. illeg.*, Art. 53.1); **incl.** *Aloe oligospila* Baker (1902); **incl.** *Aloe schimperi* G. Karsten & Schenk (1905) (*nom. illeg.*, Art. 53.1).

[5] Acaulescent or caulescent, simple or in groups; stem erect or decumbent, to 80 × 15 cm, dried **L** persisting; **L** 24 or more, densely rosulate, deltoid, 40 - 55 × 10 - 15 cm, glaucous green or grey-green, with slight bluish to reddish tinge, sometimes with spots in young shoots, with whitish to pinkish margin; marginal teeth 3 - 5 mm, 0.6 - 16 mm apart; exudate drying yellow; **Inf** 60 - 80 cm, with 5 - 12 **Br**; racemes cylindrical-conical, (6.5-) 12 - 25 × 5 - 6 cm, dense; **Bra** ovate-acuminate, 8 - 20 × 3 - 6 mm; **Ped** 11 - 20 mm; **Fl** scarlet, paler at mouth, 17 - 23 mm, base shortly attenuate, 6 mm ∅ across **Ov**, not narrowed above; **Tep** free for 5 - 7 mm. – *Cytology:* 2n = 14 (Fernandes 1930).

Natural hybrids with *A. camperi* have been reported (Reynolds 1966).

A. perrieri Reynolds (JSAB 22(3): 131, 1956). **T:** Madagascar, Fianarantsoa (*Perrier* 10995 [P]). – **D:** Madagascar; 1000 m. **I:** Reynolds (1966: 405).

Incl. *Aloe parvula* H. Perrier (1926) (*nom. illeg.*, Art. 53.1).

[6] Acaulescent, suckering to form dense groups; **R** fusiform; **L** ± 10, rosulate, linear-attenuate, to 30 × 1.5 - 2 cm, green with many pale spots, surface rough; marginal teeth ± 1 mm, firm, white, 1 - 2 mm apart; **Inf** 40 - 50 cm, simple; racemes cylindrical-acuminate, 15 - 20 cm, lax, with 20 - 30 **Fl**; **Bra** ovate-acute, 3 - 4 × 2 mm; **Ped** 15 mm; **Fl** deep rose, almost white at mouth, 16 mm, campanulate,

base attenuate, slightly enlarged above **Ov** to wide open mouth; **Tep** free for 8 mm; **St** exserted 0 - 1 mm.

Perrier (1926) misidentified this as *A. parvula* A. Berger.

A. perryi Baker (JLSB 18: 161, 1881). **T:** Socotra (*Perry* s.n. [K]). – **D:** Socotra; limestone, to 515 m. **I:** Reynolds (1966: 196).

[18] Caulescent, simple; stem erect or decumbent, ± 30 × 5 cm; **L** 12 - 30, densely rosulate, lanceolate-acute, 35 × 7.5 cm, glaucous green tinged reddish; marginal teeth ± 4 mm, pale brown, 6 mm apart; **Inf** 50 - 60 cm, usually with 2 - 3 **Br**; racemes oblong-cylindrical, 15 - 25 × 5 - 6 cm, dense; **Bra** lanceolate-deltoid, 4 - 6 mm; **Ped** 8 mm; **Fl** bright red tipped greenish, turning yellow at anthesis, 20 - 25 mm, base shortly attenuate, slightly narrowed above **Ov**, then enlarged again towards mouth; **Tep** free for 8 mm; **St** exserted 0 - 1 mm.

Some details were not mentioned in the protologue. This description is based on a description of cultivated material at Kew, which might not represent all characters as they would be in the natural habitat.

A. petricola Pole-Evans (Trans. Roy. Soc. South Afr. 5: 707, 1917). **T:** not typified. – **D:** RSA (Mpumalanga); exposed sandstone slopes and granite outcrops. **I:** Reynolds (1950: 451-452, t. 47).

[5] Acaulescent, simple or in small groups; **L** 20 - 30, densely rosulate, lanceolate-attenuate, ± 60 × 10 cm, glaucous, sometimes with a few scattered prickles, lower face usually with a few prickles along an obtuse keel near tip; marginal teeth 5 mm, dark brown, 15 mm apart; **Inf** to 1 m, with 3 - 6 **Br**; racemes cylindrical, slightly acuminate, 40 - 50 cm, very dense; **Bra** ovate-acute, ± 12 × 5 mm, pale brown, strongly deflexed near base; **Ped** 2 mm; **Fl** greenish-white to pale orange, 28 - 30 mm, slightly ventricose, base obconical, enlarging above **Ov**; **Tep** free for 19 - 20 mm; **St** exserted 10 mm. – *Cytology:* 2n = 14 (Riley 1959).

Natural hybrids with other species have been reported (Reynolds 1950).

A. petrophila Pillans (South Afr. Gard. 23: 213, 1933). **T:** RSA, Northern Prov. (*Frames* s.n. [BOL 20482]). – **D:** RSA (Northern Prov.); rock faces. **I:** Reynolds (1950: 217-218).

[5] Acaulescent or rarely shortly caulescent, simple or suckering to form small groups; stem to 8 cm; **L** 10 - 20, densely rosulate, oblong-lanceolate, acuminate, 20 - 25 × 5 - 6 cm, bright green, lineate, with scattered elongated H-shaped dull white spots; marginal teeth 3 - 5 mm, pungent, dark brown, 8 - 12 mm apart; **Inf** 50 - 75 cm, with 3 - 6 **Br**; racemes capitate-corymbose, 4 - 8 × 5 - 6 cm, dense, with 20 - 30 **Fl**; **Bra** lanceolate-acuminate, ± 7 mm, white; **Ped** to 15 mm; **Fl** coral-pink with whitish margins

on **Tep**, 28 mm, base truncate, 7 mm ∅ across **Ov**, abruptly narrowed to 4 mm above, then enlarging to mouth; **Tep** free for 8 mm.

A. peyrierasii Cremers (Adansonia, n.s., 15(4): 500-503, ill., 1976). **T:** Madagascar (*Cremers* 2495 [P]). – **D:** Madagascar; dry forest under-storey.

≡ *Lomatophyllum peyrierasii* (Cremers) Rauh (1998).

[9] Caulescent, simple; stem erect, to 4 m, with dead **L** persistent; **L** ± 20, rosulate, lanceolate, 165 × 12 cm, yellowish-green, with 3 - 4 mm cartilaginous violet-tinged margin; marginal teeth 3 - 5 mm, 12 - 27 mm apart; **Inf** 60 - 75 cm, with 2 - 4 **Br**; racemes cylindrical, 8 - 10 cm, dense; **Bra** not known; **Ped** 25 - 30 mm; **Fl** not known; **Fr** berries.

A. pictifolia D. S. Hardy (BT 12(1): 62-64, ills., 1976). **T:** RSA, Eastern Cape (*Marais* s.n. [PRE 32328]). – **D:** RSA (Eastern Cape); stony country. **I:** Wyk & Smith (1996: 179).

[15] Caulescent, branching at base; stem to 12 cm, with dead **L** bases persistent; **L** rosulate, lanceolate-acuminate, 12 - 15 × 1 - 2.5 cm, glaucous with many white spots, lower face with prickles on a keel near tip; marginal teeth to 1 mm, pungent, red-brown, 4 - 5 mm apart; **Inf** 20 cm, simple; raceme cylindrical-acuminate, 14 - 17 × 3.5 - 4 cm, sub-dense; **Bra** ovate-acute, 10 × 3 - 4 mm; **Ped** 11 - 12 mm; **Fl** scarlet with greenish mouth, 15 - 16 mm, base rounded, 3 - 4 mm ∅ across **Ov**, slightly narrowing to mouth; **Tep** free to base.

A. pillansii L. Guthrie (J. Bot. 66: 15, 1928). **T:** RSA, Northern Cape (*Pillans* 5012 [BOL]). – **Lit:** Williamson (1998). **D:** Namibia, RSA (Northern Cape); hot arid rocky slopes. **I:** Reynolds (1950: 495-497, t. 58). **Fig. XVII.d**

[17] Caulescent, dichotomously branching from ± the middle upwards; stem to 10 m or more, 1 - 2 m ∅ at base, narrowing to ± 20 cm above; **L** densely rosulate, lanceolate-attenuate, slightly falcate, 50 - 60 × 10 - 12 cm, grey-green to brownish-green, with white margin, surface smooth; marginal teeth 1 - 2 mm, white, 5 - 8 mm apart; **Inf** ± 50 cm, spreading horizontally with racemes turning upwards, with up to 50 **Br**; racemes cylindrical, to 15 cm, lax, with ± 30 **Fl**; **Bra** filiform, slightly < 10 mm; **Ped** 10 mm; **Fl** yellow, to 35 mm, base attenuate, ± 12 mm ∅ across **Ov**, slightly narrowing above to mouth; **Tep** free for 25 mm; **St** exserted 10 mm.

A. pirottae A. Berger (BJS 36: 65, 1905). **T:** Ethiopia, Gamo Gofa / Sidamo Region (*Ruspoli & Riva* 1682 [B [† ?], FT]). – **Lit:** Gilbert & Sebsebe (1992). **D:** Ethiopia, Kenya; open *Acacia* woodland, 1300 - 1820 m.

[7] Acaulescent or very shortly caulescent, suckering to form small groups; **L** rosulate, 45 - 90 × 6.5

- 13 cm, dark green with many elongated pale spots, surface smooth; marginal teeth (3-) 4 - 5.5 mm, brown-tipped, 10 - 14 mm apart; **Inf** with up to 28 **Br**; racemes cylindrical, often with secund **Fl**, 7 - 33 cm, lax; **Bra** ovate, acute or acuminate, 3 - 10 × 2 - 5 mm; **Ped** 3 - 9 mm; **Fl** yellow with green markings, orange, or sometimes red, 20 - 28 mm, cylindrical to subclavate; **Tep** free for 6 - 15 mm.

Reynolds (1966) misidentified the type locality, and his description is of a different plant, later named *A. parvidens*.

A. plicatilis (Linné) Miller (Gard. Dict., Ed. 8, no. 7, 1768). **T:** not typified. – **D:** RSA (Western Cape); rocky slopes. **I:** Reynolds (1950: 503-505). **Fig. XVII.e**

≡ *Aloe disticha* var. *plicatilis* Linné (1753) ≡ *Rhipidodendrum plicatile* (Linné) Haworth (1821); **incl.** *Aloe linguaeformis* Linné *fil.* (1782); **incl.** *Aloe tripetala* Medikus (1783); **incl.** *Aloe lingua* Thunberg (1785); **incl.** *Kumara disticha* Medikus (1786) ≡ *Rhipidodendrum distichum* (Medikus) Willdenow (1811); **incl.** *Aloe flabelliformis* Salisbury (1796); **incl.** *Aloe plicatilis* var. *major* Salm-Dyck (1817).

[16,17] Caulescent, dichotomously branched from low down; stem erect, to 5 m; **L** ± 12 - 16, distichous, linear to lorate, 30 × 4 cm, dull to glaucous green, surface smooth; marginal teeth minute in upper ⅓, sometimes almost absent; **Inf** to 50 cm, simple; racemes cylindrical, slightly acuminate, 15 - 25 cm, lax, with 25 - 30 **Fl**; **Bra** ovate-deltoid, 8 - 6 mm; **Ped** 10 mm; **Fl** scarlet, to 55 mm, base rounded, not narrowed above **Ov**; **Tep** free for ± 18 mm. – *Cytology:* 2n = 14 (Fernandes 1930).

A. plowesii Reynolds (JSAB 30(2): 71-73, t. 14, 1964). **T:** Moçambique (*Pedro & Pedrogaõ* 7393 [PRE]). – **D:** Moçambique; in grass among sandstone rocks, 1525 - 1770 m. **I:** Reynolds (1966: 16).

[1] Acaulescent; **R** fusiform; **L** ± 10, rosulate, linear-acute, 20 - 30 × 0.6 - 1 cm, dull green with a few scattered small white spots near base, lower face with spots near base; marginal teeth 0.5 mm, white, more distant becoming obsolete towards tip; **Inf** 30 - 45 cm, simple; racemes cylindrical-conical, to 10 × 6 cm, sub-dense; **Bra** ovate-acute, 15 × 7 mm, imbricate in bud stage; **Ped** 30 mm; **Fl** scarlet, paler to greenish at mouth, 30 - 35 mm, base shortly attenuate, 8 mm ∅ across **Ov**, not narrowed above; **Tep** free to base.

A. pluridens Haworth (Philos. Mag. J. 64: 299, 1824). **T** [neo]: RSA, Eastern Cape (*Reynolds* 1425 [PRE]). – **D:** RSA (Eastern Cape, KwaZulu-Natal); in bush on mountain slopes. **I:** Reynolds (1950: 415-416).

Incl. *Aloe atherstonei* Baker (1880); **incl.** *Aloe pluridens* var. *beckeri* Schönland (1903).

[9,16] Caulescent, branching; stem erect, to 5 m;

L 30 - 40, densely rosulate, lanceolate-falcate, 60 - 70 × 5 - 6 cm, pale to yellowish green, obscurely lineate, with narrow white cartilaginous margin; marginal teeth 2 - 3 mm, white or pale pink, 5 - 10 mm apart; **Inf** 80 - 100 cm, with up to 4 **Br**; racemes conical, 25 - 30 × 9 - 10 cm, dense; **Bra** ovate-acute, ± 20 × 10 - 12 mm, white; **Ped** 30 - 35 mm; **Fl** salmon-pink to dull scarlet, 40 - 45 mm, base rounded, slightly narrowed above **Ov**, then enlarging to mouth; **Tep** free to base. – *Cytology:* 2n = 14 (Ferguson 1926).

Natural hybrids with other species have been reported (Reynolds 1950).

A. polyphylla Schönland *ex* Pillans (South Afr. Gard. 24: 267, 1934). **T:** Lesotho (*Reynolds* 934 [BOL 21370]). – **D:** Lesotho; steep rocky slopes, 2370 m. **I:** Reynolds (1950: 194-195).

[5] Acaulescent or very shortly caulescent, simple or in dense groups; **L** ± 150, densely rosulate with 5 spiral rows, ovate-oblong, acuminate, 20 - 30 × 6 - 10 cm, grey-green, becoming dry and deep purplish at tip, lower face with 1 - 2 pale green to white keels near margin in upper ⅓, surface smooth; marginal teeth 5 - 8 mm, firm, pale green, 2 - 12 mm apart; **Inf** 50 - 60 cm, with 3 - 8 **Br**; racemes cylindrical, 12 - 15 × 10 cm, dense; **Bra** lanceolate-deltoid, acuminate, ± 4 - 6 mm, fleshy, tinged pink; **Ped** 4 - 6 mm; **Fl** pale red to salmon-pink, rarely yellow, 45 - 55 mm, base shortly attenuate, slightly enlarged to ± 10 mm above **Ov**; **Tep** free to base. – *Cytology:* 2n = 14 (Riley 1959).

A. porphyrostachys Lavranos & Collenette (CSJA 72(1): 18-19, ills., 2000). **T:** Saudi Arabia, Hijaz Prov. (*Collenette* 2900 [K]). – **D:** Saudi Arabia; on granite, ± 2000 m. **I:** Collenette (1999: 25). **Fig. XVIII.a**

[7] Acaulescent, usually suckering at base; **L** to 60, densely rosulate, erect, deltoid-attenuate, 50 - 55 × 7.5 cm, bluish-grey with white cartilaginous margin; marginal teeth firm, 3 - 5 mm, white, 25 - 30 mm apart; **Inf** to 110 cm, erect, with up to 6 **Br**; racemes cylindrical, 45 × 4 cm, very dense; **Bra** ovate-lanceolate, 12 - 15 mm, papery, 5-nerved; **Ped** 2 - 5 mm; **Fl** bright red, 30 - 35 mm, clavate, base rounded, 11 mm ∅ in the middle; **Tep** free for ± 20 - 21 mm; **St** exserted 12 - 15 mm.

A. powysiorum L. E. Newton & Beentje (CSJA 62(5): 252-255, ills., 1990). **T:** Kenya, Laikipia / Meru Districts (*Beentje & al.* 4124 [EA, K, MO]). – **D:** Kenya; rock faces and steep rocky slopes, 1600 - 1900 m. **Fig. XVIII.c**

[13] Caulescent, branching at base and higher, sometimes dichotomously at tip; stem pendulous, to 1.8 m × 12 cm ∅; **L** densely rosulate, triangular, to 55 × 9 cm, pale green with light bloom, surface smooth; marginal teeth to 1 mm, white, (1-) 3 - 8 mm apart; exudate colourless; **Inf** to 92 cm, slightly

ascending, almost horizontal, simple; racemes with secund **Fl**, to 48 cm, dense; **Bra** triangular, 11 × 3 mm, whitish; **Ped** 5 - 6 mm; **Fl** salmon-red, paler at mouth, 32 - 36 mm, ventricose, base obconical, 10 mm ⌀ across **Ov**, enlarged above to middle, then narrowing to 6.5 - 7 mm at mouth; **Tep** free for ⅓; **St** exserted to 10 mm.

A. pratensis Baker (JLSB 18: 156, 1880). **T** [lecto]: RSA, Eastern Cape (*MacOwen* 1896 [GRA]). — **D:** Lesotho, RSA (Eastern Cape, KwaZulu-Natal); rocky slopes, from near coast to 1830 m. **I:** Reynolds (1950: 191-192).

[4] Acaulescent, rarely simple, usually in small groups; **L** 30 - 40, densely rosulate, lanceolate or ovate-lanceolate, to 15 × 4 - 5 cm, glaucous, lineate, upper face sometimes with a few scattered reddish-brown prickles, lower face usually with a few scattered reddish-brown prickles and sometimes with a median keel towards tip armed with a few 2 - 3 mm brown prickles; marginal teeth ± 5 mm, pungent, reddish-brown, 10 mm apart; exudate drying deep orange; **Inf** to 60 cm, simple; racemes cylindrical, ± 20 × 10 cm, dense; **Bra** ovate-acuminate, to 40 × 15 - 18 mm; **Ped** 25 - 30 mm; **Fl** rose-red, 35 - 40 mm, base truncate, enlarged slightly above **Ov**; **Tep** free to base; **St** exserted 0 - 1 mm. — *Cytology:* 2n = 14 (Riley 1959).

Natural hybrids with *A. maculata* have been reported (Reynolds 1950: as *A. saponaria*).

A. pretoriensis Pole-Evans (Gard. Chron., ser. 3, 56: 106, 1914). **T:** RSA, Gauteng (*Pole-Evans* 12 [PRE]). — **D:** Zimbabwe, RSA (Gauteng, Mpumalanga, Northern Prov.), Swaziland; grassy slopes, 1370 m. **I:** Reynolds (1950: 307-308). **Fig. XVIII.b**

[8] Caulescent, simple; stem suberect or arcuate-erect, to 1 m × 25 cm, with dead **L** persistent; **L** 40 - 60, densely multifarious, lanceolate-acuminate, to 60 × 15 cm, green with greyish powdery bloom, obscurely lineate; marginal teeth ± 3 - 4 mm, pungent, reddish, 10 - 15 mm apart; **Inf** 2 - 3.5 m, with 5 - 8 **Br**; racemes conical to cylindrical-acuminate, 20 - 30 × 10 cm, dense; **Bra** ovate-deltoid, ± ½ as long as **Ped**; **Ped** 25 - 40 mm; **Fl** rose-pink to rich peach-red with a bloom, sometimes yellowish at mouth, 40 - 50 mm, base very shortly attenuate, enlarged slightly above **Ov**, narrowing to mouth; **Tep** free to base; **St** exserted 0 - 1 mm. — *Cytology:* 2n = 14 (Müller 1941).

Natural hybrids with *A. greatheadii* var. *davyana* have been reported (Reynolds 1950).

A. prinslooi I. Verdoorn & D. S. Hardy (FPA 37: t. 1453 + text, 1965). **T:** RSA, KwaZulu-Natal (*Hardy* 1907 [PRE]). — **D:** RSA (KwaZulu-Natal); grassland.

[5] Acaulescent, simple or in small groups; **L** 16 - 30, rosulate, lanceolate, 14 - 20 × 4 - 8 cm, green

with few to many irregular oblong white spots, denser on upper face, sometimes in transverse bands; marginal teeth ± 4 mm, pungent; **Inf** to 60 cm, with 2 - 4 **Br**; racemes capitate-corymbose, 6 - 12 × 6 - 7 cm, dense; **Bra** attenuate, to 30 × 5 mm; **Ped** to 30 mm; **Fl** pale whitish-green tinged with pale to deep shell-pink, ± 17 mm, base rounded, sometimes slightly narrowed above **Ov**, enlarging slightly to mouth; **Tep** free for 5 - 7 mm; **St** exserted 0 - 1 mm.

A. procera L. C. Leach (JSAB 40(2): 117-121, ills., 1974). **T:** Angola, Cuanza Sul Distr. (*Leach & Cannell* 14617 [LISC, SRGH]). — **D:** Angola; tall grass in deciduous woodland on steep hillside, ± 1230 m.

[3] Acaulescent or usually shortly caulescent, simple; stem erect, to 25 cm; **L** ± 20, densely rosulate, ovate-attenuate, apical portion soon drying, to 55 × 8 - 9.5 cm, pale green, obscurely lineate, with narrow cartilaginous pale yellow margin; marginal teeth 1.5 - 3.5 mm, orange-brown tipped, 10 - 18 mm apart; **Inf** 2.2 - 2.75 m, with 9 - 12 **Br**, lower **Br** rebranched; racemes with secund **Fl**, 25 - 40 cm, lax; **Bra** deltate-attenuate, 5 - 6 × ± 4 mm, brownish; **Ped** 1.5 - 5 mm; **Fl** dull reddish-purple, 28 - 33 mm, base rounded or truncate, not or slightly narrowed above **Ov**; **Tep** free for 9 - 11 mm.

A. propagulifera (Rauh & Razafindratsira) L. E. Newton & G. D. Rowley (Bradleya 16: 114, 1998). **T:** Madagascar (*Razafindratsira* s.n. in *Rauh* 74409 [HEID]). — **D:** C-E Madagascar. **I:** Rauh (1998). **Fig. XIX.a**

≡ *Lomatophyllum propaguliferum* Rauh & Razafindratsira (1998).

[2] Usually acaulescent, solitary; **L** 10 - 12, densely rosulate, 20 × 1.5 - 2 cm, grey-green with dark green spots and stripes; marginal teeth deltoid, 1 - 2 mm, olive-green, 5 - 10 mm apart; **Inf** < 20 cm, simple; raceme 2 - 3 cm, with 5 - 10 **Fl**, with 1 - 2 bulbils at base or above **Fl**; **Bra** triangular, acute; **Ped** 7 - 10 mm, dark red; **Fl** bright cinnabar-red, upper ⅓ pale red with green margins, 25 mm, base rounded, 7 - 9 mm ⌀ across **Ov**, constricted above the **Ov**; **Tep** free only at the tips; **Fr** berries, ± 10 mm ⌀.

A. prostrata (H. Perrier) L. E. Newton & G. D. Rowley (Excelsa 17: 61, 1997). **T:** Madagascar (*Perrier* 11014 [P]). — **D:** Madagascar.

≡ *Lomatophyllum prostratum* H. Perrier (1926).

[6] Acaulescent, stoloniferous; **L** 10 - 20, densely rosulate, lanceolate-attenuate, falcate, tip with 3 - 4 teeth, 15 - 20 × 1.5 - 2 cm, dark green to dark brown, with many confluent white spots; marginal teeth 3 mm, white, 5 - 15 mm apart; **Inf** to 12 cm, simple; racemes cylindrical, 4 - 6 cm, with few **Fl**; **Bra** longer than **Ped**; **Ped** 4 -5 mm; **Fl** bright carmine-red, green tipped, 20 - 25 mm, base truncate, 6 - 7 mm ⌀ across **Ov**, narrowed to 5 mm above, then

enlarging to 6 - 7 mm at mouth; **Tep** free for 7 - 8 mm; **Fr** berries.

A. prostrata ssp. **pallida** Rauh & Mangelsdorff (KuaS 51(6): 157-159, ills., 2000). **T:** Madagascar, Toliara (*Mangelsdorff* s.n. in *BG Heidelberg* 70588 [HEID]). – **D:** Madagascar; humus on limestone in forest.

[6] Differs from ssp. *prostrata*: **L** brown at base; **Fl** pale reddish to yellow.

A. prostrata ssp. **prostrata** – **D:** Madagascar; crevices in limestone slopes. **I:** Rauh (1995a: 328).

[6] Description as for the species.

A. pruinosa Reynolds (JSAB 2(2): 122-124, t. 17, 1936). **T:** RSA, KwaZulu-Natal (*Reynolds* 377 [PRE, BOL]). – **Lit:** Smith & Crouch (1995). **D:** RSA (KwaZulu-Natal); thorn bush. **I:** Reynolds (1950: 251-252).

[3] Caulescent, simple; stem procumbent and rooting below, to 50 cm; **L** 12 - 24, densely rosulate, lanceolate-attenuate, to 70 × 8 - 10 cm, green with many white spots, scattered or in transverse bands, more numerous on lower face; marginal teeth to 4 mm, pungent, pale pinkish-brown, 15 - 20 mm apart; exudate drying deep violet; **Inf** to 2 m, with ± 11 **Br**, lower **Br** rebranched; racemes cylindrical-acuminate, 10 - 30 × 7 cm, lax; **Bra** linear-lanceolate, acuminate, 10 - 20 mm or slightly longer; **Ped** 10 - 20 mm; **Fl** dull dark brownish-red to pinkish-white, with heavy greyish powdery bloom, 30 - 40 mm, base truncate, 8 mm ∅ across **Ov**, abruptly narrowed to 5 mm above, then enlarging to 8 mm at mouth; **Tep** free for 7 mm. – *Cytology:* 2n = 14 (Müller 1941).

A. pseudorubroviolacea Lavranos & Collenette (CSJA 72(1): 17-18, ills., 2000). **T:** Saudi Arabia, Asir Prov. (*Collenette* 4409 [K]). – **D:** Saudi Arabia; steep hillsides, ± 2000 m. **I:** Collenette (1999: 26).

[13] Caulescent, simple or less frequently sparsely suckering; stem decumbent, to 4 m with persistent dry **L** bases; **L** ± 18, densely rosulate, deltoid-attenuate, to 60 × 15 cm, bluish-grey, more vivid green in yellow-flowered plants; marginal teeth hard, triangular, 2 - 3 mm, 25 - 40 mm apart; **Inf** to 120 cm, obliquely ascending or recurved, with 4 - 20 (-52) **Br**; racemes cylindrical, 25 - 60 cm, dense; **Bra** ovate-deltoid, acute, 7 mm, papery with brown nerves; **Ped** 3 mm; **Fl** bright red, orange, or rarely golden-yellow, 32 - 40 mm, base rounded, ventricose, 12 mm ∅ in the middle; **Tep** free for 20 - 22 mm; **St** exserted 14 mm.

A. pubescens Reynolds (JSAB 23(1): 10-12, t. 10-11, 1957). **T:** Ethiopia, Arissu Prov. (*Reynolds* 8144 [PRE, K]). – **D:** Ethiopia; rocky banks, 1800 - 2550 m. **I:** Reynolds (1966: 136-137).

[4] Acaulescent or shortly caulescent, simple or more frequently in groups; **L** ± 16, rosulate, lanceolate-attenuate, 45 × 8 cm, grey-green; marginal teeth 2 - 3 mm, pungent, with white base, reddish-brown-tipped, 15 - 20 mm apart; **Inf** 0.7 - 1 m, simple or with 1 **Br**; racemes cylindrical-acuminate, ± 20 cm, sub-dense; **Bra** ovate-deltoid, 20 × 6 mm, imbricate in bud stage; **Ped** 15 mm; **Fl** coral-pink, shortly pubescent, 42 mm, base rounded, 8 mm ∅ across **Ov**, slightly narrowed above, then enlarging slightly and narrowing to mouth; **Tep** free for 16 mm. – *Cytology:* 2n = 14 (Brandham 1971).

A. pulcherrima M. G. Gilbert & Sebsebe (KB 52(1): 147-148, 1997). **T:** Ethiopia, Shewa Region (*Gilbert* 1669 [ETH, K]). – **D:** Ethiopia; sparse bush on steep basalt slopes and cliffs, 2480 - 2750 m.

[13] Caulescent, branching dichotomously; stem prostrate or pendulous, to 1 m × 8 cm; **L** 35 - 50, densely rosulate, lanceolate-attenuate, to 50 × 12 cm, pale blue-green, slightly glaucous, finely striate, with red margin, esp. in dry season; marginal teeth 0.2 - 0.3 mm, to 3 cm apart; exudate drying purple; **Inf** 50 - 80 cm, descending at base and curving upwards again with U-bend, with 3 - 6 (-11) **Br**; racemes cylindrical, 12 - 28 cm, lax; **Bra** ovate, acuminate, 8 - 9 (-15) × 7 - 8 mm, rather fleshy; **Ped** 8 - 12 mm; **Fl** red, 32 - 33 mm, 6 - 8.5 mm ∅ when pressed; **Tep** free for ± 20 mm.

A. purpurea Lamarck (Encycl. 1: 85, 1783). **T:** "Réunion" [Mauritius] (*Commerson* s.n. [P-LA]). – **Lit:** Marais (1975: 601-602). **D:** Mauritius; dry mountain slopes.

≡ *Lomatophyllum purpureum* (Lamarck) T. Durand & Schinz (1895); **incl.** *Dracaena marginata* Aiton (1789) (*nom. illeg.*, Art. 52) ≡ *Aloe marginata* (Aiton) Willdenow (1809) (*nom. illeg.*, Art. 53.1); **incl.** *Aloe marginalis* De Candolle (1800) (*nom. illeg.*, Art. 52); **incl.** *Dracaena dentata* Persoon (1805) (*nom. illeg.*, Art. 52); **incl.** *Lomatophyllum borbonicum* Willdenow (1811) (*nom. illeg.*, Art. 52.1); **incl.** *Phylloma aloiflorum* Ker Gawler (1813) (*nom. illeg.*, Art. 52.1) ≡ *Lomatophyllum aloiflorum* (Ker Gawler) G. Nicholson (1885); **incl.** *Aloe rufocincta* Haworth (1819) ≡ *Phylloma rufocinctum* (Haworth) Sweet (1827) ≡ *Lomatophyllum rufocinctum* (Haworth) Salm-Dyck ex Schultes & Schultes *fil.* (1829) **incl.** *Lomatophyllum marginatum* Hoffmannsegg (1824) (*nom. inval.*, Art. 32.1c).

[8] Caulescent; stem erect, to 3 m × 7 - 10 cm; **L** 12 - 20, densely rosulate, linear-lanceolate to ensiform, to 100 × 8 - 12 cm, dark green, with red horny margin; marginal teeth small, red; **Inf** 50 - 60 cm, with up to 10 **Br**; racemes cylindrical, 15 - 22 cm; **Bra** deltoid, 4 - 5 mm; **Ped** 20 - 33 mm; **Fl** yellowish-red, 19 mm, base attenuate, slightly narrowed above **Ov**, then enlarging to mouth; **Tep** free for 14 - 15 mm; **St** not exserted; **Fr** berries.

A. pustuligemma L. E. Newton (Brit. Cact. Succ. J. 12(2): 53-54, ills., 1994). **T:** Kenya, Rift Valley Prov. (*Newton* 3739 [K, EA]). – **D:** Kenya; grassland with scattered trees and shrubs, 1370 m. **Fig. XVII.f**

[13] Caulescent, branching at base; stem erect or ascending to ± 40 cm, becoming decumbent to 1.5 m; **L** rosulate and persisting for 30 - 40 cm below, lanceolate, to 32 × 8.5 cm, dull green, surface slightly rough; marginal teeth 2 mm, firm, brown-tipped, 7 - 11 mm apart; exudate yellow; **Inf** 56 - 65 cm, with 8 - 9 **Br**; racemes cylindrical, terminal 13 - 21 cm, laterals 4 - 17 cm, sub-dense; **Bra** triangular, 11 - 13 × 3 mm, imbricate in bud stage; **Ped** 10 mm; **Fl** pale pink, lobes with white margins, surface minutely pustulate (esp. obvious in bud), 27 - 28 mm, base obconical, 5 mm ∅ across **Ov**, narrowed to 4.5 mm above, then enlarging to 5 mm at mouth; **Tep** free for 15 - 16 mm.

A. ×qaharensis Lavranos & Collenette (CSJA 72(2): 87, ills. (p. 86), 2000). **T:** Saudi Arabia (*Collenette* 7749 [K]). – **D:** Saudi Arabia. **I:** Collenette (1999: 27).

= *A. fleurentiniorum* × *A. woodii*.

A. rabaiensis Rendle (JLSB 30: 410, 1895). **T:** Kenya, Coast Prov. (*Taylor* s.n. [BM]). – **D:** Somalia, Kenya, Tanzania; sandy soil in open bushland, 18 - 500 m.

[14,16] Caulescent, branching from base; stem erect or sprawling, often supported by surrounding shrubs, to 2 m, mostly with dead **L** persisting; **L** laxly rosulate and persistent below stem tip, lanceolate-attenuate, 30 - 45 × 3 - 8 cm, greyish-green, often tinged reddish, often with few scattered whitish spots on young shoots, surface smooth; marginal teeth 2 - 3 mm, brown-tipped, 8 - 15 mm apart; exudate yellow, drying red; **Inf** to 60 cm, with 5 - 9 **Br**, lower **Br** sometimes rebranched; racemes sub-capitate, ± 8 × 8 cm, sub-dense to dense; **Bra** lanceolate, 10 - 12 × 3 mm; **Ped** 10 - 15 (-20) mm; **Fl** shades of orange-red, yellow at mouth, sometimes entirely yellow, 20 - 25 mm, base shortly attenuate, 8 mm ∅ across **Ov**, slightly narrowed above; **Tep** free for ± ½ of their length. – *Cytology:* 2n = 14 (Cutler & al. 1980).

Plants from upland Kenya and Tanzania (1370 - 1900 m) included in this taxon by Reynolds (1966), including his plate 83, are *A. ngongensis*. Possible hybrids with *A. secundiflora* have been reported (Carter 1994).

A. ramosissima Pillans (JSAB 5(3): 66-67, t. 7, 1937). **T:** RSA, Northern Cape (*Reynolds* 2547 [BOL, PRE]). – **Lit:** Rebelo & al. (1989). **D:** Namibia, RSA (Northern Cape); hot arid mountain slopes. **I:** Reynolds (1950: 487).

≡ *Aloe dichotoma* var. *ramosissima* (Pillans) Glen & D. S. Hardy (2000).

[16] Caulescent, much branched at base and above, branched dichotomously; stem erect or ascending, to 3 m × 8 cm, smaller **Br** to 2 cm ∅, covered with waxy grey powder; **L** 10 - 14, densely rosulate, lanceolate-linear, 15 - 20 × 2.2 cm, glaucous-green with very narrow pale yellow subcartilaginous margin; marginal teeth ± 1 mm, pale brownish, 1 - 4 mm apart; **Inf** 15 - 20 cm, with 1 - 2 **Br**; racemes cylindrical, 12 - 15 cm, sub-dense; **Bra** deltoid-acuminate, 4 - 5 × 3 mm, white; **Ped** 8 mm; **Fl** greenish-yellow to canary-yellow, 35 mm, slightly ventricose, base rounded, slightly enlarged above **Ov** to about the middle, narrowing to mouth; **Tep** free for 25 mm; **St** exserted 12 mm. – *Cytology:* 2n = 14 (Riley 1959).

A. rauhii Reynolds (JSAB 29(4): 151-152, t. 24-25, 1963). **T:** Madagascar, Toliara (*Rauh* 7594 [K, HEID]). – **D:** Madagascar; sandstone rocks in dense bush. **I:** Reynolds (1966: 413). **Fig. XV.c**

≡ *Guillauminia rauhii* (Reynolds) P. V. Heath (1994).

[11] Acaulescent or very shortly caulescent, branching to form dense groups; **L** to 20, densely rosulate, lanceolate-deltoid, tip acute, 7 - 10 × 1.5 - 2 cm, grey-green, sometimes tinged brownish, with many scattered H-shaped spots, with white cartilaginous margin; marginal teeth ± 0.5 mm, white, 1 - 2 mm apart; **Inf** 30 cm, simple or rarely with 1 **Br**; racemes cylindrical, slightly acuminate, ± 7 × 4 cm, lax, with 12 - 18 **Fl**; **Bra** ovate-acute, attenuate, 4 - 5 × 2 mm, white; **Ped** 10 mm; **Fl** rose-scarlet, paler at mouth, 25 mm, base shortly attenuate, 5 mm ∅ across **Ov**, slightly narrowed above, then enlarging to mouth; **Tep** free to base; **St** exserted 0 - 1 mm. – *Cytology:* 2n = 14 (Brandham 1971).

A. reitzii Reynolds (JSAB 3(3): 135-137, t. 20, 1937). **T:** RSA, Mpumalanga (*Reynolds* 2308 [PRE, BOL]). – **D:** RSA.

A. reitzii var. **reitzii** – **D:** RSA (Mpumalanga); rocky slopes. **I:** Reynolds (1950: 453-454, t. 50).

[5] Acaulescent or caulescent, simple or rarely branched; stem decumbent, to 60 cm; **L** densely rosulate, lanceolate-ensiform, to 65 × 12 cm, the tip a pungent **Sp**, green, surface smooth, lower face sometimes with 4 - 8 brownish ± 2 mm prickles in median line near tip; marginal teeth 3 mm, pungent, brownish to reddish-brown, 7 - 15 mm apart; **Inf** 1 - 1.3 m, with 2 - 6 **Br**; racemes cylindrical, slightly acuminate, 35 - 45 × 5 - 6 cm, very dense; **Bra** lanceolate-acute, 14 × 7 mm, brownish; **Ped** 3 mm; **Fl** bright red above, lower face lemon-yellow, to 50 mm, base rounded, ± 7 mm ∅ across **Ov**, enlarged to ± 9 mm in middle, then slightly narrowing to mouth; **Tep** free for 20 mm; **St** exserted 10 mm. – *Cytology:* 2n = 14 (Müller 1941).

A. reitzii var. **vernalis** D. S. Hardy (BT 13(3/4):

451-452, ills., 1981). **T:** RSA, KwaZulu-Natal (*Hardy* 3589 [PRE]). – **D:** RSA (KwaZulu-Natal); steep well-drained granitic slopes.

[5] Differs from var. *reitzii*: **Bra** deltoid-acuminate, 6 × 4 - 5 mm; **Fr** smaller; different range and flowering season.

A. retrospiciens Reynolds & P. R. O. Bally (JSAB 24(4): 182-184, t. 25-26, 1958). **T:** Somalia (*Reynolds* 8482 [PRE, EA, K]). – **Lit:** Gilbert & Sebsebe (1992). **D:** Somalia; amongst boulders on sandy soil, 820 - 1160 m. **I:** Reynolds (1966: 356-357).

Incl. *Aloe ruspoliana* var. *dracaeniformis* A. Berger (1908).

[8] Caulescent, sometimes simple, usually 2- to 6-branched; stem erect, 100 - 125 × 3 cm; **L** ± 12, rosulate and persistent for 10 - 20 cm below stem tip, lanceolate-attenuate, tip obtuse with a few small white teeth, 25 × 5 - 6 cm, bluish-grey tinged reddish, sometimes with a few scattered white spots near base, with white cartilaginous margin; marginal teeth 1 mm, firm, white, 5 mm apart; **Inf** 45 cm, with ± 10 **Br**; terminal raceme cylindrical, laterals mostly with secund **Fl** pointing backwards, 3 × 5 cm, sub-dense; **Bra** ovate-deltoid, 5 × 2.5 mm, white; **Ped** 5 - 6 mm; **Fl** yellow, greenish at mouth, 20 mm, base rounded, 6 mm ⌀ across **Ov**, slightly narrowed above, then enlarging slightly to mouth; **Tep** free for 10 mm.

A. reynoldsii Letty (FPSA 14: t. 558 + text, 1934). **T:** RSA, Eastern Cape (*Reynolds* 140 [PRE]). – **D:** RSA (Eastern Cape); rock faces and steep grassy slopes, to 760 m. **I:** Reynolds (1950: 300-301).

[7] Acaulescent or shortly caulescent, forming groups of up to 12 **Ros**; stem 5 cm ⌀; **L** 16 - 20, densely rosulate, ovate-lanceolate, acuminate, to 35 × 11 cm, glaucous green, lineate, with many scattered oblong or H-shaped dull white spots, lower face with fewer spots, with 2 mm pink cartilaginous margin; marginal teeth minute, soft to firm, 1 - 4 mm apart; **Inf** 40 - 60 cm, with ± 4 **Br**; racemes subcapitate, 5 - 6 cm, lax; **Bra** lanceolate-deltoid, 10 mm; **Ped** 20 - 25 mm; **Fl** yellow, tinged orange towards mouth, 28 mm, base truncate, 7 mm ⌀ across **Ov**, abruptly narrowed to 5 mm above, then enlarging to mouth; **Tep** free for 5 mm; **St** exserted 0 - 2 mm.

A. rhodesiana Rendle (JLSB 40: 215, 1911). **T:** Zimbabwe, Melsetter Distr. (*Swynnerton* 6047 [BM]). – **D:** Moçambique, Zimbabwe; mostly on rocky or stony ground, 1830 - 2135 m. **I:** Reynolds (1966: 27-28).

Incl. *Aloe eylesii* Christian (1936).

[4] Acaulescent or shortly caulescent, simple or with 2 - 3 **Br** at base; **R** fusiform; stem to 10 × 3 - 4 cm; **L** 8 - 12, densely rosulate, triangular, 25 - 30 × 4 - 5 cm, dull green, lower face sometimes with a few elliptic whitish spots near base, with narrow

white cartilaginous margin; marginal teeth ± 0.5 - 1 mm, firm, white, 1 - 4 mm apart; **Inf** 40 - 45 cm, simple; raceme cylindrical-acuminate, 12 - 15 × 8 cm, sub-dense; **Bra** ovate-acute, cuspidate, 20 × 11 mm, imbricate in bud stage; **Ped** to 30 mm; **Fl** salmon-pink, 35 mm, base obconical-attenuate, 6 mm ⌀ across **Ov**, enlarging slightly to just above middle; **Tep** free to base; **St** exserted 0 - 1 mm.

A. richardsiae Reynolds (JSAB 30(2): 67-69, t. 12-13, 1964). **T:** Tanzania, Iringa Distr. (*Richards* 15604 [K, PRE]). – **D:** Tanzania; grass clearings in woodland. **I:** Reynolds (1966: 37-38).

[1] Acaulescent, simple; **R** thick, fleshy; **L** 8 - 10, rosulate, with bases expanded to form an underground bulb 3 - 4 cm ⌀, linear, 20 - 25 × 1.5 cm, green, obscurely lineate; marginal teeth 0.5 mm, firm, white, 1 - 2 mm apart; **Inf** 35 - 45 cm, simple; raceme cylindrical-acuminate, ± 25 × 5 - 6 cm, lax; **Bra** ovate-acute, 25 - 30 × 7 - 8 mm, white, imbricate in bud stage; **Ped** 5 - 7 mm; **Fl** pale orange-scarlet, to 48 mm, base rounded, 7 - 8 mm ⌀ across **Ov**, slightly narrowed above, then enlarging to mouth; **Tep** free for 15 mm.

A. rigens Reynolds & P. R. O. Bally (JSAB 24(4): 177-179, t. 20-21, 1958). **T:** Somalia (*Reynolds* 8369 [PRE, EA, K]). – **D:** Yemen, Somalia.

A. rigens var. **mortimeri** Lavranos (CSJA 39(4): 123-125, ills., 1967). **T:** Yemen (*Lavranos & Rauh* 3174 [PRE]). – **D:** Yemen; amongst boulders on limestone hills, ± 750 m.

[6] Differs from var. *rigens*: Suckering to form large groups; **L** 15 - 18, soft, 40 - 60 × 6 - 10 cm, glaucous green; marginal teeth 3 mm, whitish, 10 - 20 mm apart; **Inf** to 1.25 m, simple or with 1 - 2 **Br**; **Bra** 20 - 22 × 6 mm; **Fl** brick-red, minutely papillose, lobes with yellowish margin.

A. rigens var. **rigens** – **D:** Somalia; dry plains, 700 - 1555 m. **I:** Reynolds (1966: 124-125).

[5] Acaulescent or shortly caulescent, usually simple, sometimes in small groups; **L** ± 24, densely rosulate, lanceolate-attenuate, 60 - 80 × 12 - 15 cm, pale to darker grey-green, sometimes reddish-tinged; marginal teeth 4 - 6 mm, pungent, reddish-brown with pale base, 20 - 35 mm apart; exudate drying yellow to orange; **Inf** 1.25 - 1.75 m, with 3 - 4 **Br**; racemes cylindrical-acuminate, 20 - 30 × 6 cm, sub-dense; **Bra** ovate-deltoid, to 15 × 6 mm, white, imbricate in bud stage; **Ped** 5 - 6 mm; **Fl** rose-pink to dull scarlet, very shortly pubescent, 30 - 34 mm, base obtuse, 7 mm ⌀ across **Ov**, very slightly narrowed above and enlarging to mouth; **Tep** free for 10 - 12 mm. – *Cytology:* 2n = 14 (Brandham 1971: as *A. rigens*).

A natural hybrid with *A. megalacantha* has been reported (Reynolds 1966).

169

A. rivae Baker (FTA 7: 465, 1898). **T:** Ethiopia, Sidamo Prov. (*Ruspoli & Riva* 1509 [B]). – **D:** Ethiopia, Kenya; open wooded grassland on rocky slopes, 1000 - 1500 m. **I:** Reynolds (1966: 115-116).

[5] Acaulescent or shortly caulescent, simple or in small groups; stem erect, ascending or decumbent, to 60 cm; **L** ± 20, densely rosulate, ovate-lanceolate, 50 - 55 × 17 - 20 cm, the tip a **Sp**, dull olive-green to brownish-green, with reddish-tinged margin; marginal teeth ± 4 mm, firm, reddish-brown, 10 - 15 mm apart; exudate drying purple; **Inf** 60 - 70 cm, with ± 12 **Br**, lower **Br** rebranched; racemes cylindrical or with **Fl** slightly secund, 10 × 7 cm, lax; **Bra** ovate-acute, 2 - 4 × 2 - 3 mm; **Ped** 12 mm; **Fl** scarlet with a bloom, 33 mm, base truncate, 10 mm Ø across **Ov**, narrowed slightly to mouth; **Tep** free for 13 mm.

A. rivierei Lavranos & L. E. Newton (CSJA 49(3): 114-116, ills., 1977). **T:** Yemen, Taiz Prov. (*Lavranos & Newton* 13121 [E, K]). – **Lit:** Wood (1983). **D:** Saudi Arabia, Yemen; cliffs and rocky slopes, 1000 - 2000 m. **I:** Collenette (1999: 27).

[8] Caulescent, branching mostly at base; stem erect, to 2 m × 5 cm; **L** ± 15, rosulate, triangular-acute, 55 × 8 cm, pale green, with many pale spots on young shoots only, with white hyaline margin; marginal teeth 2 mm, pungent, white, tipped reddish-brown, 4 - 5 mm apart; **Inf** to 1.2 m, with 2 **Br**; racemes cylindrical-acuminate, lax; **Bra** 10 × 7 mm, pale green, imbricate in bud stage; **Ped** to 15 mm; **Fl** coral-red, becoming yellow to orange towards mouth, sometimes entirely yellow, 30 mm, base rounded, 4 mm Ø across **Ov**, very slightly narrowed above, enlarging slightly to mouth; **Tep** free for 11 - 13 mm. – *Cytology:* 2n = 14 (Wood 1983).

This is probably conspecific with *A. arborea* Forsskål, which is currently treated as a *nomen ambiguum* but would have priority if its identity were to be confirmed (Wood 1983).

A. rosea (H. Perrier) L. E. Newton & G. D. Rowley (Excelsa 17: 61, 1997). **T:** Madagascar (*Perrier* 13979 [K]). – **D:** Madagascar; deep shade in deciduous forest, on limestone, ± 400 - 800 m. **I:** Rauh (1995a: 100, 328).

≡ *Lomatophyllum roseum* H. Perrier (1926).

[6,7] Acaulescent, suckering; **L** 12 - 15, laxly rosulate, lanceolate-attenuate, 30 - 45 × 2.5 - 4 cm, dull green; marginal teeth 5 mm, brown, 6 - 18 mm apart; **Inf** 25 - 30 cm, simple or with 1 - 3 **Br**; racemes cylindrical, 6 - 12 cm, sub-dense, with 25 - 30 **Fl**; **Bra** acute, to 5 - 6 mm; **Ped** 5 - 6 mm; **Fl** white at base, rose-pink above, 22 - 25 mm, base rounded, narrowed above **Ov**, then enlarging to mouth; **Tep** free for 4.5 - 5 mm; **Fr** berries.

A. rubroviolacea Schweinfurth (Bull. Herb. Boissier 2(App. 2): 71, 1894). **T:** Yemen (*Schweinfurth*

1658 [K [iso]]). – **Lit:** Wood (1983). **D:** Saudi Arabia, Yemen; rocky slopes, 2500 - 2900 m. **I:** FPA 41: t. 1610, 1970; Collenette (1999: 28).

[13] Caulescent, suckering; stem decumbent, to 1 m; **L** densely rosulate, lanceolate-ensiform, 60 × 10 - 11 cm, red-violet with a bloom; marginal teeth ± 2 - 3 mm, reddish, hooked, 20 - 25 mm apart; **Inf** 1 m, arcuate-ascending, simple or with 1 **Br**; racemes cylindrical, 30 - 40 × 8 - 10 cm, very dense; **Bra** lanceolate-acute, 20 - 30 × 11 mm; **Ped** 2 - 4 mm; **Fl** bright red, 25 - 35 mm, ventricose, base rounded, enlarged above **Ov** to middle, slightly narrowing to mouth; **Tep** free for ± 15 mm; **St** exserted 10 - 15 mm. – *Cytology:* 2n = 14 (Wood 1983).

Yellow-flowered plants are also reported, but may belong with *A. pseudorubroviolacea.*

A. ruffingiana Rauh & Petignat (KuaS 50(11): 270-272, ills., 1999). **T:** Madagascar, Tolanaro (*Petignat* 671 [HEID]). – **D:** Madagascar; open bush on granite.

[18] Acaulescent or with stem to 10 cm, suckering at base; **L** 12 - 15, densely rosulate, spreading, oval-triangular, attenuate, 10 - 15 × 4 cm, both faces green with numerous ± rectangular white flecks, margin pale green with narrow reddish edge; marginal teeth deltoid, small, white or reddish; **L** exudate colourless; **Inf** to 20 cm, erect, usually simple, or with 1 **Br**; racemes cylindrical, 6 cm with 10 - 15 **Fl**, lax; **Bra** lanceolate, 10 mm; **Ped** ± 15 mm; **Fl** red at base, above whitish with green midstripe, 25 - 28 mm, base rounded, 4 mm Ø across **Ov**, slightly constricted above, widening to mouth; **OTep** free to base; **St** shortly exserted.

A. rugosifolia M. G. Gilbert & Sebsebe (KB 47(4): 652-653, 1992). **T:** Kenya, North-East Prov. (*Ruspoli & Riva* 476 [B, FT]). – **D:** Ethiopia, Kenya; dry bushland, 1060 - 1700 m. **I:** Reynolds (1966: 159, as *A. otallensis* var. *elongata*).

Incl. *Aloe otallensis* var. *elongata* A. Berger (1908).

[5] Caulescent, sometimes simple, usually in small groups; stem becoming decumbent, to 50 cm; **L** 16 - 20, densely rosulate, lanceolate-attenuate, 20 - 40 × 5.5 - 8 cm, deep green to brownish-green with many scattered dull white lenticular spots, surface finely rugose; marginal teeth 3 - 5 mm, rigid, reddish-brown with paler base, 10 - 15 mm apart; exudate drying yellow; **Inf** 1.5 - 1.8 m, with 8 - 10 **Br**; racemes cylindrical-acuminate, 10 - 20 cm, lax; **Bra** ovate-acute, 10 - 13 × 4 - 8 mm, white; **Ped** 5.5 - 7 mm; **Fl** rose-pink, 25 - 28 mm, base rounded, 5 - 6 mm Ø across **Ov**, slightly enlarged to 8 mm above; **Tep** free for 15 mm.

A. rupestris Baker (FC 6: 327-328, 1896). **T:** RSA, KwaZulu-Natal (*MacOwan* 1556 [SAM]). – **D:** Moçambique, Swaziland, RSA (KwaZulu-Natal); tall bush on rocky slopes. **I:** Reynolds (1950: 473-474).

Incl. *Aloe pycnacantha* MacOwan ms. (s.a.) (*nom. inval.*, Art. 29.1); **incl.** *Aloe nitens* Baker (1880) (*nom. illeg.*, Art. 53.1).

[9] Caulescent, mostly simple; stem erect, to 8 m × 20 cm, dead **L** persisting in upper ⅓; **L** ± 30 - 40, densely rosulate, lanceolate-attenuate, to 70 × 7 - 10 cm, dull to slightly glossy deep green, with deep pink to pale red margin; marginal teeth 4 - 6 mm, pungent, reddish-brown, 8 - 12 mm apart; **Inf** 1 - 1.25 m, with 6 - 9 **Br**, lower **Br** rebranched; racemes cylindrical, slightly acuminate, 20 - 25 × 7 cm, very dense; **Bra** ± 1 × 2 mm; **Ped** 1 mm; **Fl** lemon-yellow, becoming orange-yellow to brownish-yellow towards mouth, 20 mm, slightly ventricose, base rounded, 4 mm ∅ across **Ov**, enlarged above to ± middle, then narrowing to mouth; **Tep** free for 12 mm; **St** exserted 15 mm. – *Cytology:* 2n = 14 (Riley 1959).

Natural hybrids with *A. marlothii* have been reported (Reynolds 1950).

A. rupicola Reynolds (JSAB 26(2): 89-91, t. 10-11, 1960). **T:** Angola, Bié Distr. (*Reynolds 9243* [PRE, K, LUA]). – **D:** Angola; rocky hills, 1780 m. **I:** Reynolds (1966: 323-324).

[9] Caulescent, simple or branched at base; stem erect, to 5 m × 10 - 12 cm; **L** ± 40, densely rosulate, lanceolate, 40 - 45 × 6 cm, apical 10 cm usually drying early, green, obscurely lineate; marginal teeth 4 - 5 mm, pungent, reddish-brown, 10 mm apart; exudate drying pale yellow; **Inf** 70 - 90 cm, with 3 - 8 **Br**; racemes cylindrical, 15 - 18 × 8 - 9 cm, sub-dense; **Bra** ovate-acute, 9 × 5 mm; **Ped** 12 mm; **Fl** orange-scarlet, 42 mm, base very shortly attenuate, 7 mm ∅ across **Ov**, slightly enlarged above; **Tep** free for 21 mm. – *Cytology:* 2n = 14 (Brandham 1971).

A. ruspoliana Baker (FTA 7: 460, 1898). **T:** Ethiopia, Ogaden (*Ruspoli & Riva 918* [B]). – **D:** Ethiopia, Kenya, Somalia; arid bushland, 400 - 950 m. **I:** Reynolds (1966: 254).

Incl. *Aloe stephaninii* Chiovenda (1916); **incl.** *Aloe jex-blakeae* Christian (1942).

[7] Acaulescent or shortly caulescent, suckering to form groups, sometimes large; stem ascending or decumbent, to 50 cm; **L** ± 16, densely rosulate, lanceolate-attenuate, 50 - 60 × 12 cm, yellowish-green, sometimes with a few lenticular white spots near base, surface smooth; marginal teeth to 0.5 mm, white, 5 - 8 mm apart; exudate very pale yellow, almost colourless; **Inf** ≥ 1.5 m, with ≥ 12 **Br**; racemes capitate, 2 - 4 × 5 cm, dense; **Bra** deltoid-acute, 3 × 1.5 mm; **Ped** 5 mm; **Fl** yellow, 16 - 20 mm, base rounded, 5 mm ∅ across **Ov**, enlarging above to ± 6 - 7 mm at mouth; **Tep** free for 6 - 8 mm.

A. sabaea Schweinfurth (Bull. Herb. Boissier 2(App. 2): 74, 1894). **T:** Yemen (*Schweinfurth 941*

[K [iso]]). – **D:** Saudi Arabia, Yemen; cliffs and steep rocky slopes, 600 - 2060 m. **I:** Reynolds (1966: 313, as *A. gillilandii*); Collenette (1999: 28).

Incl. *Aloe gillilandii* Reynolds (1962).

[9] Caulescent; stem erect, simple, to 3 m × 10 cm; **L** ± 16, densely rosulate, lanceolate-acute, lowest **L** decurved, 60 - 80 × 15 cm, grey-green, with pale pink cartilaginous margin; marginal teeth 1 - 1.5 mm, soft, pale pink, 5 - 10 mm apart; **Inf** ± 90 cm, with ± 8 **Br**; racemes cylindrical-acuminate, 15 × 6 cm, sub-dense; **Bra** ovate-acute, 10 - 17 × 8 - 13 mm, imbricate in bud stage; **Ped** 12 mm; **Fl** scarlet to red-brown, paler at mouth, 22 - 30 mm, base rounded, 10 mm ∅ across **Ov**, not narrowed above except at upturned mouth; **Tep** free to base. – *Cytology:* 2n = 14 (Johnson & Brandham 1997).

A. saundersiae (Reynolds) Reynolds (JSAB 13(2): 103, ills., 1947). **T:** RSA, KwaZulu-Natal (*Reynolds 1799* [PRE]). – **D:** RSA (KwaZulu-Natal); rock crevices or in grass, ± 1525 m. **I:** Reynolds (1950: 111, t. 1).

≡ *Leptaloe saundersiae* Reynolds (1936); **incl.** *Aloe minima* Medley-Wood (1906) (*nom. illeg.*, Art. 53.1).

[1] Acaulescent, simple or in small groups; **R** fusiform; **L** 10 - 16, rosulate, linear, 4 - 8 × 0.3 cm, green, lower face sometimes with a few spots near base; marginal teeth ± 0.5 mm, soft, white, 1 mm apart; **Inf** 14 - 18 cm, simple; racemes capitate, 2 - 2.5 × 3 - 3.5 cm, sub-dense, with 12 - 16 **Fl**; **Bra** ovate-acuminate, 7 × 3 - 4 mm, white; **Ped** 9 - 10 mm; **Fl** pale cream-pink, 10 - 12 mm, base attenuate, not narrowed above **Ov**; **Tep** free to base; **St** not exserted. – *Cytology:* 2n = 14 (Müller 1945: as *Leptaloe*).

A. scabrifolia L. E. Newton & Lavranos (CSJA 62(5): 219-221, ills., 1990). **T:** Kenya, Rift Valley Prov. (*Newton 3476* [K, EA, MO]). – **D:** Kenya; open *Acacia* bushland, 1000 - 1630 m. **I:** Reynolds (1966: 225, as *A. turkanensis*).

[13] Caulescent, branching sparsely at base; stem erect for ± 30 cm, becoming decumbent to 1.2 m × 3 - 4 cm ∅; **L** to 25, laxly rosulate, lanceolate-attenuate, often slightly falcate, to 55 × 10 - 12.5 cm, dull green to grey-green with few to many scattered elliptic white spots, with narrow white margin, surface rough; marginal teeth to 2 mm, uncinate, white, 7 - 15 mm apart; exudate drying brown; **Inf** to 1.4 m, ascending, with up to 12 **Br**; racemes with secund **Fl**, 30 - 45 cm, lax; **Bra** triangular, 3 × 2 - 3 mm; **Ped** 5 mm; **Fl** dull red, with paler margins on **Tep**, 25 mm, base rounded, 6 - 7 mm ∅ across **Ov**, narrowed to 5 mm above, then enlarging to 7 mm at mouth; **Tep** free for 12 - 13 mm.

Reynolds (1966) confused this with *A. turkanensis*, which differs in several characters, notably its tighter growth habit and its very smooth leaves.

171

A. schelpei Reynolds (JSAB 27(1): 1-3, t. 1-2, 1961). **T**: Ethiopia, Shoa Prov. (*Curle & Schelpe* 61 [BM]). − **D**: Ethiopia; grassland on steep slopes, 2130 - 2350 m. **I**: Reynolds (1966: 284-286).

[6] Caulescent, branching at base and above to form dense groups; stem decumbent, to 50 × 5 - 6 cm; **L** 16 - 20, rosulate, lanceolate-attenuate, 45 × 10 - 12 cm, glaucous tinged bluish, sometimes with several pale green to creamy lenticular spots near base, lower face deeper green usually with several spots near base, with prominent reddish-pink margin; marginal teeth 2 - 3 mm, firm, reddish-pink, paler tipped, ± 15 mm apart; exudate drying dark brown; **Inf** 50 cm, simple or with 1 **Br**; racemes cylindrical-conical, 6 - 9 × 6 - 7 cm, dense; **Bra** ovate-acute, 5 × 3 mm; **Ped** 13 - 15 mm; **Fl** orange-red, paler at mouth, 28 - 30 mm, base rounded, 7 mm ∅ across **Ov**, slightly narrowed above, then slightly enlarging to mouth; **Tep** free for 12 mm. − *Cytology:* 2n = 14 (Brandham 1971).

A. schilliana L. E. Newton & G. D. Rowley (Excelsa 17: 61, 1997). **T**: Madagascar (*Perrier* 1104 [P]). − **D**: Madagascar; on gneiss, 200 m.

Incl. *Lomatophyllum viviparum* H. Perrier (1926).

[6] Caulescent, simple; stem prostrate, short; **L** 12 - 15, laxly rosulate, lanceolate-attenuate, 45 - 55 × 2.5 - 3.5 cm, green, tip with 3 **Sp**; marginal teeth ± 1 mm, green, crowded; **Inf** 50 - 80 cm, simple; bulbils developing in **Ax** of sterile **Bra** below raceme; racemes cylindrical, ± 13 cm, sub-dense; **Bra** acute, ⅓ - ½ as long as **Ped**; **Ped** 8 - 10 mm; **Fl** purple-red, green-tipped, 30 - 33 mm, 6 - 7 mm ∅ across **Ov**; **Tep** free for 10 - 11 mm; **Fr** berries.

A. schoelleri Schweinfurth (Bull. Herb. Boissier 2(App. 2): 107, 1894). **T**: Eritrea (*Schweinfurth* 158 [B, K]). − **D**: Eritrea; > 1000 m.

[18] Habit unknown; **L** lanceolate, 40 - 45 × 10 - 13 cm, surface smooth; marginal teeth minute; exudate drying brown; **Inf** 50 - 60 cm, simple; racemes ± 30 cm, very dense; **Bra** rhomboidal to obovate, acute, 14 - 17 × 9 - 10 mm, minutely papillate; **Ped** ± 10 mm mm; **Fl** at least 15 mm (mature **Fl** not seen); **Tep** free for 11 mm.

Known only from the type collection and incompletely known.

A. schomeri Rauh (KuaS 17(2): 22-24, ills., 1966). **T**: Madagascar, Toliara (*Rauh* M1382 [HEID, PRE]). − **D**: S Madagascar; gneissic rocks. **I**: Reynolds (1966: 429-430).

[4] Acaulescent or very shortly caulescent, simple or suckering to form groups; **L** ± 30, densely rosulate, lanceolate-attenuate, 20 - 30 × 3 - 5 cm, dark green, with pale, almost white, cartilaginous margin; marginal teeth 2 mm, pale, almost white, 5 - 8 mm apart; **Inf** usually simple, sometimes with 1 - 2 **Br**; racemes subcapitate or shortly cylindrical, 6

- 10 × 7 cm, dense, with 60 - 70 **Fl**; **Bra** ± 9 × 5 mm; **Ped** 12 mm; **Fl** yellow, 21 mm, base rounded, enlarging above **Ov** to mouth; **Tep** free almost to base; **St** exserted 5 - 8 mm.

A. schweinfurthii Baker (JLSB 18: 175, 1880). **T**: Sudan, Equatoria Prov. (*Schweinfurth* ser. 3, 167 [K]). − **D**: Benin, Burkina Faso, Ghana, Mali, Nigeria, Sudan, Uganda, Zaïre; granite outcrops, 600 - 1200 m. **I**: Reynolds (1966: 289-290).

Incl. *Aloe barteri* var. *lutea* A. Chevalier (1913); **incl.** *Aloe trivialis* A. Chevalier (1952) (*nom. inval.*, Art. 36.1).

[7] Acaulescent or shortly caulescent, suckering to form dense groups; **L** 16 - 20, densely rosulate, lanceolate-attenuate, 30 - 60 × 6 - 7 cm, grey-green tinged bluish, usually with a few scattered whitish spots near base, surface smooth; marginal teeth 4 mm, pungent, reddish-brown, 10 - 12 mm apart; exudate drying purplish; **Inf** 90 cm, with 8 - 10 **Br**; racemes cylindrical-acuminate, 15 × 7 cm, sub-dense; **Bra** ovate-acute, 5 × 2 - 3 mm; **Ped** 13 mm; **Fl** scarlet, becoming orange at mouth, 28 mm, base shortly attenuate, 7 mm ∅ across **Ov**, slightly narrowed above; **Tep** free for 12 mm. − *Cytology:* 2n = 14 (Newton 1970).

A natural hybrid with *A. buettneri* is known as *A. ×keayi* (Newton 1976). A natural hybrid with *A. macrocarpa* var. *major* is also known (Newton, unpubl.).

Until 1963 only 1 species was reported from West Africa, *A. barteri*. Keay (1963) showed that the type of this name was a mixture of 2 taxa, *A. buettneri* and *A. schweinfurthii*.

A. scobinifolia Reynolds & P. R. O. Bally (JSAB 24(4): 174-175, t. 17-18, 1958). **T**: Somalia (*Reynolds* 8403 [PRE, EA, K]). − **D**: Somalia; exposed gypsum soils, 1525 - 1675 m. **I**: Reynolds (1966: 197-199).

[5] Acaulescent or shortly caulescent, simple or usually forming small groups; **L** 16 - 20, densely rosulate, lanceolate-attenuate, 30 × 7 cm, the tip a **Sp**, dull green, with very narrow pale pink cartilaginous margin, surface rough; marginal teeth absent; exudate drying deep brown; **Inf** 60 - 70 cm, with 5 - 8 **Br**; racemes capitate-corymbose, 3 - 4 × 6 cm, dense; **Bra** deltoid, deflexed, 8 × 2 mm, white; **Ped** 15 - 18 mm; **Fl** yellow, orange or scarlet, 22 mm, slightly clavate, base shortly attenuate, 4 - 5 mm ∅ across **Ov**, enlarged above; **Tep** free for 9 - 10 mm. − *Cytology:* 2n = 14 (Brandham 1971).

A. scorpioides L. C. Leach (JSAB 40(2): 106-111, ills., 1974). **T**: Angola, Moçamedes Distr. (*Leach & Cannell* 14654 [LISC, BM, BR, K, LUA, LUAI, M, MO, PRE, SRGH]). − **D**: Angola; rocky slopes, often in shade of woodland.

[11] Caulescent, branching at base and above; stem usually divergent, to 50 (-100) cm; **L** laxly ro-

sulate, persistent below, ovate-attenuate, to 30×2.5 - 3.5 cm, yellowish-green, lower face darker, obscurely lineate, rarely with a few spots near base; marginal teeth 2 - 3 mm, pungent, yellowish- or brownish-tipped, 10 - 15 mm apart; sheath 1 - 2 cm, striate; **Inf** ± 15 cm, descending at base and curving upwards again with U-bend, simple or with 1 - 2 **Br**; racemes conical or cylindrical-acuminate, 11 - 25×6 cm, sub-dense; **Bra** ovate, subacute or acuminate, ± 6.5×3.5 mm, orange-brown; **Ped** 6 - 10 mm; **Fl** scarlet, yellow-striped with green patch at base, 21 - 28 mm, base very shortly attenuate, ± 7 mm ∅ across **Ov**, narrowed to ± 5.5 mm above, then enlarging to mouth; **Tep** free for 8.5 - 10 mm.

A. secundiflora Engler (Pfl.-welt Ost-Afr., Teil C, 140, 1895). **T:** Tanzania, Moshi Distr. (*Volkens* 530 [B]). – **D:** Ethiopia, Kenya, Rwanda, Tanzania.

A. secundiflora var. **secundiflora** – **D:** Ethiopia, Kenya, Rwanda, Tanzania; grassland and open woodland on sandy soil, 750 - 1980 m. **I:** Reynolds (1966: 231-233). **Fig. XIX.d**
    Incl. *Aloe engleri* A. Berger (1905); **incl.** *Aloe floramaculata* Christian (1940); **incl.** *Aloe marsabitensis* I. Verdoorn & Christian (1940).
    [5,7] Acaulescent or very shortly caulescent, usually simple, sometimes in small groups; **L** ± 20, densely rosulate, ovate-lanceolate, attenuate, 30 - 75×8 - 30 cm, glossy dull green, often with horny margin, surface smooth; marginal teeth 3 - 6 mm, pungent, dark brown, 10 - 20 mm apart; exudate drying yellow; **Inf** 1 - 1.5 m, with 10 - 12 **Br**, lower **Br** rebranched; racemes with secund **Fl**, 15 - 20 cm, lax; **Bra** ovate-acute, 3 - 7×2 - 5 mm; **Ped** 5 - 10 mm; **Fl** rose-pink to dull scarlet, paler at mouth, 25 - 35 mm, base truncate, 9 mm ∅ across **Ov**, slightly narrowed above, then enlarging slightly to mouth; **Tep** free for ½ their length.

A. secundiflora var. **sobolifera** S. Carter (Fl. Trop. East Afr., Aloaceae, 32-33, ills., 1994). **T:** Tanzania, Kilosa Distr. (*Congdon* 282 [K, EA, NHT]). – **D:** Tanzania; woodland on sandy soil, 600 - 1825 m.
    [7] Differs from var. *secundiflora*: Suckering to form groups, often large; **L** lanceolate, to 8 - 15 (-20) cm wide at base, dark green, often with bronze hue; marginal teeth not pungent, never joined by a horny rim.

A. seretii De Wildeman (Pl. Bequaert. 1: 28, 1921). **T:** Zaïre, Oriental Prov. (*Seret* 299 [BR]). – **D:** Zaïre; rock outcrops, 945 - 1770 m. **I:** Reynolds (1966: 257-258).
    [7] Acaulescent or shortly caulescent, suckering to form dense groups; **L** ± 16, densely rosulate, lanceolate-attenuate, 40×6 - 7 cm, grey bluish-green tinged reddish, sometimes with obscure dull white spots, with pinkish margin; marginal teeth 3 - 4 mm, pungent, white with reddish-brown tips, 8 - 10

mm apart; **Inf** 60 - 70 cm, with 3 **Br**; racemes cylindrical-conical, 15 - 20×5 - 6 cm, dense; **Bra** ovate-acute, 9 - 15×5 - 10 mm, pink, fleshy; **Ped** 14 - 18 mm; **Fl** dull to bright scarlet, 28 - 33 mm, base shortly attenuate, 7 mm ∅ across **Ov**, very slightly narrowed above; **Tep** free for 9 - 10 mm.

A. serriyensis Lavranos (JSAB 31(1): 76-77, t. 15, 1965). **T:** Yemen (*Lavranos* 2101 [PRE]). – **D:** Yemen; wooded valleys, 300 m. **I:** Reynolds (1966: 140).
    [5] Acaulescent, usually forming small groups; **L** rosulate, attenuate, 30 - 35×6 - 7 cm, green, sometimes tinged brownish, obscurely lineate; marginal teeth horny, blunt, dark brown, 20 - 40 mm apart; **Inf** 40 cm, with 2 - 3 **Br**; racemes conical, 20 - 25×5 cm, lax; **Bra** deltoid, 8 - 10×3 mm; **Ped** 10 mm; **Fl** scarlet-pink with powdery bloom, 27 mm, base attenuate, 6 mm ∅ across **Ov**, slightly narrowed above; **Tep** free for 7 - 9 mm.

A. shadensis Lavranos & Collenette (CSJA 72(2): 82, ills., 2000). **T:** Saudi Arabia, Hijaz Prov. (*Collenette* 6718 [K]). – **D:** Saudi Arabia; in woody vegetation, on granite, 750 - 1800 m. **I:** Collenette (1999: 29).
    [3] Acaulescent, solitary; **L** densely rosulate, spreading to ascending, lanceolate-attenuate, ± 60×15 cm, pinkish-grey; marginal teeth small, pale brown, widely spaced; **Inf** to 150 cm, erect, with 2 - 5 **Br**; racemes cylindrical, sub-dense; **Bra** 7 - 9×3 - 4 mm; **Ped** 7 - 11 mm; **Fl** pale pink, 30 - 35 mm, curved, base shortly attenuate, 6 - 7 mm ∅ across **Ov**, not constricted above; **Tep** free for 17 mm; **St** not exserted.

A. sheilae Lavranos (CSJA 57(2): 71-72, ills., 1985). **T:** Saudi Arabia (*Collenette* 3397 [K, E, MO]). – **D:** Saudi Arabia; amongst tonolite rocks, ± 1600 m. **I:** Collenette (1999: 29).
    [5] Acaulescent or sometimes shortly caulescent, simple or sometimes with 2 - 3 basal suckers; **L** laxly rosulate, deltoid-acuminate, to 55×6.5 cm, green, often with a few pale green rounded spots, with narrow white cartilaginous margin, surface rough; marginal teeth absent or few, to 1.5 mm, white, brown-tipped, 30 - 80 mm apart; **Inf** 50 - 70 cm, with 2 - 4 (-7) **Br**; racemes subcapitate or conical, sub-dense; **Bra** ovate-deltoid, 5 - 7×3 - 4 mm; **Ped** 15 - 18 mm; **Fl** coral-red becoming yellowish, 30 - 35 mm, base attenuate, 5 mm ∅ across **Ov**, slightly narrowed above, then enlarging to mouth; **Tep** free for 20 mm.

A. silicicola H. Perrier (Mém. Soc. Linn. Normandie, Bot. 1(1): 42, 1926). **T:** Madagascar (*Perrier* 13160 [P]). – **D:** Madagascar; on quartzite, ± 2000 m. **I:** Reynolds (1966: 454).
    [8] Caulescent, simple; stem to 2 m × 5 cm; **L** densely rosulate, lanceolate-attenuate, 45 - 50×7 -

8 cm, green; marginal teeth 1 - 1.5 mm, green, to 7 mm apart; **Inf** 50 - 60 cm, with 3 - 4 **Br**; racemes cylindrical, 6 - 8 cm, sub-dense, with 20 - 25 **Fl**; **Bra** lanceolate-acute, 5 - 6 × 1 - 2 mm; **Ped** 20 - 22 mm; **Fl** reddish-orange, 28 - 30 mm, narrowed above **Ov**, from middle enlarging to mouth; **Tep** free to base.

A. simii Pole-Evans (Trans. Roy. Soc. South Afr. 5: 704, 1917). **T:** RSA, Mpumalanga (*Sim* 137 [PRE]). − **D:** RSA (Mpumalanga); in grass on steep rocky slopes. **I:** Reynolds (1950: 280-281).

[5] Acaulescent, simple or rarely suckering to form small groups; **L** 15 - 20, densely rosulate, lanceolate-attenuate, 40 - 60 × 9 - 12 cm, bright to milky green, obscurely lineate, sometimes with a few obscure paler spots; marginal teeth 3 - 4 mm, horny, light brown, 10 - 15 mm apart; **Inf** 1 - 2 m, with 5 - 9 **Br**, lower **Br** sometimes rebranched; racemes cylindrical-acuminate, 30 - 65 cm, lax; **Bra** lanceolate-acuminate, ± 12 - 15 mm; **Ped** 12 - 15 mm; **Fl** strawberry-pink, 35 - 40 mm, base truncate, 12 mm ∅ across **Ov**, abruptly narrowed to 5 mm above, then enlarging to mouth; **Tep** free for 12 mm; **St** exserted 1 - 2 mm. − *Cytology:* 2n = 14 (Müller 1945).

A. sinana Reynolds (JSAB 23(1): 3-5, t. 3-4, 1957). **T:** Ethiopia, Shoa Prov. (*Reynolds* 8126 [PRE, EA, K]). − **D:** Ethiopia; mountain slopes, 1410 - 1950 m. **I:** Reynolds (1966: 210).

[15] Caulescent, branching; stem erect or divergent, to 1 m × 8 - 10 cm; **L** 12 - 16, rosulate, persistent for 20 cm below, lanceolate-attenuate, 60 - 70 × 10 - 13 cm, grey-green, usually with a few scattered pale green lenticular spots towards base, spots usually more numerous on lower face, with horny reddish margin; marginal teeth 3 - 4 mm, pungent, reddish-brown, 10 - 20 mm apart; exudate drying deep brown; **Inf** ± 1 m, with 4 - 7 **Br**; racemes conical to subcapitate, 6 - 10 × 8 cm, sub-dense; **Bra** ovate-attenuate, 5 × 3 mm; **Ped** 18 - 20 mm; **Fl** orange-scarlet, paler at mouth, 28 mm, slightly clavate, base obconical, shortly attenuate, 6 mm ∅ across **Ov**, enlarging above; **Tep** free for 14 mm.

A. sinkatana Reynolds (JSAB 23(2): 39-42, t. 14-16, 1957). **T:** Sudan, Kassala Prov. (*Reynolds* 8020 [PRE, K, KHU]). − **D:** Sudan; flat sandy ephemeral water courses, 850 - 1100 m. **I:** Reynolds (1966: 201-202).

[5] Acaulescent, simple or usually suckering to form groups; **L** 16 - 20, densely rosulate, lanceolate-attenuate, tip rounded with 3 - 5 small reddish teeth, 50 - 60 × 6 - 8 cm, dull grey-green, sometimes with scattered dull white lenticular spots, usually with reddish margin; marginal teeth 2 - 3 mm, firm, pale red, 15 - 25 mm apart; **Inf** 75 - 90 cm, with 5 - 6 **Br**; racemes capitate or subcapitate, 4 - 6 × 7 cm, dense; **Bra** ± 3 - 4 × 2 mm; **Ped** 16 - 20

mm; **Fl** scarlet, orange or yellow, 22 mm, base obconical and shortly attenuate, 5 mm ∅ across **Ov**, enlarged above; **Tep** free for 9 - 10 mm. − *Cytology:* 2n = 14 (Brandham 1971).

Gilbert & Sebsebe (1997) regard this as possibly conspecific with *A. elegans.*

A. sladeniana Pole-Evans (Ann. Bolus Herb. 3(1): 13, 1920). **T:** Namibia (*Pearson* s.n. [BOL 9000]). − **D:** Namibia; quartz hills. **I:** Reynolds (1950: 213).

Incl. *Aloe carowii* Reynolds (1938).

[6] Acaulescent, suckering to form groups; **L** 6 - 8, trifarious, lanceolate-acute, 4 - 8 × 3 - 4 cm, green with many white elongate-confluent spots scattered or in irregular transverse bands, lower face obscurely carinate towards tip, the keel with a few small white hard prickles, with narrow whitish cartilaginous margin; marginal teeth ± 1 mm, hard, white, 2 - 5 mm apart; **Inf** ± 50 cm, simple or with 1 - 2 **Br**; racemes cylindrical-acuminate, ± 18 × 7 cm, lax, with 30 - 40 **Fl**; **Bra** deltoid-acuminate, ± 6 × 4 mm; **Ped** 17 mm; **Fl** dull pink, slightly greenish at mouth, 30 mm, base truncate, 7 mm ∅ across **Ov**, abruptly narrowed to 5 mm above, then enlarging to 8 mm near mouth; **Tep** free for 7 mm; **St** exserted 0 - 1 mm.

A. socialis (H. Perrier) L. E. Newton & G. D. Rowley (Excelsa 17: 61, 1997). **T** [syn]: Madagascar (*Perrier* 1807 [P?]). − **D:** Madagascar; in forests on basalt or limestone, 600 - 900 m. **I:** Perrier (1938: 73).

≡ *Lomatophyllum sociale* H. Perrier (1926).

[15] Caulescent, branching to form dense bushes; stem prostrate or ascending, to 30 cm; **L** 14 - 16, laxly rosulate, lanceolate-attenuate, 30 - 40 × 1 - 1.5 cm, green; marginal teeth 3 mm, greenish, 12 mm apart; sheath to 2 cm; **Inf** 15 - 25 cm, usually simple, rarely with 1 **Br**; racemes ovate to subcapitate, 4 - 9 × 4 - 5 cm, dense; **Bra** acute, 2 - 4 mm; **Ped** 8 - 12 mm; **Fl** carmine-red, lobes green bordered white, 16 - 22 mm, 7 mm ∅ across **Ov**, narrowed to 4 mm above, then enlarging to 5 mm at mouth; **Tep** free for ± 5 - 7 mm; **Fr** berries.

A. somaliensis W. Watson (Gard. Chron., ser. 3, 26: 430, 1899). **T:** Somalia (*Cole* 261/1895 [K]). − **Lit:** Carter & al. (1984). **D:** Somalia.

A. somaliensis var. **marmorata** Reynolds & P. R. O. Bally (JSAB 30(4): 222-223, t. 30, 1964). **T:** Somalia (*Bally* 11793 [K, EA, PRE]). − **D:** Somalia; below shrubs on alluvial plains. **I:** Reynolds (1966: 58-59).

[5] Differs from var. *somaliensis*: **L** ± 16 - 20, to 40 × 6 - 8 cm, dark green obscurely speckled with diffuse darker green longitudinal markings giving a marbled effect; **Inf** to 85 cm, with up to 16 **Br**, lo-

wer **Br** sometimes rebranched; racemes, 4 - 6 cm; **Bra** 6 - 7 × 3 - 4 mm; **Ped** 10 mm; **Fl** bright red, 26 mm; **Tep** free for 13 mm. – *Cytology:* 2n = 14 (Carter & al. 1984).

Carter & al. (1984) concluded that this could not be maintained as a distinct variety, but it was still recognized by Lavranos (1995).

A. somaliensis var. **somaliensis** – **D:** Somalia; sandstone rocks, 1310 - 1555 m. **I:** Reynolds (1966: 56-57).

[5] Acaulescent or shortly caulescent, simple or suckering to form small groups; **L** 12 - 16, densely rosulate, lanceolate-attenuate, 20 × 7 cm, glossy brownish-green with many pale green lenticular spots, lower face paler with more spots, surface smooth; marginal teeth 4 mm, pungent, reddish-brown, 8 - 10 mm apart; exudate drying brown; **Inf** 60 - 80 cm, suberect to oblique, with 5 - 8 **Br**, lower **Br** sometimes rebranched; racemes cylindrical or with **Fl** subsecund on oblique **Br**, 15 - 20 × 5 - 6 cm, sub-dense; **Bra** ovate-attenuate, ± 8 × 4 mm; **Ped** 8 mm; **Fl** pinkish-scarlet, minutely speckled, 28 - 30 mm, base rounded, 9 mm ∅ across **Ov**, very slightly narrowed above and towards mouth; **Tep** free for 10 mm. – *Cytology:* 2n = 14 (Brandham 1971).

A. soutpansbergensis I. Verdoorn (FPA 35: t. 1381 + text, 1962). **T:** RSA, Northern Prov. (*Crundall* s.n. [PRE 27035 (= 29005)]). – **D:** RSA (Northern Prov.); S-facing rocky slopes with dense fog in wet season. **Fig. XIX.b**

[10] Caulescent, simple or suckering to form small groups; **R** fleshy; stem to 5 × 0.8 cm; **L** ± 7, distichous at first, soon becoming rosulate, linear, ± 25 × 1 cm, green with a few obscure white spots, lower face with more spots, esp. near base, with narrow translucent cartilaginous margin; marginal teeth 0.5 mm, translucent, ± 3 - 4 mm apart; **Inf** ± 20 cm, simple; raceme subcapitate, lax, with 8 or more **Fl**; **Bra** lanceolate-acuminate, ± 17 × 10 mm; **Ped** ± 25 mm; **Fl** apricot-orange, ± 27 mm, base truncate, 7 mm ∅ across **Ov**, narrowing slightly above; **Tep** free to base; **St** exserted 0 - 1 mm.

A. speciosa Baker (JLSB 18: 178, 1880). **T:** not typified. – **D:** RSA (Western Cape, Eastern Cape); rocky slopes, 760 m. **I:** Reynolds (1950: 423-425).

[9,16] Caulescent, simple or branched; stem to 4 (-6) m, dead **L** persistent; **L** densely rosulate, lanceolate-attenuate, 60 - 80 × 7 - 9 cm, dull glaucous-green, tinged bluish to reddish, with very narrow deep pink to pale reddish margin; marginal teeth 1 mm, pale red, ± 10 mm apart; **Inf** ± 50 cm, arcuate-erect, simple; racemes cylindrical, slightly acuminate, ± 30 × 12 cm, very dense; **Bra** lanceolate-obtuse, to 20 × 10 mm, brownish; **Ped** 5 - 8 mm; **Fl** white, tinged greenish, 30 - 35 mm, ventricose, base rounded, enlarged above **Ov**, narrowing slightly at mouth; **Tep** free almost to base; **St** exserted to 16 mm. – *Cytology:* 2n = 14 (Resende 1937).

Natural hybrids with other species have been reported (Reynolds 1950).

A. spicata Linné *fil.* (Suppl. Pl., 205, 1782). **T:** S Africa, sine loco (*Thunberg* 8599 [UPS]). – **Lit:** Glen & Hardy (1995). **D:** Moçambique, Zimbabwe, RSA (KwaZulu-Natal, Mpumalanga), Swaziland; granite outcrops and cliffs. **I:** Reynolds (1950: 432-433, as *A. sessiliflora*).

Incl. *Aloe sessiliflora* Pole-Evans (1917); **incl.** *Aloe tauri* L. C. Leach (1968).

[8] Caulescent, simple or branching; stem to 2 m; **L** ± 30, densely rosulate, lanceolate-attenuate, ± 50 - 60 × 7 - 9 cm, green to reddish, with reddish margin; marginal teeth 1 - 1.5 mm, deep pink to reddish, 8 - 12 mm apart; **Inf** 1 m or more, simple; racemes cylindrical, ± 30 - 40 × 4 - 5 cm, very dense; **Bra** ovate-cuspidate; **Ped** absent; **Fl** greenish-yellow, 14 - 15 mm, campanulate; **Tep** free to base; **St** exserted ± 10 mm.

Natural hybrids with other species have been reported (Reynolds 1950).

A. splendens Lavranos (JSAB 31(1): 77-80, t. 16, 1965). **T:** Yemen (*Rauh & Lavranos* 13008 [HEID, K, PRE]). – **D:** Yemen; sparse thornscrub, ± 760 m. **I:** Reynolds (1966: 160-161).

[5] Acaulescent or shortly caulescent, simple or in small groups; stem prostrate, to 40 cm; **L** 20 - 24, densely rosulate, deltoid-attenuate, to 60 × 15 cm, dull green, often tinged purplish-brown, with dark brown cartilaginous margin; marginal teeth 2 mm, blunt, dark brown, 15 - 30 mm apart; **Inf** 1.4 m, with many **Br**, lower **Br** sometimes rebranched; racemes conical-cylindrical, 30 - 35 × 6 cm, sub-dense; **Bra** deltoid, 10 × 6 mm; **Ped** 8 mm; **Fl** bright scarlet with a bloom, 30 - 33 mm, base shortly attenuate, 8 mm ∅ across **Ov**, narrowed to 6 mm above, then enlarging to 7 mm at mouth; **Tep** free for 13 mm.

A. squarrosa Baker (Proc. Roy. Soc. Edinburgh 12: 97, 1883). **T:** Socotra (*Balfour* 282 [K]). – **Lit:** Lavranos (1969). **D:** Socotra; limestone cliffs, ± 300 m. **I:** FPA 41: t. 1611, 1970.

Incl. *Aloe concinna* Baker (1898) (*nom. illeg.*, Art. 53.1); **incl.** *Aloe zanzibarica* Milne-Redhead (1947).

[13] Caulescent, branching at base; stem pendulous, to 40 cm; **L** rosulate, lanceolate-attenuate, 5 - 7 × 2 - 3 cm, light green with many whitish rounded spots, surface rough; marginal teeth 3 - 4 mm, firm, whitish, ± 5 mm apart; **Inf** 10 - 20 cm, usually pendulous and arcuate ascending, simple; racemes cylindrical, 6 × 4.5 cm, lax; **Bra** deltoid, 5 mm; **Ped** 7 - 8 mm; **Fl** scarlet, 23 - 25 mm, base shortly attenuate, 5 mm ∅ across **Ov**, slightly narrowed above, then enlarging to mouth; **Tep** free for 5 - 6 mm. – *Cytology:* 2n = 14 (Resende 1937: as *A. concinna*).

A. steffanieana Rauh (KuaS 51(3): 71-73, ills., 2000). **T:** Madagascar, Tolonaro (*Razafindratsira* s.n. in *BG Heidelberg* 73599 [HEID]). – **D:** Madagascar; bare granite rocks near the sea, ± 50 - 100 m.

[8] Caulescent, stem to 15 × 2 - 3 cm, unbranched; **L** rosulate, lanceolate, to 35 × 3.5 cm, green; marginal teeth deltoid, small, pale yellow; **Inf** to 80 cm, erect, simple; racemes subcapitate, 7 cm, subdense; **Bra** 10 × 7 mm; **Ped** 4 mm; **Fl** reddish at base, cream-white above, 35 mm, base rounded, 8 - 10 mm ⌀ across **Ov**, slightly narrowed above, widening to mouth; **Tep** free for 15.5 mm; **St** exserted ± 10 mm.

A. steudneri Schweinfurth (Bull. Herb. Boissier 2(App. 2): 73, 1894). **T:** Eritrea (*Penzig* 1424 [K]). – **D:** Eritrea, Ethiopia; mountain peaks, 2600 - 3500 m. **I:** Reynolds (1966: 277).

[6,7] Caulescent, branching at base; stem short; **L** ± 25, densely rosulate, lanceolate-attenuate, ± 60 × 12 - 15 cm, grey-green, with narrow hyaline rose margin; marginal teeth ± 2 mm, rose, 10 - 40 mm apart; **Inf** 70 - 90 cm, simple or with 3 - 5 **Br**; racemes cylindrical, 15 - 18 cm, sub-dense; **Bra** ovate-lanceolate, acuminate, 15 - 20 mm; **Ped** 15 - 30 mm; **Fl** deep red, 40 - 45 mm, base rounded, 12 - 14 mm ⌀ across **Ov**, not narrowed above; **Tep** free almost to base. – *Cytology:* 2n = 14 (Resende 1937).

A. striata Haworth (Trans. Linn. Soc. London 7: 18, 1804). **T** [neo]: RSA, Cape Prov. (*Bottomley* s.n. [PRE 27]). – **D:** Namibia, RSA.

A. striata ssp. **karasbergensis** (Pillans) Glen & D. S. Hardy (SAJB 53(6): 491, 1987). **T:** RSA, Northern Cape (*Pillans* 5848 [BOL]). – **D:** Namibia, RSA (Northern Cape); arid sandy areas and stony mountain slopes. **I:** Reynolds (1950: 297-299, as *A. karasbergensis*). **Fig. XVIII.d**

≡ *Aloe karasbergensis* Pillans (1928).

[5,13] Differs from ssp. *striata*: Usually acaulescent, sometimes caulescent, simple or with up to 19 **Br**; stem to 30 cm; **L** with 2 - 3 mm dull white margin, sometimes crenulate; **Inf** 50 - 60 cm; racemes conical, lax; **Fl** pink to pale coral-red, 25 - 27 mm. – *Cytology:* 2n = 14 (Riley 1959: as *A. karasbergensis*).

A. striata ssp. **komaggasensis** (Kritzinger & van Jaarsveld) Glen & D. S. Hardy (SAJB 53(6): 491, 1987). **T:** RSA, Northern Cape (*Kritzinger* 12 [NBG, PRE]). – **D:** RSA (Northern Cape); open xerophytic scrub on quartz slopes. **I:** SAJB 51: 287, 1985, as *A. komaggasensis*.

≡ *Aloe komaggasensis* Kritzinger & van Jaarsveld (1985).

[13] Differs from ssp *striata*: **L** 30 - 40 × 10 - 13 cm, grey-white, obscurely striate, with white to yellowish margin; marginal teeth minute; exudate dry-

ing orange-yellow; **Inf** with racemes 6 - 8 cm; **Bra** 8 × 2 - 3 mm; **Fl** yellow, rarely orange, 20 mm, 4.5 mm ⌀ across **Ov**; **Tep** free for 4 mm.

A. striata ssp. **striata** – **D:** RSA (Western Cape, Eastern Cape); in grass or bush on rocky slopes. **I:** Reynolds (1950: 295-296, t. 20).

Incl. *Aloe paniculata* Jacquin (1809); **incl.** *Aloe albocincta* Haworth (1819); **incl.** *Aloe hanburyana* Naudin (1875); **incl.** *Aloe rhodocincta* hort. *ex* Baker (1880); **incl.** *Aloe striata* var. *oligospila* Baker (1894).

[13] Caulescent, usually simple, sometimes with up to 5 **Br**; stem decumbent, to 1 m, with dead **L** persistent; **L** 12 - 20, densely rosulate, lanceolate-attenuate, to 50 × 20 cm, glaucous-green to reddish-tinged, usually striate, sometimes with obscure spots on upper face, with broad pale pink to almost red margin; marginal teeth absent; **Inf** to 1 m, with 6 - 12 **Br**, lower **Br** rebranched; racemes capitate to slightly conical, ± 6 × 6 cm, dense; **Bra** deltoid-acuminate, ± 5 mm; **Ped** 15 - 25 mm; **Fl** peach-red to coral-red, 30 mm, base truncate, 6 mm ⌀ across **Ov**, abruptly narrowed above, then enlarging to mouth; **Tep** free for 6 - 8 mm; **St** exserted to 1 mm. – *Cytology:* 2n = 14 (Vosa 1982).

Natural hybrids with other species have been reported (Reynolds 1950).

A. striatula Haworth (Philos. Mag. J. 1825: 281, 1825). **T:** not typified. – **D:** Lesotho, RSA.

A. striatula var. **caesia** Reynolds (FPSA 16: t. 633 + text, 1936). **T:** RSA, Eastern Cape (*Reynolds* 1607 [PRE, BOL]). – **D:** RSA (Eastern Cape); rocky slopes. **I:** Reynolds (1950: 365, t. 31).

Incl. *Aloe striatula* fa. *typica* Resende (s.a.) (*nom. inval.*, Art. 24.3); **incl.** *Aloe striatula* fa. *conimbricensis* Resende (1943); **incl.** *Aloe striatula* fa. *haworthii* Resende (1943).

[16] Differs from var. *striatula*: Stem to 2 m × 1.5 - 2 cm; **L** 10 - 15 × 1.5 - 2.5 cm, milky-green; sheath 5 - 15 mm, obscurely green-lineate; **Fl** yellow, tipped greenish, 30 - 33 mm, slightly narrowed above **Ov**, then enlarging to mouth.

A. striatula var. **striatula** – **D:** Lesotho, RSA (Eastern Cape); amongst rocks on mountain tops. **I:** Reynolds (1950: 362-364, t. 31). **Fig. XVIII.e**

Incl. *Aloe macowanii* Baker (1880); **incl.** *Aloe aurantiaca* Baker (1892); **incl.** *Aloe cascadensis* Kuntze (1898).

[16] Caulescent, branching; stem to 1.75 m × 2.5 cm; **L** scattered along stem for 40 - 60 cm, linear-lanceolate, acuminate, to 25 × 2.5 cm, semiglossy green, with very narrow white cartilaginous margin; marginal teeth ± 1 mm, firm, white, 3 - 8 mm apart; sheath 15 - 20 mm, prominently green-lineate; **Inf** to 40 cm, simple; racemes cylindrical-conical, 10 - 15 cm, dense; **Bra** deltoid-subulate, ± ½ as long as

Ped; **Ped** 3 - 5 mm; **Fl** reddish-orange to orange, 40 - 45 mm, base truncate, very slightly narrowed above **Ov**; **Tep** free almost to base. – *Cytology:* 2n = 14 (Fernandes 1930).

A. suarezensis H. Perrier (Mém. Soc. Linn. Normandie, Bot. 1(1): 21, 1926). **T:** Madagascar, Majunga (*Perrier* 16221 [P]). – **D:** Madagascar; limestone hills, ± 40 m. **I:** Reynolds (1966: 459-461). **Fig. XIX.c**

[3] Acaulescent or caulescent, simple; stem erect, to 30 cm, with dead **L** persistent; **L** 20 - 24, densely rosulate, lanceolate-attenuate, tip rounded with 2 - 3 short teeth, to 50 - 60 × 9 - 10 cm, dull green tinged reddish; marginal teeth ± 2 mm, dirty-white to pale pinkish, 10 mm apart; exudate drying pale yellow; **Inf** 60 - 80 cm, with 4 - 12 **Br**, lower **Br** sometimes rebranched; racemes cylindrical, 10 - 15 × 6 cm, dense; **Bra** ovate-attenuate, 10 - 12 × 4 - 6 mm, dirty-white; **Ped** 10 - 12 mm; **Fl** dull to pale scarlet, paler at mouth, minutely puberulent, 28 mm, base rounded and very shortly attenuate, 7 mm ∅ across **Ov**, narrowed to 5.5 mm above, then enlarging to mouth; **Tep** free to base.

A. subacutissima G. D. Rowley (Nation. Cact. Succ. J. 28(1): 6, 1973). **T:** Madagascar, Toliara (*Perrier* 12690 [P]). – **D:** Madagascar; flat rocks, 600 - 1000 m. **I:** Reynolds (1966: 501-502, as *A. intermedia*).

Incl. *Aloe deltoideodonta* var. *intermedia* H. Perrier (1926) ≡ *Aloe intermedia* (H. Perrier) Reynolds (1957) (*nom. illeg.*, Art. 53.1).

[15] Caulescent, branching; stem erect, divergent or decumbent, to 1 m × 3 cm; dead **L** persistent; **L** 20 - 26, rosulate and persistent for 20 cm below, lanceolate-attenuate, to 25 - 30 × 5 - 6 cm, dull green tinged reddish; marginal teeth 3 - 4 mm, reddish-brown, tips paler, 10 mm apart; **Inf** ± 60 cm, with 2 - 3 **Br**; racemes cylindrical-acuminate, 10 - 15 cm, dense; **Bra** lanceolate-acute, 12 mm, imbricate in bud stage; **Ped** 15 mm; **Fl** scarlet, 28 mm, base obconical-attenuate, 5 mm ∅ across **Ov**, narrowed to 4 mm above, then enlarging to mouth; **Tep** free for 11 mm; **St** exserted 0 - 1 mm.

A. succotrina Allioni (Auct. Syn., 13, 1773). **T:** [icono]: Commelin, Hort. Med. Amstel. 1: fig. 48, 1697. – **D:** RSA (Western Cape); between sandstone boulders. **I:** Reynolds (1950: 390-394).

Incl. *Aloe perfoliata* var. ξ Linné (1753); **incl.** *Aloe soccotrina* Garsault (1767) (*nom. inval.*, Art. 32.8); **incl.** *Aloe vera* Miller (1768) (*nom. illeg.*, Art. 53.1); **incl.** *Aloe succotrina* Lamarck (1783) (*nom. illeg.*, Art. 53.1); **incl.** *Aloe perfoliata* var. *purpurascens* Aiton (1789) ≡ *Aloe purpurascens* (Aiton) Haworth (1804); **incl.** *Aloe perfoliata* var. *succotrina* Aiton (1789); **incl.** *Aloe sinuata* Thunberg (1794); **incl.** *Aloe soccotrina* var. *purpurascens* Ker Gawler (1812) (*nom. inval.*, Art. 43.1);

incl. *Aloe soccotorina* Schultes & Schultes *fil.* (1829) (*nom. inval.*, Art. 61.1); **incl.** *Aloe succotrina* var. *saxigena* A. Berger (1908).

[4,12] Caulescent, sometimes almost acaulescent, simple or branching at base or above; stem erect or decumbent, short or to 2 m × 15 cm, dead **L** persistent; **L** densely rosulate, lanceolate-attenuate, to 50 × 10 cm, dull green to grey-green, obscurely lineate, sometimes with a few small scattered white spots, usually with dull white narrow cartilaginous margin; marginal teeth 2 - 4 mm, firm, white, to 10 mm apart; **Inf** ± 1 m, usually simple; racemes cylindrical-acuminate, 25 - 35 cm, sub-dense; **Bra** lanceolate, 20 × 10 mm; **Ped** 30 mm; **Fl** glossy red to reddish-salmon, green-tipped, 40 mm, base truncate, not narrowed above **Ov**; **Tep** free to base. – *Cytology:* 2n = 14 (Resende 1937).

A. suffulta Reynolds (JSAB 3: 151, 1937). **T:** Moçambique (*Reynolds* 2457 [PRE, BOL]). – **D:** Moçambique, RSA (KwaZulu-Natal); in the shade of shrubs on sandy soil, ± 90 m. **I:** Reynolds (1950: 343-344).

Incl. *Aloe subfulta* hort. (s.a.) (*nom. inval.*, Art. 61.1); **incl.** *Aloe subfulta* hort. (s.a.) (*nom. inval.*, Art. 61.1).

[3] Caulescent, simple; stem to 20 × 2 cm; **L** ± 16, scattered along stem, attenuate, 40 - 50 × 4 cm, green with dull white spots, the spots sometimes scattered, usually ± in transverse bands; marginal teeth 1 - 2 mm, usually uncinate, whitish, 5 - 10 mm apart; sheath 5 - 10 mm, striatulate; **Inf** ± 1.75 m, supported by shrubs, with up to 9 **Br**; racemes cylindrical, slightly acuminate, ± 8 - 15 × 5 cm, lax, with up to 20 **Fl**; **Bra** 9 mm; **Ped** 9 mm; **Fl** light jasper-red, whitish at mouth, 30 - 35 mm, base shortly attenuate, 6 mm ∅ across **Ov**, narrowed to 5.5 mm above, then enlarging to mouth; **Tep** free for 7 mm; **St** exserted 6 mm. – *Cytology:* 2n = 14 (Brandham 1971).

A. suprafoliata Pole-Evans (Trans. Roy. Soc. South Afr. 5: 603, 1916). **T:** Swaziland (*Pole Evans* 215 [PRE]). – **D:** Swaziland, RSA (KwaZulu-Natal, Mpumalanga); rocky slopes on mountains, mostly with mist or low clouds. **I:** Reynolds (1950: 303-304, t. 18).

Incl. *Aloe suprafoliolata* hort. (s.a.) (*nom. inval.*, Art. 61.1).

[4] Acaulescent or shortly caulescent, usually simple; stem erect or procumbent, rarely to 50 cm; **L** ± 30, distichous in young plants, densely rosulate later, lanceolate-acuminate, 30 - 40 × 7 cm, bluish-green to bluish-grey, becoming reddish-brown towards tip, obscurely lineate; marginal teeth 2 - 5 mm, reddish-brown, sometimes bifid, 5 - 10 mm apart; **Inf** to 1 m, simple; raceme conical to cylindrical-acuminate, to 25 × 10 cm, sub-dense; **Bra** lanceolate-acute, ± 20 mm; **Ped** to 20 mm; **Fl** rose-pink to scarlet, with a bloom, greenish at mouth, 40

- 50 mm, base rounded or very slightly attenuate, not narrowed above **Ov**; **Tep** free to base; **St** exserted 0 - 1 mm. – *Cytology:* 2n = 14 (Müller 1941).

Natural hybrids with *A. arborescens* have been reported (Reynolds 1950).

A. suzannae Decary (Bull. Econ. Madag. 18: 26, 1921). **T:** Madagascar, Toliara (*Decary* 2913 [P, BM]). – **D:** Madagascar; in dense bush, ± 30 m. **I:** Reynolds (1966: 515-517). **Fig. XX.a**

[9] Caulescent, usually simple, sometimes with 1 or 2 **Br**; stem erect, to 4 m × 30 cm; **L** 60 - 100, densely rosulate and persistent for up to 1 m below, lanceolate-attenuate, tip rounded with 5 - 7 short teeth, 1 m × 8 - 9 cm, dull green, surface very rough; marginal teeth 2 mm, pungent, pale brown, 8 - 10 mm apart; exudate drying deep brown-orange; **Inf** ± 3 m, simple; raceme cylindrical, ± 2 m × 17 cm, dense; **Bra** linear-deltoid, ± 15 × 2 mm, green with pale margin; **Ped** 28 - 30 mm; **Fl** ivory tinged with pale rose, 33 mm, base rounded, 10 mm ⌀ across **Ov**, slightly narrowed above, lobes spreading; **Tep** free for 16 - 17 mm, reflexed; **St** exserted 10 mm. – *Cytology:* 2n = 14 (Brandham 1971).

The flowers are nocturnal and fragrant, probably pollinated by bats and lemurs.

A. swynnertonii Rendle (JLSB 40: 215, 1911). **T:** Zimbabwe (*Swynnerton* 722 [BM, K]). – **D:** Zimbabwe, RSA (Northern Prov.); grassland and dry forest, 610 - 1830 m. **I:** Reynolds (1966: 85, fig. 84-85).

Incl. *Aloe chimanimaniensis* Christian (1936); **incl.** *Aloe melsetterensis* Christian (1938).

[5] Acaulescent or very shortly caulescent, simple or in small groups of 3 - 4 **Ros**; **L** ± 20, densely rosulate, lanceolate-attenuate, to 75 × 8 - 10 cm, apical 10 cm soon drying, upper face dark green with oblong to confluent H-shaped spots in irregular transverse bands, lower face paler green, lineate, usually without spots; marginal teeth 4 mm, pungent, reddish-brown, 10 - 25 mm apart; exudate drying yellow; **Inf** 1.5 - 1.75 m, with 8 - 12 **Br**, lower **Br** sometimes rebranched; racemes capitate-corymbose, 6 - 8 × 7 - 8 cm, dense; **Bra** deltoid, 8 × 3 mm; **Ped** 25 - 30 mm; **Fl** flesh-pink to dull coral-red, with a slight bloom, 30 mm, base truncate, 8 mm ⌀ across **Ov**, abruptly narrowed to 5 mm above **Ov**, then enlarging above and slightly narrowing to mouth; **Tep** free for 9 mm; **St** exserted 0 - 1 mm.

Populations in Malawi and Moçambique documented by Reynolds (1966) have been distinguished as *A. menyharthii* by Carter (1996).

A. tenuior Haworth (Philos. Mag. J. 1825: 281, 1825). **T:** not typified. – **Lit:** Wyk & Smith (1996). **D:** RSA (Eastern Cape); thorn-bush or more open ground, sometimes on steep slopes. **I:** Reynolds (1950: 347-352).

Incl. *Aloe tenuior* var. *glaucescens* Zahlbruckner

(1900); **incl.** *Aloe tenuior* var. *decidua* Reynolds (1936); **incl.** *Aloe tenuior* var. *rubriflora* Reynolds (1936); **incl.** *Aloe tenuior* var. *densiflora* Reynolds (1950).

[14,15] Caulescent, branching; **R** from a stock to ± 60 (-200) cm ⌀; stem erect, to 60 cm, or spreading to decumbent, scandent, or supported by bushes, to 3 m × 1.5 cm; **L** laxly rosulate, sometimes persistent for up to 20 cm below stem tip, linear-lanceolate, 10 - 18 × 1 - 2.2 cm, glaucous-green, with very narrow white cartilaginous margin; marginal teeth to 0.5 mm, white, 1 - 2 mm apart; sheath 5 - 25 mm, obscurely green-lineate; **Inf** 35 - 40 (-50) cm, simple or with 1 - 2 **Br**; racemes cylindrical, slightly acuminate, 10 - 20 × 4 cm, sub-dense to dense; **Bra** linear-deltoid, acuminate, ± 5 mm; **Ped** 3 - 5 mm; **Fl** yellow or red with yellow tips, 11 - 15 mm, base shortly attenuate, very slightly narrowed above **Ov**, then enlarging to mouth; **Tep** free for ± 3 - 6 mm. – *Cytology:* 2n = 14 (Müller 1941).

Some extreme variants of this very variable species have been described as varieties in the past. A possible intergeneric natural hybrid with *Bulbine aloides* has been reported (Reynolds 1950).

A. tewoldei M. G. Gilbert & Sebsebe (KB 52(1): 143, 1997). **T:** Ethiopia, Harerge Region (*Tewolde-Berhan Gebre-Egziabher* s.n. [K]). – **D:** Ethiopia; limestone cliff faces.

[13] Caulescent; stem sprawling or pendulous, to 50 × 0.6 cm; **L** scattered along stem, oblong-lanceolate, semiterete, up to 13.5 (-32) × 1.5 - 2 (-2.2) cm, grey-green, obscurely spotted; marginal teeth ± 0.5 mm, white, 2 - 3 mm apart; **Inf** simple; racemes cylindrical, ± 27 cm, lax; **Bra** ± 4 × 2 mm; **Ped** ± 12 mm; **Fl** glaucous-orange, tipped greenish, 20 mm, base truncate, 7 mm ⌀ when pressed.

A. thompsoniae Groenewald (Tydskr. Wetensk. Kuns 14: 64, 1936). **T** [lecto]: RSA, Northern Prov. (*Thompson* s.n. [PRE 274]). – **D:** RSA (Northern Prov.); fissures in quartzite, ± 1830 m. **I:** Reynolds (1950: 131).

[1] Acaulescent or very shortly caulescent, suckering and branching at base, forming dense groups; **R** fusiform; **L** 12 - 18, rosulate-multifarious, lanceolate-attenuate, 15 - 20 × ± 1.5 cm, green, obscurely lineate, with a few scattered white elongated spots near base, surface with spots more numerous and more rounded; marginal teeth ± 1 mm, firm, white, 1 - 2 mm apart; **Inf** to 20 cm, simple; raceme pyramidal-capitate, 3 - 4 × 4 - 5 cm, dense, with 10 - 15 **Fl**; **Bra** ovate-acuminate, ± 10 × 5 - 6 mm; **Ped** 15 - 20 mm; **Fl** coral-red, 25 - 28 mm, base shortly attenuate, not narrowed above **Ov**; **Tep** free to base; **St** not exserted.

A. thorncroftii Pole-Evans (Trans. Roy. Soc. South Afr. 5: 709, 1917). **T:** RSA, Mpumalanga (*Thorncroft* s.n. [PRE 247]). – **D:** RSA (Mpumalanga);

mountain slopes, ± 1525 m. **I:** Reynolds (1950: 305-306).

[2] Acaulescent or sometimes shortly caulescent, simple; **L** ± 25 - 30, densely rosulate, lanceolate, to 40 × 10 - 14 cm, dull grey-green, reddish-green towards tip, indistinctly lineate, surface rough; marginal teeth 3 - 4 mm, pungent, reddish-brown, 8 - 12 mm apart; **Inf** to 1 m, simple, rarely branched; raceme cylindrical-acuminate, 40 - 50 cm, lax; **Bra** lanceolate-ovate, 20 × 15 mm, thick and fleshy, closely imbricate in bud stage; **Ped** 20 mm; **Fl** dull rose-red to scarlet, with pale bluish-grey bloom, to 55 mm, base shortly attenuate, slightly enlarged above **Ov**, then narrowing slightly to mouth; **Tep** free for 20 mm; **St** exserted 0 - 1 mm.

A. thraskii Baker (JLSB 18: 180, 1880). **T:** not typified. – **D:** RSA (Eastern Cape, KwaZulu-Natal); almost pure sand in low coastal vegetation or taller bush, just above sea level. **I:** Reynolds (1950: 475, t. 54).

Incl. *Aloe candelabrum* Engler & Drude (1910) (*nom. illeg.*, Art. 53.1).

[9] Caulescent, simple, dead **L** persistent; stem usually to 2 m, to 4 m in dense bush; **L** densely rosulate, lanceolate-attenuate, to 1.6 m × 22 cm, dull green to glaucous, lower face sometimes with a few median prickles in upper ½, with narrow reddish or brownish-red margin; marginal teeth ± 2 mm, reddish, 10 - 20 mm apart; **Inf** with 4 - 8 **Br**; racemes cylindrical, slightly acuminate, to 25 × 10 - 12 cm, very dense; **Bra** ovate-acute, 9 × 6 mm; **Ped** 1 - 2 mm, green; **Fl** lemon-yellow to pale orange, greenish-tipped, ± 25 mm, base truncate, 6 mm Ø across **Ov**, enlarged above, narrowed at mouth; **Tep** free for ± 17 mm; **St** exserted 15 - 20 mm. – *Cytology:* 2n = 14 (Müller 1941).

Natural hybrids with *A. maculata* (as *A. saponaria*) have been reported (Reynolds 1950).

A. tomentosa Deflers (Voy. Yemen, 211, 1889). **T:** Yemen (*Deflers* 616 [P ?]). – **D:** Saudi Arabia, Yemen; rocky slopes, 1430 - 3050 m. **I:** Reynolds (1966: 128-129). **Fig. XX.b**

Incl. *Aloe tomentosa* fa. *viridiflora* Lodé (1997) (*nom. inval.*, Art. 34.1b, 36.1).

[7] Caulescent, suckering to form dense groups; stem decumbent, short; **L** 16 - 20, densely rosulate, lanceolate-deltoid, ± 35 × 9 cm, grey-green tinged reddish, with narrow pinkish-brown cartilaginous margin; marginal teeth 0.5 - 1 mm, blunt, 20 - 40 mm apart, sometimes absent; exudate drying pale yellow; **Inf** 60 - 70 cm, with 3 - 4 **Br**; racemes cylindrical-conical, to 15 × 5 - 6 cm, sub-dense; **Bra** ovate-deltoid, 7 × 4 mm; **Ped** 6 - 9 mm; **Fl** rose-pink, conspicuously tomentose, 24 - 28 mm, base rounded, 7 - 8 mm Ø across **Ov**, slightly narrowed above; **Tep** free for 9 mm.

A. tormentorii (Marais) L. E. Newton & G. D.

Rowley (Excelsa 17: 61, 1997). **T:** Mauritius (*Guého* s.n. [MAU 13060]). – **D:** Mauritius; exposed rocky hillsides, 240 - 310 m. **I:** Marais (1978: 183: 12).

≡ *Lomatophyllum tormentorii* Marais (1975).

[7] Acaulescent or shortly caulescent, sometimes forming large groups; stem decumbent; **L** densely rosulate, ovate-acuminate, ± 60 × 15 cm, pale green or bluish; marginal teeth cartilaginous; **Inf** 60 - 120 cm, with 3 - 4 **Br**; racemes to 30 cm, dense; **Bra** deltoid, 1 - 2 mm; **Ped** 13 - 20 mm; **Fl** orange-red, green-tipped, 14 - 17 mm, base shortly attenuate, slightly narrowed above **Ov**, then enlarging to mouth; **Tep** free for 9 - 12 mm; **St** exserted to 1 mm; **Fr** berries.

A. tororoana Reynolds (FPA 29: t. 1144 + text, 1953). **T:** Uganda (*Bally & Reynolds* 6594 [PRE]). – **D:** Uganda; steep rock faces, 1340 - 1465 m. **I:** Reynolds (1966: 335).

[11] Caulescent, branching; stem ± 20 × 1.5 cm; **L** ± 12, densely rosulate, lanceolate-attenuate, 15 × 3 - 5 cm, dull milky-green with few to many small oblong dull white spots, more numerous and crowded on lower face; marginal teeth 2 - 3 mm, pungent, whitish, brown-tipped, 5 - 10 mm apart; **Inf** to 40 cm, simple or with 1 - 2 **Br**; racemes cylindrical-acuminate, 8 - 10 cm, sub-dense; **Bra** ovate-deltoid, 3 × 2 mm; **Ped** 8 - 10 mm; **Fl** coral-red to scarlet, green-tipped, 20 - 22 mm, base rounded, 5 mm Ø across **Ov**, narrowed to 4 mm above, slightly enlarging to mouth; **Tep** free for 7 mm; **St** exserted 1 mm. – *Cytology:* 2n = 14 (Brandham 1971).

A. torrei I. Verdoorn & Christian (FPA 25: t. 987 + text, 1946). **T:** Moçambique (*da Torre* s.n. [PRE 27239]). – **D:** Moçambique; in grass on exposed granite slabs, ± 1525 m. **I:** Reynolds (1966: 14-15).

[10] Caulescent, branching at base forming dense clumps; stem erect, to 15 × 1.5 cm; **L** ± 10, rosulate, linear, 40 - 45 × 0.5 cm, limp and deflexed, green with a few white spots near base, lower face with many, sometimes tuberculate, spots; marginal teeth minute, ± 1 - 2 mm apart; **Inf** ± 50 cm, simple; raceme cylindrical, slightly acuminate, 9 × 4 - 5 cm, lax, with ± 10 **Fl**; **Bra** ovate-acute, 15 × 7 mm; **Ped** 15 mm; **Fl** scarlet, grey-green at mouth, 30 mm, slightly ventricose, base shortly attenuate, enlarged to 6 - 7 mm above **Ov**, then narrowing to mouth; **Tep** free to base; **St** not exserted.

A. trachyticola (H. Perrier) Reynolds (JSAB 23(2): 72-73, t. 26-27, 1957). **T:** Madagascar, Antananarivo (*Perrier* 11000 [P]). – **D:** Madagascar; on trachyte and quartzite, 1400 - 2200 m. **I:** Reynolds (1966: 462-463).

≡ *Aloe capitata* var. *trachyticola* H. Perrier (1926).

[2] Acaulescent, sometimes shortly caulescent,

simple; stem procumbent, short; **L** 6 - 10 and distichous when young, to 14 and spiral to subrosulate later, lanceolate, tip rounded with short teeth, 10 - 15 × 3 - 4 cm, bluish-grey tinged reddish; marginal teeth 1 - 1.5 mm, pungent, reddish-brown, 3 - 5 mm apart; **Inf** 65 - 90 cm, simple; raceme capitate, 2 - 3 × 7 - 8 cm, dense; **Bra** ovate-acute, 10 × 6 mm; **Ped** lowest 3 - 5 mm, uppermost 15 - 20 mm; **Fl** red, to 35 mm, base rounded, 8 mm ∅ across **Ov**, slightly enlarged above; **Tep** free almost to base; **St** exserted 1 - 2 mm.

A. trichosantha A. Berger (BJS 36: 62, 1905). **T** [lecto]: Eritrea (*Schweinfurth & Riva* 2291 [K, FT, G]). – **Lit:** Gilbert & Sebsebe (1997). **D:** Eritrea, Ethiopia.

The choice of a new lectotype by Gilbert & Sebsebe (1997) appears contrary to the rules and is not followed here.

A. trichosantha ssp. **longiflora** M. G. Gilbert & Sebsebe (KB 52(1): 142-143, 1997). **T:** Ethiopia, Harerge Region (*Burger* 3394 [K, ETH, FT]). – **D:** Ethiopia; open deciduous bushland on volcanic rocks and alluvial soils, 1000 - 1950 m. **I:** Reynolds (1966: 133, fig. 134, lower, as *A. trichosantha*).

[5] Differs from ssp. *trichosantha*: **L** marginal teeth 2 - 4 mm; **Fl** 25 - 30 mm.

This occurs further S than ssp. *trichosantha*.

A. trichosantha ssp. **trichosantha** – **D:** Eritrea, Ethiopia; rocky slopes or arid flat areas, 520 - 1700 m. **I:** Reynolds (1966: 132-133). **Fig. XX.c**

Incl. *Aloe percrassa* Schweinfurth (1894) (*nom. illeg.*, Art. 53.1); incl. *Aloe percrassa* var. *albopicta* Schweinfurth (1894) (*incorrect name*, Art. 11.4).

[5] Acaulescent or very shortly caulescent, simple or suckering to form groups; **L** 12 - 16, densely rosulate, lanceolate-attenuate, 40 - 50 × 10 cm, dull green, sometimes with a few scattered spots, rarely with many spots; marginal teeth ± 4.5 - 5.5 mm, pungent, reddish-brown, 12 - 15 mm apart; exudate drying yellow; **Inf** 1 - 1.5 m, with 2 - 3 **Br**; racemes cylindrical-acuminate, ± 30 (-50) cm, subdense; **Bra** ovate-lanceolate, acute, 14 × 6 mm; **Ped** 5 - 6 mm; **Fl** strawberry-pink or coral-pink, white-tomentose, 20 - 23 mm, base rounded, 7 - 8 mm ∅ across **Ov**, not or very slightly narrowed above; **Tep** free for 10 - 12 mm. – *Cytology:* 2n = 14 (Brandham 1971).

Plants in S Ethiopia, Kenya (near Garissa), and Somalia, mentioned by Reynolds (1966: 143) as an undescribed species, are now *A. citrina*.

A. trigonantha L. C. Leach (JSAB 37(1): 46-51, ills., 1971). **T:** Ethiopia, Begemdir Prov. (*McLeay* s.n. in *Reynolds* 11618 [PRE, SRGH]). – **D:** Ethiopia; grassland, ± 2500 m.

[7] Acaulescent, suckering at base to form dense groups; **L** ± 24, densely rosulate, lanceolate-attenuate, 30 - 45 × 5 - 10 cm, pale green, pinkish-brown towards tip, with a few elongated whitish spots, lower face bluish-green with many spots; marginal teeth ± 3 mm, pinkish brown, 10 - 20 mm apart; **Inf** 60 - 90 cm, with ± 3 **Br**, each rebranched; racemes cylindrical-acuminate, ± 20 × 7 cm, lax; **Bra** deltoid-attenuate to ovate-acuminate, 5 - 10 × 3 - 5 mm; **Ped** to 15 mm; **Fl** bright scarlet to orange-scarlet, very thick and fleshy, 35 mm, base shortly attenuate, 10 - 12 mm ∅ across **Ov**, narrowed to 7 - 8 mm above, and to 3 - 4 mm at mouth; **Tep** free for ± 10 mm; **St** exserted 0 - 1 mm.

A. tugenensis L. E. Newton & Lavranos (CSJA 62(5): 215-217, ills., 1990). **T:** Kenya, Rift Valley Prov. (*Newton* 3514 [K, EA, MO]). – **D:** Kenya; dry *Acacia* scrub, 1300 - 1325 m.

[13] Caulescent, branching at or near base; stem erect or ascending to 70 cm, becoming decumbent to 1.2 m × 3 cm, dead **L** persistent; **L** 12 - 20, rosulate, lanceolate-attenuate, to 61 × 12 cm, dull green, tinged brownish-red in sun, with several scattered whitish spots on seedlings and young shoots only, surface slightly rough; marginal teeth to 5 mm, firm, uncinate, brown-tipped, 6 - 15 mm apart; exudate yellow, drying brownish-yellow; **Inf** 95 - 130 cm, with up to 12 **Br**, lower **Br** sometimes rebranched; racemes cylindrical, 10 - 24 cm, subdense; **Bra** linear, 11 × 4 mm, densely imbricate in bud stage; **Ped** 7 - 9 mm; **Fl** pale pink, **Tep** with whitish margin, 22 mm, base shortly attenuate, 5 mm ∅ across **Ov**, narrowed to 4.5 mm above, then enlarging to 7 mm at mouth; **Tep** free for 14 - 15 mm.

A. turkanensis Christian (JSAB 8(2): 173-174, t. 6, 1942). **T:** Kenya, Rift Valley Prov. (*Erens* 1610 [SRGH, PRE]). – **Lit:** Newton & Lavranos (1990: ills.). **D:** Kenya, Uganda; usually in shade of shrubs in arid areas, 915 - 1500 m.

[12] Caulescent, branching sparsely at base forming clumps to 2 m ∅; stem ascending, to 45 cm, becoming decumbent to 70 cm; **L** 14 - 18, densely rosulate, lanceolate-attenuate, to 70 × 9 cm, dull green, sometimes with slight bluish bloom, with a few elongated pale green spots, often more numerous and in ± transverse bands on the lower face, surface smooth; marginal teeth 2 mm, whitish, 12 - 18 mm apart; exudate drying yellow; **Inf** to 1 m, with up to 8 **Br**, lower **Br** rebranched; racemes with secund **Fl**, 15 - 26 × 6 cm, sub-dense; **Bra** ovate-cuspidate, 5 - 7 × 3 mm; **Ped** 8 - 9 mm; **Fl** red to orange-red, tipped slate-grey, 25 mm, base shortly attenuate, 8 - 9 mm ∅ across **Ov**, narrowed to 6.5 - 7 mm above; **Tep** free for 9 - 11 mm.

A. turkanensis sensu Reynolds (1966) is based on *A. scabrifolia*, and his illustrations are of the latter species.

A. tweediae Christian (JSAB 8(2): 175-176, t. 7, 1942). **T**: Uganda, Karamoja Distr. (*Tweedie* 262 [SRGH, K, PRE]). – **D**: Kenya, Sudan, Uganda; dry sandy bushland, 1340 - 1800 m. **I**: Reynolds (1966: 269-270).

[5] Acaulescent or shortly caulescent, simple or suckering to form small groups; stem rarely to 50 cm; **L** ± 20, densely rosulate, lanceolate-attenuate, ± 50 × 13 cm, dull to glossy green, usually with many pale green spots, surface smooth; marginal teeth ± 4 mm, pungent, reddish-brown, 10 - 15 mm apart; exudate yellow, drying brownish; **Inf** 1.2 - 1.5 m, with 15 - 20 **Br**, lower **Br** sometimes re-branched; terminal racemes cylindrical, laterals with secund **Fl**, to 15 × 5 cm, lax; **Bra** ovate-acute, 2 × 2 mm; **Ped** ± 7 mm; **Fl** coral-pink, paler becoming yellowish towards mouth, ± 24 mm, base rounded, 8 mm ∅ across **Ov**, narrowed slightly above; **Tep** free for 16 mm.

Reynolds (1966) states that the leaf exudate dries deep purple, but this is not evident on herbarium specimens, and was not the case on specimens collected by the present author near the type locality on the Kenyan side of the border.

Natural hybrids with other species have been reported (Reynolds 1966).

A. ukambensis Reynolds (JSAB 22(1): 33-35, 1956). **T**: Kenya, Eastern Prov. (*Reynolds* 7651 [EA, K, PRE]). – **D**: Kenya; gneissic rock faces and rocky slopes on hills, 520 - 1370 m. **I**: Reynolds (1966: 264-266).

[7] Acaulescent, branching at base to form dense groups; **L** 30 - 40, densely rosulate, lanceolate-attenuate, to 50 × 10 - 12 cm, grey-green tinged reddish, with conspicuous longitudinal striations, sometimes with a few scattered oval white spots near base, on lower face the spots H-shaped when present, surface smooth; marginal teeth 3 - 4 mm, reddish-brown, 8 - 10 mm apart; exudate drying pale orange-brown; **Inf** 50 cm, with 2 - 3 **Br**; racemes capitate or subcapitate, 4 - 6 × 6 cm, dense; **Bra** ovate-deltoid, 5 × 2.5 mm; **Ped** 18 - 20 mm; **Fl** bright glossy red, ± 40 mm, base shortly attenuate, 7 mm ∅ across **Ov**, slightly narrowed above, then enlarging to mouth; **Tep** free for 20 mm. – *Cytology*: 2n = 14 (Brandham 1971).

A. umfoloziensis Reynolds (JSAB 3(1): 42-45, t. 2, 1937). **T**: RSA, KwaZulu-Natal (*Reynolds* 2011 [PRE]). – **D**: RSA (KwaZulu-Natal); low-lying subtropical parkland. **I**: Reynolds (1950: 223-224).

[5,7] Acaulescent or shortly caulescent, simple or usually suckering to form groups; stem to 30 cm; **L** ± 20, densely rosulate, lanceolate-attenuate, 20 - 25 × 8 - 9 cm, green to brownish-green, with many dull white oblong spots scattered or ± in interrupted transverse bands, lower face paler green with or without spots, usually somewhat lineate; marginal teeth 3 - 5 mm, pungent, 10 - 15 mm apart; **Inf** 1 -

1.5 m, with 5 - 8 **Br**, lower **Br** sometimes re-branched; racemes capitate, 7 - 9 × 7 - 9 cm, sub-dense; **Bra** deltoid-acuminate, shorter than **Ped**; **Ped** 10 - 15 mm; **Fl** coral-red, 33 - 38 mm, base truncate, 8 - 9 mm ∅ across **Ov**, abruptly narrowed to 5 - 6 mm above, then enlarging to mouth; **Tep** free for 8 - 9 mm.

A. vacillans Forsskål (Fl. Aegypt.-Arab., 74, 1775). **T**: Yemen (*Forsskål* s.n. [C †]). – **Lit**: Wood (1983). **D**: Saudi Arabia, Yemen; rocky slopes and eroded grassland, 700 - 2700 m. **I**: Reynolds (1966: 156); Collenette (1999: 30).

Incl. *Aloe audhalica* Lavranos & D. S. Hardy (1965); incl. *Aloe dhalensis* Lavranos (1965).

[4,5] Acaulescent or shortly caulescent, simple or rarely suckering to form small groups; stem erect or decumbent, to 50 cm; **L** 15 - 20, rosulate, ensi-form-attenuate, ± 30 - 60 × 7 - 13 cm, glaucous, lower face sometimes with a few median prickles near tip, surface rough; marginal teeth 2 - 3 mm, brown or reddish-brown, 6 - 10 mm apart; **Inf** to 1 - 2 m, simple or with up to 3 **Br**; racemes cylindrical, 35 - 40 cm, sub-dense; **Bra** ovate-acuminate, 10 - 15 × 6 mm; **Ped** 5 - 12 mm; **Fl** red or yellow, ± 30 mm, base rounded, 8 mm ∅ across **Ov**, very slightly narrowed above, then slightly enlarging to mouth; **Tep** free for 6 mm. – *Cytology*: 2n = 14 (Wood 1983).

Red-flowered plants are predominant at the N end of the range, and yellow-flowered plants are predominant in the S (Wood 1983). The same author reports a natural hybrid with *A. inermis*.

A. vallaris L. C. Leach (JSAB 40(2): 111-115, ills., 1974). **T**: Angola, Huila (*Leach & Cannell* 14651 [LISC, BM, BR, K, LUAI, M, MO, PRE, SRGH, WIND, ZSS]). – **D**: Angola; cliffs, ± 1230 m.

[15] Caulescent, branching at base; stem to 50 cm; **L** densely rosulate, ovate-attenuate, 22 - 34 × 4 - 5 cm, greyish or greenish-blue to bluish-green, with a few small oval or round whitish spots near base, lower face with more spots, with narrow yellowish margin; marginal teeth 2 - 2.5 mm, pungent, yellowish, orange- or brown-tipped, 10 - 12 mm apart; exudate frothy, drying as opaque crystalline yellow crust; **Inf** ≥ 50 - 60 cm, oblique or suberect, simple or with 1 **Br**; racemes cylindrical-acuminate, 17 - 45 × ± 4 cm, lax; **Bra** ovate-acute, to 4.5 × 2.5 mm; **Ped** 4 - 4.5 mm; **Fl** bright scarlet, yellowish at mouth, 20 - 25 mm, base rounded, ± 5 mm ∅ across **Ov**, narrowed to 4.5 mm above, then enlarging to 5.5 mm at mouth; **Tep** free for 4.5 - 6 mm; **St** exserted 0 - 1 mm.

A. vanbalenii Pillans (South Afr. Gard. 24: 25, 1934). **T**: RSA, KwaZulu-Natal (*Van Balen* s.n. [BOL]). – **D**: RSA (KwaZulu-Natal); in shade on flat rocks or rocky slopes in bushland. **I**: Reynolds (1950: 420-421, t. 41).

[7] Acaulescent or shortly caulescent, branching to form dense groups; stem to 30 cm; **L** densely rosulate, lanceolate-attenuate, strongly recurved, 70 - 80 × 12 - 15 cm, green to copper-red, usually obscurely lineate, with horny reddish to reddish-brown margin; marginal teeth ± 3 - 5 mm, pungent, reddish, 10 - 15 mm apart; **Inf** ± 1 m, with 2 - 3 **Br**; racemes conical, 25 - 30 × 8 - 10 cm, sub-dense; **Bra** ovate-acute, 15 × 7 mm; **Ped** to 20 mm; **Fl** usually buff-yellow, sometimes dull red, ± 35 mm, base shortly attenuate, not narrowed above **Ov**, enlarging slightly to mouth; **Tep** free to base; **St** exserted 10 mm.

Natural hybrids with *A. marlothii* have been reported (Reynolds 1950).

A. vandermerwei Reynolds (Aloes South Afr. [ed. 1], 268-270, ills., 1950). **T:** RSA, Northern Prov. (*van der Merwe* s.n. [PRE 21288]). − **D:** RSA (Northern Prov.); grassy clearings in bushland.

[7] Acaulescent, suckering to form dense groups; **L** 15 - 20, rosulate, linear-attenuate, to 60 × 3.5 cm, dark green with large whitish spots ± in interrupted transverse bands; marginal teeth 3 - 4 mm, firm, pale brownish, 10 - 15 mm apart; exudate drying deep purple; **Inf** ± 1 m, with 6 - 10 **Br**; racemes cylindrical, slightly acuminate, 10 - 20 cm, lax, with ± 25 **Fl**; **Bra** deltoid-acuminate, to 10 × 3 mm; **Ped** 10 mm; **Fl** flesh-pink to pale pinkish-red, 30 mm, base rounded, 7 - 8 mm Ø across **Ov**, abruptly narrowed to 5 mm above, then enlarging to 9 mm at mouth; **Tep** free for 9 - 10 mm; **St** exserted 1 - 2 mm.

A. vaombe Decorse & Poisson (Recherch. Fl. Merid. Madag., 96, 1912). **T:** Madagascar, Toliara (*Anonymus* s.n.). − **D:** Madagascar.

A. vaombe var. **poissonii** Decary (Bull. Econ. Madag. 18: 23, 1921). **T:** Madagascar, Ambovombe Distr. (*Anonymus* s.n.). − **D:** Madagascar; on gneiss.

[9] Differs from var. *vaombe*: Stem to 5 m, more slender; **L** more densely rosulate, more deflexed; **Inf** with **Br** more spreading.

A. vaombe var. **vaombe** − **D:** Madagascar; dry thorn-bush. **I:** Reynolds (1966: 509-511).

[9] Caulescent, simple; stem erect, to 3 m × 20 cm, dead **L** persistent; **L** 30 - 40, densely rosulate, lanceolate-attenuate, 80 - 100 × 15 - 20 cm, dull green; marginal teeth 5 - 6 mm, subpungent, 15 - 20 mm apart; exudate drying deep purple; **Inf** ± 90 cm, with ± 12 **Br**, lower **Br** rebranched; racemes cylindrical, slightly acuminate, to 15 × 6 cm, sub-dense; **Bra** triangular, 8 × 5 mm; **Ped** ± 12 mm; **Fl** bright crimson, ± 28 mm, base rounded, 6 - 7 mm Ø across **Ov**, narrowed above, then enlarging to mouth; **Tep** free for 14 mm; **St** exserted 0 - 1 mm. − *Cytology:* 2n = 14 (Brandham 1971).

A. vaotsanda Decary (Bull. Econ. Madag. 18: 23, 1921). **T:** Madagascar, Toliara (*Decary* s.n. [P]). − **D:** Madagascar; dry bush on limestone outcrops, to 50 m. **I:** Reynolds (1966: 512).

[9] Caulescent, simple; stem erect, to 4 m × 15 cm, dead **L** persistent; **L** 30 - 40, densely rosulate, lanceolate-attenuate, strongly deflexed, to 1 m × 15 cm, green tinged reddish; marginal teeth ± 5 - 6 mm, pungent, 15 mm apart; exudate drying yellowish-brown; **Inf** ± 50 cm, with many **Br**; racemes secund, dense, with ± 50 - 70 **Fl**; **Bra** triangular-attenuate, 7 - 8 × 3 mm; **Ped** 4 - 6 mm; **Fl** orange-yellow, ± 22 mm, base rounded, slightly narrowed above **Ov**, then enlarging to mouth; **Tep** free for ± 13 mm.

A. variegata Linné (Spec. Pl. [ed. 1], 1: 321, 1753). **T:** not typified. − **D:** Namibia, RSA (Northern Cape, Western Cape, Eastern Cape, Free State); in shade of bushes on hard or stony ground. **I:** Reynolds (1950: 206-209). **Fig. XIX.e**

Incl. *Aloe punctata* Haworth (1804); incl. *Aloe variegata* var. *haworthii* A. Berger (1908); incl. *Aloe ausana* Dinter (1931).

[6] Acaulescent, suckering to form groups; **L** to 24, trifarious, lanceolate-deltoid, ± 10 - 15 × 4 - 6 cm, green with oblong white spots in irregular transverse bands, with whitish horny crenate-dentate margin; **Inf** ± 30 cm, simple or with 1 - 2 **Br**; racemes cylindrical, 10 - 20 cm, lax, with ± 20 - 30 **Fl**; **Bra** ovate-acuminate, to 15 × 7 mm; **Ped** 4 - 7 mm; **Fl** flesh-pink to dull scarlet, 35 - 45 mm, base truncate, slightly narrowed above **Ov**, then enlarging to mouth; **Tep** free for 5 - 7 mm; **St** exserted 0 - 2 mm. − *Cytology:* 2n = 14 (Kondo & Megata 1943).

Natural hybrids with other species have been reported (Reynolds 1950).

A. vera (Linné) Burman *fil.* (Fl. Indica, 83, 1768). **T:** BM [Hort. Cliff., †]. − **Lit:** Newton (1979). **D:** Origin uncertain, probably Arabia; widely cultivated in warm countries, esp. around the Mediterranean (since ancient times), India, and the West Indies. **I:** Reynolds (1966: 146-149, as *A. barbadensis*). **Fig. XIX.f**

≡ *Aloe perfoliata* var. *vera* Linné (1753); incl. *Aloe barbadensis* Miller (1768) ≡ *Aloe perfoliata* var. *barbadensis* (Miller) Aiton (1789); incl. *Aloe vulgaris* Lamarck (1783); incl. *Aloe elongata* Murray (1789); incl. *Aloe flava* Person (1805); incl. *Aloe barbadensis* var. *chinensis* Haworth (1819) ≡ *Aloe chinensis* (Haworth) Baker (1877) ≡ *Aloe vera* var. *chinensis* (Haworth) A. Berger (1908); incl. *Aloe indica* Royle (1839); incl. *Aloe vera* var. *littoralis* Koenig *ex* Baker (1880); incl. *Aloe lanzae* Todaro (1891) ≡ *Aloe vera* var. *lanzae* (Todaro) A. Berger (1908); incl. *Aloe vera* var. *wratislaviensis* Kostecka-Madalska (1953).

[6] Acaulescent or shortly caulescent, suckering

to form dense groups; stem to 30 cm; **L** ± 16, densely rosulate, lanceolate-attenuate, 40 - 50 × 6 - 7 cm, grey-green tinged reddish, with slightly pinkish margin, surface smooth; marginal teeth ± 2 mm, firm, pale, 10 - 20 mm apart; exudate drying yellow; **Inf** 60 - 90 cm, simple or with 1 - 2 **Br**; racemes cylindrical acuminate, 30 - 40 × 5 - 6 cm, dense; **Bra** ovate-acute, deflexed, 10 × 5 - 6 mm; **Ped** ± 5 mm; **Fl** yellow, ± 28 - 30 mm, slightly ventricose, base rounded, 7 mm ∅ across **Ov**, enlarged above, then narrowing to mouth; **Tep** free for 18 mm; **St** exserted ± 5 mm. – *Cytology:* 2n = 14 (Sutaria 1932).

Linné did not publish the combination *Aloe vera* as a numbered species. Reynolds (1966) argued that the name should be *A. barbadensis*, but he had overlooked the combination published by N. L. Burman (not later than April 6, 1768), which has priority over Miller's (April 16, 1768) name (Newton 1979).

Wood (1983) suggested that this is conspecific with *A. officinalis*, which he treated as a doubtfully distinct variety. As *A. vera* is not known as naturally occurring populations anywhere, though plants from the Sumail Gap (Oman) reported by Lavranos (1965) need further investigation, it seems best to maintain it as a separate species at present. It is possible that the plants now cultivated are the result of selection over 2000 years or more, and perhaps they should have the status of cultivars.

A. indica is probably the red-flowered variant introduced to India (Reynolds 1966).

A. verecunda Pole-Evans (Trans. Roy. Soc. South Afr. 5: 703, 1917). **T:** RSA, Northern Prov. (*Pienaar* s.n. [PRE]). – **D:** RSA (North-West Prov., Northern Prov., Gauteng, Mpumalanga); rocky slopes. **I:** Reynolds (1950: 134-135).

[4] Acaulescent or shortly caulescent, simple or usually suckering to form groups; **R** fusiform; stem to 20 × 2 cm; **L** 8 - 10, usually distichous, sometimes subrosulate, linear, 25 - 35 × 0.8 - 1 cm, dull green, lower face with many small white spots near base; marginal teeth minute, soft, white, 2 - 7 mm apart; **Inf** 25 cm, simple; raceme capitate, dense; **Bra** ovate-acute, 20 × 15 mm; **Ped** 25 - 30 mm; **Fl** rich peach-red to scarlet, greenish-tipped, 26 - 30 mm, base shortly attenuate, 12 mm ∅ across **Ov**, not narrowed above; **Tep** free to base. – *Cytology:* 2n = 14 (Müller 1945).

Natural hybrids with *A. zebrina* (as *A. transvaalensis*) have been reported (Reynolds 1950).

A. versicolor Guillaumin (Bull. Mus. Nation. Hist. Nat., Sér. 2, 21: 723, 1950). **T:** Madagascar, Toliara (*Humbert* 20617 [P]). – **D:** Madagascar; soil pockets on silicate rocks, ± 50 - 70 m. **I:** Reynolds (1966: 422-423).

[6] Acaulescent or very shortly caulescent, suckering to form dense groups; **L** ± 15, densely rosu-

late, 15 × 2 cm, almost linear, tip rounded with 3 - 5 white 1 mm teeth, dull bluish-green with a bloom; marginal teeth 1.5 - 2 mm, firm, white, 5 - 6 mm apart; exudate drying deep brown; **Inf** 30 - 40 cm, simple; raceme cylindrical, subcapitate, ± 5 cm, sub-dense; **Bra** ovate-acute, 7 × 4 mm; **Ped** 15 - 20 mm; **Fl** coral-red to pale scarlet at base, paler above, yellowish at tip, ± 25 mm, base rounded, 7 mm ∅ across **Ov**, not narrowed above; **Tep** free almost to base.

A. veseyi Reynolds (JSAB 25(4): 315-317, pl. 32, 1959). **T:** Zambia, Abercorn Distr. (*Reynolds* 8659 [PRE, K, SRGH]). – **D:** Tanzania, Zambia; cliffs, 1200 m. **I:** Reynolds (1966: 168-169).

[13] Caulescent, branching at base; stem pendulous, to 40 × 2 cm; **L** ± 12, rosulate, lanceolate-falcate, attenuate, 40 - 50 × 3 cm, dull grey-green tinged reddish, with many scattered dull white spots; marginal teeth 1 - 2 mm, firm, to 20 mm apart; **Inf** ± 60 cm, pendulous with racemes ascending, with 2 - 4 **Br**; racemes cylindrical-conical, 12 × 7 cm, lax, with ± 30 **Fl**; **Bra** lanceolate-attenuate, 6 × 3 mm; **Ped** 14 mm; **Fl** pale yellow, 25 mm, base rounded, 5 mm ∅ across **Ov**, slightly narrowed above, then enlarging to 10 mm at mouth; **Tep** free for 5 - 6 mm. – *Cytology:* 2n = 14 (Brandham 1971).

A. viguieri H. Perrier (Bull. Acad. Malgache, n.s. 10: 20, 1927). **T:** Madagascar, Toliara (*Perrier* 17592 [P, K]). – **D:** Madagascar; limestone slopes and cliffs, 60 - 350 m. **I:** Reynolds (1966: 441-442).

[4] Acaulescent or caulescent, simple or branching to form small groups; stem decumbent to 30 cm or pendulous to 1 m, dead **L** persistent; **L** 12 - 16, densely rosulate, lanceolate-attenuate, ± 30 - 40 × 8 - 9 cm, light green, lineate, with 1 mm white cartilaginous margin; marginal teeth 0.5 - 1 mm, firm, white, 1 - 2 mm apart; **Inf** ± 45 cm, simple; raceme cylindrical, 20 - 25 cm, lax, with ± 22 **Fl**; **Bra** ovate-acute, 1.5 mm; **Ped** 11 mm; **Fl** scarlet, 22 mm, slightly clavate, base shortly attenuate, 4 mm ∅ across **Ov**, enlarging above; **Tep** free for 11 mm; **St** exserted 0 - 1 mm.

A. viridiflora Reynolds (JSAB 3(4): 143-145, t. 23, 1937). **T:** Namibia (*Reynolds* 1626 [PRE]). – **D:** Namibia; rocky slopes, ± 1830 m. **I:** Reynolds (1950: 322-323).

[3] Acaulescent, simple; **L** 50 - 60, densely rosulate, lanceolate-attenuate, to 40 × 8 cm, glaucous, obscurely lineate; marginal teeth 2 mm, pungent, pinkish-brown, 2 - 5 mm apart; **Inf** ± 1.5 m, with 6 **Br**; racemes capitate, 10 × 8 cm, dense, with ± 50 - 60 **Fl**; **Bra** ovate-acute, 15 × 7 mm; **Ped** 20 mm; **Fl** green, lemon-tinged about middle, 33 mm, clavate, base attenuate, enlarged above **Ov** to 9 - 10 mm near mouth; **Tep** free to base; **St** exserted 10 mm.

A. vituensis Baker (FTA 7: 458, 1898). **T:** Kenya, Coast Prov. (*Thomas* 113 [B]). − **Lit:** Leach (1970: ills.). **D:** Kenya, Sudan; usually in shade, below large shrubs or in grass on rocky slopes, 290 - 1525 m.

[11] Caulescent, branching sparingly; stem erect at first, becoming decumbent later, to 40 cm; **L** scattered along stem for 15 - 20 cm, lanceolate-attenuate, to 35 × 4 cm, yellow-green near base, becoming bronze to brown above, obscurely striatulate, with many lenticular or H-shaped whitish spots, more numerous on lower face, surface smooth; marginal teeth 3 - 4 mm, pungent, orange-brown, 5 - 10 mm apart; sheath to 3 cm, brown striatulate; exudate none, with internal fibres, esp. at base and in sheath; **Inf** ± 50 - 60 cm, simple; raceme cylindrical, slightly acuminate, 7 - 8 × 6 cm, sub-dense; **Bra** ovate-acute, usually with a small tooth on one or both sides, markedly convex and appearing fleshy, ± 7 - 9 × 6 mm; **Ped** 4 - 7 mm; **Fl** coral-pink, greenish-tipped, 27 - 29 mm, base shortly attenuate, ± 5 mm ∅ across **Ov**, narrowed above, then enlarging to 8 - 10 mm at mouth; **Tep** free for 6 - 8 mm.

The type specimen was collected on an expedition that started from Witu, but it was collected further N.

A. vogtsii Reynolds (JSAB 2(3): 118-120, t. 15, 1936). **T:** RSA, Northern Prov. (*Reynolds* 1488 [PRE, BOL]). − **D:** RSA (Northern Prov.); in long grass or among shrubs on rocky slopes in the mist belt, 1430 m. **I:** Reynolds (1950: 257-258).

[7] Caulescent, suckering to form small groups; stem to 20 cm; **L** 16 - 20, densely rosulate, lanceolate-attenuate, tip a pungent **Sp**, 20 - 25 × 5 - 6 cm, cress-green, obscurely lineate, with many small white H-shaped spots, scattered or ± in wavy interrupted transverse bands, lower face dull green, with a few pale brown median prickles near tip; marginal teeth ± 3 mm, pungent, pale brown, 10 - 15 mm apart; **Inf** ± 66 cm, with ± 7 **Br**, lower **Br** rebranched; racemes cylindrical, slightly acuminate, ± 20 × 8 cm, sub-dense, with 30 - 40 **Fl**; **Bra** ovate-acuminate, 10 - 15 mm; **Ped** to 18 mm; **Fl** scarlet, paler at mouth, 34 mm, base truncate, 9 mm ∅ across **Ov**, abruptly narrowed to 5 mm above, then enlarging to mouth; **Tep** free for 9 mm; **St** scarcely exserted. − *Cytology:* 2n = 14 (Müller 1945).

Natural hybrids with other species have been reported (Reynolds 1950).

A. volkensii Engler (Pfl.-welt Ost-Afr., Teil C, 141, 1895). **T:** Tanzania, Moshi Distr. (*Volkens* 406 [B, BM]). − **D:** Kenya, Tanzania, Uganda.

A. volkensii ssp. **multicaulis** S. Carter & L. E. Newton (Fl. Trop. East Afr., Aloaceae, 56, 1994). **T:** Tanzania, Musoma Distr. (*Greenway* 10351 [K, EA]). − **D:** Kenya, Tanzania, Uganda; rocky bushland, 1150 - 2100 m.

[9] Differs from ssp. *volkensii*: Branching at base with ≥ 3 main stems; **L** marginal teeth ± 15 mm apart; **Fl** brick-red, 30 - 35 mm.

Natural hybrids with *A. ngongensis* have been reported (Carter 1994).

A. volkensii ssp. **volkensii** − **D:** Kenya, Tanzania; dry forest on rocky slopes, 10 - 1800 m. **I:** Reynolds (1966: 328-329).

Incl. *Aloe stuhlmannii* Baker (1898).

[17] Caulescent, simple or branching above base with single basal trunk; stem erect, to 9 m × 30 cm, dead **L** persistent; **L** densely rosulate, lanceolate-attenuate, to 1 m × 10 cm, glaucous-green to olive-green, surface smooth; marginal teeth 4 mm, pungent, brown-tipped, 8 - 15 mm apart; exudate drying yellow; **Inf** 70 - 85 cm, with ± 10 **Br**; racemes subcapitate, becoming cylindrical, 8 - 12 × 8 - 9 cm, dense; **Bra** ovate-acute, ± 5 × 5 mm; **Ped** ± 15 mm; **Fl** reddish-orange to pale scarlet, yellow at mouth, ± 35 mm, base rounded, 7 - 8 mm ∅ across **Ov**, very slightly narrowed above, then slightly enlarging to mouth; **Tep** free for 15 mm.

A. vossii Reynolds (JSAB 2(2): 65-68, t. 4, 1936). **T:** RSA, Northern Prov. (*Reynolds* 557 [PRE]). − **D:** RSA (Northern Prov.); grassy slopes. **I:** Reynolds (1950: 136-137).

[6] Acaulescent or very shortly caulescent, suckering to form small groups; **L** 14 - 20, multifarious, long-attenuate, to 50 × 3 cm, deep green with several scattered elongate white spots, the spots occasionally subtuberculate and spinulescent, lower face with many spots near base, the spots frequently with a firm white prickle, margin very narrow, white, cartilaginous; marginal teeth 2 mm, firm, white, 2 - 4 mm apart; **Inf** to 50 cm, simple; raceme capitate, ± 8 × 7 cm, dense; **Bra** ovate-acute, 16 × 11 mm; **Ped** 30 mm; **Fl** scarlet, 28 mm, base rounded, 8 - 9 mm ∅ across **Ov**, slightly narrowed above towards mouth; **Tep** free to base; **St** exserted 0 - 1 mm. − *Cytology:* 2n = 14 (Müller 1941).

A. vryheidensis Groenewald (Tydskr. Wetensk. Kuns 15: 129-131, 1937). **T:** RSA, KwaZulu-Natal (*van der Merwe* 266 [PRE]). − **D:** RSA (KwaZulu-Natal); rocky slopes and dolomite outcrops. **I:** Reynolds (1950: 429-431, and 434-435 as *A. dolomitica*).

Incl. *Aloe dolomitica* Groenewald (1938).

[4] Acaulescent or caulescent, simple or sometimes 2- to 4-branched at base; stem decumbent, suberect or erect, to 2 m × 20 cm; **L** 20 - 50, densely rosulate, lanceolate-attenuate, to 65 × 13 cm, dark green to glaucous-green tinged bluish, with reddish margin, surface smooth; marginal teeth 2 - 3 mm, pungent, reddish to brownish-red, 10 - 15 mm apart; exudate drying yellow; **Inf** 60 - 150 cm, oblique with raceme erect, simple; raceme cylindrical, 30 - 40 × 5 - 7 cm, dense; **Bra** ovate-acute, 8 - 15 × 5 -

10 mm; **Ped** absent; **Fl** rose or greenish-yellow to yellowish, 12 - 20 mm, campanulate, base obconical, 5 mm ⌀ across **Ov**, enlarged above to 12 mm at mouth; **Tep** free to base; **St** exserted 12 - 15 mm.

A. vryheidensis was described as acaulescent or very shortly caulescent, and *A. dolomitica* as having a stem to 2 m high and occurring further N. Glen & Hardy (1995) reported that the 2 taxa are not disjunct, and they reported having found a variable population with every possible intermediate between the 2, and they concluded that they are conspecific.

Natural hybrids with other species have been reported (Reynolds 1950).

A. whitcombei Lavranos (CSJA 67(1): 30-33, ills., 1995). **T:** Oman, Dhofar (*Collenette* 8950 [E]). – **D:** Oman; cliffs of limestone or calcareous sandstone, 900 m.

[13] Caulescent, branching freely at base; stem mostly pendulous, to ± 30 cm; **L** 5 - 8, densely rosulate, deltoid, tip rounded, 50 - 80 × 1.5 cm, green with many rounded whitish spots; marginal teeth 0.5 mm, soft, white, 1 - 2 mm apart; **Inf** 35 - 80 cm, obliquely ascending and arched downwards, or entirely pointing obliquely downwards, simple or with 1 **Br**; racemes conical, 30 - 60 cm, dense, with 12 - 50 **Fl**; **Bra** deltoid-acute, 4 × 2.5 mm; **Ped** ± 6 mm; **Fl** white with green veins, 14 mm, base rounded, 4 mm ⌀ across **Ov**, not narrowed above; **Tep** free for 6 mm.

A. wildii (Reynolds) Reynolds (Kirkia 4: 13, 1964). **T:** Zimbabwe, Melsetter Distr. (*Wild* 3541 [SRGH, K, PRE]). – **D:** Zimbabwe; open bush on mountain slopes, to ± 2135 m. **I:** Reynolds (1966: 20).

≡ *Aloe torrei* var. *wildii* Reynolds (1961).

[1] Acaulescent, simple or branching at base to form small groups; **L** ± 6, distichous, linear, 15 - 30 × 0.5 - 1 cm, dull green tinged brownish, with a few scattered small white spots near base, lower face with spots more numerous and minutely spinulescent; marginal teeth ± 0.5 mm, soft, white, 1 - 2 mm apart; **Inf** ± 25 - 30 (-50) cm, simple; raceme cylindrical, 6 - 7 × 5 cm, lax, with 12 - 16 **Fl**; **Bra** ovate-acute, 5 × 2 - 3 mm, dull pink; **Ped** 10 - 15 mm; **Fl** bright orange-scarlet, green tipped, 30 - 40 mm, slightly ventricose, base shortly attenuate, enlarged above **Ov**, slightly narrowed at mouth; **Tep** free to base; **St** exserted 0 - 1 mm.

A. wilsonii Reynolds (JSAB 22(3): 137-140, 1956). **T:** Uganda, Karamoja Distr. (*Tweedie* 1365 [PRE, EA]). – **D:** Kenya, Uganda; rocky slopes on isolated hills, 1525 - 3000 m. **I:** Reynolds (1966: 263-264).

[13] Caulescent, simple or in small groups; stem decumbent, usually short, sometimes to 80 × 6 cm; **L** ± 24, densely rosulate, lanceolate-attenuate, to 25 × 9 cm, deep green tinged yellowish, slightly glossy, lower face sometimes with a few crowded white spots near base, with pale brown horny margin; marginal teeth ± 3 - 4 mm, pungent, reddish, 10 mm apart; exudate yellow drying brownish; **Inf** 50 - 60 cm, with ± 8 **Br**; racemes cylindrical-acuminate, 15 - 18 × 5 - 6 cm, lax; **Bra** ovate-acuminate, 5 × 3 mm; **Ped** 15 mm; **Fl** dull scarlet with a bloom, 28 mm, base shortly attenuate, 7 mm ⌀ across **Ov**, very slightly narrowed above, then slightly enlarged to mouth; **Tep** free for 14 mm; **St** exserted 1 - 2 mm. – *Cytology:* 2n = 14 (Cutler & al. 1980).

A. wollastonii Rendle (JLSB 38: 238, 1908). **T:** Uganda, Toro Distr. (*Wollaston* s.n. [BM]). – **D:** Kenya, Tanzania, Uganda, Zaïre; grassland and with grass in open woodland, 1100 - 2285 m. **I:** Reynolds (1966: 97, as *A. lateritia*; 99, as *A. lateritia* var. *kitaliensis*).

Incl. *Aloe angiensis* De Wildeman (1921); **incl.** *Aloe bequaertii* De Wildeman (1921); **incl.** *Aloe lanuriensis* De Wildeman (1921); **incl.** *Aloe angiensis* var. *kitaliensis* Reynolds (1955) ≡ *Aloe lateritia* var. *kitaliensis* (Reynolds) Reynolds (1966).

[5] Acaulescent, usually simple; **L** 12 - 15, densely rosulate, lanceolate-attenuate, 40 - 50 × 8 - 10 cm, dull green, with many elongated whitish spots in irregular transverse bands, surface smooth; marginal teeth 4 - 6 mm, pungent, red-brown, 10 - 20 mm apart; **Inf** to 1.25 m, with 4 - 6 **Br**, lower **Br** sometimes rebranched; racemes cylindrical-conical, 10 - 30 × 8 cm, lax; **Bra** linear-lanceolate, 10 - 20 × 4 mm; **Ped** 15 - 20 mm; **Fl** pinkish to orange-red, rarely yellow, 30 - 35 mm, base truncate, 8 - 10 mm ⌀ across **Ov**, abruptly narrowed to ± 6 mm above, then enlarging to mouth; **Tep** free for ± 10 - 12 mm; **St** exserted 0 - 1 mm.

A. woodii Lavranos & Collenette (CSJA 72(2): 83, ills., 2000). **T:** Saudi Arabia, Asir Prov. (*Vesey-Fitzgerald* 16076/6 [BM]). – **D:** Saudi Arabia, Yemen; open stony ground or in *Juniperus* forests, 2000 - 3000 m. **I:** Collenette (1999: 31).

[5] Usually acaulescent, sometimes with a short procumbent stem, usually solitary, occasionally with a few suckers; **L** 15 - 20, densely rosulate, spreading to ascending, very rigid, lanceolate-attenuate, 45 - 50 × 10 - 15 cm, glaucous with a yellowish tinge; marginal teeth 2 - 3 mm, brown-tipped, 8 - 20 mm apart; **Inf** 80 - 100 cm, erect, with up to 6 **Br**; racemes cylindrical, sub-dense; **Bra** lanceolate, 15 - 18 × 5 - 10 mm; **Ped** 7 - 10 mm; **Fl** cream to yellow with green longitudinal stripes, usually densely white-tomentose, 26 - 33 mm, base rounded, ± 8 mm ⌀ across **Ov**, not constricted above; **Tep** free for ± 10 mm; **St** scarcely exserted.

A. wrefordii Reynolds (JSAB 22(3): 141-143, 1956). **T:** Uganda, Karamoja Distr. (*Tweedie* 659 [PRE, EA, K]). – **D:** Kenya, Sudan, Uganda; stony thornbush and exposed rocky slopes, 950 - 1430 m. **I:** Reynolds (1966: 208-209).

[5] Acaulescent, simple or suckering to form small groups; **L** ± 24, densely rosulate, lanceolate-attenuate, ± 60 × 15 cm, dull grey-green, sometimes tinged reddish-brown, obscurely lineate, tip a reddish-brown **Sp**, surface smooth; marginal teeth 4 mm, pungent, reddish-brown, 10 - 15 mm apart; exudate drying yellow; **Inf** ± 1.2 m, with up to 16 **Br**, lower **Br** rebranched; racemes subcapitate, 5 - 8 × 6 cm, dense; **Bra** ovate-acute, 9 × 4 mm; **Ped** 16 mm; **Fl** scarlet or orange, 22 mm, clavate, base shortly attenuate, 5 - 6 mm ⌀ across **Ov**, enlarged above towards mouth; **Tep** free for 13 mm.

According to Gilbert & Sebsebe (1992) the Ethiopian material referred to this species by Reynolds (1966) is *A. otallensis*.

A. yavellana Reynolds (JSAB 20(1): 28-30, t. 4, 1954). **T:** Ethiopia, Sidamo Prov. (*Reynolds* 7063 [PRE, K]). – **D:** Ethiopia; in forest, in clearings, or on rocks, 1700 - 1950 m. **I:** Reynolds (1966: 345).

[13] Caulescent, branching at base; stem erect to 1 m, then decumbent, to 3 m × 4 cm; **L** ± 16 - 20, scattered along stem for top-most 20 cm, ensiform-attenuate, to 40 × 6 - 8 cm, bronze-brown, paler in shade, lower face paler bronze-brown to dull brownish-green and green-lineate near base, surface smooth; marginal teeth 2 - 3 mm, pungent, pale reddish-brown tipped, 10 - 15 mm apart; sheath striate; exudate drying yellow; **Inf** 60 - 90 cm, with 8 - 10 **Br**, lower **Br** rebranched; racemes capitate or subcapitate, 2 - 3 cm, dense; **Bra** deltoid, 3 × 2 mm; **Ped** 10 mm; **Fl** in bud dark scarlet with longitudinal grey stripes and white flecks, at anthesis dull scarlet at base, paler to orange above, 27 mm, base attenuate, 5 - 6 mm ⌀ across **Ov**, slightly narrowed above, slightly enlarging to the slightly upturned mouth; **Tep** free for 9 mm; **St** exserted 1 mm. – *Cytology:* 2n = 14 (Cutler & al. 1980).

A. yemenica J. R. I. Wood (KB 38(1): 20-21, ills., 1983). **T:** Yemen (*Wood* 2537 [K]). – **D:** Saudi Arabia, Yemen; sandstone cliffs, 1000 - 2000 m. **I:** Collenette (1999: 32).

[13] Caulescent, branching at base; stem pendulous, rarely erect, to ± 50 cm; **L** scattered along stem, ensiform, ± 35 × 3.5 cm, green, with spots in juvenile stage; marginal teeth short, 8 - 10 mm apart; sheath to 3 cm; **Inf** 30 cm, ascending or recurved, simple or usually with **Br**; racemes cylindrical, lax; **Bra** ovate-acute, 10 × 7 mm; **Ped** 8 - 10 mm; **Fl** red, rarely yellow, ± 22 mm. – *Cytology:* 2n = 14 (Wood 1983).

A. zebrina Baker (Trans. Linn. Soc. London, Bot. 1: 264, 1878). **T** [lecto]: Angola, Luanda Distr. (*Welwitsch* 3721 [LISU, BM, K]). – **D:** Angola, Botswana, Moçambique, Zambia, Zimbabwe, Namibia, Malawi, RSA (Gauteng, Mpumalanga, Northern Prov.); grassland and thickets on dry hills. **I:** Reynolds (1966: 90).

Incl. *Aloe platyphylla* Baker (1878); **incl.** *Aloe transvaalensis* Kuntze (1898); **incl.** *Aloe lugardiana* Baker (1901); **incl.** *Aloe bamangwatensis* Schönland (1904); **incl.** *Aloe ammophila* Reynolds (1936); **incl.** *Aloe laxissima* Reynolds (1936); **incl.** *Aloe angustifolia* Groenewald (1938) (*nom. illeg.*, Art. 53.1); **incl.** *Aloe transvaalensis* var. *stenacantha* F. S. Müller (1940).

[5,7] Acaulescent or shortly caulescent, solitary or usually suckering to form small or large groups; **L** 10 - 20, densely rosulate, lanceolate, 20 - 30 × 5 - 7 cm, dull green, striate, with oblong whitish spots scattered or usually in irregular transverse bands, lower face with few or many spots; marginal teeth 4 - 6 mm, pungent, brownish, 8 - 15 mm apart; exudate drying purplish or orange; **Inf** 1 - 1.3 m, with 5 - 10 **Br**, lower **Br** sometimes rebranched; racemes cylindrical, slightly acuminate, 20 - 40 × 6 - 7 cm, lax; **Bra** deltoid, acuminate, ± 6 - 15 mm; **Ped** 6 - 15 mm; **Fl** coral-red or dull reddish, ± 30 mm, base rounded or truncate, 7 - 10 mm ⌀ across **Ov**, abruptly narrowed to 4 - 6 mm above, then enlarging to mouth; **Tep** free for 7 - 11 mm; **St** exserted 1 - 2 mm. – *Cytology:* 2n = 14 – Fernandes (1930); Müller (1941: as *A. laxissima*); Brandham (1971: as *A. ammophila*).

Natural hybrids with *A. littoralis* (Reynolds 1966) and with *A. verecunda* (as *A. transvaalensis*) (Reynolds 1950) have been reported.

A. zombitsiensis Rauh & M. Teissier (KuaS 51(8): 201-203, ills., 2000). **T:** Madagascar, Toliara (*Anonymus* s.n. in *BG Heidelberg* 73994 [HEID]). – **D:** SW Madagascar; in humus on the forest floor, ± 1200 m.

[2] Acaulescent; **Ros** solitary, to 30 cm ⌀; **L** to 25, densely rosulate, narrowly lanceolate-attenuate, to 18 × 2.5 cm, dark green with irregular white flecks; marginal teeth deltoid, 2 - 3 mm, 3 - 15 mm apart; **L** exudate yellow; **Inf** 15 cm, erect, simple; raceme cylindrical, 5 cm, subdense; **Bra** as long as the **Ped** or longer; **Ped** 5 - 7 mm; **Fl** bright cinnabar-red at the base, base rounded, 2.7 mm ⌀ across **Ov**, slightly narrowed above and widening towards mouth; **Tep** free for ⅓ to ½ of their length, lobes green with white margins, 2.5 - 2.7 mm; **St** scarcely exserted; **Fr** berries.

The description states that the tepals are united for ± 4 mm, but it is clear from fig. 3 of the protologue that this is an error.

×ALOLIRION

U. Eggli

×**Alolirion** G. D. Rowley (Nation. Cact. Succ. J. 28(1): 7, 1973).

= *Aloe* × *Chortolirion*. No formally named hybrids are known. Rowley (1982) lists a hybrid between *Chortolirion* and *Aloe striatula*.

×ALOLOBA

U. Eggli

×Aloloba G. D. Rowley (Nation. Cact. Succ. J. 22(3): 74, 1967).
Incl. ×*Chamaeloba* D. M. Cumming (1974).
Incl. ×*Lomatoloba* D. M. Cumming (1974).

= *Aloe* × *Astroloba*. None of these hybrids have been formally named, but see Rowley (1982) for a list of known crosses.

×ALWORTHIA

U. Eggli

×Alworthia G. D. Rowley (Nation. Cact. Succ. J. 28(1): 7, 1973).

= *Aloe* × *Haworthia*. No formally named hybrids are known, and Rowley (1982: 46) lists only a single known cross between an unidentified *Aloe* and *H. cymbiformis*.

ASTROLOBA

N. L. Meyer & G. F. Smith

Astroloba Uitewaal (Succulenta 1947 (5): 53, 1947). **T:** *Aloe spiralis* var. *pentagona* Aiton. — **Lit:** Groen (1986); Groen (1987). **D:** RSA (Western Cape, Eastern Cape). **Etym:** Gr. 'aster, astros', star; and Gr. 'lobos', lobe; for the stellately spreading perianth lobes.
Incl. *Apicra* Haworth (1819) (*nom. illeg.*, Art. 53.1). **T:** *Aloe pentagona* (Aiton) Haworth [G. D. Rowley, Nation. Cact. Succ. J. 31(3): 54, 1976.].

Small herbaceous **L**-succulent slow-growing perennials, proliferating from the base or with subterranean stolons to form small or large clusters, caulescent, erect or creeping; **R** succulent, terete; **L** numerous, alternate, spirally arranged, imbricate, basally amplexicaul, erectly spreading, deltoid-acuminate, triangular, leathery, light to dark green to glaucous green, upper face plane or convex, sometimes with reddish striations, lower face convex, usually with a distinct asymmetric keel extending for ⅔ of the length, margins and keel acute or slightly rounded, smooth or denticulate, tip acute, sometimes terminating in a mucro, surface smooth or tuberculate, green or brownish; **Inf** laxly flowered racemes or panicles, 10 - 40 cm, occasionally with accessory branchlets or buds in the **Ax**; peduncle smooth, simple or branched, lower sterile parts bracteate; **Bra** membranous, persistent, triangular to long-acuminate, keeled with 1 - 3 brownish-green or pinkish central veins; **Ped** short, erect; **Fl** erect, tubular, regular, ± straight, sometimes with inflated tissue on sides of **Per**; **Tep** 6, connate below, free but contiguous above, usually monochrome with dull colours, often drab greenish-white, rarely cream, beige, yellow or pink, distinctly veined, lobes short; **St** 6, shorter than **Per**, included or barely exserted; **Fil** yellowish-white, filiform; **Anth** yellow, dorsifixed, dehiscing longitudinally, introrse; **Ov** superior, green, sessile, oblong, ± 4 × 3 mm ∅; **Sty** whitish, straight, capitate, ± 4 mm, subulate; **Sti** minute, apical; **Fr** 3-locular capsules, cylindrical, oblong to obtuse or ovoid to acuminate, dehiscing loculicidally, ± 12 × 5 - 6 mm ∅; **Se** angled, dark brown to black, laterally compressed, angles obscurely winged, ± 4 mm.

The genus has been revised in an unpublished thesis by Roberts Reinecke (1965). Another revision was done by Groen (1986) and Groen (1987), but the present authors prefer a compromise between these treatments. *Astroloba* has floral affinities with the genus *Haworthia*, esp. *H.* subg. *Robustipedunculares*. It is retained as separate mainly for its caulescent habit, regular (not bilabiate) flowers, and the leaves which are ± arranged in 5 distinct ranks.

The names *A. smutsiana* and *A. hallii* (= *Haworthia hallii* (Roberts Reineke) M. Hayashi) and the combination *A. foliolosa* ssp. *robusta* have been proposed by Roberts Reinecke (1965) for newly discovered taxa in the genus. These names are not validated here, pending completion of the revision of the genus by N. L. Meyer.

A. bullulata (Jacquin) Uitewaal (Succulenta 1947 (5): 53, 1947). **T:** [lecto – icono]: Jacquin, Fragm. Bot. t. 109, 1809. — **D:** RSA (Western Cape, Eastern Cape: Ceres, Laingsburg and Sutherland Distr.). **I:** Pilbeam (1983: 149, as *A. bicarinata*).

≡ *Aloe bullulata* Jacquin (1809) ≡ *Apicra pentagona* var. *bullulata* (Jacquin) Baker (1880) ≡ *Haworthia bullulata* (Jacquin) Parr (1971) ≡ *Apicra bullulata* (Jacquin) Willdenow (1811) (*incorrect name*, Art. 11.3); **incl.** *Astroloba bullata* hort. (s.a.) (*nom. inval.*, Art. 61.1); **incl.** *Apicra egregia* von Poellnitz (1930) ≡ *Astroloba egregia* (von Poellnitz) Uitewaal (1947) ≡ *Haworthia egregia* (von Poellnitz) Parr (1971); **incl.** *Astroloba fardeniana* Uitewaal (1948) ≡ *Haworthia egregia* var. *fardeniana* (Uitewaal) Parr (1971).

L in 5 straight rows or rarely imbricate, spiral angle usually 0° - 10°, dull greenish-brown, usually suberect, tips curved upwards, frequently to the keeled side, keel forming a margin at the apex, 23 - 40 × 13 - 26 mm, mucro 0.3 - 2 mm, vein lines absent, surface of some or all **L** tuberculate, tubercles fairly prominent, ± 1 mm ∅, few and irregularly scattered or more numerous and grouped in transverse rows; **Inf** lax racemes, 14 - 30 cm; **Fl** erect, greenish-brown with yellow lobes; **Ped** 3 - 6 mm; **Per** tube straight, 8 - 11 × ± 3 mm ∅, lobes ± 2 mm. – *Cytology:* 2n = 14.

A. congesta (Salm-Dyck) Uitewaal (Succulenta 1947(5): 54, 1947). **T:** [lecto – icono]: Salm-Dyck, Monogr. Gen. Aloes & Mesembr., Sect. 6: fig. 1,

1854. – **D:** RSA (Eastern Cape: Albany, Bedford and Cradock Distr.).

≡ *Aloe congesta* Salm-Dyck (1854) ≡ *Apicra congesta* (Salm-Dyck) Baker (1881) ≡ *Astroloba foliolosa* ssp. *congesta* (Salm-Dyck) Roberts Reinecke (1965) (*nom. inval.*, Art. 29.1) ≡ *Haworthia congesta* (Salm-Dyck) Parr (1971); **incl.** *Aloe deltoidea* Hooker *fil.* (1873) ≡ *Apicra deltoidea* (Hooker *fil.*) Baker (1881) ≡ *Astroloba deltoidea* (Hooker *fil.*) Uitewaal (1947) ≡ *Haworthia deltoidea* (Hooker *fil.*) Parr (1971); **incl.** *Apicra turgida* Baker (1889) ≡ *Apicra deltoidea* var. *turgida* (Baker) A. Berger (1908) ≡ *Astroloba deltoidea* var. *turgida* (Baker) H. Jacobsen (1954) (*nom. inval.*, Art. 33.2) ≡ *Astroloba turgida* (Baker) H. Jacobsen (1960) (*nom. inval.*, Art. 33.2?) ≡ *Haworthia deltoidea* var. *turgida* (A. Berger) Parr (1971) (*nom. inval.*, Art. 34); **incl.** *Apicra deltoidea* var. *intermedia* A. Berger (1908) ≡ *Astroloba deltoidea* var. *intermedia* (A. Berger) Uitewaal (1947) ≡ *Haworthia deltoidea* var. *intermedia* (A. Berger) Parr (1971); **incl.** *Haworthia shieldsiana* Parr (1971) (*nom. inval.*, Art. 34.1).

L in 5 straight rows or rarely imbricate, spiral angle 10° - 20°, light to dark green with a glossy sheen, erect to patent, tips curved upwards to outwards, keel not forming a margin at the apex, 20 - 47 × 14 - 28 mm, mucro 0.5 - 1.3 mm, tubercles absent, but rarely small elongated, very slightly raised concolorous patches present on some **L**; **Inf** lax racemes, 6 - 31 cm, sometimes branched; **Fl** green with a creamy tinge with white or cream lobes (never yellow); **Ped** 0.7 - 4 mm; **Per** tube straight, 6 - 9 × ± 3 mm ∅, lobes recurved, 1.4 - 3 mm. – *Cytology:* 2n = 14.

A. corrugata N. L. Meyer & G. F. Smith (BT 28(1): 61-62, ill., 1998). **T:** RSA, Western Cape (*van Jaarsveld* 13913 [PRE]). – **D:** RSA (Western Cape). **I:** Pilbeam (1983: 148-149, as *A. aspera*). **Fig. XX.d**

Incl. *Aloe aspera* Salm-Dyck (1854) (*nom. illeg.*, Art. 53.1) ≡ *Haworthia aspera* (Salm-Dyck) Parr (1971) (*nom. illeg.*, Art. 53.1); **incl.** *Astroloba rugosa* Roberts Reinecke (1965) (*nom. inval.*, Art. 29.1); **incl.** *Astroloba muricata* L. E. Groen (1987) (*nom. inval.*, Art. 36.1) ≡ *Haworthia corrugata* (N. L. Meyer & G. F. Smith) M. Hayashi (2000).

L in 5 straight rows or imbricate, spiral angle usually 0° - 10°, light to dark green, usually suberect, tips curved outwards, keel not forming a margin at the apex, 14 - 25 × 11 - 18 mm, mucro 0.4 - 1 mm, surface always tuberculate, tubercles tending to be arranged in longitudinal lines, 0.5 mm ∅, concolorous; **Inf** lax racemes, 10 - 43 cm, unbranched; **Fl** white or cream with faint pink or greenish tinge, lobes cream or whitish, midrib green with beige or pink tinge; **Ped** 2 - 9 mm; **Per** tube 7 - 12 × 2.5 - 3.5 mm ∅, lobes ± 1.5 × 2 mm, tissue on either side of the **OTep** sometimes slightly inflated. – *Cytology:* 2n = 14.

Differing from the other tuberculate-leaved species by the non-marginate leaf tips and a more even, denser distribution of the tubercles, as well as smaller leaves. The names *Astroloba aspera* (Haworth) Uitewaal and *A. aspera* var. *major* have been misapplied to this species, but seem to apply to a species of *Haworthia* on the base of the original descriptions.

A. foliolosa (Haworth) Uitewaal (Succulenta 1947 (5): 54, 1947). **T:** [neo – icono]: Curtis's Bot. Mag., 1811: t. 1352. – **D:** RSA (Eastern Cape). **I:** Pilbeam (1983: 151).

≡ *Aloe foliolosa* Haworth (1804) ≡ *Apicra foliolosa* (Haworth) Willdenow (1811) (*incorrect name*, Art. 11.3) ≡ *Haworthia foliolosa* (Haworth) Haworth (1812) ≡ *Astroloba spiralis* ssp. *foliolosa* (Haworth) Groen (1987).

L 5-ranked or imbricate, spiral angle 10° - 40°, light to grey-green with a glossy sheen, erect to patent, tips curved upwards to outwards, keel not forming a margin at the apex, 14 - 40 × 9 - 24 mm, mucro 0.4 - 1.5 mm, margins and keels concolorous, paler or whitish, surface smooth or with small elongate slightly raised concolorous patches, lower face sometimes with whitish flecks or darker green lines; **Inf** lax racemes, 5 - 29 cm, sometimes branched; **Fl** greenish-white or pale cream with white or cream lobes; **Ped** 0.8 - 3.8 mm; **Per** tube straight, 6 - 9 × ± 3 mm ∅, lobes recurved, 1.5 - 3 mm, midrib green with glaucous or beige tinge. – *Cytology:* 2n = 14.

A. herrei Uitewaal (Desert Pl. Life 20: 37-39, ills., 1948). **T:** RSA, Western Cape (*Herre* 5703 [WAG]). – **D:** RSA (Western Cape: Prince Albert and Uniondale Distr.). **I:** Aloe 13: 99, 1975.

Incl. *Astroloba dodsoniana* Uitewaal (1950) ≡ *Haworthia dodsoniana* (Uitewaal) Parr (1971) ≡ *Astroloba herrei* cv. *Dodsoniana* (Uitewaal *pro sp.*) L. E. Groen (1987); **incl.** *Haworthia harlandiana* Parr (1971).

L in 5 straight rows or imbricate, light green, usually suberect, tips curved upwards to outwards, keel not forming a margin at the apex, 18 - 32 × 9 - 16 mm, narrowly acuminate, mucro 0.7 - 1.8 mm, margins and keels concolorous or paler, lamina with very fine longitudinal lines, smooth; **Inf** lax racemes, 10 - 30 cm, unbranched; **Fl** white with yellow lobes, midribs pale green with glaucous or beige tinge; **Ped** 3.5 - 10.8 mm; **Per** tube with very marked smooth or slightly undulating inflations of tissue on either side of the 3 **OTep**, 7 - 9 × 2.5 - 4 mm ∅, lobes 1.5 - 3 × 1.5 - 3 mm. – *Cytology:* 2n = 14.

Differs from *A. spiralis* in chromosome number and in the smooth or slightly undulating nature of the inflated tissue of the perianth tube. Other differences are the very marked lines, frequently occurring as fine longitudinal ridges, and the narrowly acuminate leaf tips.

A. spiralis (Linné) Uitewaal (Succulenta 1947(5): 53, 1947). **T:** [lecto – icono]: Dillenius, Hort. Eltham., t. 16: fig. 14, 1732. – **D:** RSA (Eastern Cape: Graaff Reinet Distr.; Western Cape: Ladismith and Oudtshoorn Distr.). **I:** Pilbeam (1983: 154). **Fig. XX.g**

≡ *Aloe spiralis* Linné (1753) ≡ *Haworthia spiralis* (Linné) Duval (1809) ≡ *Apicra spiralis* (Linné) Baker (1881); **incl.** *Aloe spiralis* var. *imbricata* Aiton (1789) ≡ *Aloe imbricata* (Aiton) Haworth (1811) ≡ *Apicra imbricata* (Aiton) Willdenow (1811) (*incorrect name*, Art. 11.3) ≡ *Haworthia imbricata* (Aiton) Haworth (1812); **incl.** *Aloe spiralis* var. *pentagona* Aiton (1789) ≡ *Aloe pentagona* (Aiton) Haworth (1804) ≡ *Apicra pentagona* (Aiton) Willdenow (1811) (*incorrect name*, Art. 11.3) ≡ *Haworthia pentagona* (Aiton) Haworth (1812) ≡ *Astroloba pentagona* (Aiton) Uitewaal (1947) ≡ *Astroloba spiralis* cv. 'Pentagona' (Aiton *pro var.*) Groen (1987); **incl.** *Aloe spiralis* Haworth (1804) (*nom. illeg.*, Art. 53.1) ≡ *Aloe pentagona* var. *spiralis* (Haworth) Salm-Dyck (1836) ≡ *Astroloba pentagona* var. *spiralis* (Haworth) Uitewaal (1947) ≡ *Haworthia pentagona* var. *spiralis* (Salm-Dyck) Parr (1971); **incl.** *Haworthia spirella* Haworth (1812) ≡ *Astroloba spirella* (Haworth) hort. (s.a.) (*nom. inval.*, Art. 29.1) ≡ *Aloe spirella* (Haworth) Salm-Dyck (1817) ≡ *Apicra pentagona* var. *spirella* (Haworth) Baker (1881) ≡ *Astroloba pentagona* var. *spirella* (Haworth) Uitewaal (1947) ≡ *Haworthia pentagona* var. *spirella* (Haworth) Parr (1971); **incl.** *Apicra pentagona* var. *torulosa* Haworth (1821) ≡ *Astroloba pentagona* var. *torulosa* (Haworth) Uitewaal (1947) ≡ *Haworthia pentagona* var. *torulosa* (Haworth) Parr (1971); **incl.** *Haworthia gweneana* Parr (1971) (*nom. inval.*, Art. 33.2).

L in 5 straight rows or imbricate, erect to suberect, grey-green to blue-green, tips following the axis of the **L** or curved outwards, keel not forming a margin at the apex, 19 - 42 × 10 - 18 mm, mucro 0.4 - 1.6 mm, margins and keels concolorous or darker, **L** tips slightly reddish-brown, surface occasionally with visible lines, smooth; **Inf** lax racemes, 16 - 39 cm, very rarely branched; **Fl** with a marked rugose inflation of tissue on either side of the 3 **OTep**, white with yellow lobes, midrib pale green with glaucous or beige tinge; **Ped** 1.5 - 8 mm; **Per** tube straight, 7 - 13 × 2.5 - 4 mm ∅, lobes 1.5 × 2 mm. – *Cytology:* 2n = 28.

×ASTROWORTHIA

N. L. Meyer & G. F. Smith

×**Astroworthia** G. D. Rowley (Nation. Cact. Succ. J. 22(3): 74, 1967).

Incl. ×*Apworthia* von Poellnitz (1943).
= *Astroloba* × *Haworthia*.

×**A. bicarinata** (Haworth) G. D. Rowley (Nation. Cact. Succ. J. 28(1): 7, 1973). – **D:** RSA (Western Cape: Montagu District). **I:** Groen (1987: as *A. skinneri*).

≡ *Apicra bicarinata* Haworth (1819) ≡ *Aloe bicarinata* (Haworth) Schultes & Schultes *fil.* (1829) ≡ *Astroloba bicarinata* (Haworth) Uitewaal (1947) ≡ *Haworthia bicarinata* (Haworth) Parr (1971); **incl.** *Apicra skinneri* A. Berger (1908) ≡ *Haworthia skinneri* (A. Berger) Resende (1943) ≡ *Astroloba skinneri* (A. Berger) Uitewaal (1947) ≡ ×*Astroworthia bicarinata* nvar. *skinneri* (A. Berger) G. D. Rowley (1973) ≡ ×*Astroworthia skinneri* (A. Berger) Groen (1987); **incl.** *Apicra bullulata* Uitewaal (1938) (*nom. illeg.*, Art. 53.1); **incl.** *Haworthia olivettiana* Parr (1971).

To 30 cm tall; **L** in 5 straight rows to irregularly imbricate, 33 - 51 × 17 - 25 mm, dull dark green, erect to suberect, tips curved upwards to outwards, tip marginate, mucro 0.3 - 2 mm, keel not always distinct, occasionally doubly keeled, vein lines usually absent, tubercles present, evenly distributed or sometimes in groups, ± 0.5 mm ∅, few and irregularly scattered or more numerous and grouped in transverse rows; **Inf** lax racemes, 12 - 36 cm; **Fl** erect, white to yellowish with pale green midribs; **Ped** 3 - 9 mm; **Per** tube straight, 7 - 13 × 2.5 - 4.5 mm ∅, lobes 1.5 - 3 mm. – *Cytology:* 2n = 14.

This is the naturally occurring hybrid *Astroloba corrugata* × *Haworthia margaretifera*.

×BAYERARA

U. Eggli

×**Bayerara** D. M. Cumming (Haworthiad 13(1): 20, 1999). – **Lit:** Cumming (1999). **Etym:** For M. B. (Bruce) Bayer (*1935), South African succulent plant enthusiast and gardener, and former curator of the Karoo National Botanic Gardens, Worcester, RSA; plus the suffix '-ara' indicating plurigeneric hybrids.

= *Aloe* × *Gasteria* × *Haworthia*. The only known combination (with an interspecific *Aloe* hybrid as one parent) was not formally named.

CHORTOLIRION

G. F. Smith

Chortolirion A. Berger (in Engler, A. (ed.), Pflanzenr. IV. 38 (Heft 33): 72, 1908). **T:** *Haworthia angolensis* Baker. – **Lit:** Smith (1995a). **D:** S Angola, Zimbabwe, Namibia, Botswana, RSA, Lesotho; inland above the Great Escarpment, sparse to dense grasslands, to 2000 m. **Etym:** Gr. 'chortos', feeding place; and Gr. 'lirion', lily; alluding to the usual occurrence of the genus in grassland.

Herbaceous acaulescent perennial; **L** from a short subterranean butt; **R** fusiform, fleshy; bulb ovoid-oblong, formed by slightly fleshy **L**-bases, 3 - 4 cm long, 2 cm ∅; **L** rosulate, slightly succulent, grass-like, flaccid, light green, usually once or twice twisted, ± 15 cm, ± 2 mm ∅, margins denticulate, uppermost 5 - 10 mm of tips often dry; **Inf** racemose, simple, ± 36 cm, lower sterile parts bracteate; **Fl** erect, zygomorphic, **Tep** greenish, brownish or pinkish-white with greenish keels, base obtuse, bud decurved and with pinkish tip; **Ped** short, erect, persistent; **Per** tube straight, 14 mm, ± 2 mm ∅; **Tep** basally united, closely adhering for ⅔, limb bilabiate; **St** 6, inserted at the base of the **Per** tube, ± 7 mm; **Fil** white, tapering towards apex; **Anth** yellow, dorsifixed, dehiscing longitudinally, introrse; **Ov** green, sessile, 3 mm, 2 mm ∅; **Sty** white, straight, capitate, 4 mm; **Fr** 3-locular capsules, cylindrical, apically acute, dehiscing loculicidally, ± 15 mm long, 5 - 6 mm ∅; **Se** angled, dark brown to black, shortly winged, ± 3 mm. – *Cytology:* 2n = 14.

Following a recent revision by Smith (1995a) *Chortolirion* is now a monotypic genus. Morphologically it is quite distinct from *Haworthia*, esp. with regard to the presence of an underground bulbous rootstock, and in its acuminate capsules. Furthermore, it is the only haworthioid species of which the leaves are deciduous and die back to ground-level after fires or frost.

The (invalid) name ×*Gastrolirion orpetii* E. Walther 1933 was erected for a suposed hybrid between *Gasteria sp.* and *Chortolirion tenuifolium*.

C. angolense (Baker) A. Berger (in Engler, A. (ed.), Pflanzenr. IV.38 (Heft 33): 73, 1908). **T:** Angola, Huila (*Welwitsch* 3756 [BM, PRE [photo]]). – **D:** As for the genus. **I:** Smith (1991: 90). **Fig. XX.e**

≡ *Haworthia angolensis* Baker (1878); **incl.** *Haworthia tenuifolia* Engler (1888) ≡ *Chortolirion tenuifolium* (Engler) A. Berger (1908); **incl.** *Haworthia stenophylla* Baker (1891) ≡ *Chortolirion stenophyllum* (Baker) A. Berger (1908); **incl.** *Haworthia saundersiae* Baker (1891) (*nom. nud.*); **incl.** *Haworthia subspicata* Baker (1904) ≡ *Chortolirion subspicatum* (Baker) A. Berger (1908); **incl.** *Chortolirion bergerianum* Dinter (1914).

Description as for the genus.

×CUMMINGARA

U. Eggli

×**Cummingara** G. D. Rowley (Haworthiad 13(3): 115, 1999). – **Lit:** Cumming (1999). **Etym:** For David M. Cumming (fl. 1999), Australian succulent plant enthusiast and breeder; plus the suffix '-ara', indicating plurigeneric hybrids.

Incl. ×*Smithara* D. M. Cumming (1999) (*nom. illeg.*, Art. 53.1).

= *Gasteria* × *Haworthia* × *Poellnitzia*. The 2 known combination have not been named formally.

×GASTERALOE

L. E. Newton

×**Gasteraloe** Guillaumin (Bull. Mus. Nation. Hist. Nat., Sér. 2, 3: 339, 1931). – **Lit:** Walther (1930: as *Gastrolea*); Newton (1998c).
Incl. ×*Gastrolea* E. Walther (1930) (*nom. inval.*, Art. H6.2).
Incl. ×*Lomateria* Guillaumin (1931).
Incl. ×*Chamaeteria* D. M. Cumming (1974).
Incl. ×*Gaslauminia* P. V. Heath (1994).

= *Aloe* × *Gasteria.* Mostly acaulescent or almost so, mostly caespitose; **L** rosulate, variously spotted or tuberculate, margins with teeth; **Inf** lateral, simple or branched; racemes lax; **Fl** bracteate, pedicellate, nutant or spreading, zygomorphic; **Per** tube sometimes inflated at base, often curved above **Ov**, usually reddish at base, greenish towards tip, lobes often longer than tube; **St** and **Sty** scarcely exserted. – *Cytology:* Most of the taxa for which chromosomes have been counted have a somatic number of 2n = 14, but one count of 2n = 28 and two aneuploids have also been reported. These are summarized by Riley & Majumdar (1979).

Rowley (1982) included × *Poellneria* (= *Poellnitzia* × *Gasteria*) as a synonym here, but *Poellnitzia* is an accepted genus distinct from *Aloe*.

In addition to the nothotaxa reported below, numerous other crosses are listed in the literature.

×**G. bedinghausii** (Radl) Guillaumin (Bull. Mus. Nation. Hist. Nat., Sér. 2, 3: 339, 1931).
≡ *Aloe* ×*bedinghausii* Radl (1896) ≡ ×*Gastrolea bedinghausii* (Radl) E. Walther (1930) (*nom. inval.*, Art. 43.1).
= *Aloe aristata* × *Gasteria disticha.* Habit approaching *A. aristata*; **L** 6 × 2 cm, shape and ornamentation approaching *G. disticha.* – *Cytology:* 2n = 14 (Resende 1937: as *Aloe*).

×**G. beguinii** (hort. *ex* Radl) Guillaumin (Bull. Mus. Nation. Hist. Nat., Sér. 2, 3: 339, 1931).
≡ *Aloe* ×*beguinii* hort. *ex* Radl (1896) ≡ ×*Gastrolea beguinii* (hort. *ex* Radl) E. Walther (1930) (*nom. inval.*, Art. 43.1).
= *Aloe aristata* (as *A. longiaristata*) × *Gasteria carinata* var. *verrucosa* (as *G. verrucosa*). Larger and coarser than *A. aristata*, usually simple; **L:** 10 × 2.5 cm, dark green, with prominent scattered white pearly tubercles; **Inf** to 60 cm; raceme lax; **Fl** pink at base, primrose-yellow with green mid-stripe above, 38 mm, tube slightly inflated at base, straight. – *Cytology:* 2n = 14 (Resende 1937: as *Aloe*).

×G. beguinii nvar. **beguinii**

×G. beguinii nvar. **chludowii** (Radl) G. D. Rowley (Nation. Cact. Succ. J. 37(2): 49, 1982).

≡ *Aloe ×chludowii* Radl (1896) ≡ ×*Gastrolea chludowii* (Radl) E. Walther (1930) (*nom. inval.*, Art. 43.1) ≡ ×*Gasteraloe chludowii* (Radl) Guillaumin (1931) ≡ ×*Gastrolea beguinii* nvar. *chludowii* (Radl) G. D. Rowley (1973) (*nom. inval.*, Art. 43.1).

= *Aloe aristata × Gasteria carinata* var. *verrucosa* (as *G. verrucosa* var. *asperrima*). Differs from nvar. *beguinii*: **L** 12 - 16 cm.

×G. beguinii nvar. **perfectior** (Radl) Guillaumin (Bull. Mus. Nation. Hist. Nat., Sér. 2, 3: 339, 1931).

≡ *Aloe ×beguinii* var. *perfectior* Radl (1896) ≡ *Aloe ×perfectior* (Radl) A. Berger (1908) ≡ ×*Gastrolea perfectior* (Radl) E. Walther (1930) (*nom. inval.*, Art. 43.1) ≡ ×*Gastrolea beguinii* var. *perfectior* (Radl) G. D. Rowley (1955) (*nom. inval.*, Art. 43.1).

= *Aloe aristata × Gasteria carinata* var. *verrucosa* (as *G. verrucosa*). Differs from nvar. *beguinii*: To 20 cm tall; **L** longer and brighter green.

×G. derbetzei (hort. *ex* A. Berger) Guillaumin (Bull. Mus. Nation. Hist. Nat., Sér. 2, 3: 339, 1931).

≡ *Aloe ×derbetzei* hort. *ex* A. Berger (1908) ≡ ×*Gastrolea derbetzei* (hort. *ex* A. Berger) E. Walther (1930) (*nom. inval.*, Art. 43.1).

= *Aloe striata × Gasteria acinacifolia.*

×G. gloriosa (Radl) G. D. Rowley (Nation. Cact. Succ. J. 37(2): 49, 1982).

≡ *Aloe hybrida* var. *gloriosa* Radl (1896) ≡ ×*Lomateria gloriosa* (Radl) Guillaumin (1931).

= *Aloe purpurea* (as *Lomatophyllum purpureum*) × *Gasteria bicolor* var. *bicolor* (as *G. maculata*). **L** 40 × 3 cm, dark green with large whitish spots, margin scarcely toothed.

×G. lapaixii (Radl) Guillaumin (Bull. Mus. Nation. Hist. Nat., Sér. 2, 3: 339, 1931). – **Lit:** Rowley (1968).

≡ *Aloe ×lapaixii* Radl (1896) ≡ ×*Gastrolea lapaixii* (Radl) H. Jacobsen (1954) (*nom. inval.*, Art. 43.1).

= *Aloe aristata* (as *A. longiaristata*) × *Gasteria bicolor* var. *bicolor* (as *G. maculata*). Larger and coarser than *A. aristata*, usually simple; **L** rosulate, deltoid-lanceolate, 10 × 2 cm, dark green with a few large whitish spots in irregular transverse bands, margin with blunt white teeth; **Inf** to 1.2 m, simple or with 1 - 3 **Br**; racemes 20 - 25 cm, lax; **Fl** pink with green longitudinal stripes, 25 - 27 mm, base attenuate; **St** not exserted. – *Cytology:* 2n = 14 (Resende 1937: as *Aloe*).

×G. lapaixii nvar. **lapaixii**

×G. lapaixii nvar. **latifolia** (Radl) Guillaumin (Bull. Mus. Nation. Hist. Nat., Sér. 2, 3: 339, 1931).

Incl. *Aloe ×lapaixii* var. *latifolia* Radl (1896).

Differs from nvar. *lapaixii*: **L** broader, slightly acute, margin strongly toothed.

×G. lynchii (Baker) G. D. Rowley (Nation. Cact. Succ. J. 37(2): 49, 1982).

≡ *Aloe ×lynchii* Baker (1881) ≡ ×*Gastrolea lynchii* (Baker) E. Walther (1930) (*nom. inval.*, Art. 43.1).

= *Aloe striata × Gasteria carinata* var. *verrucosa.*

×G. mortolensis (A. Berger) Guillaumin (Bull. Mus. Nation. Hist. Nat., Sér. 2, 3: 339, 1931).

≡ *Aloe ×mortolensis* A. Berger (1908) ≡ ×*Gastrolea mortolensis* (A. Berger) E. Walther (1930) (*nom. inval.*, Art. 43.1).

= *Aloe ?variegata × Gasteria acinacifolia.* Caulescent; stem procumbent, 40 - 60 cm, suckering at base; **L** rosulate, deltoid-lanceolate, 25 × 9 cm, glossy green, with rounded white spots in irregular transverse bands, more numerous on lower face, margins with teeth 2 mm, 10 - 18 mm apart; **Inf** simple or with up to 3 **Br**; racemes 30 cm, lax; **Bra** cuspidate, 2.5 cm; **Ped** 3 - 3.5 cm; **Fl** 5.8 - 6 cm, base orange, pale brownish-green or grey towards tip, base not inflated; **St** not exserted.

×G. nowotnyi (Radl) G. D. Rowley (Nation. Cact. Succ. J. 37(2): 49, 1982).

≡ *Aloe ×nowotnyi* Radl (1896) ≡ ×*Gastrolea nowotnyi* (Radl) E. Walther (1930) (*nom. inval.*, Art. 43.1).

= *Aloe aristata* (as *A. longiaristata*) × *Gasteria sp.* **L** rosulate, deltoid, 3.5 × 2 - 2.5 cm, bright green with spots, fewer on upper face. – *Cytology:* 2n = 20 (Brandham 1969).

×G. peacockii (Baker) G. D. Rowley (Nation. Cact. Succ. J. 37(2): 49, 1982).

≡ *Gasteria ×peacockii* Baker (1880) ≡ ×*Gastrolea peacockii* (Baker) E. Walther (1930) (*nom. inval.*, Art. 43.1).

= *Aloe sp.* (as *A. heteracantha*) × *Gasteria acinacifolia* (as *G. acinacifolia* var. *ensifolia*).

×G. pethamensis (Baker) G. D. Rowley (Nation. Cact. Succ. J. 37(2): 49, 1982). – **I:** Walther (1930: as *Gastrolea*).

≡ *Gasteria pethamensis* Baker (1880) ≡ ×*Gastrolea pethamensis* (Baker) E. Walther (1930) (*nom. inval.*, Art. 43.1).

= *Aloe variegata × Gasteria carinata* var. *verrucosa.*

×G. pfrimmeri Guillaumin (Bull. Mus. Nation. Hist. Nat., Sér. 2, 3: 339-340, 1931).

≡ ×*Gastrolea pfrimmeri* (Guillaumin) E. Walther (1933) (*nom. inval.*, Art. 43.1).

= *Aloe variegata × Gasteria sp.* **L** 10 in dense **Ros** 18 cm ∅ × 7 cm high, deltoid, upper face concave, lower face obliquely keeled, tip acute, to 10 × 4 cm, green with numerous irregularly confluent white spots, denser on upper face; **Inf** to 30 cm, branching; racemes 8 cm, dense; **Bra** 5 mm; **Fl** red with green markings on lobes, 3 cm; **St** not exserted.

×G. prorumpens (A. Berger) G. D. Rowley (Nation. Cact. Succ. J. 37(2): 49, 1982).
≡ *Aloe* ×*prorumpens* A. Berger (1908) ≡ ×*Gastrolea prorumpens* (A. Berger) E. Walther (1930) (*nom. inval.*, Art. 43.1).
= *Aloe sp.* × *Gasteria sp.*

×G. quehlii (Radl) G. D. Rowley (Nation. Cact. Succ. J. 37(2): 49, 1982).
≡ *Aloe* ×*quehlii* Radl (1896) ≡ ×*Gastrolea quehlii* (Radl) E. Walther (1930) (*nom. inval.*, Art. 43.1).
= *Aloe sp.* × *Gasteria bicolor* var. *bicolor* (as *G. maculata* or *G. picta* var. *formosa*). **L** deltoid, 4 cm wide, with a few spots; **Inf** 60 - 70 cm, branching; racemes 30 - 40 cm; **Fl** 25 mm.

×G. radlii L. E. Newton (Brit. Cact. Succ. J. 15(1): 34, 1997).
Incl. *Aloe* ×*imbricata* hort. *ex* A. Berger (1908) (*nom. illeg.*, Art. 53.1) ≡ ×*Gastrolea imbricata* (hort. *ex* A. Berger) E. Walther (1930) (*nom. inval.*, Art. 43.1) ≡ ×*Gasteraloe imbricata* (hort. *ex* A. Berger) G. D. Rowley (1982).
= *Aloe variegata* or *A. serrulata* × *Gasteria sp.* **L** rosulate, in 5 rows, lanceolate-deltoid with shortly mucronate tip, 6 × 2 cm, green, margin cartilaginous, warty.

×G. rebutii (hort. *ex* A. Berger) Guillaumin (Bull. Mus. Nation. Hist. Nat., Sér. 2, 3: 340, 1931).
≡ *Aloe rebutii* hort. *ex* A. Berger (1908) ≡ ×*Gastrolea rebutii* (hort. *ex* A. Berger) E. Walther (1930) (*nom. inval.*, Art. 43.1).
= *Aloe variegata* × *Gasteria sp.* **L** spiral in 5 rows, 13 - 14 × 2.5 - 3.5 cm, green with numerous spots in irregular transverse bands, margin with minute teeth; **Inf** with raceme 35 cm; **Fl** subsecund, 33 - 35 mm, base slightly inflated to 9 mm ∅; **St** not exserted. – *Cytology:* 2n = 14 (Brandham 1969: as *Gasteria*).

×G. sculptilis G. D. Rowley *ex* L. E. Newton (Haseltonia 5: 94, 1998). **T:** [icono]: Cact. Succ. J. (US) 7)9): 136, 1936. – **I:** Poindexter (1936).
Incl. ×*Gastrolea sculptilis* Poindexter (1936) (*nom. inval.*, Art. 36.1, 43.1).
= *Aloe variegata* × *Gasteria* ×*cheilophylla*. Acaulescent or shortly caulescent, suckering sparsely; **L** with groove on upper face, with keel on lower face, ± 18 × 8.5 cm, olive-green with numerous whitish spots, mostly in irregular transverse bands, margin white, denticulate; **Inf** ± 50 cm, branched; **Fl** pale red with yellowish tip, 3 cm. – *Cytology:* 2n = 14 (Cutler & Brandham 1977).

The plant was described as sterile by Poindexter (1936).

×G. simoniana (Deleuil) Guillaumin (Bull. Mus. Nation. Hist. Nat., Sér. 2, 3: 340, 1931).
≡ *Aloe* ×*simoniana* Deleuil (1893) ≡ ×*Gastrolea simoniana* (Deleuil) H. Jacobsen (1954) (*nom. inval.*, Art. 43.1).
= *Aloe aristata* × *Gasteria disticha*. Plants to 40 cm tall; **L** 15 - 20 × 5 - 6 cm, green with numerous spots.

×G. smaragdina (hort. *ex* A. Berger) Guillaumin (Bull. Mus. Nation. Hist. Nat., Sér. 2, 3: 340, 1931).
≡ *Aloe* ×*smaragdina* hort. *ex* A. Berger (1908) ≡ ×*Gastrolea smaragdina* (hort. *ex* A. Berger) E. Walther (1930) (*nom. inval.*, Art. 43.1).
= *Aloe variegata* × *Gasteria acinacifolia* (as *G. ?candicans*). Suckering freely at base; **L** ± 15, rosulate, lanceolate-deltoid, 20 × 7 cm, pale green, glossy, with whitish-green spots in irregular transverse bands, margin with minute teeth; **Inf** simple or with few **Br**; raceme 20 - 30 cm, lax; **Bra** deltoid, cuspidate; **Ped** ± 1.2 cm; **Fl** bright red, 4 cm, base not inflated; **St** not exserted. – *Cytology:* 2n = 14 (Brandham 1969).

×GASTERHAWORTHIA

U. Eggli

×Gasterhaworthia Guillaumin (Bull. Mus. Nation. Hist. Nat., Sér. 2, 3: 339, 1931).
= *Gasteria* × *Haworthia*. Rowley (1982: 76) lists 5 known combinations (1 cultivar, 4 formally named).

GASTERIA

E. van Jaarsveld

Gasteria Duval (Pl. Succ. Horto Alencon., 6, 1809). **T:** *Aloe angustifolia* Aiton [Lectotype, selected by Maire, Fl. Afr. Nord. 5: 71, 1958]. – **Lit:** Jaarsveld (1992a); Jaarsveld & Ward-Hilhorst (1994). **D:** S Namibia, RSA (mostly below the escarpment, mainly in the Eastern Cape); mainly in subtropical thickets. **Etym:** Gr. 'gaster', stomach; for the stomach-shaped basally inflated perianth.
Incl. *Papilista* Rafinesque (1840). **T:** *Aloe verrucosa* Miller.

Slow-growing perennial glabrous **L** succulents; **R** mostly terete, slightly succulent to somewhat clavate; acaulescent or **Br** short, rarely pendent and long-stemmed, solitary or proliferating from base to form dense groups; **L** ± brittle, mottled, often with dense white spots in transverse bands, dimorphic, distichous in juvenile phase becoming rosulate (sometimes remaining distichous); juvenile **L** lorate, flat above, convex below, adult **L** linear to lorate-lanceolate to triangular, margin acute, rounded, wavy or rugulose, sometimes tuberculate-crenulate with white tubercles ± merging towards the apex, forming a continuous margin which may be entire, serrulate, crenulate or rarely denticulate; epidermis asperulous, tuberculate, rugulose or smooth; **L** tip truncate, obtuse, acute or acuminate, rarely retuse, mucronate; **Inf** terminal but pushed aside by the **L** and thus appearing lateral, 1 - 5 per plant, racemose or paniculate, ± secund; **Fl** 6-merous, pendent, pink to reddish (rarely nearly white), apically white with green striae, laxly arranged in the upper ½; **Tep** fused for the greater part of their length forming a tubular curved basally asymmetrically gasteriform **Cl**, gasteriform part variously shaped, extending for up to ⅔ of **Per** length; **St** 6, included or rarely exserted, **Fil** becoming contracted after anthesis; **Ov** oblong-ovoid, with 3 locules, with 6 grooves on the outside; **Sty** becoming shortly exserted after anthesis; **Sty** apical, minute, 3-lobed; **Fr** erect oblong 3-angled woody capsules, dehiscing longitudinally; **Se** black, compressed, irregularly angled and obscurely oblong in outline, 2 - 8 mm.

The genus as recently monographed by Jaarsveld & Ward-Hilhorst (1994) is divided into 2 sections, each with 2 series:

[-] Sect. *Gasteria*: **L** lorate, smoooth or asperulous (rarely tuberculate), distichous or spirally distichous; **Per** 12 - 50 mm, gasteriform portion globose or globose-ellipsoid (rarely narrowly ellipsoid):

[1] Ser. *Gasteria*: **Per** 12 - 22 (-25) mm, gasteriform part > ½ of **Per** length. 6 species from the Lower Karoo regions and the Eastern Cape.

[2] Ser. *Namaquana* van Jaarsveld 1992: **Per** 25 - 45 (-50) mm, gasteriform part ≤ ⅓ of **Per** length, globose but often ± indistinct from upper part of **Per** and of ± the same ∅. Only *G. pillansii* from the Namaqualand.

[-] Sect. *Longiflorae* Haworth 1827: **L** tapering, smooth or tuberculate (rarely asperulous), rosulate; **Per** 18 - 45 (-50) mm, gasteriform portion narrowly ellipsoid:

[3] Ser. *Longifoliae* (Haworth) van Jaarsveld 1992: **Per** 35 - 50 mm, hardly gasteriform. 3 species from the Eastern Cape, Transkei, Swaziland and Mpumalanga.

[4] Ser. *Multifariae* (Haworth) van Jaarsveld 1992: **Per** 18 - 33 mm, gasteriform part ± well defined. 6 species from the Western Cape, Eastern Cape and the Transkei.

Gasteria are popular indoor horticultural subjects. Most require some form of shading and they are easily propagated by off-shoots, leaf cuttings or seed which should be sown when fresh.

The following names are of unresolved application but are referred to this genus: *Aloe linguiformis* De Candolle (s.a.); *Aloe trigona* Salm-Dyck (s.a.) ≡ *Gasteria trigona* (Salm-Dyck) Haworth (s.a.); *Gasteria acinacifolia* var. *spathulata* A. Berger (s.a.); *Gasteria angulata* var. *truncata* (Willdenow) A. Berger (s.a.); *Gasteria brevifolia* Haworth (s.a.); *Gasteria brevifolia* var. *laetevirens* Haworth (s.a.); *Gasteria brevifolia* var. *perviridis* Haworth (s.a.); *Gasteria crassifolia* Haworth (s.a.); *Gasteria dicta* N. E. Brown (1876); *Gasteria elongata* Baker (s.a.); *Gasteria fasciata* (Salm-Dyck) Haworth (s.a.); *Gasteria fasciata* var. *laxa* Haworth (s.a.); *Gasteria formosa* Haworth (s.a.); *Gasteria gracilis* Baker (s.a.); *Gasteria laevis* (Salm-Dyck) Haworth (s.a.); *Gasteria linita* Haworth (s.a.); *Gasteria nigricans* var. *fasciata* (Salm-Dyck) Haworth (1821); *Gasteria nigricans* var. *guttata* (Salm-Dyck) Baker (s.a.); *Gasteria nigricans* var. *polyspila* Baker (1880) ≡ *Gasteria fasciata* var. *polyspila* (Baker) A. Berger (1908); *Gasteria prolifera* Lemaire (s.a.); *Gasteria subnigricans* Haworth (1827) ≡ *Aloe subnigricans* (Haworth) Sprengel (1828) ≡ *Gasteria nigricans* var. *subnigricans* (Haworth) Baker (1880); *Gasteria subnigricans* var. *canaliculata* Salm-Dyck (s.a.); *Gasteria subnigricans* var. *glabrior* Haworth (1827) ≡ *Gasteria pseudonigricans* var. *glabrior* (Haworth) H. Jacobsen (1955); *Gasteria transvaalensis* Hort. de Smet *ex* Baker (s.a.).

G. acinacifolia (Jacquin) Haworth (Suppl. Pl. Succ., 49, 1819). **T:** [lecto − icono]: Jacquin, Eclog. Pl. Rar. 49, t. 31, 1811-1816. − **D:** RSA (Eastern Cape); coastal dune thickets, flowers spring to mid-summer. **I:** Jaarsveld & Ward-Hilhorst (1994: 40).

≡ *Aloe acinacifolia* Jacquin (1811); **incl.** *Gasteria acinacifolia* var. *acinacifolia*; **incl.** *Gasteria nitens* Haworth (1819) ≡ *Gasteria acinacifolia* var. *nitens* (Haworth) Baker (s.a.) ≡ *Aloe nitens* (Haworth) Roemer & Schultes (1829); **incl.** *Gasteria candicans* Haworth (1821) ≡ *Aloe candicans* (Haworth) Roemer & Schultes (1829); **incl.** *Gasteria ensifolia* Haworth (1825) ≡ *Aloe ensifolia* (Haworth) Roemer & Schultes (1829) ≡ *Gasteria acinacifolia* var. *ensifolia* (Haworth) Baker (1880); **incl.** *Gasteria pluripuncta* Haworth (1827) ≡ *Aloe pluripuncta* (Haworth) Roemer & Schultes (1829) ≡ *Gasteria acinacifolia* var. *pluripuncta* (Haworth) Baker (1896); **incl.** *Gasteria venusta* Haworth (1827) ≡ *Aloe venusta* (Haworth) Roemer & Schultes (1829) ≡ *Gasteria acinacifolia* var. *venusta* (Haworth) Baker (1896); **incl.** *Gasteria inexpectata* von Poellnitz (1938).

[4] Acaulescent, decumbent to erect, 25 - 75 × 65 cm, solitary or proliferating to form small groups; **L** rosulate, 22 - 60 × 4.5 - 10 cm, linear-lanceolate to lorate, erectly spreading and sometimes falcate, keeled, both surfaces dark green, with dense white spots arranged in transverse bands, epidermis smooth, rarely slightly tuberculate, margin cartilaginous, serrulate, rarely entire, tip acute, rarely obtuse, mucronate; juvenile **L** distichous, lorate,

patent to erectly spreading, tuberculate, rarely smooth; **Inf** variable, racemose, usually flat-topped panicles to 1 m; side **Br** horizontal to erectly spreading; **Per** 35 - 45 (-50) mm, gasteriform part narrowly ellipsoid, ½ of the **Per** length, 5 - 9 mm ∅, often not constricted and indistinct, **Per** pink, upper ½ white with green striations; **Sty** included or exserted for up to 5 mm; **Fr** 35 - 43 mm, truncate or obtuse; **Se** 6 - 8 × 5 - 6 mm.

G. batesiana G. D. Rowley (Nation. Cact. Succ. J. 10(2): 32, 1955). **T** [neo]: RSA, Mpumalanga (*Rowley* s.n. [RNG]). – **D**: RSA (Mpumalanga, KwaZulu-Natal); flowers spring to mid-summer. **Fig. XXI.a**

[4] Acaulescent, decumbent to erect, 3 - 10 × 8 - 30 cm, proliferating from the base to form small to large groups, rarely solitary; **L** distichous at first, becoming rosulate, 5 - 18 × 1.5 - 4 cm, triangular-lanceolate to linear, erectly spreading, becoming recurved, dark green with white spots densely arranged in transverse bands, densely rugulose-tuberculate, margin cartilaginous, serrulate (rarely denticulate), tip acute, rarely obtuse, mucronate; juvenile **L** lorate, densely tuberculate, tip obtuse, mucronate; **Inf** racemose, 30 - 45 cm; **Bra** 6 - 12 × 2 - 5 mm; **Ped** 9 mm; **Per** 35 - 40mm long, stipitate for 3 - 5 mm, gasterifom part narrowly ellipsoid, ½ of **Per** length, 6 - 9 mm ∅, light pink, upper ½ of **Per** white with green striations, inflated to the same ∅ as the lower portion, with a slight constriction in the middle, **Per** tips obtuse, white with green median stripes; **Sti** included or exserted for up to 5 mm; **Fr** 16 - 20 mm; **Se** 4 - 6 × 2 - 3 mm.

G. batesiana var. **batesiana** – **D**: RSA (Mpumalanga, KwaZulu-Natal); S-facing cliff faces in river valleys, Bushveld. **I**: Jaarsveld & Ward-Hilhorst (1994: 36).

[4] Description as for the species.

G. batesiana var. **dolomitica** van Jaarsveld & E. A. van Wyk (Aloe 36(4): 74, 2000). **T**: RSA, Mpumalanga (*van Jaarsveld & Hankey* 15081 [NBG]). – **D**: RSA (Mpumalanga); sheer dolomite cliffs.

[4] Differs from var. *batesiana*: **L** linear, 10 × 1 - 2 cm, becoming biconvex when turgid, tip obtuse.

G. baylissiana Rauh (JSAB 43(3): 187-191, 1977). **T**: RSA, Eastern Cape (*Bayliss* s.n. [HEID 30517, PRE]). – **D**: RSA (Eastern Cape); thickets on quartzitic sandstones, spring-flowering. **I**: Jaarsveld & Ward-Hilhorst (1994: 72).

[1] Acaulescent, decumbent to erect, 0.5 - 4 cm tall, proliferating from the base to form small dense groups to 8 cm ∅, rarely solitary; **L** distichous, 2.5 - 5.5 × 2 - 2.3 cm, lorate, erectly spreading, often becoming patent or recurved, epidermis with dense white cartilaginous tubercles, these very dense, domed to globose and confluent forming a dense

reticulation, margin crenulate, becoming continuous towards tip, tip obtuse, truncate or retuse, mucronate; juvenile **L** lorate, tuberculate and slightly asperulous; **Inf** racemose, 8 - 35 cm, erectly spreading, occasionally with a pair of side **Br**; **Per** 14 - 16 mm, stipitate for 1 mm, gasteriform part ⅔ of **Per** length, red-pink, 6 - 7.5 mm ∅, then abruptly constricted into a tube 3 - 4 mm ∅; **Cl** tube white with green striations; **Sty** included; **Fr** 14 - 20 mm; **Se** oblong, 4 × 3 mm.

G. bicolor Haworth (Philos. Mag. J. 1826: 275, 1826). **T**: [neo – icono]: Salm-Dyck, Monogr. Gen. Aloes & Mesembr., 29, t. 5, 1836-1863. – **D**: RSA.

G. bicolor var. **bicolor** – **D**: RSA (Eastern Cape); subtropical thickets, flowers in spring and summer. **I**: Jaarsveld & Ward-Hilhorst (1994: 68). **Fig. XXI.b, XXI.c**

Incl. *Gasteria maculata* var. *maculata*; **incl.** *Gasteria spiralis* Baker (s.a.); **incl.** *Aloe maculata* Thunberg (1785) (*nom. illeg.*, Art. 53.1) ≡ *Gasteria maculata* (Thunberg) Haworth (1827); **incl.** *Aloe maculata* var. *obliqua* Aiton (1789); **incl.** *Aloe obliqua* Haworth (1802) ≡ *Gasteria obliqua* (Haworth) Duval (1809); **incl.** *Aloe lingua* Ker (1807); **incl.** *Gasteria maculata* var. *fallax* Haworth (1827) ≡ *Aloe obliqua* var. *fallax* (Haworth) Roemer & Schultes (1829); **incl.** *Gasteria picta* Haworth (1827); **incl.** *Gasteria retata* Haworth (1827); **incl.** *Aloe dictyodes* Roemer & Schultes (1829); **incl.** *Aloe zeyheri* Salm-Dyck (1836) ≡ *Gasteria zeyheri* (Salm-Dyck) Baker (1880); **incl.** *Aloe planifolia* Baker (1870) (*nom. illeg.*, Art. 53.1) ≡ *Gasteria planifolia* (Baker) Baker (1880); **incl.** *Gasteria variolosa* Baker (1871); **incl.** *Gasteria colubrina* N. E. Brown (1877); **incl.** *Gasteria marmorata* Baker (1880); **incl.** *Gasteria spiralis* var. *tortulata* Baker (1880); **incl.** *Gasteria maculata* var. *dregeana* A. Berger (1908); **incl.** *Gasteria caespitosa* von Poellnitz (1937); **incl.** *Gasteria chamaegigas* von Poellnitz (1937); **incl.** *Gasteria herreana* von Poellnitz (1938); **incl.** *Gasteria longiana* von Poellnitz (1938); **incl.** *Gasteria longibracteata* von Poellnitz (1938); **incl.** *Gasteria salmdyckiana* von Poellnitz (1938); **incl.** *Gasteria biformis* von Poellnitz (1940); **incl.** *Gasteria kirsteana* von Poellnitz (1940); **incl.** *Gasteria loeriensis* von Poellnitz (1940); **incl.** *Gasteria multiplex* von Poellnitz (1940).

[1] Decumbent to erect, 8 - 50 cm tall, with a short foliated stem to 20 cm, proliferating from the base to form small groups; **L** distichous or rosulate, 8 - 40 × 1.5 - 6 cm, lorate to linear, erectly spreading, slightly falcate and twisted sideways when distichous, with an asymmetrical keel when spirally arranged, dark green and with dense white spots arranged in obscure transverse bands, epidermis smooth, rarely slightly asperulous, margin entire, cartilaginous, serrulate, tubercles becoming conflu-

ent towards tip, tip obtuse, rarely acuminate, with an asymmetrical mucro; juvenile **L** patent or erectly spreading, lorate, asperulous, obtuse, mucronate; **Inf** 0.1 - 1.5 m, rarely simple or mostly branched from the middle, with up to 8 erectly spreading side **Br**; **Per** 12 - 20 mm, stipitate for 1 - 2 mm, gasteriform part > ½ of the **Per** length, light pink (rarely white), globose to globose-ellipsoid, 6 - 9 mm ∅, then abruptly constricted into a tube 3 - 4 mm ∅; **Cl** tube white with green striations; **Fr** 10 - 25 × 6 - 10 mm; **Se** oblong to rectangular, 2 - 4 mm.

A very variable taxon with many local forms. The varieties grade into each other.

G. bicolor var. **liliputana** (von Poellnitz) van Jaarsveld (Aloe 29(1): 21, 1992). **T:** RSA, Eastern Cape (*Dyer & Britten* 508 [PRE]). – **D:** RSA (Eastern Cape); subtropical thickets, spring-flowering. **Fig. XXI.d**

≡ *Gasteria liliputana* von Poellnitz (1938).

[1] Differs from var. *bicolor*: **L** distichous or rosulate, 1.5 - 10 × 0.8 - 1.4 cm, epidermis smooth, rarely slightly tuberculate or somewhat asperulous, tip obtuse or acute; **Inf** simple racemes, 16 - 40 cm, occasionally with a pair of side **Br**; **Per** 12 - 15 mm, ∅ of the gasteriform part variable.

G. brachyphylla (Salm-Dyck) van Jaarsveld (Aloe 29(1): 19, 1992). **T:** [lecto – icono]: Salm-Dyck, Monogr. Gen. Aloes & Mesembr. 29, t. 8. – **D:** RSA.

≡ *Aloe brachyphylla* Salm-Dyck (1840).

G. brachyphylla var. **bayeri** van Jaarsveld (Aloe 29(1): 20, fig. 21 (p. 19), 1992). **T:** RSA, Western Cape (*Bayer* 1751 [NBG]). – **D:** RSA (Western Cape: Little Karoo); Succulent Karoo, flowers late spring to mid-summer.

[1] Differs from var. *brachyphylla*: **L** shorter, 1.5 - 5 × 2.2 - 2.8 cm, at first erect, becoming patent and recurved, tips often incurved, truncate, epidermis smooth or slightly asperulous; **Inf** racemose, 25 - 28 cm; **Per** 18 mm.

G. brachyphylla var. **brachyphylla** – **D:** RSA (Western Cape); Succulent Karoo, spring-flowering. **I:** Jaarsveld & Ward-Hilhorst (1994: 64).

Incl. *Aloe nigricans* var. *marmorata* Salm-Dyck (1821); **incl.** *Gasteria nigricans* var. *marmorata* Haworth (1821); **incl.** *Gasteria nigricans* var. *platyphylla* Baker (1880); **incl.** *Gasteria angustiarum* von Poellnitz (1937); **incl.** *Gasteria triebneriana* von Poellnitz (1938); **incl.** *Gasteria joubertii* von Poellnitz (1940); **incl.** *Gasteria vlaaktensis* von Poellnitz (1940).

[1] Acaulescent, decumbent to erect, 9 - 23 × 7.5 - 23 cm, proliferating from the base to form small groups; **L** distichous, 8.5 - 23 × 2.2 - 8 cm, lorate, rarely triangular-lanceolate, epidermis smooth, dark green, both faces with dense white spots arranged in obscure transverse bands, margin crenulate, becoming continuous towards tip, tip acute, obtuse or truncate; juvenile **L** lorate, asperulous, densely spotted, obtuse; **Inf** racemose, simple or with a pair of side **Br**, 0.2 - 1.1 m; **Per** 12 - 22 mm, stipitate for 2 - 3 mm, gasteriform part > ½ of **Per** length, pink, 5 - 7 mm ∅, globose or globose-ellipsoid, then constricted into a tube 3 - 4 mm ∅, white with green striations; **Sty** 7 - 10 mm, included; **Fr** oblong, 15 - 23 × 7 mm; **Se** 3 - 4 × 2 - 3 mm.

G. carinata (Miller) Duval (Pl. Succ. Horto Alencon., [], 1809). **T:** [lecto – icono]: Commelin, Hort. Med. Amstel., t. 9, 1701. – **Lit:** Jaarsveld (1998). **D:** RSA (Western Cape).

≡ *Aloe carinata* Miller (1768).

[4] Acaulescent, decumbent to erect, 3 - 18 cm tall, proliferating from the base to form small dense groups 15 - 80 cm ∅; **L** distichous at first, becoming rosulate, 3 - 12 × 1 - 5 cm, triangular to triangular-lanceolate, erectly spreading, lower **L** spreading, both faces tuberculate or smooth, rarely asperulous, spotted with raised or immersed white domed tubercles in obscure transverse bands, margin cartilaginous, tuberculate-crenulate, rarely denticulate, tip acute, rarely obtuse, truncate or retuse, mucronate; juvenile **L** distichous, erectly spreading, tuberculate or smooth, lorate; **Inf** racemose, 15 - 90 cm, occasionally with a pair of side **Br**; **Per** variable, 16 - 27 mm, gasteriform part > ½ of **Per** length, pink, narrowly ellipsoid to rarely globose-ellipsoid, above constricted into a tube 3 - 5 mm ∅; **Per** tips light pink to white with central green stripes; **St** oblong, included; **Ov** 6 - 7 × 2.5 mm; **Sty** 14 mm; **Fr** 19 - 23 × 7 mm; **Se** oblong, 3 - 4 × 2 mm.

G. carinata var. **carinata** – **D:** RSA (Western Cape); Renosterveld and thickets on shale or quartz, spring-flowering. **I:** Jaarsveld & Ward-Hilhorst (1994: 54).

Incl. *Aloe tristicha* Medikus (1786); **incl.** *Aloe carinata* var. *subglabra* Haworth (1804); **incl.** *Aloe lingua* var. *angulata* Haworth (1804); **incl.** *Aloe lingua* var. *multifaria* Haworth (1804); **incl.** *Aloe angulata* Willdenow (1811) ≡ *Gasteria angulata* (Willdenow) Haworth (1827); **incl.** *Aloe excavata* Willdenow (1811) ≡ *Gasteria excavata* (Willdenow) Haworth (1827); **incl.** *Aloe obscura* var. *truncata* Salm-Dyck (1817); **incl.** *Aloe subcarinata* Salm-Dyck (1817) ≡ *Gasteria subcarinata* (Salm-Dyck) Haworth (1819); **incl.** *Aloe sulcata* Salm-Dyck (1821) ≡ *Gasteria sulcata* (Salm-Dyck) Haworth (1827); **incl.** *Gasteria laetepuncta* Haworth (1827) ≡ *Aloe laetepuncta* (Haworth) Roemer & Schultes (1829); **incl.** *Gasteria parva* Haworth (1827) ≡ *Gasteria carinata* var. *parva* (Haworth) Baker (1896); **incl.** *Gasteria strigata* Haworth (1827) ≡ *Gasteria carinata* var. *strigata* (Haworth) Baker (1896); **incl.** *Aloe carinata* var. *laevior*

Salm-Dyck (1836); **incl.** *Gasteria pallescens* Baker (1880); **incl.** *Gasteria parvifolia* Baker (1880); **incl.** *Gasteria porphyrophylla* Baker (1880); **incl.** *Gasteria carinata* var. *falcata* A. Berger (1908); **incl.** *Gasteria carinata* var. *latifolia* A. Berger (1908); **incl.** *Gasteria trigona* var. *kewensis* A. Berger (1908); **incl.** *Gasteria humilis* von Poellnitz (1929); **incl.** *Gasteria bijliae* von Poellnitz (1937); **incl.** *Gasteria schweickerdtiana* von Poellnitz (1938); **incl.** *Gasteria patentissima* von Poellnitz (1940).

[4] **L** first distichous, becoming rosulate or spirally arranged, 3 - 18 × 2.5 - 10 cm, triangular to triangular-lanceolate with a distinct keel, epidermis with raised white tubercles, tip acute or obtuse.

Differs from the other varieties by the triangular to triangular-lanceolate tuberculate and keeled leaves.

G. carinata var. **glabra** (Salm-Dyck) van Jaarsveld (CSJA 70(2): 70, ill. (p. 68), 1998). **T:** [neo − icono]: Salm-Dyck, Monogr. Gen. Aloes & Mesembr., t. 29. − **D:** RSA (Western Cape: E of the Gouritz River, around Mossel Bay etc.). **Fig. XXI.g**
≡ *Aloe glabra* Salm-Dyck (1817) ≡ *Gasteria glabra* (Salm-Dyck) Haworth (1827).
[4] Differs from var. *carinata*: **L** 3 - 18 × 2.5 - 10 mm, without tubercles.

G. carinata var. **retusa** van Jaarsveld (Aloe 29(1): 15, ill., 1992). **T:** RSA, Western Cape (*van Jaarsveld & Stayner* 4656 [NBG]). − **D:** RSA (Western Cape); Succulent Karoo on shale, spring-flowering. **Fig. XXI.f**
[4] Differs from var. *carinata*: **L** distichous, 5 - 9 × 2.5 - 3.5 cm, lorate, with raised semitranslucent white dome-shaped tubercles, tip retuse; **Inf** 30 - 45 cm; **Per** variable, gasteriform part narrowly to globosely ellipsoid.

G. carinata var. **thunbergii** (N. E. Brown) van Jaarsveld (CSJA 70(2): 4, ill., 1998). **T:** RSA, Western Cape (*Thunberg* 8595 [UPS]). − **D:** RSA (Western Cape, Gouritz River Valley); subtropical thickets on conglomerate, flowers mainly in autumn.
≡ *Gasteria thunbergii* N. E. Brown (1923); **incl.** *Aloe disticha* Thunberg (1785) (*nom. illeg.*, Art. 53.1).
[4] Differs from var. *carinata*: **L** remaining distichous, linear, acuminate, deeply channelled, 5 - 18 × 1 - 1.5 cm, with white and green tubercles in transverse bands.

G. carinata var. **verrucosa** (Miller) van Jaarsveld (Aloe 29(1): 15, 1992). **T:** [lecto − icono]: Boerhaave, Index Alter Pl., t. 2, p. 131 (no. 36), 1720. − **D:** RSA (Western Cape); Fynbos on limestone, mainly spring-flowering. **I:** Jaarsveld & Ward-Hilhorst (1994: 56).

≡ *Aloe verrucosa* Miller (1768) ≡ *Aloe linguiformis* var. *verrucosa* (Miller) De Candolle (1799) ≡ *Gasteria verrucosa* (Miller) Duval (1809); **incl.** *Gasteria intermedia* var. *longior* Haworth (s.a.); **incl.** *Gasteria subverrucosa* var. *marginata* Baker (s.a.); **incl.** *Aloe racemosa* Lamarck (1783); **incl.** *Aloe carinata* De Candolle (1799) (*nom. illeg.*, Art. 53.1); **incl.** *Aloe intermedia* Haworth (1804); **incl.** *Aloe lingua* Ker (1810) (*nom. illeg.*, Art. 52.1); **incl.** *Gasteria intermedia* Haworth (1812) ≡ *Gasteria verrucosa* var. *intermedia* (Haworth) Baker (1880); **incl.** *Aloe subverrucosa* Salm-Dyck (1817) ≡ *Gasteria subverrucosa* (Salm-Dyck) Haworth (1827); **incl.** *Aloe verrucosa* var. *striata* Salm-Dyck (1817) ≡ *Gasteria verrucosa* var. *striata* (Salm-Dyck) von Poellnitz (1938); **incl.** *Aloe subverrucosa* var. *grandipunctata* Salm-Dyck (1821) ≡ *Gasteria subverrucosa* var. *grandipunctata* (Salm-Dyck) Haworth (1827); **incl.** *Aloe subverrucosa* var. *parvipunctata* Salm-Dyck (1821) ≡ *Gasteria subverrucosa* var. *parvipunctata* (Salm-Dyck) Haworth (1827); **incl.** *Aloe verrucosa* var. *latifolia* Salm-Dyck (1821) ≡ *Gasteria verrucosa* var. *latifolia* (Salm-Dyck) Haworth (1821); **incl.** *Gasteria repens* Haworth (1821); **incl.** *Aloe intermedia* var. *asperrima* Salm-Dyck (1821) (*nom. inval.*, Art. 43.1) ≡ *Gasteria intermedia* var. *asperrima* (Salm-Dyck) Haworth (1821) (*nom. inval.*, Art. 43.1) ≡ *Gasteria verrucosa* var. *asperrima* (Salm-Dyck) von Poellnitz (1938) (*nom. inval.*, Art. 43.1); **incl.** *Gasteria intermedia* var. *laevior* Haworth (1827); **incl.** *Aloe scaberrima* Salm-Dyck (1834) ≡ *Gasteria verrucosa* var. *scaberrima* (Salm-Dyck) Baker (1896); **incl.** *Gasteria radulosa* Baker (1889).

[4] Differs from var. *carinata*: **L** distichous, rarely becoming rosulate, 3 - 28 × 1.5 - 3.5 cm, linear-lanceolate to lorate, distinctly tuberculate due to raised semitranslucent white or green domed tubercles, green to glaucous-green; **Inf** 12 - 30 cm; **Per** 20 - 25 mm.

G. croucheri (Hooker *fil.*) Baker (JLSB 18: 196, 1880). **T:** RSA (*Cooper* s.n. [K]). − **D:** RSA (KwaZulu-Natal, N Eastern Cape); subtropical thickets, quartzitic sandstone cliff faces, flowers in mid-summer. **I:** Jaarsveld & Ward-Hilhorst (1994: 38).
≡ *Aloe croucheri* Hooker *fil.* (1869); **incl.** *Gasteria disticha* var. *natalensis* Baker (s.a.).
[4] Acaulescent, decumbent to erect, 25 - 40 × 60 cm ∅, solitary or dividing to form dense groups; **L** rosulate, 20 - 36 × 3 - 10 cm, triangular to linear-lanceolate, erectly spreading, rarely patent or recurved, both surfaces dark green, often glaucous, with dense white spots arranged in transverse bands, epidermis smooth, rarely slightly asperulous, margin tuberculate, serrulate, rarely denticulate, tip obtuse or acute, mucronate; juvenile **L** distichous, lorate, patent to erectly spreading, smooth or asperulous, tip acute, rarely obtuse, mucronate; **Inf** vari-

able, flat-topped panicles to 50 cm, or racemose, with or without a pair of side **Br**; **Bra** 5 mm, piliferous; **Ped** 6 - 7 mm, pink; **Per** 28 - 40 mm, stipitate for up to 6 mm, gasteriform part narrowly ellipsoid, ½ of **Per** length, 5 - 9 mm ∅, pink, upper ½ of **Per** white with green striations, inflated to same ∅ as lower part; **Per** tips erect becoming erectly spreading, obtuse; **Sti** included or shortly exserted; **Fr** 18 - 25 mm, obtuse; **Se** 3 - 4 × 2 - 3 mm.

G. disticha (Linné) Haworth (Philos. Mag. Ann. Chem. 1827: 352, 1827). **T**: [lecto – icono]: Commelin, Hort. Med. Amstel., t. 8, 1701. – **D**: RSA (Western Cape); Succulent Karoo, Nama Karoo, Renosterveld, flowers spring to late spring. **I**: Jaarsveld & Ward-Hilhorst (1994: 62).

≡ *Aloe disticha* Linné (1753); **incl.** *Aloe linguiformis* Miller (1768); **incl.** *Aloe lingua* var. *angustifolia* Aiton (1789) ≡ *Gasteria angustifolia* (Aiton) Duval (1809) ≡ *Gasteria disticha* var. *angustifolia* (Aiton) Baker (1880); **incl.** *Aloe lingua* var. *crassifolia* Aiton (1789) ≡ *Gasteria nigricans* var. *crassifolia* (Aiton) Haworth (1821) ≡ *Aloe crassifolia* (Aiton) Roemer & Schultes (1829); **incl.** *Aloe lingua* var. *latifolia* Haworth (1804) ≡ *Gasteria latifolia* (Haworth) Haworth (1812); **incl.** *Aloe lingua* var. *longifolia* Haworth (1804) ≡ *Gasteria longifolia* (Haworth) Duval (1809); **incl.** *Aloe nigricans* Haworth (1804) ≡ *Gasteria nigricans* (Haworth) Duval (1809); **incl.** *Aloe obscura* Willdenow (1811) (*nom. illeg.*, Art. 53.1); **incl.** *Aloe conspurcata* Salm-Dyck (1817) ≡ *Gasteria conspurcata* (Salm-Dyck) Haworth (1827) ≡ *Gasteria disticha* var. *conspurcata* (Salm-Dyck) Baker (1880); **incl.** *Aloe nigricans* var. *denticulata* Salm-Dyck (1817) ≡ *Gasteria denticulata* (Salm-Dyck) Haworth (1819); **incl.** *Aloe obtusifolia* Salm-Dyck (1821) ≡ *Gasteria obtusifolia* (Salm-Dyck) Haworth (1827); **incl.** *Gasteria mollis* Haworth (1821) ≡ *Aloe mollis* (Haworth) Roemer & Schultes (1829); **incl.** *Aloe angustifolia* Salm-Dyck (1821) (*nom. illeg.*, Art. 53.1); **incl.** *Gasteria disticha* var. *major* Haworth (1827); **incl.** *Gasteria disticha* var. *minor* Haworth (1827); **incl.** *Aloe angustifolia* Salm-Dyck (1849) (*nom. illeg.*, Art. 53.1).

[1] Acaulescent, decumbent to erect, 2.5 - 23 × 7.5 - 23 cm, proliferating from the base to form small groups; **L** distichous, 6 - 17 × 3 - 4.5 cm, lorate, erectly spreading, both sides with dense white spots arranged in irregular transverse bands, epidermis asperulous, margin irregularly undulating, tuberculate-crenulate, tip obtuse, rarely truncate, mucronate; juvenile **L** often patent or recurved, asperulous, truncate; **Inf** racemose, simple or with a pair of side **Br**, 20 - 90 cm; **Per** 12 - 20 mm, stipitate for 2 - 3 mm, gasteriform part > ½ of **Per** length, globose-ellipsoid or narrowly ellipsoid, 6 mm ∅, pink to reddish-pink, then constricted into a tube of 3 - 4 mm ∅, white with green striations; **Fr** oblong, 15 - 23 × 7 mm; **Se** 3 - 4 × 2 - 3 mm.

G. ellaphieae van Jaarsveld (CSJA 63(1): 3-7, ills., 1991). **T**: RSA, Eastern Cape (*van Jaarsveld & al.* 9904 [NBG]). – **D**: RSA (Eastern Cape); subtropical thickets, in rock crevices; flowers early to mid-summer.

[4] Acaulescent, decumbent to erect, 1.5 - 4 × 5 - 16 cm, solitary or proliferating from the base to form small groups; **L** first distichous, becoming rosulate, 2 - 5 × 1 - 2 cm, triangular to triangular-lanceolate and falcate, the inner erectly spreading, the outer recurved, lower face with an asymmetrical keel, both faces dark green, with dense white tubercles arranged in irregular transverse bands, epidermis tuberculate, margin tuberculate-denticulate, tip acute or acuminate, recurved, mucronate; juvenile **L** distichous, lorate, ascending at first, becoming patent or recurved; **Inf** erectly spreading racemes, 25 - 40 cm, unbranched or branched in the upper ⅓ and with 8 - 15 **Fl**; **Per** 22 - 27 mm, reddish-pink, stipitate for 2.5 mm, gasteriform for slightly > ½ of **Per** length, 7.5 mm ∅, narrowly ellipsoid, above constricted into a tube 4.5 mm ∅; **Per** tips erectly spreading, obtuse, white with central green striae; **Sty** included; **Fr** 12 - 15 × 5 - 6 mm; **Se** 3 × 2 mm.

G. excelsa Baker (JLSB 18: 195, 1880). **T**: RSA, Eastern Cape (*Cooper* s.n. [K]). – **D**: RSA (Eastern Cape); subtropical thickets, rocky river valleys, flowers mid-summer. **I**: Jaarsveld & Ward-Hilhorst (1994: 42).

Incl. *Gasteria fuscopunctata* Baker (1880); **incl.** *Gasteria huttoniae* N. E. Brown (1908); **incl.** *Gasteria lutzii* von Poellnitz (1933).

[4] Acaulescent, robust, decumbent to erect, solitary, 30 - 60 × 60 - 75 cm; **L** in a dense **Ros**, 10 - 40 × 10 - 18 cm, triangular to triangular-lanceolate, erectly spreading to somewhat recurved, both faces dark green with indistinct white spots arranged in transverse bands, which (rarely) may be barely visible, rarely striate, epidermis smooth, margin often very sharp, cartilaginous, serrulate, tip acute or obtuse, mucronate; juvenile **L** distichous, densely white-spotted, patent, lorate, tuberculate; **Inf** very large spreading panicles, 1 - 3 per plant, 1 - 1.9 m, side **Br** erectly (rarely horizontally) spreading; **Per** 22 - 26 mm, stipitate for 2 - 3 mm, gasteriform part > ½ of **Per** length, narrowly ellipsoid, tube constricted to 5 mm ∅, light pink (rarely white), 6 - 7 mm ∅ at widest point; **Per** tips white with green midstripes; **Sty** included; **Fr** 17 - 20 × 8 - 12 mm; **Se** 4 - 5 × 2 - 3 mm.

G. glauca van Jaarsveld (CSJA 70(2): 65-66, ills., 1998). **T**: RSA, Eastern Cape (*van Jaarsveld & Welsh* 14670 [PRE]). – **D**: RSA (Eastern Cape: Kouga); sheer S-facing cliff-faces, flowers in mid-summer.

[4] Acaulescent, decumbent to erect, proliferating from the base to form dense clusters to 30 cm

⌀; **L** rosulate, 5 - 7 × 1.5 - 1.8 cm, lorate-lanceolate, upper face slightly canaliculate in the upper ⅓, lower face somewhat convex with a distinct asymmetrical keel, both surfaces glaucous, epidermis tuberculate-asperulous, margins tuberculate-dentate, tip acute, mucronate; **Inf** simple racemes to 25 cm with 10 - 20 **Fl** in the upper ½; **Per** 30 - 34 mm, reddish-pink, gasteriform basally for ± ½ of its length or slightly less, globose-ellipsoid, then constricted into a tube 5 mm wide; **St** to 32 mm; **Anth** yellow, included; **Ov** yellowish-green, 6 × 2.5 mm; **Fr** oblong, 20 × 8 mm; **Se** 3 × 2 mm, black.

Closely related to *G. ellaphieae* which also occurs along the Kouga River.

G. glomerata van Jaarsveld (Bradleya 9: 100-104, ills., SEM-ills., 1991). **T:** RSA, Eastern Cape (*van Jaarsveld & Sardien* 11054 [NBG]). – **D:** RSA (Eastern Cape); thickets on sheer sandstone cliff faces, S aspects, spring-flowering.

[1] Acaulescent, decumbent to erect, 1.5 - 4 × 2 - 8 cm, proliferating from the base to form dense globose clusters up to 20 cm ⌀; **L** distichous, 1.5 - 5 × 1.5 - 2.5 cm, lorate to widely ovate, the inner erectly spreading, the outer patent or recurved, biconvex in cross-section to almost terete, becoming flattened during the dry season, upper face often retuse in the upper ⅓, both faces glaucous, unspotted, epidermis minutely tuberculate, asperulous, margin entire, minutely crenulate-tuberculate in the upper ¼, tip truncate or obtuse, mucronate; juvenile **L** distichous, lorate, ascending at first, becoming patent or recurved, asperulous, only slightly and obscurely tuberculate; **Inf** erectly spreading racemes, 12 - 20 cm; **Per** 20 - 27 mm, gasteriform part slightly > ½ of the **Per** length, reddish-pink, globose to globose-ellipsoid, 6 - 9 (-10) mm ⌀, variable, then constricted into a tube 4 mm ⌀; **Fr** 16 × 8 mm; **Se** 3 × 2 mm.

G. nitida (Salm-Dyck) Haworth (Philos. Mag. Ann. Chem. 1827: 359, 1827). **T:** [neo – icono]: Salm-Dyck, Monogr. Gen. Aloes & Mesembr., 29, t. 17. – **D:** RSA.

≡ *Aloe nitida* Salm-Dyck (1817).

G. nitida var. **armstrongii** (Schönland) van Jaarsveld (Aloe 29(1): 12, 1992). **T:** RSA, Eastern Cape (*Anonymus* s.n. [K]). – **D:** RSA (Eastern Cape); Renosterveld, conglomerate or quartzitic sandstone, flowers in mid-summer. **I:** Jaarsveld & Ward-Hilhorst (1994: 48).

≡ *Gasteria armstrongii* Schönland (1912).

[4] Differs from var. *nitida*: **L** remaining distichous, patent, lorate, epidermis tuberculate, rarely smooth, tip somewhat retuse, obtuse or truncate; **Inf** racemose, 4 - 50 cm; **Per** 20 mm.

G. nitida var. **nitida** – **D:** RSA (Eastern Cape); Grassland and Renosterveld, flowers in mid-sum-

mer. **I:** Jaarsveld & Ward-Hilhorst (1994: 46).

Incl. *Aloe nitida* var. *major* Salm-Dyck (1817); incl. *Aloe nitida* var. *minor* Salm-Dyck (1817); incl. *Aloe nitida* var. *obtusa* Salm-Dyck (1817) ≡ *Gasteria obtusa* (Salm-Dyck) Haworth (1827); incl. *Aloe nitida* var. *grandipunctata* Salm-Dyck (1821) ≡ *Gasteria nitida* var. *grandipunctata* (Salm-Dyck) A. Berger (1908); incl. *Aloe nitida* var. *parvipunctata* Salm-Dyck (1821) ≡ *Gasteria nitida* var. *parvipunctata* (Salm-Dyck) A. Berger (1908); incl. *Aloe trigona* var. *obtusa* Salm-Dyck (1821); incl. *Aloe nitida* Ker-Gawler (1822) (*nom. illeg.*, Art. 53.1); incl. *Haworthia nigricans* Haworth (1824); incl. *Gasteria decipiens* Haworth (1827) ≡ *Aloe decipiens* (Haworth) Roemer & Schultes (1829); incl. *Gasteria beckeri* Schönland (1907); incl. *Gasteria stayneri* von Poellnitz (1938).

[4] Acaulescent, decumbent to erect, 6 - 20 × 5 - 28 cm, solitary or proliferating from the base to form small groups; **R** somewhat fusiform, succulent, to 12 mm ⌀; **L** first distichous, becoming rosulate, 1.6 - 18 × 2.5 - 8 cm, triangular-lanceolate, rarely lanceolate-acuminate, erectly spreading with a distinct asymmetrical keel, both faces dark green, with faint to dense white spots arranged in irregular transverse bands, epidermis smooth and shiny, margin entire or indistinctly tuberculate, tip acute, mucronate; juvenile **L** distichous, lorate, 2 - 4 × 2 - 3.5 cm, epidermis tuberculate, rarely smooth, dark green, not or rarely spotted; **Inf** erectly spreading lax panicles 0.2 - 1.2 m; **Per** 20 - 25 mm, bright reddish-pink, stipitate for 2 - 3 mm, gasteriform for slightly > ½ of the **Per** length, 5 - 8 mm ⌀, narrowly ellipsoid, constricted above into a tube 4 - 5 mm ⌀, **Per** tips erectly spreading, obtuse, yellowish; **Sty** included; **Fr** oblong, 24 - 30 × 8 mm; **Se** 3 - 4 × 2 mm.

G. pillansii Kensit (Trans. Roy. Soc. South Afr. 1: 163, 1909). **T:** RSA, Northern Cape (*Pillans* 833 [BOL, PRE]). – **D:** Namibia, RSA.

G. pillansii var. **ernesti-ruschii** (Dinter & von Poellnitz) van Jaarsveld (Aloe 29(1): 17, 1992). **T:** [lecto – icono]: Kakt. and. Sukk. (Berlin), 1938(2): fig., p. 36. – **D:** S Namibia, RSA (Northern Cape); Succulent Karoo, rocks, flowers mid-summer to autumn. **I:** Jaarsveld & Ward-Hilhorst (1994: 60). **Fig. XX.f**

≡ *Gasteria ernesti-ruschii* Dinter & von Poellnitz (1938).

[2] Differs from var. *pillansii*: **L** 2 - 7 cm; **Inf** 6 - 30 cm; **Per** 25 - 30 mm; **St** shortly exserted.

G. pillansii var. **pillansii** – **D:** RSA (Northern Cape: Namaqualand); Succulent Karoo, flowers in mid-summer. **I:** Jaarsveld & Ward-Hilhorst (1994: 58). **Fig. XX.h, XXI.e**

Incl. *Gasteria neliana* von Poellnitz (1930).

[2] Acaulescent, decumbent to erect, 5 - 20 × 6 -

40 cm, proliferating from subterraneous stolons and forming dense groups; **L** distichous, 2 - 20 × 1.5 - 5 cm, lorate, erectly spreading or patent, both faces spotted with immersed tubercles in obscure transverse bands, epidermis asperulous, rarely with few domed tubercles, margin cartilaginous, tuberculate or crenulate, tip obtuse to acute, mucronate; **Inf** racemose, 1 - 3 per plant, spreading and slightly curved, 0.6 - 1.2 (-1.65) m, rarely with a pair of side **Br**; **Fl** laxly arranged in the upper ½; **Per** 25 - 50 × 6 - 8 mm, stipitate for 3 mm, obscurely gasteriform basally for ⅓ or less of the **Per** length, globose-ellipsoid to slightly constricted above the **Ov**, above gradually enlarging and rarely clavate, gasteriform part pink, tube white with green striations; **St** included or exserted to 5 mm; **Fr** oblong, 15 - 23 × 7 mm; **Se** 4 - 5 × 2.5 mm.

G. pulchra (Aiton) Haworth (Synops. Pl. Succ., 86, 1812). **T:** [lecto – icono]: Miller, Gard. Dict., 1759, t. 292. – **D:** RSA (Eastern Cape); subtropical thickets on conglomerate, spring-flowering. **I:** Jaarsveld & Ward-Hilhorst (1994: 44).

 ≡ *Aloe maculata* var. *pulchra* Aiton (1789) ≡ *Aloe pulchra* (Aiton) Haworth (1804); **incl.** *Aloe obliqua* De Candolle (1802) (*nom. illeg.*, Art. 53.1); **incl.** *Aloe pulchra* Jacquin (1805) (*nom. illeg.*, Art. 53.1); **incl.** *Gasteria poellnitziana* H. Jacobsen (1954) (*nom. inval.*, Art. 36.1).

 [4] Acaulescent, decumbent to erect, 20 - 36 × 20 - 36 cm, solitary or proliferating from the base to form small groups; **L** in a dense **Ros**, 24 - 36 × 2.5 - 4 cm, erectly spreading, often falcate, linear-ensiform to linear-acuminate, with an indistinct asymmetrical keel, both faces dark green with dense white spots in transverse bands, epidermis smooth, margin cartilaginous, serrulate, tip acute or acuminate, mucronate; juvenile **L** distichous, erectly spreading, lorate, acuminate, distinctly tuberculate, mucronate; **Inf** erect to erectly spreading lax panicles 0.35 - 1.5 m, rarely simple or side **Br** erectly spreading; **Per** 18 - 25 mm, stipitate for 2 mm, gasteriform part > ½ of the **Per** length, globose-ellipsoid, reddish-pink, 6 - 7 mm ∅, then constricted into a tube of 4 - 4.5 mm ∅, light to dark pink, **Per** tips white, rarely pink, erectly spreading, obtuse, with green median stripes; **Sty** included, **Fr** oblong, 12 - 27 × 7 mm; **Se** 2 - 5 × 2 - 3 mm.

G. rawlinsonii Obermeyer (FPA 43(3/4): t. 1701 + text, 1976). **T:** RSA, Eastern Cape (*Rawlinson* s.n. [PRE 34421]). – **D:** RSA (Eastern Cape); sheer shady quartzitic sandstone cliff faces.

 [1] Caulescent, pendulous, proliferating from the base; stems leafy, to 1 m, rarely branched; **Int** 1 - 2 cm; **L** distichous or rosulate, 3 - 8 × 1 - 2.5 cm, linear, lorate, slightly falcate, tips recurved, lower face convex, without keel, both faces green, unspotted or with faint white spots, epidermis asperulous, margin sparingly denticulate, or sometimes unarmed, prick-

les turning black with age, tip obtuse, mucronate; **Inf** racemose, 10 - 50 cm; **Bra** 5 × 2 mm at base; **Per** reddish-pink, variable, 16 - 25 mm, stipitate for 1 - 3 mm, gasteriform part > ½ of the **Per** length, globose-ellipsoid or globose, pink, then constricted into a tube 4 - 6 mm ∅, tube pink or white, occasionally with green striations; **Fr** 18 mm, oblong-ovoid; **Se** 3 - 4 mm.

 A slow growing cliff-dweller with 2 forms grading into each other: One has the leaves remaining distichous and short inflorescences; the other is with rosulate leaves and longer inflorescences and globose gasteriform parts of the perianth.

G. vlokii van Jaarsveld (CSJA 58(4): 170-174, ills., 1987). **T:** RSA, Western Cape (*Vlok* 880 [NBG]). – **D:** RSA (Western Cape); quartzitic sandstones, Fynbos, flowers in mid-summer.

 [4] Acaulescent, proliferating from the base to form small groups to 14 cm ∅; **R** succulent, to 8 mm ∅, fusiform; **L** distichous, ultimately becoming rosulate, 5 - 9 × 2 - 3 mm broad at the base, patent, lorate, lanceolate to triangular, falcate, with an asymmetrical keel in the upper ⅓, both faces green with dense white spots in obscure transverse bands, epidermis asperulous, margin acute, tuberculate, crenulate, becoming continuous towards tip; tip acute or obtuse with an asymmetrical mucro; juvenile **L** lorate, tip obtuse, mucronate; **Inf** racemose, curved and spreading, 30 - 84 cm, rarely with a pair of side **Br**; **Per** 29 - 33 mm, stipitate for 2 - 3 mm, gasteriform part slightly > ½ of the **Per** length, 6 - 7 mm ∅, dark reddish-pink, above constricted into a tube of 4 mm ∅; **Per** tips white with green median striations; **Fr** 15 - 18 × 6 mm; **Se** oblong, 3 - 5 × 2 mm.

×GASTROLOBA

U. Eggli

×**Gastroloba** D. M. Cumming (Bull. Afr. Succ. Pl. Soc. 9: 36, 1974).

 = *Gasteria* × *Astroloba*. Rowley (1982: 76) lists 2 known combinations, one of them formally named as *G. apicroides* (Baker) G. D. Rowley (≡ *Gasteria apicroides* Baker).

HAWORTHIA

M. B. Bayer & E. van Jaarsveld

Haworthia Duval (Pl. Succ. Horto Alencon., [7], 1809). **T:** *Aloe pumila* var. *arachnoidea* Linné. – **Lit:** Bayer (1982); Scott (1985); Breuer & Metzing (1997); Breuer (1998a); Breuer (1998b); Bayer (1999). **D:** S Namibia, RSA (predominantly Western Cape, Eastern Cape, but also Northern Cape, Free State, KwaZulu-Natal and Mpumalanga); mainly Succulent Karoo and Nama Karoo, mostly in rocky places in the shade of grasses or shrubs.

Etym: For Adrian H. Haworth (1768 - 1833), English zoologist and botanist and specialist for succulent plants.

Incl. *Catevala* Medikus (1786) (*nomen rejiciendum*, Art. 56.1). **T:** *Aloe retusa* Linné [Pfeiffer, Nomencl. Bot., 1871-74 (cf. P. V. Heath, Calyx 4(3): 83, 1994).].

Incl. *Apicra* Willdenow (1811). **T:** not typified.

Incl. *Aprica* D. Dietrich (1840). **T:** not typified.

Dwarf rosulate perennials, acaulescent to caulescent, solitary or proliferating to form dense tight clusters; **R** fibrous to succulent, sometimes fusiform; **Ros** sessile or caulescent, 2 - 15 cm ∅, flat to globose or elongate; **L** erect, spreading to recurved, sometimes highly reduced, usually densely spirally arranged, rarely distichous, triangular-lanceolate to linear, usually firm, often drying back from the tip and becoming papery, surfaces often tuberculate and variously marked, glabrous to rarely pubescent, often with translucent windows, and with various patterns, markings and reticulations, very pale green, blue-green to dark (blackish-) green, upper face flat, or channelled to convex, lower face convex, often keeled towards the tip, margin entire, denticulate, ciliate, with bristly **Sp** or tuberculate, tip acute, or acuminate to mucronate, sometimes retuse or truncate; **Inf** 1 - 5 racemes from **Ros** centre, erect, simple or branched, firm to wiry, bracteate, peduncle brownish-green, often with a powdery bloom; **Bra** small; **Fl** small, zygomorphic, bilabiate, 5 - 50 per **Inf**, ± 15 mm, tube curved or straight, slightly swollen towards base; **Tep** 6, white to pink, often with central green or brownish striations, tips spreading to recurved with green, brownish or occasionally yellow median colouration; **Fil** 6, included; **Ov** oblong to oblong-ovoid, 3-chambered, with 6 grooves; **Sty** subulate; **Sti** apical, minute, trilobate or round; **Fr** erect oblong-ovoid woody capsules, dehiscing longitudinally; **Se** black to grey, compressed, irregularly angled or flattish.

Many species are highly variable and phenotypically plastic. Here, species definitions are based on geographical distribution and sympatrical occurrence, and the concepts of Bayer (1982) and Bayer (1999) are followed. Based on flower morphology, 3 subgenera can be recognized:

[1] Subgen. *Haworthia*: Base of **Fl** triangular or rounded-triangular, tube obclavate, curved; **OTep** free; **Sty** upcurved; **Se** irregularly angled. – Here belongs the majority of the species.

[2] Subgen. *Hexangulares* Uitewaal *ex* M. B. Bayer 1971: Base of **Fl** 6-angular, gradually narrowing to junction with **Ped**; tube obcapitate, curved; **OTep** partly fused to **ITep**; **Sty** straight; **Se** irregularly angled. – 15 species.

[3] Subgen. *Robustipedunculares* Uitewaal *ex* M. B. Bayer 1971: Base of **Fl** 6-angular, rounded, abruptly joined to **Ped**; **OTep** partly fused to **ITep**, tube obcapitate, straight; **Sty** straight; **Se** flattish. – 4 species.

Haworthias are popular amongst succulent plant collectors, and many can tolerate relatively shaded growing conditions. Propagation is usually from offsets, but also possible from detached whole leaves or in some taxa even from detached roots, as well as from seed or tissue culture.

In the following list of names of unresolved application, many of the names are of horticultural derivation.

The following names are of unresolved application but are referred to this genus: *Aloe anomala* Haworth (1804) ≡ *Apicra anomala* (Haworth) Willdenow (1811) (*incorrect name*, Art. 11.3); *Aloe arachnoidea* var. *pumila* Aiton (1789) ≡ *Aloe pumila* (Aiton) Haworth (1804) (*nom. illeg.*, Art. 53.1); *Aloe aspera* Haworth (1804) ≡ *Apicra aspera* (Haworth) Willdenow (1811) (*incorrect name*, Art. 11.3) ≡ *Astroloba aspera* (Haworth) Uitewaal (1947); *Aloe cylindrica* var. *rigida* Lamarck (1783) ≡ *Aloe rigida* (Lamarck) De Candolle (1799) ≡ *Apicra rigida* (Lamarck) Willdenow (1811) (*incorrect name*, Art. 11.4) ≡ *Haworthia rigida* (Lamarck) Haworth (1821); *Aloe expansa* Haworth (1804) ≡ *Haworthia expansa* (Haworth) Haworth (1812) ≡ *Aloe rigida* var. *expansa* (Haworth) Salm-Dyck (1836) ≡ *Haworthia rigida* var. *expansa* (Haworth) Baker (1880); *Aloe expansa* var. *major* Haworth (1804); *Aloe glabrata* var. *concolor* Salm-Dyck (1849) ≡ *Haworthia glabrata* var. *concolor* (Salm-Dyck) Baker (1880); *Aloe glabrata* var. *perviridis* Salm-Dyck (1849) ≡ *Haworthia glabrata* var. *perviridis* (Salm-Dyck) Baker (1880); *Aloe hybrida* Salm-Dyck (1817) ≡ *Haworthia hybrida* (Salm-Dyck) Haworth (1821); *Aloe pseudorigida* Salm-Dyck (1817) ≡ *Apicra pseudorigida* (Salm-Dyck) Haworth (1819) ≡ *Haworthia tortuosa* var. *pseudorigida* (Salm-Dyck) A. Berger (1908); *Aloe pumila* var. ε Linné (1753); *Aloe radula* var. *major* Salm-Dyck (1817); *Aloe radula* var. *media* Salm-Dyck (1817); *Aloe radula* var. *minor* Salm-Dyck (1817); *Aloe rigida* var. *paulo-major* Salm-Dyck (1817); *Aloe rugosa* Salm-Dyck (1834) ≡ *Haworthia rugosa* (Salm-Dyck) Baker (1880); *Aloe rugosa* var. *laetevirens* Salm-Dyck (1834); *Aloe rugosa* var. *perviridis* Salm-Dyck (1834) ≡ *Haworthia rugosa* var. *perviridis* (Salm-Dyck) A. Berger (1908); *Aloe semimargaritifera* var. *glabrata* Salm-Dyck (1834); *Aloe semimargaritifera* var. *major* Salm-Dyck (1817) ≡ *Haworthia semimargaritifera* var. *major* (Salm-Dyck) Haworth (1819) ≡ *Haworthia margaritifera* subvar. *major* (Haworth) A. Berger (1908); *Aloe semimargaritifera* var. *minor* Salm-Dyck (1817) ≡ *Haworthia semimargaritifera* var. *minor* (Salm-Dyck) Haworth (1819); *Aloe semimargaritifera* var. *multipapillosa* Salm-Dyck (1834) ≡ *Haworthia margaritifera* subvar. *multipapillosa* (Salm-Dyck) A. Berger (1908); *Aloe subalbicans* Salm-Dyck (1854) ≡ *Haworthia margaritifera* var. *subalbicans* (Salm-Dyck) A. Berger (1908); *Aloe* subalbicans var. *acuminata* Salm-Dyck (1854) ≡ *Haworthia margaritifera* subvar. *acuminata* (Salm-Dyck) A. Berger (1908); *Aloe subalbicans* var. *laevior* Salm-Dyck (1854) ≡ *Haworthia margaritifera* subvar. *laevior* (Salm-Dyck) A. Berger (1908); *Aloe subattenuata* Salm-Dyck (1834) ≡ *Haworthia subattenuata* (Salm-Dyck) Baker (1880); *Aloe subfasciata* Salm-Dyck (1825) (*nom. illeg.*, Art. 52.1) ≡ *Haworthia subfasciata* (Salm-Dyck) Baker (1880); *Aloe subrigida* Roemer & Schultes (1829) ≡ *Haworthia subrigida* (Roemer & Schultes) Baker (1880); *Aloe subulata* Salm-Dyck (1822); *Aloe subulata* Salm-Dyck (1829) ≡ *Haworthia subulata* (Salm-Dyck) Baker (1880); *Aloe tortuosa* var. *major* Salm-Dyck (1837) ≡ *Haworthia tortuosa* var. *major* (Salm-Dyck) Baker (1896); *Apicra aspera* var. *major* Haworth (1819) ≡ *Astroloba aspera* var. *major* (Haworth) Uitewaal (1947) ≡ *Haworthia aspera* var. *major* (Haworth) Parr (1971); *Apicra patula* Willdenow (1811) (*incorrect name*, Art. 11.3); *Haworthia affinis* Baker (1880) ≡ *Haworthia bilineata* var. *affinis* (Baker) von Poellnitz (1938); *Haworthia altilinea* Haworth (1824) ≡ *Aloe altilinea* (Haworth) Roemer & Schultes (1829) ≡ *Haworthia mucronata* var. *altilinea* (Haworth) Halda (1997); *Haworthia altilinea* fa. *acuminata* von Poellnitz (1938) ≡ *Haworthia mucronata* fa. *acuminata* (von Poellnitz) von Poellnitz (1940); *Haworthia altilinea* fa. *minor* Triebner (1938) ≡ *Haworthia mucronata* fa. *minor* (Triebner) von Poellnitz (1940); *Haworthia altilinea* fa. *subglauca* von Poellnitz (1938) ≡ *Haworthia mucronata* fa. *subglauca* (von Poellnitz) von Poellnitz (1940); *Haworthia altilinea* var. *bicarinata* Triebner (1938) ≡ *Haworthia mucronata* var. *bicarinata* (Triebner) von Poellnitz (1940); *Haworthia altilinea* var. *brevisetata* von Poellnitz (1937); *Haworthia altilinea* var. *setulifera* Triebner & von Poellnitz (1938) ≡ *Haworthia mucronata* var. *setulifera* (Triebner & von Poellnitz) von Poellnitz (1940); *Haworthia argyrostigma* Baker (1896) (*nom. inval.*, Art. 32.1) ≡ *Haworthia attenuata* var. *argyrostigma* (Baker) A. Berger (1908) (*nom. inval.*, Art. 32.1); *Haworthia aristata* var. *helmiae* (von Poellnitz) Pilbeam (1983) (*nom. inval.*, Art. 33.2); *Haworthia aspera* Haworth (1812); *Haworthia asperula* Haworth (1824) ≡ *Aloe asperula* (Haworth) Roemer & Schultes (1829) ≡ *Haworthia retusa* ssp. *asperula* (Haworth) Halda (1997) ≡ *Haworthia retusa* var. *asperula* (Haworth) Halda (1997); *Haworthia baccata* G. G. Smith (1944); *Haworthia bijliana* von Poellnitz (1929) ≡ *Haworthia setata* var. *bijliana* (von Poellnitz) von Poellnitz (1938); *Haworthia bilineata* Baker (1880); *Haworthia broteriana* Resende (1941) ≡ *Haworthia sampaiana* fa. *broteriana* (Resende) Resende & Pinto Lopes (1946); *Haworthia cassytha* Baker (1896); *Haworthia coarctata* var. *sampaiana* Resende (1938) ≡ *Haworthia sampaiana* (Resende) Resende (1940); *Haworthia ×coarctatoides* Resende & Viveiros (1948); *Haworthia columnaris* Baker

(1889) ≡ *Haworthia pilifera* var. *columnaris* (Baker) von Poellnitz (1938) ≡ *Haworthia obtusa* var. *columnaris* (Baker) Uitewaal (1948); *Haworthia confusa* von Poellnitz (1933) ≡ *Haworthia minima* var. *confusa* (von Poellnitz) von Poellnitz (1938) (*incorrect name*, Art. 11.4) ≡ *Haworthia tenera* var. *confusa* (von Poellnitz) Uitewaal (1948); *Haworthia curta* Haworth (1819) ≡ *Aloe curta* (Haworth) Sprengel (1825) ≡ *Haworthia tortuosa* var. *curta* (Haworth) Baker (1896); *Haworthia cuspidata* Haworth (1819) ≡ *Aloe cuspidata* (Haworth) Roemer & Schultes (1829); *Haworthia fergusoniae* von Poellnitz (1930); *Haworthia ferox* von Poellnitz (1933); *Haworthia ferox* var. *armata* von Poellnitz (1937); *Haworthia gracilidelineata* von Poellnitz (1933) ≡ *Haworthia bilineata* var. *gracilidelineata* (von Poellnitz) von Poellnitz (1938) ≡ *Haworthia cymbiformis* fa. *gracilidelineata* (von Poellnitz) Pilbeam (1983); *Haworthia henriquesii* Resende (1941); *Haworthia icosiphylla* Baker (1880); *Haworthia janseana* Uitewaal (1940); *Haworthia krausiana* Hort. Haage & Schmidt (s.a.); *Haworthia ×krausii* hort. *ex* J. R. Brown (1957); *Haworthia lisbonensis* Resende (1946); *Haworthia longifolia* Farden (1939); *Haworthia multifaria* Haworth (1824) ≡ *Aloe multifaria* (Haworth) Roemer & Schultes (1829); *Haworthia pearsonii* C. H. Wright (1907) ≡ *Haworthia arachnoidea* ssp. *pearsonii* (C. H. Wright) Halda (1997) ≡ *Haworthia arachnoidea* var. *pearsonii* (C. H. Wright) Halda (1997); *Haworthia pellucens* var. *delicatula* A. Berger (1908) ≡ *Haworthia translucens* var. *delicatula* (A. Berger) von Poellnitz (1938); *Haworthia perplexa* von Poellnitz (1938); *Haworthia polyphylla* Baker (1880) ≡ *Haworthia altilinea* var. *polyphylla* (Baker) von Poellnitz (1938) ≡ *Haworthia mucronata* var. *polyphylla* (Baker) von Poellnitz (1940); *Haworthia pseudogranulata* von Poellnitz (1937); *Haworthia radula* var. *asperior* Haworth (1821); *Haworthia radula* var. *laevior* Haworth (1821); *Haworthia radula* var. *magniperlata* Haworth (1821); *Haworthia reinwardtii* var. *minor* Baker (1880); *Haworthia resendeana* von Poellnitz (1938); *Haworthia revendettii* Uitewaal (1940); *Haworthia rubrobrunea* von Poellnitz (1940); *Haworthia ryderiana* von Poellnitz (1937); *Haworthia semiglabrata* Haworth (1819) ≡ *Aloe semiglabrata* (Haworth) Roemer & Schultes (1829); *Haworthia semimargaritifera* var. *multiperlata* Haworth (1819) ≡ *Aloe semimargaritifera* var. *multiperlata* (Haworth) Roemer & Schultes (1829) ≡ *Haworthia margaritifera* subvar. *multiperlata* (Haworth) von Poellnitz (1938); *Haworthia sessiliflora* Baker (1896); *Haworthia setata* var. *subinermis* von Poellnitz (1936); *Haworthia stiemiei* von Poellnitz (1938); *Haworthia tauteae* Archibald (1946); *Haworthia tisleyi* Baker (1880); *Haworthia tortella* Haworth (1819) ≡ *Haworthia tortuosa* var. *tortella* (Haworth) Baker (1896); *Haworthia triebneriana* var. *lanceolata* Triebner & von Poellnitz (1938); *Haworthia trieb-*

neriana var. *nitida* von Poellnitz (1940); *Haworthia uitewaaliana* von Poellnitz (1939); *Haworthia walmsleyi* Hort. Haage & Schmidt (s.a.).

H. angustifolia Haworth (Philos. Mag. J. 66: 283, 1825). **T** [neo]: RSA, Eastern Cape (*Bruyns* 1653 [NBG 120 017]). – **D:** RSA (Eastern Cape).
≡ *Haworthia chloracantha* ssp. *angustifolia* (Haworth) Halda (1997) ≡ *Haworthia chloracantha* var. *angustifolia* (Haworth) Halda (1997); **incl.** *Aloe stenophylla* Roemer & Schultes (1829); **incl.** *Haworthia albanensis* Schönland (1912) ≡ *Haworthia angustifolia* var. *albanensis* (Schönland) von Poellnitz (1937); **incl.** *Haworthia angustifolia* var. *grandis* G. G. Smith (1943) ≡ *Haworthia angustifolia* fa. *grandis* (G. G. Smith) Pilbeam (1983).

H. angustifolia var. **altissima** M. B. Bayer (Haworthia Revisited, 26, ill. (p. 27), 1999). **T:** RSA, Eastern Cape (*Smith* 5220 [NBG]). – **D:** RSA (Eastern Cape: Grahamstown Distr.).
≡ *Haworthia altissima* (M. B. Bayer) M. Hayashi (2000).
[1] Differs from var. *angustifolia*: **Ros** proliferous; **L** slender, erect, to 15 × 1 cm, finely denticulate along margins and keel, colour tending to greyish-green rather than darkish green.

H. angustifolia var. **angustifolia** – **D:** RSA (Eastern Cape). **I:** Bayer (1982: fig. 1a); Scott (1985: 55).
[1] **Ros** to 4 cm ∅, proliferous, stemless; **L** 10 - 40, slender, erect, to 10 × 1 cm, lanceolate-acuminate, somewhat flaccid, brownish to dark green, margins and keel finely denticulate; **Inf** to 20 cm, lax; **Fl** 8 - 10, white to dull pinkish-white.

H. angustifolia var. **baylissii** (C. L. Scott) M. B. Bayer (Haworthia Revisited, 27, ill., 1999). **T:** RSA (*Scott* 796 [PRE, NBG]). – **D:** RSA (Eastern Cape: Oudekraal). **I:** Bayer (1982: fig. 1b, as fa.); Scott (1985: 102).
≡ *Haworthia baylissii* C. L. Scott (1968) ≡ *Haworthia angustifolia* fa. *baylissii* (C. L. Scott) M. B. Bayer (1982) ≡ *Haworthia chloracantha* ssp. *baylissii* (C. L. Scott) Halda (1997).
[1] Differs from var. *angustifolia*: **L** deltoid, recurving.

H. angustifolia var. **paucifolia** G. G. Smith (JSAB 14: 48, ill. (p. 58), 1948). **T:** RSA, Eastern Cape (*Smith* 6819 [NBG]). – **D:** RSA (Eastern Cape: Frazer's Camp to Kaffirdrift).
≡ *Haworthia angustifolia* fa. *paucifolia* (G. G. Smith) Pilbeam (1983).
[1] Differs from var. *angustifolia*: **L** 3 - 4 , to 3 cm.

H. arachnoidea (Linné) Duval (Pl. Succ. Horto Alencon., 7, 1809). **T:** [lecto – icono]: Commelin,

Praeludia Bot., t. 27, 1703. – **D:** RSA (Western Cape, Eastern Cape).
≡ *Aloe pumila* var. *arachnoidea* Linné (1753) ≡ *Aloe arachnoidea* (Linné) Burman *fil.* (1768) ≡ *Catevala arachnoidea* (Linné) Medikus (1786) ≡ *Apicra arachnoidea* (Linné) Willdenow (1811) (*incorrect name*, Art. 11.3).

H. arachnoidea var. **arachnoidea** – **D:** RSA (Western Cape: Little Karoo). **I:** Bayer (1982: fig. 2). **Fig. XXII.b**
Incl. *Haworthia arachnoidea* var. *minor* Haworth (1819).
[1] **R** slender, succulent; **Ros** stemless, variable in size from 6 to exceptionally 12 cm ∅, solitary or forming small clusters; **L** 25 - 80, dense, incurving, uniformly light to dark green, not translucent and only occasionally faintly reticulate, flattened and often drying grey-white to brownish at the tips forming a protective cover, triangular- to ovate-lanceolate, 2 - 7 × 1 - 1.5 cm, keeled, margin and keels with translucent bristly **Sp** to 12 mm, tip acuminate-aristate; **Inf** to 30 cm; **Fl** 20 - 30, white.

H. arachnoidea var. **aranea** (A. Berger) M. B. Bayer (Haworthia Revisited, 30, ill. (p. 31), 1999). **T:** [lecto – icono]: Pflanzenr. 4(38 = Heft 33): 114, fig. 39A-E. – **D:** RSA (Western Cape: Little Karoo); dry Mountain Fynbos. **I:** Bayer (1982: fig. 3, as *H. aranea*).
≡ *Haworthia bolusii* var. *aranea* A. Berger (1908) ≡ *Haworthia aranea* (A. Berger) M. B. Bayer (1976).
[1] Differs from var. *arachnoidea*: **Ros** smaller, to 6 cm ∅; **L** softer in texture.

H. arachnoidea var. **namaquensis** M. B. Bayer (Haworthia Revisited, 31, ills., 1999). **T:** RSA, Northern Cape (*Bayer* 1674 [NBG]). – **D:** RSA (Northern Cape, Western Cape).
[1] Differs from var. *arachnoidea*: **Ros** to 6 cm ∅; **L** paler green.

H. arachnoidea var. **nigricans** (Haworth) M. B. Bayer (Haworthia Revisited, 32, ill., 1999). **T** [neo]: RSA (*Bayer* 2419 [NBG]). – **D:** RSA (Western Cape: Little Karoo). **I:** Scott (1985: 100, as *H. helmiae*).
≡ *Haworthia setata* var. *nigricans* Haworth (1821) ≡ *Aloe setata* var. *nigricans* (Haworth) Roemer & Schultes (1829); **incl.** *Haworthia venteri* von Poellnitz (1939) ≡ *Haworthia unicolor* var. *venteri* (von Poellnitz) M. B. Bayer (1976).
[1] Differs from var. *arachnoidea*: **L** with dark purplish coloration.

H. arachnoidea var. **scabrispina** M. B. Bayer (Haworthia Revisited, 34, ills., 1999). **T:** RSA, Western Cape (*Bayer* 2105 [NBG]). – **D:** RSA (Western Cape).

[1] Differs from var. *arachnoidea*: **Ros** roundish, raised above ground-level; **L** with firm rigid brownish spines.

H. arachnoidea var. **setata** (Haworth) M. B. Bayer (Haworthia Revisited, 34, ills. (p. 35), 1999). **T:** RSA, Cape Prov. (*Makrill* s.n. [[lecto – icono]: K [drawing]]). – **D:** RSA (Western Cape: Little Karoo). **I:** Scott (1985: 69, as *H. setata*).

≡ *Haworthia setata* Haworth (1819) ≡ *Aloe setata* (Haworth) Roemer & Schultes (1829) ≡ *Haworthia arachnoidea* ssp. *setata* (Haworth) Halda (1997); **incl.** *Haworthia setata* var. *major* Haworth (1821) ≡ *Aloe setata* var. *major* (Haworth) Roemer & Schultes (1829); **incl.** *Haworthia setata* var. *media* Haworth (1821) ≡ *Aloe setata* var. *media* (Haworth) Roemer & Schultes (1829); **incl.** *Aloe setosa* Roemer & Schultes (1829); **incl.** *Haworthia gigas* von Poellnitz (1933) ≡ *Haworthia setata* var. *gigas* (von Poellnitz) von Poellnitz (1938) ≡ *Haworthia arachnoidea* var. *gigas* (von Poellnitz) M. Hayashi (2000); **incl.** *Haworthia minima* var. *major* von Poellnitz (1938) (*incorrect name*, Art. 11.4) ≡ *Haworthia tenera* var. *major* (von Poellnitz) Uitewaal (1948).

[1] Differs from var. *arachnoidea*: **Ros** variable; **L** white-spined.

H. arachnoidea var. **xiphiophylla** (Baker) Halda (Acta Mus. Richnov. Sect. Nat. 4(2): 43, 1997). **T:** RSA, Eastern Cape (*Howlett* s.n. [K]). – **D:** RSA (Eastern Cape). **I:** Bayer (1982: fig. 44, as *H. xiphiophylla*).

≡ *Haworthia xiphiophylla* Baker (1896) ≡ *Haworthia setata* var. *xiphiophylla* (Baker) von Poellnitz (1938); **incl.** *Haworthia longiaristata* von Poellnitz (1937).

[1] Differs from var. *arachnoidea*: **L** bright green, long, slender, with broad but short marginal **Sp**.

H. aristata Haworth (Suppl. Pl. Succ., 51, 1819). **T:** [lecto – icono]: K. – **D:** RSA (Eastern Cape).

≡ *Aloe aristata* (Haworth) Roemer & Schultes (1829) (*nom. illeg.*, Art. 53.1); **incl.** *Haworthia denticulata* Haworth (1821) ≡ *Aloe denticulata* (Haworth) Roemer & Schultes (1829) ≡ *Aloe altilinea* var. *denticulata* (Haworth) Salm-Dyck (1842) ≡ *Haworthia altilinea* var. *denticulata* (Haworth) von Poellnitz (1938).

[1] **Ros** stemless, proliferating slowly, to 6 cm ∅; **L** slender, erect, incurved, dark green, hardly translucent and faintly reticulated, margins and keel entire or minutely spined; **Inf** to 15 cm, lax; **Fl** 10 - 15, white.

H. attenuata (Haworth) Haworth (Synops. Pl. Succ., 92, 1812). **T:** RSA (*Perry* 660 [NBG 144 672]). – **D:** RSA (Eastern Cape).

≡ *Aloe attenuata* Haworth (1804) ≡ *Apicra atte-nuata* (Haworth) Willdenow (1811) (*incorrect name*, Art. 11.4) ≡ *Haworthia pumila* ssp. *attenuata* (Haworth) Halda (1997).

H. attenuata var. **attenuata** – **D:** RSA (Eastern Cape). **I:** Bayer (1982: fig. 46). **Fig. XXII.d**

Incl. *Haworthia clariperla* Haworth (1826) ≡ *Aloe clariperla* (Haworth) Roemer & Schultes (1829) ≡ *Aloe attenuata* var. *clariperla* (Haworth) Salm-Dyck (1836) ≡ *Haworthia attenuata* var. *clariperla* (Haworth) Baker (1880) ≡ *Haworthia attenuata* fa. *clariperla* (Haworth) M. B. Bayer (1976); **incl.** *Haworthia fasciata* var. *caespitosa* A. Berger (1908) ≡ *Haworthia attenuata* var. *caespitosa* (A. Berger) R. S. Farden (1939) ≡ *Haworthia attenuata* fa. *caespitosa* (A. Berger) Pilbeam (1983) (*nom. inval.*, Art. 33.2); **incl.** *Haworthia britteniana* von Poellnitz (1937) ≡ *Haworthia attenuata* var. *britteniana* (von Poellnitz) von Poellnitz (1937) ≡ *Haworthia attenuata* fa. *britteniana* (von Poellnitz) M. B. Bayer (1982); **incl.** *Haworthia attenuata* var. *deltoidea* R. S. Farden (1939); **incl.** *Haworthia attenuata* var. *inusitata* R. S. Farden (1939); **incl.** *Haworthia attenuata* var. *linearis* R. S. Farden (1939); **incl.** *Haworthia attenuata* var. *minissima* R. S. Farden (1939); **incl.** *Haworthia attenuata* var. *odonoghueana* R. S. Farden (1939); **incl.** *Haworthia attenuata* var. *uitewaaliana* R. S. Farden (1939).

[2] **Ros** stemless, proliferous, to 13 cm tall and 10 cm ∅; **L** to 13 × 1.5 cm, attenuate, spreading, lanceolate-deltoid, surfaces scabrid with distinct raised non-confluent tubercles; **Inf** sparsely branched, 24 - 28 cm, lax; **Fl** tube obcapitate, **Tep** revolute.

H. attenuata var. **radula** (Jacquin) M. B. Bayer (Haworthia Revisited, 167, ill., 1999). **T:** [lecto – icono]: Jacquin, Pl. Hort. Schoenbr. 4: t. 422, 1804. – **D:** RSA (Eastern Cape: Baviaanskloof). **I:** Bayer (1982: fig. 56, as *H. radula*).

≡ *Aloe radula* Jacquin (1804) ≡ *Apicra radula* (Jacquin) Willdenow (1811) (*incorrect name*, Art. 11.3) ≡ *Haworthia radula* (Jacquin) Haworth (1812) ≡ *Haworthia pumila* ssp. *radula* (Jacquin) Halda (1997); **incl.** *Haworthia radula* var. *pluriperlata* Haworth (1821).

[2] Differs from var. *attenuata*: **L** with many minute crowded white tubercles.

H. bayeri J. D. Venter & S. A. Hammer (CSJA 69(2): 75-76, ill., 1997). **T:** RSA, Western Cape (*Stayner* s.n. in *Karoo Garden* 164/69 [NBG]). – **D:** RSA (Western Cape: Little Karoo).

Incl. *Haworthia uniondalensis* hort. (s.a.) (*nom. inval.*, Art. 29.1).

[1] **Ros** stemless, to 8 cm ∅; **L** 15 - 20, retuse, dark brownish-green to blackish-green, slightly scabrid, keels and margins with minute **Sp** or smooth, tip rounded and not pointed, the end-area opaque, cloudy-transparent, with sparse reticulate patterning or longitudinal lines; **Inf** to 30 cm; **Fl** 15 - 25.

Here belong *H. willowmorensis* and *H. correcta* in the sense of Scott.

H. blackburniae W. F. Barker (JSAB 3: 93, 1937). **T:** RSA (*Reynolds* 1842 [NBG, BOL, PRE]). – **D:** RSA (Western Cape).

    Incl. *Haworthia blackburniana* hort. (s.a.) (*nom. inval.*, Art. 61.1).

H. blackburniae var. **blackburniae** – **D:** RSA (Western Cape: mountains of the C Little Karoo). **I:** Bayer (1982: fig. 6).

    R fusiform; **Ros** stemless, basally 1 - 1.5 cm ∅, clumping; **L** 10 - 15, long, slender, to 40 cm × 3 - 5 mm, bright green to brownish-green or dark greyish-green, upper face channelled, margins glabrous or finely spined; **Inf** to 30 cm; **Fl** 15 - 20, white with green veins.

H. blackburniae var. **derustensis** M. B. Bayer (Haworthia Revisited, 41, ill., 1999). **T:** RSA, Western Cape (*Vlok* s.n. in *Venter* 93/24 [NBG]). – **D:** RSA (Western Cape: Oudtshoorn Distr.).

    ≡ *Haworthia derustensis* (M. B. Bayer) M. Hayashi (2000).

    [1] Differs from var. *blackburniae*: **Ros** robust, to 1.8 cm ∅; **L** very long, to 45 cm and 3 mm wide, brownish-green at the base, green above.

H. blackburniae var. **graminifolia** (G. G. Smith) M. B. Bayer (Haworthia Revisited, 42, ills., 1999). **T:** RSA (*Smith* 5222 [NBG]). – **D:** RSA (Western Cape: Swartberg Mts.). **I:** Bayer (1982: fig. 16, as *H. graminifolia*).

    ≡ *Haworthia graminifolia* G. G. Smith (1942) ≡ *Haworthia blackburniae* ssp. *graminifolia* (G. G. Smith) Halda (1997).

    [1] Differs from var. *blackburniae*: **L** < 3 mm broad, dark greyish-green.

H. bolusii Baker (JLSB 18: 215, 1880). **T:** RSA (*Bolus* 158 [K, BOL]). – **D:** RSA (Eastern Cape).

    ≡ *Haworthia arachnoidea* var. *bolusii* (Baker) Halda (1997).

H. bolusii var. **blackbeardiana** (von Poellnitz) M. B. Bayer (New Haworthia Handb., 31, 1982). **T:** B. – **D:** RSA (interior of Eastern Cape). **I:** Bayer (1982: fig. 7b). **Fig. XXII.c**

    ≡ *Haworthia blackbeardiana* von Poellnitz (1933); **incl.** *Haworthia inermis* von Poellnitz (1933) ≡ *Haworthia altilinea* var. *inermis* (von Poellnitz) von Poellnitz (1937) ≡ *Haworthia altilinea* fa. *inermis* (von Poellnitz) von Poellnitz (1940) ≡ *Haworthia mucronata* fa. *inermis* (von Poellnitz) von Poellnitz (1940); **incl.** *Haworthia blackbeardiana* var. *major* von Poellnitz (1937); **incl.** *Haworthia batteniae* C. L. Scott (1979) ≡ *Haworthia cooperi* ssp. *batteniae* (C. L. Scott) Halda (1997).

[1] Differs from var. *bolusii*: **Ros** generally larger, to 15 cm ∅; **L** with sparse **Sp** > 2 mm.

H. bolusii var. **bolusii** – **D:** RSA (interior of Eastern Cape). **I:** Scott (1985: 72).

    [1] **Ros** 4 - 8 cm ∅, slowly proliferating; **L** oblong-lanceolate, incurved, translucent bluish-green, margins and keel with **Sp** > 2 mm; **Inf** robust, to 30 cm; **Fl** broad and flat across the base of the tube.

H. bruynsii M. B. Bayer (JSAB 47: 789, 1981). **T:** RSA (*Rossouw* 456 [NBG]). – **D:** RSA (Eastern Cape: Little Karoo). **I:** Bayer (1982: fig. 47).

    ≡ *Haworthia retusa* var. *bruynsii* (M. B. Bayer) Halda (1997).

    [2] **Ros** stemless, solitary, to 6 cm ∅; **L** 5 - 11, brownish-green, with flat retuse end-area, opaque and slightly scabrid with small raised tubercles; **Inf** simple, slender; **Fl** distant, slender, tube obcapitate; **Tep** fused, tips revolute.

H. chloracantha Haworth (Revis. Pl. Succ., 57, 1821). **T** [lecto]: RSA, Cape Prov. (*Bowie* s.n. [[lecto – icono]: K [drawing]]). – **D:** RSA (Western Cape).

    ≡ *Aloe chloracantha* (Haworth) Roemer & Schultes (1829).

H. chloracantha var. **chloracantha** – **D:** RSA (Western Cape). **I:** Bayer (1982: fig. 8a).

    [1] **Ros** 2.5 - 4 cm ∅, proliferous; **L** 18 - 25, 0.5 - 1.5 cm long, erect spreading, light green, faintly reticulate, firm to slightly scabrid, triangular in cross-section, margins and keel with translucent **Sp** to 0.3 mm; **Inf** lax, to 25 cm; **Fl** small.

H. chloracantha var. **denticulifera** (von Poellnitz) M. B. Bayer (Haworthia Handb., 112, 1976). **T:** B [lecto: unpubl. ill.]. – **D:** RSA (Western Cape: Mosselbay). **I:** Bayer (1982: fig. 8b).

    ≡ *Haworthia angustifolia* var. *denticulifera* von Poellnitz (1937) ≡ *Haworthia denticulifera* (von Poellnitz) M. Hayashi (2000); **incl.** *Haworthia angustifolia* var. *liliputana* Uitewaal (1953).

    [1] Differs from var. *chloracantha*: **L** 2.5 - 3.5 cm long, dark green.

H. chloracantha var. **subglauca** von Poellnitz (Kakteenkunde 9: 135, 1937). **T** [neo]: RSA (*Hurling & Neil* s.n. [BOL]). – **D:** RSA (Western Cape: Mosselbay). **I:** Bayer (1982: fig. 8c).

    ≡ *Haworthia subglauca* (von Poellnitz) M. Hayashi (2000).

    [1] Differs from var. *chloracantha*: **L** bluish-green, **Sp** larger and sparser.

H. coarctata Haworth (Philos. Mag. J. 64: 301, 1824). **T** [neo]: RSA (*Smith* 7092 [NBG 68473]). – **D:** RSA (Eastern Cape).

    ≡ *Aloe coarctata* (Haworth) Roemer & Schultes

(1829) ≡ *Haworthia reinwardtii* ssp. *coarctata* (Haworth) Halda (1997) ≡ *Haworthia reinwardtii* var. *coarctata* (Haworth) Halda (1997).

H. coarctata fa. **greenii** (Baker) M. B. Bayer (Haworthia Revisited, 172, ill., 1999). **T**: RSA (*Cooper* s.n. [K]). – **D**: RSA (Eastern Cape: Grahamstown). **I**: Bayer (1982: fig. 48b, as var.).

≡ *Haworthia greenii* Baker (1880) ≡ *Haworthia coarctata* var. *greenii* (Baker) M. B. Bayer (1973) ≡ *Haworthia reinwardtii* var. *greenii* (Baker) Halda (1997); **incl.** *Haworthia peacockii* Baker (1880); **incl.** *Haworthia greenii* fa. *minor* Resende (1943); **incl.** *Haworthia greenii* fa. *bakeri* Resende (1943) (*nom. illeg.*, Art. 52.1, 26.1).

[2] Differs from typical *H. coarctata*: **L** without tubercles.

H. coarctata var. **adelaidensis** (von Poellnitz) M. B. Bayer (Haworthia Revisited, 172, ills., 1999). **T**: RSA, Eastern Cape (*Armstrong* s.n. [B [lecto: unpubl. ill.]]). – **D**: RSA (Eastern Cape: Grahamstown). **I**: Bayer (1982: fig. 48d, as ssp.).

≡ *Haworthia reinwardtii* var. *adelaidensis* von Poellnitz (1940) ≡ *Haworthia coarctata* ssp. *adelaidensis* (von Poellnitz) M. B. Bayer (1973); **incl.** *Haworthia reinwardtii* var. *riebeekensis* G. G. Smith (1944); **incl.** *Haworthia reinwardtii* var. *bellula* G. G. Smith (1945) ≡ *Haworthia coarctata* fa. *bellula* (G. G. Smith) Pilbeam (1983).

[2] Differs from var. *coarctata*: **Ros** smaller, stems to 15 cm, 3 cm ∅; **L** 3.2 × 1 cm.

H. coarctata var. **coarctata** – **D**: RSA (Eastern Cape). **I**: Bayer (1982: fig. 48a).

Incl. *Haworthia chalwinii* Marloth & A. Berger (1906) ≡ *Haworthia reinwardtii* var. *chalwinii* (Marloth & A. Berger) Resende (1943) ≡ *Haworthia coarctata* fa. *chalwinii* (Marloth & A. Berger) Pilbeam (1983); **incl.** *Haworthia fallax* von Poellnitz (1933) ≡ *Haworthia reinwardtii* var. *fallax* (von Poellnitz) von Poellnitz (1937); **incl.** *Haworthia reinwardtii* var. *conspicua* von Poellnitz (1937) ≡ *Haworthia coarctata* fa. *conspicua* (von Poellnitz) Pilbeam (1983); **incl.** *Haworthia reinwardtii* var. *pseudocoarctata* von Poellnitz (1940) ≡ *Haworthia coarctata* fa. *pseudocoarctata* (von Poellnitz) Resende (1943) ≡ *Haworthia greenii* fa. *pseudocoarctata* (von Poellnitz) Resende & Pinto-Lopes (1946); **incl.** *Haworthia coarctata* fa. *major* Resende (1943); **incl.** *Haworthia coarctata* var. *kraussii* Resende (1943); **incl.** *Haworthia fulva* G. G. Smith (1943); **incl.** *Haworthia greenii* var. *silvicola* G. G. Smith (1943); **incl.** *Haworthia reinwardtii* var. *committeesensis* G. G. Smith (1943); **incl.** *Haworthia coarctata* var. *haworthii* Resende (1943) (*nom. illeg.*, Art. 52.1, 26.1); **incl.** *Haworthia reinwardtii* var. *huntsdriftensis* G. G. Smith (1944); **incl.** *Haworthia musculina* G. G. Smith (1948).

[2] **Ros** to 12 cm ∅, caulescent, proliferating; **L** numerous, to 7 × 2 cm, ratio stem diameter : **L** width = 1:1.7, erect-spreading or incurved, scabrid, both sides brownish-green, usually with rounded tubercles; **Inf** simple or occasionally compound, to 60 cm; **Fl** tube obcapitate; **Tep** revolute.

Not always easily separated from *H. reinwardtii* and its forms. The ratio between stem diameter and leaf width is a good help.

H. coarctata var. **tenuis** (G. G. Smith) M. B. Bayer (Nation. Cact. Succ. J. 28: 80, 1973). **T**: RSA, Eastern Cape (*Smith* 3420 [NBG]). – **D**: RSA (Eastern Cape: Alexandria). **I**: Bayer (1982: fig. 48c).

≡ *Haworthia reinwardtii* var. *tenuis* G. G. Smith (1948).

[2] Differs from var. *coarctata*: **Ros** with elongated narrow stems to 40 cm, 2.5 cm ∅; **L** 3.5 × 0.8 cm.

H. cooperi Baker (Refug. Bot. 4: t. 233 + text, 1871). **T**: RSA (*Cooper* s.n. [K]). – **D**: RSA (Eastern Cape).

H. cooperi var. **cooperi** – **D**: RSA (Eastern Cape). **I**: Bayer (1982: fig. 10a).

Incl. *Haworthia vittata* Baker (1871).

[1] **Ros** to 12 cm ∅, often proliferous, stemless; **L** 20 - 40, fleshy, swollen, erect, oblong-lanceolate, quickly tapering, acuminate or truncate, bluishgreen, slightly translucent, with veins usually reddening and **L** becoming purplish in exposed situations, marginal **Sp** < 2 mm when present; **Inf** compact, firm, to 20 cm; **Fl** 20 - 30, closely arranged, white.

H. cooperi var. **dielsiana** (von Poellnitz) M. B. Bayer (Haworthia Revisited, 51, ill., 1999). **T** [neo]: RSA, Eastern Cape (*van der Merwe* s.n. in *Smith* 1140 [NBG]). – **D**: RSA (Eastern Cape: Cookhouse). **I**: Bradleya 13: 80-81, as *H. joeyae*.

≡ *Haworthia dielsiana* von Poellnitz (1930) ≡ *Haworthia cooperi* fa. *dielsiana* (von Poellnitz) hort. (s.a.) (*nom. inval.*, Art. 29.1) ≡ *Haworthia pilifera* var. *dielsiana* (von Poellnitz) von Poellnitz (1940) ≡ *Haworthia obtusa* var. *dielsiana* (von Poellnitz) Uitewaal (1948); **incl.** *Haworthia joeyae* C. L. Scott (1995).

[1] Differs from var. *cooperi*: **L** obtuse, truncate, not acuminate, conspicuously veined.

H. cooperi var. **gordoniana** (von Poellnitz) M. B. Bayer (Haworthia Revisited, 52, ills., 1999). **T** [neo]: RSA, Eastern Cape (*Smith* 3028 [NBG]). – **D**: RSA (Eastern Cape: Baviaanskloof).

≡ *Haworthia gordoniana* von Poellnitz (1937) ≡ *Haworthia pilifera* var. *gordoniana* (von Poellnitz) von Poellnitz (1938) ≡ *Haworthia obtusa* var. *gordoniana* (von Poellnitz) Uitewaal (1948).

[1] Differs from var. *cooperi*: **Ros** small, to 6 cm ∅; **L** erect, acuminate, not truncate.

H. cooperi var. **leightonii** (G. G. Smith) M. B. Bayer (Haworthia Handb., 128, 1976). **T:** RSA (*Smith 6938* [NBG]). – **D:** RSA (Eastern Cape: between Chalumna and East London). **I:** Bayer (1982: fig. 10b).

≡ *Haworthia leightonii* G. G. Smith (1950).

[1] Differs from var. *cooperi*: **Ros** very proliferous; **L** lanceolate-acuminate.

H. cooperi var. **pilifera** (Baker) M. B. Bayer (Haworthia Revisited, 54, ills., 1999). **T:** [lecto – icono]: Refug. Bot. 4: t. 234, 1871. – **D:** RSA (Eastern Cape). **I:** Scott (1985: 105, as *H. pilifera*).

≡ *Haworthia pilifera* Baker (1871) ≡ *Haworthia obtusa* var. *pilifera* (Baker) Uitewaal (1948) ≡ *Haworthia cooperi* fa. *pilifera* (Baker) Pilbeam (1983); **incl.** *Haworthia stayneri* von Poellnitz (1937) ≡ *Haworthia pilifera* var. *stayneri* (von Poellnitz) von Poellnitz (1938) ≡ *Haworthia obtusa* var. *stayneri* (von Poellnitz) Uitewaal (1948); **incl.** *Haworthia stayneri* var. *salina* von Poellnitz (1937) ≡ *Haworthia pilifera* var. *salina* (von Poellnitz) von Poellnitz (1938) ≡ *Haworthia obtusa* var. *salina* (von Poellnitz) Uitewaal (1948); **incl.** *Haworthia pilifera* fa. *acuminata* von Poellnitz (1940) ≡ *Haworthia obtusa* fa. *acuminata* (von Poellnitz) Uitewaal (1948).

[1] Differs from var. *cooperi*: **L** obtusely acuminate, margins and keel acute, with small bristly **Ha**, tips becoming truncate, contracting and flattening as exposure to light increases.

H. cooperi var. **truncata** (H. Jacobsen) M. B. Bayer (Haworthia Revisited, 55, ill. (p. 56), 1999). **T:** [neo – icono]: H. Jacobsen, Handb. Sukk. Pfl. 2: 724, fig. 644, 1956.. – **D:** RSA (Eastern Cape). **I:** Jacobsen (1960: 2: fig. 756, as *H. obtusa* fa.).

≡ *Haworthia obtusa* fa. *truncata* H. Jacobsen (1955).

[1] Differs from var *cooperi*: **Ros** very proliferous, to 7 cm ∅; **L** 20 - 25, 20 - 25 × 8 mm, pale blue-green, erect, truncate, translucent and lightly veined above.

H. cooperi var. **venusta** (C. L. Scott) M. B. Bayer (Haworthia Revisited, 56, ills., 1999). **T:** RSA, Eastern Cape (*Britten 781* [GRA]). – **D:** RSA (Eastern Cape: Alexandria). **I:** Bradleya 14: 87, 1996, as *H. venusta*.

≡ *Haworthia venusta* C. L. Scott (1996).

[1] Differs from var. *cooperi*: **L** shortly pilose.

H. cymbiformis (Haworth) Duval (Pl. Succ. Horto Alencon., 7, 1809). **T** [neo]: RSA (*Smith 2844* [NBG 68015]). – **D:** RSA (Eastern Cape).

≡ *Aloe cymbiformis* Haworth (1804); **incl.** *Aloe cymbaefolia* Schrader (1807) ≡ *Apicra cymbaefolia* (Schrader) Willdenow (1811) (*incorrect name*, Art. 11.4); **incl.** *Haworthia concava* Haworth (1821); **incl.** *Aloe hebes* Roemer & Schultes (1829); **incl.**

Haworthia planifolia fa. *alta* Triebner & von Poellnitz (1938); **incl.** *Haworthia planifolia* fa. *olivacea* Triebner & von Poellnitz (1938); **incl.** *Haworthia planifolia* fa. *robusta* Triebner & von Poellnitz (1938); **incl.** *Haworthia planifolia* subvar. *agavoides* Triebner & von Poellnitz (1938).

H. cymbiformis var. **cymbiformis** – **D:** RSA (Eastern Cape). **I:** Bayer (1982: fig. 11a). **Fig. XXII.a**

Incl. *Haworthia planifolia* Haworth (1825) ≡ *Aloe planifolia* (Haworth) Roemer & Schultes (1829) ≡ *Haworthia cymbiformis* var. *planifolia* (Haworth) Baker (1880) ≡ *Haworthia cymbiformis* fa. *planifolia* (Haworth) Pilbeam (1983); **incl.** *Haworthia cymbiformis* fa. *subarmata* von Poellnitz (1938) ≡ *Haworthia cymbiformis* subvar. *subarmata* (von Poellnitz) M. B. Bayer (1982) (*nom. inval.*, Art. 33.2); **incl.** *Haworthia cymbiformis* var. *angustata* von Poellnitz (1938); **incl.** *Haworthia cymbiformis* var. *compacta* Triebner (1938); **incl.** *Haworthia planifolia* fa. *agavoides* Triebner & von Poellnitz (1938); **incl.** *Haworthia planifolia* fa. *calochlora* Triebner & von Poellnitz (1938); **incl.** *Haworthia planifolia* var. *exulata* von Poellnitz (1938); **incl.** *Haworthia planifolia* var. *incrassata* von Poellnitz (1938); **incl.** *Haworthia planifolia* var. *longifolia* Triebner & von Poellnitz (1938); **incl.** *Haworthia planifolia* var. *sublaevis* von Poellnitz (1938); **incl.** *Haworthia planifolia* var. *poellnitziana* Resende (1940); **incl.** *Haworthia lepida* G. G. Smith (1944).

[1] **Ros** to 13 cm ∅, usually stemless, proliferous; **L** broadly ovate to lanceolate, flat to slightly concave, generally < ⅓ as thick as wide, usually opaque, green turning yellowish or developing a pink hue in full sun; **Inf** to 25 cm, lax; **Fl** 10 - 15, white.

A variegated form ("fa. *variegata*") is known in cultivation.

H. cymbiformis var. **incurvula** (von Poellnitz) M. B. Bayer (Haworthia Handb., 124, 1976). **T:** RSA (*Britten* s.n. [BOL 71307]). – **D:** RSA (Eastern Cape: E Grahamstown). **I:** Bayer (1982: fig. 11c).

≡ *Haworthia incurvula* von Poellnitz (1933).

[1] Differs from var. *cymbiformis*: **Ros** smaller, to 5 cm ∅; **L** narrow, acuminate, incurved, tip obtuse.

H. cymbiformis var. **obtusa** (Haworth) Baker (JLSB 18: 209, 1880). **T:** K [lecto – icono, publ. Succulenta 1948: 49]. – **D:** RSA (Eastern Cape: Fort Beaufort). **I:** Bayer (1982: fig. 11e, as *H. umbraticola*).

≡ *Haworthia obtusa* Haworth (1825); **incl.** *Haworthia hilliana* von Poellnitz (1937) ≡ *Haworthia umbraticola* var. *hilliana* (von Poellnitz) von Poellnitz (1938); **incl.** *Haworthia umbraticola* von Poellnitz (1937) ≡ *Haworthia cymbiformis* var. *umbraticola* (von Poellnitz) M. B. Bayer (1976).

[1] Differs from var. *cymbiformis*: **Ros** prolifer-

ous, not contracting into the soil; **L** obtuse, venation subdued.

H. cymbiformis var. **ramosa** (G. G. Smith) M. B. Bayer (Haworthia Revisited, 60, ill., 1999). **T:** RSA (*Smith* 3168 [NBG, PRE]). – **D:** RSA (Eastern Cape: Peddie). **I:** Bayer (1982: fig. 11b, as fa.).

≡ *Haworthia ramosa* G. G. Smith (1940) ≡ *Haworthia cymbiformis* fa. *ramosa* (G. G. Smith) M. B. Bayer (1976).

[1] Differs from var. *cymbiformis*: **Ros** developing long leafy stems.

H. cymbiformis var. **reddii** (C. L. Scott) M. B. Bayer (Haworthia Revisited, 61, ills., 1999). **T:** RSA, Eastern Cape (*Scott* 8968 [PRE]). – **D:** RSA (interior of Eastern Cape). **I:** CSJA 66: 182, 1994, as *H. reddii*.

≡ *Haworthia reddii* C. L. Scott (1994).

[1] Differs from var. *cymbiformis*: **L** subdeltoid, surface with reticulate pattern.

H. cymbiformis var. **setulifera** (von Poellnitz) M. B. Bayer (Haworthia Revisited, 62, ills., 1999). **T:** RSA (*Anonymus* s.n. in *SUG* 3332 [[lecto – icono]: Kakteenkunde 1938: 54]). – **D:** RSA (SE Eastern Cape).

≡ *Haworthia planifolia* var. *setulifera* von Poellnitz (1938); **incl.** *Haworthia cymbiformis* var. *obesa* von Poellnitz (1938) ≡ *Haworthia cymbiformis* fa. *obesa* (von Poellnitz) Pilbeam (1983).

[1] Differs from var. *cymbiformis*: **L** deltoid, margins spiny.

H. cymbiformis var. **transiens** (von Poellnitz) M. B. Bayer (Haworthia Handb., 162, 1976). **T:** B [lecto: unpubl. ill.]. – **D:** RSA (Eastern Cape: Baviaanskloof, Langkloof). **I:** Bayer (1982: fig. 11d).

≡ *Haworthia planifolia* var. *transiens* von Poellnitz (1938) ≡ *Haworthia transiens* (von Poellnitz) M. Hayashi (2000); **incl.** *Haworthia cymbiformis* var. *brevifolia* Triebner & Poellnitz (1938); **incl.** *Haworthia cymbiformis* var. *multifolia* Triebner (1938) ≡ *Haworthia cymbiformis* fa. *multifolia* (Triebner) Pilbeam (1983); **incl.** *Haworthia cymbiformis* var. *translucens* Triebner & von Poellnitz (1938).

[1] Differs from var. *cymbiformis*: **L** obtuse, surface reticulate with translucent areas between the veins.

H. decipiens von Poellnitz (RSN 28: 103, 1930). **T** [neo]: RSA (*Fourcade* 4637 [BOL]). – **D:** RSA (Western Cape, Eastern Cape: S Karoo).

H. decipiens var. **cyanea** M. B. Bayer (Haworthia Revisited, 65, ills. (pp. 65-66), 1999). **T:** RSA, Eastern Cape (*Bayer* 4180 [NBG]). – **D:** RSA (Western Cape, Eastern Cape).

≡ *Haworthia cyanea* (M. B. Bayer) M. Hayashi (2000).

[1] Differs from var. *decipiens*: **Ros** smaller with more **L**; **L** slender, incurving, bluish-green.

H. decipiens var. **decipiens** – **D:** RSA (Western Cape, Eastern Cape: S Karoo). **I:** Bayer (1982: fig. 12).

[1] **Ros** stemless, slowly proliferous, to 20 cm ⌀; **L** ascending, broadly ovate, sometimes acuminate, relatively thin, bright green, marginal **Sp** sparse but broad at their base; **Inf** robust, to 40 cm; **Fl** numerous, densely arranged, broad and flat across the upper base of the tube.

H. decipiens var. **minor** M. B. Bayer (Haworthia Revisited, 66, ills., 1999). **T:** RSA, Eastern Cape (*Smith* 3588 [NBG]). – **D:** RSA (Eastern Cape).

[1] Differs from var. *decipiens*: **Ros** smaller, to 6 cm ⌀; **L** broad, incurved, light green.

H. decipiens var. **pringlei** (C. L. Scott) M. B. Bayer (Haworthia Revisited, 67, ill., 1999). **T:** RSA, Eastern Cape (*Pringle* s.n. in *Scott* 8970 [PRE]). – **D:** RSA (Eastern Cape). **I:** CSJA 66: 104, 1994, as *H. pringlei*.

≡ *Haworthia pringlei* C. L. Scott (1994).

[1] Differs from var. *decipiens*: **L** ± lanceolate, incurved and erect to suberect, bright green, margins and keel with white **Sp**.

H. emelyae von Poellnitz (RSN 42: 271, 1937). **T:** B lecto: unpubl. ill.]. – **D:** RSA (Western Cape, Eastern Cape).

≡ *Haworthia retusa* ssp. *emelyae* (von Poellnitz) Halda (1997) ≡ *Haworthia retusa* var. *emelyae* (von Poellnitz) Halda (1997).

H. emelyae var. **comptoniana** (G. G. Smith) J. D. Venter & S. A. Hammer (CSJA 69(2): 77, 1997). **T:** RSA, Eastern Cape (*Malherbe* s.n. in *Smith* 3433 [NBG]). – **D:** RSA (Eastern Cape: E Little Karoo). **I:** Bayer (1982: fig. 9, as *H. comptoniana*). **Fig. XXII.e**

≡ *Haworthia comptoniana* G. G. Smith (1945) ≡ *Haworthia retusa* var. *comptoniana* (G. G. Smith) Halda (1997); **incl.** *Haworthia comptoniana* fa. *brevifolia* Hort. Sheilam (s.a.) (*nom. inval.*, Art. 29.1); **incl.** *Haworthia comptoniana* fa. *major* Pilbeam (1983) (*nom. inval.*, Art. 37.1).

[1] Differs from var. *emelyae*: **Ros** to 12 cm ⌀; upper face of **L** smooth and markedly reticulate.

H. emelyae var. **emelyae** – **D:** RSA (Western Cape: Little Karoo). **I:** Bayer (1982: fig. 14a).

Incl. *Haworthia blackburniae* von Poellnitz (1937) (*nom. illeg.*, Art. 53.1); **incl.** *Haworthia correcta* von Poellnitz (1938); **incl.** *Haworthia picta* von Poellnitz (1938).

[1] **Ros** to 10 cm ⌀, rarely proliferous; **L** 15 - 20, distinctly retuse, pointed, barely translucent, dark green, with scattered elongate small flecks, with ob-

scure raised tubercles, lined and with reddish-brown hue; **Inf** to 30 cm; **Fl** 15 - 20, white.

H. emelyae var. **major** (G. G. Smith) M. B. Bayer (Aloe 34(1-2): 6, 1997). **T:** RSA (*Smith 5370* [NBG]). – **D:** RSA (Western Cape: S Little Karoo). **I:** Bayer (1982: fig. 22c, as *H. magnifica* var.).

≡ *Haworthia schuldtiana* var. *major* G. G. Smith (1946) ≡ *Haworthia maraisii* var. *major* (G. G. Smith) M. B. Bayer (1976) ≡ *Haworthia magnifica* var. *major* (G. G. Smith) M. B. Bayer (1977); **incl.** *Haworthia wimii* M. Hayashi (2000).

[1] Differs from var. *emelyae*: **L** acuminate, tubercles with an apical **Sp**.

H. emelyae var. **multifolia** M. B. Bayer (Nation. Cact. Succ. J. 34(2): 28-31, ills., 1979). **T:** RSA (*Bayer* 1558 [NBG]). – **D:** RSA (Western Cape: S Little Karoo). **I:** Bayer (1982: fig. 14b).

≡ *Haworthia multifolia* (M. B. Bayer) M. Hayashi (2000).

[1] Differs from var. *emelyae*: **L** 25 to 30, erect.

H. fasciata (Willdenow) Haworth (Revis. Pl. Succ., 54, 1821). **T** [neo]: RSA (*Stayner* s.n. [NBG 110 360]). – **D:** RSA (Eastern Cape). **I:** Bayer (1982: fig. 49a).

≡ *Apicra fasciata* Willdenow (1811) (*incorrect name*, Art. 11.3) ≡ *Aloe fasciata* (Willdenow) Salm-Dyck (1834) ≡ *Haworthia pumila* ssp. *fasciata* (Willdenow) Halda (1997); **incl.** *Haworthia fasciata* fa. *fasciata*; **incl.** *Aloe fasciata* var. *major* Salm-Dyck (1817) ≡ *Haworthia fasciata* var. *major* (Salm-Dyck) Haworth (1819) ≡ *Haworthia fasciata* fa. *major* (Salm-Dyck) von Poellnitz (1938); **incl.** *Aloe fasciata* var. *minor* Salm-Dyck (1834); **incl.** *Haworthia browniana* von Poellnitz (1937) ≡ *Haworthia fasciata* fa. *browniana* (von Poellnitz) M. B. Bayer (1976) ≡ *Haworthia fasciata* var. *browniana* (von Poellnitz) C. L. Scott (1985); **incl.** *Haworthia fasciata* var. *subconfluens* von Poellnitz (1937) ≡ *Haworthia fasciata* fa. *subconfluens* (von Poellnitz) von Poellnitz (1938); **incl.** *Haworthia fasciata* fa. *ovato-lanceolata* von Poellnitz (1938); **incl.** *Haworthia fasciata* fa. *sparsa* von Poellnitz (1938); **incl.** *Haworthia fasciata* fa. *vanstaadensis* von Poellnitz (1938); **incl.** *Haworthia fasciata* fa. *variabilis* von Poellnitz (1938).

[2] **Ros** to 15 cm ∅ and 18 cm tall, stemless, proliferating; **L** 60 - 80, erect, to 6 × 1.5 cm, incurved, scabrid, with white tubercles on the lower face only; **Inf** simple or occasionally compound, to 30 cm; **Fl** tube obcapitate, curved; **Tep** revolute.

H. floribunda von Poellnitz (RSN 40: 149, 1936). **T:** B [lecto: unpubl. ill.]. – **D:** RSA (Western Cape). **I:** Bayer (1982: fig. 15); Scott (1985: 58).

≡ *Haworthia chloracantha* var. *floribunda* (von Poellnitz) Halda (1997).

[1] **Ros** stemless, to 3 cm ∅, slowly proliferating;

L 20 - 30, ovate-lanceolate, spreading, twisted with flattened rounded tip, dark green, opaque, margins scabrid to dentate; **Inf** to 25 cm; **Fl** 10 - 15, greenish-white, few open together.

H. floribunda var. **dentata** M. B. Bayer (Haworthia Revisited, 73, ills., 1999). **T:** RSA, Western Cape (*Dekenah 90* in *Smith 5502* [NBG]). – **D:** RSA (Western Cape: Bredasdorp and Riversdale Distr.).

≡ *Haworthia dentata* (M. B. Bayer) M. Hayashi (2000).

[1] Differs from var. *floribunda*: **Ros** to 4 cm ∅; **L** very dark green, slightly scabrid, margins spiny.

H. floribunda var. **floribunda** – **D:** RSA (Western Cape: Bredasdorp Distr.).

[1] **L** glabrous.

H. floribunda var. **major** M. B. Bayer (Haworthia Revisited, 74, ill., 1999). **T:** RSA, Western Cape (*De Kok* s.n. [NBG]). – **D:** RSA (Western Cape: Bredasdorp Distr.).

Incl. *Haworthia kondoi* M. Hayashi (2000).

[1] Differs from var. *floribunda*: Plants more robust (esp. in cultivation); **L** very green with darker colouration at the basal margins, relatively glabrous.

H. glabrata (Salm-Dyck) Baker (JLSB 18: 206, 1880). **T:** [neo – icono]: Salm-Dyck, Monogr. Gen. Aloes & Mesembr. 3: Aloe t. 7 [Sect. 6: 13], 1840. – **D:** RSA (Eastern Cape: former Transkei). **I:** Bayer (1982: fig. 50).

≡ *Aloe glabrata* Salm-Dyck (1834).

[2] **Ros** stemless, proliferous, to 12 cm tall; **L** to 8 × 1.5 cm, lanceolate-deltoid, attenuate, spreading, scabrid with or without distinct raised non-confluent excrescences; **Inf** sparsely branched, lax; **Fl** obcapitate, curved; **Tep** revolute.

H. glauca Baker (JLSB 18: 203, 1880). **T:** RSA (*Cooper* s.n. [K]). – **D:** RSA (Eastern Cape).

≡ *Haworthia reinwardtii* ssp. *glauca* (Baker) Halda (1997) ≡ *Haworthia reinwardtii* var. *glauca* (Baker) Halda (1997).

H. glauca var. **glauca** – **D:** RSA (Eastern Cape: Zuurberg Mts.). **I:** Bayer (1982: fig. 51a).

Incl. *Haworthia carrissoi* Resende (1941).

[2] **Ros** to 8 cm ∅, caulescent, proliferating; **L** numerous, to 6 × 1.5 cm, erectly spreading or incurved, scabrid, glaucous grey-green, surfaces without tubercles; **Inf** simple or occasionally compound, to 30 cm; **Fl** tube obcapitate, curved; **Tep** revolute.

H. glauca var. **herrei** (von Poellnitz) M. B. Bayer (Haworthia Handb., 122, 1976). **T** [neo]: RSA (*Barker* 5069 [NBG 68132]). – **D:** RSA (Eastern Cape: Little Karoo). **I:** Bayer (1982: fig. 51b).

≡ *Haworthia herrei* von Poellnitz (1929) ≡ *Haworthia reinwardtii* var. *herrei* (von Poellnitz) Halda (1997); **incl.** *Haworthia herrei* var. *depauperata* von Poellnitz (1933); **incl.** *Haworthia armstrongii* von Poellnitz (1937) ≡ *Haworthia glauca* fa. *armstrongii* (von Poellnitz) M. B. Bayer (1976); **incl.** *Haworthia eilyae* von Poellnitz (1937); **incl.** *Haworthia jacobseniana* von Poellnitz (1937) ≡ *Haworthia glauca* fa. *jacobseniana* (von Poellnitz) Pilbeam (1983); **incl.** *Haworthia jonesiae* von Poellnitz (1937) ≡ *Haworthia glauca* fa. *jonesiae* (von Poellnitz) Pilbeam (1983); **incl.** *Haworthia herrei* var. *poellnitzii* Resende (1941) (*nom. illeg.*, Art. 52.1, 26.1); **incl.** *Haworthia eilyae* var. *zantneriana* Resende (1943); **incl.** *Haworthia eilyae* var. *poellnitziana* Resende (1943) (*nom. illeg.*, Art. 52.1, 26.1).

[2] Differs from var. *glauca*: **L** lanceolate, spreading, tuberculate.

H. gracilis von Poellnitz (RSN 27: 133, 1930). **T** [neo]: RSA (*Britten* s.n. [PRE]). – **D:** RSA (Eastern Cape).

≡ *Haworthia arachnoidea* var. *gracilis* (von Poellnitz) Halda (1997).

This taxon was previously treated as a synonym of *H. translucens* by M. B. Bayer.

H. gracilis var. **gracilis** – **D:** RSA (Eastern Cape).

[1] **Ros** stemless, to 6 cm ∅, proliferous; **L** 30 - 40, lanceolate-acuminate, incurved, pale greyish-green, upper face translucent between the veins, margins with slender short **Sp**; **Inf** to 20 cm; **Fl** 15 - 20, white.

H. gracilis var. **isabellae** (von Poellnitz) M. B. Bayer (Haworthia Revisited, 77, ills., 1999). **T** [neo]: RSA, Eastern Cape (*Hall* s.n. in *NBG* 68799 [NBG]). – **D:** RSA (Eastern Cape: Baviaanskloof). **I:** Bayer (1982: fig. 38a, as *H. translucens*).

≡ *Haworthia isabellae* von Poellnitz (1938).

[1] Differs from var. *gracilis*: **Ros** more proliferous; **L** spreading, greyish-green, margins and keel finely spined.

H. gracilis var. **picturata** M. B. Bayer (Haworthia Revisited, 78, ills. (pp. 78-79), 1999). **T:** RSA, Eastern Cape (*Thode* 21507 [NBG]). – **D:** RSA (Eastern Cape: Baviaanskloof).

≡ *Haworthia picturata* (M. B. Bayer) M. Hayashi (2000).

[1] Differs from var. *gracilis*: **L** glabrous, bright green, translucent areas contrasting with the dark green opaque reticulation.

H. gracilis var. **tenera** (von Poellnitz) M. B. Bayer (Haworthia Revisited, 77, ill. (p. 78), 1999). **T** [neo]: RSA (*Smith* 5416 [NBG 115210]). – **D:** RSA (Eastern Cape: Fish River Valley). **I:** Bayer (1982: fig. 38b, as *H. translucens* ssp.).

≡ *Haworthia tenera* von Poellnitz (1933) ≡ *Haworthia translucens* ssp. *tenera* (von Poellnitz) M. B. Bayer (1976) ≡ *Haworthia arachnoidea* ssp. *tenera* (von Poellnitz) Halda (1997) ≡ *Haworthia arachnoidea* var. *tenera* (von Poellnitz) Halda (1997); **incl.** *Haworthia minima* Baker (1880) (*nom. illeg.*, Art. 53.1).

[1] Differs from var. *gracilis*: **Ros** smaller, more compact; **L** closely incurved.

H. gracilis var. **viridis** M. B. Bayer (Haworthia Revisited, 79, ills., 1999). **T:** RSA, Eastern Cape (*Smith* 6867 [NBG]). – **D:** RSA (Eastern Cape: Winterhoek Mts.).

[1] Differs from var. *gracilis*: **L** with brighter green colouration.

H. heidelbergensis G. G. Smith (JSAB 14: 42, 1948). **T:** RSA (*Dekenah* 230 in *Smith* 6566 [NBG]). – **D:** RSA (Western Cape). **I:** Bayer (1982: fig. 18).

≡ *Haworthia retusa* var. *heidelbergensis* (G. G. Smith) Halda (1997).

[1] **Ros** stemless, proliferous, to 8 cm ∅; **L** numerous, erect to recurved, 3 - 5 × 0.5 - 1 cm, turgid and with semiretuse face, usually dark green with reddish hues, end-area slightly translucent, margins and keel generally with small **Sp**; **Inf** to 20 cm; **Fl** 10 - 15, white with brownish veins.

H. heidelbergensis var. **heidelbergensis** – **D:** RSA (Western Cape: Bredasdorp Distr.).

[1] **Ros** to 8 cm ∅; **L** usually dark green with reddish hues.

H. heidelbergensis var. **minor** M. B. Bayer (Haworthia Revisited, 82, ill., 1999). **T:** RSA, Western Cape (*Bayer* s.n. in *Karoo Garden* 36/70 [NBG]). – **D:** RSA (Western Cape: Bredasdorp Distr.).

[1] Differs from var. *heidelbergensis*: **Ros** to 3 cm ∅; **L** light yellowish-green, well-spined.

H. heidelbergensis var. **scabra** M. B. Bayer (Haworthia Revisited, 82, ills., 1999). **T:** RSA, Western Cape (*Bayer* 1700 [NBG]). – **D:** RSA (Western Cape: Bredasdorp and Montagu Distr.).

[1] Differs from var. *heidelbergensis*: **Ros** to 3 cm ∅; **L** very dark green, erect or suberect, slightly scabrid along margins and keel.

H. heidelbergensis var. **toonensis** M. B. Bayer (Haworthia Revisited, 83, ill., 1999). **T:** RSA, Western Cape (*Smith* 6797 [NBG]). – **D:** RSA (Western Cape: Bredasdorp Distr.).

≡ *Haworthia maraisii* var. *toonensis* (M. B. Bayer) M. Hayashi (2000).

[1] Differs from var. *heidelbergensis*: **L** recurved, with a distinct transparent end-area.

H. herbacea (Miller) Stearn (Cact. J. (Croydon) 7:

40, 1938). **T**: [lecto – icono]: Boerhaave, Ind. Alter Hort. Lugd.-Bat., 2: ad. p. 131, 1720. – **D**: RSA (Western Cape: Worcester-Robertson Karoo). **I**: Bayer (1982: fig. 19).

≡ *Aloe herbacea* Miller (1768); **incl.** *Haworthia translucens* ssp. *translucens*; **incl.** *Aloe pumila* Linné (1753) ≡ *Haworthia pumila* (Linné) Duval (1809); **incl.** *Catevala atroviridis* Medikus (1786); **incl.** *Aloe arachnoidea* var. *pumila* Willdenow (1799) (*nom. illeg.*, Art. 53.1); **incl.** *Aloe atrovirens* De Candolle (1799) (*nom. illeg.*, Art. 52.1) ≡ *Apicra atrovirens* (De Candolle) Willdenow (1811) (*nom. illeg.*, Art. 52.1) ≡ *Haworthia atrovirens* (De Candolle) Haworth (1821) (*nom. illeg.*, Art. 52.1); **incl.** *Aloe bradlyana* Jacquin (1804); **incl.** *Aloe translucens* Haworth (1804) ≡ *Aloe arachnoidea* var. *translucens* (Haworth) Ker Gawler (1811) ≡ *Apicra translucens* (Haworth) Willdenow (1811) ≡ *Haworthia translucens* (Haworth) Haworth (1819) ≡ *Haworthia arachnoidea* var. *translucens* (Haworth) Halda (1997); **incl.** *Haworthia pellucens* Haworth (1812); **incl.** *Aloe arachnoidea* var. *pellucens* Salm-Dyck (1817); **incl.** *Aloe papillosa* Salm-Dyck (1817) ≡ *Haworthia papillosa* (Salm-Dyck) Haworth (1819); **incl.** *Haworthia pallida* Haworth (1821) ≡ *Aloe pallida* (Haworth) Roemer & Schultes (1829); **incl.** *Haworthia papillosa* var. *semipapillosa* Haworth (1821) ≡ *Aloe papillosa* var. *semipapillosa* (Haworth) Roemer & Schultes (1829); **incl.** *Haworthia aegrota* von Poellnitz (1939); **incl.** *Haworthia luteorosea* Uitewaal (1939); **incl.** *Haworthia submaculata* von Poellnitz (1939).

[1] **Ros** stemless, proliferating, to 8 cm ∅; **L** erect, incurved, ovate-lanceolate, to 6 × 1 cm, greenish-yellow, with reticulate pattern with translucent areas between the veins, scabrid, margins and keel with firm **Sp**; **Inf** to 30 cm; **Fl** 30 - 40, large, beige with pinkish tips, buds with an S-bend.

Here belongs *H. arachnoidea* in the sense of Scott.

H. herbacea var. **flaccida** M. B. Bayer (Haworthia Revisited, 86, ills., 1999). **T**: RSA, Western Cape (*Bruyns* 7114 [NBG]). – **D**: RSA (Western Cape: Worcester-Robertson Karoo).

≡ *Haworthia pallida* var. *flaccida* (M. B. Bayer) M. Hayashi (2000).

[1] Differs from typical *H. herbacea*: **Ros** small and delicate.

H. herbacea var. **herbacea** – **D**: RSA (Western Cape: Worcester-Robertson Karoo). **Fig. XXII.f**

[1] Description as for *H. herbacea*.

H. herbacea var. **lupula** M. B. Bayer (Haworthia Revisited, 86, ills. (pp. 86-87), 1999). **T**: RSA, Western Cape (*Esterhuysen* s.n. [NBG]). – **D**: RSA (Western Cape: Worcester-Robertson Karoo).

≡ *Haworthia lupula* (M. B. Bayer) M. Hayashi (2000).

[1] Differs from typical *H. herbacea*: **L** broader, shorter, finely flecked, less scabrid; **Fl** larger, pink.

H. herbacea var. **paynei** (von Poellnitz) M. B. Bayer (Haworthia Revisited, 87, ill., 1999). **T**: RSA, Cape Prov. (*Payne* s.n. [B [lecto, unpubl. ill.]]). – **D**: RSA (Western Cape: Worcester-Robertson Karoo).

≡ *Haworthia paynei* von Poellnitz (1937) ≡ *Haworthia pallida* var. *paynei* (von Poellnitz) von Poellnitz (1937).

[1] Differs from typical *H. herbacea*: Plants small-sized; **Fl** bicoloured, pink above and white below.

H. kingiana von Poellnitz (RSN 41: 203, 1937). **T** [neo]: RSA (*Dekenah* 201 [NBG 68719]). – **D**: RSA (Western Cape). **I**: Bayer (1982: fig. 64).

≡ *Haworthia subfasciata* var. *kingiana* (von Poellnitz) von Poellnitz (1938) ≡ *Haworthia pumila* var. *kingiana* (von Poellnitz) Halda (1997).

[3] **Ros** stemless, slowly proliferating, to 18 cm tall; **L** to 16 × 1.8 cm, nearly as thick as wide, yellowish-green, lanceolate-deltoid, attenuate, spreading, surfaces scabrid with raised rounded nonconfluent tubercles; **Inf** sparsely branched, lax; **Fl** 30 - 40, tube obcapitate; **Tep** lobes short, veins pinkish.

H. koelmaniorum Obermeyer & D. S. Hardy (FPA 1967: t. 1502 + text, 1967). **T**: RSA, Mpumalanga (*Hardy & Mauve* 2267 [PRE]). – **D**: RSA (Mpumalanga).

≡ *Haworthia limifolia* ssp. *koelmaniorum* (Obermeyer & D. S. Hardy) Halda (1997).

H. koelmaniorum var. **koelmaniorum** – **D**: RSA (Mpumalanga). **I**: Bayer (1982: fig. 52).

[2] **Ros** stemless, slowly proliferating, 5 - 7 cm ∅; **L** 14 - 20, ovate, 7 × 2 cm, dark brownish-green, opaque, somewhat recurved, scabrid with small raised tubercles, margins and keel with small **Sp**; **Inf** slender, to 35 cm; **Fl** 10 - 15, slender, **Tep** tips revolute.

H. koelmaniorum var. **mcmurtryi** (C. L. Scott) M. B. Bayer (Haworthia Revisited, 181, ills., 1999). **T**: RSA, Mpumlanga (*McMurtry* 5247 [PRE]). – **D**: RSA (Mpumlanaga). **I**: Scott (1985: 141, as *H. mcmurtryi*).

≡ *Haworthia mcmurtryi* C. L. Scott (1984); **incl.** *Haworthia macmurtryi* hort. (s.a.) (*nom. inval.*, Art. 61.1).

[2] Differs from var. *koelmaniorum*: **Ros** smaller, 4 - 5 cm ∅; **L** with more prominent surface markings.

H. limifolia Marloth (Trans. Roy. Soc. South Afr. 1: 409, 1908). **T**: RSA (*Marloth* 4678 [PRE]). – **D**: Moçambique, Swaziland, RSA (Mpumalanga, KwaZulu-Natal).

H. limifolia var. **gigantea** M. B. Bayer (JSAB 28: 215-216, 1962). **T:** RSA (*Bayer* 112 [PRE]). – **D:** RSA (Mpumalanga, N KwaZulu-Natal?). **I:** Bayer (1982: fig. 53b).

≡ *Haworthia gigantea* (M. B. Bayer) M. Hayashi (2000).

[2] Differs from var. *limifolia*: **Ros** larger, to 20 cm ∅; **L** to 13 × 5 cm, finely tuberculate.

H. limifolia var. **limifolia** – **D:** Moçambique, Swaziland, RSA (Mpumalanga, KwaZulu-Natal). **I:** Bayer (1982: fig. 53a).

Incl. *Haworthia limifolia* fa. *diploidea* Resende (1940); **incl.** *Haworthia limifolia* fa. *schuldtiana* Resende (1940) ≡ *Haworthia limifolia* var. *schuldtiana* (Resende) Resende (1943); **incl.** *Haworthia limifolia* fa. *tetraploidea* Resende (1940); **incl.** *Haworthia limifolia* var. *diploidea* Resende (1940); **incl.** *Haworthia limifolia* var. *tetraploidea* Resende (1940); **incl.** *Haworthia limifolia* fa. *marlothiana* Resende (1941) ≡ *Haworthia limifolia* var. *marlothiana* (Resende) Resende (1943); **incl.** *Haworthia limifolia* fa. *major* Resende (1943); **incl.** *Haworthia limifolia* fa. *pimentelii* Resende (1943); **incl.** *Haworthia limifolia* var. *stolonifera* Resende (1943); **incl.** *Haworthia limifolia* var. *keithii* G. G. Smith (1950) ≡ *Haworthia keithii* (G. G. Smith) M. Hayashi (2000); **incl.** *Haworthia limifolia* var. *striata* Pilbeam (1983) (*nom. inval.*, Art. 37.1).

[2] **Ros** stemless, slowly proliferating, with or without stolons, 5 - 7 cm ∅; **L** 12 - 30, ovate-lanceolate, to 6 × 2 cm, spreading, light to very dark green and even brownish-green, opaque, scabrid with white or concolorous tubercles or with confluent transverse ridges, margins and keel scabrid; **Inf** slender, to 35 cm; **Fl** 15 - 20, slender, **Tep** tips revolute.

A variegated form ("fa. *variegata*") is known in cultivation.

H. limifolia var. **ubomboensis** (I. Verdoorn) G. G. Smith (JSAB 16: 4, 1950). **T:** RSA (*Keith* s.n. [PRE]). – **D:** E Swaziland. **I:** Bayer (1982: fig. 53c).

≡ *Haworthia ubomboensis* I. Verdoorn (1941).

[2] Differs from var. *limifolia*: **L** bright green, glabrous.

H. lockwoodii Archibald (FPSA 20: t. 792 + text, 1940). **T:** RSA (*Lockwood Hill* 215 [GRA]). – **D:** RSA (Western Cape: SW Great Karoo). **I:** Bayer (1982: fig. 20).

≡ *Haworthia mucronata* ssp. *lockwoodii* (Archibald) Halda (1997).

[1] **Ros** stemless, to 10 cm ∅, slowly proliferating, withdrawn into the soil in habitat; **L** many, ovate, 7 × 2 cm, incurved, smooth, usually **Sp**-less and dying back at the tips, pale green, translucent towards tips; **Inf** robust; **Fl** numerous, large, appressed to **Inf** axis, broad across and flat at the base of the tube.

H. longiana von Poellnitz (RSN 41: 203, 1937). **T:** B [lecto: unpubl. ill.]. – **D:** RSA (Eastern Cape: Baviaanskloof). **I:** Bayer (1982: fig. 54).

≡ *Haworthia pumila* ssp. *longiana* (von Poellnitz) Halda (1997); **incl.** *Haworthia longiana* var. *albinota* G. G. Smith (1948).

[2] **Ros** stemless, slowly proliferating, to 30 cm tall and 6 cm ∅; **L** to 30 × 2 cm, narrowly long attenuate, erect, incurving, surfaces minutely scabrid with small indistinctly raised tubercles (occasionally with small white tubercles); **Inf** sparsely branched, lax; **Fl** tube obcapitate, curved; **ITep** tips revolute.

H. maculata (von Poellnitz) M. B. Bayer (Haworthia Handb., 130, 1976). **T** [lecto]: RSA, Cape Prov. (*Venter* 6 [BOL]). – **D:** RSA (Western Cape).

≡ *Haworthia schuldtiana* var. *maculata* von Poellnitz (1940).

H. maculata var. **intermedia** (von Poellnitz) M. B. Bayer (Haworthia Revisited, 91, ills., 1999). **Incorrect name**, Art. 11.5. **T** [neo]: RSA, Western Cape (*Bayer* 4461 [NBG]). – **D:** RSA (Western Cape: Worcester).

≡ *Haworthia intermedia* von Poellnitz (1937).

[1] Differs from var. *maculata*: **L** slightly reticulate.

H. maculata var. **maculata** – **D:** RSA (Western Cape: Worcester-Robertson Karoo). **I:** Bayer (1982: fig. 21).

[1] **Ros** stemless, proliferous, to 8 cm ∅; **L** numerous, ovate-lanceolate, 6 × 1 cm, suberect to spreading, purplish-green, spotted with colourless dots, margins and keel with short **Sp**; **Inf** slender; **Fl** 15 - 20, only a few open together, white with yellowish throat and green veins.

H. magnifica von Poellnitz (RSN 33: 239-240, 1933). **T:** RSA (*Ferguson* s.n. [BOL]). – **D:** RSA (Western Cape).

≡ *Haworthia maraisii* var. *magnifica* (von Poellnitz) M. B. Bayer (1976) (*incorrect name*, Art. 11.5) ≡ *Haworthia retusa* var. *magnifica* (von Poellnitz) Halda (1997).

H. magnifica var. **acuminata** (M. B. Bayer) M. B. Bayer (Aloe 34(1-2): 6, 1997). **T:** RSA (*Bayer* s.n. [NBG †?]). – **D:** RSA (Western Cape: Albertinia). **I:** Bayer (1982: fig. 33b, as *H. retusa* var.).

≡ *Haworthia retusa* fa. *acuminata* M. B. Bayer (1976) ≡ *Haworthia retusa* var. *acuminata* (M. B. Bayer) M. B. Bayer (1982) ≡ *Haworthia acuminata* (M. B. Bayer) M. Hayashi (2000).

[1] Differs from var. *magnifica*: **L** with long acuminate truncate end-area.

H. magnifica var. **atrofusca** (G. G. Smith) M. B. Bayer (Nation. Cact. Succ. J. 32(1): 18, 1977). **T:** RSA (*Smith* 6169 [NBG]). – **D:** RSA (Western Cape: Riversdale). **I:** Bayer (1982: fig. 22b).

≡ *Haworthia atrofusca* G. G. Smith (1948).

[1] Differs from var. *magnifica*: **L** brownish-green, tips bluntly rounded.

H. magnifica var. **dekenahii** (G. G. Smith) M. B. Bayer (Aloe 34(1-2): 6, 1997). **T:** RSA (*Smith* 5489 [NBG, PRE]). – **D:** RSA (Western Cape: Albertinia). **I:** Bayer (1982: fig. 33c, as *H. retusa* var.).

≡ *Haworthia dekenahii* G. G. Smith (1944) ≡ *Haworthia retusa* var. *dekenahii* (G. G. Smith) M. B. Bayer (1982).

[1] Differs from var. *magnifica*: **L** with raised tubercles and silvery flecks, tips bluntly rounded.

H. magnifica var. **magnifica** – **D:** RSA (Western Cape). **I:** Bayer (1982: fig. 22a).

[1] **Ros** stemless, slowly proliferating, to 8 cm ⌀; **L** spreading, retuse with the end-area flush with ground level in habitat, dark green to purplish, margins scabrid to finely spined, end-area slightly translucent between the veins, surfaces with small slightly raised tubercles; **Inf** slender, to 40 cm; **Fl** 15 - 20, only few open together, white with brownish veins, **OTep** with pinched tips.

H. magnifica var. **splendens** S. A. Hammer & D. J. Venter (CSJA 70(4): 180-182, ills., 1998). **T:** RSA, Western Cape (*Venter* 93/57 [NBG]). – **D:** RSA (Western Cape: Albertinia).

≡ *Haworthia splendens* (S. A. Hammer & D. J. Venter) M. Hayashi (2000).

[1] Differs from var. *magnifica*: **L** with shiny raised tubercles.

H. ×mantelii Uitewaal (Succulenta 1947(4): 37-38, ill., 1947).

The garden hybrid *H. truncata × H. cuspidata* according to the protologue. *H. cuspidata* is treated as unresolved name in this treatment. The often asymmetrical rosettes with truncate leaves clearly show the influence of *H. truncata* as one parent. – [U. Eggli]

H. maraisii von Poellnitz (RSN 38: 194, 1935). **T:** RSA, Western Cape (*Marais* s.n. [B lecto: unpubl. ill.]]). – **D:** RSA (Western Cape: Robertson).

≡ *Haworthia magnifica* var. *maraisii* (von Poellnitz) M. B. Bayer (1977).

H. maraisii var. **maraisii** – **D:** RSA (Western Cape: Robertson). **I:** Bayer (1982: fig. 22d, as *H. magnifica* var.).

Incl. *Haworthia schuldtiana* von Poellnitz (1937); **incl.** *Haworthia sublimpidula* von Poellnitz (1937); **incl.** *Haworthia whitesloaneana* von Poell-

nitz (1937) ≡ *Haworthia schuldtiana* var. *whitesloaneana* (von Poellnitz) von Poellnitz (1940); **incl.** *Haworthia triebneriana* var. *diversicolor* Triebner & von Poellnitz (1938); **incl.** *Haworthia schuldtiana* var. *minor* Triebner & von Poellnitz (1940); **incl.** *Haworthia schuldtiana* var. *robertsonensis* von Poellnitz (1940); **incl.** *Haworthia schuldtiana* var. *simplicior* von Poellnitz (1940); **incl.** *Haworthia schuldtiana* var. *sublaevis* von Poellnitz (1940); **incl.** *Haworthia schuldtiana* var. *subtuberculata* von Poellnitz (1940); **incl.** *Haworthia schuldtiana* var. *unilineata* von Poellnitz (1940); **incl.** *Haworthia angustifolia* var. *subfalcata* von Poellnitz *ex* Zantner (1951) (*nom. inval.*, Art. 36.1).

[1] **Ros** stemless, slowly proliferating, 4 - 7 cm ⌀; **L** few to many, ovate-lanceolate, to 4 × 1 cm, very dark green, opaque, usually retuse, scabrid with small raised tubercles, tubercles occasionally spined, margins and keel with small **Sp**; **Inf** slender, to 30 cm; **Fl** frequently with yellow throat; tips of **OTep** pinched.

H. maraisii var. **meiringii** M. B. Bayer (Haworthia Handb., 134, 1976). **T:** RSA (*Bayer* s.n. [NBG]). – **D:** RSA (Western Cape: Bonnievale). **I:** Bayer (1982: fig. 22e, as *H. magnifica* var.).

≡ *Haworthia magnifica* var. *meiringii* (M. B. Bayer) M. B. Bayer (1977) ≡ *Haworthia divergens* var. *meiringii* (M. B. Bayer) M. Hayashi (2000).

[1] Differs from var. *maraisii*: **L** incurved, green, spiny.

H. maraisii var. **notabilis** (von Poellnitz) M. B. Bayer (Haworthia Handb., 141, 1976). **T:** B [lecto: unpubl. ill.]. – **D:** RSA (Western Cape: Robertson). **I:** Bayer (1982: fig. 22f, as *H. magnifica* var.).

≡ *Haworthia notabilis* von Poellnitz (1938) ≡ *Haworthia magnifica* var. *notabilis* (von Poellnitz) M. B. Bayer (1977); **incl.** *Haworthia schuldtiana* var. *erecta* Triebner & von Poellnitz (1940); **incl.** *Haworthia nitidula* var. *opaca* von Poellnitz (1948).

[1] Differs from var. *maraisii*: **L** erect, green, turgid.

H. marginata (Lamarck) Stearn (Cact. J. (Croydon) 7(2): 39, 1938). **T:** [lecto – icono]: Commelin, Praeludia Bot., t. 30, 1703. – **D:** RSA (Western Cape: Heidelberg). **I:** Bayer (1982: fig. 65).

≡ *Aloe marginata* Lamarck (1783); **incl.** *Aloe albicans* Haworth (1804) ≡ *Apicra albicans* (Haworth) Willdenow (1811) (*incorrect name*, Art. 11.3) ≡ *Haworthia albicans* (Haworth) Haworth (1812); **incl.** *Haworthia laevis* Haworth (1821) ≡ *Haworthia marginata* var. *laevis* (Haworth) H. Jacobsen (1955); **incl.** *Haworthia ramifera* Haworth (1821) ≡ *Aloe ramifera* (Haworth) Roemer & Schultes (1829) ≡ *Haworthia marginata* var. *ramifera* (Haworth) H. Jacobsen (1955); **incl.** *Haworthia virescens* Haworth (1821) ≡ *Aloe virescens* (Haworth) Roemer & Schultes (1829) ≡ *Haworthia al-*

bicans var. *virescens* (Haworth) Baker (1896) ≡ *Haworthia marginata* var. *virescens* (Haworth) Uitewaal (1939); **incl.** *Haworthia virescens* var. *minor* Haworth (1821).

[3] **Ros** stemless, slowly proliferating, to 20 cm tall; **L** 50 - 60, to 18 × 2 cm, lanceolate-deltoid, attenuate, spreading, pale brownish-green, smooth without tubercles; **Inf** sparsely branched, lax; **Fl** tube straight; **Tep** short, veins pinkish.

H. marumiana Uitewaal (Cact. Vetpl. (Amsterdam) 9: 20, 1940). **T:** RSA (*Anonymus* s.n. in *SUG* 6610 [AMD]). – **D:** RSA (Western Cape, Eastern Cape: Great Karoo).

≡ *Haworthia arachnoidea* var. *marumiana* (Uitewaal) Halda (1997).

H. marumiana var. **archeri** (W. F. Barker *ex* M. B. Bayer) M. B. Bayer (Haworthia Revisited, 104, ills. (pp. 104-105), 1999). **T:** RSA (*Archer* s.n. [NBG]). – **D:** RSA (Western Cape: SW Great Karoo). **I:** Bayer (1982: fig. 4a, as *H. archeri*).

≡ *Haworthia archeri* W. F. Barker *ex* M. B. Bayer (1981); **incl.** *Haworthia archeri* var. *archeri.*

[1] Differs from var. *marumiana*: **Ros** small, to 6 cm ∅; **L** brownish-green.

H. marumiana var. **batesiana** (Uitewaal) M. B. Bayer (Haworthia Revisited, 105, ill. (p. 106), 1999). **T:** RSA (*Ferguson* s.n. [AMD]). – **D:** RSA (Western Cape, Eastern Cape). **I:** Bayer (1982: fig. 5, as *H. batesiana*).

≡ *Haworthia batesiana* Uitewaal (1948) ≡ *Haworthia reticulata* ssp. *batesiana* (Uitewaal) Halda (1997).

[1] Differs from var. *marumiana*: **Ros** smaller, to 5 cm ∅; **L** smooth, bright green with pale reticulation.

H. marumiana var. **dimorpha** (M. B. Bayer) M. B. Bayer (Haworthia Revisited, 106, ills., 1999). **T:** RSA (*Bayer* 2092 [NBG]). – **D:** RSA (Western Cape: S Great Karoo, Touws River).

≡ *Haworthia archeri* var. *dimorpha* M. B. Bayer (1981) ≡ *Haworthia dimorpha* (M. B. Bayer) M. Hayashi (2000).

[1] Differs from var. *marumiana*: **Ros** to 12 cm ∅; **L** to 25, curving outwards (in cultivation).

H. marumiana var. **marumiana** – **D:** RSA (Eastern Cape: Great Karoo). **I:** Bayer (1982: fig. 23).

[1] **Ros** stemless, very proliferous, to 7 cm ∅; **L** erect, incurved, ± soft, purplish-green, opaque, with reticulate pattern, margins and keel with **Sp**; **Inf** to 20 cm; **Fl** smallish, white.

H. marumiana var. **viridis** M. B. Bayer (Haworthia Revisited, 107, ills., 1999). **T:** RSA, Western Cape (*Bayer* 3620 [NBG]). – **D:** RSA (Western Cape: S Karoo).

[1] Differs from var. *marumiana*: **L** light green, narrow, more erect.

H. maxima (Haworth) Duval (Pl. Succ. Horto Alencon., 7, 1809). **T:** [lecto – icono]: Commelin, Hort. Amstel. t. 10, 1701. – **D:** RSA (Western Cape: Worcester-Robertson Karoo, W Little Karoo). **I:** Bayer (1982: fig. 68, as *H. pumila*).

≡ *Aloe margaritifera* var. *maxima* Haworth (1804) ≡ *Apicra margaritifera* var. *maxima* (Haworth) Willdenow (1811) (*incorrect name*, Art. 11.3) ≡ *Aloe semimargaritifera* var. *maxima* (Haworth) Salm-Dyck (1817) ≡ *Haworthia semimargaritifera* var. *maxima* (Haworth) Haworth (1819) ≡ *Apicra maxima* (Haworth) Steudel (1821) (*incorrect name*, Art. 11.3) ≡ *Haworthia margaritifera* subvar. *maxima* (Haworth) A. Berger (1908) ≡ *Haworthia margaritifera* var. *maxima* (Haworth) Uitewaal (1947); **incl.** *Aloe pumila* var. *margaritifera* Linné (1753) (*nom. inval.*, Art. 43.1) ≡ *Aloe margaritifera* (Linné) Miller (1768) ≡ *Apicra margaritifera* (Linné) Willdenow (1811) (*incorrect name*, Art. 11.3) ≡ *Haworthia margaritifera* (Linné) Haworth (1819); **incl.** *Aloe pumila* Burman *fil.* (1768) (*nom. illeg.*, Art. 53.1); **incl.** *Aloe semimargaritifera* Salm-Dyck (1817) ≡ *Haworthia semimargaritifera* (Salm-Dyck) Haworth (1819) ≡ *Haworthia margaritifera* var. *semimargaritifera* (Salm-Dyck) Baker (1880).

[3] **Ros** stemless, slowly proliferating, to 25 cm tall; **L** 14 × 2 cm, almost as thick as wide, lanceolate-deltoid, attenuate, spreading, brownish-green to olive-green, surfaces scabrid with raised rounded non-confluent tubercles; **Inf** sparsely branched, lax, to 60 cm; **Fl** tube straight; **Tep** short, veins brownish-green.

The nomenclature is confused beause *H. pumila* (Miller) Duval is a synonym of *H. herbacea* (Miller) Stearn, and *H. margaritifera* (Linné) Haworth is a synonym for *H. minima* (Aiton) Haworth. Therefore, *H. maxima* (Haworth) Duval includes *H. pumila* and *H. margaritifera* in the sense of Bayer (1982) and Scott (1985).

H. minima (Aiton) Haworth (Synops. Pl. Succ., 92, 1812). **T:** [lecto – icono]: Dillenius, Hort. Eltham., t. 16: f. 18, 1732. – **D:** RSA (Western Cape: Swellendam to Mosselbay).

≡ *Aloe margaritifera* var. *minima* Aiton (1789) ≡ *Haworthia margaritifera* var. *minima* (Aiton) Uitewaal (1947) ≡ *Haworthia pumila* ssp. *minima* (Aiton) Halda (1997); **incl.** *Aloe pumila* var. β Linné (1753); **incl.** *Aloe pumila* var. γ Linné (1753); **incl.** *Aloe margaritifera* var. *major* Aiton (1789) ≡ *Haworthia major* (Aiton) Duval (1809) ≡ *Apicra margaritifera* var. *major* (Aiton) Willdenow (1811) (*incorrect name*, Art. 11.3); **incl.** *Aloe margaritifera* var. *minor* Aiton (1789) ≡ *Haworthia minor* (Aiton) Duval (1809) ≡ *Apicra minor* (Aiton) Steudel (1821) (*incorrect name*, Art. 11.3) ≡ *Aloe minor*

(Aiton) Roemer & Schultes (1829) ≡ *Haworthia margaritifera* var. *minor* (Aiton) Uitewaal (1947); **incl.** *Aloe margaritifera* var. *media* De Candolle (1799); **incl.** *Apicra granata* Willdenow (1811) (*incorrect name*, Art. 11.3) ≡ *Haworthia granata* (Willdenow) Haworth (1819) ≡ *Aloe granata* (Willdenow) Roemer & Schultes (1829) ≡ *Haworthia margaritifera* var. *granata* (Willdenow) Baker (1880); **incl.** *Haworthia brevis* Haworth (1819) ≡ *Aloe brevis* (Haworth) Roemer & Schultes (1829); **incl.** *Haworthia erecta* Haworth (1819) ≡ *Aloe erecta* (Haworth) Salm-Dyck (1829) ≡ *Haworthia margaritifera* var. *erecta* (Haworth) Baker (1880); **incl.** *Haworthia granata* var. *polyphylla* Haworth (1821) ≡ *Haworthia margaritifera* subvar. *polyphylla* (Haworth) von Poellnitz (1938); **incl.** *Aloe erecta* var. *laetivirens* Salm-Dyck (1834) ≡ *Haworthia margaritifera* subvar. *laetivirens* (Salm-Dyck) A. Berger (1908); **incl.** *Aloe granata* var. *major* Salm-Dyck (1834); **incl.** *Aloe granata* var. *minor* Salm-Dyck (1834) ≡ *Haworthia margaritifera* subvar. *minor* (Salm-Dyck) A. Berger (1908); **incl.** *Haworthia margaritifera* var. *corallina* Baker (1880); **incl.** *Haworthia mutabilis* von Poellnitz (1938).

H. minima var. **minima** – **D:** RSA (Western Cape). **I:** Bayer (1982: fig. 66).
[3] **Ros** stemless, slowly proliferating, to 15 cm tall; **L** to 13 × 1.5 cm, nearly as thick as wide, lanceolate-deltoid, attenuate, spreading, blue-green, surfaces scabrid with raised flattened non-confluent tubercles; **Inf** sparsely branched, lax, 30 - 40 cm; **Fl** tube straight; **Tep** short, veins pinkish.

H. minima var. **poellnitziana** (Uitewaal) M. B. Bayer (Haworthia Revisited, 213, ills., 1999). **T:** RSA (*Anonymus* s.n. in *SUG* 7796 [AMD]). – **D:** RSA (Western Cape: Worcester-Robertson Karoo). **I:** Bayer (1982: fig. 67, as *H. poellnitziana*).
≡ *Haworthia poellnitziana* Uitewaal (1939).
[3] Differs from var. *minima*: **L** slender, longer, to 18 cm, grey-green; **Tep** tips yellowish.

H. mirabilis (Haworth) Haworth (Synops. Pl. Succ., 95, 1812). **T:** [neo – icono]: Curtis's Bot. Mag. t. 1354, 1811. – **D:** RSA (Western Cape).
≡ *Aloe mirabilis* Haworth (1804) ≡ *Apicra mirabilis* (Haworth) Willdenow (1811) (*incorrect name*, Art. 11.3) ≡ *Haworthia retusa* var. *mirabilis* (Haworth) Halda (1997).

H. mirabilis var. **badia** (von Poellnitz) M. B. Bayer (Haworthia Revisited, 109, ills., 1999). **T:** [lecto – icono]: Kakteenkunde 1938: 76, ill, 1938. – **D:** RSA (Western Cape: Napier). **I:** Bayer (1982: fig. 25b).
≡ *Haworthia badia* von Poellnitz (1938) ≡ *Haworthia mirabilis* ssp. *badia* (von Poellnitz) M. B. Bayer (1976).
[1] Differs from var. *mirabilis*: **Ros** robust; **L** attenuate, retuse, smooth, deep shiny brown.

H. mirabilis var. **beukmanii** (von Poellnitz) M. B. Bayer (Haworthia Revisited, 110, ills., 1999). **T** [neo]: RSA (*Smith* 3969 [NBG]). – **D:** RSA (Western Cape: N Caledon).
≡ *Haworthia emelyae* var. *beukmanii* von Poellnitz (1940) ≡ *Haworthia mirabilis* fa. *beukmanii* (von Poellnitz) Pilbeam (1983) ≡ *Haworthia beukmanii* (von Poellnitz) M. Hayashi (2000).
[1] Differs from var. *mirabilis*: **Ros** very robust, to 12 cm ∅; **L** scabrid, strongly retuse, margins with short **Sp**.

H. mirabilis var. **calcarea** M. B. Bayer (Haworthia Revisited, 110, ill. (p. 111), 1999). **T:** RSA, Western Cape (*Burgers* 1648 [NBG]). – **D:** RSA (Western Cape: Bredasdorp Distr.).
≡ *Haworthia calcarea* (M. B. Bayer) M. Hayashi (2000).
[1] Differs from var. *mirabilis*: **Ros** proliferous; **L** short, erect, with short retuse end-area.

H. mirabilis var. **consanguinea** M. B. Bayer (Haworthia Revisited, 111, ills., 1999). **T:** RSA, Western Cape (*Bayer* s.n. [NBG]). – **D:** RSA (Western Cape: Riviersonderend Mts.).
≡ *Haworthia consanguinea* (M. B. Bayer) M. Hayashi (2000).
[1] Differs from var. *mirabilis*: **Ros** small, proliferous; **L** soft, turgid.

H. mirabilis var. **mirabilis** – **D:** RSA (Western Cape: Bredasdorp). **I:** Bayer (1982: fig. 25a following p. 16).
Incl. *Haworthia mundula* G. G. Smith (1946) ≡ *Haworthia mirabilis* ssp. *mundula* (G. G. Smith) M. B. Bayer (1976).
[1] **Ros** stemless, proliferous, to 7 cm ∅; **L** 10 - 15, 3 - 4 × 1.5 cm, dark green, markedly retuse, tips acute, face translucent and lined, marginal **Sp** turning reddish in sun; **Inf** slender, to 25 cm; **Fl** narrowly elongate, bud S-shaped, **lTep** pinched at tips.

H. mirabilis var. **paradoxa** (von Poellnitz) M. B. Bayer (Aloe 34(1-2): 6, 1997). **T** [neo]: RSA (*Ferguson* s.n. [BOL]). – **D:** RSA (Western Cape: Riversdale). **I:** Bayer (1982: fig. 22g, as *H. magnifica* var.).
≡ *Haworthia paradoxa* von Poellnitz (1933) ≡ *Haworthia maraisii* var. *paradoxa* (von Poellnitz) M. B. Bayer (1976) (*nom. inval.*, Art. 33.2) ≡ *Haworthia magnifica* var. *paradoxa* (von Poellnitz) M. B. Bayer (1977).
[1] Differs from var. *mirabilis*: **L** more densely maculate with opaque light coloured dots on both sides.

H. mirabilis var. **sublineata** (von Poellnitz) M. B. Bayer (Haworthia Revisited, 113, ills., 1999). **T** [neo]: RSA (*Smith* 3966 [NBG]). – **D:** RSA (Western Cape: Bredasdorp).

≡ *Haworthia triebneriana* var. *sublineata* von Poellnitz (1938) ≡ *Haworthia mirabilis* fa. *sublineata* (von Poellnitz) Pilbeam (1983).

[1] Differs from var. *mirabilis*: **L** long and slender.

H. mirabilis var. **triebneriana** (von Poellnitz) M. B. Bayer (Haworthia Revisited, 113, ills. (p. 114), 1999). **T:** RSA (*Helm s.n.* in *Triebner 841* [B lecto: unpubl. ill.]]). – **D:** RSA (Western Cape: Overberg). **I:** Bayer (1982: fig. 25c, as *H. mirabilis* ssp. *mundula*).

≡ *Haworthia triebneriana* von Poellnitz (1936); **incl.** *Haworthia willowmorensis* von Poellnitz (1937); **incl.** *Haworthia rossouwii* von Poellnitz (1938); **incl.** *Haworthia triebneriana* var. *depauperata* von Poellnitz (1938); **incl.** *Haworthia triebneriana* var. *multituberculata* von Poellnitz (1938); **incl.** *Haworthia triebneriana* var. *napierensis* Triebner & von Poellnitz (1938) ≡ *Haworthia mirabilis* fa. *napierensis* (Triebner & von Poellnitz) Pilbeam (1983); **incl.** *Haworthia triebneriana* var. *rubrodentata* Triebner & von Poellnitz (1938) ≡ *Haworthia mirabilis* fa. *rubrodentata* (Triebner & von Poellnitz) Pilbeam (1983); **incl.** *Haworthia triebneriana* var. *subtuberculata* Triebner & von Poellnitz (1938); **incl.** *Haworthia triebneriana* var. *turgida* Triebner (1938); **incl.** *Haworthia nitidula* von Poellnitz (1939); **incl.** *Haworthia triebneriana* var. *pulchra* von Poellnitz (1940).

[1] Differs from var. *mirabilis*: **Ros** slowly proliferating; **L** suberect to erect.

H. monticola Fourcade (Trans. Roy. Soc. South Afr. 21: 78, 1937). **T:** RSA (*Fourcade 2498* [K]). – **D:** RSA (Western Cape, Eastern Cape). **I:** Scott (1985: 57).

≡ *Haworthia chloracantha* var. *monticola* (Fourcade) Halda (1997); **incl.** *Haworthia divergens* M. B. Bayer (1976).

[1] **Ros** stemless, proliferous, 2 - 4 cm ∅; **L** 30 - 40, elongate lanceolate, 2 - 6 cm, tips incurving, margins and keel with small **Sp**, upper face often with pellucid spots; **Inf** slender, to 30 cm; **Fl** laxly arranged, 15 - 20, white.

H. monticola var. **asema** M. B. Bayer (Haworthia Revisited, 117, ills., 1999). **T:** RSA, Western Cape (*Venter & de Vries 12(85/83)* [NBG]). – **D:** RSA (Western Cape: Ladismith Distr.)

≡ *Haworthia asema* (M. B. Bayer) M. Hayashi (2000).

[1] Differs from typical *H. monticola*: **L** smoother, more turgid, generally shorter, more uniformly grey-green; flowering very much earlier.

H. monticola var. **monticola** – **D:** RSA (Western Cape, Eastern Cape: E Little Karoo).
[1] Description as for *H. monticola*.

H. mucronata Haworth (Suppl. Pl. Succ., 50, 1819). **T:** K [lecto – icono, publ. Brit. Cact. Succ. J. 1: 98, 1983]. – **D:** RSA (Western Cape: Little Karoo).

≡ *Aloe mucronata* (Haworth) Roemer & Schultes (1829) ≡ *Haworthia altilinea* var. *mucronata* (Haworth) von Poellnitz (1937); **incl.** *Haworthia limpida* Haworth (1819) ≡ *Aloe limpida* (Haworth) Roemer & Schultes (1829) ≡ *Haworthia altilinea* var. *limpida* (Haworth) von Poellnitz (1937) ≡ *Haworthia mucronata* var. *limpida* (Haworth) von Poellnitz (1940); **incl.** *Haworthia altilinea* fa. *typica* von Poellnitz (1938) (*nom. inval.*, Art. 24.3) ≡ *Haworthia mucronata* fa. *typica* (von Poellnitz) von Poellnitz (1940) (*nom. inval.*, Art. 24.3).

H. mucronata var. **habdomadis** (von Poellnitz) M. B. Bayer (Haworthia Revisited, 120, ills., 1999). **T** [neo]: RSA (*Barker & Lewis s.n.* ex cult. *NBG 2764/32* [BOL]). – **D:** RSA (Western Cape: Little Karoo). **I:** Bayer (1982: fig. 17a, as *H. habdomadis*). **Fig. XXIII.b**

≡ *Haworthia habdomadis* von Poellnitz (1938) ≡ *Haworthia inconfluens* var. *habdomadis* (von Poellnitz) M. B. Bayer (1976) (*incorrect name*, Art. 11.4); **incl.** *Haworthia habdomadis* var. *habdomadis*.

[1] Differs from var. *mucronata*: **L** tips rounded, margins and keel spiny.

H. mucronata var. **inconfluens** (von Poellnitz) M. B. Bayer (Haworthia Revisited, 121, ills. (p. 122), 1999). **T:** RSA (*Triebner 1031* [B lecto: unpubl. ill.]]). – **D:** RSA (Western Cape: Little Karoo). **I:** Bayer (1982: fig. 17b, as *H. habdomadis* var.).

≡ *Haworthia altilinea* fa. *inconfluens* von Poellnitz (1938) ≡ *Haworthia mucronata* fa. *inconfluens* (von Poellnitz) von Poellnitz (1940) ≡ *Haworthia inconfluens* (von Poellnitz) M. B. Bayer (1976) ≡ *Haworthia habdomadis* var. *inconfluens* (von Poellnitz) M. B. Bayer (1977); **incl.** *Haworthia bijliana* var. *joubertii* von Poellnitz (1937) ≡ *Haworthia setata* subvar. *joubertii* (von Poellnitz) von Poellnitz (1938) ≡ *Haworthia setata* var. *joubertii* (von Poellnitz) H. Jacobsen (1960) (*nom. inval.*, Art. 33.2).

[1] Differs from var. *mucronata*: **L** pale green, often without **Sp**.

H. mucronata var. **morrisiae** (von Poellnitz) von Poellnitz (RSN 49(1-4): 29, 1940). **T:** B [lecto: unpubl. ill.]. – **D:** RSA (Western Cape: Little Karoo). **I:** Bayer (1982: fig. 17c, as *H. habdomadis* var.).

≡ *Haworthia altilinea* var. *morrisiae* von Poellnitz (1938) ≡ *Haworthia inconfluens* var. *morrisiae* (von Poellnitz) M. B. Bayer (1976) ≡ *Haworthia habdomadis* var. *morrisiae* (von Poellnitz) M. B. Bayer (1977); **incl.** *Haworthia sakaii* M. Hayashi (2000).

[1] Differs from var. *mucronata*: **L** bright to emerald-green with brownish tips.

H. mucronata var. **mucronata** – **D:** RSA (Western Cape: Little Karoo). **I:** Bayer (1982: fig. 41a, as *H. unicolor*). **Fig. XXIII.a**

Incl. *Haworthia unicolor* var. *unicolor*; **incl.** *Haworthia integra* von Poellnitz (1933) ≡ *Haworthia reticulata* var. *integra* (von Poellnitz) Halda (1997); **incl.** *Haworthia helmiae* von Poellnitz (1937) ≡ *Haworthia unicolor* var. *helmiae* (von Poellnitz) M. B. Bayer (1976) ≡ *Haworthia aristata* ssp. *helmiae* (von Poellnitz) Halda (1997); **incl.** *Haworthia unicolor* von Poellnitz (1937); **incl.** *Haworthia mclarenii* von Poellnitz (1939) ≡ *Haworthia chloracantha* var. *mclarenii* (von Poellnitz) Halda (1997).

[1] **Ros** stemless, proliferous, 6 - 12 cm ∅; **L** 30 - 45, soft, incurved, broadly ovate-lanceolate, slightly pellucid, with translucent margins and keel, both often spiny; **Inf** robust, to 40 cm; **Fl** numerous, closely arranged, broad and flat across the base of the tube, white with green venation.

Here belongs *H. aristata* in the sense of von Poellnitz and Scott (1985).

H. mucronata var. **rycroftiana** (M. B. Bayer) M. B. Bayer (Haworthia Revisited, 124, ills., 1999). **T:** RSA, Western Cape (*Bayer* 1701 [NBG]). – **D:** RSA (Western Cape: Little Karoo). **I:** Bayer (1982: fig. 34, as *H. rycroftiana*).

≡ *Haworthia rycroftiana* M. B. Bayer (1981).

[1] Differs from var. *mucronata*: **Ros** compact; **L** shorter, broader and more turgid.

H. mutica Haworth (Revis. Pl. Succ., 55, 1821). **T:** K [lecto: ill. publ. Excelsa 8: 50, 1978]. – **D:** RSA (Western Cape). **I:** Bayer (1982: fig. 26).

≡ *Aloe mutica* (Haworth) Roemer & Schultes (1829) ≡ *Haworthia retusa* var. *mutica* (Haworth) Baker (1896); **incl.** *Haworthia otzenii* G. G. Smith (1945).

[1] **Ros** stemless, solitary, 6 - 8 cm ∅; **L** 12 - 15, 6 × 1.5 cm, retuse, brownish-green, in habitat developing a purplish cloudiness, barely pellucid with several longitudinal lines, tip blunt; **Inf** to 20 cm; **Fl** white with brownish veins.

H. mutica var. **mutica** – **D:** RSA (Western Cape: Overberg).

[1] Description as for *H. mutica*.

H. mutica var. **nigra** M. B. Bayer (Haworthia Revisited, 126, ills., 1999). **T:** RSA, Western Cape (*Smith* 5753 [NBG]). – **D:** RSA (Western Cape: Heidelberg area).

≡ *Haworthia silviae* var. *nigra* (M. B. Bayer) M. Hayashi (2000).

[1] Differs from typical *H. mutica*: **L** dark green.

H. nigra (Haworth) Baker (JLSB 18: 203, 1880). **T:** RSA, Cape Prov. (*Bowie* s.n. [[lecto – icono]: K [drawing]]). – **D:** RSA (Northern Cape, Western Cape, Eastern Cape).

≡ *Apicra nigra* Haworth (1825) ≡ *Aloe nigra* (Haworth) Roemer & Schultes (1829) ≡ *Haworthia venosa* ssp. *nigra* (Haworth) Halda (1997) ≡ *Haworthia viscosa* ssp. *nigra* (Haworth) Halda (1998).

H. nigra var. **diversifolia** (von Poellnitz) Uitewaal (Succulenta 1948: 51, 1948). **T** [neo]: RSA (*Bruyns* s.n. in *Karoo Garden* 435/75 [NBG]). – **D:** RSA (Western Cape, Northern Cape, Eastern Cape). **I:** Scott (1985: 30, as *H. nigra*).

≡ *Haworthia diversifolia* von Poellnitz (1937) ≡ *Haworthia schmidtiana* var. *diversifolia* (von Poellnitz) von Poellnitz (1938); **incl.** *Haworthia schmidtiana* fa. *nana* von Poellnitz (1938) ≡ *Haworthia nigra* fa. *nana* (von Poellnitz) Uitewaal (1948).

[2] Differs from var. *nigra*: **L** more tightly appressed to the stem, greyish-green, tubercles pale and confluent in transverse bands.

H. nigra var. **nigra** – **D:** RSA (Eastern Cape: Karoo). **I:** Bayer (1982: fig. 55).

Incl. *Haworthia schmidtiana* von Poellnitz (1929) ≡ *Haworthia nigra* var. *schmidtiana* (von Poellnitz) Uitewaal (1948); **incl.** *Haworthia schmidtiana* var. *angustata* von Poellnitz (1937) ≡ *Haworthia nigra* var. *angustata* (von Poellnitz) Uitewaal (1948) ≡ *Haworthia nigra* fa. *angustata* (von Poellnitz) Pilbeam (1983); **incl.** *Haworthia schmidtiana* var. *suberecta* von Poellnitz (1937) ≡ *Haworthia nigra* var. *suberecta* (von Poellnitz) Uitewaal (1948); **incl.** *Haworthia schmidtiana* var. *elongata* von Poellnitz (1938) ≡ *Haworthia nigra* var. *elongata* (von Poellnitz) Uitewaal (1948); **incl.** *Haworthia schmidtiana* var. *pusilla* von Poellnitz (1938) ≡ *Haworthia nigra* var. *pusilla* (von Poellnitz) Uitewaal (1948); **incl.** *Haworthia ryneveldii* von Poellnitz (1939).

[2] **Ros** usually caulescent, slowly proliferating or often stoloniferous, to 5 cm tall or occasionally taller; **L** to 3 × 1.5 cm, ovate-deltoid, erect or recurved-spreading, blackish to grey-green, opaque, surfaces scabrid with distinct raised non-confluent concolorous tubercles; **Inf** simple, lax, to 40 cm; **Fl** erect, tube obcapitate, **ITep** revolute.

H. nortieri G. G. Smith (JSAB 12: 13, 1946). **T:** RSA, Western Cape (*Smith* 1676a [NBG]). – **D:** RSA (Northern Cape, Western Cape).

≡ *Haworthia mucronata* var. *nortieri* (G. G. Smith) Halda (1997).

H. nortieri var. **globosiflora** (G. G. Smith) M. B. Bayer (Haworthia Handb., 119, 1976). **T:** RSA (*Smith* 7198 [NBG]). – **D:** RSA (Western Cape: N Ceres Karoo). **I:** Bayer (1982: fig. 27b).

≡ *Haworthia globosiflora* G. G. Smith (1950).

[1] Differs from var. *nortieri*: **Ros** compact; **L** ovate, with translucent dots; **Fl** ovoid to globose.

H. nortieri var. **nortieri** – **D:** RSA (Northern Cape,

N Western Cape: Namaqualand). **I:** Bayer (1982: fig. 27a).

Incl. *Haworthia nortieri* var. *giftbergensis* G. G. Smith (1950); **incl.** *Haworthia nortieri* var. *montana* G. G. Smith (1950).

[1] **Ros** stemless, proliferous, 3 - 5 cm ⌀; **L** 25 - 45, soft, suberect, 6 × 0.75 cm, ovate-lanceolate to obovate, pale to purplish-green, with translucent spots, margins and keel with small **Sp**; **Inf** slender, to 30 cm; **Fl** 15 - 20, greyish-white, throat yellowish.

H. nortieri var. **pehlemanniae** (C. L. Scott) M. B. Bayer (Haworthia Revisited, 130, ills., 1999). **T:** RSA (*Scott* 7450 [PRE]). – **D:** RSA (Western Cape: Laingsburg).

≡ *Haworthia pehlemanniae* C. L. Scott (1982) ≡ *Haworthia arachnoidea* var. *pehlemanniae* (C. L. Scott) Halda (1997).

[1] Differs from var. *nortieri*: **L** greyish-green, unspotted.

H. outeniquensis M. B. Bayer (Haworthia Revisited, 130-131, ills., 1999). **T:** RSA, Western Cape (*Venter & al.* 94/61 [NBG]). – **D:** RSA (Western Cape: Oudtshoorn Distr.).

[1] **Ros** stemless, 4 - 6 cm ⌀, proliferous; **L** erect to suberect, to 60 × 6 - 10 mm, 2 - 3 mm thick, incurving tips with a 2 mm long terminal awn, surfaces with pellucid anastomosing dots, yellowish-green, upper face convex with 4 - 5 prominent rows of conspicuous pellucid dots, lower face convex with 3 - 6 rows of pellucid dots, with a sharper keel with spines to 1 mm, margins similarly spined; **Inf** with 7 - 15 **Fl**; **Ped** 4 - 6 mm; **Per** white and yellowish-green, 15 mm.

H. parksiana von Poellnitz (RSN 41: 205, 1937). **T:** RSA (*Helms* s.n. in *Parks* 636/32 [B [lecto: ill. publ. Desert Pl. Life 10: 48, 1938]]). – **D:** RSA (Western Cape: Great Brak). **I:** Bayer (1982: fig. 28).

[1] **Ros** stemless, proliferous, 3 - 4 cm ⌀; **L** 25 - 35, triangular sublanceolate, 1.5 - 3 cm, blackish-green, sharply recurved, minutely tubercled, tip barely pointed; **Inf** slender, to 20 cm; **Fl** few, narrow, whitish with dull greenish venation.

H. pubescens M. B. Bayer (JSAB 38: 129-130, ills. (pp. 126-127), 1973). **T:** RSA, Western Cape (*Bayer* s.n. in *Karoo Garden* 112/70 [NBG]). – **D:** RSA (Western Cape: Worcester-Robertson Karoo). **I:** Bayer (1982: fig. 29).

[1] **Ros** stemless, rarely proliferating, to 4 cm ⌀; **L** 20 - 35, ovate-lanceolate, 5 × 0.8 cm, shortly incurved, opaque grey-green, covered with minute **Sp**; **Inf** to 20 cm; **Fl** 10 - 15, white with pinkish venation, upper **Tep** tips flaring.

H. pubescens var. **livida** M. B. Bayer (Haworthia

Revisited, 134, ills., 1999). **T:** RSA, Western Cape (*Bayer* 1128 [NBG]). – **D:** RSA (Western Cape: Worcester Distr.).

≡ *Haworthia maraisii* var. *livida* (M. B. Bayer) M. Hayashi (2000).

[1] Differs from var. *pubescens*: Plants less pubescent; **L** slightly broader and fewer, partly with pellucid spots.

H. pubescens var. **pubescens** – **D:** RSA (Western Cape: Worcester Distr.).

[1] **L** without pellucid spots.

H. pulchella M. B. Bayer (JSAB 39: 232, 1973). **T:** RSA (*Bayer* 162 in *Karoo Garden* 43/71 [NBG]). – **D:** RSA (Western Cape: Touws Rivier). **I:** Bayer (1982: fig. 30).

≡ *Haworthia chloracantha* var. *pulchella* (M. B. Bayer) Halda (1997).

[1] **Ros** stemless, occasionally proliferous, to 5 cm ⌀; **L** 30 - 45, narrowly triangular, incurved, coriaceous, dark to emerald-green, with block-patterned reticulation, margins and keel with pronounced whitish **Sp**; **Inf** slender, to 30 cm; **Fl** 15 - 20, white.

H. pulchella var. **globifera** M. B. Bayer (Haworthia Revisited, 136, ill., 1999). **T:** RSA, Western Cape (*Bruyns* 7338 [BOL]). – **D:** RSA (Western Cape: Anysberg).

≡ *Haworthia globifera* (M. B. Bayer) M. Hayashi (2000).

[1] Differs from var. *pulchella*: **Ros** glabrous, slightly caulescent, forming clusters.

H. pulchella var. **pulchella** – **D:** RSA (Western Cape: Touws Rivier).

[1] **Ros** stemless, occasionally proliferous.

H. pungens M. B. Bayer (Haworthia Revisited, 188-189, ills., 1999). **T:** RSA, Eastern Cape (*Bruyns* 7090 [BOL]). – **D:** RSA (Eastern Cape: Willowmore). **Fig. XXIII.c**

[2] **Ros** to 6 cm ⌀, caulescent, proliferating; **L** many, usually in 5 (rarely 3) rows, to 5 × 1.8 cm, spreading, smooth, rigid, sharp-pointed, green in shade, darkening and reddening in sun; **Inf** simple, to 30 cm; **Fl** with straight tube; **ITep** revolute.

Similar to *H. viscosa* and superficially reminiscent of the genus *Astroloba*.

H. pygmaea von Poellnitz (RSN 27: 132, 1929). **T** [neo]: RSA (*Fourcade* 4759 [BOL]). – **D:** RSA (Western Cape).

H. pygmaea var. **argenteo-maculosa** (G. G. Smith) M. B. Bayer (Aloe 34(1-2): 6, 1997). **T:** RSA (*Emett* s.n. [NBG 68037, PRE]). – **D:** RSA (Western Cape: Albertinia). **I:** Bayer (1976: 68, as *H. dekenahii* var.).

≡ *Haworthia dekenahii* var. *argenteo-maculosa*
G. G. Smith (1945) ≡ *Haworthia retusa* fa. *argen-teo-maculosa* (G. G. Smith) M. B. Bayer (1976);
incl. *Haworthia silviae* M. Hayashi (2000).

[1] Differs from var. *pygmaea*: **L** almost smooth,
conspicuously white-spotted.

H. pygmaea var. **pygmaea** – **D:** RSA (Western
Cape: Mossel Bay). **I:** Bayer (1982: fig. 31).

Incl. *Haworthia pygmaea* fa. *crystallina* Pilbeam
(1983); **incl.** *Haworthia pygmaea* fa. *major* Pilbeam
(1983) (*nom. inval.*, Art. 37.1).

[1] **Ros** stemless, slowly proliferating, 6 - 10 cm
⌀; **L** 12 - 15, 6 × 1.8 cm, retuse, round-tipped, surface pellucid with obscure raised tubercles, sometimes intensely papillose; **Inf** robust, to 30 cm; **Fl**
15 - 20, white with greenish veins.

H. reinwardtii (Salm-Dyck) Haworth (Revis. Pl.
Succ., 53, 1821). **T:** [neo – icono]: Salm-Dyck,
Monogr. Gen. Aloes & Mesembr. 1: Aloe t. 12
(Sect. 6: 16), 1836. – **D:** RSA (Eastern Cape).

≡ *Aloe reinwardtii* Salm-Dyck (1821).

H. reinwardtii fa. **chalumnensis** (G. G. Smith) M.
B. Bayer (Haworthia Handb., 106, 1976). **T:** RSA
(*Smith* 513 [NBG, PRE]). – **D:** RSA (Eastern Cape:
Chalumna). **I:** Bayer (1982: fig. 57b).

≡ *Haworthia reinwardtii* var. *chalumnensis* G. G.
Smith (1943).

[2] Differs from var. *reinwardtii*: **L** elongate, incurved, conspicuously tuberculate.

H. reinwardtii fa. **kaffirdriftensis** (G. G. Smith)
M. B. Bayer (Haworthia Handb., 126, 1976). **T:**
RSA (*Smith* 3364 [NBG, PRE]). – **D:** RSA (Eas-
tern Cape: Fish River). **I:** Bayer (1982: fig. 57c).

≡ *Haworthia reinwardtii* var. *kaffirdriftensis* G.
G. Smith (1943).

[2] Differs from var. *reinwardtii*: Outer side of **L**
with tubercles in longitudinal rows.

H. reinwardtii fa. **olivacea** (G. G. Smith) M. B.
Bayer (Haworthia Handb., 142, 1976). **T:** RSA
(*Smith* 5260 [NBG, PRE]). – **D:** RSA (Eastern
Cape: Fish River). **I:** Bayer (1976: 66).

≡ *Haworthia reinwardtii* var. *olivacea* G. G.
Smith (1944).

[2] Differs from var. *reinwardtii*: **L** olive-green,
relatively smooth, tubercles rounder and sparser.

H. reinwardtii fa. **zebrina** (G. G. Smith) M. B. Bayer (Nation. Cact. Succ. J. 32(1): 18, 1977). **T:**
RSA, Eastern Cape (*Smith* 5258 [NBG, PRE]). –
D: RSA (Eastern Cape: Fish River). **I:** Bayer
(1976: 65).

≡ *Haworthia reinwardtii* var. *zebrina* G. G. Smith
(1944).

[2] Differs from var. *reinwardtii*: Outside of **L**
with tubercles conspicuous in prominent transverse
white bands.

H. reinwardtii var. **brevicula** G. G. Smith (JSAB
10: 11, 1944). **T:** RSA, Eastern Cape (*Smith* 3138
[NBG, PRE]). – **D:** RSA (Eastern Cape: Grahamstown). **I:** Bayer (1982: fig. 57f).

≡ *Haworthia reinwardtii* fa. *brevicula* (G. G.
Smith) Pilbeam (1975) (*nom. inval.*, Art. 33.3); **incl.**
Haworthia reinwardtii var. *diminuta* G. G. Smith
(1948) ≡ *Haworthia reinwardtii* fa. *diminuta* (G. G.
Smith) Pilbeam (1975) (*nom. inval.*, Art. 33.3).

[2] Differs from var. *reinwardtii*: **Ros** small, to
10 cm tall and 4 cm ⌀.

H. reinwardtii var. **reinwardtii** – **D:** RSA (Eastern
Cape). **I:** Bayer (1982: fig. 57a). **Fig. XXIII.d**

Incl. *Haworthia reinwardtii* var. *major* Baker
(1880); **incl.** *Haworthia reinwardtii* var. *archibaldi-
ae* von Poellnitz (1937); **incl.** *Haworthia reinwardtii* var. *pulchra* von Poellnitz (1937); **incl.** *Haworthia reinwardtii* var. *peddiensis* G. G. Smith (1943);
incl. *Haworthia reinwardtii* var. *triebneri* Resende
(1943); **incl.** *Haworthia reinwardtii* var. *valida* G.
G. Smith (1943); **incl.** *Haworthia reinwardtii* var.
haworthii Resende (1943) (*nom. illeg.*, Art. 52.1,
26.1); **incl.** *Haworthia reinwardtii* var. *grandicula*
G. G. Smith (1944).

[2] **Ros** to 10 cm ⌀, caulescent, to 20 cm tall,
proliferating; **L** numerous, to 7 × 2 cm, ratio stem
diameter : **L** width = 1:1.2, **L** erectly spreading or
incurved, scabrid, brownish-green, usually with
flattened **Sc**-like tubercles; **Inf** simple or occasionally compound, to 30 cm; **Fl** 15 - 20, tube obcapitate, curved, **ITep** revolute.

See comment for *H. coarctata* var. *coarctata*.

H. reticulata (Haworth) Haworth (Synops. Pl.
Succ., 94, 1812). **T:** [neo – icono]: Curtis's Bot.
Mag. t. 1314, 1810. – **D:** RSA (Western Cape).

≡ *Aloe reticulata* Haworth (1804) ≡ *Aloe arachnoidea* var. *reticulata* (Haworth) Ker Gawler
(1811) ≡ *Apicra reticulata* (Haworth) Willdenow
(1811) (*incorrect name*, Art. 11.3).

H. reticulata var. **attenuata** M. B. Bayer (Hawor-
thia Revisited, 140, ills., 1999). **T:** RSA, Western
Cape (*Smith* 3979 [NBG]). – **D:** RSA (Western
Cape: Bonnievale).

[1] Differs from var. *reticulata*: **L** longer, more
slender.

H. reticulata var. **hurlingii** (von Poellnitz) M. B.
Bayer (New Haworthia Handb., 52, 1982). **T:** RSA,
Western Cape (*Hurling* s.n. [B [lecto: ill. publ. De-
sert Pl. Life 10: 125, 1938]]). – **D:** RSA (Western
Cape: W of Bonnievale). **I:** Bayer (1982: fig. 32b).

≡ *Haworthia hurlingii* von Poellnitz (1937).

[1] Differs from var. *reticulata*: **Ros** small, to 3
cm ⌀; **L** obtuse.

H. reticulata var. **reticulata** – **D:** RSA (Western
Cape: Worcester-Robertson Karoo). **I:** Bayer
(1982: fig. 32a).

Incl. *Aloe pumilio* Jacquin (1804) ≡ *Apicra pumilio* (Jacquin) Willdenow (1811) (*incorrect name*, Art. 11.3); **incl.** *Haworthia hurlingii* var. *ambigua* Triebner & von Poellnitz (1938); **incl.** *Haworthia reticulata* var. *acuminata* von Poellnitz (1938); **incl.** *Haworthia guttata* Uitewaal (1947).

[1] **Ros** proliferous, sometimes with a short stem, to 8 cm ⌀; **L** 25 - 40, lanceolate-acuminate, 6 × 1 cm, firmly suberect, incurved, yellowish-green, opaque with reticulate to mottled patterning, reddening in the sun, margins and keel frequently with short **Sp**; **Inf** to 25 cm; **Fl** 20 - 30, large, white to pinkish, buds arcuate with flattened tips.

H. reticulata var. **subregularis** (Baker) Pilbeam (Haworthia and Astroloba Coll. Guide, 116, 1983). **T:** [lecto – icono]: Refug. Bot. 4: t. 232, 1871. – **D:** RSA (Western Cape: Worcester).

≡ *Haworthia subregularis* Baker (1871); **incl.** *Haworthia haageana* von Poellnitz (1930); **incl.** *Haworthia haageana* var. *subreticulata* von Poellnitz (1937).

[1] Differs from var. *reticulata*: **Ros** > 10 cm ⌀; **L** more suberect to spreading.

H. retusa (Linné) Duval (Pl. Succ. Horto Alencon., 7, 1809). **T:** [lecto – icono]: Commelin, Horti Med. Amstelod. 2: t. 6, 1701. – **D:** RSA (Western Cape: Riversdale). **I:** Bayer (1982: fig. 33a).

≡ *Aloe retusa* Linné (1753) ≡ *Catevala retusa* (Linné) Medikus (1786) ≡ *Apicra retusa* (Linné) Willdenow (1811) (*incorrect name*, Art. 11.3); **incl.** *Haworthia fouchei* von Poellnitz (1940) ≡ *Haworthia retusa* fa. *fouchei* (von Poellnitz) Pilbeam (1983); **incl.** *Haworthia retusa* var. *densiflora* G. G. Smith (1946); **incl.** *Haworthia retusa* var. *multilineata* G. G. Smith (1946) ≡ *Haworthia retusa* fa. *multilineata* (G. G. Smith) Pilbeam (1983) ≡ *Haworthia multilineata* (G. G. Smith) C. L. Scott (1985); **incl.** *Haworthia retusa* var. *solitaria* G. G. Smith (1946) ≡ *Haworthia solitaria* (G. G. Smith) C. L. Scott (1973); **incl.** *Haworthia geraldii* C. L. Scott (1965) ≡ *Haworthia retusa* fa. *geraldii* (C. L. Scott) Pilbeam (1983) (*nom. inval.*, Art. 33.2).

[1] **Ros** stemless, slowly or rarely proliferating, to 12 cm ⌀; **L** 10 - 15, turgid, rigid, with pronouncedly retuse end-areas, 8 × 2 cm, brownish or green and rarely with purplish hue, variously lined and windowed, surface and usually also margins and keel without **Sp** or tubercles, tips pointed; **Inf** robust, to 30 cm; **Fl** 20 - 30, closely spaced, white with greenish-brown veins.

H. scabra Haworth (Suppl. Pl. Succ., 58, 1819). **T:** K [lecto: ill. publ. Cact. Succ. J. (US) 52: 274, 1980]. – **D:** RSA (Western Cape, Eastern Cape).

≡ *Aloe scabra* (Haworth) Roemer & Schultes (1829).

H. scabra var. **lateganiae** (von Poellnitz) M. B. Bayer (Haworthia Revisited, 195, ills., 1999). **T:** B [lecto: ill. publ. Desert Pl. Life 9: 103, 1937]. – **D:** RSA (Western Cape: Little Karoo, E of Oudtshoorn). **I:** Bayer (1982: fig. 60b, as *H. starkiana* var.).

≡ *Haworthia lateganiae* von Poellnitz (1937) ≡ *Haworthia starkiana* var. *lateganiae* (von Poellnitz) M. B. Bayer (1976).

[2] Differs from var. *scabra*: **L** long and slender, smooth and shiny, dark green.

H. scabra var. **morrisiae** (von Poellnitz) M. B. Bayer (Haworthia Handb., 137, 1976). **T:** B [lecto: ill., publ. Kakteenkunde 1937: 132]. – **D:** RSA (Western Cape: Little Karoo, Schoemanspoort). **I:** Bayer (1982: fig. 58b).

≡ *Haworthia morrisiae* von Poellnitz (1937).

[2] **L** surfaces minutely scabrid, tubercles small and confluent.

H. scabra var. **scabra** – **D:** RSA (Eastern Cape: Baviaanskloof). **I:** Bayer (1982: fig. 58a).

Incl. *Haworthia tuberculata* von Poellnitz (1931) ≡ *Haworthia scabra* var. *tuberculata* (von Poellnitz) Halda (1997); **incl.** *Haworthia tuberculata* var. *acuminata* von Poellnitz (1938); **incl.** *Haworthia tuberculata* var. *subexpansa* von Poellnitz (1938); **incl.** *Haworthia tuberculata* var. *sublaevis* von Poellnitz (1938); **incl.** *Haworthia tuberculata* var. *angustata* von Poellnitz (1940).

[2] **Ros** stemless, slowly proliferating, to 16 cm tall, 6 cm ⌀; **L** 12 - 25, to 1.6 × 2.2 cm, triangular-lanceolate, attenuate, almost as thick as wide, dull green covered with dust, incurved, surfaces scabrid or smooth, with or without distinct raised non-confluent concolourous tubercles; **Inf** sparsely branched, lax, to 48 cm; **Fl** 15 - 20, tube obcapitate, curved; **ITep** revolute.

H. scabra var. **starkiana** (von Poellnitz) M. B. Bayer (Haworthia Revisited, 197, ills., 1999). **T:** RSA, Western Cape (*Taylor* s.n. [B [lecto: unpubl. ill.]]). – **D:** RSA (Western Cape: Little Karoo, Schoemanspoort). **I:** Bayer (1982: fig. 60a, as *H. starkiana*).

≡ *Haworthia starkiana* von Poellnitz (1933) ≡ *Haworthia scabra* ssp. *starkiana* (von Poellnitz) Halda (1997); **incl.** *Haworthia starkiana* var. *starkiana*; **incl.** *Haworthia taylorii* W. F. Barker *ms.* (s.a.) (*nom. inval.*, Art. 29.1); **incl.** *Haworthia smitii* von Poellnitz (1938) ≡ *Haworthia pumila* var. *smitii* (von Poellnitz) Halda (1997).

[2] Differs from var. *scabra*: **Ros** larger, forming clumps; **L** smooth, shiny, without tubercles, yellowish-green.

H. semiviva (von Poellnitz) M. B. Bayer (Haworthia Handb., 153, 1976). **T:** [lecto – icono]: B (ill., publ. Succulenta 22: 25, 1940). – **D:** RSA (Northern Cape, Western Cape). **I:** Bayer (1982: fig. 35).

≡ *Haworthia bolusii* var. *semiviva* von Poellnitz (1938) ≡ *Haworthia arachnoidea* var. *semiviva* (von Poellnitz) Halda (1997).

[1] **Ros** stemless, rarely proliferating, 5 - 6 cm ∅; **L** 30 - 40, broadly ovate, 6 × 1.5 cm, thin, incurved, pale green, translucent and usually drying up from the tips; **Inf** 20 - 30 cm; **Fl** 30 - 35, white with green venation, broad and flat across the upper base of the tube.

H. serrata M. B. Bayer (JSAB 39: 249, 1973). **T:** RSA, Western Cape (*Bayer* 166 [NBG]). – **D:** RSA (Western Cape: Heidelberg). **I:** Bayer (1982: fig. 36).

≡ *Haworthia chloracantha* var. *serrata* (M. B. Bayer) Halda (1997).

[1] **Ros** stemless, rarely proliferating, to 7 cm ∅; **L** 20 - 30, narrow, 6 × 1 cm, acuminate, bright yellowish-green with translucent lines above, margins and keel spiny; **Inf** robust, to 40 cm; **Fl** 20 - 30, white with green venation.

H. sordida Haworth (Revis. Pl. Succ., 51, 1821). **T:** [neo – icono]: Salm-Dyck, Monogr. Gen. Aloes & Mesembr. 7: Aloe t. 1 (Sect. 7: 2), 1863. – **D:** RSA (Eastern Cape).

≡ *Aloe sordida* (Haworth) Roemer & Schultes (1829) ≡ *Haworthia scabra* ssp. *sordida* (Haworth) Halda (1997) ≡ *Haworthia scabra* var. *sordida* (Haworth) Halda (1997).

H. sordida var. **lavrani** C. L. Scott (CSJA 53(3): 124-126, ill., 1981). **T:** RSA, Eastern Cape (*Hechter* s.n. [PRE 61124]). – **D:** RSA (Eastern Cape: Little Karoo). **I:** Scott (1985: 8).

≡ *Haworthia scabra* var. *lavrani* (C. L. Scott) Halda (1997).

[2] Differs from var. *sordida*: **Ros** smaller, to 5 cm ∅; **L** shorter, recurved.

H. sordida var. **sordida** – **D:** RSA (Eastern Cape: E Little Karoo). **I:** Bayer (1982: fig. 59).

Incl. *Haworthia agavoides* Zantner & von Poellnitz (1938) ≡ *Haworthia sordida* var. *agavoides* (Zantner & von Poellnitz) G. G. Smith (1950).

[2] **Ros** stemless, rarely proliferating, to 15 cm tall, 8 cm ∅; **L** 6 - 15, to 15 × 2 cm, lanceolate-deltoid, attenuate, erect, dark grey to blackish-green, surfaces scabrid with indistinct slightly raised nonconfluent tubercles, margins obtuse; **Inf** sparsely branched, lax, to 40 cm; **Fl** tube straight; **ITep** revolute.

H. springbokvlakensis C. L. Scott (JSAB 36: 287-288, 1970). **T:** RSA, Eastern Cape (*Scott* 245 [PRE]). – **D:** RSA (Eastern Cape: E Little Karoo). **I:** Bayer (1982: fig. 37).

≡ *Haworthia retusa* var. *springbokvlakensis* (C. L. Scott) Halda (1997).

[1] **Ros** stemless, solitary, to 10 cm ∅; **L** 8 - 12,

turgid, 6 × 1.5 cm, very rounded and retuse with translucent end-area and several short longitudinal lines, smooth; **Inf** 20 - 25 cm; **Fl** white with brownish venation.

H. truncata Schönland (Trans. Roy. Soc. South Afr. 1: 391, 1910). **T:** RSA, Western Cape (*Britten* s.n. [K]). – **D:** RSA (Western Cape: Little Karoo).

H. truncata var. **maughanii** (von Poellnitz) B. Fearn (Nation. Cact. Succ. J. 21(1): 28-29, 1966). **T** [neo]: RSA, Western Cape (*Malherbe* s.n. in *NBG* 307/40 [NBG 68307]). – **D:** RSA (Western Cape: Little Karoo, Calitzdorp). **I:** Bayer (1982: fig. 24, as *H. maughanii*).

≡ *Haworthia maughanii* von Poellnitz (1933).

[1] Differs from var. *turgida*: **L** multifarious, round in cross-section.

H. truncata var. **truncata** – **D:** RSA (Western Cape: C Little Karoo). **I:** Bayer (1982: fig. 39). **Fig. XXIII.e**

Incl. *Haworthia truncata* fa. *crassa* von Poellnitz (1938); **incl.** *Haworthia truncata* fa. *normalis* von Poellnitz (1938); **incl.** *Haworthia truncata* fa. *tenuis* von Poellnitz (1938) ≡ *Haworthia truncata* var. *tenuis* (von Poellnitz) M. B. Bayer (1976).

[1] **Ros** stemless, slowly proliferating; **L** distichous, 10 - 12, abruptly truncate with flat to slightly corrugated and subpellucid end-areas, 1.2 - 4 cm wide, 0.3 - 1 cm thick, dark grey-green, scabrid with minute tubercles; **Inf** to 20 cm; **Fl** 20 - 30, white with brownish veins.

H. turgida Haworth (Suppl. Pl. Succ., 52, 1819). **T** [neo]: RSA, Western Cape (*Bayer* 2420 [NBG 132378]). – **D:** RSA (Western Cape).

≡ *Aloe turgida* (Haworth) Roemer & Schultes (1829); **incl.** *Haworthia rodinii* hort. (s.a.) (*nom. inval.*, Art. 29.1).

H. turgida var. **longibracteata** (G. G. Smith) M. B. Bayer (Haworthia Revisited, 154, ills. (pp. 154-155), 1999). **T:** RSA, Western Cape (*Dekenah* 18 in *Smith* 5378 [NBG, PRE]). – **D:** RSA (Western Cape). **I:** Scott (1985: 127, as *H. longibracteata*).

≡ *Haworthia longibracteata* G. G. Smith (1945) ≡ *Haworthia retusa* fa. *longibracteata* (G. G. Smith) Pilbeam (1983).

[1] Differs from var. *turgida*: **L** erect to suberect, ovate-lanceolate.

H. turgida var. **suberecta** von Poellnitz (RSN 45: 134, 1938). **T** [neo]: RSA, Western Cape (*Bayer* s.n. in *Karoo Garden* 631/69 [NBG]). – **D:** RSA (Western Cape). **I:** Scott (1985: 126, as *H. dekenahii*).

≡ *Haworthia turgida* fa. *suberecta* (von Poellnitz) Pilbeam (1983); **incl.** *Haworthia turgida* var. *subtuberculata* von Poellnitz (1938); **incl.** *Haworthia* *turgida* var. *pallidifolia* G. G. Smith (1946) ≡ *Haworthia turgida* fa. *pallidifolia* (G. G. Smith) Pilbeam (1983).

[1] Differs from var. *turgida*: **L** strongly mottled, tips slightly truncate and rounded.

H. turgida var. **turgida** – **D:** RSA (Western Cape: Swellendam to Riversdale). **I:** Bayer (1982: fig. 40).

Incl. *Haworthia laetevirens* Haworth (1819) ≡ *Aloe laetevirens* (Haworth) Link (1822); **incl.** *Haworthia caespitosa* von Poellnitz (1937) ≡ *Haworthia turgida* fa. *caespitosa* (von Poellnitz Pilbeam (1983); **incl.** *Haworthia caespitosa* fa. *subplana* von Poellnitz (1938); **incl.** *Haworthia caespitosa* fa. *subproliferans* von Poellnitz (1938).

[1] **Ros** partially stemless, proliferous, 5 - 10 cm ∅; **L** 20 - 40, ovate-lanceolate, 4 × 1.2 cm, turgid, often as thick as broad, recurved or slightly retuse, yellow-green to pink in sun, generally mottled, margins and keel lightly spined; **Inf** 15 - 20 cm; **Fl** 20 - 30, slender, brownish-white with darker venation.

H. variegata L. Bolus (J. Bot. 67: 137, 1929). **T:** RSA (*Ferguson* s.n. [BOL]). – **D:** RSA (Western Cape). **I:** Scott (1985: 57).

≡ *Haworthia chloracantha* ssp. *variegata* (L. Bolus) Halda (1997) ≡ *Haworthia chloracantha* var. *variegata* (L. Bolus) Halda (1997).

[1] **Ros** stemless, proliferous, to 4 cm ∅; **L** 30 - 40, erect, slender lanceolate, dark green, variegated, margins and keel spined; **Inf** slender, to 35 cm, lax; **Fl** 15 - 20, greenish-white with brownish venation.

H. variegata var. **hemicrypta** M. B. Bayer (Haworthia Revisited, 158, ills., 1999). **T:** RSA, Western Cape (*Burgers* 2582 [NBG]). – **D:** RSA (Western Cape: Bredasdorp).

≡ *Haworthia hemicrypta* (M. B. Bayer) M. Hayashi (2000).

[1] Differs from var. *variegata*: Plants moderately variegated to plain; **L** long and slender, tending to arch out and then curve inwards.

H. variegata var. **modesta** M. B. Bayer (Haworthia Revisited, 159, ills., 1999). **T:** RSA, Western Cape (*Bayer* 2551 [NBG]). – **D:** RSA (Western Cape: Bredasdorp).

≡ *Haworthia modesta* (M. B. Bayer) M. Hayashi (2000).

[1] Differs from var. *variegata*: **L** broader and shorter and with less conspicuous spination.

H. variegata var. **petrophila** M. B. Bayer (Haworthia Revisited, 159, ills. (pp. 159-160), 1999). **T:** RSA, Western Cape (*Burgers* 2158 [NBG]). – **D:** RSA (Western Cape: Bredasdorp).

≡ *Haworthia petrophila* (M. B. Bayer) M. Hayashi (2000).

[1] Differs from var. *variegata*: **Ros** very proliferous; **L** slender, shorter and incurved, strongly spined.

H. variegata var. **variegata** – **D:** RSA (Western Cape: Riversdale).

[1] Plants highly variegated.

H. venosa (Lamarck) Haworth (Revis. Pl. Succ., 51, 1821). **T:** [lecto – icono]: Commelin, Praeludia Bot. t. 29, 1703. – **D:** Namibia, RSA (Northern Cape, Western Cape, Eastern Cape).

≡ *Aloe venosa* Lamarck (1783).

H. venosa ssp. **granulata** (Marloth) M. B. Bayer (Haworthia Handb., 120, 1976). **T:** RSA (*Marloth* 4217 [BOL]). – **D:** RSA (Northern Cape, Western Cape: Tanqua (Ceres) Karoo). **I:** Bayer (1982: fig. 61b).

≡ *Haworthia granulata* Marloth (1910) ≡ *Haworthia scabra* ssp. *granulata* (Marloth) Halda (1997).

[2] Differs from var. *venosa*: **Ros** caulescent to 15 cm; **L** scabrid.

H. venosa ssp. **tessellata** (Haworth) M. B. Bayer (New Haworthia Handb., 76, 1982). **T:** K [lecto: ill. publ. in Cact. Succ. J. (US): 50: 75, 1978]. – **D:** S Namibia, RSA (Northern Cape, Free State). **I:** Bayer (1982: fig. 61c). **Fig. XXIII.f**

≡ *Haworthia tessellata* Haworth (1824) ≡ *Aloe tessellata* (Haworth) Roemer & Schultes (1829) ≡ *Haworthia venosa* var. *tessellata* (Haworth) Halda (1997); **incl.** *Haworthia parva* Haworth (1824) ≡ *Aloe parva* (Haworth) Roemer & Schultes (1829) ≡ *Haworthia tessellata* var. *parva* (Haworth) Baker (1880); **incl.** *Haworthia tessellata* var. *inflexa* Baker (1880); **incl.** *Haworthia engleri* Dinter (1914) ≡ *Haworthia tessellata* var. *engleri* (Dinter) von Poellnitz (1938); **incl.** *Haworthia pseudotessellata* von Poellnitz (1929); **incl.** *Haworthia tessellata* var. *tuberculata* von Poellnitz (1936); **incl.** *Haworthia minutissima* von Poellnitz (1939) ≡ *Haworthia tessellata* var. *minutissima* (von Poellnitz) Viveiros (1949); **incl.** *Haworthia tessellata* var. *elongata* van Woerden (1940); **incl.** *Haworthia tessellata* fa. *brevior* Resende & von Poellnitz (1942); **incl.** *Haworthia tessellata* fa. *longior* Resende & von Poellnitz (1942); **incl.** *Haworthia tessellata* var. *coriacea* Resende & von Poellnitz (1942); **incl.** *Haworthia tessellata* var. *luisierii* Resende & von Poellnitz (1942); **incl.** *Haworthia tessellata* var. *obesa* Resende & von Poellnitz (1942); **incl.** *Haworthia tessellata* var. *palhinhiae* Resende & von Poellnitz (1942); **incl.** *Haworthia tessellata* var. *simplex* Resende & von Poellnitz (1942); **incl.** *Haworthia tessellata* var. *stepheniana* Resende & von Poellnitz (1942); **incl.** *Haworthia tessellata* var. *velutina* Resende & von Poellnitz (1942); **incl.** *Haworthia tessellata* fa. *major* J. R. Brown (1947).

[2] Differs from var. *venosa*: **Ros** stemless; **L** 8 - 10, subdeltoid, variously patterned.

Here belongs *H. venosa* ssp. *recurva* in the sense of Bayer (1976).

H. venosa ssp. **venosa** – **D:** RSA (Northern Cape, Western Cape). **I:** Bayer (1982: fig. 61a).

Incl. *Aloe recurva* Haworth (1804) ≡ *Apicra recurva* (Haworth) Willdenow (1811) (*incorrect name*, Art. 11.3) ≡ *Haworthia recurva* (Haworth) Haworth (1812) ≡ *Haworthia venosa* ssp. *recurva* (Haworth) M. B. Bayer (1976); **incl.** *Aloe tricolor* Haworth (1804) ≡ *Apicra tricolor* (Haworth) Willdenow (1811) (*incorrect name*, Art. 11.3); **incl.** *Haworthia distincta* N. E. Brown (1876); **incl.** *Haworthia venosa* var. *oertendahlii* Hjelmquist (1943).

[2] **Ros** usually stemless, slowly proliferating with offset or stolons, to 3 cm tall and 5 cm ∅; **L** 12 - 20, to 10 × 1.5 cm, ovate-deltoid, spreading to recurved, upper surface smooth, reticulate, lower surface usually slightly scabrid; **Inf** sparsely branched, lax, to 35 cm; **Fl** 15 - 20, tube obcapitate, **ITep** revolute.

H. venosa ssp. **woolleyi** (von Poellnitz) Halda (Acta Mus. Richnov. Sect. Nat. 4(2): 40, 1997). **T:** B [lecto, ill. publ. Cact. J. (Croydon) 7: 3, 1938]. – **D:** RSA (Eastern Cape: E Little Karoo: Steytlerville). **I:** Bayer (1982: fig. 63, as *H. woolleyi*).
≡ *Haworthia woolleyi* von Poellnitz (1937).

[2] Differs from var. *venosa*: **Ros** stemless; **L** 20 - 25, slender, attenuate.

H. viscosa (Linné) Haworth (Synops. Pl. Succ., 90, 1812). **T:** [lecto – icono]: Commelin, Praeludia Bot. t. 31, 1703. – **D:** RSA (Western Cape, Eastern Cape: Little Karoo). **I:** Bayer (1982: fig. 62). **Fig. XXIII.g**
≡ *Aloe viscosa* Linné (1753) ≡ *Apicra viscosa* (Linné) Willdenow (1811) (*incorrect name*, Art. 11.3); **incl.** *Aloe triangularis* Medikus (1784); **incl.** *Aloe tortuosa* Haworth (1804) ≡ *Apicra tortuosa* (Haworth) Willdenow (1811) (*incorrect name*, Art. 11.3) ≡ *Haworthia tortuosa* (Haworth) Haworth (1812); **incl.** *Aloe pseudotortuosa* Salm-Dyck (1817) ≡ *Haworthia pseudotortuosa* (Salm-Dyck) Haworth (1819) ≡ *Haworthia viscosa* var. *pseudotortuosa* (Salm-Dyck) Baker (1880) ≡ *Haworthia viscosa* fa. *pseudotortuosa* (Salm-Dyck) Pilbeam (1983) (*nom. inval.*, Art. 33.2); **incl.** *Haworthia asperiuscula* Haworth (1819) ≡ *Aloe asperiuscula* (Haworth) Salm-Dyck (1840) ≡ *Haworthia viscosa* fa. *asperiuscula* (Haworth) Pilbeam (1983) (*nom. inval.*, Art. 33.2); **incl.** *Haworthia concinna* Haworth (1819) ≡ *Aloe concinna* (Haworth) Roemer & Schultes (1829) ≡ *Haworthia viscosa* var. *concinna* (Haworth) Baker (1880); **incl.** *Haworthia cordifolia* Haworth (1819) ≡ *Aloe cordifolia* (Haworth) Roemer & Schultes (1829); **incl.** *Haworthia indurata* Haworth (1821) ≡ *Aloe indurata* (Haworth) Roemer & Schultes (1829) ≡ *Aloe viscosa* var. *indurata* (Haworth) Salm-Dyck (1836) ≡ *Haworthia viscosa* var.

indurata (Haworth) Baker (1880); **incl.** *Haworthia viscosa* var. *major* Haworth (1821) ≡ *Aloe viscosa* var. *major* (Haworth) Roemer & Schultes (1829); **incl.** *Haworthia viscosa* var. *minor* Haworth (1821) ≡ *Aloe viscosa* var. *minor* (Haworth) Roemer & Schultes (1829); **incl.** *Haworthia viscosa* var. *parvifolia* Haworth (1821) ≡ *Aloe viscosa* var. *parvifolia* (Haworth) Roemer & Schultes (1829); **incl.** *Haworthia torquata* Haworth (1827) ≡ *Aloe torquata* (Haworth) Salm-Dyck (1836) ≡ *Haworthia viscosa* var. *torquata* (Haworth) Baker (1880) ≡ *Haworthia viscosa* fa. *torquata* (Haworth) Pilbeam (1983) (*nom. inval.*, Art. 33.2); **incl.** *Aloe subtortuosa* Roemer & Schultes (1829); **incl.** *Haworthia viscosa* var. *caespitosa* von Poellnitz (1938); **incl.** *Haworthia viscosa* var. *subobtusa* von Poellnitz (1938) ≡ *Haworthia viscosa* fa. *subobtusa* (von Poellnitz) Pilbeam (1983) (*nom. inval.*, Art. 33.2); **incl.** *Haworthia beanii* G. G. Smith (1944) ≡ *Haworthia viscosa* fa. *beanii* (G. G. Smith) Pilbeam (1983); **incl.** *Haworthia beanii* var. *minor* G. G. Smith (1944); **incl.** *Haworthia asperiuscula* var. *subintegra* G. G. Smith (1945); **incl.** *Haworthia viscosa* var. *cougaensis* G. G. Smith (1945); **incl.** *Haworthia viscosa* var. *viridissima* G. G. Smith (1945); **incl.** *Haworthia asperiuscula* var. *patagiata* G. G. Smith (1946); **incl.** *Haworthia viscosa* var. *quaggaensis* G. G. Smith (1948); **incl.** *Haworthia viscosa* ssp. *dereki-clarki* Halda (1998).

[2] **Ros** caulescent, proliferous, to 8 cm ∅, to 30 cm tall; **L** 20 - 60, to 5 × 1.5 cm, deltoid, closely arranged in 3 rows, spreading, surfaces scabrid, tips spreading, pungent; **Inf** sparsely branched, lax, 15 - 20 cm; **Fl** 8 - 15, tube obcapitate, curved; **ITep** revolute.

H. vlokii M. B. Bayer (Haworthia Revisited, 160-161, ills., 1999). **T:** RSA, Western Cape (*Vlok* s.n. in *Venter* 91/2 [NBG]). – **D:** RSA (Western Cape: Little Karoo, Swartberg Mts.).

[1] **Ros** acaulescent, 4 - 5 cm ∅, proliferous; **L** spreading to suberect, surface opaque, with inconspicuous whiter dots towards the tips, margins and keel with short spines; **Inf** simple, 30 - 45 cm; **Fl** white and brownish-pink.

H. wittebergensis W. F. Barker (JSAB 8: 245, 1942). **T:** RSA, Western Cape (*Pieterse* s.n. [NBG 68214]). – **D:** RSA (Western Cape: Laingsburg: Witteberg Mts.). **I:** Bayer (1982: fig. 43). **Fig. XXIII.h**

[1] **Ros** stemless, slowly proliferating, to 3 cm ∅; **L** 20 - 30, long and slender, 15 × 0.8 cm, attenuate, conspicuously amplexicaul, grey-green with white **Sp** on margins and keel, coriaceous; **Inf** slender, to 30 cm, lax; **Fl** 15 - 20, white with green venation, sparsely arranged.

H. zantneriana von Poellnitz (Cact. J. (Croydon) 5: 35, in clavi, 1935). **T:** B [lecto, ill. publ. Desert Pl.

Life 9: 90, 1937]. – **D:** RSA (Eastern Cape). **I:** Bayer (1982: fig. 45).

≡ *Haworthia chloracantha* var. *zantneriana* (von Poellnitz) Halda (1997).

[1] **Ros** stemless, proliferous, 5 - 6 cm ∅; **L** 20 - 40, lanceolate, acuminate, 6 × 1.2 cm, attenuate, soft, glabrous, spreading, pale green, usually with pellucid white longitudinal marks; **Inf** slender, to 25 cm; **Fl** sparsely arranged, 20 - 30, white with green venation.

H. zantneriana var. **minor** M. B. Bayer (Haworthia Revisited, 164, ills., 1999). **T:** RSA, Eastern Cape (*Bayer* 1702 [NBG]). – **D:** RSA (Eastern Cape: Willowmore, S Karoo).

[1] Differs from var. *zantneriana*: **Ros** small, to 5 cm ∅; **L** unmarked.

H. zantneriana var. **zantneriana** – **D:** RSA (Eastern Cape: Willowmore, Steytlerville).

[1] Plants to 8 cm tall; **L** with markings.

×MAYSARA

U. Eggli

×**Maysara** D. M. Cumming (Haworthiad 13(3): 115, 1999). – **Lit:** Cumming (1999). **Etym:** For Harry Mays (fl. 1999), English succulent plant enthusiast and editor of the journal "Haworthiad"; plus the suffix '-ara', indicating plurigeneric hybrids.

Incl. ×*Rowleyara* D. M. Cumming (1999) (*nom. illeg.*, Art. 53.1).

= *Astroloba* × *Gasteria* × *Haworthia*. The only known combination has not been named formally.

×POELLNERIA

U. Eggli

×**Poellneria** G. D. Rowley (Nation. Cact. Succ. J. 28(1): 7, 1973).

= *Poellnitzia* × *Gasteria*. No formally named taxa are known, and Rowley (1982: 49) lists a single combination with an unidentified *Gasteria* as the other parent.

POELLNITZIA

G. F. Smith

Poellnitzia Uitewaal (Succulenta 22: 61, 1940). **T:** *Apicra rubriflora* L. Bolus. – **Lit:** Smith (1995b). **D:** RSA (Western Cape). **Etym:** For Joseph Karl L. A. von Poellnitz (1896 - 1945), German agriculturist and botanist in Thüringen, strongly interested in succulent plant systematics.

Herbaceous perennials, caulescent, proliferous, stems to 25 cm, ± 1 cm ∅; **Ros** of 4-ranked **L** in spiralling rows; **L** thick, hard, squarrose-imbricate, dark green, 2 - 4 × ± 2 cm broad near base, to 5 mm thick, margin minutely scabrid, apex pungent-acuminate; **Inf** racemose, simple, born horizontally, to 50 cm, sterile part bracteate, with secund erect **Fl**; **Ped** short, erect, persistent; **Tep** orange to red with dark green tips, basally united into a narrow, elongate tube ± 2 cm, 3 mm ∅ with the upper ⅓ slightly decurved, free parts of **Tep** closely adhering, apically spoon-shaped, connivent; **St** 6, inserted in the **Per** tube, ± 18 mm; **Fil** light green; **Anth** yellow, dorsifixed, dehiscing longitudinally, introrse; **Ov** green, sessile, 6 - 7 mm, 3 mm ∅; **Sty** white, straight, capitate, 12 mm; **Fr** 3-locular capsules, cylindrical, apically retuse, dehiscing loculicidally, ± 16 mm, 3 - 4 mm ∅; **Se** angled, dark brown to black, shortly winged, ± 4 mm. – *Cytology:* 2n= 14.

On vegetative morphological grounds, this monotypic genus shows affinities with some representatives of *Aloe, Astroloba* and *Haworthia*. However, the flower morphology is unique in the family in that the dark green free portions of the tepals are connivent and reduplicate-valvate with the very tips scarcely separated.

P. rubriflora (L. Bolus) Uitewaal (Succulenta 22: 61, 1940). **T:** RSA, Western Cape (*Smith* s.n. [BOL 45213]). – **D:** RSA (Robertson Karoo [Robertson, Bonnievale and McGregor districts]); predominantly winter rainfall areas, karroid shrublands, ± 150 - 200 m. **I:** Smith (1994: 74). **Fig. XXIV.b, XXIV.c**

≡ *Apicra rubriflora* L. Bolus (1920) ≡ *Astroloba rubriflora* (L. Bolus) E. Lamb (1955) (*nom. inval.*, Art. 33.2) ≡ *Haworthia rubriflora* (L. Bolus) Parr (1971) ≡ *Aloe rubriflora* (L. Bolus) G. D. Rowley (1981); **incl.** *Apicra jacobseniana* von Poellnitz (1939) ≡ *Poellnitzia rubriflora* var. *jacobseniana* (von Poellnitz) Uitewaal (1955) ≡ *Haworthia rubriflora* var. *jacobseniana* (von Poellnitz) Parr (1972) (*nom. inval.*, Art. 33.2).

Description as for the genus.

Amaryllidaceae

Perennial or biennial bulbous herbs with fleshy tunics, rarely rhizomatous (*Clivia, Scadoxus*); **R** contractile; **L** linear to oblong, sometimes ovate or tapering; **Inf** umbellate (rarely **Fl** solitary) with an involucre (= spatha) of 2 (-8) **Bra**; **Fl** bisexual, actinomorphic; **Tep** 6 in 2 series; **St** 6 (rarely more); **Fil** free or united at base, filiform; **Anth** dorsifixed or basifixed; **Ov** inferior, 3-locular, each with 1 to many ovules, placentation axile (rarely parietal); **Sty** terete; **Sti** capitate or tricuspidate or 3-branched; **Fr** 3-valved capsules, rarely fleshy; **Se** few to many, voluminous or flattish, angled or winged; endosperm fleshy.

Distribution: Cosmopolitan, confined mainly to temperate, subtropical and tropical regions, more common in mediterranean and savanna regions.

Literature: Meerow & Snijman (1998).

A family with ± 59 genera and ± 850 species, of which many are frequently cultivated (e.g. *Clivia, Hippeastrum*). There are a few xerophytic succulent species amongst the genera *Boophane, Brunsvigia, Cyrtanthus* and *Haemanthus* from the savanna regions of RSA (Eastern Cape, KwaZulu-Natal and Mpumalanga). These have mainly epigeous bulbs and are occasionally cultivated. *C. flammosus* is esp. striking with its large solitary red flower.

Rauhia is the only American genus which merits inclusion for its thickly succulent leaves.

Key to genera with succulents:

1 **L** usually 2, opposite, hairy or glabrous; **Fr** fleshy berries: **Haemanthus**
– **L** 2 or more, glabrous; **Fr** dry capsules: **2**
2 Spathe valves 2 - 4; **Se** flattened, usually appearing winged: **Cyrtanthus**
– Spathe valves 2 (rarely more); **Se** globose: **3**
3 **Fl** ± 100 per **Inf**; **Per** 1 - 2 (-3) cm, actinomorphic: **Boophane**
– **Fl** ≤ 30 per **Inf**; **Per** 4 - 15 cm: **4**
4 **L** tongue-shaped; **Per** pink or red; peduncle caducous at **Fr** maturity: **Brunsvigia**
– **L** oval to oblong, very fleshy, pseudopetiolate; **Per** greenish-whitish; peduncle persistent: **Rauhia**

[E. van Jaarsveld]

BOOPHANE

E. van Jaarsveld

Boophane Herbert (Appendix, 18, 1821). **T:** *Amaryllis disticha* Linné *fil.* [following ING, following E. Phillips, Gen. South. Afr. Pl., ed. 2, 201, 1951.]. – **D:** S Africa. **Etym:** Gr. 'bouphonos', killing cat-

tle; for the possibly poisonous nature of some of its species.

Incl. *Boophone* Herbert (1821) (*nom. inval.*, Art. 61.1). **T:** *Amaryllis disticha* Linné *fil.*.
Incl. *Buphane* Herbert (1825) (*nom. inval.*, Art. 61.1). **T:** *Amaryllis disticha* Linné *fil.*.
Incl. *Buphone* Herbert (1825) (*nom. inval.*, Art. 61.1). **T:** *Amaryllis disticha* Linné *fil.*.

Perennials with thickly tunicated bulbs to 30 cm ∅, above or below ground-level; **L** distichous, younger erect becoming spreading with age, ensiform, lingulate, leathery, sometimes undulate, glaucous-green to green; **Inf** densely umbellate, many-flowered; peduncle short, firm, somewhat laterally compressed; **Bra** 2, triangular; **Per** consisting of a short subcylindrical tube and spreading slender **Tep** lobes; **St** from the throat of the **Per** tube; **Fil** straight, filiform; **Anth** oblong, dorsifixed, versatile; **Ov** inferior, turbinate, 3-locular, each locule with 1 to few ovules; **Sty** simple, obscurely 3-lobed; **Fr** dry capsules, triquetrous, obtriangular, indehiscent or loculicidally 3-valved; **Se** globose.

A small genus of 5 species, widespread in S Africa. The spelling of the generic name is much in dispute and various spellings are encountered in the literature. Only the following species with epigeous bulbs are sometimes referred to as succulents:

B. disticha (Linné *fil.*) Herbert (Appendix, 18, 1821). **T:** UPS. – **D:** RSA; widespread in grassland and Karoo. **I:** Wyk & al. (1997: 60). **Fig. XXIV.a**
 ≡ *Amaryllis disticha* Linné *fil.* (1782).

Bulbs solitary, epigeous, subglobose, to 30 cm ∅, thickly tunicated, outer tunics brown and firm; **R** terete, fleshy; **L** 30 - 45 × 2.5 - 3 cm, to 16, ensiform, ± leathery, glaucous-green, with closely set ribs, young erect, later spreading, margins sometimes undulate; peduncle firm, glaucous, to 30 cm; **Bra** subtending the dense umbels triangular, 5 - 7 cm; **Fl** numerous, bright red; **Ped** 5 - 10 cm; **Per** infundibuliform with subcylindrical tube 1 - 1.3 cm, free parts of **Tep** linear, 2 - 2.5 cm; **St** ± as long as **Tep**; **Anth** yellow, oblong; **Ov** turbinate, green, to 4 mm ∅; **Sty** red, curved, slightly longer than **St**; **Fr** turbinate, ± 2 × 1.3 cm ∅.

B. haemanthoides F. M. Leighton (JSAB 13: 59-61, fig. 4, 1947). **T:** RSA, Western Cape (*Leighton* 2361 [BOL]). – **D:** RSA (Northern Cape, Western Cape); coastal regions, autumn-flowering.

Bulb solitary, glabrous, epigeous, ovoid to globose, to 18 cm ∅, thickly tunicated, outer tunics brown and firm; **R** terete, fleshy; **L** 15 - 30 × 5 - 10 cm, oblong, coriaceous, inner erect becoming spreading, margin entire, tip obtuse; peduncle firm, to 25 cm, to 3 cm ∅; **Fl** numerous, in dense umbels, creamy-yellow becoming reddish with age; **Bra** suberect, pink or red, to 14 × 8.5 cm; **Ped** 5 - 10 cm; **Per** tube 5 - 7 mm, 6-angled with deep grooves between the angles, limb segments to 3.5 cm, cucullate, to 4 mm wide near tips narrowed to 2 mm at the base; **St** exserted to 1 cm beyond the **Per**; **Fil** slender, erect, attached to the **Tep**; **Sty** to 1 cm longer than the **St**; **Sti** small, inconspicuous, minutely papillose; **Ov** obconical, sharply angled.

BRUNSVIGIA

E. van Jaarsveld

Brunsvigia Heister (Beschr. neu. Geschl., 3, 1755). **T:** *Brunsvigia orientalis* Heister. – **Lit:** Dyer (1950). **D:** S Africa. **Etym:** Honouring the House of Braunschweig [Brunswick]-Lüneburg.

Perennial herbs with hypo- or epigeous bulbs to 20 cm ∅, usually thickly tunicated; **L** 2 to many per season, variously shaped, usually broad, appearing after the **Fl**; **Inf** few- to many-flowered umbels; peduncle firm, to 35 cm, deciduous at **Fr** time; spathe valves 2 (= **Bra**); **Ped** elongating and spreading after anthesis; **Per** zygomorphic or almost actinomorphic with short tube, segments spreading-recurved; **St** arising from the **Per** tube, ± declinate or erect; **Ov** turbinate, with many superposed ovules in each locule; **Sty** filiform, declinate; **Sti** capitate; **Fr** dry capsules, obtuse or acute, 3-angled, dehiscent loculicidally or breaking unevenly; **Se** subglobose.

A genus of ± 20 species, mainly in S Africa. The species with epigeous succulent bulbs are considered succulent and are treated below. They are all from arid winter rainfall regions and occur in Succulent Karoo.

B. herrei F. M. Leighton *ex* W. F. Barker (JSAB 29: 165, t. 32, 1963). **T:** RSA, Northern Cape (*Herre* 3368 [NBG]). – **D:** RSA (Northern Cape); Succulent Karoo, mountains, autumn-flowering.

Bulb to ± 15 cm ∅, epigeous to partly underground, globose and covered with brownish cartilaginous tunics; **L** 6, hysteranthous, to 20 × 9 cm, erect at first becoming spreading, oblong, smooth, glaucous, margins narrowly reddish, tip obtuse; peduncle green, slightly laterally compressed, to 23 cm, 15 mm ∅; **Bra** ovate, pale greenish-fawn, to 6 × 2.5 cm; umbel to 40-flowered, globose and head-like; **Ped** pale green, to 18 cm, 5 mm ∅ at base; **Per** pale pink, actinomorphic; **Tep** to 5.5 × 1.2 cm, oblong, obtuse, tube 3 mm long, deep pink; **St** declinate, the longest somewhat shorter than the **Per**, the rest ½ as long; **Fil** without appendages, adnate to the **Per** tube; **Sty** becoming longer than the **St**; **Ov** to 6 mm ∅, obtusely angeled when young; **Fr** and **Se** not seen.

B. radula (Jacquin) Aiton (Hort. Kew., ed. 2 2: 230, 1811). **T:** [icono]: Jacquin, Hort. Schönbr. 1: 35, t. 68, 1797. – **D:** RSA (Western Cape); limestone outcrops, Succulent Karoo.

≡ *Amaryllis radula* Jacquin (1797).

Bulb to ± 25 cm ∅, globose and with short neck, covered with dry brownish tunics; **L** 2 - 3, to 8 × 3.5 cm, lingulate to oblong, appressed to the ground, hysteranthous, obtuse, thick, upper face and margin with papillae with enlarged bases, lower face smooth, pale green, tip obtuse; peduncle 1 - 2, to 10 cm, 5 mm ∅, slightly laterally compressed; **Bra** ovate-oblong, reddish, to 2 cm; umbel 3- to 5-flowered; **Ped** to 5 cm, 5 mm ∅; **Per** flesh-coloured to pink; **Tep** to 3 cm, linear-lanceolate, spreading, recurved, undulate, the lower often subtending **St** and **Sty**; **St** declinate, ± as long as the **Tep**; **Sty** slightly longer than the **St**; **Ov** roundish, obtusely angled; **Fr** 3-angled; **Se** not seen.

CYRTANTHUS

E. van Jaarsveld

Cyrtanthus Aiton (Hort. Kew. 1: 414, 1789). **T:** *Crinum angustifolium* Linné *fil.* – **D:** S Africa; mainly summer rainfall regions. **Etym:** Gr. 'kyrtos', curved; and Gr. 'anthos', flower; referring to the curved flower tube.

Incl. *Vallota* Salisbury *ex* Herbert (1821). **T:** *Vallota purpurea* Herbert [*nom. illeg.*, ≡ *Crinum speciosum* Linné *fil.*)].

Incl. *Anoiganthus* Baker (1878). **T:** *Cyrtanthus breviflorus* Harvey.

Plants bulbous, evergreen or deciduous; bulbs tunicate, epi- or hypogaeous, solitary, dividing or offsetting to form dense groups; **L** ascending, straight or curved, oblong, lorate or linear, green to glaucous; **Inf** umbellate, 1- to many-flowered with up to 4 spathe valves; peduncle ascending, usually hollow; **Fl** tube narrow, shorter than the lobes, widening towards mouth; **St** inserted in the throat of the tube; **Anth** dorsifixed, versatile; **Ov** 3-chambered, with numerous ovules in each locule; **Sty** filiform; **Sti** 3-lobed; **Fr** oblong loculicidal capsules; **Se** flat.

Mainly a S African genus with ± 50 species, some of them frequently cultivated. 4 taxa from arid to semi-arid sheer cliff-faces have succulent epigeous bulbs; flowers are attractively red.

C. flammosus Snijman & van Jaarsveld (FPA 54: 100-103, t. 2120, 1995). **T:** RSA, Eastern Cape (*van Jaarsveld & al.* 13803 [NBG, PRE]). – **D:** RSA (Eastern Cape); sheer cliff-faces in arid savanna, summer-rainfall region. **Fig. XXIV.e**

Evergreen, bulbous, to 25 cm; bulb solitary, ovoid, partly exposed or completely epigeous, to 4 cm ∅, tunics fleshy, withering papery, brown; **L** 2 - 4, ascending, 13 - 29 × 1.5 - 2 cm, glaucous, fleshy; **Inf** 1-flowered, scape erect, to 17 cm; spathe valves suberect, lanceolate, 5.5 × 1 cm, **Per** tube 4 - 5 cm, **Tep** spreading, broadly obovate, 4 - 5 × 3 - 3.5 cm, fire-red; **Fil** free for 2 - 2.5 cm, cream to pale green; **Anth** dorsifixed, 1.5 cm, yellow; pollen yellow; **Sty** declinate; **Sti** branches 5 mm; **Ov** 1 - 1.5 × 0.6 cm; **Fr** not seen.

C. herrei (F. M. Leighton) R. A. Dyer (FPA 33: t. 1281 + text, 1959). **T:** RSA, Northern Cape (*Herre* s.n. [BOL]). – **D:** RSA (Northern Cape); winter-rainfall region. **Fig. XXIV.d**

≡ *Cryptostephanus herrei* F. M. Leighton (1932).

Bulbs large, epigeous, clustering, obclavate, to 6 cm ∅; **R** succulent, terete; **L** synanthous, distichous, lorate, to 45 × 5 cm, leathery, glaucous, apex obtuse; **Inf** to 40 cm, scape glaucous, to 28-flowered; **Bra** to 8 × 1.3 cm, linear-lanceolate, soon withering; **Ped** to 4 cm; **Per** pendulous, zygomorphic, to 5.5 cm, tube to 4 cm, red, lobes yellowish-green; **St** fused to **Per**, free part of **Fil** to 6 mm; **Anth** yellow; **Ov** to 8 mm; **Se** black, winged.

C. labiatus R. A. Dyer (BT 13(1-2): 135, ills., 1980). **T:** RSA, Eastern Cape (*Bayliss* 5660 [PRE]). – **D:** RSA (Eastern Cape); sheer cliff-faces in arid savanna: summer-rainfall region.

Bulb epigeous, globose, 4 - 7.5 × 6 - 7.5 cm, purplish-green, solitary or forming groups, bulbiferous from base, withering with papery brown tunics; **L** 2 - 4, synanthous, evergreen, lorate-elliptic to strap-shaped, 18 - 30 × 1.4 - 2 cm, apex obtuse; **Inf** 12 - 30 cm, scape to 2.3 cm ∅, glacous, umbel to 8-flowered; **Bra** 5 × 0.5 cm, triangular-lanceolate, soon withering; **Ped** 2 - 2.5 cm; **Per** red, zygomorphic, bilabiate, 5 - 6 cm, tubular-curved, tube 1 cm, infundibuliform, with 4 upper linear-oblanceolate lobes forming a hood and 2 lower lobes; **St** fused to **Per**, free for the last 1 cm; **Anth** yellow, 3 mm, oblong; **Sty** 4 cm, **Sti** 3-lobed; **Ov** oblong-triangular, 6 × 4 mm.

C. montanus R. A. Dyer (FPA 44: t. 1756 + text, 1977). **T:** RSA, Eastern Cape (*Skinner* s.n. [PRE 37061]). – **D:** RSA (Eastern Cape); sheer cliff-faces in arid savanna, summer-rainfall region.

Bulb epigeous, globose, 6.5 × 7 cm, purplish-greenish, solitary or forming groups, bulbiferous from base, tunics withering grey-brown, papery; **L** 2 - 4, synanthous, evergreen, lorate-elliptic to 30 × 2 cm, ascending, glaucous; **Inf** to 10 cm; **Bra** 2, linear-lanceolate, to 5 cm, with smaller bracteoles subtending the **Fl**; umbels to 10-flowered; **Ped** to 3 cm; **Per** red, erect, to 5 cm, tube infundibuliform, to 1.5 cm; **OTep** linear lanceolate, to 9 mm wide; **ITep** to 11 mm wide; **St** 2-seriate; **Fil** 9 - 11 mm; **Ov** oblong, to 6 mm; **Sty** filiform, **Sti** tricuspiate; **Fr** oblong; **Se** compressed, black.

HAEMANTHUS

E. van Jaarsveld

Haemanthus Linné (Spec. Pl. [ed. 1], 325, 1753). **T:** *Haemanthus coccineus* Linné [Lectotype, desig-

nated by Hitchcock & Green, Stand. Spec. Linnean Gen., 1929.]. – **Lit:** Snijman (1984). **D:** S Africa, Namibia. **Etym:** Gr. 'haima', blood; and Gr. 'anthos', flower; for the dark red flowers of some species.

Incl. *Leucodesmis* Rafinesque (1838). **T:** *Haemanthus pubescens sensu* Rafinesque.

Incl. *Perihema* Rafinesque (1838). **T:** *Haemanthus coarctatus* Jacquin.

Incl. *Serena* Rafinesque (1838). **T:** *Haemanthus carneus* Ker Gawler [Lectotype, designated by Snijman, J. South Afr. Bot. Suppl. Vol. 12: 17, 1984.].

Incl. *Diacles* Salisbury (1866). **T:** *Haemanthus pubescens sensu* Ker Gawler [Lectotype, designated by Snijman, J. South Afr. Bot. Suppl. Vol. 12: 17, 1984.].

Incl. *Melicho* Salisbury (1866). **T:** *Haemanthus amarylloides* Jacquin [Lectotype, designated by Bjørnstad & Friis, Norweg. J. Bot. 19: 187-206, 1972.].

Deciduous bulbous geophytes (rarely evergreen); bulbs usually hypogeous, rarely epigeous or half exposed, variable, usually ovoid to pyriform with fleshy tunics in successive horizontal layers; **L** few, usually 2 or more, distichous, fleshy, strap-shaped, patent to ascending; **Fl** in dense umbels to 10 cm ∅; scape compressed; spathe valves firm, fleshy, white to red, broadly ovate; **Per** with short cylindrical tube, lobes spreading to ascending, longer than the tube; **St** 6, fused to the throat of the tube; **Fil** filiform; **Anth** oblong, dorsifixed; **Ov** inferior, globose; **Sty** filiform, **Sti** tricuspidate; **Fr** soft fleshy globose berries; **Se** black, shiny.

A genus of 22 species confined mainly to the winter rainfall regions in S Africa with a concentration in Namaqualand. Only 3 taxa are somewhat succulent. They are native to the summer rainfall region in E RSA and are easily cultivated in containers, but only *H. albiflos* is seen frequently in cultivation.

H. albiflos Jacquin (Pl. Hort. Schoenbr. 1: 31, t. 59, 1797). **T:** [icono]: l.c., t. 59. – **D:** RSA (Eastern Cape to KwaZulu-Natal); subtropical thickets and dry savanna, autumn- to spring-flowering. **Fig. XXIV.f**

≡ *Haemanthus virescens* var. *albiflos* (Jacquin) Herbert (1837); **incl.** *Haemanthus pubescens* auct. pl. (s.a.) (*nom. illeg.*, Art. 53.1); **incl.** *Haemanthus virescens* var. *intermedius* Herbert (1837) ≡ *Haemanthus intermedius* (Herbert) M. Roemer (1847); **incl.** *Haemanthus leucanthus* Miquel (1861); **incl.** *Haemanthus ciliaris* Salisbury (1866) (*nom. inval.*, Art. 32.1c); **incl.** *Haemanthus albomaculatus* Baker (1878); **incl.** *Haemanthus albiflos* var. *brachyphyllus* Baker (1888); **incl.** *Haemanthus albiflos* var. *burchellii* Baker (1888).

Evergreen bulbous geophytes; bulbs ovoid, to 8 cm broad, clustering, epi- to semi-hypogeous; tunics truncate at top, green when exposed to light; **L** strap-shaped to elliptic, appressed to the ground or spreading, flat or canaliculate, smooth or slightly pubescent; apex obtuse to acute; **Inf** 5 - 35 cm; scape compressed, to 14 mm wide; umbel compact, laterally compressed, to 7 cm wide with 4 - 8 spathe valves; **Fl** to 50, white; **Ped** to 1 cm; **Per** funnel-shaped, to 23 mm, tube to 7 mm with spreading oblong segments 10 - 18 × 1 - 2.5 mm; **Ov** globose, to 3 mm ∅; **Fr** ovoid, to 1 cm ∅, white to red.

H. deformis Hooker *fil.* (CBM 97: t. 5903 + text, 1871). **T:** RSA, KwaZulu-Natal (*McKen* s.n. [[lecto − icono]: l.c., t. 5903]). – **D:** RSA (KwaZulu-Natal); coastal sheltered rocky regions to 1000 m, Bushveld. **I:** Snijman (1984: 65).
Incl. *Haemanthus baurii* Baker (1885); **incl.** *Haemanthus mackenii* Baker (1888).

Evergreen bulbous geophytes; bulbs solitary or in small clusters, epi- to semi-hypogeous, laterally compressed, to 16 × 10 cm, tunics truncate at the top, green when exposed; **L** 2 - 4, broadly oblong, often broader than long, to 24 × 30 cm, appressed to the ground, glabrous to pubescent, margin ciliate, tip obtuse; **Inf** 1 - 3, almost sessile to 6 cm, umbel compact, with 6 - 7 spathe valves; **Fl** to 45, white; **Ped** to 7 mm; **Per** funnel-shaped, to 31 mm, tube 7 - 9 mm, segments to 23 mm; **Fr** ovoid, to 15 mm ∅, orange to red.

Closely related to *H. albiflos*.

H. pauculifolius Snijman & van Wyk (SAJB 59(2): 247-250, ills., 1993). **T:** RSA, Mpumalanga (*Matthews* 624 [NBG, K, MO, PRE]). – **D:** RSA (Mpumalanga); Bushveld, mountaineous rocky terrain.

Evergreen bulbous geophytes; bulbs ovoid, 4 - 5 cm ∅, clustering, epi- to semi-hypogeous, green, smooth; **L** 1 - 2, fleshy, strap-shaped to linear-lanceolate, canaliculate, tomentose, 7 - 10 (- 30) × 2 - 4.5 cm, apex acute; **Inf** 5 - 19 cm; scape compressed, to 7 mm wide; umbel compact, compressed, to 3 cm wide, with 4 spathe valves; **Fl** to 19, white; **Ped** to 3 mm; **Per** funnel-shaped, to 35 mm, tube to 13 mm with spreading lanceolate segments 17 - 20 × 3 - 4 mm; **Fr** globose, to 15 mm ∅, orange; **Se** ovoid, 10 mm.

RAUHIA

S. Arroyo-Leuenberger

Rauhia Traub (Pl. Life 13: 73, 1952). **T:** *Rauhia peruviana* Traub. – **Lit:** Ravenna (1969). **D:** Peru; Andean rocky slopes. **Etym:** For Werner Rauh (1913 - 2000), German botanist in Heidelberg, and specialist of Madagascan succulents.

Bulbous perennials; bulbs tunicate, hypogeous, solitary; **L** ascending, hysteranthous, distichous, oval to oblong, glaucous, fleshy, pseudopetiolate; **Inf** lateral, umbellate; scape erect, solid; **Bra** sub-

erect, lanceolate; **Fl** declinate or horizontal, greenish-whitish, tube funnel-shaped, **Tep** oblanceolate, spreading; **St** inserted in the throat of the **Per** tube, without staminal cup; **Anth** dorsifixed; **Ov** with numerous ovules in each locule; **Sty** slightly declinate; **Sti** capitate; **Fr** loculicidal capsules.

This small genus consists of 3 species and is known only from Peru. Only *R. multiflora* is encountered in cultivation; the other 2 are insufficiently known and perhaps conspecific. All appear to be evergreen in cultivation.

R. decora Ravenna (Pl. Life 37: 77, 1981). **T:** Peru, Amazonas (*Ravenna* 3060 [Herb. Ravenna, K]). – **D:** Peru; 500 - 1000 m.

Bulb ovoid, ± 5.5 cm ∅, tunics fleshy, withering papery, brown; **L** 2 - 6, 20 - 30 × 8 - 12 cm; **Inf** 8- to 25-flowered; scape to 20 - 34 cm; **Bra** 2, free to the base, broadly ovate, ± 3.8 cm; **Ped** 6.5 - 7.5 cm; **Per** tube ± 2.2 cm; **Tep** 4 - 5 cm, greenish-whitish; shorter **Fil** ± 6.2 cm, longer ± 6.7 cm; **Ov** 0.9 - 1.5 × 0.5 cm; **Sty** declinate, longer than the **St**; **Sti** capitate-trilobate.

R. multiflora (Kunth) Ravenna (Pl. Life 25: 61, 1969). **T:** Peru, Jaén (*Herb. Humboldt* 3582 [B]). – **D:** Peru (Jaén); rocky slopes, 500 - 1500 m. **Fig. XXV.b**
≡ *Phaedranassa multiflora* Kunth (1850); **incl.** *Phaedranassa megistophylla* Kraenzlin (1917) ≡ *Rauhia megistophylla* (Kraenzlin) Traub (1966); **incl.** *Rauhia peruviana* Traub (1957).

Bulb ovoid, ± 10 - 15 cm ∅, tunics fleshy, withering papery, brown; **L** 2 - 4, 20 - 30 × 9 - 15 cm, spreading; **Inf** 8- to 25-flowered; scape to 50 - 120 cm; **Bra** 2 (-3), free to the base, broadly ovate, ± 3 cm; **Ped** 3 - 9 cm; **Per** tube ± 2.2 cm; **Tep** 4 - 5 cm, greenish-whitish; **Fil** 2 shorter ± 3.1 cm, 4 longer ± 3.6 cm; **Sty** declinate, as long as or longer than the **St**; **Sti** capitate; **Ov** 0.5 - 1.2 × 0.5 cm.

R. staminosa Ravenna (Pl. Life 34: 68, 1978). **T:** Peru, Amazonas (*Ravenna* 2091 [Herb. Ravenna]). – **D:** Peru; 500 - 1000 m.

Bulb ovoid, 8 - 10 cm ∅, tunics fleshy, withering papery, brown; **L** 2, 15 - 30 × 9 - 15 cm; **Inf** 8- to 24-flowered; scape 40 - 80 cm; **Bra** 2, free to the base, broadly ovate, 3.2 - 4 cm; **Ped** 4 - 9 cm; **Per** tube 1.4 - 1.7 cm; **OTep** 2 - 2.3 cm, greenish-whitish; **ITep** 1.8 - 2.15 cm; shorter **Fil** ± 4.6 cm, longer ± 6.5 cm; **Ov** 0.9 - 1.1 × 0.5 cm; **Sty** declinate, longer than the **St**; **Sti** capitate.

Anthericaceae

Perennial herbaceous evergreen or geophytic herbs; **R** variable, mostly succulent, sometimes tuberous; rhizome creeping, often covered with old **L** bases, or rarely with (semi-) erect succulent stems; **L** rosulate, rarely reduced to 1, or distichous, lamina flat, terete or triangular, oblong, linear, glabrous to hairy, rarely succulent and xeromorphic, rigid, sheathing basally, margin often fimbriate; **Inf** terminal, racemose or paniculate, or condensed into clusters; peduncle bracteate; **Fl** 3-merous, actinomorphic; **Per** stellate, rarely urceolate or campanulate; **Tep** 6, oblong, white, yellow or blue; **St** 6; **Fil** free or connate basally, glabrous or papillate; **Anth** basifixed or dorsifixed-epipeltate, dehiscing longitudinally; **Ov** superior, 3-locular; **Sty** simple, terete, smooth; **Sti** 3-lobed; placentation axile; **Fr** loculicidal capsules; **Se** sometimes with an elaiosome, flat to rounded.

Distribution: Worldwide, mainly tropics and subtropics.

Literature: Conran (1998).

A small family of 9 genera and ± 200 species with a worldwide but mainly subtropical and tropical distribution. *Chlorophytum* is the largest genus with ± 100 species, followed by the closely related *Anthericum* with less than 10 species. In traditional classifications, the family is included within *Liliaceae* s. lat. The 2 genera mentioned were formerly delimited on account of the shape of the fruits and seeds. Recent investigations show that *Anthericum* should be limited to species with unibracteate nodes and compact seeds, while *Chlorophytum* species have multibracteate nodes and thin seeds (Kativu & Nordal 1993).

A few species of both genera mentioned are occasionally cultivated as ornamentals, but only *Chlorophytum* is here included. Propagation is by division or from seed.   [E. van Jaarsveld]

CHLOROPHYTUM

E. van Jaarsveld

Chlorophytum Ker Gawler (CBM 27: t. 1071 + text, 1807). **T:** *Chlorophytum inornatum* Ker Gawler. – **Lit:** Obermeyer (1962: 690-711). **D:** Africa, Madagascar, Asia; forests, savanna and Karoo region. **Etym:** Gr. 'chloros', yellowish-green, pale green; and Gr. 'phyton', plant; for the leaves of some taxa.

Incl. *Hartwegia* Nees (1831). **T:** *Hartwegia comosa* Nees [*nom. illeg.*, = *Anthericum sternbergianum* J. A. & J. H. Schultes].

Incl. *Asphodelopsis* Steudel *ex* Baker (1876). **T:** not typified.

Incl. *Acrospira* Welwitsch *ex* Baker (1878) (*nom. illeg.*, Art. 53.1). **T:** *Acrospira asphodeloides* Welwitsch ex Baker.

Incl. *Dasystachys* Baker (1878) (*nom. illeg.*, Art. 53.1). **T:** *Dasystachys colubrina* Baker [Lectotype, designated by Kativu & Nordal, Nordic J. Bot. 13(1): 62, 1993.].

Incl. *Debesia* Kuntze (1891). **T:** *Acrospira asphodeloides* Welwitsch ex Baker.

Incl. *Verdickia* De Wildemann (1902). **T:** *Verdickia katangensis* De Wildemann.

Perennial evergreen or deciduous and geophytic herbs; **R** variable, mostly succulent, sometimes tuberous; rhizome creeping, often covered with old **L** bases; **L** rosulate, tufted, rarely reduced to 1, or subdistichous, lamina flat, oblong to linear, glabrous to hairy, margin often fimbriate; **Inf** terminal, racemose or paniculate; peduncle bracteate; **Ped** articulated near the middle; **Per** stellate, rarely urceolate or campanulate, white to blue; **Tep** 6, oblong, marcescent, midrib darkening, **OTep** narrower than **ITep**; **St** 6; **Fil** glabrous or papillate; **Anth** basifixed, introrse; **Ov** sessile or shortly stipitate, trigonous, with 6 - 30 biseriate ovules; **Sty** terete, smooth; **Sti** apical, minute; **Fr** loculicidal trilobate or 3-angled capsules; **Se** flat, rounded, black, shiny.

A large genus with ± 100 species occurring mainly in Africa, Madagascar and Asia. *C. comosum* ("Spider Plant") from South Africa is amongst the World's most commonly cultivated house plants. The genus differs from *Anthericum* by its triangular capsules (rounded in *Anthericum*) and flat (vs. angular) seeds.

Several taxa could be regarded as weakly developed succulents due to their fusiformly thickened roots (incl. *C. comosum*); some also have slightly fleshy leaves, but none of the latter appears to be in cultivation. *C. suffruticosum* is notable for its stem succulence.

C. comosum (Thunberg) Jacques (J. Soc. Imp. Centr. Hort. 8: 345, 1862). **T:** RSA, Eastern Cape (*Thunberg* s.n. [UPS]). – **D:** RSA (Eastern Cape, KwaZulu-Natal, Mpumalanga, Northern Prov.); coastal forest, savanna and thickets, summer-flowering.

≡ *Anthericum comosum* Thunberg (1794) ≡ *Phalangium comosum* (Thunberg) Poiret (1804); **incl.** *Anthericum planifolium* Thunberg (1818); **incl.** *Anthericum sternbergianum* Roemer & Schultes (1829); **incl.** *Chlorophytum burchellii* Baker (1876) ≡ *Chlorophytum elatum* var. *burchelii* (Baker) Baker (1897); **incl.** *Chlorophytum delagoense* Baker (1897); **incl.** *Phalangium viviparum* Baker (1897); **incl.** *Chlorophytum longituberosum* von Poellnitz (1942); **incl.** *Chlorophytum vallistrappii* von Poellnitz (1942).

Perennial evergreen herbs to 80 cm tall and 1 m ∅; **R** fusiform, succulent, to 1 cm ∅; rhizome creeping, 1 - 1.5 cm ∅, often covered with old **L** bases; **L** 21 - 65 × 1 - 3.2 cm, in a dense basal **Ros**, lamina flat, linear, glabrous, channelled, bright green, margins entire, base sheathing, tips acute; **Inf** terminal spreading lax racemes, 30 - 100 cm; peduncle bracteate, 2 - 4 mm ∅; **Bra** linear-lanceolate, 2 - 4 cm, acuminate; **Fl** often replaced by propagules, which root and serve for vegetative reproduction; **Fl** in axillary fascicles; **Per** stellate, white; **Tep** oblong, to 1 cm, becoming reflexed; **St** 6; **Fil** glabrous; **Sty** terete, smooth; **Sti** minute; **Fr** 3-angled emarginate capsules; **Se** flat, rounded, black, shiny.

A very variable complex of plants. For a longer discussion and synonymy, see Nordal & al. (1997: 57-58).

C. suffruticosum Baker (Gard. Chron., ser. nov. 24: 230, 1885). **T:** Kenya (*Wakefield* s.n. [K]). − **D:** Kenya. **I:** Nordal & al. (1997: 41). **Fig. XXV.a, XXV.d**

≡ *Anthericum suffruticosum* (Baker) Milne-Redhead (1936); **incl.** *Chlorophytum rhizomatosum* Baker (1885); **incl.** *Anthericum acuminatum* Rendle (1895); **incl.** *Anthericum campestre* Engler (1895); **incl.** *Dasystachys polyphylla* Baker (1898) ≡ *Chlorophytum polyphyllum* (Baker) von Poellnitz (1946); **incl.** *Anthericum inexpectatum* von Poellnitz (1942); **incl.** *Anthericum longisetosum* von Poellnitz (1942).

Rstock fleshy; stems solitary or little-branched, to 1.5 cm ∅, with semicircular ± horizontal but often irregularly arranged **L** scars, dull greenish, sometimes partly covered by fibrous remains of old **L** bases; **L** 45 - 60 × 1.5 - 2 cm, in apical **Ros**, ascending to erect, basally sheathing, linear, firm, glabrous, pale greyish-green, slightly succulent and canaliculate; **Inf** erect to drooping or irregular, normally unbranched, to 60 cm; peduncle terete, glabrous, to 30 cm; **Ped** short, ascending, articulate near the middle; **Bra** lanceolate, acuminate, lowermost to 1.3 cm; **Per** to 2 cm ∅, stellate; **Tep** white, outside with green midvein, 3-veined; **Fil** much shorter than **Tep**, **Anth** narrowly oblong, yellow; **Fr** and **Se** not described.

This interesting stem-succulent is without parallel in the family. It is mostly encountered in collections under the name *Anthericum suffruticosum.* − [E. van Jaarsveld & U. Eggli]

Araceae

Dwarf to gigantic herbs, terrestrial, hemi-epiphytic or epiphytic, often geophytes or lithophytes, rarely floating (*Pistia*) or submerged aquatic (*Jasarum, Cryptocoryne*); stems climbing, arborescent, erect, repent or subterranean, varying from subglobose tubers to very long rhizomes with elongate **Int**; all tissues always with raphides, laticifers commonly present but absent in certain groups; **L** flat with petiole and lamina of very variable shape (except *Gymnostachys* with linear **L**), sometimes pinnatifid to pinnatisect, rarely pinnate, venation reticulate or parallel-pinnate (rarely strictly parallel); **Inf** pseudanthia consisting of a spadix subtended and/or enveloped or surrounded by a spathe (= last **L** of shoot, usually specialized in form and colour); spadix usually a dense fleshy spike with sessile (except *Pedicellarum*) minute ebracteate **Fl**; **Fl** bisexual or unisexual, when unisexual the female **Fl** forming the basal part and the male **Fl** the intermediate or apical part of the spadix (except *Spathicarpa*); **Fl** with or without **Per**; **St** opposite to the **Tep**, free or connate into a synandrium; **Anth** extrorse (except *Zamioculcas, Pedicellarum*); **Ov** syncarpous, usually 1- to 3-locular (1- to 8-locular in *Spathicarpeae*, 2- to 47-locular in *Philodendron*), ovules anatropous to orthotropous or intermediate; **Sty** present or absent, often inconspicuous; **Sti** various, discoid to capitate, or ± distinctly lobed, always wet at anthesis; **Fr** berries (basally dehiscent in *Lagenandra*), usually free and borne in ± cylindrical dense infructescences (rarely syncarpous, *Syngonium, Cryptocoryne*); **Se** small to very large, globose to ellipsoid, sometimes reniform, testa very thin to very thick, smooth, ribbed, warty or rarely spiny; embryo minute to large, endosperm present or absent.

Distribution: Worldwide, but mostly (> 90 %) in the tropics, esp. in rainforests.

Literature: Engler (1905); Mayo & al. (1997).

A family of 105 genera with over 3300 species. Several genera such as *Alocasia, Amorphophallus, Colocasia* ("Taro"), *Cyrtosperma* and *Xanthosoma* produce edible underground tubers or stems and have great importance as food plants in the tropics. *Monstera deliciosa* produces edible fruits, and some species are valued as house-plants, together with species from several other genera (*Anthurium, Dieffenbachia, Philodendron, Zantedeschia*, etc.).

Succulence is uncommon in the family. Several species of *Anthurium* have leathery leaves, and many tuberous (e.g. *Biarum, Eminium*) or rhizomatous members (e.g. *Stylochaeton*, some of its species with very fleshy roots) have adaptations to seasonally dry conditions. The only taxon with undisputed succulence is *Zamioculcas zamiifolia*.

The genera *Zamioculcas* and *Gonatopus* form together the tribe *Zamioculcadeae*, which is distributed in tropical E and SE Africa. It has an isolated position and is included in the expanded subfamily *Aroideae* today. Tribe *Zamioculcadeae* and tribe *Stylochaetoneae* (with the only genus *Stylochaeton*) are the only *Araceae* with unisexual flowers with a perigone.   [J. Bogner]

ZAMIOCULCAS

J. Bogner

Zamioculcas Schott (Syn. Aroid., 71, 1856). **T:** *Zamioculcas loddigesii* Schott [*nom. illeg.*, = *Z. zamiifolia*]. – **D:** Tropical E and SE Africa. **Etym:** For the genus *Zamia* (*Zamiaceae*); and from Arab. 'qolqas', 'kulkas', the name of the Taro plant (*Colocasia*); for the leaves which resemble those of *Zamia*.

Seasonally dormant or evergeeen herbs with a short very thick rhizome 2 - 4 cm ⌀; **L** few to many, erect, 40 - 80 cm, pinnate, leaflets deciduous leaving the persistent succulent petiole; petiole 10 - 35 cm, terete, basally thickened, 1 - 2 cm ⌀, geniculate at apex; sheath ligulate, free almost to the base, very short and inconspicuous; leaflets ± opposite, oblong-elliptic, 5 - 12 × 1.5 - 6 cm, coriaceous and thickish, dark glossy green, midvein thick, primary lateral veins pinnate, higher order venation reticulate; **Inf** 1 - 2 in each floral sympodium at ground level; peduncle 3 - 20 cm, erect but recurved at **Fr** time; spathe entirely persistent to **Fr** stage, 5 - 10 cm, coriaceous, slightly constricted between tube and lamina, outside greenish, inside whitish, tube convolute, lamina longer than tube, first expanded horizontally, later reflexed; spadix sessile, 5 - 8 cm, female zone subcylindrical, 1 - 2 × 0.7 - 1.5 cm ⌀, separate from the male zone by a short constriction with sterile **Fl**; male zone cylindrical, ellipsoid to clavate, fertile to apex, 4 - 6 × 1 - 1.5 cm ⌀, falling off after anthesis; **Fl** unisexual; **Tep** 4 in 2 series, ± 3 mm, whitish, decussate, thickish; **male Fl** with 4 **St** shorter than **Tep**; **Fil** free, oblong, thick, somewhat flattened; **Anth** introrse, pollen extruded in strands; pistillode clavate, as long as **Tep**; **sterile Fl** with 4 **Tep** surrounding a clavate pistillode; **female Fl**: **Gy** equalling the **Tep**; **Ov** ovoid, 2-locular with 1 ovule per locule, placentation axile near base of septum; **Sti** large, discoid-capitate, ± 1.5 mm ⌀; staminodes lacking; **Fr** a subglobose to ellipsoid infructescence at ground level, to 3 cm ⌀, berries depressed-globose, white, to 1.2 cm ⌀, furrowed at the septum, 1- to 2-seeded, surrounded by persistent **Tep**; **Se** ellipsoid, ± 8 × 4 - 5 mm, testa smooth, brown. – *Cytology:* 2n = 34.

The only species of this monotypic genus is highly variable and is notable for the pronounced development of 'leaf petiole succulence'. The upper part of the leaf abscisses at a pre-formed layer at the top of the petiole. Interestingly, the plants can be propagated from individual leaflets which easily break off.

The leaves are reported to be poisonous to goats (P. R. O. Bally, pers. comm.). The plants are used in the traditional medicine in Tanzania and Malawi.

Z. zamiifolia (Loddiges) Engler (Pflanzenr. IV. 23B: 305, fig. 85, 1905). **T:** [icono]: Bot. Cab. 15: t. 1408, 1829. – **D:** Kenya, Tanzania (incl. Pemba and Zanzibar), Malawi, Zimbabwe, Moçambique, RSA (KwaZulu-Natal); tropical moist forests or savannas, often in stony ground, to 600 m. **I:** FPA 40: t. 1562, 1969. **Fig. XXV.c**

≡ *Caladium zamiifolium* Loddiges (1829); **incl.** *Zamioculcas loddigesii* Schott (1856) (*nom. illeg.*, Art. 52.1); **incl.** *Zamioculcas lanceolata* Peter (1930).

Description as for the genus.

Asparagaceae

Shrubs or subshrubs, often scrambling or climbing, with herbaceous to woody perennial or annual shoots; rhizome short, sympodial; **R** sometimes succulent and fusiform; **Br** sometimes green and assimilating, sometimes transformed into **L**-like cladodes (phylloclades), other modified stems (assimilatory structures) fasciculate, green, slender and needle-like; **L** often reduced to **Sc**-like structures; **Fl** small, actinomorphic, solitary or in umbellate or racemose **Inf**, hermaphrodite or unisexual (then plants dioecious); **Tep** 3 + 3, white, yellow or green, free, spreading or basally fused to form a campanulate **Per**; male and hermaphrodite **Fl** with 3 + 3 **St**; **Fil** free from each other; **Anth** introrse, dorsifixed; **Ov** superior, 3-locular, placentation axile with 2 - 12 ovules per locule; **Sty** short; **Sti** capitate or lobed; **Fr** globose red, blue or white berries; **Se** black.

Distribution: Widely distributed in the Old World.

Literature: Obermeyer & al. (1992: 11-82); Kubitzki & Rudall (1998).

The *Asparagaceae*, only recently recognized as a family separate from the *Liliaceae*, are widely distributed in the Old World. They occur mainly in semi-arid to arid mediterranean climates. Plants with xerophmic adaptations such as succulent tuberous roots are common. Photosynthesis occurs mainly with green branches and branchlets. The small family embraces only 3 genera (Obermeyer & al. 1992): *Protoasparagus, Asparagus* and *Myrsiphyllum*, with the first two embracing the majority of the ± 300 species of the family. Based on the minor differences between the 3 genera, Fellingham & Meyer (1995) recently proposed to re-unite them all under *Asparagus*, and this view is also followed by Kubitzki & Rudall (1998). Economically important plants include *Asparagus officinalis* Linné ("Garden Asparagus", used as vegetable) as well as *A. setaceus* (Kunth) Jessop ("Fern Asparagus", used by florists for decoration). Only a few species of *Myrsiphyllum* can be considered succulent due to their succulent roots, but it should be noted that similarly thickened lateral roots occur also in some species of *Protoasparagus*, which differs from *Myrsiphyllum* in completely free tepals. Both have hermaphrodite flowers, while the closely related *Asparagus* has unisexual flowers. [E. van Jaarsveld]

MYRSIPHYLLUM

E. van Jaarsveld

Myrsiphyllum Willdenow (Ges. Naturf. Freunde Berlin Mag. Neuesten Entdeck. Gesammten Naturk. 2: 25, 1808). **T:** *Medeola asparagoides* Linné. – **Lit:** Obermeyer & al. (1992: 71-81). **D:** S Africa; 1 species naturalized in Australia as troublesome weed. **Etym:** Gr. 'myrsine', a myrtle branch; and Gr. 'phyllon', leaf; for the leaf-like phyllocladia ('leaves') which resemble those of myrtle (*Myrtus communis*).

Incl. *Hecatris* Salisbury (1866) (*nom. illeg.*, Art. 52.1). **T:** *Medeola asparagoides* Linné.

Glabrous perennial unarmed climbers; rhizome cylindrical, often not lignified, scale **L** small or vestigial; **R** orientated radially on long creeping rhizomes or irregularly secund on compact rhizomes, forming fusiform tubers crowded on the rhizome or at a distance, inner tissue softly watery; stems twining or erect; phylloclades solitary or 2- to 3-nate, from the **Ax** of scale **L**, the latter not forming **Sp**; **Fl** in groups of 1 - 3, hermaphrodite, pendulous on short to long **Ped**; **Tep** white, usually with a green central band, connate at base, forming a short cup or tube, free tips recurved; **St** erect, usually connivent around the **Ov**; **Fil** flattened, attenuate above and widened below and sometimes with 2 extended spurs; **Anth** introrse, yellow, orange or red; **Ov** 3-locular with 6 - 12 biseriate ovules in each locule; **Sty** 1 or 3; **Sti** 3, papillate; **Fr** globose or ovoid-apiculate berries, red, yellow or orange; **Se** globose, black.

A small genus with 12 species, predominantly in the winter rainfall region of S Africa, but *M. asparagoides* and *M. ramosissimum* extending throughout E Africa to S Europe. All species show a moderate degree of succulence due to the fusiform lateral roots, but only a small selection is here included. *M. asparagoides* is widely used in floristic arrangements (and erroneously referred to as 'Smilax').

M. asparagoides (Linné) Willdenow (Ges. Naturf. Freunde Berlin Mag. Neuesten Entdeck. Gesammten Naturk. 2: 25, 1808). **T:** [icono]: Tilli, Cat. Pl. Horti Pisani, t. 12: fig. 1, 1723. – **D:** Tropical Africa to Namibia and RSA; Succulent Karoo, thickets, savanna, flowers mid-winter to spring; naturalized in Australia and a troublesome weed. **I:** Obermeyer & al. (1992: 72, fig. 14: 2).

≡ *Medeola asparagoides* Linné (1753) ≡ *Hecatris asparagoides* (Linné) Salisbury (1866) ≡ *Asparagus asparagoides* (Linné) W. Wight (1909); **incl.** *Medeola angustifolia* Miller (1768) (*nom. illeg.*, Art. 52.1) ≡ *Myrsiphyllum angustifolium* (Miller) Willdenow (1808) (*nom. illeg.*, Art. 52.1) ≡ *Asparagus medeoloides* var. *angustifolius* (Miller) Baker (1896) (*nom. illeg.*, Art. 52.1); **incl.** *Dracaena medeoloides* Linné *fil.* (1782) ≡ *Asparagus medeoloides* (Linné *fil.*) Thunberg (1794); **incl.** *Myrsiphyllum falciforme* Kunth (1850) ≡ *Asparagus medeoloides* var. *falciformis* (Kunth) Baker (1896); **incl.** *Asparagus kuisibensis* Dinter (1931).

Scandent much-branched perennials with shiny green ovate phylloclades, deciduous or semideciduous; rhizome cylindrical, with numerous lateral fusiform radially arranged **R** tubers, these to 6×2 cm but variable in size, close to the rhizome but not overlapping; stems wiry, twisting, smooth or ridged, to 2 m; **Br** short, with beaded ridges; phylloclades variable in size and shape, ovate-acuminate, to 4×2 cm, flat or folded and curved, many-veined but mostly with 3 more pronounced veins on each face, margin smooth or minutely denticulate; **Ped** to 1 cm, articulated below **Per**; **Tep** to 1 cm, forming a tube in the lower ½, reflexed above; **St** erect, connivent; **Fil** expanded below into 2 small spreading teeth; **Anth** red; **Ov** pear-shaped, stipitate, narrowed into a **Sty** as long as the **Ov**, ovules to ± 6 in each locule; **Sti** short, spreading, ciliate; **Fr** globose, ± 1 cm ∅, usually many-seeded.

M. multituberosum (R. A. Dyer) Obermeyer (BT 15: 77, 1984). **T:** RSA, Cape Prov. (*Marloth* 9006 [PRE, STE]). – **D:** RSA (Northern Cape, Western Cape); Renosterveld and Fynbos, flowers in winter and spring. **I:** Obermeyer & al. (1992: 72, fig. 14: 1).

≡ *Asparagus multituberosus* R. A. Dyer (1954).

Scandent much-branched perennials to 40 cm tall; rhizome thin, horizontal, to 50 cm, with densely overlapping small fusiform tubers 10×3 mm; phylloclades ovate to cordate, to 2.5 cm, many-veined, tip apiculate, margin papillate; **Fl** 1 - 3, axillary from membranous scale-**L**; **Ped** 0.5 cm, curved, articulated below **Per**; **Tep** ± 7 mm, forming a wide tube below, spreading above; **St** as long as **Tep**; **Fil** basally broadened, flat; **Ov** ovoid, with up to 12 ovules per locule; **Sty** 3, curved outwards; **Sti** apical, papillate; **Fr** not seen.

M. ovatum (Salter) Obermeyer (BT 15: 79, 1984). **T:** RSA, Cape Prov. (*Salter* 8214 [BOL, NBG, PRE]). – **D:** RSA (Western Cape, Eastern Cape). **I:** Obermeyer & al. (1992: 72, fig. 14: 3,5).

≡ *Asparagus ovatus* Salter (1940).

Scandent deciduous twiners to 1.5 m tall; rhizome compact, scaly, woody, with long spreading **R**, tubers numerous, hard, swollen, fusiform, 5 - 10×2 - 4 mm, developing at some distance from the rhizome; phylloclades deciduous, ovate, 3×1.5 cm, many-veined, shiny; **Fl** 2 - 3, behind the base of the phylloclades; **Ped** 1 cm, articulated near the base of the **Fl**; **Tep** ± 6 mm, reflexed; **Fil** flattened, erect, with small basal spur on each side; **Ov** oblong, with up to 10 ovules per locule; **Sty** and **Sti** just exserted from the staminal column; **Fr** 1 cm ∅, globose.

Asphodelaceae

Perennial rhizomatous mesophytic to xerophytic herbs, acaulescent geophytes or small herbaceous to rarely erect shrubs; rhizome often fleshy; **R** fibrous, fleshy or fusiform; basal stem parts rarely swollen (caudiciform); **L** rosulate, often succulent, mostly oblong, linear to triangular; **Inf** simple or racemose panicles; **Fl** with 6 **Tep** in 2 series, stellate; **Tep** oblong, white, pink or yellow; **St** 6 in 2 whorls of 3; **Fil** smooth or occasionally hairy; **Anth** dorsifixed, dehiscing longitudinally; **Ov** superior, with 3 locules with 2 ovules each; placentation axile; **Sty** filiform; **Sti** trilobed; **Fr** loculicidal capsules; **Se** elongate, ovoid or winged, sometimes arillate.

Distribution: Europe, Africa, Asia, Australia.

Literature: Smith & Wyk (1998).

A family of ± 10 genera and about 350 species, mainly confined to Europe, Africa, Asia and Australia. Distribution centres are the Mediterranean region and S Africa. According to Dahlgren & al. (1985), the *Asphodelaceae* fall into the 2 subfamilies *Asphodeloideae* and *Alooideae*, but the latter are here treated as separate family *Aloaceae*.

The family in its strict sense counts with 2 genera with succulent taxa, *Bulbine* and *Trachyandra*. Both have their main centres of diversity in the winter rainfall region of the Cape Provinces. *Bulbine* with ± 80 species is the larger genus and is easily recognized on account of the softly bearded filaments and yellow flowers.

The family *Asphodelaceae* has no recognized economic importance except that many species are of horticultural interest, and some are locally used in various ways. Roots of the mediterranean species *Asphodelus albus* are used in the fermentation of alcohol, *A. ramosus* provides gums and *A. aestivus* is used as dye for carpets. Young inflorescences of *Trachyandra falcata* and *T. divaricata* from the Cape are cooked as a stew and have moderate local economical potential; they are locally known as "Veldkool". The sap of the leaves of *Bulbine frutescens* is placed on wounds because of its antiseptic properties (locally known as "Kopieva"), or is used against burns and insect bites.

Key to genera with succulents:

1 **Fl** shades of yellow (rarely white or orange); **Fil** softly bearded: **Bulbine**
- **Fl** white to mauve; **Fil** not softly bearded, at most scabrid: **Trachyandra**

[E. van Jaarsveld]

BULBINE

E. van Jaarsveld & P. I. Forster

Bulbine Wolf (Gen. Pl., 84, 1776). **T:** *Bulbine frutescens* (Linné) Willdenow. – **Lit:** Watson (1987). **D:** S Africa (mostly RSA), Australia. **Etym:** Lat., an onion-like plant (from Lat. 'bulbus', bulb).

Incl. *Blephanthera* Rafinesque (1837). **T:** not designated.
Incl. *Nemopogon* Rafinesque (1837). **T:** *Nemopogon glaucum* Rafinesque.
Incl. *Bulbinopsis* Borzi (1897). **T:** not designated.

Plants perennial (rarely annual), caulescent or acaulescent, deciduous to evergreen, dwarf succulent shrublets, often geophytic with tuberous or rhizomatous base (caudex), solitary or dividing to form dense groups; tuber very variable, depressed to rounded or oblong, often lobed; **L** very variable, terete to flattened, firm to softly succulent, glabrous or hairy, rosulate or rarely distichous, green to glaucous, sometimes with translucent lines or windows, often surrounded at the base by **Bri**, margin entire, ciliate or minutely denticulate; **Inf** racemose, few- to many-flowered, lax or dense, cylindrical to corymbose, erect; peduncle ebracteate, smooth; fertile **Bra** small, membranous; **Fl** actinomorphic; **Per** stellate, spreading, rarely ± connivent, **Tep** 6 in 2 series, free, often becoming reflexed, mostly yellow, the outer smaller; **St** 6; **Fil** bearded, yellow, rarely orange; **Anth** dorsifixed, introrse; **Ov** ovoid to globose, 3-locular; **Sty** terete; **Sti** capitate; **Fr** 3-locular loculicidal globose capsules; **Se** blackish, flat, angular, oblong to rounded in outline.

A genus of ± 80 species occuring in Africa and Australia with the greatest diversity in the SW Cape (most species here are deciduous geophytes). A very variable group easily propagated by seeds or divisions; the plants are fast-growing (though somewhat delicate at least in the N hemisphere) attractive garden and container subjects. Some species show very attractive leaf markings and ± 30 species are found cultivated. Flowering is hysteranthous or synanthous. *B. frutescens* is caulescent and frequently cultivated as a ground cover in S Africa. *Bulbine* is related to *Bulbinella* and *Trachyandra*, but is distinguished from these by the bearded filaments.

In the following descriptions of the succulent taxa, the flower colour is yellow when not specifically mentioned.

B. abyssinica A. Richard (Tent. Fl. Abyss. 2: 334, t. 97, 1851). – **D:** NE Africa to RSA; grassland, savanna, flowering in early summer.

Evergreen, acaulescent, to 50 cm tall; **R** yellow, fleshy, terete; **L** rosulate, green, 15 - 50, linear, soft, 8 - 20 cm × 4 - 6 mm, upper face flat, lower face convex, apex acuminate; **Inf** 1 - 5, densely flowered to 35 cm tall with **Fl** crowded in the upper ¼; peduncle terete, 3 mm ∅; **Bra** linear-lanceolate, 11

- 12 × 1.5 mm; **Ped** terete, 1.5 - 3 mm; **Per** spreading, stellate; **OTep** lanceolate, cymbiform, 10 × 5 mm; **ITep** 9 × 6 mm, obtuse; **St** 6 mm; **Fil** densely bearded in the upper ⅓; **Anth** 1 mm, oblong; **Ov** ovoid, 3 mm long; **Sty** erect, terete, 5 mm; **Fr** ovoid, 5 mm; **Se** grey-black, 3 mm. – [E. van Jaarsveld]

B. alata Baijnath (Brunonia 1(1): 117-120, ills., 1978). **T:** Australia, South Australia (*Hill 714* [BM]). – **D:** Australia.

Annual succulent herbs without caudex; **R** fibrous; **L** 3 - 10, linear to subulate, channelled, 1 - 2.5 mm wide, glaucous; **Inf** several to many, erect, peduncle terete; **Tep** 3 - 5 mm; **St** 3 short, 3 long, ± erect and closely bunched; **Fil** with clavate **Ha** below apex; **Anth** yellow, horizontally arched after dehiscence; ovules 2 per locule; **Sty** erect, 1 - 1.5 mm; **Fr** globose, 4 - 9 mm; **Ped** 4 - 17 mm, erect; **Se** with membranous wing. – [P. I. Forster]

B. alooides (Linné) Willdenow (Enum. Pl. Hort. Reg. Berol., 372, 1809). **T:** [icono]: Dillenius, Hort. Eltham., 312, t. 232: fig. 100, 1732. – **D:** RSA (Western Cape); Renosterveld, winter-grower flowering in winter and spring.

≡ *Anthericum alooides* Linné (1753); **incl.** *Anthericum praemorsum* Jacquin (1793) ≡ *Bulbine praemorsa* (Jacquin) Roemer & Schultes (1817); **incl.** *Bulbine zeyheri* Baker (1876).

Solitary shortly caulescent geophytes 6 - 25 cm tall; tuber depressed, 5- to 7-lobed, brownish, to 4 cm ∅, lobes to 9 mm ∅ at base, tapering; **R** terete; stem erect to 3 cm with a ring of **Bri** at base surrounding the **L**; **L** 2 - 4, unequal, 6 -25 cm × 8 - 12 mm, opposite, erect, light green, amplexicaul, linear-lanceolate, faintly striated, upper side channelled, lower face convex, apex acute; **Inf** 1 - 2, to 40 cm, synanthous, with up to 35 **Fl**; peduncle biconvex to 5 mm ∅ at the base; **Fl** stellate, spreading, ± 20 mm ∅; **Bra** to 4 mm, ovate-acuminate, 3 - 4 mm wide at base, clasping; **Ped** 10 - 15 mm; **Tep** becoming reflexed; **OTep** elliptic, 10 × 2 mm, acute; **ITep** obovate, 9 × 5 mm, obtuse; **St** 7 mm; **Anth** yellow, oblong, 1 mm; **Ov** elliptic-oblong, 2 mm; **Sty** curved, yellow, 5 mm; **Fr** oblong, 10 × 4 mm; **Se** oblong, 3 × 1.5 mm, grey-black. – [E. van Jaarsveld]

B. angustifolia von Poellnitz (Feddes Repert. Spec. Nov. Regni Veg. 54: 45, 1944). **T:** Namibia, Hereroland (*Dinter 4404* [B]). – **D:** Namibia, RSA (Northern Cape); grassland and savanna, spring-flowering. **I:** FPA 26: t. 1019, 1947, as *B. tortifolia*.

Incl. *Bulbine tortifolia* I. Verdoorn (1947).

Plants solitary, to 35 cm tall; **R** fleshy, terete; **L** rosulate, amplexicaul, to 20 per **Ros**, shiny green, ascending, 35 × 5 mm, linear-terete and twisted, flattened; **Inf** solitary, to 80 cm with **Fl** in the upper ⅓; peduncle terete, 6 mm ∅ at the base; **Bra** 12 - 15 mm, membranous, linear-acuminate, clasping; **Ped** 5 - 10 mm; **Per** stellate, 12 - 14 mm $\emptyset$; all **Tep** ovate-lanceolate, 6 × 4 mm, obtuse; **St** 5 mm; **Anth** cream; **Ov** oblong-globose, 2.5 × 2 mm; **Sty** erect, yellow, 2.5 mm; **Fr** subglobose, 10 mm $\emptyset$; **Se** grey-black, 3 mm. – [E. van Jaarsveld]

B. annua (Linné) Willdenow (Enum. Pl. Hort. Reg. Berol., 372, 1809). – **D:** RSA (Western Cape); Strandveld and coastal Fynbos, spring-flowering.

≡ *Anthericum annuum* Linné (1753); **incl.** *Bulbine caespitosa* Baker (1897).

Acaulescent rosulate annuals or weak perennials, 6 - 20 cm; **R** terete to 1 mm $\emptyset$; **L** 4 - 35, linear, subterete, amplexicaul at base, ascending to spreading, 1.8 - 18 cm, softly succulent, upper face flat, lower face convex, apex acute; **Inf** 11 - 25 cm, ascending, **Fl** in the upper ½; peduncle terete, 3 mm $\emptyset$ at the base; **Bra** ovate-acuminate, 5 × 2 mm, deltoid-acuminate, membranous; **Per** stellate, 12 - 13 mm $\emptyset$; **OTep** elliptic, 6 × 2 mm; **ITep** 5 × 3 mm, obtuse; **Ov** globose, 1.5 mm $\emptyset$; **Sti** 3 mm; **Fr** globose, 5 - 6 mm $\emptyset$; **Se** angular, 2 mm $\emptyset$. – [E. van Jaarsveld]

B. asphodeloides (Linné) Willdenow (Enum. Pl. Hort. Reg. Berol., 311, 1809). **T:** LINN 432.11. – **D:** RSA (Western Cape); Renosterveld, flowers in mid-winter and spring.

≡ *Anthericum asphodeloides* Linné (1753).

Perennials, dividing to form dense groups 20 cm high and $\emptyset$ with up to 15 heads; **R** fleshy, brownish-orange, terete, 1.5 mm $\emptyset$, tips yellow; stem short, to 2 cm; **L** rosulate, to 10, amplexicaul, green to reddish-green, ascending, linear, firm, 8 - 20 cm × 3 - 4 mm, flattened above, convex below, surface with fine longitudinal grooves, margin minutely dentate, apex acute; **Inf** solitary, 30 - 45 cm, ascending to spreading with **Fl** in the upper ½ and 3 - 10 mm apart; peduncle 2 mm $\emptyset$ at the base, terete; **Bra** 6 - 7 × 1.5 mm, membranous, deltoid-acuminate, clasping; **Ped** 10 - 13 mm; **Per** stellate; **OTep** elliptic-lanceolate, 8 × 3 mm, obtuse; **ITep** broadly ovate, 8 × 5 mm, obtuse; **St** 5 - 6 mm; **Anth** yellow, oblong; **Ov** globose, 2 mm $\emptyset$, yellowish-green, grooved; **Sty** erect, yellow, 5 mm; **Fr** ovoid-globose, 6 × 5 mm; **Se** grey-black, 2 × 1.5 mm. – [E. van Jaarsveld]

B. brunsvigiifolia Baker (FC 6: 366, 1896). **T:** RSA, Cape Prov. (*Drège* 2674 [K]). – **D:** RSA (Northern Cape, Western Cape); Succulent Karoo and Renosterveld.

Incl. *Bulbine stenophylla* I. Verdoorn (1948).

Rosulate geophytes, solitary or forming small groups; tuber depressed, to 4.5 cm $\emptyset$ and 3 cm high with numerous **Bri** to 3 cm surrounding the **L**; **R** terete to 4 mm $\emptyset$; **L** oblong-lorate, attenuate, to 10 × 4.5 cm, flat, margin ciliate, apex obtuse, minutely cuspidate; **Inf** 30 - 75 cm with **Fl** in the upper ⅓; peduncle 1 cm wide at the base; **Bra** membranous,

ovate-lanceolate, 8 - 15 × 2 mm; **Ped** 1 - 2 cm; **OTep** lanceolate, 10 - 12 × 2.5 mm; **ITep** ovate, 7 × 4 mm; **St** 7 mm; **Ov** 3 mm, ovoid; **Sty** 5 mm; **Fr** oblong, 15 × 5 mm; **Se** oblong, 4.5 × 1.5 mm. – [E. van Jaarsveld]

B. bruynsii S. A. Hammer (Cactus & Co. 2(4): 6-7, ills., 1998). **T:** RSA, Northern Cape (*Bruyns* 6126 [BOL]). – **D:** RSA (Northern Cape: Namaqualand); Succulent Karoo.

Deciduous erect soft succulent herbs 6 - 12 cm tall; tuber napiform, to 4 × 1 cm, with brown skin, producing offshoots from secondary tubers; **L** 2, soft, erect, semitranslucent, transversely banded red and green, 1 - 6 × 1 - 1.8 cm, oblong-ovate and shaped like a Chinese lantern (horizontally ridged) and tapering at the tip; second **L** subulate; **Inf** erect; peduncle slender, to 20 cm × 1 - 2 mm; **Fl** 8 - 20, sweetly scented; **Bra** triangular, membranous, < 2 mm; **Ped** to 15 mm; **OTep** narrowly linear, 1.2 - 1.5 mm, bright yellow; **ITep** ovate, to 4 mm wide, yellow with a bright green midstripe, strongly crinkled, shiny; **Fil** to 6 mm; **Anth** 1 mm; **Ov** broadly pear-shaped, bright green; **Sty** 6 mm; **Sti** knob-like; **Fr** globose; **Se** 1.6 × 0.8 mm, dull black, narrowly triangular. – [E. van Jaarsveld]

B. bulbosa (R. Brown) Haworth (Revis. Pl. Succ., 33, 1821). **T:** Australia, New South Wales (*Brown* s.n. [BM]). – **D:** Australia (Queensland, New South Wales). **I:** Cunningham & al. (1981: 183).

≡ *Anthericum bulbosum* R. Brown (1810) ≡ *Phalangium bulbosum* (R. Brown) Kuntze (1891) ≡ *Bulbinopsis bulbosa* (R. Brown) Borzi (1897); **incl.** *Bulbine australis* Sprengel (1825); **incl.** *Blephanthera depressa* Rafinesque (1837); **incl.** *Blephanthera hookeri* Rafinesque (1837); **incl.** *Bulbine fraseri* Kunth (1843).

Perennial succulent herbs with bulb-shaped caudex; **R** succulent; **L** 3 - 10, linear, channelled, 1.5 - 1.8 mm wide, green, rarely glaucous; **Inf** 1 - 2, terete, erect; **Tep** 9 - 22 mm; **St** ± equal, closely bunched or spreading; **Fil** with clavate or acute **Ha** at tip; **Anth** yellow, ± basifixed, all or some remaining erect after dehiscence and sometimes twisting about vertical axis; ovules 3 - 8 per locule; **Sty** decumbent, 2.5 - 9 mm; **Fr** globose to obovate, 4 - 7 mm; **Ped** 10 - 28 mm, not drooping at end; **Se** smooth, slightly tuberculate or ridged. – [P. I. Forster]

B. canaliculata G. Williamson (Bradleya 18: 37, ills. (p. 39), 2000). **Nom. illeg.**, Art. 53.1. **T:** RSA, Eastern Cape (*Williamson* 5928 [NBG]). – **D:** RSA (Eastern Cape: near Jeffreys Bay); coastal grassland, flowers early summer.

Small solitary geophytes; tuber ovoid, 10 × 15 mm; **R** orange-brown; **L** up to 5, to 180 × 8 mm, rosulate, erect, linear-lanceolate, bright and shiny green, acute from a broad base, upper face canaliculate, margins hyaline, finely ciliate in the lower part; **Inf** solitary, erect, to 22 cm, to 20-flowered; peduncle 1.2 mm ∅ at the base, terete; **Bra** deltoid-acuminate, 5 × 3 mm; **Tep** recurved, shiny yellow; **OTep** elliptic, subacute, 6 × 1.8 mm; **ITep** ovate, 6 × 3.2 mm, tips obtuse; **St** projecting forwards in a cluster; **Ov** ovoid, 1.8 mm, bright green; **Sty** erect, 6 mm; **Sti** papillate; **Fr** and **Se** not seen.

This name is unfortunately an illegitimate later homonym. − [E. van Jaarsveld]

B. capitata von Poellnitz (Feddes Repert. Spec. Nov. Regni Veg. 53: 37, 1944). **T:** RSA, Cape Prov. (*Wilms* 1508 [B]). − **D:** Namibia, RSA; grassland and savanna, spring-flowering.

Evergreen, rosulate, clustering, short-stemmed; **R** yellow, fleshy, terete; stems to 5 cm; **L** green, 20 per stem, linear, subterete, 7 - 20 cm × 2 mm, upper face flat to slightly chanelled, lower face convex, attenuate, apex acute; **Inf** densely flowered to 20 cm, flowering head subcorymbose; peduncle terete, 3 mm ∅ at the base; **Bra** triangular-lanceolate, 3 × 2 mm; **Ped** terete, 15 mm; **Fl** scented, spreading, stellate; **OTep** oblong-ovate, 7 × 3 mm; **ITep** 7 × 4 mm, obtuse; **St** 6 mm; **Fil** densely bearded in the upper ⅓; **Anth** 1 mm, oblong; **Ov** ovoid, 2 mm; **Sty** erect, terete, 3.5 mm; **Fr** ovoid. − [E. van Jaarsveld]

B. caput-medusae G. Williamson (Aloe 32(3-4): 80-83, ills., 1995). **T:** Namibia, Witputz (*Williamson* 4208 [NBG]). − **D:** S Namibia; desert, flowers mid-winter to late spring.

Perennials, dividing to form clusters with up to 8 heads; **R** dense, fibrous, matted, to 20 cm; **L** up to 15 in dense clusters arising from dense white fibrous sheaths, 4 cm × 2 mm, channelled to 1 mm below, terete in the upper ½ to ⅔, spreading, strongly circinate and contorted, shiny dark green, minutely sparsely granulate; **Inf** 1 - 2 per head; peduncle green, suberect to spreading, to 8 cm; raceme 4 - 6 cm, crowded, with up to 20 **Fl**; **Bra** membranous, triangular, 1.2 mm; **Ped** 1.5 cm; **Per** stellate, **OTep** light canary yellow, 9 × 1.8 mm; **ITep** elliptic, 8 × 2 mm; **Fil** bearded with **Ha** 5 mm long; **Ov** globose, dark green; **Sty** slightly curved, 5 mm; **Fr** rounded, to 12 mm long; **Se** black, oblong with contorted projecting ridges, 5 × 4 mm. − [E. van Jaarsveld]

B. cataphyllata von Poellnitz (Feddes Repert. Spec. Nov. Regni Veg. 53: 37-38, 1944). **T:** RSA, Western Cape (*Bachmann* 1797 [B]). − **D:** RSA (Western Cape); Strandveld.

Geophytes to 6 cm tall; tuber oblong cone-shaped, 2 - 4 cm ∅; **L** numerous, older basal **L** persistent; lower **L** 3 - 6 × 2 cm, ovate, other **L** ascending, linear, 8 - 15 cm × 2 mm, apex mucronate; **Inf** densely flowered in the upper ½, to 56 cm; peduncle 25 - 30 cm, terete, 2 mm ∅ at the base; **Bra** triangular-lanceolate, 5 - 6 mm; **Ped** terete, 6 - 10 mm; **Fl** scented, spreading, stellate, 11 - 13 mm ∅;

OTep oblong-ovate, 6 mm; **ITep** 5 mm, obtuse; **St** 3 mm; **Anth** 1 mm, oblong; **Ov** globose, 1 mm ∅; **Sty** erect, terete, 3 mm; **Fr** and **Se** not seen. − [E. van Jaarsveld]

B. cepacea (Burman *fil.*) Wijnands (BT 21(2): 157, 1991). **T:** [icono]: Hermann, Horti Acad. Lugd.-Bat. Cat. t. 467, 1687. − **D:** RSA (Western Cape); well-drained Renosterveld, autumn-flowering.

≡ *Ornithogalum cepaceum* Burman *fil.* (1768); **incl.** *Ornithogalum tuberosum* Miller (1768) ≡ *Bulbine tuberosa* (Miller) Obermeyer (1976); **incl.** *Anthericum pugioniforme* Jacquin (1793) ≡ *Bulbine pugioniformis* (Jacquin) Link (1821).

Hysteranthous tuberous geophytes; tuber depressed, globose on top, up to 5 cm high and 10 cm ∅; **R** fleshy, terete, to 8 mm ∅, radiating from lower margin of tuber, tapering; stem very short with **Bri** at base; **L** subterete, rosulate, 4 - 15, unequal, ± linear, 6 - 25 × 0.3 - 1 cm ∅, spirally arranged, erect, green, amplexicaul, faintly striated, upper face of lower **L** flat, lower face convex, apex acute; **Inf** 1 - 2, 40 - 70 cm; peduncle 5 mm ∅ at the base, **Fl** numerous in the upper ½, stellate, spreading, ± 20 mm ∅; **Bra** 7 × 3 - 4 mm, triangular-acuminate, clasping; **Ped** 1 - 1.5 cm; **OTep** elliptic, 9 × 2.5 mm, acute; **ITep** obovate, 8 × 3.5 mm, obtuse; **St** 7 mm; **Anth** yellow, oblong, 1 mm; **Ov** globose, 2 mm ∅; **Sty** 5 mm; **Fr** narrowly ovoid, 8 - 10 × 4 mm; **Se** dark brown, oblong, winged, to 7 mm. − [E. van Jaarsveld]

B. coetzeei Obermeyer (BT 9(2): 343-344, 1967). **T:** RSA, Mpumalanga (*Coetzee* s.n. [PRE 30026]). − **D:** RSA (KwaZulu-Natal, Mpumalanga), Swaziland; grassland.

Base tuberous, 2.5 × 1.5 cm; **R** terete, succulent, spreading; **L** rosulate, ± 7, ascending, linear, terete, to 50 × 1 cm, amplexicaul and sheathing below forming a short neck 2 cm long, upper face grooved, tip acuminate; **Inf** 1, to 50 cm; **Fl** in the upper ⅓, dense; peduncle terete; **Bra** to 25 mm, membranous, linear-acuminate, aristate; **Ped** to 25 mm; **Per** stellate, to 23 mm ∅, reflexed; **Tep** linear-lanceolate, ± 6 × 1 mm, obtuse; **Ov** ovoid, 7 mm, truncate at apex; **Sty** erect; **Fr** obovoid, 4 mm ∅; **Se** black, angled, 2 mm. − [E. van Jaarsveld]

B. cremnophila van Jaarsveld (Aloe 36(4): 72, ills., 2000). **T:** RSA, Eastern Cape (*van Jaarsveld* 7238 [NBG]). − **D:** RSA (Eastern Cape); shady quartzitic sandstone cliff-faces, midsummer-flowering.

Plants dwarf, rosulate, clustering, to 8 cm high and 10 cm ∅ with 3 - 8 heads; **R** grey, fleshy, terete; **L** 5 - 7 in **Ros**, curving downwards, linear-triangular to linear-lanceolate, 6 - 10 × 1 -1.5 cm, upper face channelled above, cymbiform below, glaucous, apex acute, mucronate; **Inf** 1, to 30 cm, 17- to 35-flowered in the upper ½; peduncle 2 mm ∅ at the base, terete; **Bra** deltoid, acuminate, 5 × 1 mm, clasping; **Fl** drooping; **Per** stellate, becoming reflexed, ± 8 - 10 mm ∅; **Ped** 15 - 18 mm; **Tep** pale orange-yellow; **OTep** elliptic, 7 × 2 mm; **ITep** ovate to ovate-elliptic, 6 × 2.5 mm, obtuse; **St** 5 mm; **Ov** globose, 1.5 mm ∅; **Sty** erect, 6 mm; **Fr** ovoid, 3 × 2.5 mm; **Se** 2 mm. – [E. van Jaarsveld]

B. crocea L. Guthrie (Ann. Bolus Herb. 4: 30, 1928). **T:** RSA, Eastern Cape (*Page* s.n. [BOL 16393]). – **D:** RSA (Eastern Cape).

Acaulescent, tuberous, tuber to 8 mm ∅; **L** rosulate, ± 7, amplexicaul, ascending, linear, terete, to 17 cm, 5 mm ∅; **Inf** 1, to 16 cm; peduncle 5 mm ∅ at the base, terete; **Fl** in upper ⅓; **Bra** membranous, deltoid-cuspidate, to 5 mm; **Ped** 13 mm; **Per** stellate; **Tep** linear-ovate, 5 × 2 mm, obtuse; **St** 5 mm; **Ov** oblong, 1.5 × 1 mm; **Sty** erect, 3 mm; **Fr** and **Se** not seen. – [E. van Jaarsveld]

B. densiflora Baker (JLSB 15: 347, 1876). **T:** RSA, Eastern Cape (*Burke* s.n. [K]). – **D:** RSA (Eastern Cape).

Plants evergreen, rosulate, to 25 cm; **R** terete, 2 mm ∅; **R**stock tuberous; **L** 10 - 14 per stem, linear-lorate, 14 - 24 cm × 5 - 7 mm, apex obtuse to subacute; **Inf** densely flowered, to 38 cm; peduncle terete, 3 mm ∅; **Bra** triangular-lanceolate, 12 - 15 × 4 mm, margins membranous; **Ped** terete, 5 - 7 mm; **Fl** spreading, stellate, 12 - 13 mm ∅; **OTep** oblong-ovate, 6 × 2 mm; **ITep** 5.5 × 2 mm, obtuse; **St** 6 mm; **Fil** bearded in upper ⅓; **Ov** ovoid, 2 × 1 mm; **Sty** erect, terete, 2 mm; **Fr** ovoid, 4 × 2 mm; **Se** black, 2 × 1 mm. – [E. van Jaarsveld]

B. diphylla Schlechter *ex* von Poellnitz (Feddes Repert. Spec. Nov. Regni Veg. 53: 40-41, 1944). **T:** RSA, Western Cape (*Schlechter* 8202 [B]). – **D:** RSA (Western Cape); Succulent Karoo, quartz gravel flats, flowers in mid-winter.

Dwarf solitary caulescent geophytes to 7 cm tall; tuber 3- to 7-lobed, brownish, lobes ovoid, 5 - 10 mm, obtuse; **R** terete; stem erect to 5 - 20 mm tall with distinct maroon striations; **L** 2, unequal, subopposite, erect, light green, amplexicaul; lower **L** subterete, ovate-lanceolate, 3 - 5 × 1.5 cm, reticulate, upper face slightly channelled, lower face convex, apex acute; upper **L** linear-terete to ovate, 1.5 - 3 × 1 - 1.5 cm, channelled; **Inf** 1, 12 - 17 cm, 5- to 10-flowered; **Fl** spreading, ± 15 mm ∅; **Bra** to 2 × 1 mm, ovate-acuminate, clasping; **Ped** 1 - 1.5 cm; **Per** stellate, becoming reflexed; **OTep** lanceolate, 8 × 2 mm; **ITep** elliptic, 8 × 4.2 mm, obtuse; **St** 5 mm; **Anth** yellow, oblong, 0.5 mm; **Ov** ellipsoid-oblong, 1 - 2.5 mm, **Sty** erect, yellow, 5 mm; **Fr** and **Se** not seen. – [E. van Jaarsveld]

B. dissimilis G. Williamson (Aloe 34(3-4): 70-72, ills., 1997). **T:** RSA, Northern Cape (*Williamson & Hammer* 5710 [NBG]). – **D:** RSA (Northern Cape: Springbok Distr.).

Solitary acaulescent perennial geophytes; tuber to 3 × 2 cm, oblong to turnip-shaped, covered with brownish to blackish imbricate papery sheaths; **R** short, fleshy, tapering downwards, deltoid, to 1.5 × 1 cm; **L** subtended by 3 triangular acute sheathing papery **Bra** to 25 × 7 mm; **L** 3, erect, 13 × 1 cm, dark glaucous-green with powdery bloom, linear-triangular, upper face grooved; **Inf** erect, solitary, to 24 cm, raceme 8 × 2 cm; **Fl** 14 - 20, sweetly scented; **Ped** 7 mm; **Tep** spreading, becoming reflexed, dark orange-yellow with shiny luminous surface; **OTep** obovate, 7 × 4.5 mm; **St** 7 mm; **Ov** globose, green, 1 mm ∅; **Sty** slender, 6 mm; **Fr** ovoid, 2.5 mm; **Se** shiny black, oblong and triangular in cross-section, 2 × 1.2 mm. – [E. van Jaarsveld]

B. esterhuyseniae Baijnath (SAJB 53(6): 427-430, ills., map, 1987). **T:** RSA, Western Cape (*Esterhuysen* 25487 [BOL, K, PRE]). – **D:** RSA (Western Cape); Fynbos, quartzitic sandstones, flowers in mid-summer.

Dwarf solitary geophytes to 2.5 cm tall; tuber depressed, 4- to 5-lobed, ± 5 mm tall and 10 - 15 mm ∅; **R** terete; **L** rosulate, 5 - 10, 5 - 35 × 0.5 mm, ascending, apex obtuse, surrounded at base by 2 - 3 **Sc** 2 × 4 mm; **Inf** 1 - 2, 5 - 12 mm, 2- to 8-flowered; peduncle to 2 mm ∅, terete; **Fl** spreading, stellate, ± 8 - 10 mm ∅; **Bra** 2.5 × 3 mm, ovate; **Ped** 10 - 14 mm, recurved; **OTep** elliptic, 6 × 2 mm; **ITep** elliptic, 5 × 2.5 mm, obtuse; **St** 4 mm; **Anth** yellow, oblong; **Ov** globose, 1 mm; **Sty** erect, yellow, 2 mm; **Fr** ovoid, 2 mm; **Se** 1.5 mm, blackish. – [E. van Jaarsveld]

B. fallax von Poellnitz (Feddes Repert. Spec. Nov. Regni Veg. 53: 41, 1944). **T:** RSA, Northern Cape (*Schlechter* 11019 [B]). – **D:** RSA (Northern Cape, Western Cape); Succulent Karoo, quartz gravel flats. **Fig. XXV.f**

Dwarf deciduous acaulescent rosulate solitary geophytes; tuber oblong, 3.5 × 1.5 cm, brown, with a ring of **Bri** around base of **Ros**; **R** fleshy, terete; **L** densely arranged, up to 20, linear-lanceolate, to 9 × 1.6 cm, green, with reticulate pattern, flat above, convex below, margin minutely ciliate, apex acute; **Inf** 1, to 25 - 33 cm; peduncle terete, 3 mm ∅ at the base; **Bra** 1 cm, ovate-lanceolate, clasping; **Ped** 1.5 - 3 cm; **Fl** stellate, to 26 mm ∅; **Tep** lanceolate, 6 - 7 mm, obtuse; **St** 6 - 8 mm; **Sty** erect, terete, yellow; **Fr** oblong-ovoid, 8 - 12 mm; **Se** angled, black, 3 mm. – [E. van Jaarsveld]

B. favosa (Thunberg) Roemer & Schultes (Syst. Veg. 7: 444, 1829). **T:** RSA, Western Cape (*Thunberg* s.n. [K]). – **D:** RSA (Western Cape); flats and lower slopes in Fynbos, autumn-flowering.

≡ *Anthericum favosum* Thunberg (1800); **incl.** *Bulbine dubia* Schlechter (s.a.).

Hysteranthous geophytes, solitary, to 10 cm tall

(excl. **Inf**); tuber oblong, 3 - 3.5 × 1 cm ∅; **R** 1 mm ∅, terete; **L** 1 - 3, subterete, 2 - 10 cm × 1 - 1.5 mm, linear, attenuate, upper face flat, lower face convex, apex acute; **Inf** laxly flowered in the upper ⅓, 12 - 27 cm; peduncle terete, 1 mm ∅; **Bra** triangular-lanceolate, 2 × 2.5 mm; **Ped** 6 - 7 mm; **Per** stellate, to 1.6 cm ∅; **OTep** oblong-ovate, 8 × 2 mm; **ITep** 8 × 3 mm, obtuse; **Fil** 6 - 7 mm, bearded in the upper ⅓; **Anth** 0.75 mm, oblong; **Ov** globose, 2.5 mm; **Sty** erect, terete, 5.5 mm; **Fr** globose, 4 mm ∅; **Se** 1.5 mm ∅. – [E. van Jaarsveld]

B. filifolia Baker (JLSB 15: 344, 1876). **T:** RSA, Eastern Cape (*Bolus* 762 [K]). – **D:** RSA (Eastern Cape, KwaZulu-Natal, Free State, former Transvaal); shallow quartzitic sandstone in rock pockets, flowers in spring and summer.

Incl. *Bulbine trichophylla* Baker (1876).

Dwarf solitary tuberous geophytes with slender solitary stem to 4.5 cm; tuber globose, lobed, ± 1 cm tall and wide, subterranean, occasionally exposed, with greyish bark; **R** terete; stem 2 - 3 mm ∅, with grey papery sheaths; **L** 2, linear-lanceolate, erect, 1 - 12 cm × 1 - 4 mm ∅, grey-green, withering in the dry season, basally amplexicaul, upper face flat, lower face convex, apex acute; **Inf** 1, 10- to 25-flowered, 11 - 27 cm, spreading; **Bra** 3 × 1 mm, membranous, deltoid, acuminate, basally clasping; **Ped** 10 - 12 mm; **Fl** 5 - 18 mm apart, 6 - 15 mm ∅; **Tep** stellate, becoming reflexed, pale yellow; **OTep** lanceolate, 5 - 8 × 4 - 5 mm; **ITep** ovate to ovate-elliptic, 5 - 6 × 2 mm, obtuse; **St** 7 mm; **Ov** globose, 1 mm; **Sty** 6 mm; **Fr** globose, 2 mm; **Se** 1.5 mm, black. – [E. van Jaarsveld]

B. flexicaulis Baker (FC 6: 365, 1896). **T:** RSA, Eastern Cape (*Pappe* 30 [K]). – **D:** RSA (Eastern Cape).

Acaulescent geophytes; tuber depressed, **R** terete; **L** 4 - 6, linear, succulent, ascending, 6 - 10 × 1.2 cm; **Inf** lax, flexuose to 15 cm; **Bra** triangular, minute; **Ped** 6 - 9 mm; **Fl** stellate, ± 12 - 13 mm ∅; **OTep** oblong-ovate, 5 - 6 × 1.5 mm; **ITep** 4.5 × 2 mm, obtuse; **St** 6 mm; **Fil** bearded in upper ⅓; **Ov** globose, 0.75 mm; **Sty** erect, terete, 3.5 mm; **Fr** globose, 4 mm ∅; **Se** not seen. – [E. van Jaarsveld]

B. flexuosa Schlechter (J. Bot. 36: 27, 1889). **T:** RSA, Western Cape (*Leipoldt* s.n. [BOL]). – **D:** RSA (Western Cape); dry Fynbos, flowers in autumn.

Geophytes with 1 - 3 heads, to 13 cm tall; **R** 1 mm ∅, terete; tuber depressed, to 3 cm ∅ and 1 cm tall; **L** hysteranthous, filiform, 5.5 - 13 cm × 1 mm; **Inf** laxly flowered, flexuose to 14 cm; peduncle terete, 0.75 mm ∅; **Bra** triangular, clasping, 1.5 × 1 mm; **Ped** 12 - 35 mm; **Fl** stellate, to 12 mm ∅; **OTep** oblong-ovate, 6 × 2 mm; **ITep** 6 × 4 mm, obtuse; **St** 3 - 4 mm; **Fil** bearded in upper ⅓; **Ov** globose, 1 mm ∅; **Sty** erect, terete; **Fr** globose, 3 mm

∅; **Se** oblong, 3 × 1 mm, grey-black. – [E. van Jaarsveld]

B. foleyi E. Phillips (Ann. South Afr. Mus. 1917: 352, 1917). **T:** RSA, Western Cape (*Kensit* 10612 [BOL]). – **D:** RSA (Western Cape); flowers in spring.

Plants short-stemmed, to 15 cm tall; tuber depressed-globose, to 1.4 cm ∅ and 1 cm tall; **R** terete, 1 mm ∅; stem to 4 cm; **L** 2 - 5 per stem, linear, 5.5 - 10 cm × 1.5 - 2 mm ∅, upper face flat, lower face convex, apex mucronate; **Inf** laxly flowered in the upper ½, 9 - 23 cm; peduncle terete, 1 mm ∅; **Bra** triangular-lanceolate, 5 × 2 mm; **Ped** 6 - 7 mm; **Fl** spreading, stellate, 10 - 11 mm ∅; **OTep** oblong-ovate, 5 × 1.5 mm; **ITep** 4 × 1.5 mm, obtuse; **St** 3 - 5 mm; **Fil** bearded in upper ⅓; **Anth** 0.5 mm, oblong; **Ov** globose, 0.5 mm ∅; **Sty** erect, terete, 2.5 mm; **Fr** globose, 2.5 mm ∅; **Se** grey-black, 3 × 2 mm, flattened. – [E. van Jaarsveld]

B. fragilis G. Williamson (Haseltonia 4: 13-15, 17, ills., 1996). **T:** RSA, Northern Cape (*Williamson & Hammer* 5411 [NBG]). – **D:** RSA (Northern Cape: Richtersveld); Succulent Karoo, flowers mid-winter to spring.

Deciduous erect soft acaulescent herbs to 10 cm tall; tuber depressed-pyramidal, to 3 cm tall with up to 7 thick and fleshy decurrent **R** tapering from the tuber; **L** to 3, erect, light green, basal portion subtended by an elongated papyraceous sheath to 1 cm long; **L** linear-lanceolate, channelled or subterete to terete, to 10 cm × 4 mm, tip acute; **Inf** erect; peduncle slender, 7 - 28 cm, to 1.5 mm ∅; raceme to 15 cm; **Fl** 3 - 13; **Bra** triagular-acuminate, membranous, to 1.5 mm; **Ped** to 15 mm; **Tep** luminous cadmium-yellow with green midvein, stellately spreading; **OTep** elliptic, 7 × 3 mm; **ITep** obovate, 7 × 5 mm; **Fil** slender-bearded; pollen green, later brown; **Ov** globose, orange, 1.8 mm; **Sty** 5 mm, decurved; **Se** flattened, deltoid, to 1.2 mm broad. – [E. van Jaarsveld]

B. francescae G. Williamson & Baijnath (SAJB 61(6): 312-314, ills., 1995). **T:** Namibia (*Williamson* 4711 [NBG]). – **D:** S Namibia; flowers mid-winter.

Dwarf solitary geophytes to 10 cm tall; tuber dorsiventrally flattened-deltoid, ± 1.5 cm ∅ and 1 cm tall; **R** succulent, terete, to 1.5 cm long; **L** 1 - 2, ascending, slightly dissimilar, first **L** lanceolate-acuminate, 4 - 10 × 0.5 - 1 cm, translucent, pale green, upper face channelled, with 16 dark green longitudinal veins and 16 transverse constrictions; second **L** similar to first but lanceolate with acute tip, 1 - 4 cm × 1 - 2 mm, curved outwards; **Inf** 1, 6- to 12-flowered, to 11 cm, ascending; **Bra** 1 × 0.5 mm, membranous, deltoid-acuminate; **Ped** 8 - 10 mm; **Per** stellate, bright yellow; **OTep** narrowly elliptic, 10 × 3 mm; **ITep** elliptic, 10 × 6 mm, tips obtuse; **St** 7 mm; **Ov** globose, 2 mm ∅, yellowish-green, grooved; **Sty** erect, 9 mm, slightly curved; **Sti** capitate, papillate; **Fr** ellipsoid, 5 × 3 mm; **Se** deltoid, 2 × 1.5 mm, greyish-black, surface reticulate-alveolate. – [E. van Jaarsveld]

B. frutescens (Linné) Willdenow (Enum. Pl. Hort. Reg. Berol., 372, 1809). **T:** [icono]: Dillenius, Hort. Eltham. t. 231: fig. 298. – **D:** RSA (Northern Cape, Western Cape, Eastern Cape), arid savanna, spring-flowering. **Fig. XXVI.a** (cultivar 'Hallmark')

≡ *Anthericum frutescens* Linné (1753); **incl.** *Bulbine caulescens* Linné (1753) (*nom. inval.*, Art. 34.1c); **incl.** *Anthericum rostratum* Jacquin (1796) ≡ *Bulbine caulescens* var. *rostrata* (Jacquin) von Poellnitz (s.a.) ≡ *Bulbine rostrata* (Jacquin) Willdenow (1809) ≡ *Bulbine frutescens* var. *rostrata* (Jacquin) G. D. Rowley (1973); **incl.** *Bulbine curvifolia* von Poellnitz (1944).

Plants caulescent to 15 cm tall, ± shrubby, dividing to form dense clumps to 30 cm ∅ with up to 30 stems; **R** terete, brownish, apex yellow; stem short, often with stilt-**R**; **L** up to 8 per stem, very variable in length, alternate, basally amplexicaul, ascending, slightly falcate, linear, 4 - 13 cm × 4 - 5 mm, lamina obscurely striate, upper face flat, lower face convex, apex acute; **Inf** racemose, solitary, 30 - 60 cm, ascending, densely flowered in the upper ½, **Fl** 2 - 25 mm apart; peduncle 3 mm ∅ basally and triangular, terete upwards, **Bra** 5 × 1 mm, membranous, deltoid-acuminate, clasping; **Ped** 8 - 12 mm; **Fl** stellate, 19 mm ∅; **OTep** elliptic-lanceolate, 8 - 9 × 3 mm, obtuse, **ITep** broadly ovate, 8 × 4 - 5 mm, obtuse; **St** 5 - 6 mm; **Anth** yellow, oblong; **Ov** globose, 2 mm ∅; **Sty** erect, yellow, 5 mm; **Fr** ovoid, 6 × 5 mm; **Se** grey-black, 2 × 1.5 mm.

Frequently cultivated as a ground-cover in RSA. – [E. van Jaarsveld]

B. glauca (Rafinesque) E. M. Watson (in George, A. S. (ed.), Fl. Austral. 45: 469, 1987). **T:** Australia, Tasmania (*Anonymus* s.n. [K]). – **D:** Australia (New South Wales, Victoria, Tasmania). **I:** CBM 59: t. 3129, 1832, as *Anthericum semibarbatum*.

≡ *Nemopogon glaucum* Rafinesque (1837); **incl.** *Bulbine suavis* Lindley (1838); **incl.** *Bulbine hookeri* Kunth (1843); **incl.** *Bulbinopsis terrae-victoriae* von Poellnitz (1945).

Perennial succulent herbs without caudex; **R** succulent; **L** 6 - 16, subulate, terete towards apex, otherwise channelled, 2 - 10 mm wide, glaucous; **Inf** 2 or more, erect, peduncle terete; **Tep** 9 - 17 mm; **St** subequal, spreading or loosely grouped; **Fil** with clavate **Ha**; **Anth** yellow, dorsifixed, forming a convex horizontal arch after dehiscence; ovules 3 - 4 per locule; **Sty** decumbent, 3 - 7 mm; **Fr** globose to obovate, 4 - 6 mm, **Ped** 12 - 22 mm, erect; **Se** smooth, slightly tuberculate or ridged. – [P. I. Forster]

B. hallii G. Williamson (Aloe 32(3-4): 80-81, ill. (p. 82), 1995). **T:** RSA, Northern Cape (*Williamson & Hammer* 5421 [NBG]). – **D:** RSA (Northern Cape: Namaqualand); flowers early spring.

Dwarf perennial deciduous acaulescent rosulate solitary geophytes; tuber deltoid-depressed, 2 cm tall, 1.5 cm ∅, with a ring of **Bri** surrounding the **Ros**, tunics brown; **R** fleshy, terete; **L** dense, up to 20, oblong-acute, ascending, 45 × 2 mm, flat to slightly channelled in the lower part, convex below, terete in upper ⅓, dull silvery green, surface minutely granulate, margin entire, tip acute; **Inf** solitary, to 25-flowered, to 7 cm; peduncle wiry, terete, 1 mm ∅ at the base; **Bra** 1.5 mm, deltoid; **Ped** to 11 mm; **Per** stellate becoming reflexed, to 16 mm ∅, light yellow with dark green central veins; **OTep** lanceolate, 8 × 2.5 mm; **ITep** 8 × 6 mm, tips obtuse; **St** 7 mm; **Ov** oblong, 2 mm; **Sty** oblique, terete, yellow, 5 mm; **Sti** capitate; **Fr** broadly ellipsoid, 3 × 2 mm; **Se** black, deltoid with contorted ridges, 1.2 × 1.2 mm, surface granular. – [E. van Jaarsveld]

B. haworthioides B. Nordenstam (Bot. Not. 117: 183-187, 1964). **T:** RSA, Western Cape (*Nordenstam* 807 [LD]). – **D:** RSA (Western Cape); Succulent Karoo, quartz flats, flowers in late spring and early summer. **Fig. XXVI.b**

Dwarf deciduous acaulescent rosulate solitary geophytes; tuber oblong to globose, 5- to 7-lobed, to 1.5 cm high and 2 cm broad, yellowish-brown; **R** yellow, fleshy, terete; **L** dense, 8 - 14, linear-lanceolate, cymbiform, spreading, grey-green, lamina 7 - 10 × 2.5 - 3 mm, upper face flat, lower face convex, margin ciliate, apex obtuse to acute; **Inf** 1, ± 10-flowered, to 15 cm; peduncle wiry; **Bra** 1 - 3 mm, triangular to lanceolate, clasping; **Ped** 8 - 12 mm; **Fl** to 15 mm ∅, 5 - 10 mm apart; **Tep** stellate, becoming reflexed; **OTep** lanceolate, 5 - 6 × 1.2 - 2 mm; **ITep** 6.5 - 7 × 3 - 4 mm, obtuse; **St** 4 - 5 mm; **Ov** oblong-globose, 1 - 1.5 mm; **Sty** erect, terete, yellow, 4 - 6 mm; **Fr** globose, 3 mm ∅; **Se** 1.5 mm, blackish. – [E. van Jaarsveld]

B. inflata Obermeyer (BT 9(2): 342-343, pl. 3, 1967). **T:** RSA, Mpumalanga (*Codd* 9503 [PRE]). – **D:** RSA (KwaZulu-Natal, Mpumalanga), Swaziland; grassland.

Geophytes with compact rhizome; **R** spreading, terete; **L** 10 - 15, rosulate, ascending, linear, terete, to 50 cm, to 1 cm ∅, sheathing below forming a short neck, upper face grooved, apex subulate; **Inf** 1, to 1.25 m, densely flowered in the upper ⅓; peduncle terete; **Bra** to 2.5 cm, membranous, linear-acuminate, aristate; **Ped** to 2.5 cm; **Fl** stellate, to 23 mm ∅; **Tep** becoming reflexed, linear-lanceolate, 10 × 3 mm, obtuse; **St** 7 mm; **Ov** oblong, 7 mm, tip truncate; **Sty** erect, 7 mm; **Fr** globose, 13 mm ∅; **Se** black, angled, 2 mm. – [E. van Jaarsveld]

B. lagopus (Thunberg) N. E. Brown (BMI 1931:

195, 1931). **T:** RSA, Western Cape (*Thunberg* s.n. [K]). – **D:** RSA (Western Cape); among quartzitic sandstones in Fynbos, ± 1000 m, flowers in late spring.

≡ *Anthericum lagopus* Thunberg (1800).

Evergreen, rosulate, clustering with 3 - 5 growths per plant, shortly caulescent to 12 cm tall and 15 cm ∅; **R** yellow, 2 mm ∅, terete; stems short, 2 - 3 × 1 cm ∅; **L** 12 - 15 per growth, green, curved inwards, firm, linear to linear-lanceolate, 4 - 16 × 0.5 cm, upper face flat, lower face convex, attenuate, margin entire, apex acute, mucronate; **Inf** 30 - 60 cm, densely flowered in the upper ⅓, peduncle terete, 3 - 6 mm ∅; **Bra** triangular-lanceolate, 5 × 2 mm; **Ped** 15 - 25 mm; **Fl** stellate, 14 mm ∅; **OTep** lanceolate, 5 × 2 mm; **ITep** ovate, 5 × 3 mm, obtuse; **St** 6 mm; **Ov** globose; **Sty** erect, terete; **Fr** globose, 4 - 5 mm ∅; **Se** 2.5 × 2 mm, angled, blackish. – [E. van Jaarsveld]

B. lamprophylla G. Williamson (Bradleya 14: 84-86, ills., 1996). **T:** RSA, Northern Cape (*Williamson & Hammer* 5707 [NBG]). – **D:** RSA (Northern Cape); Succulent Karoo.

Dwarf deciduous acaulescent rosulate solitary geophytes; tuber pinkish, deltoid, 7 - 13 × 5 - 8 mm, with several basal fleshy terete **R**; **L** 4 - 6, crowded and forming a tight basal cylinder to 8 × 2 mm, lamina 15 - 40 × 1.2 - 3.5 mm, oblong-acute, subterete, ascending, shiny luminescent, upper face flat to slightly channelled, lower face convex, epidermis finely tuberculate with minute trichomes towards the base, tip acute; **Inf** 1, 4 - 14 cm; raceme to 6 cm, 6- to 14-flowered; peduncle wiry, terete; **Fl** sweetly scented; **Ped** 1 cm; **OTep** elliptic, 7 × 2.5 mm, light orange-green, lower face apricot; **ITep** 8 × 5 mm, obovate, orange-yellow; **St** to 6 mm; **Ov** oblong, 2 mm; **Sty** 4 mm; **Fr** broadly ellipsoidal, 3 × 2 mm; **Se** pyramidal, dark brown, rugose, 1.4 × 1 mm. – [E. van Jaarsveld]

B. latifolia (Linné *fil.*) Roemer & Schultes (Syst. Veg. 7: 477, 1829). **T:** [icono]: Jacquin, Icon. Pl. Rar. 2: t. 406, 1793. – **D:** E RSA; pioneer in arid savannas, spring-flowering. **Fig. XXVI.c**

≡ *Anthericum latifolium* Linné *fil.* (1781).

Plants rosulate, solitary, to 20 cm tall; **R** fleshy, terete; **L** in a dense **Ros**, green, obscurely striate, firm, ascending, older **L** becoming recurved, lamina triangular-lanceolate, 19 - 40 × 3 - 6 cm, upper face flat, slightly channelled towards apex, lower face flat or somewhat convex, margin acute, minutely ciliate, apex acuminate; **Inf** 1 - 4, densely flowered in the upper ½, 40 -100 cm, ascending to spreading; **Fl** up to 8 open at the same time, ± 7 - 12 mm ∅, crowded; peduncle flattened below, biconvex, to 15 mm wide and tapering to 6 - 8 mm; **Bra** membranous, withering, linear-lanceolate, 5 - 8 × 1 mm; **Ped** 12 - 14 mm; **OTep** lanceolate, 7 × 3 mm, obtuse; **ITep** 7 × 3.5 mm, obtuse; **St** 7 mm; **Anth** yellow, 1 mm,

oblong; **Ov** globose, 1.5 mm, yellowish-green, grooved; **Sty** erect, 6 mm; **Fr** rounded, 4 × 3 mm; **Se** 2.5 × 1.5 mm, ellipsoid, grey-black. – [E. van Jaarsveld]

B. lavrani G. Williamson & Baijnath (Aloe 36(2-3): 28-30, ills., 2000). **T:** RSA, Northern Cape (*Williamson* 5283 [NBG]). – **D:** RSA (Northern Cape); steep quartzitic sandstone ridges.

Dwarf, solitary acaulescent single-headed geophytes; tuber oblong, tapering, black, 3 × 2 cm; **R** fibrous; **L** 1, erect, basally amplexicaul, linear, subterete, glaucous, 10 - 15 × 2 - 4 mm, upper face channelled; **Inf** 1, erect, to 11 - 22 cm, up to 15-flowered; peduncle 1.5 mm ∅ basally, terete; **Bra** deltoid, acuminate, 2 × 1 mm, clasping; **Fl** spreading, stellate, ± 15 mm ∅, 7 - 12 mm apart; **Ped** 5 - 7 mm; **OTep** elliptic, 7 × 2 mm; **ITep** ovate to ovate-elliptic, 6 × 2.5 mm, obtuse to subacute; **St** 6 mm; **Ov** globose, 1 mm ∅; **Sty** erect, 7 mm; **Fr** globose, 2.5 mm; **Se** angular, grey-black, 1 mm. – [E. van Jaarsveld]

B. longifolia Schinz (Bull. Herb. Boissier, sér. 2, 11: 939, 1902). **T:** RSA, Western Cape (*Schlechter* 8715 [BOL]). – **D:** RSA (Western Cape); dry Fynbos, flowers in winter and spring.

Small solitary geophytes to 4.5 - 20 cm tall; tuber depressed, ± 1 cm ∅; **R** terete; **L** linear to linear-lanceolate, 2 - 5 per stem, 3 - 20 cm × 2 - 5 mm, flat to channelled above, convex below, attenuate, apex acute; **Inf** laxly flowered, to 15 - 25 cm; peduncle terete, 1 mm ∅ at the base; **Bra** triangular-lanceolate, 3 - 6 × 1 - 3 mm; **Ped** terete, 8 - 9 mm; **Fl** stellate, 14 - 15 mm ∅; **OTep** oblong-ovate, 7 × 2 mm; **ITep** 7 × 5 mm, obtuse; **St** 5 mm; **Anth** 0.5 mm, oblong; **Ov** globose, 1.5 mm; **Sty** erect, terete, 4 mm; **Fr** and **Se** not seen. – [E. van Jaarsveld]

B. longiscapa (Jacquin) Willdenow (Enum. Pl. Hort. Reg. Berol., 372, 1809). **T:** [icono]: Jacquin, Icon. Pl. Rar. 2: 17, t. 404, 1793. – **D:** RSA (Eastern Cape); flowers in mid-summer and early autumn.

≡ *Anthericum longiscapum* Jacquin (1787).

Plants acaulescent, rosulate, 20 - 34 cm tall; **R** terete, 1 mm ∅; **L** up to 15, 20 - 30 × 1.8 cm, softly succulent, ascending, linear-lanceolate, dorsiventrally flattened, apex tapering, acuminate; **Inf** 1 - 2, erect to 40 - 70 cm, densely flowered in the upper ½; peduncle 4 mm ∅ at the base; **Fl** stellate, to 12 mm ∅; **Bra** 4 - 10 × 1 mm; **Ped** 10 - 18 mm; **OTep** elliptic, 6 × 1.5 mm; **ITep** 5 × 1.5 mm; **St** 5 mm; **Ov** oblong-globose, 1.5 mm; **Sty** 4 mm; **Fr** oblong-globose, 4 × 3 mm. – [E. van Jaarsveld]

B. louwii L. I. Hall (SAJB 3(6): 356-357, 1984). **T:** RSA, Western Cape (*Hall* 4718 [NBG]). – **D:** RSA (Western Cape); Succulent Karoo, quartz flats.

Dwarf deciduous acaulescent rosulate solitary

geophytes; tuber depressed, globose above, flat below, with radiating tap**R**, 1 cm tall, 1.5 cm broad; **R** terete; **L** dense, to 30, spatulate, ascending, 4 × 0.8 - 1 cm, striate, margin ciliate, apex obtuse, mucronate; **Inf** 1 - 2, to 12 cm, 30- to 40-flowered; peduncle reddish-brown, wiry, 1.5 mm ∅ at the base; **Bra** to 3 mm, deltoid-cuspidate; **Ped** 10 - 12 mm; **Per** stellate, becoming reflexed, to 2 cm ∅; **OTep** lanceolate, 10 × 3 mm; **ITep** 10 × 7 mm, obtuse; **St** 5 mm; **Ov** oblong, 1.5 mm; **Sty** oblique, terete, yellow, 7 mm; **Fr** 4 mm; **Se** not seen. − [E. van Jaarsveld]

B. margarethae L. I. Hall (SAJB 3(6): 357, 1984). **T:** RSA, Western Cape (*Hall* 5083 [NBG]). − **D:** RSA (Western Cape); Succulent Karoo, on limestone outcrops, flowers in late spring and early summer. **Fig. XXVI.d**

Dwarf deciduous acaulescent rosulate clustering geophytes; tuber depressed, globose above, flat below, 1 × 1.5 cm, lobed; **R** fibrous, terete; **L** dense, to 25, linear, 5 × 0.5 cm, biconvex, upper face distinctly tessellate, apex acute; **Inf** 1, ± 25-flowered, to 15 cm; peduncle wiry, terete, to 1.5 mm ∅ at the base; **Bra** 2 - 3 mm, triangular-lanceolate, clasping; **Ped** 8 - 10 mm; **Per** stellate, becoming reflexed, to 18 mm ∅; **OTep** oblong, 9 × 2.5 mm; **ITep** oblong, 8 × 6 mm, obtuse; **St** 6 mm; **Anth** 1 mm; **Ov** globose, 1.5 mm; **Sty** erect, 6 mm; **Fr** ovoid, 4 × 3 mm; **Se** 1.25 mm, angular, rough. − [E. van Jaarsveld]

B. meiringii van Jaarsveld ([in press], 2001). − **D:** RSA (Western Cape: Little Karoo, Swartberg Mtn.); quartzitic sandstones, rock crevices, flowers in spring.

Dwarf rosulate succulents to 10 cm tall, dividing to form clusters ± 20 cm ∅ with up to 12 heads; **R** grey, fleshy, terete; stem short, 2 cm, thickening towards the base; **L** 4 - 7, curving downwards, striate, linear-lanceolate, 10 - 15 cm × 6 - 8 mm, upper face flat, lower face convex, apex acute; **Inf** 1, 21 - 26 cm, ascending to spreading, **Fl** in the upper ⅓, 4 - 6 mm apart; peduncle 3 mm wide at the base, biconvex, terete upwards; **Bra** 3 mm, membranous, deltoid-acuminate, 1 mm, clasping; **Ped** 10 - 12 mm; **Per** stellate, becoming reflexed; **Tep** pale yellow; **OTep** elliptic, 7 × 2 mm, obtuse; **ITep** ovate to ovate-elliptic, 6 × 2.5 - 3 mm, obtuse; **St** 5 mm; **Anth** yellow, oblong; **Ov** globose, 1.5 mm ∅; **Sty** erect, 5 mm; **Fr** ovoid, 3 × 4 mm; **Se** grey-black, 1.5 × 1 mm. − [E. van Jaarsveld]

B. mesembryanthoides Haworth (Philos. Mag. J. 66: 31-32, 1825). **T:** RSA, Cape Prov. (*Bowie* s.n. [K [Duncanson-drawing, status?]]). − **D:** RSA (Northern Cape, Western Cape).

Incl. *Bulbine mesembryanthemoides* Baker (1896) (*nom. inval.*, Art. 61.1).

Dwarf solitary summer-deciduous geopytes 1 - 2.5 cm tall; tuber globose to deltoid, depressed, 3-

to 5-lobed, ± 1.2 - 2 cm ∅ and 1 cm tall, lobes ovoid, tapering, flesh grey-white where exposed; **R** light yellowish-grey, fleshy, terete; **L** 1 - 4, erect, amplexicaul, dissimilar; larger **L** 1.5 - 3 × 1 - 2 cm, oblong, tip acute or truncate to concave (appearing scalped), pinkish-grey-green, striate, upper face grooved; younger **L** linear-lanceolate to lanceolate, 1.5 × 0.3 cm, almost terete, acute at first, withering from the top and becoming truncate or remaining acute, upper face grooved, lower face convex; **Inf** 1, 17.5 - 20 cm, 1- to 7-flowered, erect to ascending, **Fl** spreading and ± 8 - 10 mm ∅, 5 - 10 mm apart; peduncle 1 mm ∅ at the base, terete; **Bra** 1.5 - 2 × 1 mm, membranous, deltoid-acuminate, clasping; **Ped** 8 - 10 mm; **Per** stellate, 10 mm ∅ when fully spread; **OTep** lanceolate, 6 - 7 × 1.5 - 2 mm; **ITep** ovate to ovate-elliptic to oblanceolate, 5 - 6 × 2 - 3 mm, obtuse; **St** 4 - 5 mm; **Anth** yellow, oblong, 0.75 mm; **Ov** globose, 1 mm ∅; **Sty** erect, yellow, 5 mm; **Fr** globose, 3 mm ∅; **Se** 1.5 mm. − [E. van Jaarsveld]

B. mesembryanthoides ssp. **mesembryanthoides** − **D:** RSA (Northern Cape, Western Cape); Succulent Karoo, rocky regions, summer-flowering.

Incl. *Bulbine orchioides* Drège *ex* von Poellnitz (1944).

Tuber globose, elongating into the **R**; **L** 2 - 3, terete, acute even when withering; **Inf** usually with 3 - 6 **Fl**; **Fil** with a single tuft of **Ha**. − [E. van Jaarsveld]

B. mesembryanthoides ssp. **namaquensis** G. Williamson (Aloe 36(1): 14-15, ills., 1999). **T:** RSA, Northern Prov. (*Le Roux* 3116 [STE in NBG]). − **D:** RSA (Northern Cape); Succulent Karoo, rocky regions.

Differs from ssp. *mesembryanthoides*: Tubers deltoid; **L** 1 normal and 1 vestigial, normal **L** truncate, terete, vestigial **L** falcate, acute; **Inf** shorter, 1- to 6-flowered; **Fil** with a double tuft of **Ha**. − [E. van Jaarsveld]

B. minima Baker (JLSB 15: 344, 1876). **T:** RSA, Western Cape (*Drège* 953 [K]). − **D:** RSA (Western Cape); Renosterveld, summer-flowering.

Dwarf rosulate evergreen acaulescent geophytes; tuber globose-depressed, 1 - 1.5 cm tall and 1.5 - 2 cm broad with numerous **Bri** to 15 mm surrounding the **Ros**; **R** terete, 1 mm ∅; **L** filiform, curved, 15 - 20 per **Ros**, 2.5 - 4.5 cm × 0.75 mm; **Inf** 1 - 3, 13 - 16 cm, laxly flowered in the upper ½; peduncle terete, 1 mm ∅; **Bra** triangular, 1 × 0.75 mm; **Ped** terete, 6 - 7 mm; **Per** 7 - 8 mm ∅, stellate becoming reflexed; **OTep** oblong-ovate, 4 × 1 mm; **ITep** 3.5 × 1.5 mm, obtuse; **Fr** globose, 3 mm ∅; **Se** 1.5 × 1 mm, flattened, semicircular in outline. − [E. van Jaarsveld]

B. monophylla von Poellnitz (Feddes Repert. Spec.

Nov. Regni Veg. 53: 45, 1944). **T:** RSA, Western Cape (*Schlechter* 4912 [B]). – **D:** RSA (Western Cape); Renosterveld, spring-flowering.

Solitary geophytes to 45 cm tall; tuber depressed-globose, 1 - 1.4 cm tall, 1.2 - 1.5 cm ∅; **R** terete; **L** 1, terete, synanthous, 30 - 45 cm × 3 - 7 mm ∅, apex tapering, acute; **Inf** 30 - 40 cm, laxly flowered in the upper ½; peduncle terete, 2 - 3 mm ∅ at the base; **Bra** triangular-lanceolate, 5 × 2 mm; **Ped** terete, 10 - 14 mm; **Per** stellate, 16 - 18 mm ∅; **OTep** oblong-elliptic, 9 × 2 mm; **ITep** oblong-ovate, 8 × 3 mm, obtuse; **St** 5 mm; **Ov** ellipsoid, 2 mm; **Sty** erect, terete, 6 mm; **Fr** oblong-globose, 4 × 5 mm; **Se** not seen. – [E. van Jaarsveld]

B. muscicola G. Williamson (Bradleya 18: 36, ills. (p. 32, 38), 2000). **T:** RSA, Western Cape (*Williamson* 5930 [NBG]). – **D:** RSA (Western Cape: lower foothills of the Cederberg); shallow quartzitic sandstone outcrops, among mosses, flowers late spring.

Dwarf solitary single-headed geophytes; tuber dome-shaped, 10 × 8 mm, with spreading wedge-shaped **R**; **L** 1, erect, amplexicaul, linear-terete becoming lanceolate, translucently light green, 35 × 4 mm, later in the season becoming shorter and truncate (25 × 5 mm); **Inf** solitary, erect, to 8 cm, raceme 4 cm, to 10-flowered; peduncle terete, 1.5 mm ∅ at the base; **Bra** deltoid-acuminate, 2.5 × 0.8 mm, clasping; **Ped** 8 mm; **Tep** pale yellowish-green; **OTep** recurved, narrowly ovate, 6 × 2 mm; **ITep** ovate, 6 × 3 mm, obtuse; **Ov** ovoid, 1 mm ∅; **Sty** erect, 6 mm; **Sti** capitate; **Fr** 3 × 2 mm; **Se** narrowly pyramidal, brownish-black, 1.3 × 0.5 mm. – [E. van Jaarsveld]

B. namaensis Schinz (Bull. Herb. Boissier, sér. 2, 2: 939, 1902). **T:** Namibia (*Schinz* 934 [W ?]). – **D:** Namibia, RSA (Northern Cape); Strandveld, flowers in July.

Plants rosulate, clustering with 2 - 5 heads per plant, to 14 cm tall; tuber small, oblong 1 × 0.7 cm; **R** terete, 1 - 2 mm ∅; **L** 10 - 15, ascending, to 14 cm × 1 mm, curved at the top, filiform with basal membranous sheaths of 2.5 cm; **Inf** 18 - 43 cm, densely flowered in the upper ⅓; peduncle terete, 3 - 6 mm ∅ at the base; **Bra** triangular-acuminate, 7 × 1.5 - 2 mm; **Ped** terete, 12 - 15 mm; **Fl** scented, spreading, stellate, 16 mm ∅; **OTep** oblong-ovate, 8 × 2 mm; **ITep** 7 × 3 mm, obtuse; **St** 5 mm; **Anth** 1 mm, oblong; **Ov** globose, 1 mm; **Fr** globose, 1 cm; **Se** angled, oblong, 4 mm ∅. – [E. van Jaarsveld]

B. narcissifolia Salm-Dyck (Hort. Dyck., 334, 1834). – **D:** RSA (Eastern Cape, KwaZulu-Natal); grassland, flowers from spring to autumn.

Plants acaulescent, clustering, 8 - 15 cm tall; tuber depressed, 1 × 2 cm; **R** yellow, fusiform, terete; **L** 4 - 5, distichous, with slight spiral twist, amplexicaul, green, ascending, linear, 6.5 - 15 × 0.4 - 2.2

cm, lamina obscurely striate, flat, apex acute, mucronate; **Inf** 1, racemose, to 38 cm, densely flowered, capitate, **Fl** ± 2 - 3 mm apart; peduncle 2 - 8 mm ∅ at the base, terete to flattened; **Bra** 8 - 9 × 3 mm, membranous, lanceolate-acuminate; **Ped** 15 - 25 mm; **Per** stellate, 15 mm ∅, becoming reflexed; **OTep** elliptic-lanceolate, 7 × 2 mm; **ITep** lanceolate-ovate, 6 × 2.5 mm, obtuse; **St** 5 mm; **Anth** yellow, oblong; **Ov** oblong-globose, 2 mm; **Sty** erect, yellow, 3 mm; **Fr** ovoid, 6 × 5.5 mm; **Se** grey-black, 1.5 mm. – [E. van Jaarsveld]

B. natalensis Baker (FC 6: 366, 1896). **T:** RSA, KwaZulu-Natal (*Wood* 553 [K]). – **D:** RSA (Eastern Cape, KwaZulu-Natal); sandstone cliff faces, flowers in early spring and summer. **I:** Wyk & al. (1997: 64).

Plants evergreen, rosulate, with short stem, without tuber; **R** yellow, fleshy, terete; **L** 10 - 14, triangular-lanceolate, grey-green, striate, soft, spreading, 8.5 - 13 × 3.5 - 4.5 cm, upper face flat to broadly channelled, lower face flat to rounded, margin densely ciliate, apex acuminate, cilia 2 mm; **Inf** 1 - 3, densely flowered, to 55 cm; peduncle flattened at the base, 6 - 7 mm ∅; **Bra** linear-lanceolate, 10 × 1 mm; **Ped** 9 - 11 mm; **Fl** spreading; **Per** stellate, to 15 mm ∅; **OTep** lanceolate, 6 × 3 mm; **ITep** 6 × 4 mm, obtuse; **St** 5 mm; **Ov** globose, 1.5 mm; **Sty** 6 mm; **Fr** globose, 3 mm; **Se** 1.5 mm, grey-black, ellipsoid. – [E. van Jaarsveld]

B. nutans (Jacquin) Roemer & Schultes (Syst. Veg. 7(1): 447, 1829). **T:** [icono]: Jacquin, Collectanea 2: 17, t. 407, 1787. – **D:** RSA (Western Cape); dry Fynbos.

≡ *Anthericum nutans* Jacquin (1787); **incl.** *Bulbine bachmanniana* Schinz (1902).

Solitary rosulate geophytes to 30 cm tall (without **Inf**); tuber oblong, 4 × 3.5 cm with few radiating succulent **R** to 12 mm ∅ at the base and tapering, crowned with a ring of **Bri** and withered **L** to 4.5 cm; **L** linear to ovate-lanceolate, 5 - 30 × 0.6 - 2.2 cm, flat, keeled, spreading, amplexicaul at the base, margins ciliate, apex acuminate; **Inf** 1 - 2, 25 - 40 cm, **Fl** in the upper ⅓; peduncle terete, 2 - 5 mm ∅ at the base; **Bra** triangular to triangular-lanceolate, 3 × 1 mm; **Ped** terete, 11 - 15 mm; **Fl** spreading, stellate, 12 - 13 mm ∅; **Tep** 6 × 1.5 mm, oblong-ovate, obtuse; **Fr** oblong-ovate, 6 mm; **Se** oblong, 2 - 3 × 1 mm. – [E. van Jaarsveld]

B. pendens G. Williamson & Baijnath (SAJB 61 (6): 316-319, ills., 1995). **T:** RSA, Northern Cape (*Williamson* 4738 [NBG]). – **D:** RSA (Northern Cape); sheer shady quartzitic sandstone cliff faces.

Dwarf acaulescent solitary or 2-headed geophytes, pendent from cliff faces; tuber oblong, 3 × 1.5 cm, stellately lobed, lobes tapering into **R**; **L** 1 - 2, pendent, amplexicaul with basal sheath 1 - 2 cm, lamina linear, terete, striate, 15 - 18 × 2 - 5 mm; **Inf**

1, erect, 8 - 18 cm, up to 4-flowered; peduncle 1 mm ∅ at base, terete; **Bra** deltoid-acuminate, 3 - 4 × 1 mm, clasping; **Ped** to 25 mm; **Fl** spreading, stellate, ± 18 mm ∅, 10 - 15 mm apart; **OTep** 8 × 3 mm; **ITep** ovate to ovate-elliptic, 8 × 4 mm, obtuse to subacute; **St** 5 mm; **Ov** globose, 1.5 - 2 mm ∅; **Sty** erect, 5 mm; **Fr** and **Se** not seen. – [E. van Jaarsveld]

B. pusilla von Poellnitz (Feddes Repert. Spec. Nov. Regni Veg. 53: 46-47, 1944). **T:** RSA, Western Cape (*Schlechter* 9035 [B]). – **D:** RSA (Western Cape); quartzitic sandstone rocks, in crevices; midsummer-flowering.

Dwarf solitary summer-deciduous geophytes to 2.5 cm tall; tuber globose, 3- to 5-lobed, ± 1 cm tall, 1.5 cm ∅, with greyish papery tunics, tissue pinkish; **R** grey, fleshy, terete; **L** 2, erectly spreading, greygreen, amplexicaul, linear-lanceolate, 1 - 3 cm × 2 - 4 mm, upper face flat, channelled towards the tip, lower face convex, tip acute; **Inf** 1, 8- to 12-flowered, 5 - 8 cm; peduncle 1 mm ∅ at the base, terete; **Bra** 1.5 × 1 mm, deltoid-acuminate, clasping; **Ped** 7 - 9 mm; **Fl** spreading and slightly drooping, ± 8 - 10 mm ∅; **Per** stellate, becoming reflexed, pale yellow; **OTep** lanceolate, 7 × 2 mm; **ITep** ovate to ovate-elliptic, 6 × 3.5 mm, obtuse; **St** 6 mm; **Anth** yellow, oblong; **Ov** globose, 1 mm ∅; **Sty** erect, yellow, 6 mm; **Fr** ovoid, 4 × 3 mm; **Se** 1.5 mm, grey-black. – [E. van Jaarsveld]

B. quartzicola G. Williamson (Haseltonia 4: 19-21, ills. (p. 16), 1996). **T:** RSA, Northern Cape (*Williamson & Hammer* 5416 [NBG]). – **D:** RSA (Northern Cape: Namaqualand); Succulent Karoo, flowers late spring and early summer.

Deciduous erect softly succulent herbs to 7 cm tall; tuber deltoid to round, 1.5 - 2.5 cm ∅ and 1.5 cm tall, decurrent into tapering thick and fleshy **R**; **L** 3, erect, light green and many-veined, circinate in the upper ½, basal part subtended by an elongated papyraceous sheath to 4 cm long; **L** flattened, subterete to terete, 4 - 8 cm × 2.5 - 4 mm, tip acute; **Inf** erect; peduncle slender, to 12 cm; raceme to 6 cm; **Fl** 6 - 16; **Bra** triangular-acuminate, membranous, to 5 mm; **Ped** to 1.2 mm; **Tep** light yellow, ovate; **OTep** strongly reflexed, 7 × 2.5 mm; **ITep** reflexed at an angle of 45°, 7 × 4 mm; **Fil** slender-bearded; **Ov** oblong, green, 1 mm; **Sty** 6 mm, decurved; **Fr** and **Se** not known. – [E. van Jaarsveld]

B. ramosa van Jaarsveld ([in press], 2001). – **D:** RSA (Western Cape: Little Karoo); Succulent Karoo.

Plants rosulate, dividing to form dense groups; **R** fleshy, terete; stem short, 2.5 cm ∅; **L** 6 - 8, green, obscurely striate, soft, ascending to slightly falcate, older **L** spreading, lamina linear to triangular-lanceolate, 8 - 11 × 1.5 - 2.3 cm, upper face flat, lower face convex, apex acute; **Inf** 1 - 2, 3 - 4.7 cm, laxly

20- to 35-flowered in the upper ⅓, slightly drooping; peduncle flattened, 3 - 4 mm wide at the base, terete upwards and tapering to 2 mm ∅; **Bra** membranous, withering, lower **Bra** triangular-acuminate, 2 - 3 × 1 mm; **Ped** 11 - 15 mm, terete; **Fl** spreading, stellate, ± 18 mm ∅; **OTep** elliptic, 9 × 3 mm, obtuse; **ITep** 9 × 6 mm, obtuse; **St** 6 mm; **Anth** yellow, 1 mm, oblong; **Ov** globose, 1.5 mm ∅; **Sty** erect, terete, yellow, 6.5 mm; **Fr** rounded, 4 × 3 mm; **Se** 1.5 mm, oblong-elliptic, grey-black. – [E. van Jaarsveld]

B. rhopalophylla Dinter (RSN 29: 271, 1931). **T:** Namibia (*Dinter* 6494 [LUS]). – **D:** S Namibia, RSA (Northern Cape: Richtersveld); Succulent Karoo, flowers in mid-winter.

Dwarf rosulate succulent geophytes to 5 cm tall; tuber depressed-globose, 1.5 - 2 cm tall and wide; **R** terete; **L** 2 - 4, erect, green, club-shaped, 2 - 3 × 0.4 - 1 cm, apex obtuse, mucronate; **Inf** 4 - 4.5 cm, laxly flowered; peduncle terete, 0.75 mm ∅; **Bra** triangular-lanceolate, 2 × 0.75 mm; **Ped** terete, 2 - 5 mm; **Fl** spreading, stellate, 12 mm ∅; **OTep** oblong-ovate, 6 × 1.5 mm; **ITep** 5 × 2 mm, obtuse; **St** 3 mm; **Ov** 2 × 1 mm, oblong; **Sty** erect, terete, 4 mm; **Fr** and **Se** not seen. – [E. van Jaarsveld]

B. rupicola G. Williamson (Bradleya 18: 35-36, ills., 2000). **T:** RSA, Eastern Cape (*Williamson* 5927 [NBG]). – **D:** RSA (Eastern Cape); crevices in rocky kloofs, flowers mid-summer.

Dwarf geophytes, proliferating to form chains of ovoid tubers, tubers 8 - 12 × 10 - 12 mm, with up to 7 horizontal incremental growth lines; **R** 6, spreading, to 40 × 1.2 mm; **L** 3 - 4, rosulate, erect, arising from papery sheathing remains of old **L**, lanceolate, terete to subterete, acute, 30 - 60 × 3 - 7 mm, transparently green, margins finely toothed towards the base; **Inf** solitary, erect, to 22 cm, to 8-flowered; peduncle to 1.5 mm ∅ at the base, terete; **Bra** deltoid-acuminate, 3 × 2 mm; **Ped** 10 mm; **Tep** recurved, light yellow; **OTep** ovate with narrow obtuse tip, 7 × 2 mm; **ITep** broadly ovate, 7 × 3 mm, tip emarginate; **St** projecting forwards in a tight bunch; **Ov** ovoid, 1.3 mm; **Sty** erect, 5.5 mm; **Sti** minute, rugulose; **Fr** and **Se** not seen. – [E. van Jaarsveld]

B. sedifolia Schlechter *ex* von Poellnitz (Feddes Repert. Spec. Nov. Regni Veg. 53: 47, 1944). **T:** RSA, Western Cape (*Schlechter* 11003 [B]). – **D:** RSA (Western Cape); Fynbos on quartzitic sandstone, in rock crevices.

Rosulate evergreen solitary succulents 2.5 - 4 cm tall; tuber depressed-globose, 1 - 1.2 cm wide and tall; **R** terete; **L** 6 - 10, linear-lanceolate, acuminate, 1.6 - 3 cm × 3 - 4 mm, upper face flat, lower face convex, tip acute; **Inf** 5 - 17 cm, laxly flowered in the upper ½; peduncle terete, 1 mm ∅ at the base; **Bra** triangular-lanceolate, 2 × 1 mm; **Ped** terete, 7 mm; **Fl** spreading, stellate, 12 - 15 mm ∅; **OTep** oblong-ovate, 7 × 2 mm; **ITep** ovate, 6 × 4 mm, obtuse; **St** 4 mm; **Ov** ovate to ellipsoid, 1 - 1.5 mm; **Sty** erect, 5 mm, terete; **Fr** ellipsoid, 3 - 4 mm; **Se** angular, 1 mm ∅. – [E. van Jaarsveld]

B. semibarbata (R. Brown) Haworth (Revis. Pl. Succ., 33, 1821). **T:** Australia, South Australia (*Brown* s.n. [BM]). – **D:** Australia; widespread. **I:** Cunningham & al. (1981: 183).

≡ *Anthericum semibarbatum* R. Brown (1810) ≡ *Phalangium semibarbatum* (R. Brown) Kuntze (1891) ≡ *Bulbinopsis semibarbata* (R. Brown) Borzi (1897).

Annual succulent herbs without caudex; **R** fibrous; **L** 6 - 16, linear to subulate, channelled, 1 - 5 mm wide, green, rarely glaucous; **Inf** erect, 1 to many; peduncle terete; **Tep** 3 - 7 mm, yellow; **St** 3 short, 3 long, closely bunched; **Anth** reddish, orange or yellow-brown, dorsifixed, erect to horizontal after dehiscence; ovules 2 per locule; **Sty** straight, 0.5 - 1.5 mm; **Fr** globose, 2 - 4.5 mm, **Ped** 8 - 25 mm, erect to recurved; **Se** smooth. – [P. I. Forster]

B. striata van Jaarsveld & Baijnath (SAJB 53(6): 424-426, ills., 1987). **T:** RSA, Northern Cape (*van Jaarsveld & Patterson* 6640 [NBG]). – **D:** RSA (Northern Cape); Succulent Karoo, quartz rocks, spring-flowering.

Solitary geophytes; tuber oblong, ± 2.5 × 1.8 cm, rounded at the top with dark brown tunics overtopped with a ring of **Bri**; **R** in appressed bundles; **L** 4 - 6, rosulate, spreading, ovate-lanceolate, 5 - 15 × 2.5 - 4 cm, upper face striate, translucent; **Inf** 1, erect, to 57 cm; peduncle 3.5 mm ∅, terete; **Bra** membranous, deltoid-acuminate, 2 × 1 mm; **Ped** 1 - 2 cm, ascending; **Fl** stellate, 14 - 16 mm ∅, **Tep** pale yellow, reflexed when fully open; **OTep** ovate-elliptic, 8 × 1.5 mm; **ITep** 8 × 2.5 mm; **St** 4 - 5 mm; **Ov** globose; **Sty** 5 mm; **Fr** not seen; **Se** black, oblong-angular, 3 × 1 mm. – [E. van Jaarsveld]

B. succulenta Compton (Trans. Roy. Soc. South Afr. 14: 275, 1931). **T:** RSA, Western Cape (*Compton* 2910 [NBG]). – **D:** RSA (Northern Cape, Western Cape); Succulent Karoo, spring-flowering.

Acaulescent, rosulate, up to 13 cm tall, solitary; tuber oblong-ovoid, 3 × 1.5 cm with black tunics and a ring of **Bri** surrounding the **L**; **R** brownish, tips yellow, terete; **L** 8 - 10, amplexicaul, glaucous-green, variable, ascending, linear-clavate, 2 - 13 × 0.3 - 1 cm, widening towards the tip, lamina obscurely striate, upper face flat, lower face convex, tip obtuse to acute, mucronate; **Inf** 1, 10 - 20 cm, ascending to spreading, laxly flowered in the upper ⅙, **Fl** 3 - 7 mm apart; peduncle 1.5 - 4 mm ∅ at the base, biconvex, terete upwards; **Bra** 4 × 1 mm, membranous, deltoid-acuminate, clasping; **Ped** 8 - 10 mm; **Per** stellate, 15 - 20 mm ∅, **Tep** becoming reflexed; **OTep** elliptic-lanceolate, 7 - 10 × 2

mm; **ITep** broadly ovate, 7 - 9 × 3 - 4 mm, obtuse; **St** 5 - 6 mm; **Anth** yellow, oblong; **Ov** globose, 2 mm ∅; **Sty** erect, yellow, 5 mm; **Fr** ovoid, 10 × 5 mm; **Se** grey-black, 2 × 4 mm. – [E. van Jaarsveld]

B. tetraphylla Dinter (RSN 29: 272, 1931). **T:** Namibia (*Dinter* 5201 [LUS]). – **D:** S Namibia; Karoo, flowers in mid-winter.

Plants rosulate, to 14 cm tall; tuber depressed with few swollen radiating **R**; **L** with brownish basal papery sheaths 1 - 2 cm, 4 per plant, dark blue-green, linear-lanceolate, tapering, 10 - 14 × 1 - 1.5 cm; **Inf** 20 -30 cm, floriferous part cylindrical with 15 - 30 **Fl**; peduncle terete, 2 - 3 mm ∅; **Bra** triangular, 2 × 2 mm; **Ped** terete, 5 - 6 mm; **Fl** spreading, stellate, to 16 mm ∅; **Tep** 8 mm; **Anth** 1 mm, oblong; **Ov**, **Fr** and **Se** not seen. – [E. van Jaarsveld]

B. torsiva G. Williamson (Haseltonia 4: 21, ills. (p. 17, 20), 1996). **T:** RSA, Northern Cape (*Williamson & Hammer* 5420 [NBG]). – **D:** RSA (Northern Cape: Richtersveld); Succulent Karoo, flowers late spring to mid-summer.

Deciduous erect softly succulent herbs to 11 cm tall; tuber depressed-pyramidal, 2 - 3 cm ∅ and 1.5 cm tall, tapering into fleshy decurrent **R**; **L** 3, erect, light green, multiveined, circinate in the upper ½, basal part subtended by an elongated papyraceous sheath 4 - 15 mm long; **L** subterete, forming tight spirals, 5 - 8 × 1.5 - 2 mm, tip acute; **Inf** erect; peduncle slender, 5 - 15 cm; raceme 2.5 - 6 cm with 6 - 12 **Fl**; **Bra** narrowly triangular-acuminate, membranous, to 1 mm; **Ped** to 8 mm; **Tep** strongly reflexed; **OTep** light yellow, subacute, 5 - 7 × 1.3 - 2.5 mm; **ITep** elliptic, 7 × 4 mm; **Fil** stout, bearded; **Ov** oblong to globose, dark green, 1 mm; **Sty** 4 mm, yellow; **Fr** subglobose to oblong, ± 2 mm; **Se** black, pyramidal, with undulate ridges, 1 - 1.5 mm ∅. – [E. van Jaarsveld]

B. torta N. E. Brown (BMI 1908: 409, 1908). **T:** RSA, Western Cape (*Wiers* s.n. [K]). – **D:** RSA (Western Cape); dry Fynbos, spring-flowering.

Small rosulate geophytes to 5 - 12 cm tall; tuber depressed, 2 cm ∅, 1 cm tall; **R** terete; **L** numerous, 3.5 - 12 cm × 1 mm, tortuose, filiform; **Inf** 8 - 10 cm, laxly flowered; peduncle 1 mm ∅ at the base; **Bra** triangular-lanceolate, 3 × 1 mm; **Ped** terete, 8 - 10 mm; **Fl** spreading, stellate, 13 - 14 mm ∅; **OTep** oblong-ovate, 7 × 1.5 mm; **ITep** 6 × 4 mm, obtuse; **St** 5 mm; **Anth** 0.5 mm, oblong; **Ov** globose, 1.5 mm; **Sty** erect, terete, 3 mm; **Fr** globose; **Se** 2 mm. – [E. van Jaarsveld]

B. truncata G. Williamson (Haseltonia 4: 19, ills. (p. 15, 20), 1996). **T:** RSA, Northern Cape (*Williamson & Hammer* 5415 [NBG]). – **D:** RSA (Northern Cape: Namaqualand); Succulent Karoo, flowers in mid-summer.

Deciduous erect soft acaulescent herbs to 3.5 cm

tall; tuber dome-shaped, to 1.1 cm wide with up to 6 thick and fleshy **R** spreading downwards and decurrent and tapering from the tuber; **L** up to 3, erect, dull green, surface minutely asperulous with 3 longitudinal veins, outer 2 **L** slightly falcate, 3 cm × 4 mm, distinctly truncate, inner **L** smaller and acute, all **L** subtended by an elongated papyraceous sheath to 4 mm long; **Inf** erect; peduncle purple, slender, to 5 cm; raceme to 2.5 cm; **Fl** 6 - 8; **Bra** triangular-acuminate, membranous, to 1.5 mm; **Ped** to 7 mm; **Tep** brownish-orange with darker midvein, reflexed; **OTep** elliptic-acuminate, obtuse, 7 × 1.8 mm; **ITep** obovate, 7 × 4 mm; **Fil** slender, 6.5 mm; **Ov** ellipsoid, green, 2 mm; **Sty** 5 mm; **Se** up to 3 per **Fr**, deltoid, ruminate, to 2 mm wide. – [E. van Jaarsveld]

B. undulata G. Williamson (Bradleya 18: 32-34, ills., 2000). **T:** RSA, Western Cape (*Williamson* 5931 [NBG]). – **D:** RSA (Western Cape: Cederberg); moss-covered quartzitic sandstone rocks, flowers late spring.

Dwarf solitary single-headed geophytes; tuber dome-shaped, to 15 × 10 mm, with several horizontally spreading wedge-shaped **R** from the flat base; **L** 4, erect, linear, terete, unequal in length, translucent and shiny green, longest **L** 80 × 3 mm, flattish at the base with a broader sheathing base, shorter **L** mainly terete and 40 × 1.2 mm, with a sheathing base, withering sheaths clasping the young **Inf**; **Inf** solitary lax racemes to 14.5 cm, to 20-flowered; **Bra** deltoid-acuminate, 1.2 × 1 mm, clasping; **Ped** 7 mm; **Tep** pale yellow with very fine green midveins; **OTep** recurved, lanceolate, obtuse, 7 × 2 mm; **ITep** narrowly ovate, obtuse, 7 × 4 mm; **Ov** globose, 1 mm ∅; **Sty** erect, 6 mm; **Sti** capitate, minute; **Fr** ovoid, 2 × 1.8 mm; **Se** narrowly pyramidal, black, 1 × 0.8 mm. – [E. van Jaarsveld]

B. vagans E. M. Watson (in George, A. S. (ed.), Fl. Austral. 45: 470, ill., 1987). **T:** Australia, Queensland (*Watson* 152 [CBG, AD, BRI, K]). – **D:** Australia (Queensland, New South Wales). **I:** Forster (1993).

Perennial succulent herbs without caudex; **R** succulent; **L** many, linear to subulate, channelled, 2 - 8 mm wide, green, not glaucous; **Inf** 1 to many, straggling; peduncle angular; **Tep** 8 - 13 mm, yellow; **St** 3 short, 3 long, closely bunched; **Anth** yellow, ± basifixed, inclining over the **Sty** after dehiscence; ovules 2 - 4 per locule; **Sty** erect or down-curved, 2 - 3 mm; **Fr** 3-lobed, 3 - 5 mm, **Ped** 18 - 24 mm, often twisting and drooping at the tip; **Se** heavily ridged. – [P. I. Forster]

B. vitrea G. Williamson & Baijnath (SAJB 61(6): 314-316, ills., 1995). **T:** RSA, Northern Cape (*Williamson* 4471 [NBG]). – **D:** RSA (Northern Cape: Richtersveld); Succulent Karoo, flowers in mid-summer.

Dwarf solitary summer-deciduous geophytes to 1 cm tall; tuber globose, 3- to 5-lobed, ± 1 - 1.5 cm ∅, 1 cm tall, with greyish papery tunics, flesh pinkish where exposed; **R** grey, succulent, terete; **L** 1 - 4, ascending, linear-lanceolate, 5 - 10 × 2 - 4 mm, tessellate, grey-green tinged pink, amplexicaul, surface undulating, older **L** becoming truncate, upper face flat, lower face convex, apex acute; **Inf** 1 - 3, 2.5 - 6 cm, 3- to 7-flowered in the upper ⅓, ascending; peduncle 1 mm ∅ at the base, terete; **Bra** 3 × 1 mm, membranous, deltoid-acuminate, amplexicaul; **Ped** 4 mm; **Fl** 2 - 6 mm apart; **Per** stellate, becoming reflexed, ± 10 - 14 mm ∅; **OTep** ovate-lanceolate, 5 × 1.5 - 2 mm; **ITep** ovate to ovate-elliptic, 5 × 3 mm, obtuse; **St** 4.5 mm; **Anth** yellow, oblong, 1 mm; **Ov** globose, 1 mm ∅, yellowish-green, grooved; **Sty** erect, 5 mm; **Fr** ovoid, 2 mm long; **Se** 1 mm. – [E. van Jaarsveld]

B. vittatifolia G. Williamson (Haseltonia 4: 21-22, ills. (pp. 18, 20), 1996). **T:** RSA, Northern Cape (*Williamson* 5527 [NBG]). – **D:** RSA (Northern Cape: Namaqualand); Succulent Karoo, flowers in mid-summer.

Small rosulate deciduous geophytes to 8 cm tall; tuber depressed-ovoid, 3 - 4.5 × 3 - 4 cm ∅; **L** 15 - 20 per plant, 20 - 35 × 2 - 3 mm, erect, upper ½ curved horizontally towards the light, softly succulent with translucent greenish-white base below ground-level, upper face flat to slightly convex, translucent and with 3 darker green striations, lower face convex, not translucent, tip acute, mucronate; **Inf** 1 - 4, to 12 - 19 cm with 25 - 45 spreading **Fl**, 1 - 1.5 mm ∅ at the base; **Bra** triangular-acuminate, 1.5 × 1 mm; **Ped** terete, 4 - 8 mm; **Per** stellately spreading, yellow, 12 mm ∅; **OTep** oblong-lanceolate, 7 × 1.5 mm, tips acute; **ITep** 6 × 2 mm, tips obtuse; **St** 5 - 6 mm; **Fil** bearded in upper ⅓; **Ov** globose, 1 mm; **Sty** erect, terete, 4.5 mm; **Sti** capitate; **Fr** globose, 7 mm ∅; **Se** 3 × 4 mm, broadly ellipsoid, angled, shiny and black. – [E. van Jaarsveld]

B. wiesei L. I. Hall (SAJB 3(6): 357, 1984). **T:** RSA, Western Cape (*Wiese* s.n. [NBG 126754, K, PRE]). – **D:** RSA (Western Cape); Succulent Karoo, quartz flats, flowers from late spring to early summer.

Dwarf deciduous acaulescent rosulate solitary geophytes; tuber depressed, globose above, flat below, 1 cm tall and 1.5 cm wide, with a ring of **Bri** surrounding the **Ros**; tunics brown; **R** fleshy, terete; **L** to 20, crowded, oblong-acute, ascending, 3.5 - 4 × 0.5 cm, upper face flat to slightly convex, lower face convex, dull silvery-green, apex acute; **Inf** 1, to 10 cm, 20-flowered; peduncle wiry, terete, 1 mm ∅ at the base; **Bra** 3 mm, deltoid-cuspidate; **Ped** 8 - 12 mm; **Per** stellate, becoming reflexed, to 16 mm ∅; **OTep** lanceolate, 8 × 3 mm; **ITep** 8 × 6 mm, obtuse; **St** 6 mm; **Ov** oblong, 1.5 mm; **Sty** oblique, terete, yellow, 7 mm; **Fr** pyriform, 4 × 3 mm; **Se** black, oblong, 2 × 1 mm. − [E. van Jaarsveld]

TRACHYANDRA

E. van Jaarsveld

Trachyandra Kunth (Enum. Pl. 4: 573, 1843). **T:** *Anthericum hispidum* Linné. − **Lit:** Obermeyer (1962). **D:** Africa (mostly SW RSA). **Etym:** Gr. 'trachys', rough; and Gr. 'aner, andros', male; for the rough filaments.
Incl. *Dilanthes* Salisbury (1866). **T:** not typified.
Incl. *Liriothamnus* Schlechter (1924). **T:** *Anthericum involucratum* Baker.

Plants variable, perennial, glabrous to glandular-pubescent rosulate **L** succulents, acaulescent, geophytic or shrubby to 2 m tall; **R**stock tuberous or as erect rhizome; **R** fusiform to fibrous, terete; stem erect, woody and covered with **L** bases; **L** dimorphous or uniform, lamina flat or triangular to terete, often linear-lorate, rarely caniculate, straight or undulating; **Inf** racemose, occasionally paniculate; peduncle terete, bracteose; **Fl** solitary, spreading to pendulous; **Per** rotate, often becoming recurved; **Tep** 6 in 2 whorls, white to mauve, **OTep** narrower than **ITep**; **St** 6, adnate to base of **Tep**; **Fil** scabrid; **Anth** dorsifixed, introrse; **Ov** globose, superior, 3-locular; **Sty** terete, **Sti** capitate; **Fr** loculicidal globose capsules; **Se** grey-brown, smooth to papillate, becoming glutinous.

A genus of ± 50 species, most of which are geophytic succulents. *Trachyandra* has a wide distribution in S Africa but shows a concentration of species in the winter rainfall regions of the former Cape Prov., RSA. They are seldom cultivated but are easy and should be kept dry during the summer months.

The genus is divided into 3 sections according to Obermeyer (1962: 717):
[1] Sect. *Liriothamnus* (Schlechter) Obermeyer 1962: Plants glabrous; outer **L** not reduced to **Sc**; **Inf** mostly simple.
[2] Sect. *Trachyandra*: Plants glabrous; outer **L** reduced to **Sc**; **Inf** mostly branched.
[3] Sect. *Glandulifera* Obermeyer 1962: Plants glandular-pubescent.

T. adamsonii (Compton) Obermeyer (BT 7(4): 720, 1962). **T:** RSA, Cape Prov. (*Compton in NBG* 318/22 [NBG]). − **D:** RSA (Northern Cape, Western Cape); dry Fynbos and Succulent Karoo on quartzitic sandstones.
≡ *Liriothamnus adamsonii* Compton (1931).
[1] Erect sparingly branched woody rosulate shrubs to 2 m; caudex depressed, globose above and tapering into stem, flat below with grey bark; **R** succulent, terete, 5 mm ∅; **Br** 1 - 2 cm ∅, grey with peeling **L** remains, becoming smooth at base; **L** glaucous, striate, crowded at stem tips, linear-lance-

olate, flat, lower side slightly keeled, upper face basally channelled, margin denticulate, apex acuminate; **Inf** axillary, ascending, racemose, to 50 cm, with a pair of side **Br** at the base; **Bra** triangular to ovate-acuminate, 2 - 10 × 1 - 3 mm; **Ped** erect, 6 - 11 mm; **Per** white with yellow eye at base of **Tep**, to 3 cm ∅; **OTep** 11 × 2.5 mm, linear-lanceolate, **ITep** 10 × 3.5 mm; **St** 8 mm; **Ov** oblong, 2 mm; **Fr** erect, ovate, 12 mm; **Se** grey, oblong, 3 mm long.

T. aridimontana J. C. Manning (SAJB 56(1): 1-5, ills., 1990). **T:** RSA, Northern Cape (*Oliver & al.* 478 [PRE]). − **D:** RSA (Northern Cape: Richtersveld); Succulent Karoo.
[1] Erect sparingly branched rosulate deciduous succulent herbs to 35 cm; caudex depressed, globose above and tapering into stem, flat below with grey bark; **R** succulent, terete, 3 mm ∅; **Br** 4 - 15 mm ∅, bark gey; **L** glaucous, crowded at stem tips, 3 - 6, linear-lanceolate, 3 - 12 cm × 3 - 7 mm ∅, subterete but dorsiventrally flattened, lower face convex, upper face flat to caniculate, margin denticulate, apex mucronate; **Inf** axillary, ascending, racemose, to 22 cm; **Bra** ovate to lanceolate-acuminate, to 8 mm; **Ped** patent, 1 - 5 mm; **Per** white, 1.5 - 2 cm ∅; **OTep** linear, 11 - 13 × 2 - 3.5 mm; **ITep** ovate, 11 - 13 × 3 - 5.5 mm; **St** 8 - 9 mm, scabrid; **Ov** ovoid, 1.5 mm long; **Sty** 7 - 8 mm; **Fr** erect, narrowly ovate, 8 - 12 × 4 mm; **Se** black, oblong, 3 × 1.5 mm.

T. ciliata (Linné *fil.*) Kunth (Enum. Pl. 4: 585, 1843). **T:** RSA, Cape Prov. (*Thunberg* s.n. [UPS]). − **D:** RSA (Western Cape, Eastern Cape); Fynbos, Renosterveld. **Fig. XXV.e**
≡ *Anthericum ciliatum* Linné *fil.* (1781) ≡ *Bulbine ciliata* (Linné *fil.*) Link (1821); **incl.** *Anthericum longifolium* Jacquin (1786) ≡ *Phalangium longifolium* (Jacquin) Poiret (1804) ≡ *Trachyandra longifolia* (Jacquin) Kunth (1843); **incl.** *Anthericum caniculatum* Aiton (1789) ≡ *Phalangium caniculatum* (Aiton) Poiret (1804) ≡ *Bulbine caniculata* (Aiton) Sprengel (1825) ≡ *Trachyandra caniculata* (Aiton) Kunth (1843); **incl.** *Anthericum vespertinum* Jacquin (1804) ≡ *Phalangium vespertinum* (Jacquin) Poiret (1804) ≡ *Trachyandra vespertina* (Jacquin) Kunth (1843); **incl.** *Anthericum blepharophoron* Roemer & Schultes (1829) ≡ *Trachyandra blepharophora* (Roemer & Schultes) Kunth (1843); **incl.** *Trachyandra bracteosa* Kunth (1843); **incl.** *Anthericum recurvatum* Dinter (1931); **incl.** *Anthericum hamatum* von Poellnitz (1942); **incl.** *Anthericum maculatum* von Poellnitz (1942); **incl.** *Anthericum pilosiflorum* von Poellnitz (1942); **incl.** *Anthericum pilosiflorum* var. *subpapillosum* von Poellnitz (1942); **incl.** *Anthericum spongiosum* von Poellnitz (1942).
[2] Acaulescent fast-growing geophytes, rosulate; **R** swollen, terete, spreading; **L** linear, to 1 m, to 4 mm ∅, upper face flat, lower side keeled, softly

succulent, margin ciliate; **Inf** racemose, ascending-spreading to 50 cm, with a pair of side **Br** at the base; peduncle pubescent, becoming glabrous; **Bra** to 1 cm, cymbiform, auriculate, younger **Bra** imbricate, subulate; **Fl** white, translucent, 2 cm ∅, **Tep** recurved; **OTep** linear, 10 × 2 mm; **ITep** ovate, 10 × 3 mm; **St** 8 - 9 mm, scabrid; **Ov** globose, 1.5 mm long; **Sty** 7 - 8 mm; **Fr** globose, to 14 mm ∅; **Se** black, rough.

T. involucrata (Baker) Obermeyer (BT 7(4): 721, 1962). **T:** RSA, Northern Cape (*Drège* 2681 [K, L, PRE [photo]]). – **D:** RSA (Northern Cape); Succulent Karoo, rocks.

≡ *Anthericum involucratum* Baker (1876) ≡ *Liriothamnus involucratus* (Baker) Schlechter (1924).

[1] Erect sparingly branched rosulate deciduous succulent gnarled shrublets to 60 cm; caudex depressed, globose above and tapering into stem, flat below with grey bark; **R** succulent, terete, 3 mm ∅; **Br** 5 - 15 mm ∅, bark grey; **L** glaucous, crowded at the stem tips, 3 - 6, linear-lanceolate, 17 cm × 6 mm ∅, subterete, canaliculate, denticulate at the base, apex mucronate; **Inf** axillary, ascending, racemose, to 26 cm; **Bra** ovate to lanceolate-acuminate, to 7 mm, clasping; **Ped** recurved in **Fr**, 5 - 10 mm; **Per** white, 1.5 - 2 cm ∅; **OTep** linear, 10 - 14 × 2 - 2.5 mm; **ITep** ovate, 10 - 14 × 3 - 4 mm; **St** 8 - 9 mm, scabrid; **Ov** oblong to globose, 1.5 mm; **Sty** 7 - 10 mm; **Fr** erect, narrowly ovate, 10 - 19 × 2 - 3 mm; **Se** black, ovoid, 3 × 1.5 mm.

T. tortilis (Baker) Obermeyer (BT 7(4): 745, 1962). **T:** RSA, Western Cape (*Schlechter* 4846 [Z, PRE]). – **D:** RSA (Western Cape); Succulent Karoo, rocks.

≡ *Anthericum tortile* Baker (1904); **incl.** *Anthericum salteri* Leighton (1938); **incl.** *Anthericum oocarpum* Schlechter *ex* von Poellnitz (1942).

[2] Acaulescent geophytes to 15 cm tall with a subterranean tuber crowned with broad scale **L**; **R** swollen, fused to the elongate tuber, spreading, terete; **L** 3 - 6, linear, glaucous, widely spreading, transversely and plicately folded, 6 - 10 cm × 2 mm, glabrous or sparingly pubescent, flat, margin entire; **Inf** lax ascending divaricate panicles to 9.5 cm, with up to 5 pairs of side **Br**; peduncle pubescent becoming glabrous, basally arcuate; **Bra** to 3 mm, ovate-lanceolate, cymbiform, auriculate; **Ped** to 5 mm; **Fl** white to pale pink, 1.5 cm ∅; all **Tep** linear-obovate, 5 × 2 mm; **St** 2 - 3 mm, scabrid; **Ov** globose, 0.75 mm ∅; **Sty** 2 - 3 mm; **Fr** linear-ovoid, 7 mm long; **Se** ridged.

Bromeliaceae

Perennial small to large herbs or shrubby, terrestrial or often epiphytic with caulescent or sessile small to large **Ros**; **L** spiral (rarely distichous), simple, entire or serrate, basally often broadened and ± sheathing, often leathery or succulent, usually with peltate shortly stalked whitish multicellular water-absorbing scales; **Inf** terminal (and **Ros** monocarpic) or lateral, pedunculate or sessile, simple or compound panicles, racemes, spikes or congested heads; **Bra** often brightly coloured; **Fl** bisexual (sometimes functionally unisexual), regular or almost so; **Per** various, with 3 outer **Tep** (usually referred to as **Sep**, often petaloid) and 3 inner free or basally united **Tep** (usually referred to as **Pet**); **St** 6 in 2 series; **Fil** united to each other or to **Tep**, or free; **Ov** of 3 **Ca**, superior to inferior, with septal **Nec** and axile placentation; **Sty** 3-parted; **Fr** longitudinally dehiscing 3-parted capsules, or fleshy berries, rarely the individual **Fr** of an **Inf** united into a fleshy collective **Fr** (*Ananas*); **Se** various, often winged or plumose, embryo small with copious mealy endosperm.

Distribution: Tropical and subtropical America, 1 species in tropical W Africa.

Literature: Smith & Downs (1974-1979); Rauh (1990); Baensch (1994); Smith & Till (1998).

This large family (± 56 genera, 2600 species) is well-known in cultivation, where its species are colloquially referred to as Bromeliads. Numerous species are cultivated as house and foliage plants, and a multitude of hybrids and cultivars is generally available. *Tillandsia* ("Air Plants") are popular among specialized collectors, and numerous hobby clubs and societies exist worldwide. *Ananas* ("Pineapple") is an important fruit crop, and several species have been valued for their strong fibres (*Aechmea, Bromelia, Neoglaziovia*).

Ecological adaptations are numerous. Especially epiphytic taxa are often densely covered in greyish-white scales which absorb water from the atmosphere. The large tank bromeliads from humid climates harbour an often unique micro-fauna (frogs, insects etc.) and -flora (specialized *Utricularia* species) in their water-filled rosettes. Species of *Brocchinia* growing in acid peat bogs have long been suspected to be partly carnivorous.

Succulence is well-developed in many taxa, predominantly in the form of leaf succulence, and bromeliads are an important constituent of the flora of semi-arid and arid regions. CAM photosynthesis has been found in several species.

Systematically, the family can be divided into 3 subfamilies as follows:

Pitcairnioideae: Usually terrestrial; **Ov** superior or almost so, or half-inferior; **Fr** dehiscent capsules;

Se winged but without a hairy crown. – This is probably the most primitive group within the family. Important genera: *Dyckia, Hechtia, Navia, Puya.* *Hechtia* is notable for being dioecious.

Tillandsioideae: Mostly epiphytic; **Ov** superior to half-inferior; **Fr** dehiscent capsules; **Se** winged and with a hairy plumose crown. – Important genera: *Catopsis, Guzmania, Tillandsia, Vriesea.* Some species of *Catopsis* are dioecious.

Bromelioideae: Terrestrial or epiphytic; **Ov** inferior or almost so; **Fr** juicy berries or all **Fr** of an **Inf** united into a fleshy collective **Fr** (*Ananas*); **Se** unwinged and without a hairy crown. – Here belongs the bulk of the family. Important genera: *Aechmea, Ananas, Billbergia, Bromelia, Cryptanthus, Neoregelia, Nidularium.*   [U. Eggli]

Commelinaceae

Herbs, mostly perennial, often with a rhizomatous base or **R** tubers, more rarely annual, often grasslike when not in flower; stems usually succulent or subsucculent, often swollen at the nodes; **Int** usually marked with a line of **Ha**; **L** simple, entire, often longitudinally striped with whitish, silvery or purplish bands, arranged spirally or distichous, or sometimes in mainly terminal or basal **Ros**; L-base forming a closed (i.e. tubular) sheath around the stem (as in many grasses, but without a projecting ligule), margins of young **L** involute or convolute, venation clear or obscure; **Fl** actinomorphic or zygomorphic, bisexual or rarely polygamous, in terminal or axillary 1- to many-flowered cincinni (helicoid cymes) often aggregated in thyrses or fused in pairs; cincinni usually bracteate and bracteolate; **Bra** often spathaceous or beak-like; **Sep** usually 3, free (rarely connate, very rarely 2); **Pet** 3 (very rarely 2), free, sometimes clawed and/or connate at base, equal or unequal; **St** typically 6, in 2 whorls of 3, but all or some of one or other whorl sometimes missing, variously differentiated or staminodal; **Ov** superior, 3- or rarely 2-locular with several to 2, or rarely 1, biseriate or uniseriate, ovules per locule; **Fr** capsules, splitting loculicidally, or rarely berrylike and indehiscent; **Se** with mealy endosperm, position of micropyle and embryo marked by a callosity (operculum or embryotega) on the testa.

Distribution: Mostly tropics and subtropics worldwide.

Literature: Clarke (1881); Woodson (1942); Faden (1985); Faden & Hunt (1991); Hunt (1993a); Hunt (1993b); Faden (1998).

A family with ± 40 genera and 650 species in 2 very unequal subfamilies: *Cartonematoideae*, with 2 genera and a handful of species in Australia and Africa, and *Commelinoideae*, widely distributed, with all the rest. Many are cultivated, primarily for their ornamental foliage. Apart from this the family is of little economic value.

The following account is confined to the most xerophytic species with succulent or subsucculent leaves in the genera *Aneilema, Callisia, Cyanotis, Tradescantia* and *Tripogandra* and treats only those of potential interest to collectors of succulents. Most of the species are easily cultivated and readily propagated by cuttings.

Many members of the family exhibit a degree of succulence in the stems and/or leaves, but even in some of those occasionally seen in collections of succulent plants, such as *Callisia navicularis* and *Tradescantia sillamontana* (2 of the 3 species listed by Jacobsen), the normal stems and leaves die down at the commencement of the winter dry season, i.e. the plants are hemicryptophytes. Many other members of the family in seasonally dry and/or cold habitats are hemicryptophytes or geophytes with tubers or thickened roots.

Identification of individual genera is usually not possible by vegetative features alone and depends in the first instance on features of the inflorescence, which is composed of one to many helicoid cymes or cincinni. The cincinnus may be relatively lax, with visible internodes between the flowers, as usually in species of *Aneilema*, or very condensed, so that the flowers appear crowded, as in a small umbel. The diagnostic feature of the large American genus *Tradescantia* and its immediate allies, *Callisia* and *Tripogandra*, is that the cincinni are fused in pairs, back to back. This may not be immediately apparent, because of the crowding of the flowers, but is often indicated by the presence of paired bracts beneath the flower-heads, or else by an open flower on each side (normally each cincinnus opens one flower at a time). *Callisia* includes several of the most succulent species in the family, but its characters are more obscure (seeds, chromosomes), with affinities to both *Tradescantia* and *Tripogandra. Cyanotis* is the Old World equivalent of *Tradescantia*, but the cincinni are single, not fused in pairs.

Key to genera with succulents:

1 Cincinni single, not fused in pairs: **2**
– Cincinni fused in bifacial pairs: **3**
2 **St** dimorphic: **Aneilema**
– **St** all similar: **Cyanotis**
3 **St** dimorphic: **Tripogandra**
– **St** all similar: **4**
4 **Bra** paired or very rarely 3 or more:
                                                                                              Tradescantia

– **Bra** solitary or absent: **5**
5 Paired cincinni stipitate, or if sessile then **Fl**
   also sessile: **Callisia**
– Paired cincinni sessile, **Fl** pedicellate:
                                                                                              Tradescantia

[D. R. Hunt]

ANEILEMA

D. R. Hunt

Aneilema R. Brown (Prodr., 270, 1810). **T:** *Aneilema biflorum* R. Brown. – **Lit:** Faden (1991). **D:** Africa (55 species + 1 extending to the Arabian Peninsula and 1 to tropical America), Yemen (1 species), Madagascar (1 species), Australia (5 species). **Etym:** Gr. 'a-, an-', without; and Gr. 'eilema', involucre; because the inflorescences are without conspicuous subtending bracts.
Incl. *Lamprodithyros* Hasskarl (1864). **T:** not designated.
Incl. *Ballya* Brenan (1964). **T:** *Aneilema zebrinum* Chiovenda.

Annual and perennial herbs; **R** sometimes tuberous; **L** various; **Inf** single cincinni or thyrses of cincinni, without conspicuous **Bra**; **Fl** bisexual, zygomorphic; **Sep** 3, free, equal; **Pet** 3, free, often unequal and differing in colour, upper pair clawed, white to lilac or lavender, or yellow to orange, lower much reduced (rarely enlarged), often inconspicuous; **St** (5-) 6, the 3 anterior fertile (sometimes the median sterile), the (2-) 3 posterior staminodal; **Fil** equal, or more commonly the median shorter, all glabrous or the lateral bearded; **Anth** of median **St** usually different from those of the lateral, antherodes usually bilobed; **Ov** 3- or 2-locular, ventral locules with 1 - 6 uniseriate ovules, dorsal not developed or with usually 1 ovule only; **Fr** capsules, usually 2-valved; **Se** with linear hilum and dorsal embryotega.

7 sections are recognized. The only species treated here belongs to Sect. *Lamprodithyros* (Hasskarl) C. B. Clarke 1881.

Ballya was separated on the grounds of 3 characters, but 2 have been shown by Faden (1991) to be found in other species of *Aneilema* Sect. *Lamprodithyros*, whilst the third was atypical of *A. zebrinum* itself.

A. zebrinum Chiovenda (Webbia 8: 38, t. 12, 1951). **T** [syn]: Ethiopia, Gemu-Gofa Prov. (*Corradi 2154* [FT]). – **D:** Ethiopia, Kenya, Tanzania, RSA (KwaZulu-Natal); 10 - 1150 m. **I:** Faden (1991: 140-143). **Fig. XXVI.e**
   ≡ *Ballya zebrina* (Chiovenda) Brenan (1964).

Creeping perennials; **R** not tuberous; **L** distichous, lamina to 4 × 2 cm, shiny, greyish-green, striped paler and sometimes with maroon lines; **Inf** lateral, piercing the **L**-sheath, usually a single cincinnus; **Fl** 4 (-6), to 1 cm ∅; **Pet** pale lilac, lower **Pet** cup-like; **Ov** 3-locular; **Fr** 3 - 4 × 2 - 3 mm, puberulous.

CALLISIA

D. R. Hunt

Callisia Loefling (Iter Hispan., 305, 1758). **T:** *Callisia repens* Linné. – **Lit:** Moore (1958); Hunt (1986b). **D:** SE USA, Mexico, tropical America. **Etym:** Gr. 'kallos', beauty.
Incl. *Aploleia* Rafinesque (1837). **T:** *Aploleia diffusa* Rafinesque [*Nom. illeg.*, = *Callisia monandra* (Swartz) Schultes *fil.*].
Incl. *Phyodina* Rafinesque (1837). **T:** *Tradescantia gracilis* Kunth.
Incl. *Spironema* Lindley (1840) (*nom. illeg.*, Art. 53.1). **T:** *Spironema fragrans* Lindley.
Incl. *Cuthbertia* Small (1903). **T:** *Tradescantia rosea* Ventenat.
Incl. *Leptocallisia* (Bentham & Hooker *fil.*) Pichon (1946). **T:** *Callisia umbellulata* Lamarck.
Incl. *Hadrodemas* H. E. Moore (1963). **T:** *Tradescantia warszewicziana* Kunth & Bouché.

Perennial or rarely annual or short-lived herbs; **L** succulent; **Inf** various, cincinni typically fused in pairs, without conspicuous **Bra**; **Fl** usually actinomorphic and bisexual, sessile or pedicellate; **Sep** 3 or very rarely 2, sometimes hyaline; **Pet** 3 or very rarely 2, free, usually white or pink; **St** 6, all similar or subsimilar, or reduced to 3 or 1; **Fil** typically **Ha**-less, sometimes bearded; **Anth** versatile, connectives usually broad; **Ov** (2- to) 3-locular; ovules 2 or very rarely 1 per locule; **Sti** usually penicillate or minutely capitate; **Fr** small capsules; **Se** with punctiform (dot-like) hilum.

Closely related to *Tradescantia*, but mostly lacking paired bracts below the inflorescence, and having different seeds and chromosomes. The smaller-growing species die back to dwarf shoots or to tubers during the resting-period. As in *Tradescantia*, the cincinni are fused in pairs (in 1 species associated in pairs or threes), and the flowers are regular with the stamens all similar (though sometimes reduced in number). Various species have been placed in separate genera, now treated as sections. Those including succulent species treated here are distinguished as follows:

[1] Sect. *Hadrodemas* (H. E. Moore) D. R. Hunt 1986: Cincinni incompletely fused in pairs or threes; **Fl** pedicellate, **Sep** fleshy, **Pet** pink, **St** 6; robust *Aloe*-like plant.

[2] Sect. *Brachyphylla* D. R. Hunt 1986: Cincinni strictly fused in pairs; **Fl** pedicellate, **Sep** boat-shaped, **Pet** pink, **Sti** capitate or dot-like, **St** 6, bearded.

[3] Sect. *Leptocallisia* Bentham & Hooker *fil.* 1883: Cincinni strictly fused in pairs; **Fl** pedicellate, **Sep** boat-shaped, **Pet** white, **Sti** capitate or dot-like; **St** (1-) 3, glabrous (in treated taxa).

[4] Sect. *Callisia*: Cincinni strictly fused in pairs; **Fl** sessile or nearly so; **Sep** narrowly acute; **Sti** penicillate.

C. elegans Alexander *ex* H. E. Moore (Baileya 6: 140, fig. 28, 1958). **T:** NY. – **D:** Guatemala, Honduras.

[4] Decumbent perennials; **L** 3.5 - 10 × 1.5 - 3 cm, distichous, dark green above with paler longitudinal stripes, commonly purplish beneath, velutinous; **Inf** 6 - 15 cm or more; paired cincinni mostly sessile, some pedunculate; **Fl** sessile; **Pet** white.

Allegedly found in Mexico but not known there in the wild. Treated by Hunt (1986b: 411) as a variety of *C. gentlei* Matuda from Belize, which has pink flowers and unstriped leaves.

C. fragrans (Lindley) Woodson (Ann. Missouri Bot. Gard. 29: 154, 1942). **T:** [lecto – icono]: Edward's Bot. Reg., 26: t. 47, 1840. – **D:** Mexico (Tamaulipas to Yucatán).

≡ *Spironema fragrans* Lindley (1840).

[4] Robust perennials to 1.5 m; **Fl** stems bromeliad-like and stout, with subrosulate **L**, sparsely

branched but producing long, relatively slender stolons with distichous **L** from the lower nodes; **L** of flowering stems to 30 × 7 cm, narrowly elliptic-lanceolate, acute, subamplexicaul, usually glabrous, bright light green; **Inf** ample terminal panicles, **Br** crowded with sessile paired cincinni subtended by papery **Bra** to 2 cm; **Fl** almost sessile, small, fragrant; **Sep** 3.5 - 5 × 1.5 - 2 mm, bristly; **Pet** 5 - 6 × 2.5 - 3.5 mm, spreading, lacking an expanded blade, lanceolate to ovate, white; **St** 6 and long-exserted, **Anth**-connectives membranous, white, more conspicuous than the **Pet**; **Sti** penicillate.

A cultivar with pale-striped leaves has been named 'Melnickoff'.

C. micrantha (Torrey) D. R. Hunt (KB 38: 131, 1983). **T:** USA, Texas (*Schott* s.n. [US]). – **D:** USA (SE Texas).

≡ *Tradescantia micrantha* Torrey (1859).

[2] Tufted or creeping, sometimes trailing to 1 m or more, succulent; **L** narrowly lanceolate to oblong-lanceolate, ± falcate and canaliculate, acute, mostly 1.5 - 2.5 × 0.5 - 0.6 cm, succulent, green with purplish striation below, virtually glabrous; **Inf** of paired cincinni mostly terminal and sessile, closely subtended by the uppermost 1 - 2 **L**; **Ped** 8 - 12 mm; **Fl** 1 - 1.5 cm ∅, bright purplish-pink; **St** bearded, connective yellow; **Sti** capitate.

Resembling *C. navicularis*, but smaller in all parts and lacking overwintering dwarf shoots. The karyotype is also different.

C. multiflora (M. Martens & Galeotti) Standley (J. Washington Acad. Sci. 15: 457, 1925). **T:** Mexico, Veracruz (*Galeotti* 4964 [BR, K, W]). – **D:** Mexico to Nicaragua. **I:** CBM 81: t. 4849, 1855.

≡ *Commelina multiflora* M. Martens & Galeotti (1842) ≡ *Aploleia multiflora* (M. Martens & Galeotti) H. E. Moore (1961); **incl.** *Tradescantia martensiana* Kunth (1843) ≡ *Callisia martensiana* (Kunth) C. B. Clarke (1881).

[3] Perennials with ascending or procumbent stems to 80 cm, rooting and branching at lower nodes; **L** 3 - 9 × 1 - 2.5 cm, ovate or oblong to elliptic-lanceolate, acute or acuminate, base rounded to subcordate, pale green, succulent, densely and minutely hairy; **Inf** slender terminal panicles to 30 cm with numerous paired cincinni; **Ped** 6 - 7 mm, usually minutely glandular-pubescent; **Fl** 6 - 8 mm ∅, white, fragrant; **St** 3, opposite to **Sep**; **Fil** ± 1.5 mm, glabrous or bearded; **Anth** linear, yellow; **Ov** 3-locular; **Sty** very short; **Sti** shallowly 3-lobed.

C. navicularis (Ortgies) D. R. Hunt (KB 38(1): 132, 1983). **T:** [lecto – icono]: Gartenflora, 26: t. 901, 1877. – **D:** E and NE Mexico. **Fig. XXVI.g**

≡ *Tradescantia navicularis* Ortgies (1877); **incl.** *Tradescantia brachyphylla* Greenman (1898).

[2] Stems tufted or trailing, succulent, of 2 intergrading types, bulbil-like short shoots with ± tightly

imbricate **L**, and stolons with long **Int** rooting and producing short shoots or **Inf**; **L** lanceolate to broadly ovate, canaliculate and somewhat falcate, tip acute, mostly 2 - 3 × 1 - 2 cm, very succulent, green above, purple striate beneath, virtually glabrous; **Inf** terminal, closely subtended by the uppermost **L** or more rarely by 2 **L**; **Ped** 1 - 2 cm; **Sep** lanceolate, canaliculate and strongly keeled, ± 7 × 3 - 4 mm, keel ciliate, margins hyaline; **Pet** very broadly ovate, ± 10 × ± 9 mm, bright purplish-pink; **St** equal, 5 - 6 mm; **Fil** bearded; connectives yellow; **Sti** capitate.

The original material was allegedly collected in Peru and was cultivated in the Zürich Botanical Garden. The taxon is not known outside Mexico, however.

C. repens Linné (Spec. Pl., ed. 2, 62, 1762). **T:** Martinique (*Jacquin* s.n. [not located]). – **D:** USA (Texas) and West Indies to Argentina. **Fig. XXVI.h**

Incl. *Hapalanthus repens* Jacquin (1763).

[4] Variable perennials with slender creeping stems, rooting at the nodes and forming mats; **L** 1 - 4 × 1 - 2 cm, variable, narrowly to broadly ovate, acute, rounded to subcordate at the base, virtually glabrous; **Inf** typically spiciform, ascending; paired cincinni sessile, typically subtended by **L** reduced to a mucro on the membranous sheath; **Ped** 0.5 - 1.5 mm; **Sep** 2 - 3.5 mm, becoming scarious; **Pet** 3 - 5 × 1 - 1.5 mm, narrowly oblong acute, translucent white; **St** 3 or 6 (1 or more often staminodal), long-exserted; **Fil** 6 - 10 mm, connective ± 1 × 1.5 mm, broadly reniform, thin; **Ov** 2-locular; **Sti** penicillate.

The form of *C. repens* in commercial cultivation is distinctive. It has small rounded leaves and inconspicuous inflorescences in the axils of normal leaves. Its wild origin is uncertain. It has been confused with *C. cordifolia* (Sect. *Leptocallisia*) which has a different inflorescence and thin leaves.

C. tehuantepecana Matuda (Anales Inst. Biol. UNAM 27: 356, 1957). **T:** Mexico, Oaxaca (*MacDougall* s.n. [MEXU]). – **D:** S Mexico.

Incl. *Callisia nizandensis* Matuda (1976).

[4] Closely related to *C. gentlei* and *C. elegans*, but differing in the longer and narrower **L**, somewhat laxer habit etc. Stems decumbent or erect to 60 - 80 cm; **L** lanceolate, acuminate, 7 - 9 × 1.5 - 2 cm, glabrous, not striate, margins white-ciliate, sheaths papery; **Inf** spiciform, paired cincinni terminal and lateral, sessile or shortly pedunculate, 6- to 8-flowered; **Ped** short; **Fl** 1 - 1.5 cm ∅, pink; **St** glabrous, connective broadly quadrate; **Sti** capitate.

C. gentlei, though succulent, is of less interest than the closely related *C. elegans* and *C. tehuantepecana*, and is not treated here.

C. warszewicziana (Kunth & Bouché) D. R. Hunt (KB 38: 132, 1983). **T:** Guatemala (*Warszewicz* s.n.

[B]). – **D:** S Mexico, Guatemala. **I:** Moore (1963: 134).

≡ *Tradescantia warszewicziana* Kunth & Bouché (1847) ≡ *Tripogandra warszewicziana* (Kunth & Bouché) Woodson (1942) ≡ *Hadrodemas warszewiczianum* (Kunth & Bouché) H. E. Moore (1963).

[1] Robust bromeliad-like herbs; stems eventually 1 m or more, stout; **L** to 30 × 6.5 cm, spirally arranged and densely imbricate towards the apical **Ros**, narrowly oblong, acuminate, sessile, succulent, glabrous except for the sometimes ciliate, often purple margin; **Inf** to 35 cm overall, appearing axillary, branched below and sometimes viviparous; cincinni fused at the base in pairs or threes, sessile or subsessile, many-flowered; **Ped** 10 - 13 mm; **Sep** 4 - 5 mm, succulent, persistent, glabrous; **Pet** 6 × 6 mm, purplish-pink; **St** 6, ± 5 mm; **Fil** usually glabrous, connectives broad, yellow; **Ov** 3-locular, ovules 2 per locule.

CYANOTIS

D. R. Hunt

Cyanotis D. Don (Prodr. Fl. Nepal., 45, 1825). **T:** *Cyanotis barbata* D. Don. – **D:** Old World tropics. **Etym:** Gr. 'kyanos', dark blue; and Gr. 'ous, otos', ear; for the flowers.

Incl. *Belosynapsis* Hasskarl (1871). **T:** *Belosynapsis kewensis* Hasskarl.

Usually perennial, often tuberous-rooted herbs; **L** narrow, often distichous, ± succulent and sometimes ciliate; **Fl** violet-blue or purplish-pink, bisexual, actinomorphic or nearly so, generally almost stalkless in dense single axillary cincinni subtended by a **L**-like or reduced **Bra** and with conspicuous sickle-shaped bracteoles; **Sep** 3, free or joined below; **Pet** 3, free or joined with the **Fil**-bases to form a short tube; **St** 6, equal, fertile; **Fil** nearly always bearded, often swollen subterminally, connectives narrow; **Ov** 3-locular with 2 ovules per locule; **Fr** dry dehiscent capsules; **Se** with punctiform (dot-like) hilum, embryotega terminal.

Most species of *Cyanotis* are of little horticultural interest, but the following are valued for their attractive succulent foliage. The majority of its ± 30 species are only subsucculent, perennating by tubers.

C. kewensis (Hasskarl) C. B. Clarke (in A. & C. de Candolle, Monogr. Phan. 3: 243, 1881). **T:** not located. – **D:** India (Madras).

≡ *Belosynapsis kewensis* Hasskarl (1871); **incl.** *Erythrotis beddomei* Hooker *fil.* (1875).

Prostrate perennials creeping from its initial **Ros**, **Int** and **L**-sheaths densely brown-hairy; **L** of **Ros** lanceolate, acute, rounded at the base, to 5 × 2 cm, **L** of side-shoots overlapping in 2 ranks, often smaller, succulent, dark green above, deep purple beneath, densely velvety; **Inf** lax, to 3 cm, with up to 8

Fl ± 8 mm ∅; **Ped** 3 - 4 mm; **Pet** free, purplish-pink; **St** ± 6 mm, **Fil** bearded with violet **Ha**; **Anth** yellow with orange pollen.

C. somaliensis C. B. Clarke (BMI 1895: 229, 1895). **T** [syn]: Somalia (*Cole* s.n. [K]). – **D**: Somalia. **I**: KuaS 28(4): 98, 1977. **Fig. XXVI.f**

Succulent perennials with non-flowering basal **Ros** (usually lacking in cultivated specimens) and stolon-like creeping shoots to ± 25 cm which can produce **Inf** or **R** and form new **Ros**; **L** oblong-linear, acute and with a small point, sessile, to 12 × 1.5 cm in **Ros**, 1.5 - 4 × 0.5 - 1 cm on lateral shoots, these **L** U-shaped in section, and often recurved, succulent, densely hairy below, margins long-ciliate, sheaths inflated, persistent, becoming papery; **Inf** short, scarcely exceeding the **Bra**, with several **Fl**; **Fl** ± 5 mm ∅, purplish-blue.

The origin and status of the widespread cultivated form of *C. somaliensis* (see e.g. the illustration cited above), which generally lacks basal rosettes, is uncertain.

C. speciosa (Linné *fil.*) Hasskarl (Commelin. Ind., 108, 1870). **T**: RSA, Western Cape (*Thunberg* s.n. [UPS]). – **D**: S Africa.

≡ *Tradescantia speciosa* Linné *fil.* (1782); **incl.** *Tradescantia nodiflora* Lamarck (1786) ≡ *Cyanotis nodiflora* (Lamarck) Kunth (1843).

Variable perennial succulent herbs with basal **Ros** and stolons to 50 cm, often forming loose mats; **L** narrowly oblong, acute, channelled, to ± 10 × 2 cm in the best forms, ± densely clothed with white silky **Ha**, or nearly glabrous; **Inf** spiciform, arching to 30 cm with sessile cincinni subtended by reduced **L**; **Fl** numerous, bracteoles ± 8 mm, falcate; **Pet** blue to mauve or pink.

TRADESCANTIA

D. R. Hunt

Tradescantia Linné (Spec. Pl. [ed. 1], 288, 1753). **T**: *Tradescantia virginiana* Linné. – **Lit**: Anderson & Woodson (1935); Hunt (1980); Hunt (1986a). **D**: America (from USA to N Argentina). **Etym**: For John Tradescant (± 1570 - 1638), gardener to Charles I of England.

Incl. *Campelia* L. Richard (1808). **T**: *Commelina zanonia* Linné.

Incl. *Zebrina* Schnizlein (1849). **T**: *Zebrina pendula* Schnizlein.

Incl. *Rhoeo* Hance (1853). **T**: *Tradescantia discolor* L'Héritier.

Incl. *Treleasea* Rose (1899) (*nom. illeg.*, Art. 53.1). **T**: *Tradescantia brevifolia* (Torrey) Rose.

Incl. *Setcreasea* K. Schumann & Sydow (1901). **T**: *Tradescantia brevifolia* (Torrey) Rose.

Perennial or very rarely annual or short-lived herbs of diverse habit; **R** fibrous or tuberous; **Inf** of bifacially fused and paired sessile cincinni subtended and ± enclosed by paired boat-shaped **Bra** similar to or ± differentiated from the **L**; **Fl** actinomorphic or almost so, bisexual; **Ped** often recurved after anthesis; **Sep** 3, usually equal, free, rarely somewhat unequal or fused or in 1 species accrescent (*T. andrieuxii*) and fleshy in **Fr**; **Pet** 3, equal, usually free, sometimes clawed at the base, rarely united into a slender tube, blue-violet, purple, pink or white; **St** 6, all fertile, all similar and equal or subequal; **Fil** free or the antipetalous rarely connate with the **Pet** bases; **Anth** versatile, connectives broad; **Ov** 3-locular; ovules 2 or very rarely 1 per locule; **Sti** minutely capitate; **Fr** capsules; **Se** with linear or very rarely punctiform (dot-like) hilum and dorsal or lateral embryotega.

The genus, which numbers 70 species, is currently divided into 12 sections by Hunt (1980) and Hunt (1986a), some of which (see synonymy) have previously been treated as independent genera. The sections relevant to this account can be recognized as follows:

[1] Sect. *Zebrina* (Schnizlein) D. R. Hunt 1986: **Pet** united below into a slender tube; **L** striped silvery (*T. zebrina* only).

[2] Sect. *Rhoeo* (Hance) D. R. Hunt 1986: **Pet** free; plants evergreen, bromeliad-like with **Ros** of fleshy linear-ensiform **L**; **Bra** of **Inf** well-differentiated from foliage **L**; **Fl** mostly white (only *T. spathacea*).

[3] Sect. *Austrotradescantia* D. R. Hunt 1980: **Pet** free; plants evergreen; **Bra** of **Inf** well-differentiated from foliage **L**; **Fl** mostly white; creeping or decumbent herbs with relatively small thin or fleshy **L**; **Cal** not fleshy or accrescent.

[4] Sect. *Tradescantia*: **Pet** free; **Bra** of **Inf** similar to foliage **L**; **Fl** usually blue, violet or purple; **Inf** appearing mainly terminal; plants dying back annually to the shortly rhizomatous **R**stock.

[5] Sect. *Setcreasea* (K. Schumann & Sydow) D. R. Hunt 1975: **Pet** united, but at base only; **L** not striped.

[6] Sect. *Mandonia* D. R. Hunt 1980: **Pet** free; **Bra** of **Inf** similar to foliage **L**; **Fl** usually blue, violet or purple; **Inf** terminal and lateral, forming a compound spike; plants dying back annually to the tuberous **R**stock.

The type species, *T. virginiana*, and its allies, native to the USA, are reliably hardy in temperate gardens, perennating as hemicryptophytes. A few other species will survive mild winters out-of-doors in the cool-temperate N hemisphere, but the deciduous species require protection from damp conditions, while the evergreen taxa need cool or (in a few cases) warm greenhouse treatment.

T. cerinthoides Kunth (Enum. Pl. 4: 83, 1843). **T**: Brazil, Rio Grande do Sul (*Sello* 3033a [B, K [fragment]]). – **D:** SE Brazil. **I:** CBM 170: t. 247, 1955, as *T. blossfeldiana.*

Incl. *Tradescantia blossfeldiana* Mildbraed (1940).

[3] Similar to *T. crassula*, but more hairy, with broader **L**, and **L** and **Fl** usually suffused with purple. Stems decumbent, rooting at the nodes and becoming shortly rhizomatous; **L** elliptic-oblong to ovate, to 15 × 3.5 cm, sessile, somewhat fleshy, glossy dark green and hairy to glabrous above, green or often purple and densely hairy beneath; **Inf** terminal and lateral; peduncles to 5 cm; **Bra** 2 - 2.5 × 1 - 1.5 cm; **Ped** to 2.5 cm, hairy; **Fl** 1.5 - 2 cm ∅, purplish-pink in upper ½, white below, or all white.

Better known as *T. blossfeldiana*, which was introduced via Argentina by Blossfeld, who offered it in his 1939 catalogue but all wild plants seen by the present writer were collected in SE Brazil.

T. crassula Link & Otto (Icon. Pl. Rar. Hort. Reg. Bot. Berol. 13, t. 7, 1828). **T:** Brazil, Rio Grande do Sul (*Sello* s.n. [B]). – **D:** SE Brazil.

[3] Stem decumbent, rooting at the nodes and becoming rhizomatous at the base; **L** oblong-ellipic, obtuse, channelled, to 15 × 3 cm, somewhat fleshy, shiny green, glabrous except the minutely ciliate margins and sometimes minutely pubescent beneath; **Inf** terminal and lateral; peduncle to 6 cm; **Bra** unequal, the longer to ± 4 × 1.5 cm (unfolded); **Ped** to 2 cm; **Fl** 1 - 2 cm ∅, white.

T. fluminensis Vellozo (Fl. Flumin., 140, 1829). **T:** [lecto – icono]: Vellozo, Icones 3: t. 153, 1835. – **D:** SE Brazil (Minas Gerais to Rio Grande do Sul), Paraguay, Uruguay, NE Argentina (Misiones).

Incl. *Tradescantia albiflora* Kunth (1843).

[3] Stems weakly ascending and decumbent or pendent, rooting at the nodes; **L** variable, broadly ovate to oblong-lanceolate, acute to acuminate, ± unequal-sided at the base and rounded to subcordate, 1.5 - 12 × 1 - 3.5 cm, subpetiolate, usually glabrous, green or purplish beneath, variegated whitish or yellowish in the more decorative forms; **Inf** terminal and lateral on peduncles ± 1 - 5 cm; **Bra** folded but not keeled, 1 - 2 × 0.7 - 1.5 cm (unfolded); **Ped** ± 1 cm, usually glabrous; **Fl** 1.2 - 1.8 cm ∅, white.

The true *T. fluminensis* is a rhizomatous perennial from the region of Rio de Janeiro with subpetiolate leaves tending to the larger end of the range quoted. In different clones the leaves are all green, or variegated whitish or yellowish. The popular cultivar with sessile leaves longitudinally striped whitish, known as *T. albiflora* 'Albovittata', is intermediate in habit between *T. fluminensis* and *T. crassula*. The decorative, non-rhizomatous, annual or short-lived plant with small, yellow-variegated leaves and slender, often purplish, stems, known as *T. fluminensis* 'Argenteo-variegata' is not this species. It may be *T. mundula* Kunth or *T. anagallidea* Seubert from S

Brazil to N Argentina. Individual plants are self-fertile, but most of their seedlings have normal green leaves.

T. hirta D. R. Hunt (CBM 180(3): 121-123, ills., t. 686, 1975). **T:** M. – **D:** NE Mexico (Coahuila, Nuevo León, San Luis Potosí).

Incl. *Setcreasea hirsuta* Markgraf (1952).

[5] Stem-base shortly rhizomatous and branching; **L** annual, linear, thickly succulent, concave above and rounded below, long-attenuate at the tip, to 20 × 1 cm, with soft tissue, glaucous green, sparingly to densely hairy with long weak **Ha** esp. towards the sheath; **Inf** terminal, solitary, scapiform, to 20 cm or more; **Bra** foliaceous, unequal, to 1.8 cm, membranous, glabrous, greenish or hyaline, veiny; **Ped** ± 3 - 5 mm, glabrous or pilose at the tip; **Sep** narrowly ovate-elliptic, 10 - 15 × 3 - 5 mm, membranous, whitish, glabrous; **Pet** spatulate, claw 8 - 9 mm, whitish, coherent with the **Fil** of the antisepalous **St** to form a slender tube 2 - 8 mm, **Pet** blade ovate to suborbicular, 10 - 15 × 8 - 10 mm, deep purplish-pink; **St** subequal, inserted on the **Cl** at the tube mouth; **Fil** 8 - 12 mm, bearded with purplish-pink **Ha**; connectives broadly triangular and apiculate, 2 × 2 mm, yellow; **Anth** curved-oblong, 2 mm; **Ov** ellipsoid-cylindrical, 2 - 3 mm, glabrous; **Sty** 13 - 18 mm; **Sti** 3-lobed; **Fr** and **Se** not described.

T. pallida (Rose) D. R. Hunt (KB 30: 452, 1975). **T:** Mexico, Tamaulipas (*Palmer* s.n. [US]). – **D:** E Mexico (Tamaulipas to Yucatán).

≡ *Setcreasea pallida* Rose (1911); **incl.** *Setcreasea purpurea* Boom (1955).

[5] Stems to 40 cm, ascending or decumbent; **L** oblong-elliptic to elliptic-lanceolate, usually 7 - 15 × 2 - 3.5 cm, trough-shaped, acute and usually apiculate, tapering into the sheath, slightly succulent, glaucous green, edged red, or reddish to violet overall, usually glabrous, rarely pilose with long soft **Ha**; **Inf** mostly terminal, solitary, on peduncles to 6 - 11 cm; **Bra** unequal, ovate-acuminate, folded and ± keeled, outer to 7 × 2.5 cm (unfolded), inner smaller; bracteoles thin, sheathing the buds; **Ped** short, with long soft **Ha** at the tip; **Fl** 2 - 3 cm ∅; **Sep** free, hyaline; **Pet** clawed, connivent at the very base, pink or pink with white midline, rarely white; **St** slightly epipetalous, with weak **Ha** to 3 mm.

A purple-leaved cultivar ('Purpurea' or 'Purple Heart') is widely cultivated under glass and useful for summer bedding.

T. sillamontana Matuda (Bol. Soc. Bot. México 18: 1, fig. 1, 1955). **T:** Mexico, Nuevo León (*White & Chatters* 30 [MICH, GH, MEXU]). – **D:** NE Mexico. **I:** CBM 181: t. 706, 1976. **Fig. XXVII.c**

Incl. *Tradescantia pexata* H. E. Moore (1960).

[4] Stems ascending or decumbent, to 30 cm, mainly branched near the base, normally dieing back in autumn to the rhizomatous base; **L** distich-

ous, elliptic-ovate to broadly ovate-lanceolate, acute, rounded to amplexicaul at the base, mostly 3 - 7 × 2 - 2.5 cm, somewhat succulent, green or becoming purplish in strong light, densely villose-lanate, esp. below with **Ha** to 1 cm; **Inf** terminal and usually solitary; peduncles usually < 3.5 cm; **Bra** similar to **L** but more strongly channelled or folded, somewhat unequal, 2.5 - 5 × 2 - 4 cm (unfolded); **Fl** 1.5 - 2.5 cm ∅, purplish-pink; **Ped** 1 - 1.5 cm; **Sep** and **Pet** free; **St** glabrous.

T. spathacea Swartz (Prodr., 57, 1788). **T:** Jamaica [cult.] (*Swartz* s.n. [S]). – **Lit:** Stearn (1957). **D:** S Mexico, Belize, Guatemala (Petén).

≡ *Rhoeo spathacea* (Swartz) Stearn (1957); **incl.** *Tradescantia discolor* L'Héritier (1788) ≡ *Rhoeo discolor* (L'Héritier) Hance *ex* Walpers (1853).

[2] Erect bromeliad-like herbs; stem usually short, rarely to 1 m; **L** 20 - 35 × 3 - 5.5 cm, imbricate, subrosulate, narrowly linear-lanceolate, acuminate, scarcely narrowed above the sheath, succulent, glabrous, green on both sides or more commonly dark bluish-green above and purple beneath; **Inf** axillary; peduncle 2 - 4.5 cm, simple or branching; **Bra** 2 - 4.5 × 2.5 - 5 cm, deeply boat-shaped, broadly ovate; **Fl** numerous, 1 - 1.5 cm ∅, white, scarcely exserted; **Ped** ± 1.5 cm, recurved in **Fr**; **Sep**, **Pet** and **St** free; **Fr** capsules; **Se** 1 per locule.

The predominance of the form with purple pigmentation, and the common occurrence of the plant around ancient Maya sites, may be due to the use of the plant as source of a cosmetic decoction (Standley & Steyermark 1952: 23). The species is widely cultivated, world-wide, as an ornamental. 'Vittata' ('Variegata') is a cultivar with attractive leaves, cream-striped above and purple below. 'Concolor' has the leaves green throughout.

T. zebrina Heynhold (Alph. Aufz. Gew. 735, 1847). – **D:** S Mexico (?, not definitely known in the wild state).

Incl. *Zebrina pendula* Schnizlein (1849); **incl.** *Zebrina purpusii* Brückner (1928).

[1] Stems decumbent or creeping, rooting at the nodes; **L** 2.5 - 10 × 1.5 - 3.5 cm, ovate-oblong to broadly ovate, acute, rounded at the base, somewhat succulent, green and/or purple above, often striped with silver, usually purplish beneath; **Inf** solitary, terminal or opposite the **L**; peduncle 1.5 - 11 cm; **Bra** 2, unequal, the outer larger, 1.5 - 6 cm, the inner 0.8 - 3 cm, usually glabrous except for a conspicuous band of **Ha** on the sheath; **Ped** to 3 mm; **Sep** 5 - 8 mm, hyaline, connivent, or the posterior ± free; **Pet** united below into a slender tube to 10 × 1.3 mm, white, lobes free, 5 - 10 × 3 - 7 mm, ovate, purplish pink or violet-blue; **St** epipetalous, bearded below.

This species is very commonly cultivated and is naturalized in many warm countries.

TRIPOGANDRA

D. R. Hunt

Tripogandra Rafinesque (Fl. Tellur. 2(2): 16, 1837). **T:** *Tradescantia multiflora* Swartz. – **Lit:** Handlos (1975). **D:** Tropical America. **Etym:** Gr. 'tri-', three; Gr. 'pogon', beard; and Gr. 'aner, andros', man, [bot.] anther; because the type species has 3 bearded and 3 glabrous stamens.

Incl. *Neodonnellia* Rose (1906). **T:** *Callisia grandiflora* J. D. Smith.

Annual or perennial herbs, erect or trailing, rarely scandent; tubers absent; **L** ovate, narrowly ovate, oblong-lanceolate or rarely linear; **Inf** bifacially paired and fused cincinni not subtended by evident **Bra** but borne on a common peduncle, the paired cincinni solitary or variously clustered; **Fl** bisexual, zygomorphic; **Sep** 3, free; **Pet** 3, free, white or pink; **St** 6, dimorphic; outer whorl usually fertile, shorter, **Fil** glabrous or bearded, **Anth**-connectives narrow; inner whorl fertile or staminodal, longer, **Fil** curved to be erect in front of the upper **Pet**, glabrous or variously bearded, **Anth**-connectives various; **Ov** 3-locular; ovules 2 per locule; **Sti** capitellate or capitate; **Fr** capsules; **Se** with a punctiform (dot-like) to linear hilum and dorsal embryotega.

A genus of 22 species.

T. glandulosa (Seubert) Rohweder (Abh. Auslandsk., Reihe C, Naturwiss. 61(C18): 156, 1956). **T:** Brazil, Paraná (*Sello* 995 [B]). – **D:** Trinidad?, Brazil, Paraguay, Uruguay, N Argentina.

≡ *Tradescantia glandulosa* Seubert (1855); **incl.** *Tradescantia radiata* C. B. Clarke (1903) ≡ *Tripogandra radiata* (C. B. Clarke) Bacigalupo (1964).

Perennials, decumbent and rooting at the nodes, **Fl** stems erect to 40 cm; **L** ovate, acute, rounded at the base, to 8 × 3 cm, glabrous except towards base of midrib below and at mouth of sheath; **Inf** terminal and in the uppermost **L**-axils; peduncles to 1 - 5 cm; cincinni with up to 5 - 6 **Fl**; **Ped** 4 - 5 mm, glandular-pilose; **Sep** ovate, 3 - 4 × 1.2 - 2 mm, glandular-pilose; **Pet** ovate, 3.5 - 5 × 2.5 - 3.5 mm, white or pink; outer **St** fertile, **Fil** to 1.5 mm, bearded; inner **St** fertile, **Fil** to 3.5 mm, glabrous, dilated below tip; **Ov** globose, to 1 mm ∅; **Sty** very short; **Sti** simple or capitellate; **Fr** globose capsules to 2.7 × 3.1 mm; **Se** rounded-triangular, 0.8 - 1.4 mm, light grey to brown, ribbed, reticulate-foveate, hilum punctiform.

T. grandiflora (J. D. Smith) Woodson (Ann. Missouri Bot. Gard. 29: 153, 1942). **T:** Guatemala (*Tuerckheim* 7864 [US]). – **D:** S Mexico, Belize, Guatemala.

≡ *Callisia grandiflora* J. D. Smith (1901).

Subscandent perennials to 3 m; **L** to 15 × 4.5 cm, distichous, narrowly ovate to elliptic, acute, oblique at the base, glabrous; **Inf** paniculate; cincinni to 7-

flowered; **Ped** to 1.4 cm, glabrous, erect in **Fr**; **Fl** fragrant, to 2 cm $\varnothing$; **Sep** dull mauve; **Pet** white; outer **St** fertile, **Anth** blackish; inner **St** staminodal, S-shaped, with 2 tufts of whitish **Ha** in the upper ½, **Anth** orange.

T. ionantha (Diels) Macbride (Revista Univ. (Cuzco) 33(87): 142, 1945). **T:** Peru, Puno (*Weberbauer* 588 [B]). – **D:** Colombia, Peru, Bolivia; mountains.
≡ *Tradescantia ionantha* Diels (1906).

Succulent perennials; flowering stems erect to 35 cm, often containing a purplish pigment; **L** clustered at the stem-base, scattered above, 1.5 - 3.8 × 1 - 1.5 cm, rather fleshy, slightly hairy above, more so below; **Inf** ± closely subtended by 1 - 3 upper **L**, of which 1 - 2 are folded in a manner reminiscent of *Callisia navicularis*; **Fl** 5 - 6 mm $\varnothing$; **Sep** 2.5 × 2 mm, hairy; **Pet** 2 - 3 mm, lilac to deep magentapurple; longer **St** 3 mm, bearded, shorter **St** 1.5 mm.

Handlos (1975: 284) regarded this plant as only a mountain ecotype of *T. multiflora*, as only the flower colour, always bright purple or magenta, appeared to him to separate it from the bulk of *T. multiflora*. I have not found this difficulty myself, since the plant is also much more succulent, with shiny, often folded, leaves, giving living plants and even herbarium specimens a distinct facies.

T. multiflora (Swartz) Rafinesque (Fl. Tellur. 2: 16, 1837). **T:** Jamaica (*Swartz* s.n. [B?, M]). – **D:** Tropical America (Jamaica, Trinidad and Tobago, Costa Rica, Venezuela, Colombia, Peru, Bolivia).
≡ *Tradescantia multiflora* Swartz (1788).

Perennials, trailing and rooting at the nodes; **Fl** stems to 80 cm, erect; **L** narrowly to broadly ovate, usually glabrous, acute, to 8 × 3 cm, oblique at the base; **Inf** numerous, terminal and in the upper **L**-axils; **Fl** 3 - 8 mm $\varnothing$, white or pink; **Sep** often red-spotted at the base; outer **St** very short, glabrous, inner **St** longer, S-shaped, bearded.

Dioscoreaceae

Dioecious or rarely monoecious herbaceous perennial climbers, rarely ± woody, rarely spiny, often completely deciduous tropophytes from underground or above-ground tubers (sometimes with fissured corky bark) or rhizomes; sometimes with aerial tubers from **L** axils; **L** alternate or rarely opposite, petiole mostly well-developed, lamina entire or (palmately) lobed or more often heart-shaped, with few to many curved-ascending and again converging main veins; **Inf** paniculate or spicate from **L** axils; **Fl** small and often greenish, almost always unisexual, regular; **Tep** 6 in 2 series, basally usually united, often with **Nec**; **male Fl** with 6 **St** in 2 whorls (sometimes 1 whorl absent or as staminodes); **Anth** longitudinally dehiscent; **female Fl** without staminodes; **Ov** inferior of 3 united **Ca** with normally 2 ovules per locule, placentation mostly axile; **Sty** 3 or 1; **Sti** 3; **Fr** often 3-winged capsules, or fleshy berries; **Se** often winged.

Distribution: Tropics worldwide, with a few taxa (incl. *Tamus*, "Black Bryony") extending into N temperate regions.

Literature: Knuth (1924); Huber (1998).

The family has only ± 3 genera with about 600 species of which 95% belong to *Dioscorea*. Apart from several species of *Dioscorea* which are important food plants, the family is notable for its diverse phytochemistry (primarily lactone alkaloids and steroidal saponins). Underground storage tubers (derived from lowermost internodes and/or the hypocotyl) are present in *Tamus* ("Black Bryony") and many species of *Dioscorea*, but only few taxa of the latter can be regarded as succulents and are covered below.　　　　　　　　　　　[U. Eggli & G. D. Rowley]

DIOSCOREA

G. D. Rowley

Dioscorea Linné (Spec. Pl. [ed. 1], 1032, 1753). **T:** *Dioscorea sativa* Linné [Lectotype, designated by McNeill & al., Taxon 36: 368, 1987; and proposed for conservation by Jarvis, Taxon 41: 562, 1992.]. – **Lit:** Burkill (1952); Archibald (1967). **D:** Pantropical. **Etym:** For Pedanios Dioscorides, most influential Greek physician and herbalist of the first century A.D.
Incl. *Testudinaria* Salisbury (1824). **T:** *Tamus elephantipes* L'Héritier.

Dioecious perennials or very rarely annuals, typically with long weak annual stems twining anticlockwise from underground storage organs (swollen rhizomes, potato-like tubers or caudices derived from an enlarged epicotyl), rarely with aboveground caudices with strongly fissured and structured bark, or rarely with persistent woody or succulent stems; stems sometimes spiny, sometimes with aerial tubers from **L** axils; **L** alternate, usually simple and cordate with a well-developed petiole having a jointed pulvinate base; stipules absent; **Inf** panicles, racemes or spikes, axillary; **Fl** unisexual (but sometimes with staminodes or pistillodium); actinomorphic, small and greenish-yellow or greenish-whitish, or purple; **Tep** 6 in 2 series, shortly united into a basal tube; **St** 6 in 2 whorls; **Ov** inferior of 3 **Ca**, each locule with 2 ovules; **Fr** 3-lobed or -winged capsules; **Se** large, flattened, with a symmetrical or 1-sided papery wing. – *Cytology:* n = 10.

The large genus with ± 600 species has a predominantly tropical distribution. The tubers have a diverse morphological origin, and many species are further notable for their extrafloral nectaries (esp. found on leaves). Some 10 % of its species are economically important, esp. for their starchy tubers ("Yams", with numerous cultivars of several ploidy levels of several species cultivated throughout the tropics) or are exploited as sources of steroidal hormones used as oral contraceptives (esp. the Mexican *D. floribunda*). The turnip-like flesh of *D. elephantipes* and *D. sylvatica* is occasionally cooked and eaten by the Bantu in S Africa. Similar use is reported for *D. remotiflora* Kunth from Mexico, whose tuberous roots are used locally and taste like potatoes when cooked.

Succulence has developed in the form of enlarged and fleshy caudices in several taxa from arid regions in Africa and Mexico (caudex normally ± above ground) and one from arid Chile (caudex underground). A notable addition to the succulent species is *D. basiclavicaulis* from NE Brazil with perennial thickened virgate prickly stems. Propagation is by seed except for the last-mentioned, which can also be multiplied by division.

Knuth (1924) recognizes 4 subgenera, of which 2 are relevant here:

[1] Subgen. *Dioscorea*: **Se** winged all round.

[2] Subgen. *Testudinaria* (Salisbury) Uline 1898: **Se** wing oblique or 1-sided.

D. basiclavicaulis Rizzini & Mattos-Filho (Revista Brasil. Biol. 46(2): 317-319, ill., 1986). **T:** Brazil, Minas Gerais (*Rizzini & Mattos-Filho s.n.* [RB 215.073]). – **D:** Brazil (Minas Gerais, Bahia); in Caatinga scrub. **Fig. XXVII.f**

[1?] **R** fascicled, tuberous-elongate; stems several from a common rhizomatous **R** stock, perennial, basal parts (0.5 - 1 m) fleshy-fibrous, narrowly fusiform to 8 cm ⌀, olive-green, terete or angled or ± longitudinally ridged, densely spiny, tip elongating whip-like, thin, to several m long, with pendent to semi-twining laxly spiny side **Br**; **L** narrowly triangular, to 11 × 6 cm, base auriculate, tip long-acuminate, thin-textured, grass-green, young flushed reddish-brown; petiole to 4 cm; **Inf** axillary, spic-

ate, pendent; **Fl** 0.5 - 2 cm apart, greenish, 8 - 10 mm ⌀; **Fr** 3.5 - 4.5 cm, alate; **Se** not described.

Unique in the large genus for its thickened succulent perennial stems. Propagated by division, probably also by rooting sections of individual stems. The original publication notes the surprising similarity of the stems with those of *Smilax papyracea* Poiret (*Smilacaceae*), also native to Brazil. – [U. Eggli]

D. elephantipes (L'Héritier) Engler (Veg. Erde 9(2): 267, 1908). – **D:** RSA (Eastern Cape). **I:** Rowley (1987: 2, 11, 43-45, 47). **Fig. XXVII.a, XXVII.d**

≡ *Tamus elephantipes* L'Héritier (1788) ≡ *Testudinaria elephantipes* (L'Héritier) Lindley (1825); **incl.** *Testudinaria montana* Burchell (1824) ≡ *Dioscorea montana* (Burchell) Sprengel (s.a.) ≡ *Testudinaria elephantipes* var. *montana* (Burchell) G. D. Rowley (1953) ≡ *Testudinaria elephantipes* fa. *montana* (Burchell) G. D. Rowley (1973); **incl.** *Dioscorea elephantopus* Sprengel (1827).

[2] Caudex ± completely above ground with marginal fibrous **R**, massive, 60 (-100) × 60 cm in age with 1 or more irregular domed growing points, covered in hard grey corky bark divided into irregularly polygonal (4- to 7-sided) plates 8 × 3 - 6 cm; stems few, stiffly erect, basally woody, **Sp**-less, partly deciduous, with spirally set horizontal **Br** that twine at the tips; **L** broadly ovate to reniform with acute tip, 2 - 2.5 × 2 - 6 cm, grass-green; petiole 0.5 - 1 cm; **Fl** ± 4 mm ⌀; **Fr** to 2 × 1.8 cm, obovoid, with 3 wings; **Se** lenticular, 5 mm ⌀ with a one-sided wing 10 × 7 mm, dark brown to black.

Unique for the massive tubercled caudex and branch system in which, as Darwin noted, only the lateral shoots twine. The characteristic polygonally fissured bark starts to develop after some years; caudices of seedlings are ± smooth. The tubers are said to be edible and have been used as emergency food. Vernacular names: "Elephant's Foot", "Turtleback Plant", "Hottentot Bread".

D. fastigiata Gay (Fl. Chil. 6: 54, 1853). **T:** Chile, Atacama (*Gay s.n.* [P]). – **D:** Chile (from Antofagasta to Talca); dry coastal regions, often in sand. **I:** Navas B. & Erba V. (1968: t. 2: a-c). **Fig. XXVII.b**

Incl. *Dioscorea axilliflora* Philippi (1896); **incl.** *Dioscorea gayi* Philippi (1896); **incl.** *Dioscorea geissei* Philippi (1896); **incl.** *Dioscorea paupera* Philippi (1896); **incl.** *Dioscorea thinophila* Philippi (1896); **incl.** *Dioscorea cylindrostachya* I. M. Johnston (1929).

[1] Caudex underground, globose with flat sometimes lobed bottom, 2 - 4 cm ⌀, bark pale brown, smooth or slightly fissured; stem 1 to few, completely deciduous, creeping, contracted, 5 - 20 cm, herbaceous, spineless; **L** few, ± cordate, to 2 × 2 cm, dark green, slightly thickish; petiole to 2 cm; **Inf**

paniculate, 1 - 2 cm; **male Fl** numerous in fascicles, 1.5 mm ∅, whitish, **St** 6; **female Fl** 3 - 4 only; **Fr** roundish to obovate, 1 - 1.5 cm ∅.

Monoecious specimens are reported for this species. – [U. Eggli]

D. hemicrypta Burkill (JSAB 18: 187, 1952). – **D:** RSA (Western Cape). **I:** Rowley (1987: 48-49). **Fig. XXVII.e**

Incl. *Testudinaria glauca* Marloth (s.a.) (*nom. inval.*, Art. 32).

[2] Caudex half-underground, dome-shaped and taller than wide, underground part amorphously lobed like molten lava, above-ground part with thick corky irregular ridges and plates; stems few, ± completely deciduous or basally woody, **Sp**-less, side **Br** only twining; **L** to 6 × 4 cm, always longer than broad, bluish-green; **Fl** to 6 mm ∅; otherwise similar to *D. elephantipes*.

D. mexicana Scheidweiler (Hort. Belge 4: 99, 1837). – **D:** S Mexico; extending to C America (El Salvador, Panama). **I:** Rowley (1987: 49, as *D. macrostachya*).

Incl. *Dioscorea macrostachya* Bentham (1841) ≡ *Testudinaria macrostachya* (Bentham) G. D. Rowley (1973); **incl.** *Dioscorea macrophylla* Martius & Galeotti (1842); **incl.** *Dioscorea deppei* Schiede *ex* Schlechtendal (1843); **incl.** *Dioscorea billbergiana* Kunth (1850); **incl.** *Dioscorea leiboldiana* Kunth (1850); **incl.** *Dioscorea composita* Hemsley (1884); **incl.** *Testudinaria cocolmeca* Procopp (1892); **incl.** *Dioscorea astrostigma* Uline (1896); **incl.** *Dioscorea macrostachya* var. *sessiliflora* Uline (1897) ≡ *Dioscorea mexicana* var. *sessiliflora* (Uline) Matuda (1954); **incl.** *Dioscorea palmeri* R. Knuth (1917) ≡ *Dioscorea macrostachya* var. *palmeri* (R. Knuth) Morton (1936).

[1] Caudex ± half-underground, semiglobose, to 25 cm ∅ or more, upper surface with polygonally fissured corky bark much like *D. elephantipes*; stems few, to 6 m, ± completely deciduous, twining, spineless or with **Sp** to 1 cm from persistent petiole bases; **L** broadly rounded-cordate, long-acuminate, 10 - 17 × 15 cm, much smaller on flowering shoots, long petiolate, bright green; **male Inf** narrow spikes to 25 cm, sometimes fascicled; **female Inf** always simple; **Fl** in groups of 2 - 5, dark purple, 2.5 - 3.5 mm ∅; **Fr** oblong-elliptic; **Se** 7 - 8 mm overall including 2 ± equal wings.

Despite the wide distribution ascribed to this species, it is poorly known, esp. as to its variability. – [G. D. Rowley & U. Eggli]

D. sylvatica Ecklon (South Afr. Quart. J. 1: 363, 1830). – **D:** S Zimbabwe, RSA; widespread among trees. **I:** Rowley (1987: 48). **Fig. XXVII.g**

≡ *Testudinaria sylvatica* (Ecklon) Kunth (1850); **incl.** *Dioscorea hederifolia* Grisebach (1842); **incl.** *Tamus sylvestris* Kunth (1850); **incl.** *Testudinaria*

sylvestris Kunth (1850); **incl.** *Dioscorea montana* var. *glauca* R. Knuth (1924); **incl.** *Dioscorea sylvatica* ssp. *lydenbergensis* Blunden & al. (1971) ≡ *Testudinaria sylvatica* var. *lydenbergensis* (Blunden & al.) G. D. Rowley (1974).

[2] Caudex variable, above or below ground, conical, slab-like or obconical, often amorphously lobed like lava, 30 - 60 (-100) cm ∅; stem 1, thin, wiry, spineless, twining, branched above; **L** triangular-cordate, 2 - 8 × 1.5 - 6 cm, green, petiole to 2.5 cm; **Fl** 5 mm ∅; **Fr** to 2.7 × 1.5 cm, obovoid; **Se** 7 × 4 mm with a 1-sided wing to 16 × 6 mm.

Readily propagated when caudex lobes can be divided. This is a variable species which can be divided into a number of varieties. See Burkill (1952) for information on the complicated nomenclatural status of *D. montana* and its var. *glauca*, the latter here included provisionally as synonym.

D. sylvatica var. **brevipes** (Burtt Davy) Burkill (JSAB 18: 189, 1952). – **D:** S Zimbabwe, RSA (Cape Prov., KwaZulu-Natal, Free State, Northern Prov.).

≡ *Dioscorea brevipes* Burtt Davy (1924) ≡ *Testudinaria paniculata* var. *brevipes* (Burtt Davy) G. D. Rowley (1953) ≡ *Testudinaria sylvatica* var. *brevipes* (Burtt Davy) G. D. Rowley (1973).

[2] **L** 6 - 8 cm long with divergent auricles; **Ped** to 2 mm; **Fr** ± 2 cm long.

D. sylvatica var. **multiflora** (Marloth) Burkill (JSAB 18: 189, 1952). – **D:** RSA (Northern Prov., Mpumalanga).

≡ *Testudinaria multiflora* Marloth (1913) ≡ *Testudinaria sylvatica* var. *multiflora* (Marloth) G. D. Rowley (1973); **incl.** *Dioscorea marlothii* R. Knuth (1924).

[2] **L** thin, without divergent auricles; **Fr** 1.2 - 1.4 cm long.

D. sylvatica var. **paniculata** (Dummer) Burkill (JSAB 18: 189, 1952). – **D:** RSA (Cape Prov., KwaZulu-Natal).

≡ *Testudinaria paniculata* Dummer (1912) ≡ *Testudinaria sylvatica* var. *paniculata* (Dummer) G. D. Rowley (1973); **incl.** *Testudinaria montana* var. *paniculata* Kuntze (1898) ≡ *Dioscorea montana* var. *paniculata* (Kuntze) R. Knuth (1924); **incl.** *Dioscorea paniculata* Dummer (1912); **incl.** *Dioscorea montana* var. *duemmeri* R. Knuth (1924).

[2] Caudex domed or slab-like, often convex below; **L** 6 - 8 cm long, reniform with divergent auricles; **Ped** to 4 mm; **Fr** ± 2 cm long.

D. sylvatica var. **rehmannii** (Baker) Burkill (JSAB 18: 189, 1952). – **D:** RSA (KwaZulu-Natal, Free State, Northern Prov.).

≡ *Dioscorea rehmannii* Baker (1896) ≡ *Testudinaria rehmannii* (Baker) G. D. Rowley (1953) ≡ *Testudinaria sylvatica* var. *rehmannii* (Baker) G. D. Rowley (1973).

[2] **L** thin, with convergent auricles; **Fr** 2 - 2.5 cm long.

D. sylvatica var. **sylvatica** – **D:** RSA (Eastern Cape).
[2] **L** 2 × 4 cm with divergent auricles; **Fr** ± 1.5 cm long.

Doryanthaceae

Giant monocarpic **Ros** plants; stems very short; **R** thick, fleshy; **L** radical, numerous, flat, ± succulent, with a brown tubular tip withering to leave a ragged apex; **Inf** large scapose terminal oblong thyrses or globular compound racemes, subtended by numerous sheathing **Bra**; **Fl** bisexual, actinomorphic, 3-merous, each subtended by a conspicuous **Bra**; **Per** segments united at the base into a tube, free parts spreading; **St** 6, **Fil** linear-subulate, somewhat enlarged at the base, adnate to the **Per** for ± ½ of their length, inserted into a pit at the base of the **Anth**; **Anth** longitudinally dehiscent, bilocular; **Ov** inferior, 3-locular; ovules 40 - 50 per locule; **Sty** 1, **Sti** 3-angled; **Fr** large ellipsoidal to ovoid loculicidal capsules 7 - 10 cm with short apical beak 6 - 7 mm; **Se** winged, flattened.

Distribution: Endemic to Australia.

Literature: Clifford (1998a).

Embracing a single genus with 2 species only. Traditionally included in *Agavaceae*, *Amaryllidaceae* or *Liliaceae* s.l.   [P. I. Forster]

DORYANTHES

P. I. Forster

Doryanthes M. P. Corrêa (Trans. Linn. Soc. London 6: 211, 1802). **T:** *Doryanthes excelsa* M. P. Corrêa. – **Etym:** Gr. 'dory', wood, trunk, lance, spear; and Gr. 'anthos', flower.
Description as for the family.

D. excelsa M. P. Corrêa (Trans. Linn. Soc. London 6: 211, 1802). **T:** Australia, New South Wales (*Bass* s.n. [not traced]). – **D:** Australia (New South Wales). **I:** Elliot & Jones (1984: 330).
Incl. *Furcraea australis* Haworth (1812) ≡ *Agave australis* (Haworth) Steudel (1821).
Ros to 2 m tall and 4 m ∅; **L** numerous, spreading and recurved, ensiform, 1.5 - 2.5 m × 8 - 10 cm, tip 6 - 7 × 0.7 - 1.5 cm, brown; **Inf** scape 3 - 5 m; **Fl** 10 - 16 × 1.5 - 2 cm ∅, red, pink-red, or white; **Ped** 4 - 5 cm; **Sep** lanceolate-oblong, cucullate, 6 - 12 × 0.6 - 0.9 cm; **Pet** lanceolate-oblong, 6 - 12 × 0.6 - 0.9 cm, recurved; **Fil** 54 - 72 mm; **Fr** 4.5 - 5 cm ∅; **Se** 15 - 23 × 12 - 13 mm.
Widely cultivated in gardens and parks in Australia and elsewhere in temperate climates. Vernacular names (Australia): "Gymea Lily", "Giant Lily".

D. palmeri W. Hill *ex* Bentham (Fl. Austral. 6: 452, 1873). **T:** K, MEL. – **D:** Australia (Queensland, New South Wales). **I:** Pedley (1986: 87); CSJA 67: 344-345, 1995. **Fig. XXVII.h**
≡ *Doryanthes excelsa* var. *palmeri* (W. Hill *ex*

Bentham) F. M. Bailey (1883); **incl.** *Doryanthes larkinii* C. Moore (1885) ≡ *Doryanthes palmeri* var. *larkinii* (C. Moore) C. Moore & E. Betche (1893); **incl.** *Doryanthes guilfoylei* F. M. Bailey (1893) ≡ *Doryanthes excelsa* var. *guilfoylei* (F. M. Bailey) F. M. Bailey (1902).

Ros to 2.5 m tall and 3 m ∅; **L** numerous, spreading and recurved, ensiform, 2 - 3 m × 15 - 20 cm, tip acuminate-tubular, 8 - 12 × 1.5 - 3 cm, brown; **Inf** scape 2 - 5 m; **Fl** 5 - 5.5 × 4 - 5 cm ∅, scarlet-red to red-brown; **Ped** 5 - 5.7 cm; **Sep** lanceolate-oblong, cucullate, 58 - 62 × 14 - 15 mm; **Pet** lanceolate-ovate, 50 - 53 × 14 - 15 mm, erect to slightly spreading; **Fil** 25 - 35 mm; **Fr** 5 - 6 cm ∅; **Se** 15 - 17 × 10 - 11 mm wide.

Occasionally cultivated in gardens and parks, esp. in Australia. Vernacular name (Australia): "Spear Lily".

Dracaenaceae

Herbs to large trees; stems occasionally absent, or woody and partly or wholly subterranean and rhizomatous, or more rarely pachycaul and enormous; **L** often clustered at **Br** tips or tips of subterranean rhizomes, spirally arranged or occasionally distichous, simple, narrowly linear to ovate and sessile, sometimes conspicuously succulent and terete, always fibrous, venation parallel (and branching in *Cordyline*); **Inf** racemes or panicles, axillary and pedunculate, emerging either from **Ros** near the ground or on the ends of **Br**; **Fl** small but numerous, 3-merous, pedicellate, generally highly fragrant; **Tep** 3 + 3, elongate, all equal, usually basally connate and tubular with free tips, brownish, purple-violet or white; **St** 3 + 3; **Sty** often long and simple; **Sti** trilobate or capitate; **Ov** 3-locular; **Fr** globose berries, red or orange; **Se** globose or elongate.

Distribution: Subtropical and tropical Africa, Asia, Australasia but rare in the American tropics; centre of distribution in Africa.

Literature: Dahlgren & al. (1985); Bos (1998).

This is a very small family with ± 150 species in 3 genera (all treated below). The family placement of these genera has been in a state of flux and until recently they had a home in the *Agavaceae*. The recognition of the family *Dracaenaceae* as presented here is probably no more than an interim measure. *Dracaena* and *Sansevieria* are undoubtedly closely related and according to Bos (1984) they are so close as to be congeneric. *Cordyline* is more distantly related to *Dracaena* (see notes under *Cordyline* for details) and may not even belong in the same family. True leaf succulence is found only in *Sansevieria*; the other 2 genera are better described as woody xerophytes, with the pachycaul habit exhibited in a very few species.

A number of species are widely cultivated ornamentals, often as variegated cultivars, most notably of *Sansevieria trifasciata*, the "Mother-in-Law's Tongue", which is one of the most commonly grown houseplants because of its tolerance of dry and shady conditions. Other common cultivars occur also in *Dracaena* and *Cordyline*. Most species have fibrous leaves and a few species have been grown commercially as fibre sources. Resin from the pachycaul species of *Dracaena* has had various uses ranging from varnishes to medicines. Additionally, *Cordyline* has been a source of high fructose syrup.

Key to the genera:
1 **L** usually succulent; stems short, often horizontal, subterranean and rhizomatous, or absent:
                                           Sansevieria

– **L** never truly succulent, often narrowly linear-lanceolate, and clustered at stem tips; stems woody, occasionally massively pachycaul: **2**

2 **L** venation parallel and branching; **Ped** with 3 basal **Bra**: **Cordyline**

– **L** venation exclusively parallel; **Ped** without basal **Bra**: **Dracaena**

[C. C. Walker]

CORDYLINE

C. C. Walker

Cordyline Commerson *ex* Jussieu (Gen. Pl., 41, 1789). **T:** *Asparagus terminalis* Linné [fide ING]. – **Lit:** Pedley (1986: 81-86). **D:** SE Asia, Australasia, tropical America. **Etym:** Gr. 'kordyle', club, pestle; for the club-like roots of some taxa.
Incl. *Taetsia* Medikus (1786) (*nom. illeg.*, Art. 56.1). **T:** *Dracaena ferrea* Linné.
Incl. *Charlwoodia* Sweet (1827). **T:** *Charlwoodia congesta* Sweet.
Incl. *Calodracon* Planchon (1850). **T:** not typified.

Woody perennials, often tree-like, overall very similar to *Dracaena*, but with creeping rhizomes; **L** venation parallel and branching (parallel only in *Dracaena*); **Ped** with 3 **Bra** at the base (none in *Dracaena*); **Ov** with 3 locules, each locule with 6 or more ovules (1 in *Dracaena*).

Succulence is very borderline and the leaves are merely tough and fibrous. Some species have massive pachycaul stems. The genus has ± 20 species. It was recently classified in the small segregate family *Lomandraceae* and was previously also placed in *Asteliaceae*.

C. australis (G. Forster) Hooker *fil.* (Gard. Chron. 1860: 792, 1860). – **D:** New Zealand. **I:** Everett (1981-1982: 3: 866). **Fig. XXVIII.d**
≡ *Dracaena australis* G. Forster (1786).

Trees to 20 m, sparingly branched below, well branched above; **L** clustered at **Br** tips, lanceolate-linear, arching, 30 - 100 × 3 - 6 cm; **Inf** paniculate, much branched, to 1.5 m; **Fl** numerous, creamy-white or bluish, fragrant, 5 - 6 mm; **Fr** globose berries, white or pale blue, 4 mm ⌀.

Vernacular name: "New Zealand Cabbage Tree". Many variegated cultivars are available.

DRACAENA

C. C. Walker

Dracaena Vandelli *ex* Linné (Syst. Nat., ed. 12, 2: 246, 1767). **T:** *Asparagus draco* Linné. – **Lit:** Bos (1984); Marrero & al. (1998); Walker (1999). **D:** Predominantly tropical Africa, Macaronesia, S Arabia, Socotra, Madagascar, SE Asia, C America and West Indies. **Etym:** Lat., female dragon (from Gr. 'drakon', dragon); from the vernacular name of *D.*

draco, "Dragon's Blood Tree", which is based on the red exudate of the bruised stems.
Incl. *Draco* Crantz (1768). **T:** not typified.
Incl. *Pleomele* Salisbury (1796). **T:** not typified.
Incl. *Nemampsis* Rafinesque (1838). **T:** *Nemampsis ternifolia* Rafinesque [*nom. illeg.*, = *Dracaena surculosa* Lindley, fide ING].

Woody shrubs or trees, sometimes with massive trunks to 40 m tall; bark smooth, often with prominent **L** scars; **R** usually bright orange; **L** usually spirally arranged, often in dense **Ros** at **Br** tips, tough and fibrous to coriaceous, smooth, often variegated, sessile or petiolate, ensiform or oblong, entire; **Inf** terminal, racemose, often paniculate, **Ped** articulated at the middle; **Fl** 1 to few in fascicles, nocturnal and usually highly fragrant; **Per** to 5 cm long, white or greenish with purple tinges, basally tubular with 6 free lobes; **St** 6, inserted at the throat of the **Per** tube; **Sti** capitate; **Ov** superior, ovoid, sessile, 3-locular; **Fr** globose, baccate, fleshy; **Se** 1 - 3, testa thick, sometimes pulpy.

A genus of ± 60 species, but apparently with hundreds of redundant synonyms. According to Bos (1984), *Dracaena* is close to and possibly not separable from *Sansevieria*. Several species are extremely widely cultivated as houseplants, particularly attractive as variegated cultivars. The genus is of minor interest to succulent plant growers since small plants of the few xerophytic species are relatively uninspiring and slow-growing, in contrast to the impressive bulky specimens of mature trees. Only a few species with pachycaul stems of limited interest are covered here, and of these, only *D. draco* is frequently cultivated. *D. draco* and *D. cinnabari* are the "Dragon's Blood Trees", the dried red resinous sap (used in varnishes etc.) of which was of economic importance and at one time Socotra's major export item.

D. cinnabari Balfour *fil.* (Trans. Roy. Soc. Edinburgh 30: 623, 1882). – **D:** Socotra; endemic. **I:** NCSJ 26: 110, 1971; Palmengarten 59: 140-145, 1995. **Fig. XXVIII.c**

Trees with stout trunks to 10 m, dichotomously branched with regular semiglobose crown; **L** 30 - 60 × 2 - 3 cm, erect, rigid, ensiform with broadened base, sessile; **Inf** paniculate, well-branched; **Fl** in groups of 2 - 4; **Ped** 5 mm; **Per** to 5 mm long, cup-shaped; **St** slightly shorter than the **Per**; **Sty** filiform; **Sti** capitate.

This species belongs to a distinct group within the genus. Possibly all except *D. draco* comprise a single species complex. They are all thick-stemmed trees with dichotomous branching. Leaves are crowded into dense rosettes at branch tips. *D. draco* differs from *D. cinnabari* in its compressed ensiform leaves, smaller bracteoles, greenish perianth segments, and shorter anthers. *D. ombet* and *D. serrulata* have a less robust habit, more slender panicle branches, and longer pedicels.

259

D. draco (Linné) Linné (Syst. Nat., ed. 12, 2: 246, 1767). – **D:** Canary Islands, Madeira, Cape Verde Islands, Morocco.

≡ *Asparagus draco* Linné (1762); **incl.** *Yucca draco* Carrière (1859).

D. draco ssp. **ajgal** Benabid & Cuzin (Compt. Rend. Acad. Sci. Paris, Sér. 3, Sci. Vie 320: 270, 1997). **T:** Morocco (*Benabid & Cuzin* s.n. [RAB]). – **Lit:** Audissou (1999). **D:** Morocco; mixed open woodland, quartzite cliffs in gorges. **I:** Pl. Talk 12: 18, 1998.

Differs from ssp. *draco*: **L** smaller, 60 × 3 cm; **Ped** of **Fl** shorter, 1 - 4 mm; **Per** yellowish-white, tube campanulate, shorter, 1 - 2 mm; **Tep** shorter, 7 - 8 mm; **Anth** yellow.

This taxon was only discovered in 1996 as a population consisting of thousands of trees on quartzite cliffs in inaccessible gorges in the W part of the Anti-Atlas mountains in Morocco E of Tiznit. This discovery is highly significant, in view of the severely endangered status of ssp. *draco* in the Canary Islands.

D. draco ssp. **draco** – **D:** Canary Islands (Tenerife, Gran Canaria, La Palma), Madeira, Cape Verde Islands. **I:** Bramwell & Bramwell (1974: t. 322); Palmengarten 59: 70-74, 1995.

Trees to ± 20 m tall; trunk silvery-grey, smooth; **Br** dichotomously branched forming a semiglobose crown; **L** linear-lanceolate, sessile, glaucous, coriaceous, to 1.1 m × 4 cm, in dense terminal **Ros**; **Inf** paniculate; **Fl** in groups of 4 - 5; **Ped** 5 - 10 mm; **Per** white, pink, crimson to greenish-white, tube campanulate, 1.5 - 4 mm; **Tep** 7 - 11 mm; **Anth** greenish; **Fr** globose berries, to 1.5 cm ∅, red-orange.

This is the "Dragon Tree" of the Canary Islands, now extensively cultivated in gardens in frost-free climates, but endangered in the wild. Slow growing as a pot plant. Mature specimens are presumed to be of great age and a famous specimen on Tenerife, blown down by a storm in 1868, was 21 m tall and 15 m ∅ at the base of trunk, and was estimated to be 6,000 years old. Mägdefrau (1975) has shown, however, that an age of a few hundred years is much more resonable.

D. ellenbeckiana Engler (BJS 32: 95, 1903). **T:** Ethiopia, Harar Prov. (*Ellenbeck* 1232 [B]). – **D:** Ethiopia, Kenya, Uganda. **I:** Bally (1967: figs. 1-2).

Incl. *Dracaena ellenbeckii* hort. (s.a.) (*nom. inval.*, Art. 61.1).

Large shrubs 3 - 6 m tall, dichotomously branching to form many-stemmed clumps; **Br** 4 - 8 cm ∅; **L** 35 - 55 × 1 - 2.2 cm, often restricted to **Br** tips, ensiform, coriaceous, glabrous, entire, base amplexicaul, sessile; **Inf** terminal, paniculate; **Fl** in 2 - 7 cymes; **Ped** 3 - 5 mm; **Per** to 1 cm, **Tep** linear-

oblong, 10 × 1.8 - 2 mm; **St** slighter shorter than the **Per**; **Sty** cylindrical, 3 × 0.5 mm; **Sti** capitate; **Ov** longitudinally ribbed; **Fr** subglobose, 4 - 10 × 4 - 10 mm, turning from green over red to dark purple; **Se** globose, 4 - 5 mm ∅, ivory-white.

Vernacular name: "Ol Kedong". The Kedong Valley in Kenya is named after this locally abundant *Dracaena*. It is best described as a woody xerophyte rather than a succulent. The stems are used by the Masai in arrow-making.

D. ombet Kotschy & Peyritsch (Pl. Tinn., 47, 1867). – **D:** Sudan, Ethiopia, Eritrea, Djibouti; bushland and woodland, usually on limestone, 1000 - 1800 m. **I:** Thulin (1995: t. 1E).

Incl. *Dracaena schizantha* Baker (1877).

Trees to 8 m; **Br** thick, dichotomously branched, forming a semiglobose crown; **L** crowded, to 60 × 3 cm, often restricted to **Br** tips, rigid, glabrous, ensiform, margins smooth, entire, tip acute, base amplexicaul; **Inf** terminal, paniculate, much branched, glabrous or pubescent, to 0.5 m; **Fl** in 2 - 7 cymes; **Ped** glabrous or pubescent, 2 - 4 mm; **Per** to 1 cm, **Tep** linear-oblong, almost free, whitish, to 10 × 2 mm; **St** somewhat shorter than the **Per**; **Fil** flattened; **Sty** cylindrical; **Sti** capitate; **Ov** oblong, shortly stipitate; **Fr** subglobose berries, 10 - 12 mm ∅.

The Somali plant was previously named as *D. schizantha*, but Thulin (1995) was unable to separate these taxa. The red resin is used in traditional medicine.

D. rhabdophylla Chiovenda (Webbia 8(1): 12, fig. 2, 1952). **T:** Ethiopia (*Corradi* 4585 [FI]). – **D:** S Ethiopia.

Differs from *D. ellenbeckiana* in the **L** which are longer (to 70 cm), narrower (to 5 mm), more rigid, minutely scaberulous, not glabrous.

Described from a single sterile specimen. Requires collection of non-sterile material for confirmation of its specific status. Apparently not in cultivation.

D. serrulata Baker (BMI 1894: 342, 1894). – **D:** Yemen, Saudi Arabia, Oman. **I:** Miller & Morris (1988: 17); EJ 9: 54, 56, 1994. **Fig. XXVIII.a**

Trees to 5 m (but mostly smaller) with a single stem, branched above but without well-defined semiglobose crown; **L** up to 50 - 100 × 2 - 5 cm, linear-lanceolate, margins serrulate, tip acute; **Per** 4 mm long; **Fr** fleshy, globose, ± 6 mm ∅.

Similar to *D. cinnabari* (see there for details).

D. tamaranae A. Marrero & al. (Bot. J. Linn. Soc. 128: 294-297, ills., 1998). **T:** Canary Islands, Gran Canaria (*Marrero & al.* 18525 [MA, K, LPA, TFC]). – **D:** Canary islands (SW Gran Canaria); slopes and cliffs, 400 - 900 m.

Trees 6 - 10 m tall, bark yellow-grey, slightly

glossy; **Br** trichotomous; **L** subulate, canaliculate, rather falcate, glaucous, margins hyaline-white, entire, basally swollen with a subamplexicaul pseudosheath, 40 - 80 (-110) × 3 - 4.5 cm; **Inf** paniculate, 80 - 100 cm long; **Fl** in groups of 2 - 5 ; **Ped** 2.25 - 3.25 mm; bracteoles minute; **Per** bright greenish-white, 9.5 - 11 mm; **Tep** oblong-linear, **ITep** narrower than **OTep**, joined at the base to form a very short tube; **St** shorter than the **Tep**; **Anth** yellow-greenish, 2 mm; **Ov** trilocular, 3.6 × 2.4 mm; **Sty** filiform, 5.8 mm; **Sti** capitate, trilobulate; **Fr** globose berries, greenish, glaucous, orange when ripe; **Se** globose to broadly ovoid, 6 - 7 mm.

D. tamaranae seems closely related to the species from the Horn of Africa and Arabia (viz. *D. ombet* and *D. serrulata*). All these species have glaucous leaves, minute bracteoles and are not densely branched. In contrast to *D. tamaranae*, *D. draco* and *D. cinnabari* have flat, not glaucous, ensiform leaves and bipinnate robust inflorescences.

D. tamaranae is extremely rare and known from a few localities only, meriting critically endangered status.

SANSEVIERIA

L. E. Newton

Sansevieria Thunberg (Prodr. Fl. Cap., 65, 1794). **T:** *Aloe hyacintoides* Linné [≡ *Sanseverinia thyrsiflora* Petagna, *nom. illeg.*]. − **Lit:** Brown (1915). **D:** Africa, Arabian Peninsula (Yemen), Comoro Islands, India, Sri Lanka, Myanmar. **Etym:** For Count Pietro Antonio Sanseverino, Italian patron of horticulture in Naples around 1785.
Incl. *Acyntha* Medikus (1786) (*nomen rejiciendum*, Art. 56.1). **T:** *Aloe hyacinthoides* Linné.
Incl. *Sanseverinia* Petagna (1787) (*nomen rejiciendum*, Art. 56.1). **T:** *Sanseverinia thyrsiflora* Petagna.

Acaulescent or caulescent perennials, sometimes branching near base, with subterranean rhizomes or runners above ground; **L** solitary, few or many, distichous or in **Ros**, succulent or leathery, lanceolate, linear or lorate and flat, or cylindrical or semicylindrical and usually with a groove on the upper face, sessile, sometimes narrowed at the base to resemble a petiole, green, often with lighter blotches or transverse bands; **Inf** terminal, paniculate or simple and spike-like, sometimes capitate, dense or lax, with extrafloral **Nec** associated with the **Bra**; **Fl** solitary or in clusters, bracteate, pedicellate, actinomorphic, often nocturnal and sweetly scented; **Ped** articulated; **Tep** united at the base to form a tube with 6 free lobes that curl back at anthesis, mostly whitish; **St** 6, extended beyond the **Per** tube, exposed at anthesis by curling back of **Tep**; **Sty** simple, as long as the **St** or slightly longer; **Fr** berries with 1 - 3 **Se**. − *Cytology:* The few published

chromosome numbers suggest that there is some polyploidy, with base numbers of x = 20 or 21. Early counts were summarized by Darlington & Wylie (1955).

Seedlings of species with cylindrical leaves, as well as young plants raised from cuttings, have short flat leaves, and they look different from mature plants. In the descriptions below, leaf dimensions are length and width, plus thickness from front to back in the case of cylindrical and laterally compressed leaves. Although usually referred to as racemes, the inflorescences are spike-like compound structures with flower clusters arranged along an axis, and not opening from the base upwards.

Many of the species included in Brown's account were described from cultivated plants of undocumented wild origin, and in some cases even the country of origin was unknown.

Some species have been used at a local level as a source of cordage fibre, giving rise to such vernacular names as "Bowstring Hemp" and "African Sisal". Hybrids with improved fibre quality have been produced in the USA, see Joyner & al. (1951) and Wilson & al. (1962).

An infrageneric classification was suggested by Pfennig (1977a), but it was not presented formally with names for the groups. The following synopsis is based on Pfennig's scheme, with slight modification:
[1] **Inf** branching.
[2] **Inf** simple, capitate.
[3] **Inf** simple, spike-like.
   [3a] Plants with runners above ground.
   [3b] Plants with underground rhizomes; **L** cylindrical or semicylindrical, upper face often grooved.
   [3c] Plants with underground rhizomes; **L** flattened or folded.

The following names are of unresolved application but are referred to this genus: *Acyntha polyrhitis* Chiovenda (1932) ≡ *Sansevieria polyrhitis* (Chiovenda) Cufodontis (1971); *Sansevieria aubrytiana* Carrière (1861); *Sansevieria cylindrica* Schweinfurth (1894) (*nom. illeg.*, Art. 53.1); *Sansevieria cylindrica* Baum (1903) (*nom. illeg.*, Art. 53.1); *Sansevieria ehrenbergii* K. Schumann (1900) (*nom. illeg.*, Art. 53.1); *Sansevieria fulvocincta* Haworth (1819); *Sansevieria glauca* Haworth (1812); *Sansevieria guineensis* var. γ Schultes (1829); *Sansevieria laetevirens* Haworth (1812); *Sansevieria polyphylla* Haworth (1812); *Sansevieria pumila* Spin (1812) (*nom. illeg.*, Art. 53.1); *Sansevieria stenophylla* Link (1821); *Sansevieria striata* G. Don *ex* Steudel (1841) (*nom. inval.*, Art. 32.1c); *Sansevieria venosa* G. Don *ex* Steudel (1841) (*nom. inval.*, Art. 32.1c).

S. aethiopica Thunberg (Prodr. Fl. Cap., 65, 1794).

T: RSA, Cape Prov. (*Thunberg* s.n. [UPS]). – **D:** Botswana, Namibia, Zimbabwe, RSA (Northern Cape, North-West Prov., Eastern Cape, Gauteng); dry open places or bushland on well-drained soil. **I:** Obermeyer & al. (1992: 6). **Fig. XXVIII.e**

Incl. *Aletris hyacinthoides* var. *zeylanica* Aiton (1789) (*nom. illeg.*, Art. 53.1); incl. *Sansevieria zeylanica* Redouté (1809) (*nom. illeg.*, Art. 53.1); incl. *Sansevieria glauca* Gérôme & Labroy (1903) (*nom. illeg.*, Art. 53.1); incl. *Sansevieria caespitosa* Dinter (1926); incl. *Sansevieria scabrifolia* Dinter (1932).

[3c] Acaulescent, rhizomatous; rhizome ± 1 cm ∅; **L** 13 - 30, rosulate, ascending-spreading, linear or linear-lanceolate, 13 - 43 × 1 - 2 cm, upper face concave, base sometimes slightly narrowed, gradually narrowed above the middle to a subulate green tip 1.6 - 2.5 mm, soon becoming white, otherwise dark green with bluish hue, ± glaucous, sometimes with paler green transverse bands, margin red or whitish, surface slightly rough; **Inf** 35 - 75 cm, simple, spike-like; raceme dense, 4 - 6 **Fl** per cluster; **Bra** ovate-lanceolate, acute, 0.5 - 1 mm; **Ped** to 5 mm; **Fl** white, sometimes purple or cream, tube 2 - 3.5 cm, lobes 1.5 - 2 cm.

S. arborescens Cornu *ex* Gérôme & Labroy (Bull. Mus. Hist. Nat. (Paris) 9: 170, 172-173, ills., 1903). – **D:** Kenya, Tanzania; lowland *Acacia* bushland, 40 - 375 m. **I:** Sansevieria J. 3(1): 36-37, 1994. **Fig. XXVIII.b**

[1] Caulescent, rhizomatous; stem erect, to 20 - 150 × 2 - 2.5 cm; **L** many, densely spiralled, spreading or recurved, lanceolate or linear-lanceolate, upper face concave, 20 - 45 × 2 - 4.5 cm, base scarcely or slightly narrowed, green with whitish or reddish slightly wavy margin, surface smooth, with a 8.5 - 25.4 mm pale brown stout pungent subulate tip; **Inf** 50 cm, paniculate; raceme sub-dense, 4 - 6 **Fl** per cluster; **Bra** triangular, 2 mm; **Ped** 3 mm; **Fl** white, tube 6 mm, lobes 9 mm. – *Cytology:* 2n = 76 (Sharma & Chaudhuri 1964).

Originally described with flowers unknown.

S. bagamoyensis N. E. Brown (BMI 1913: 306, 1913). **T:** Tanzania, Bagamoyo Distr. (*Sacleux* 672 [P]). – **D:** Tanzania; half-shade of coastal bushland, 100 - 200 m.

[1] Caulescent; stem erect; **L** spiralled, recurved or recurved-spreading, linear-lanceolate or linear, upper face concave, 18 - 38 × to 1.7 cm, dark green, narrowed to a ≥ 4.2 mm brown hard **Sp**-like tip, margin narrowly red-brown with white membranous edge, surface smooth; **Inf** 40 - 55 cm, paniculate; raceme lax, ± 6 **Fl** per cluster; **Bra** lanceolate-acute, 2 - 3 mm; **Ped** to 6 mm; **Fl** cream-white, tube 6 - 7 mm, lobes ± 6.5 mm.

S. bella L. E. Newton (CSJA 72(4): 224-226, ills., 2000). **T:** Kenya, Rift Valley Prov. (*Newton* 3945

[K, EA]). – **D:** Kenya; under shrubs or amongst rocks in dry bushland, 1700 - 2030 m.

[2] Caulescent, with runners, branching freely above ground; runners spreading or ascending, to 15 cm; **L** up to 8, ± distichous, ascending, cylindrical with a groove on the upper face from the base to ± ¼ of their length, to 70 cm, to 3.5 cm thick, with distinct dark and light green transverse bands and darker green narrow longitudinal lines, tip a red-brown **Sp** to 5 mm, pungent or relatively blunt, margins with a red-brown line and a narrow colourless flange, surface very rough; **Inf** to 60 cm, simple, spike-like; peduncle light green with a slight bloom; raceme to 55 cm, dense, to 7 **Fl** per cluster; **Bra** triangular, 3 × 2 mm; **Ped** to 5 mm; **Fl** white, tube 10 - 15 mm, lobes 13 - 18 mm.

S. bracteata Baker (Trans. Linn. Soc. London, Bot. 1: 253, 1880). – **D:** Angola; dry areas.

[3c] Acaulescent, rhizomatous; **L** erect, lanceolate, 38 - 60 × 5 - 7 cm, base narrowed from the middle or below into a channelled petiole, dark green with wide pale green irregular bands or blotches on both faces, glaucous, narrowed above to a ≥ 6.4 mm tip, margin hard, brownish-red, sometimes with whitish edge; **Inf** 45 - 60 cm, simple, spike-like; raceme 5 - 6 cm, dense; **Bra** lanceolate or oblong-lanceolate, acute, 1.3 - 1.9 mm; **Ped** 2 - 3 mm; **Fl** white, tube 8.8 - 11.4 cm, lobes ± 2.5 - 3.2 cm.

S. braunii Engler & K. Krause (BJS 45: 153-154, 1911). **T:** Tanzania, Kigoma (*Braun* s.n. [EA]). – **D:** SW Tanzania; woodland in deep shade, 500 - 1160 m.

[3c] Acaulescent, rhizomatous, rhizome 3 cm ∅, brownish-orange; **L** usually 2, lanceolate-oblong, 50 - 70 × 7 - 11 cm, base slightly narrowed, green with few whitish bands or blotches, tip acuminate, margin hardened, red-brown; **Inf** 45 cm, simple, cylindrical, peduncle reddish; raceme 15 cm ∅, dense; **Bra** lanceolate-acute, 20 × 6 mm; **Ped** 1 cm; **Fl** white, tube ± 8.2 cm, lobes ± 2.5 - 3 cm.

The identity of this species is still uncertain. Plants in Kenya treated as this species by Mbugua (1998) are illustrated as *S. forskaoliana* by Teketay (1995), though Teketay's description states "leaves usually 2". The protologue of *S. forskaoliana* gives no details of leaves. Living material from the 2 type localities is required, but for the present, I am following Teketay in treating the Kenyan plants as *S. forskaoliana*.

S. burdettii Chahinian (Brit. Cact. Succ. J. 18(3): 132-133, ills., 2000). **T:** Malawi, Southern Region (*Burdett* s.n. in *Chahinian* 316 [K, MO]). – **D:** Malawi.

[3b] Acaulescent, rhizomatous; rhizome to 4 cm ∅, orange; **L** 3 - 6, distichous, slightly spreading, cylindrical with a groove on the upper face from the

base to 0.2 - 0.5 of the **L** length, to 90 cm, base 2.5 cm wide, dark green with several longitudinal lines, sometimes with faint transverse bands on young **L**, margins green with a withered non-fibrous flange, tip a withered **Sp**, surface smooth, waxy; **Inf** 16 - 20 cm, simple, spike-like; peduncle green with small white spots; raceme 12 - 14.75 cm, dense, 4 - 6 **Fl** per cluster; **Bra** 1 - 2 mm; **Ped** 4 mm; **Fl** white tinged pinkish-brown, tube 20 - 25 mm, lobes 18 - 20 mm.

S. burmanica N. E. Brown (BMI 1915(5): 48, fig. 12, 1915). **T:** Myanmar, Mandalay Prov. (*Clayton* s.n. [K]). − **D:** Myanmar; scrub forest, ± 100 m.

Incl. *Sansevieria roxburghiana* Hooker *fil.* (1896) (*nom. illeg.*, Art. 53.1).

[3c] Acaulescent, rhizomatous; rhizome 1.25 - 1.7 cm ∅; **L** 8 - 13, rosulate, erect, linear or linear-lanceolate, flat or upper face concave, 45 - 76 × 1.25 - 3.2 cm, green with paler transverse bands, upper face with 1 - 3 longitudinal striations, lower face with 6 - 9 striations, narrowed above to a 2.5 - 10 cm green soft subulate tip, margin green, becoming whitish with age, surface smooth; **Inf** 60 - 76 cm, simple, spike-like; raceme lax, 2 - 5 **Fl** per cluster; **Bra** lanceolate, acute, 2 - 4 mm; **Ped** 7.5 - 8.5 mm; **Fl** greenish-white, tube ± 8.5 mm, lobes 8.5 mm.

S. canaliculata Carrière (Rev. Hort. 1861: 449, 1861). **T:** sine loco (*Bojer* s.n. [P]). − **D:** Madagascar.

Incl. *Sansevieria sulcata* Baker (1887); **incl.** *Sansevieria schimperi* Baker (1898).

[3b] Acaulescent, rhizomatous; rhizome 1 - 1.25 cm ∅; **L** 1 - 2, to 5 cm apart along the rhizome, erect or slightly curved, cylindrical or slightly laterally compressed, with 5 - 6 shallow longitudinal grooves, 15 - 76 cm, 1 - 2 cm thick, dark green, narrowed shortly to a whitish hard acute tip, surface almost smooth; **Inf** 5 - 16 cm, simple, spike-like; raceme lax, **Fl** usually 3 per cluster, solitary near the tip; **Bra** ovate-acute, ± 4 mm; **Ped** 1 - 1.6 mm; **Fl** white tinged green, tube 2.5 cm, lobes 1.7 cm. − *Cytology:* 2n = 42 (Sharma & Chaudhuri 1964).

Described from a cultivated plant of unknown origin. Brown (1915) cited a specimen from Somalia, but Thulin (1995) found no material agreeing with the description. The species was reported by Perrier (1938) to be naturalized in Madagascar.

S. caulescens N. E. Brown (BMI 1915(5): 20, fig. 2, 1915). **T:** Kenya (*Powell* s.n. [K]). − **D:** Kenya.

[3a] Caulescent; stem erect, to 60 × 2.5 - 3.8 cm, branching at the base; **L** many, densely spiralled, spreading and slightly recurved, cylindrical with a groove along the upper face almost to the tip, 45 - 84 × 2 - 3.4 cm, 1.3 - 1.8 thick, dark green, lower face with 9 - 12 dark longitudinal striations, with whitish edges, when young with indistinct transverse dark green bands, gradually narrowed to a

firm green tip becoming pale brown or whitish, surface slightly rough; **Inf** 60 - 70 cm, simple, spike-like; raceme ± 40 - 46 cm, dense, 6 - 12 **Fl** per cluster; **Bra** very small, inconspicuous; **Ped** 7.4 - 9.5 mm; **Fl** whitish, tube 1.5 cm, lobes 1.9 cm.

S. chinensis L. Gentil *ex* N. E. Brown (BMI 1915 (5): 62, fig. 18, 1915). **T:** sine loco (*Gentil* s.n. [K]). − **D:** Unknown.

[3c] Acaulescent, rhizomatous; rhizome 2.5 - 3.2 cm ∅; **L** 3 - 6, rosulate, erect or ascending-spreading, lorate to lanceolate, 45 - 68 × 3 - 10 cm, base narrowed to a 5 - 17.8 cm channelled petiole, with dark and light green transverse bands, slightly or distinctly glaucous, with 4.2 - 12.7 mm pale brownish or whitish soft subulate tip, margin wavy, reddish-brown or whitish, surface almost smooth; **Inf** 60 - 84 cm, simple, spike-like, peduncle light green; raceme 30 - 38 cm, sub-dense, 2 - 3 **Fl** per cluster; **Bra** linear-lanceolate, acute, 6.25 - 8.5 mm; **Ped** 5.3 - 7.4 mm; **Fl** pale greenish-white, tube 2.5 cm, lobes 2.1 - 2.5 cm.

Described from a cultivated plant received in the Brussels Botanic Garden from another garden with the name *S. chinensis*, thought possibly to be a gardener's corruption of *S. guineensis*, a name applied indiscriminately to several sansevierias in cultivation (Brown 1915).

S. concinna N. E. Brown (BMI 1915(5): 53, fig. 14, 1915). **T:** Moçambique, Sofala Prov. (*Dawe* 1 [K]). − **D:** Moçambique, RSA; in shade of coastal forest in sandy soil. **I:** Jaarsveld (1994).

[3c] Acaulescent, rhizomatous; rhizome ± 1.25 cm ∅; **L** ± 5, rosulate, ascending-spreading, lanceolate, 15 - 25 × 1.25 - 3 cm, base shortly narrowed to a 3.8 - 8.9 cm channelled petiole, green with pale green transverse bands, with 4.2 - 8.5 mm green subulate tip, margin scarcely hardened, green, surface smooth; **Inf** 15 - 30 cm, simple, spike-like, peduncle green, tinged and dotted purple; raceme sub-dense, 1 or 2 **Fl** per cluster; **Bra** lanceolate-acuminate, 6 - 8.5 mm; **Ped** 3 - 4 mm; **Fl** white, tube 4.5 cm, lobes 2.1 cm.

S. conspicua N. E. Brown (BMI 1913: 306, 1913). **T:** Kenya, Coast Prov. (*Powell* s.n. [K]). − **D:** Kenya, Malawi, Tanzania; coastal thicket and grassland, to 375 m. **I:** Sansevieria J. 5(1): 3-6, 1996. **Fig. XXVIII.f**

[3c] Acaulescent, rhizomatous; rhizome 1.7 - 1.9 cm ∅; **L** 3 - 5, rosulate, ascending-spreading, lanceolate, 23 - 76 × 5 - 8.3 cm, base narrowed from below the middle, dull green, upper face with darker longitudinal striations, with 2.1 - 3.2 mm hard tip, margin hardened, reddish-brown and edged white, surface smooth; **Inf** to 60 cm, simple, spike-like, peduncle greyish-green tinged dull purple; raceme 25.5 - 30 cm, dense, 1 - 3 **Fl** per cluster; **Bra** linear-lanceolate, 0.6 - 1.3 mm; **Ped** 0.4 - 0.6 mm; **Fl**

greenish-white below, white above, tube 3.8 - 4.2 mm, lobes 2.5 - 3.2 mm.

S. cylindrica Bojer (Hort. Maurit., 349, 1837). – **D:** Angola, Zambia.

S. cylindrica var. **cylindrica** – **D:** Angola, Zambia.
Incl. *Sansevieria angolensis* Welwitsch *ex* Hooker (1856); **incl.** *Sansevieria guineensis* Weiner (1887) (*nom. illeg.*, Art. 53.1); **incl.** *Sansevieria livingstoniae* Rendle (1932).

[3b] Acaulescent, rhizomatous; rhizome 2.5 - 3.8 cm ∅; **L** 3 - 4, distichous, erect, cylindrical or slightly laterally compressed, 60 - 150 cm, 2 - 3 cm thick, green or whitish-green, with dark green transverse bands, gradually narrowed to a 4.2 - 6.4 mm hard whitish acute tip, surface slightly rough; **Inf** 60 - 90 cm, simple, spike-like; raceme 38 - 75 cm, 5 - 6 **Fl** per cluster; **Bra** lanceolate or ovate-lanceolate, 4 - 10 mm; **Ped** 4 - 8.5 mm; **Fl** white or tinted, tube 1.7 - 2.5 cm, lobes 1.7 - 1.9 cm. – *Cytology:* 2n = 40 (Roy 1956), 92 (Sharma & Chaudhuri 1964), 102 - 104 (Darlington & Wylie 1955).

S. cylindrica var. **patula** N. E. Brown (BMI 1915 (5): 38, 1915). **T:** Angola (*Anonymus* s.n. [K]). – **D:** Angola.
[3b] Differs from var. *cylindrica*: **L** 3 - 6, diverging, recurved or spreading; **Fl** tube 1.15 - 1.9 cm, lobes 1.5 - 1.9 cm.

S. dawei Stapf (JLSB 37: 529, 1906). **T:** Uganda, Central Distr. (*Dawe* 109 [K]). – **D:** Kenya, Uganda, Burundi; open woodland, 600 - 1300 m.
[3c] Acaulescent, rhizomatous; rhizome to ≥ 2.5 cm ∅; **L** 2 - 3, ascending or suberect, lanceolate, 60 - 150 × 5.7 - 11 cm, base narrowed from below middle to a long or short channelled petiole, dull dark green and glaucous, narrowed above the middle to an acute tip, margin reddish-brown, upper face smooth, lower face slightly rough; **Inf** 45 - 75 cm, simple, spike-like, dense, 3 - 4 **Fl** per cluster; **Bra** ovate or ovate-oblong, acute or subobtuse, 8.5 - 17 mm; **Ped** 4 - 6 mm; **Fl** white, tube 1.9 - 2.5 cm, lobes 1.7 - 2.2 cm.

S. dooneri N. E. Brown (BMI 1915(5): 51, fig. 13, 1915). **T:** Kenya, Rift Valley Prov. (*Dooner* s.n. [K]). – **D:** Kenya; shade in dry forest, ± 2000 - 2100 m.
[3c] Acaulescent or shortly caulescent, rhizomatous or with runners; rhizome 0.6 - 0.85 cm ∅; stem to 5 cm, concealed by **L** bases; **L** to 20, in lax **Ros**, lanceolate or lorate, base erect or ascending, recurved-spreading above, 10 - 43 × 1.5 - 3 cm, base gradually narrowed from near or above the middle, dark green, lower face slightly paler, both faces with faint and irregular pale green transverse bands, shortly narrowed to a 6 - 50 mm green soft subulate tip, margin green, surface smooth; **Inf** 30 - 38 cm,

simple, spike-like, peduncle green; raceme 15 - 19 cm, lax, 2 - 3 **Fl** per cluster; **Bra** ovate-lanceolate, acute, 2 - 4 mm; **Ped** 3 - 4 mm; **Fl** dull pink or pale purplish, whitish inside, tube ± 1.2 cm, lobes 1.16 - 1.27 cm.

Similar to, and possibly conspecific with, *S. parva*, whose type locality is only ± 55 km away from that of this species.

S. downsii Chahinian (Brit. Cact. Succ. J. 18(3): 133-135, ills., 2000). **T:** Malawi, Northern Region (*Downs* 1/75 [K, MO]). – **D:** Malawi.
[3a] Acaulescent, with runners, runners to 2 cm ∅, covered with **L** sheaths; **L** 6 - 14, rosulate, spreading-recurved, cylindrical with a groove on the upper face from the base to 0.25 - 0.8 of the **L** length, 14 - 45 cm, base 3.2 cm wide, medium bluish-green with sparse grey-green transverse bands and dark green longitudinal lines, margins with a chest-nut brown line and a narrow colourless flange, tip a **Sp**, sometimes chestnut-brown, surface slightly rough, somewhat shiny, waxy; **Inf** to 1.6 m, simple, spike-like; peduncle medium green, speckled; raceme < ⅔ of the peduncle length, lax, 3 - 5 **Fl** per cluster; **Bra** 1 - 2 mm; **Ped** 3 mm; **Fl** white tinged green, tube ± 10 mm, lobes ± 12 mm.

The plants reported from Malawi as possibly being *S. gracilis* by Thiede (1993) are *S. downsii*.

S. ebracteata (Cavanilles) C. R. Suresh (in Nicolson, Interpret. Van Rheede's Hort. Malab., 271, 1988). – **D:** India; sandy places.
≡ *Salmia ebracteata* Cavanilles (1795); **incl.** *Aletris zeylanica* var. β Lamarck (1789); **incl.** *Sansevieria lanuginosa* Willdenow (1799) (*nom. illeg.*, Art. 52.1).

[3b] Acaulescent, rhizomatous; rhizome stout; **L** ± 3 - 4, erect, semiterete with concave channel down upper face and several grooves down lateral and lower faces, 45 - 90 × 1.7 - 1.9 cm, narrowed to an acute tip, green with "woolly" grooves; **Inf** ± 60 cm, simple, spike-like, density unknown, 2 - 5 **Fl** per cluster; **Ped** ± 4.2 mm; **Fl** white, tube 0.85 - 1 cm, lobes 1.27 - 1.48 cm.

S. ehrenbergii Schweinfurth *ex* Baker (JLSB 14: 549, 1875). **T:** Sudan (*Schweinfurth* 31 [B?]). – **D:** Yemen, Djibouti, Eritrea, Ethiopia, Kenya, Somalia, Sudan, Tanzania; rocky ground, usually in the shade of thickets or small trees, 400 - 1100 m. **I:** Teketay (1995).
[1] Acaulescent or caulescent, rhizomatous; rhizome 3 cm ∅; stem erect, to 25 cm; **L** 5 - 9, distichous, erect or spreading, laterally compressed with a groove along the upper face, 76 - 180 cm, 3 - 4.5 cm thick, narrowed to an abrupt 6.5 - 20 mm hard **Sp**-like tip, dark green with 5 - 12 blackish-green shallow longitudinal grooves, margin reddish-brown with white membranous edge, surface slightly rough; **Inf** to 2 m, paniculate; raceme with 4 - 7 **Fl**

per cluster; **Ped** 2 - 4 mm; **Fl** purple to white, tube 5 - 6.5 mm, lobes 7 - 18 mm. – *Cytology:* 2n = 40 (Sharma & Chaudhuri 1964).

Often confused with *S. robusta.* For differences see Chahinian (1993).

S. eilensis Chahinian (Sansevieria J. 4(1): 9-11, ills., 1995). **T:** Somalia, Nugaal Distr. (*Lavranos* 10179 [MO, UPS]). – **D:** Somalia; shaded limestone, ± 120 m. **I:** Sansevieria J. 3(3): 51-52, 1994, as *S. sp.* **Fig. XXVIII.g**

[3b] Acaulescent, rhizomatous; **L** 2 - 3, mostly distichous, cylindrical, sometimes with a groove on the upper face, 7 - 18 cm, 1.9 - 2.5 cm thick, narrowed from middle to base, narrowed abruptly to a 5 mm acute **Sp**-like tip, medium grey-green, with light grey-green transverse bands and up to 12 medium green longitudinal lines, margin green becoming brown edged white, surface rough; **Inf** 34 cm, simple, spike-like, peduncle light green; raceme 23 cm, lax, 2 - 4 **Fl** per cluster; **Bra** 2 - 6 × 1 - 3 mm; **Ped** 4 - 5 mm; **Fl** greenish-white, tube ± 8 mm, lobes ± 14 mm.

S. erythraeae Mattei (Boll. Reale Orto Bot. Giardino Colon. Palermo 4: 170, 1918). **T:** not preserved. – **D:** Eritrea, Ethiopia, Sudan; semi-arid areas along rivers, 1200 - 2100 m.

Incl. *Sansevieria schweinfurthii* Täckholm & Drar (1954) (*nom. inval.*, Art. 36.1).

[3b] Acaulescent, rhizomatous; **L** 6 - 8, erect, cylindrical with a short groove on the upper face and 5 furrows on the lower face, 40 - 50 cm; **Inf** to 50 cm, simple, spike-like; raceme dense, 3 - 5 **Fl** per cluster; **Bra** ovate-lanceolate, acute; **Ped** 7 - 8 mm; **Fl** white, tube 5 - 7 mm, lobes > 7 mm.

S. fasciata Cornu *ex* Gérôme & Labroy (Bull. Mus. Hist. Nat. (Paris) 9: 170, 173-173, fig. 3, 1903). **T:** Congo Free State [Congo/Zaïre] (*Anonymus* s.n. [P]). – **D:** Congo or Zaïre.

Incl. *Sansevieria lasciata* L. Gentil (1907) (*nom. inval.*, Art. 61.1).

[3c] Acaulescent, rhizomatous; **L** 2 - 5, re-curved-spreading, lanceolate, 38 - 84 × 3.8 - 11.5 cm, narrowed from about the middle to the channelled petiole, with a 2.1 - 6.4 mm green acute tip, upper face pale green broken into patches by irregular transverse zigzag dark green bands, lower face whitish-green with irregular narrow transverse dark green bands, margin green, becoming reddish or whitish, surface smooth; **Inf** and **Fl** not known.

S. fischeri (Baker) Marais (KB 41(1): 58, 1986). **T:** Tanzania, Kilimanjaro Distr. (*Fischer* 9 [B?, K [fragment]]). – **D:** Ethiopia, Kenya, Somalia, Tanzania; sandy soils at edges of thickets and in dense bush, 260 - 900 m. **I:** Rauh (1963: 126, as *S. singularis*).

≡ *Buphane fischeri* Baker (1898); **incl.** *Sansevieria singularis* N. E. Brown (1911).

[2] Acaulescent, rhizomatous; rhizome to 4.5 cm ∅; **L** solitary, erect, cylindrical with 4 - 6 longitudinal furrows, 45 - 240 cm, narrowed slightly upwards to near tip, then narrowing shortly to a whitish stout acute tip, dull greyish-green or bluish-green, brighter green with pale green transverse bands when young, surface slightly rough; **Inf** to 10 cm, simple, capitate, peduncle subterranean; raceme dense; **Bra** acute, 3 - 5 × 3 mm; **Ped** 1.27 cm; **Fl** tube 2 - 5 cm, whitish sometimes tinged with violet, lobes 5 - 10 mm, white with violet venation outside.

In growth habit and leaves this is very similar to *S. stuckyi,* and as both were described without inflorescences and flowers unknown, the 2 taxa have been confused in cultivation. Rauh (1963) first described the inflorescence and flowers of *S. fischeri,* and those of *S. stuckyi* were described by Jumelle (1923) and Pfennig (1981), confirming that the 2 species are distinct.

S. forskaoliana (Schultes *fil.*) Hepper & Wood (KB 38(1): 83, 1983). **T:** Yemen (*Forskål* 9 [C]). – **Lit:** Hepper & Friis (1994); Friis (1995). **D:** Yemen, Sudan, Djibouti, Congo, Eritrea, Ethiopia, Somalia; dry rocky slopes and river beds, usually in shade, 550 - 2000 m. **I:** Teketay (1995: 53).

≡ *Smilacina forskaoliana* Schultes *fil.* (1829); **incl.** *Convallaria racemosa* Forsskål (1775); **incl.** *Sansevieria guineensis* Schweinfurth (1894) (*nom. illeg.*, Art. 53.1); **incl.** *Sansevieria guineensis* var. *angustior* Engler (1902) ≡ *Sansevieria abyssinica* var. *angustior* (Engler) Cufodontis (1971); **incl.** *Sansevieria abyssinica* N. E. Brown (1913) ≡ *Acyntha abyssinica* (N. E. Brown) Chiovenda (1916); **incl.** *Acyntha abyssinica* var. *sublaevigata* Chiovenda (1932) ≡ *Sansevieria abyssinica* var. *sublaevigata* (Chiovenda) Cufodontis (1971); **incl.** *Acyntha elliptica* Chiovenda (1932) ≡ *Sansevieria elliptica* (Chiovenda) Cufodontis (1971).

[3c] Acaulescent, rhizomatous; rhizome ≥ 2 cm ∅; **L** erect, lanceolate, ≥ 60 × 6.3 - 7.5 cm, narrowed from the middle to a channelled petiole ⅕ - ⅓ as long as the lamina, green, with a ≥ 2.1 mm brown hardened tip, margin wavy, hard, reddish-brown, 0.1 mm, surface rough, lower face finely transversely rugose; **Inf** to ≥ 95 cm, simple, spike-like; raceme with 4 - 5 **Fl** per cluster; **Ped** to 10 mm; **Fl** white.

Originally described from a fruiting specimen without leaves, bracts and flowers. See note after *S. braunii.*

S. francisii Chahinian (Sansevieria J. 4(1): 12-14, ills., 1995). **T:** Kenya, Coast Prov. (*Horwood* 432 [MO, UPS, ZSS]). – **D:** Kenya. **I:** BCSJ 2(3): 81-82, 1984, as *S. sp.*

[3a] Caulescent, with runners; stem erect, to ≥ 30

cm; runners to 16 × 0.8 - 1.4 cm; **L** to 40, spiralled in 5 rows, cylindrical with a groove on the upper face for ¼ to ¾ of the length, 8 - 15 cm, narrowed gradually to a 5 mm acute **Sp**-like tip, dark green with grey-green transverse bands and 4 - 6 dark green longitudinal lines, margin brownish-red with a white edge on the basal ½, green above, surface slightly rough; **Inf** 12 - 25 cm, simple, spike-like; raceme dense, 1 - 2 **Fl** per cluster; **Bra** triangular, ± 6 × 3 mm; **Ped** ± 2 mm; **Fl** greenish-white to brownish-green, tube 1.6 - 1.9 cm, lobes 0.8 - 1 cm.

S. frequens Chahinian (CSJA 72(3): 130-132, ills., 2000). **T:** Kenya, Laikipia Distr. (*Chahinian* 785 [MO, NYBG]). – **Lit:** Mbugua (1998: as *S. braunii*). **D:** Ethiopia, Kenya, Uganda; mostly rocky areas in grassland or open bush, 600 - 1750 m.

[3c] Acaulescent, rhizomatous; rhizome to 5 cm ∅; **L** 4 - 8 (usually 6), rosulate, erect, leathery, oblanceolate, 90 or more × 15 cm, dull grass-green, sometimes with a bloom, base slightly narrowed, tip obtuse, margins fibrous, chestnut-brown, surface smooth; **Inf** 60 - 90 cm, simple, spike-like; peduncle green; raceme dense; **Bra** 3.35 - 8 × 2.5 - 3 mm; **Ped** shorter or longer than the **Bra**; **Fl** greenish-white, tube 18 - 20 mm, lobes 26 - 28 mm.

S. gracilis N. E. Brown (BMI 1911: 96, fig. 4, 1911). **T:** Kenya, Coast Prov. (*Powell* 11 [K]). – **D:** Kenya, Tanzania; dry bushland and thicket, 30 - 600 m. **I:** Brown (1915: 24).

[3a] Caulescent, with runners; stem erect, to 8 cm; runners slightly ascending, to 90 × 0.8 cm; **L** 8 - 12, densely spiralled, ascending or spreading, cylindrical with a groove on the upper face for up to 12.5 cm from the base, 23 - 80 cm, to 0.6 - 1 cm thick, deep grass-green, sometimes with inconspicuous narrow transverse darker bands and slightly darker longitudinal lines, narrowed gradually to a 2.1 - 6.4 mm brown or whitish **Sp**-like tip, margin membranous, white, surface smooth; **Inf** to 30 cm, simple, spike-like, peduncle light green; raceme lax, 2 **Fl** per cluster; **Bra** lanceolate or linear-lanceolate, acute, 2.1 - 3.2 mm; **Ped** 1 - 1.6 mm; **Fl** white, tube 1.9 - 2.5 cm, lobes 1.05 - 1.27 cm.

S. grandicuspis Haworth (Synops. Pl. Succ., 67, 1812). – **D:** Unknown.

Incl. *Sansevieria ensifolia* Haworth (1812); **incl.** *Sansevieria pumila* Haworth (1812).

[3c] Acaulescent, rhizomatous; **L** 5 - 15, erect or ascending-spreading, linear-lanceolate, 18 - 50 × 1.25 - 3.8 cm, sometimes narrowed to a channelled petiole to 15 cm, alternating transverse bands of dull dark and lighter green, with 5 - 7 longitudinal dark green impressed lines, shortly narrowed above to a 17 - 50 mm green flexible subulate tip, margin green, surface smooth; **Inf** and **Fl** not known.

Originally described with flowers not seen and still little-known.

S. hallii Chahinian (Sansevieria J. 5: 7-10, ills., 1996). **T:** Zimbabwe, East Region (*Hall* 67/799 [MO, UPS]). – **D:** Zimbabwe; ± 300 m.

[2] Acaulescent, rhizomatous; rhizome 1.8 - 3 cm ∅, grey-orange; **L** 1 - 3, erect, cylindrical with a deep groove on the upper face, to 60 × 5 cm, dark greyish-green with numerous longitudinal lines, with inconspicuous transverse bands when young, narrowed to a 6 mm obtuse or rounded tip, margin chestnut-brown edged white, surface rough; **Inf** to 18 cm, capitate; raceme to 16 cm ∅, dense; **Bra** triangular, 18 × 6 mm; **Fl** white tinged purple, tube 4.5 - 7.8 cm, lobes 2.5 - 2.9 cm.

S. hyacinthoides (Linné) Druce (Bot. Exch. Club Soc. Brit. Isles 1913(3): 423, 1914). **T:** [lecto – icono]: Commelin, Praeludia Bot. t. 33, 1703. – **Lit:** Wijnands (1973); Obermeyer & al. (1992). **D:** Widespread in E parts of S Africa into tropical E Africa, RSA (Eastern Cape, KwaZulu-Natal, Mpumalanga); forming dense stands usually in the shade of trees or shrubs. **I:** Wijnands (1973: 112).

≡ *Aloe hyacinthoides* Linné (1753) ≡ *Aletris hyacinthoides* (Linné) Linné (1762); **incl.** *Aloe hyacinthoides* var. *guineensis* Linné (1753) ≡ *Aletris hyacinthoides* var. *guineensis* (Linné) Linné (1762) ≡ *Aloe guineensis* (Linné) Jacquin (1762) (*nom. illeg.*, Art. 52.1) ≡ *Aletris guineensis* (Linné) Jacquin (1770) (*nom. illeg.*, Art. 52.1) ≡ *Sansevieria guineensis* (Linné) Willdenow (1799) (*nom. illeg.*, Art. 52.1); **incl.** *Sanseverinia thyrsiflora* Petagna (1787) (*nom. illeg.*, Art. 52.1) ≡ *Sansevieria thyrsiflora* (Petagna) Thunberg (1794) (*nom. illeg.*, Art. 52.1); **incl.** *Salmia spicata* Cavanilles (1795) ≡ *Sansevieria spicata* (Cavanilles) Haworth (1812); **incl.** *Pleomele aloifolia* Salisbury (1796); **incl.** *Sansevieria latifolia* Bojer (1837); **incl.** *Sansevieria guineensis* var. β Kunth (1850) (*nom. inval.*, Art. 43.1); **incl.** *Sansevieria nobilis* Godefroy-Lebeuf (1861) (*nom. inval.*, Art. 32.1c); **incl.** *Sansevieria angustiflora* Lindberg (1875); **incl.** *Sansevieria rufocincta* Baker (1875); **incl.** *Sansevieria angustifolia* Baker (1875) (*nom. inval.*, Art. 61.1); **incl.** *Sansevieria grandis* Hooker *fil.* (1903); **incl.** *Sansevieria grandis* var. *zuluensis* N. E. Brown (1915).

[3c] Acaulescent, rhizomatous; rhizome stout; **L** 2 - 8, rosulate, erect, lanceolate to broadly linear, 15 - 60 × 2.5 - 9 cm, narrowed from the middle or below to the channelled petiole, dull green with numerous closely placed transverse pale green bands that fade with age, usually with a white withered acute or obtuse tip to 17 mm, margin brownish-red, surface smooth; **Inf** 45 - 75 cm, simple, spike-like; raceme 22 - 30 cm, dense, 2 - 6 **Fl** per cluster; **Bra** ovate-lanceolate to narrowly lanceolate, acute, 4 - 12.5 × 1.4 - 3.2 mm; **Ped** 3.2 - 6.3 mm; **Fl** whitish, tube ± 1.9 cm, lobes 1.9 cm. – *Cytology:* 2n = 100 (Darlington & Wylie 1955: as *S. grandis*).

S. kirkii Baker (BMI 1887(May): 3, 8, 1887). **T:** Tanzania, Pangani Distr. (*Kirk* s.n. [K]). – **D:** Malawi, Tanzania, Zanzibar.

S. kirkii var. **kirkii** – **D:** Malawi, Tanzania, Zanzibar; coral cliffs near sea level. **I:** Brown (1915: 75).

Incl. *Sansevieria aubryana* De Wildeman (1903); **incl.** *Sansevieria aubrytiana* Gérôme & Labroy (1903) (*nom. illeg.*, Art. 53.1).

[2] Acaulescent, rhizomatous; rhizome stout; **L** 1 - 3, erect or ascending-spreading, sometimes the terminal part recurved, elongate-lanceolate or broadly lorate, 75 - 275 × 6 - 9 cm, narrowed gradually from about the middle to the stout channelled petiole, whitish-brown, greyish-green, mottled or transversely banded light green, with 3 - 9 dark green longitudinal lines, with a 8.5 - 12.7 mm pale whitish-brown firm acute tip, margin wavy, reddish-brown, surface smooth; **Inf** to 60 cm, simple, capitate, peduncle dull purplish-brown, speckled pale green or dull whitish; raceme 3.8 - 10 cm, dense; **Bra** ovate or oblong-ovate, acute or subobtuse, 25 - 38 × 8.5 - 19 mm; **Ped** 6.4 - 10.5 mm; **Fl** tube 11.4 - 12.7 cm, pale purplish or dull pink, lobes 3.2 - 4.5 cm, white.

S. kirkii var. **pulchra** N. E. Brown (BMI 1915(5): 76, 1915). **T:** Zanzibar (*Last* s.n. [K]). – **D:** Zanzibar. **I:** Pfennig (1977b: 552). **Fig. XXVIII.h**

Incl. *Sansevieria longiflora* Gérôme & Labroy (1903) (*nom. illeg.*, Art. 53.1).

[2] Differs from var. *kirkii*: **L** more conspicuously marked with whitish-green, buff or almost reddish spots or irregular bands, margin red-brown with white membranous edge; **Bra** lanceolate.

Populations of *S. kirkii* seen on Pemba Island (*Newton* 5611) and Misali Island (*Newton* 5616) have plants varying from strongly marked with spots or bands to almost uniformly green and unmarked, and so the var. *pulchra* is doubtfully distinct.

S. liberica Gérôme & Labroy (Bull. Mus. Hist. Nat. (Paris) 9: 170, 173, fig. 4, 1903). **T:** Liberia (*Julien* s.n. [K]). – **D:** Central African Republic, Ivory Coast, Ghana, Liberia, Nigeria, Sierra Leone, Togo; dry shady places by streams and rock outcrops.

[3c] Acaulescent, rhizomatous; rhizome ± 1.9 cm ∅, pale greyish; **L** 1 - 6, erect or suberect, spreading with age, lanceolate, 46 - 106 × 5 - 12.5 cm, narrowed downwards from below the middle, dark green, usually with indistinct transverse paler green bands, narrowed upwards from above the middle to a 2 - 12 mm green flexible subulate acute tip becoming whitish, margin cartilaginous, pale reddish-brown, surface smooth; **Inf** 60 - 80 cm, simple, spike-like; raceme lax; **Bra** lanceolate; **Fl** white, ± 5 cm.

Originally described with flowers not seen.

S. longiflora Sims (CBM 53: t. 2634 + text, 1826). **T:** [icono]: l.c. t. 2634. – **D:** Angola, Namibia, Zaïre, Equatorial Guinea (Bioko Island).

S. longiflora var. **fernandopoensis** N. E. Brown (BMI 1915(5): 76, 1915). **T:** Equatorial Guinea, Bioko Island (*Anonymus* s.n. [not located]). – **D:** Equatorial Guinea (Bioko Island).

[3c] Differs from var. *longiflora*: **Ped** 5.3 - 6.4 mm; **Fl** tube 6.35 - 7.6 cm.

S. longiflora var. **longiflora** – **D:** Angola, Namibia, Zaïre; sandy soil, usually in shade.

[3c] Acaulescent, rhizomatous; rhizome 2.5 cm ∅; **L** ± 4 - 6, rosulate, spreading, lanceolate, 30 - 150 × 4 - 9 cm, narrowed to a 7.6 cm channelled petiole, dark green with paler green spots scattered or in irregular transverse bands, with a 3.2 - 6.3 mm brown hard **Sp**-like tip, margin hardened, red-brown or yellowish, surface smooth; **Inf** 33 - 68 cm, simple, spike-like; raceme 7.5 - 38 cm, dense, 2 - 3 **Fl** per cluster; **Bra** lanceolate, acute, 1.27 - 2.54 cm; **Ped** 1.6 - 3.2 mm; **Fl** tube 8.9 - 10.2 cm, greenish-white, lobes 2.5 - 3.8 cm, white.

S. masoniana Chahinian (CSJA 72(1): 31, ill., 2000). **T:** Zaïre (*Mason* s.n. in *Chahinian* 258 [MO, NYBG]). – **D:** Zaïre.

[3c] Acaulescent, rhizomatous; rhizome ± 4 cm ∅; **L** 1 - 2 per shoot, erect, leathery, oblanceolate, to 100 × 18 cm, tip acute, dull greyish-green with lighter mottling, margins fibrous, chestnut-brown, surface rough, basal sheath with purplish transverse bands; **Inf** to 53 cm, simple, spike-like; peduncle light green with purple lines; raceme sub-dense, 1 - 2 **Fl** per cluster; **Fl** greenish-white, tube 26 - 30 mm, 3 mm ∅ at base, widening to 4 mm at mouth, lobes 24 - 28 × 3 mm.

Original locality not known. Cultivated for many years under the cultivar name 'Mason Congo'.

S. metallica Gérôme & Labroy (Bull. Mus. Hist. Nat. (Paris) 9: 170, 173, fig. 2, 1903). **T:** P. – **D:** Tropical Africa, RSA.

S. metallica var. **longituba** N. E. Brown (BMI 1915(5): 67, 1915). **T:** K. – **D:** Tropical Africa; exact provenance unknown.

[3c] Differs from var. *metallica*: **Inf** peduncle brownish-green or dull purplish; **Fl** tube 2.96 cm, lobes 2.96 cm.

S. metallica var. **metallica** – **D:** Tropical Africa, RSA.

Incl. *Sansevieria guineensis* var. β Schultes (1829) (*nom. inval.*, Art. 43.1); **incl.** *Sansevieria guineensis* Baker (1875) (*nom. illeg.*, Art. 53.1).

[3c] Acaulescent, rhizomatous; rhizome 1.9 - 2.5 cm ∅, bright red, becoming pale brown in light; **L** 1

- 4, erect or upper part spreading or recurved, lanceolate, 45 - 152 × 5 - 12.7 cm, narrowed from about the middle to a 10 - 60 cm channelled petiole, dull dark green, upper side with obscure irregular transverse bands, lower face more distinctly marked, with a 3.2 - 6.4 mm green soft subulate tip, margin soft, green, becoming whitish or pale reddish-brown, surface smooth; **Inf** 45 - 122 cm, simple, spike-like, peduncle light green; raceme lax, 2 - 4 **Fl** per cluster; **Bra** lanceolate, acuminate, 6.4 - 12.7 mm; **Ped** 4.2 - 7.4 mm; **Fl** white, tube 1.27 - 1.7 cm, lobes 1.7 - 2.33 cm. − *Cytology:* 2n = 40 (Darlington & Wylie 1955).

S. metallica var. **nyasica** N. E. Brown (BMI 1915 (5): 67, fig. 20, 1915). **T:** Malawi (*Buchanan* s.n. [K]). − **D:** Malawi.

[3c] Differs from var. *metallica*: **Ped** 4.2 - 6.4 mm; **Fl** tube 1.48 - 1.69 cm, greenish-white or tinged red, lobes 1.9 cm, white.

S. nilotica Baker (JLSB 14: 548, 1875). **T:** Sudan, White Nile Prov. (*Murie* s.n. [K]). − **D:** Central African Republic, Ethiopia, Sudan, Uganda.

Incl. *Acyntha massae* Chiovenda (1940) ≡ *Sansevieria massae* (Chiovenda) Cufodontis (1971).

S. nilotica var. **nilotica** − **D:** Central African Republic, Ethiopia, Sudan, Uganda; riverine woodland, 900 - 1450 m.

[3c] Acaulescent, rhizomatous; rhizome ± 1.9 cm ∅; **L** 2 - 3, lorate, 91 - 122 × 2.5 - 5.7 cm, narrowed gradually to a 30.5 - 61 cm channelled petiole, conspicuously marked with numerous closely placed irregular zigzag transverse narrow bands of dark green and paler green, with a 4.2 - 16.9 mm green soft subulate tip, margin green, surface smooth; **Inf** 53 - 76 cm, simple, spike-like; raceme 30 - 46 cm, lax, 4 - 10 **Fl** per cluster in the lower part, 2 - 3 in the upper part; **Bra** lanceolate, acute, 4.2 - 10.6 mm; **Ped** 7.4 - 12.7 mm; **Fl** white, tube 0.95 - 1.06 cm, lobes 1.16 - 1.27 cm. − *Cytology:* 2n = 36, 40 (Roy 1956).

S. nilotica var. **obscura** N. E. Brown (BMI 1915 (5): 58, 1915). **T:** Uganda (*Dawe* s.n. [K]). − **D:** Uganda.

[3c] Differs from var. *nilotica*: **L** 4 - 5, lanceolate or lorate, 60 - 84 × 3.8 - 7 cm, narrowed to a 15.2 - 30.4 cm channelled petiole, grass-green, sometimes with a few inconspicuous transverse bands of slightly paler green 25.4 - 38 mm apart, with a 12.7 - 32 mm green soft subulate tip; **Inf** 61 - 91 cm, peduncle dull mottled green; raceme with 3 - 6 **Fl** per cluster; **Bra** ovate-lanceolate, acute or acuminate, 4.2 - 8.5 mm; **Ped** 5.8 - 7.4 mm; **Fl** tube 0.85 cm, dull greenish-white with 6 longitudinal dull purplish lines, lobes 1.48 cm, whitish.

Brown (1915) suggested that this variety might be specifically distinct from var. *nilotica*.

S. parva N. E. Brown (BMI 1915(5): 53, 1915). **T:** Kenya, Rift Valley Prov. (*Powell* 15 [K]). − **D:** Kenya, Uganda; shade in forest, 1660 - 2135 m. **I:** Pfennig (1977b: 552).

[3c] Acaulescent or shortly caulescent, rhizomatous; rhizome 0.85 cm ∅, brownish-orange; stem to 12.7 cm, sometimes concealed by **L** bases; **L** 6 - 14, rosulate, ascending, becoming spreading, linear to lanceolate, 20 - 45 × 0.85 - 3 cm, narrowed to a channelled petiole to 5 cm, with distinct irregular bands of dark green and paler green, becoming nearly uniform green or obscurely marked with age, narrowed to a 38 - 76 mm green soft stout subulate tip, margin green, surface smooth; **Inf** ± 30 cm, simple, spike-like, peduncle light green; raceme lax, 1 - 2 **Fl** per cluster; **Bra** lanceolate, acute, 3.2 - 4.2 mm; **Ped** 4.2 - 5.3 mm; **Fl** tube 1.06 - 1.16 cm, pale pinkish-white, lobes 0.85 - 0.95 cm, mauve-tinted.

See comment for *S. dooneri.*

S. patens N. E. Brown (BMI 1915(5): 30, fig. 5, 1915). **T:** K [ex cult.]. − **D:** Origin unknown, probably Kenya. **I:** CSJA 64: 232, 1992.

[3b] Acaulescent, rhizomatous; rhizome 1.9 - 2.54 cm ∅; **L** 5 - 10, distichous, spreading and recurved, cylindrical, laterally compressed with a channel along the upper face, 45 - 91 cm, 1.7 - 4.2 cm thick, narrowed to an abrupt 6.4 - 12.7 mm whitish hard acute tip, indistinctly marked with dark green and paler green transverse bands, becoming bluish-green with age, with numerous longitudinal blackish-green lines, margin acute, green, sometimes whitish along the basal 2.54 - 15 cm, surface slightly rough; **Inf** 38 (- 76?) cm, simple, spike-like, peduncle pale green; raceme ≥ 25 cm, lax, 2 - 3 **Fl** per cluster; **Bra** lanceolate, acute, 3.2 - 6.4 mm; **Ped** 5.3 - 6.4 mm; **Fl** white, tube 0.95 - 1.06 cm, lobes 1.27 cm.

S. pearsonii N. E. Brown (BMI 1911: 97, 1911). **T:** Angola, Cunene Prov. (*Pearson* 2073 [K]). − **D:** Angola, Botswana, Namibia, Zimbabwe, RSA (Gauteng, KwaZulu-Natal, North-West Prov., Northern Cape); dry sandy or rocky soil in savanna or open forest. **I:** Brown (1915: 36).

Incl. *Sansevieria deserti* N. E. Brown (1915).

[3b] Acaulescent, rhizomatous; rhizome stout; **L** 3 - 7, distichous, erect but gradually diverging towards tip, cylindrical, slightly compressed laterally, with a groove on the upper face, 50 - 100 cm, ± 3.4 - 3.8 cm thick, narrowed gradually to a 8.5 - 25 mm whitish rigid terete-subulate acute tip, slightly glaucous-green or bluish-green, when young with faint paler green bands, margin greenish-white, becoming red-brown edged whitish, surface smooth; **Inf** ± 1 m, simple, spike-like; raceme with 6 - 10 **Fl** per cluster; **Ped** 8.5 mm; **Fl** tube 1.2 - 2.5 cm, white, greyish or bluish mauve, red-streaked above, lobes 0.6 - 1 cm, white or cream with pale pink or mauve.

Originally described with flowers unknown. Waidhofer (1996) treated *S. deserti* as a separate species.

S. perrotii Warburg (Tropenpflanzer 5: 190, ills., 1901). **T:** Tanzania, Lindi Prov. (*Perrot* s.n. [not located]). – **D:** Tanzania; among bushes on coral.

[1] Caulescent; stem erect, to 20 × 2.5 cm, covered by **L** bases; **L** 8 - 12, distichous, ascending or spreading, cylindrical, laterally slightly compressed, with a deep and wide groove on the upper face, 91 - 152 cm, 1.1 - 1.5 cm thick, narrowed gradually to a whitish hard acute tip, margin reddish-brown, edged white; **Inf** ≥ 1.2 m, paniculate; raceme lax, 2 - 4 **Fl** per cluster; **Fl** tube ± 1.27 cm, pale greenish, lobes ± 0.95 cm, whitish inside, purplish outside.

S. phillipsiae N. E. Brown (HIP 30: t. 3000 + text, 1913). **T:** Somalia (*Lort Phillips* s.n. [K]). – **D:** Ethiopia, Somalia; in shade of trees, 1250 - 1450 m.

   Incl. *Sansevieria hargeisana* Chahinian (1994).

[3a] Caulescent, in clumps to 38 cm tall, with runners; stem erect, branching at or above the base; runners to 20 × 1.3 cm; **L** 5 - 10, rosulate, ascending, becoming spreading, cylindrical with a deep groove on the upper facee for 5 - 8.9 cm at base, 10.2 - 46 cm, 1.3 - 1.9 cm thick, narrowed gradually to a 2.1 - 3.2 mm brown hard acute or obtuse tip, dark and slightly bluish-green, with faint transverse paler green bands when young, margin white, surface smooth; **Inf** 35 - 46 cm, simple, spike-like; raceme 23 - 30 cm, 3 - 6 **Fl** per cluster; **Bra** ovate-lanceolate, acute, 3.2 - 6.4 mm; **Ped** 2.65 - 3.18 mm; **Fl** white, tube 1 cm, lobes 1.16 - 1.27 cm.

S. pinguicula P. R. O. Bally (Candollea 19: 145-147, ills., 1964). **T:** Kenya, Coast Prov. (*Bally* 4275 [K]). – **D:** Kenya; sandy plains with open bushland, 120 - 230 m. **I:** Pfennig (1977a: 510). **Fig. XXIX.d**

[1] Shortly caulescent, with runners; stem erect; runners to 8 cm; **L** 5 - 7, rosulate, cylindrical with a deep groove on the upper face and 2 - 7 narrow grooves on the lower face, 12 - 30 cm, 2.8 - 3.5 cm thick, narrowing to a horny acute tip, green, margin brown, surface slightly rough; **Inf** 15 - 32 cm, paniculate; racemes dense, 4 - 6 **Fl** per cluster; **Ped** 1.5 - 2 mm; **Cl** tube 4 - 5 mm, lobes 3 - 4 mm, white with brown mid-stripe.

S. powellii N. E. Brown (BMI 1915(5): 18, fig. 1, 1915). **T:** Kenya, Coast Prov. (*Powell* 5 [K]). – **D:** Kenya, Somalia; shade in thickets. **I:** Sansevieria J. 3(1): 12-15, 1994. **Fig. XXIX.a**

   ≡ *Acyntha powellii* (N. E. Brown) Chiovenda (1932).

[1] Caulescent, rhizomatous; stem erect, to ≥ 1.2 m × 2.5 cm; **L** distichous, the ranks becoming twisted around the stem, spreading, semicylindrical, 30 - 69 × 2.2 - 2.9 cm, ± 1.3 cm thick, narrowed gradually to a pale brown hard **Sp**-like acute tip,

faintly glaucous grass-green becoming dark bluish-green, margin red-brown edged white, surface slightly rough; **Inf** ± 46 cm, paniculate; raceme sub--dense, 4 - 6 **Fl** per cluster; **Bra** convex, fleshy, 1.1 - 2.1 mm; **Ped** ± 3.2 mm; **Fl** dull greenish-white, with dull brownish-purple slender lines outside, tube 6.4 mm, lobes 9.5 mm.

   Pfennig (1977a) has suggested that this might be the natural hybrid *S. arborescens* × *S. robusta*.

S. raffillii N. E. Brown (BMI 1915(5): 72, fig. 22, 1915). **T:** Kenya (*Powell* 7 [K]). – **D:** Kenya.

S. raffillii var. **glauca** N. E. Brown (BMI 1915(5): 72, 1915). **T:** Kenya (*Powell* 8 [K]). – **D:** Kenya.

[3c] Differs from var. *raffillii*: **L** very dark bluish-green with distinct, but not very conspicuous, irregular spots or wavy transverse bands of lighter green 2.5 - 5 cm apart, distinctly bluish-glaucous; **Inf** peduncle bluish-glaucous below, lighter green speckled with pale green above; **Bra** linear-lanceolate, acute, 8.5 - 25.4 mm; **Ped** slightly glaucous, 5.3 - 6.4 mm.

S. raffillii var. **raffillii** – **D:** Kenya; *Acacia* bushland, 900 - 1700 m. **I:** Sansevieria J. 2(2): 35-37, 1993.

[3c] Acaulescent, rhizomatous; rhizome 2 - 5 cm ∅, whitish; **L** 1 - 2, erect, lanceolate or lorate, 68 - 152 × 5.5 - 12.5 cm, narrowed below the middle to a sessile base or short petiole, with a short reddish-brown hard tip, with yellowish-green closely placed blotches or irregular transverse bands on darker background, sometimes paler on the lower face, slightly glaucous, with age with less conspicuous markings, margin hard, reddish-brown, surface smooth; **Inf** 90 - 115 cm, simple, spike-like; raceme 61 - 76 cm, dense, 2 - 5 **Fl** per cluster; **Bra** ovate-lanceolate, acuminate, 5.3 - 17 mm; **Ped** 4.2 - 6.4 mm; **Fl** tube 2.5 - 2.86 cm, greenish-white, lobes 2.86 - 2.96 cm, white.

S. rhodesiana N. E. Brown (BMI 1915(5): 32, fig. 7, 1915). – **D:** Zambia or Zimbabwe.

[3b] Acaulescent, rhizomatous; rhizome stout; **L** 3 - 4, distichous, erect, cylindrical, with a groove on the upper face, 91 - 168 × 2.9 - 3.2 cm, 1.9 - 3.2 cm thick, narrowed gradually to a 6.4 - 12.7 mm whitish-brown hard acute tip, deep green with numerous darker green longitudinal lines, margin dark red-brown, sometimes edged white, surface slightly rough or nearly smooth; **Inf** ± 53 cm, simple, spike-like, peduncle pale green mottled darker green; raceme ± 38 cm, dense, 8 - 9 **Fl** per cluster; **Bra** acuminate, 3.2 - 4.2 mm; **Ped** 5.3 - 6.4 mm; **Fl** white, tube 0.85 - 1.27 cm, lobes 1.6 - 1.9 cm.

S. robusta N. E. Brown (BMI 1915(5): 26, 1915). **T:** K. – **D:** Kenya; dry bushland, 600 - 1500 m. **I:** Mbugua (1994); Newton (1994).

[1] Caulescent, rhizomatous; rhizome to 2 cm ∅, yellowish-white; stem erect, to 60 × ≥ 2.5 cm; **L** 6 - 14, distichous, erect or slightly spreading, cylindrical with a wide groove along the upper face, to 2.13 m × 3.4 cm, to 4.5 cm thick, narrowed gradually to a 6.4 - 12.7 mm brown hard **Sp**-like tip, dark green, faintly glaucous, with 14 - 30 longitudinal darker green lines, margin narrowly bordered red-brown, edged white, surface smooth; **Inf** 80 - 140 cm, paniculate, peduncle greenish-grey; raceme 12 - 20 cm, lax, 4 - 6 **Fl** per cluster; **Ped** 1.1 - 19 mm; **Fl** white or greenish, tube 10 - 25 mm, lobes 5 - 10 mm.

Originally described with flowers unknown.

S. rorida (Lanza) N. E. Brown (BMI 1915(5): 25, 1915). **T:** Somalia (*Macaluso* 177 [not located]). − **D:** Somalia; sandy places on the coast near sea-level. **I:** Sansevieria J. 5(1): 13-14, 1996.

≡ *Sanseverinia rorida* Lanza (1910) ≡ *Acyntha rorida* (Lanza) Chiovenda (1916).

[1] Caulescent; stem to 23 cm; **L** 11 - 15, distichous, ascending-spreading, cylindrical with groove along the upper face, 30 - 53 × 2.5 - 3.4 cm, 1.9 - 2.5 cm thick, narrowed gradually to a **Sp**-like tip, green, somewhat glaucous, with numerous darker longitudinal lines on the sides and lower face, margin reddish-brown, edged white; **Inf** ± 1 m, paniculate; raceme 7.6 - 18 cm, lax, 3 - 6 **Fl** per cluster; **Bra** deltoid, acute; **Ped** ± 2 mm; **Fl** tube ± 6.4 mm, lobes 1.06 - 1.27 cm, whitish-yellow with reddish mid-stripe.

Thulin (1995) lists this as a synonym of *S. ehrenbergii*.

S. roxburghiana Schultes (Syst. Veg. 7: 357, 1829). − **D:** India.

Incl. *Sansevieria zeylanica* Roxburgh (1805) (*nom. illeg.*, Art. 53.1).

[3c] Acaulescent, rhizomatous; **L** 6 - 24, rosulate, ascending, slightly recurved, linear with a deep groove on the upper face, 20 - 60 × 1.3 - 2.5 cm, narrowed gradually to a 6.4 - 50 mm green soft tip, green with darker green irregular transverse bands and dark green longitudinal lines, margin green, becoming whitish with age, upper face smooth, lower face slightly rough; **Inf** 30 - 76 cm, simple, spike-like; raceme 25 - 38 cm, ± 4 **Fl** per cluster; **Bra** lanceolate, attenuate, 3.2 - 4.2 mm; **Ped** 5.3 - 8.5 mm; **Fl** tube 6.4 - 7.4 mm, lobes 8.5 - 9.5 mm. − *Cytology:* 2n = 40 (Darlington & Wylie 1955).

S. sambiranensis H. Perrier (Notul. Syst. (Paris) 5(2): 154, 1935). **T:** P. − **D:** Madagascar; shaded rocks in humid forest. **I:** Perrier (1938: 5).

[2] Acaulescent, rhizomatous; rhizome to 30 × 1 - 2 cm; **L** 15 - 20, laxly rosulate, oblanceolate, 120 × 2.5 - 9 cm, narrowed below the middle to a 10 - 20 cm channelled petiole, narrowed to an acute tip, green; **Inf** 2 - 8 cm, capitate; raceme 4 - 5 cm ∅, dense; **Bra** lanceolate, acute, 6 - 7 mm; **Ped** 7 - 20 mm; **Fl** carmine-red, tube ± 1 cm, lobes 2 - 2.2 cm.

S. senegambica Baker (JLSB 14: 548, 1875). **T:** Senegal (*Anonymus* s.n. [K]). − **D:** Gambia, Guinea, Guinea-Bissau, Ivory Coast, Senegal, Sierra Leone; shady places. **I:** Brown (1915: 56).

Incl. *Sansevieria cornui* Gérôme & Labroy (1903).

[3c] Acaulescent, rhizomatous; rhizome 1.3 - 1.9 cm ∅, bright red, turning pale brownish in light; **L** 2 - 4, rosulate, suberect, recurved to spreading nearer the tip, linear-lanceolate to lanceolate, 30 - 69 × 3 - 6.4 cm, narrowed at the base to a 2.5 - 7.6 cm channelled petiole, narrowed gradually from the middle to a 4.2 - 12.7 mm green soft subulate tip, upper face dark green sometimes with indistinct paler green transverse bands, lower face slightly paler with more distinct transverse bands, margin green, surface smooth; **Inf** 30 - 50 cm, simple, spike-like, peduncle light or dark green or mottled purplish; raceme lax, 3 - 6 **Fl** per cluster; **Bra** ovate-lanceolate or oblong-lanceolate, acute, 6.4 - 8.5 × 2.1 - 3.2 mm; **Ped** 5.3 - 8.5 mm; **Fl** white, tinged purple in sun, tube 0.64 - 1.27 cm, lobes 1.06 - 1.9 cm. − *Cytology:* 2n = 40 (Sharma & Chaudhuri 1964).

S. sordida N. E. Brown (BMI 1915(5): 34, fig. 8, 1915). **T:** K. − **D:** Unknown.

[3b] Acaulescent or shortly caulescent, rhizomatous; **L** 4 - 12, distichous, slightly spreading, cylindrical, slightly compressed laterally with a channel down the upper face, 69 - 107 × 0.9 - 1.3 cm, 1.3 - 1.9 cm thick, becoming flattened towards the base, narrowed to a 0.7 - 1.1 mm whitish or grey brown-based acute **Sp**-like tip, dull bluish-green with numerous darker longitudinal lines, margin hardened, dark brown edged white, surface very rough; **Inf** 30 - 60 cm, simple, spike-like, peduncle dull light green, with minute white dots; raceme 20 - 45 cm, lax, 7 - 14 **Fl** per cluster; **Bra** subulate, 3.2 - 6.4 mm; **Ped** 8.5 - 12.7 mm; **Fl** tube 0.74 - 1.06 cm, white or greenish, lobes 1.48 - 1.69 cm, white inside, green with minute dull purplish dots outside.

S. stuckyi Godefroy-Lebeuf (Sansev. Gigant. Afr. Orient., 13, 17, 33, ills., 1861). − **D:** Moçambique. **I:** Brown (1915: 40). **Fig. XXIX.e**

Incl. *Sansevieria andradae* Godefroy-Lebeuf (1861).

[2] Acaulescent, rhizomatous; rhizome to 5 cm ∅; **L** 1 - 2 (rarely 3), erect, cylindrical with a groove along the upper face, 122 - 275 × 3.8 - 6.4 cm, narrowing gradually to a pale brown hard acute subulate tip, dull green, slightly glaucous, with transverse paler green bands and 6 - 20 longitudinal lines, margin green, surface slightly rough; **Inf** 21 - 32 cm, simple, capitate, peduncle purple, speckled green; raceme 7 × 34 cm, dense; **Bra** ovate-lanceolate, 8 - 20 × 4 - 10 mm; **Fl** tube 9 - 10 cm, lobes 4 cm. − *Cytology:* 2n = 116 (Sharma & Chaudhuri 1964).

Originally described with flowers unknown. See comment for *S. fischeri*.

S. subspicata Baker (Gard. Chron., ser. 3, 6: 436, 1889). **T:** Moçambique, Gaza Prov. (*Monteiro* s.n. [K]). – **D:** Moçambique. **I:** Brown (1915: 55).

[3c] Acaulescent, rhizomatous; rhizome 1.3 - 2.5 cm ⌀, pale yellowish-brown; **L** 4 - 10, erect or re-curved-spreading, lanceolate, 23 - 60 × 2.5 - 5.7 cm, narrowed from the middle or above to a 3.8 - 23 × 0.4 - 0.6 cm channelled petiole, with a 4.2 - 6.4 mm green soft subulate tip, deep green, sometimes faintly glaucous, margin green becoming whitish, surface smooth; **Inf** 30 - 40 cm, simple, spike-like; raceme lax, 1 - 2 **Fl** per cluster; **Bra** lanceolate, acute, 2.1 - 6.4 mm; **Ped** 1.1 - 2.1 mm; **Fl** tube 2.3 - 3 cm, greenish-white, lobes 1.7 - 1.9 cm, white.

S. subtilis N. E. Brown (BMI 1915(5): 57, fig. 17, 1915). **T:** Uganda (*Dawe* s.n. [K]). – **D:** Uganda.

[3c] Acaulescent, rhizomatous; rhizome 0.85 - 1.1 cm ⌀; **L** 2 - 4, erect or slightly recurved, linear-lanceolate, 53 - 69 × 2.5 - 4.5 cm, narrowed to a 5 - 30 cm channelled petiole, narrowed gradually from the middle or above to a 12.7 - 25.4 mm green soft subulate tip, green, lower face sometimes with faint transverse bands, margin green, surface smooth; **Inf** 38 - 53 cm, simple, spike-like; raceme lax, 2 - 3 **Fl** per cluster; **Bra** lanceolate, acuminate, 3.2 - 4.2 mm; **Ped** 4.2 - 7.4 mm; **Fl** tube 0.64 - 0.85 cm, lobes 1.06 - 1.27 cm, white.

S. suffruticosa N. E. Brown (BMI 1915(5): 22, fig. 3, 1915). **T:** Kenya (*Evans* s.n. [K]). – **D:** Kenya.

S. suffruticosa var. **longituba** Pfennig (BJS 102 (1-4): 178, 1981). **T:** (*Pfennig* 1336 [EA]). – **D:** Kenya; edges of thickets and on sandy cliffs, 1450 m. **I:** Sansevieria J. 1(3): 33, 45, 1992.

[3a] Differs from var. *suffruticosa*: **L** surface very smooth; **Fl** tube 2 cm.

S. suffruticosa var. **suffruticosa** – **D:** Kenya; edges of thickets, 1700 - 1900 m.

[3a] Caulescent, with runners; stem erect, to 30 cm, branching freely 0.8 - 7.6 cm above ground; runners spreading or ascending, to 25 × 1.9 cm; **L** 7 - 18, ± distichous or irregular, ascending or spread-ing, cylindrical with a groove on the upper face from the base to ¼ - ½ of the **L**-length, 15 - 60 cm, 1.3 - 1.9 cm thick, narrowed gradually to a 3.2 - 4.2 mm brown hard acute **Sp**-like tip, dark green with faint paler green transverse bands when young, with darker green longitudinal lines, surface rough, sometimes smooth near the base; **Inf** 30 - 38 cm, simple, spike-like, peduncle green with numerous whitish minute linear dots; raceme 4.5 - 6.4 cm ⌀, dense, 2 - 5 **Fl** per cluster; **Bra** ovate, acute or acu-minate, 1.1 - 4.2 mm; **Ped** 2.8 - 4.2 mm; **Fl** whitish or greenish-white, sometimes slightly red-tinged outside, tube ± 1 cm, lobes 1.27 - 1.48 cm.

S. sulcata Bojer *ex* Baker (JLSB 14: 549, 1875). **T:** Moçambique (*Bojer* s.n. [K]). – **D:** Comoro Is-lands, Moçambique, Tanzania; sea shore. **I:** Pfen-nig (1977a: 509).

[3b] Acaulescent, rhizomatous; rhizome 1.7 - 2.5 cm ⌀, reddish; **L** 1, erect, cylindrical with 8 - 9 broadly rounded ribs separated by furrows, 45 - 60 cm, 1 - 1.9 cm thick, narrowed to a ± 3.2 mm pale brown hardened abruptly acute tip, surface smooth; **Inf** 12.7 - 22.9 cm, simple, spike-like; raceme with 3 - 6 **Fl** per cluster; **Bra** ovate or ovate-lanceolate, acute, 2.5 - 3.2 mm; **Ped** 2.1 - 3.2 mm; **Fl** tube ≥ 1.9 cm, lobes 1.48 cm.

S. trifasciata Prain (Bengal Pl. 2: 1054, 1903). **T:** Nigeria (*Anonymus* s.n. [not located]). – **Lit:** Cha-hinian (1986). **D:** Nigeria, Zaïre; naturalized else-where (e.g. India).

Widely cultivated in numerous cultivars, see Chahinian (1986). A dwarf cultivar (leaves to 50 × 7.5 cm) is 'Hahnii' and several derivatives with different leaf sizes, different degrees of variegation, and different shades of green exist.

S. trifasciata var. **laurentii** (De Wildeman) N. E. Brown (BMI 1915(5): 60, 1915). **T:** Zaïre (*Anony-mus* s.n. [K]). – **D:** Zaïre. **I:** Chahinian (1986: 21).

≡ *Sansevieria laurentii* De Wildeman (1904).

[3c] Differs from var. *trifasciata*: **L** with yellow margin to 1 cm wide. – *Cytology:* 2n = 40 (Sharma & Chaudhuri 1964: as *S. laurentii*).

A periclinal chimaera.

S. trifasciata var. **trifasciata** – **D:** Nigeria. **I:** Cha-hinian (1986: 20).

Incl. *Aletris hyacinthoides* Miller (1768) (*nom. illeg.*, Art. 53.1); **incl.** *Acyntha guineensis* Medikus (1786); **incl.** *Salmia guineensis* Cavanilles (1795); **incl.** *Sansevieria guineensis* Gérôme & Labroy (1903) (*nom. illeg.*, Art. 53.1); **incl.** *Sansevieria zebrina* L. Gentil (1907) (*nom. inval.*, Art. 32.1c); **incl.** *Sansevieria jacquinii* N. E. Brown (1911).

[3c] Acaulescent, rhizomatous; rhizome 1.3 - 2.5 cm ⌀; **L** 1 - 2 (-6), erect, linear-lanceolate, 30 - 122 × 2.5 - 7 cm, narrowed gradually from ± or above the middle to a channelled petiole, with a 3.2 - 3.8 mm green subulate tip, with alternating transverse bands of light dull green or clear whitish-green and deep grass-green to almost blackish-green, with slight glaucous bloom, margin green, surface smooth; **Inf** 30 - 76 cm, simple, spike-like, peduncle green with pale green dots; raceme lax, 3 - 8 **Fl** per cluster; **Bra** ovate or ovate-lanceolate, acuminate, 3.2 - 12.7 mm; **Ped** 5.3 - 8.5 mm; **Fl** white, tube 0.64 - 1.27 cm, lobes 1.48 - 1.9 cm. – *Cytology:* 2n = 36 (Sharma & Chaudhuri 1964).

S. varians N. E. Brown (BMI 1915(5): 29, 1915). **T:** K. – **D:** Known from cultivation only.

[3b] Acaulescent, rhizomatous; **L** 4 - 8, distich-

ous, erect or ascending, cylindrical with a groove on the upper face or semicylindrical with a shallow groove on the upper face, 38 - 114 cm, 1.3 - 2.2 cm thick, narrowed gradually to a 4.2 - 8.5 mm whitish hard acute tip, dull dark grass-green with numerous dark green longitudinal lines, when young with indistinct lighter green transverse bands, margin green or red-brown often edged white, surface slightly rough; **Inf** 60 - 76 cm, simple, spike-like, peduncle light glaucous-green or greyish-green; raceme with 6 - 10 **Fl** per cluster; **Bra** linear or filiform, acute, 4.2 - 6.4 mm; **Ped** 4.2 - 5.3 mm; **Fl** white, speckled with purple at the tips, tube 1.06 - 1.27 cm, lobes 1.48 - 1.69 cm.

S. volkensii Gürke (in Engler, Pfl.-welt Ost-Afr., Teil C, 144, 1895). **T:** Tanzania, Kilimanjaro Distr. (*Volkens* 1779 [B]). – **D:** Kenya, Tanzania; edge of thickets on sandy soil, 500 - 900 m. **I:** Brown (1915: 32, as *S. intermedia*).

Incl. *Sansevieria intermedia* N. E. Brown (1914); **incl.** *Sansevieria humbertiana* Guillaumin (1940).

[3b] Acaulescent, rhizomatous; **L** 2 - 7, erect or ascending, semicylindrical, 45 - 122 cm, 1.3 - 1.9 cm thick, narrowed gradually to a whitish acute **Sp**-like tip, dull deep green, becoming slightly bluish-green, sometimes with faint whitish transverse bands, margin green or whitish, surface slightly rough; **Inf** 20 - 46 cm, simple, spike-like, peduncle light greyish-green; raceme ± 30 - 40 cm, dense, 3 - 6 **Fl** per cluster; **Bra** ovate or ovate-lanceolate, acute, 2.1 - 4.2 mm; **Ped** 2.1 mm; **Fl** white, tube 1.48 - 1.9 cm, pale greenish, lobes 1.27 - 1.9 cm, white or greenish-white with minute purplish spots on the outside.

S. zanzibarica Gérôme & Labroy (Bull. Mus. Hist. Nat. (Paris) 9: 170, 172-173, fig. 19, 1903). **T:** Tanzania / Zanzibar? (*Sacleux* s.n. [P]). – **D:** Tanzania or Zanzibar.

Incl. *Sansevieria ehrenbergii* Gérôme & Labroy (1903) (*nom. illeg.*, Art. 53.1); **incl.** *Sansevieria ehrenbergii* De Wildeman (1905) (*nom. illeg.*, Art. 53.1).

[-] Very shortly caulescent; **L** distichous, recurved-spreading, linear-lanceolate, 15 - 30 × 1.9 - 2.5 cm, dull dark green with bluish-grey bloom, margin reddish-brown; **Inf** and **Fl** not known.

Originally described with flowers unknown, and still a little-known species.

S. zeylanica (Linné) Willdenow (Spec. Pl. 2: 159, 1799). **T:** [icono]: Commelin, Hort. Med. Amstel. Rar. Pl. 2: t. 21, 1701. – **D:** Sri Lanka; rocky or sandy places in dry regions. **I:** Wijnands (1973: 110).

≡ *Aloe hyacinthoides* var. *zeylanica* Linné (1753) ≡ *Aletris hyacinthoides* var. *zeylanica* (Linné) Linné (1762) ≡ *Aloe zeylanica* (Linné) Jacquin (1762) ≡ *Aletris zeylanica* (Linné) Miller (1768); **incl.** *Aletris zeylanica* Lamarck (1789) (*nom. illeg.*, Art. 53.1);

incl. *Sansevieria indica* Herter (1956) (*nom. illeg.*, Art. 52.1).

[3b] Acaulescent, rhizomatous; rhizome 1.3 cm ∅; **L** 5 - 11, erect below, slightly recurved above, linear to semicylindrical with a groove on the upper face, 45 - 76 × 0.9 - 2.1 cm, 5 - 8.5 mm thick, narrowed gradually to a 12.7 - 38 mm long green soft subulate acute tip, dark green with lighter green transverse bands, with 4 - 7 darker green longitudinal lines, margin green, surface almost smooth; **Inf** and **Fl** not known. – *Cytology:* 2n = 40, 42 (Darlington & Wylie 1955); 2n = 42 Janaki-Ammal (1945).

Described with flowers not seen, and still a little-known species.

Eriospermaceae

Perennial geophytes with solitary or stoloniferous tubers; tubers hypocotyledonary, very variable in size and shape, globose to oblong with an apical or dorsal growing point, tissue white to red; **L** petiole persistent forming a sheath through which the next season's **L** appears; **L** deciduous, synanthous or hysteranthous, usually solitary, flat, ovate to linear, smooth, hairy or with finger-like outgrowths or dissected appendages on the upper surface; **Inf** racemose, simple, usually appearing during summer or autumn; scape erect, without or with minute **Bra**; **Fl** pedicellate, actinomorphic, rotate to campanulate, diurnal; **Tep** and **St** 3+3 in 2 whorls; **Tep** somewhat connate basally, white, pink or yellow; **St** basally adnate to the **Tep**; **Fil** filiform to lanceolate; **Anth** dorsifixed, peltate, introrse, dehiscing longitudinally; **Ov** superior, globose, sessile, 3-locular; ovules few, placentation axillary; **Sty** terete; **Sti** 3-lobed; **Fr** oblong to ovate loculicidal capsules; **Se** few (to 12), pear- to comma-shaped, densely pilose with 1-celled white **Ha**.

Distribution: Africa S of the Sahara, with a concentration in the Succulent Karoo regions of the Western Cape (RSA).

Literature: Dahlgren & al. (1985: 168-170).

Eriospermum is the only genus of this monotypic family.   [E. van Jaarsveld]

ERIOSPERMUM

E. van Jaarsveld

Eriospermum Jacquin *ex* Willdenow (Spec. Pl. 2(1): 110, 1799). **T:** *Eriospermum lanceifolium* Jacquin ex Willdenow [Lectotype, designated by Phillips, Gen. South Afr. Pl., ed. 2, 115, 1951.]. – **Lit:** Perry (1994). **D:** Africa S of the Sahara with a diversity centre in the Western Cape (RSA). **Etym:** Gr. 'erion', wool; and Gr. '-spermus', -seeded; for the hairy seeds.
Incl. *Phylloglottis* Salisbury (1866). **T:** *Eriospermum folioliferum* Andrews.
Incl. *Thaumaza* Salisbury (1866). **T:** *Ornithogalum paradoxum* Jacquin.
    Description as for the family.
The monogeneric family *Eriospermaceae* is confined to Africa S of the Sahara and shows centres of endemism in the semi-arid winter rainfall regions of the Western Cape, RSA. They are ± easily cultivated from tubers, known as "Bobbejaanui" [Baboon's Onion] in South Africa. The tubers are eaten by various animals, including porcupines and rodents. The genus embraces 102 species, classified into 3 subgenera (and a total of 9 sections, not shown here) by Perry (1994):

[1] Subgen. *Ligulatum* P. L. Perry 1994: **Tep** all alike or almost so, spreading to recurved and ligulate; **Fil** filiform; tuber globose or irregularly shaped (31 species, mainly from tropical and subtropical Africa).
[2] Subgen. *Cyathiflorum* P. L. Perry 1994: **Tep** all alike or almost, **Fl** rotate or cup-shaped; tuber irregularly shaped (17 species, widespread but more common in summer rainfall regions of RSA, reaching Malawi and Tanzania in the N).
[3] Subgen. *Eriospermum*: **Tep** dimorphic; tuber irregularly shaped with lateral to basal growing point (54 species, mainly in the winter rainfall region of the Western Cape).

    In addition to the characters cited, the colour of the internal tissue of the tubers is of diagnostic value.
Because of the ± succulent tubers present in most taxa, most could be included here. Some species with peculiar succulent leaf appendages are particularly attractive and worthy of cultivation (in small containers). Since none are frequently seen in cultivation, the following few taxa must be sufficient as examples.

E. armianum P. L. Perry (Contr. Bolus Herb. 17: 238, fig. 139, 1994). **T:** RSA, Northern Cape (*Mitchell* 1179 [NBG]). – **D:** RSA (Northern Cape: Namaqualand escarpment); Succulent Karoo in quartz-gravel flats.
[3] Tuber irregular-globose, 28 × 22 mm; **L** hysteranthous (in winter in habitat), solitary, erect, petiole 12 - 15 mm, lamina cordate, 18 × 12 mm, striate, glaucous, base of upper surface with terete linear club-shaped outgrowths to 15 mm, glaucous, base subpeltate, apex obtuse; **Inf** to 10 cm; **Ped** to 1 cm; **Fl** ascending, white with green midveins, to 13 mm ∅; **OTep** oblong oblanceolate, 7 × 2 mm; **ITep** widening to apex, 5.5 × 1.6 mm; **Fil** connate to base of **Tep**, ovate to triangular-oblong; **Ov** ovoid, 2 mm ∅; **Sty** 1.8 mm.

E. bowieanum Baker (JLSB 15: 267, 1876). **T:** RSA, Western Cape (*Bowie* s.n. [K [lecto - icono]]). – **D:** RSA (Western Cape: Robertson Karoo); semi-arid regions; autumn-flowering.
    Incl. *Eriospermum coralliferum* Marloth (1929).
[3] Tuber irregular-globose, 5.5 × 3.8 cm; **L** hysteranthous, solitary, erect, petiole to 4 cm, lamina cordate, 1 × 1 cm, glaucous, base of upper surface with terete linear club-shaped outgrowths to 5 cm, base subpeltate, apex retrorse; **Inf** to 3 cm, dense-flowered, spicate; **Ped** to 2.5 cm; **Fl** spreading, to 7 mm ∅; **Tep** white with green midvein, **OTep** lanceolate, 5 × 1.8 mm, **ITep** widening to apex, 4 × 2.3 mm; **Fil** connate to base of **Tep**, lanceolate; **Ov** globose, 1.3 mm ∅; **Sty** 1.5 mm.

E. paradoxum (Jacquin) Ker Gawler (CBM 1811:

t. 1382, 1811). **T:** [lecto – icono]: Jacquin, Collectanea 5: t. 1, 1796. – **D:** RSA (Western Cape, W Eastern Cape); semi-arid Renosterveld and Succulent Karoo, winter rainfall regions (flowering in autumn).

≡ *Ornithogalum paradoxum* Jacquin (1796) ≡ *Thaumaza paradoxa* (Jacquin) Salisbury (1866); **incl.** *Eriospermum cylindricum* Marloth (1929); **incl.** *Eriospermum arenicola* Von Poellnitz (1943); **incl.** *Eriospermum vallisgratiae* Schlechter *ex* von Poellnitz (1944).

[1] Tuber oblong, to 5 × 2.5 cm; **L** hysteranthous, solitary, erect, ovate-cordate, 7 × 6 cm, adaxial surface hairy with central-stemmed erect oblong tree-like branched appendage 11 × 3 cm; **Inf** dense conical spikes 9 × 3.5 cm; **Ped** to 6 mm; **Fl** spreading, to 1.7 cm ∅; **Tep** white with green midvein, **OTep** lorate, 12 × 2 mm, **ITep** linear-spatulate, to 10 × 2.5 mm; **Fil** 9 × 1 mm; **Ov** globose, 2 mm ∅; **Sty** to 6 mm, white.

E. titanopsoides P. L. Perry (Contr. Bolus Herb. 17: 281, fig. 163, 1994). **T:** RSA, Western Cape (*Bruyns* 3151 [NBG]). – **D:** RSA (Western Cape: Knersvlakte); Succulent Karoo in quartz gravel flats. **I:** Desmet (2000).

[3] Tuber oblong, to 15 × 10 mm; **L** hysteranthous (in winter in habitat), solitary, prostrate, ovate, 9 × 5 mm, folded, succulent, glaucous, margin conspicuously undulate and wrinkled, epidermis papillate, base subpeltate, apex obtuse; **Inf** 2.5 cm; **Ped** to 12 mm; **Fl** ascending, white with green midveins, to 9 mm ∅; **OTep** elliptic, 5 × 1.5 mm; **ITep** widening to apex, 5 × 2 mm; **Fil** connate to base of **Tep**, ovate to triangular-oblong; **Ov** ovoid, 2 mm ∅; **Sty** to 1.8 mm.

Hyacinthaceae

Perennial bulbous herbs bearing fleshy tunics, rarely rhizomatous; **R** fleshy, contractile; **L** basal, numerous or few or solitary, lamina linear to oblong, mostly flat, strap-shaped, occasionally softly succulent, sometimes broadly ovate or tapering, rarely early deciduous; **Inf** racemose, paniculate or spicate, bracteate (rarely climbing if the only assimilating organ of the plant); **Fl** bisexual, actinomorphic; **Tep** and **St** 6 in 2 whorls of 3; **Tep** free or connate, sometimes forming a tube, often white but colour variable; **St** free or often basally adnate to **Tep**; **Fil** flat, often lanceolate; **Anth** dorsifixed, introrse, dehiscing longitudinally; **Ov** superior, 3-locular; ovules few, placentation axillary; **Sty** terete; **Sti** tricuspidate; **Fr** loculicidal capsules; **Se** variable, often ovoid to orbicular, sometimes winged.

Distribution: Africa, Mediterranean to SW Asia, but concentrated in RSA.

Literature: Dahlgren & al. (1985: 168-170).

A large family with some 900 species in 30 genera, with a concentration of diversity in the winter rainfall regions of RSA. Succulents from several genera are occasionally cultivated; most of these are from the semi-arid winter-rainfall regions of RSA and should be kept dry during summer. Readily propagated from seeds or daughter bulbs, some also from bulb scales.

Several genera have considerable horticultural importance, e.g. *Ornithogalum*, *Hyacinthus*, *Lachenalia*, *Veltheimia* and *Eucomis*. *Ornithogalum* is the largest genus with more than 100 species, of which several are here treated as succulents.

Key to genera with succulents, adopted from Dyer (1975-76):
1   Adult plants without **L**; **Inf** erect or twining, richly branched, green: **2**
–   Adult plants with **L** in various shapes; **Inf** racemes or capitate: **3**
2   **Inf** twining-climbing, succulent: **Bowiea**
–   **Inf** erect, wiry-sturdy to wiry-filiform: **Schizobasis**
3   Lower **Bra** spurred, early deciduous; **Fr** circumscissile below; **Se** flattened: **4**
–   **Bra** not spurred, persistent; **Se** globose, angular or flat; **L** normally contemporaneous with **Inf**: **8**
4   Plants 10 - 150 cm tall; **St** exserted: **5**
–   Plants to 6 cm tall; **Fl** solitary; **St** not exserted: **Litanthus**
5   **Tep** usually free or fused only near the base; **St** spreading: **Urginea**
–   **Tep** usually fused and forming a tube: **6**
6   **Tep** lobes reflexed: **Drimia**
–   **Tep** lobes spreading and **Fl** campanulate: **7**

7 **Tep** lobes fused for ½ of their length, thin-
   textured: **Rhadamanthus**
— **Tep** lobes fused for > ⅔ of their length, ± thick-
   textured: **Rhodocodon**
8 **Tep** free or fused only near the base; **Fl** neither
   tubular nor campanulate: **9**
— **Tep** distinctly fused; **Fl** tubular or campanulate;
   L in **Ros** or only 2 large basal **L**:   **11**
9 **OTep** spreading, **ITep** erect and cucullate; **St**
   often 3 fertile and 3 sterile: **Albuca**
— All **Tep** spreading: **10**
10 **Inf** apparently terminal; ovules many; **L** always
   unspotted: **Ornithogalum**
— **Inf** axillary; **Tep** tips reflexed; ovules 2 per
   locule; **L** often blotched with different colours:

                                        Ledebouria
11 **OTep** shorter than **ITep**, with distinct appendage; **Se** flat, discoid: **Dipcadi**
— If **OTep** shorter than **ITep** then without appendage; **Se** globose to pear-shaped: **12**
12 Upper **Fl** of each **Inf** rudimentary; **OTep** shorter
   than the **ITep**:   **Lachenalia**
— Upper **Fl** normally developed; all **Tep** of ± the
   same length: **13**
13 **Inf** spikes between the **L** or raised above the **L**;
   Per campanulate: **14**
— **Inf** capitate, sessile between the **L** at ground-
   level: **Massonia**
14 **L** only 2, spreading on the ground:
                                        Whiteheadia
— **L** several in a **Ros**, ascending: **Hyacinthus**

                                  [E. van Jaarsveld]

ALBUCA

E. van Jaarsveld

Albuca Linné (Spec. Pl., ed. 2, 438, 1762). **T:**
Albuca major Linné. – **D:** Africa to Arabia, centred
in RSA. **Etym:** Lat. 'albucus', "Asphodel"
(*Asphodelus sp.*, from Lat. 'albus', white, because of
white flowers); for the similarity of some species
with Asphodel.

Perennial bulbous geophytes; bulbs variable, tunicate, underground or above-ground, mostly globose, rarely depressed; tunics fleshy, imbricate; **L**
very variable, deciduous and hysteranthous, rarely
evergreen, rosulate, terete to flattened, straight or
curved; **Inf** racemose; peduncle bracteate, **Bra** acuminate; **Ped** erect, variable in length; **Fl** actinomorphic, erect or drooping; **Tep** 6, lorate, green, white
or yellow with green midstripe; **OTep** 3, free and
spreading; **ITep** covering **St** and **Ov**, cucullate with
large apical gland; **St** 6, all fertile or the outer 3
sterile; **Fil** dilated at the base and covering the **Ov**;
Anth introrse, versatile, oblong; **Ov** superior, 3-lo-
cular, oblong, placentation axile; **Sty** obconical-
prismatic; **Sti** capitate or conical with 3 deltoid

fimbriate lobes; **Fr** 3-locular loculicidal capsules,
ovoid to trigonous; **Se** flat, black, shiny.

A genus of ± 100 species occurring in Africa and
Arabia with its main centre in S Africa. About 50
species are native to RSA, and are widespread in
grassland, savanna, Karoo and Fynbos. Many have
xeromorphic features, and the few species described
below are horticultural curiosities sometimes cultivated. Albucas are easily grown from seed.

A. batteniae Hilliard & B. L. Burtt (Notes Roy.
Bot. Gard. Edinburgh 42(2): 247-249, fig. 1, 1985).
T: RSA, Eastern Cape (*Hilliard & Burtt* 12454 [E,
NU, PRE]). – **D:** RSA (Eastern Cape); Valley
Bushveld, sheer cliff faces, flowers spring to aut-
umn.

Evergreen, bulbs above-ground (rarely underground), solitary, ovoid, 5 × 3 cm, tunics fleshy,
truncate at the top, green; **R** fleshy, white, to 2 mm
∅; **L** in a **Ros**, 12 - 30 × 2 - 3 cm, oblong, linear-
attenuate, soft, succulent, caniculate, dark green,
faintly lineate, glabrous, apex acute; **Inf** spreading
racemes to 80 cm; scape to 25 cm, basally 6 - 10
mm ∅; **Bra** acuminate, 4 × 1.3 cm, green with
white, margin translucent; **Ped** erect, to 12 cm at
the base becoming shorter upwards (to 3.5 cm); **Fl**
erect; **Tep** white with green midstripes, oblong, cucullate; **OTep** 3 - 4.2 × 0.7 cm; **ITep** 2.5 - 3 × 0.7
cm; **Fil** 1.5 - 2 cm, white, flattened at the base;
Anth oblong, versatile, outer 4 × 1 mm, inner 7 ×
2.5 mm; **Ov** 8 - 12 mm, obtusely trigonous; **Sty** 1 -
1.3 cm; **Sti** trilobate, white; **Se** flat, shiny.

A. cremnophila van Jaarsveld & A. E. van Wyk
(Aloe 36(4): 72-73, ills., 2000). **T:** RSA, Eastern
Cape (*van Jaarsveld* 12171 [NBG]). – **D:** RSA
(Eastern Cape); Valley Bushveld, on sheer cliff
faces, in summer- and winter-rainfall regions.

Evergreen; bulbs above-ground (rarely underground), solitary, ovoid, 9 × 5 - 6 cm, tunics fleshy,
imbricate, truncate at the top, green-grey; **R** fleshy,
white, to 3 mm ∅; **L** in a **Ros**, 30 - 70 × 2 - 3 cm,
linear-attenuate, firm and succulent, caniculate,
drooping, dark green, glabrous, tips acute; **Inf** pendulous racemes to 2 m; scape to 25 cm; **Bra** acuminate, membranous, margin translucent, lower **Bra** to
45 × 5 mm, diminishing in size upwards to 15 × 3
mm; lower **Ped** 6 mm becoming shorter upwards; **Fl**
erect; **Tep** white with green midstripes; **OTep** linear-obovate, 20 × 8 mm; **ITep** ovate with hooded
yellowish apex, 18 × 10 mm; **Fil** 13 mm, 2.5 mm ∅
at the base, all fertile; **Ov** 6 mm, stipitate for 1 mm,
4 mm ∅ at the base narrowing to 3 mm, triangular,
basally each angle with a raised twin tubercle; **Sty**
linear, trigonous, 10 × 2 mm.

A. juncifolia Baker (Gard. Chron., ser. nov. 5: 534,
1876). **T:** RSA, Western Cape (*Hutton* s.n. [K]). –
D: RSA (Western Cape); coastal Renosterbosveld,
flowers in late spring. **Fig. XXIX.c**

                                            275

Bulbs partly above-ground, depressed-globose, to 4.5 × 3 cm, solitary or dividing to form small groups; tunics succulent, brownish-green, drying to grey-white; **R** fleshy, white, to 3 mm ∅; **L** 2 - 5, ascending to spreading, 5.5 - 20 cm × 1 - 5 mm, linear-attenuate, succulent, appearing terete but with strongly inrolled margins, base flat and channelled, apex acute; **Inf** 1 - 2, erect, racemose, to 28 cm; peduncle 3 mm ∅ at base, sterile for lower ½; lower **Bra** lanceolate-acuminate, 25 × 7 mm, becoming smaller upwards, white, soon becoming brown; lower **Ped** to 7 cm; **Fl** erect, to 25 mm; **Tep** yellow with broad green midstripes; **OTep** 25 × 7 mm, erectly spreading, oblong, lorate, apex cucullate; **ITep** 20 × 7 mm, elliptic-ovate, cymbiform, apex cucullate and notched, truncate; **Fil** 12 - 14 mm, white, flat and 3 mm ∅ at the base, inner **Fil** basally constricted; **Anth** oblong, 3 × 1.2 mm, versatile; outer **Fil** and **Anth** smaller; **Ov** 6 - 7 mm, obtusely trigonous; **Sty** prismatic, 7 mm, slightly widening at the tip, yellowish; **Sti** trilobate, yellow.

A. nelsonii N. E. Brown (Gard. Chron., ser. nov. 14: 198, t. 41, 1880). **T:** RSA, KwaZulu-Natal (*Nelson s.n.* [K]). – **D:** RSA (Eastern Cape, KwaZulu-Natal); Bushveld, summer rainfall regions, flowers midsummer to autumn.

Bulbs pear-shaped, to 12 × 9.5 cm, above-ground, solitary or dividing to form small groups; tunics succulent, green, imbricate and truncate at the apex, becoming shorter towards the base; **R** fleshy, white, to 5 mm ∅; **L** 4 - 6 in a **Ros**, 20 - 100 × 1.7 - 5 cm, ascending, curved, lamina oblong, linear-attenuate, succulent, firm, dark green, glabrous and shiny, channelled for most of the length, apex acute; **Inf** racemose, erect, to 1.2 m; peduncle sterile for 75 cm, 1.5 cm ∅ at the base; **Bra** lanceolate-acuminate, 4 - 15 cm, green with white translucent margin; **Ped** erect, 8 - 16 cm; **Fl** erect, to 3.5 cm long; **Tep** white with green midstripes; **OTep** 3.3 - 3.5 × 0.9 cm, oblong to ovate, cucullate; **ITep** 3 × 1.3 cm, elliptic-oblong, cucullate, truncate; **Fil** white, outer 24 × 1 mm, inner 28 mm × basally 3.5 mm; outer **Anth** 4 × 1 mm, inner 6 - 7 × 2.5 mm, all fertile; **Ov** 8 - 12 mm, obtusely trigonous; **Sty** prismatic, 5 mm; **Sti** trilobate, white.

A. spiralis Linné (Suppl. Pl., 196, 1782). – **D:** RSA (Western Cape); Renosterveld, winter-rainfall regions, spring-flowering.

Dwarf winter-growing geophytes with the top of the bulbs exposed, solitary or dividing to form small groups; bulb ovoid, 8 - 25 × 5 - 25 mm ∅, tunics green and fleshy, withering grey and membranous; **R** fibrous, terete; **L** numerous (15 - 20) in a dense basal **Ros**, 3 - 11 cm × 1 mm, terete, green, glandular-pubescent, withering in summer, erect and spirally curved at the tip, upper face slightly channelled at the base; **Inf** 4 - 5 cm with 1 - 5 diurnal drooping **Fl**; **Bra** linear-lanceolate, to 18 ×

4 mm, glandular-hairy; **Ped** cernuous, 2 - 2.5 cm; **Per** to 1.8 cm long, yellow; **St** all fertile; **Sty** prismatic, as long as the **Ov**; **Sti** tricuspidate; **Ov** obovoid-ellipsoid to globose, 6-grooved, with light green tubercles along the middle part; **Fr** not seen.

BOWIEA

E. van Jaarsveld

Bowiea Harvey *ex* Hooker *fil.* (CBM 43: t. 5619 + text, 1867). **T:** *Bowiea volubilis* Hooker *fil.* – **Lit:** Jessop (1977: 312-314); Jaarsveld (1992b). **D:** E and S Africa from Kenya to RSA and Namibia. **Etym:** For James Bowie (1789 - 1869), English horticulturist and botanical collector in S Africa.

Incl. *Schizobasopsis* Macbride (1918) (*nomen rejiciendum*, Art. 56.1). **T:** *Bowiea volubilis* Hooker *fil.*

Perennial geophytes; **R** fleshy, white, to 5 mm ∅; bulbs depressed-globose, subterranean to almost fully exposed; exposed parts green; tunics thickly fleshy withering to paper-like; **L** only present in immature young plants, fleshy, linear-lanceolate, canaliculate, short-lived; stem (= **Inf**) annual, scandent, twining or scrambling, much branched, softly succulent, green or glaucous, branchlets subulate; **Bra** lanceolate, spurred; **Ped** arched; **Fl** 6-lobed, diurnal, scented; **Tep** free to the base, white or green to yellowish-green, patent to reflexed, oblong to lanceolate, margins revolute towards base, apex subacute; **St** 6, **Fil** suberect; **Ov** 3-chambered, broadly conical, light green, glutinous on the upper surface; **Sty** terete, **Sti** apical, 3-lobed; **Fr** 3-locular erect capsules with emarginate to acuminate valves, dehiscing longitudinally; **Se** black, angular-oblong, shiny.

Both species are popular medicinal plants in RSA and wild populations are becoming very rare. Propagation is by bulb scales or seeds.

B. gariepensis van Jaarsveld (JSAB 49(4): 343-346, ills., 1983). **T:** RSA, Western Cape (*van Jaarsveld 6650* [NBG, PRE]). – **D:** RSA (W Northern Cape), S Namibia; along cool S-facing screes in the Orange River valley, autumn- and winter-growing / flowering. **I:** Jaarsveld (1992b). **Fig. XXIX.b**

≡ *Bowiea volubilis* ssp. *gariepensis* (van Jaarsveld) Bruyns (1987).

Bulb to 14 cm ∅; stem (= **Inf**) to 1.2 m, glaucous to glaucous-green, branchlets subulate, to 7 cm × 2 - 5 mm ∅; **Bra** lanceolate, spurred, 3 - 5 mm; **Fl** 14 - 23 mm ∅; **Tep** 12 × 3.5 mm, white, patent to somewhat reflexed, oblong to lanceolate, apex subacute; **Fr** globose, depressed, 4 - 6 mm with emarginate valves; **Se** 4 - 5 mm.

B. volubilis Harvey *ex* Hooker *fil.* (CBM 43: t. 5619 + text, 1867). **T:** RSA, KwaZulu-Natal

(*Cooper* 3263 [K]). – **D:** E and SE Africa; dry savanna and forest, spring- and summer-growing/flowering. **I:** Jaarsveld (1992b).

≡ *Schizobasopsis volubilis* (Harvey *ex* Hooker *fil.*) Macbride (1918) (*incorrect name*, Art. 11.3) ≡ *Schizobasis volubilis* (Harvey *ex* Hooker *fil.*) van Jaarsveld (1992) (*nom. inval.*, Art. 34.1c); **incl.** *Bowiea kilimandscharica* Mildbraed (1936) ≡ *Schizobasopsis kilimandscharica* (Mildbread) Barschus (1954) (*incorrect name*, Art. 11.3).

Bulb to 16 cm ∅; stem (= **Inf**) 3 - 4 (-10) m, green, branchlets subulate, to 7 cm × 2 - 5 mm ∅; **Bra** lanceolate, spurred, 3 - 5 mm; **Fl** 10 - 16 mm ∅; **Tep** 12 × 3.5 mm, green to yellowish-green, reflexed, oblong to lanceolate, apex acute to subacute; **Ov** broadly conical, 5 mm ∅, light green; **Fr** conical, 8 - 30 mm with acuminate valves; **Se** 5 - 10 mm.

DIPCADI

U. Eggli

Dipcadi Medikus (Acta Acad. Theod.-Palat. 6: 431, 1790). **T:** *Hyacinthus serotinus* Linné. – **D:** S Europe, Arabian Peninsula, tropical and S Africa, Madagascar, S India. **Etym:** Perhaps the ancient oriental name for some species today classified as *Muscari* ("Grape Hyacinth").

Incl. *Zuccangnia* Thunberg (*nomen rejiciendum*, Art. 56.1). **T:** *Hyacinthus viridis* Linné.
Incl. *Uropetalon* Burchell *ex* Ker Gawler (1816). **T:** *Hyacinthus viridis* Linné.
Incl. *Uropetalum* Burchell (1822) (*nom. inval.*, Art. 61.1). **T:** *Hyacinthus viridis* Linné.
Incl. *Polemannia* Bergius *ex* Schlechtendal (1826). **T:** *Polemannia hyacinthiflora* Schlechtendal.
Incl. *Tricharis* Salisbury (1866). **T:** not typified.

Perennials with underground bulbs with papery or subfleshy tunics; **L** synanthous or hysteranthous, linear to lorate, rarely ± succulent; **Inf** lax long-pedunculate spikes; **Per** white, greenish, yellowish or brownish, often glaucous; **Tep** in 2 distinct series, basally united to form a tube; tips of **OTep** spreading-ascending, with a filiform to stout conical appendage; **ITep** continuing the **Per** tube, tips ascending or spreading; **Fil** basally united with the **Per** tube; **Ov** sessile, with numerous ovules; **Fr** membranous capsules; **Se** compressed, winged.

The genus is widely distributed esp. in Africa and Arabia and counts ± 55 species. They are easily recognizable on the base of the continuation of the perianth tube formed by the inner tepals, and the long-appendiculate tips of the outer tepals. Many taxa are notable for the glaucous dark green or brownish coloured flowers. Moderate leaf succulence is found in the following species only:

D. hyacinthoides (Berg) Baker (JLSB 11: 398, 1871). – **D:** SW coastal Madagascar, RSA.

Incl. *Polemannia hyacinthoides* Berg (1826) ≡ *Uropetalon hyacinthoides* (Berg) Sprengel (1827).

Bulbs shallowly seated or at ground-level, ovoid, ± 1.5 cm ∅; **L** 3 - 10, semi-erect or spreading, strap-shaped, 15 - 40 × 0.4 - 1 cm, succulent and coriaceous, slightly roughened, with parallel narrow channels, margins finely white-toothed; **Inf** 1 - 2 from **L** axils; 25 - 35 cm, scape ± ⅔ of overall **Inf** length, naked; **Bra** all fertile, papery, 4 - 12 mm, longer than the **Ped**; **Ped** 2 - 10 mm, elongating at fruiting time; **Fl** pendent; **Per** 16 - 25 mm, brownish-grey, slightly glaucous; appendix of **OTep** short and thick, 3 - 4 mm; tube of **ITep** 6 mm longer than the outer tube; free part of **Fil** ± 0.5 mm, flattened; **Ov** basally narrowed.

Perrier (1938) gives the taxon as endemic for Madagascar, while Baker (l.c.) lists only localities in RSA.

DRIMIA

E. van Jaarsveld

Drimia Jacquin *ex* Willdenow (Spec. Pl. 2: 165, 1799). **T:** *Drimia elata* Jacquin *ex* Willdenow [Type according to ING from E. Phillips, Gen. South Afr. Pl., ed. 2, 190, 1951.]. – **Lit:** Jessop (1977). **D:** RSA; widespread. **Etym:** Gr. 'drimys', sharp, cutting; for the pointed capsules (Genaust 1983).

Perennial geophytes with fleshy tunics forming a lax to compact bulb, solitary or proliferating to form groups; **L** strap-shaped, flat, variable, withering during the dry season; **Inf** racemose; **Per** campanulate, **Tep** reflexed; **St** inserted at the throat (above the **Ov**) of the **Per** tube; **Fil** terete to flat; **Anth** versatile, introrse; **Ov** sessile, ovoid to pyramidal, 3-chambered with many ovules; **Sty** terete; **Sti** capitate; **Fr** 3-angled ovoid loculicidal capsules; **Se** black, angled.

A genus of ± 25 species, of which only the following are succulent:

D. anomala (Baker) Baker (FC 6: 442, 1897). **T:** RSA (*Cooper* s.n. [K]). – **D:** RSA (Western Cape, KwaZulu-Natal); Succulent Karoo and Bushveld.

≡ *Ornithogalum anomalum* Baker (1870); **incl.** *Urginea eriospermoides* Baker (1887).

Bulbs above-ground, solitary, globose to pear-shaped, 5 - 8 × 5 - 8 cm ∅, with fleshy green tunics, withering grey-translucent; **R** fleshy, 2 mm ∅, terete; **L** 1 (occasionally 2), synanthous, dark green, ascending to spreading, firmly succulent, terete, to 30 cm, 4 - 6 mm ∅, surrounded by grey papery sheaths at the base, withering from the tip; **Inf** to 60 cm, up to 80-flowered, erect; lower **Bra** broadly ovate, 1 cm, spurred; upper **Bra** 1 mm, early deciduous; **Ped** to 8 mm; **Fl** spreading, yellowish-green; **Tep** reflexed, ± 5 mm, fused at the base; **OTep** 1 mm broad; **ITep** 2 mm broad; **St** to 4 mm; **Ov** 2 mm, ovoid; **Sty** erect, 1.5 - 2 mm; **Fr** ellipsoid, 4 - 6 mm, oblong; **Se** angled, 1.5 - 2 mm.

D. haworthioides Baker (Gard. Chron., ser. nov. 3: 366, 1875). **T:** RSA, Cape Prov. (*Bolus* 40 [K]). – **D:** RSA (Western Cape, Eastern Cape); Succulent Karoo and arid savanna, usually in the shade of shrubs, midsummer-flowering. **Fig. XXX.a, XXX.b, XXX.c**

Geophytes, solitary or forming small groups; **R** succulent, spreading, to 3 mm ⌀, white; bulb with loose incurved succulent pink-purplish club-shaped **Sc**, outer **Sc** 5 - 25 mm petiolate, flat with a swollen semiglobose apical part 15 × 12 mm; inner **Sc** becoming smaller towards the centre, ovate-attenuate, 25 mm; **L** in a **Ros**, linear-lanceolate, 35 - 50 × 6 - 10 mm, ciliate, spreading and appressed against the ground (in autumn and winter), green, channelled and tubular at the base; **Inf** 20 - 24 cm; scape erect, purple, 1 mm ⌀; **Bra** 4 × 1 mm, basally spurred for 2 mm; **Fl** horizontally spreading; closed **Per** 7 - 13 mm, ventricose in the lower part and 3 mm ⌀ at the base; **Tep** free for 7 - 9 mm, distinctly reflexed, whitish, green towards apex; **Fil** 7 - 8 mm, white, flat and broadening towards the base, appressed against the **Sty**; **Anth** 0.5 mm, dorsifixed, versatile, dirty brown; **Ov** conical, green, 2 × 2 mm; **Sty** 10 mm, elongating to 11 mm after anthesis; **Fr** 3-angled, 8 × 4 mm; **Se** oblong, 5 × 2 mm, black, shiny.

D. media Jacquin *ex* Willdenow (Spec. Pl. 2: 166, 1799). **T:** [icono]: Jacquin, Icones 2: t. 1375, 1795. – **D:** RSA (Western Cape); near the sea or on slopes in Renosterveld.

Underground geophytes; bulbs ovoid, 3 - 9.5 × 2.5 - 8 cm, with tight reddish tunics; **L** to 20, synanthous, to 8 - 40 cm, green, ascending to spreading, lamina succulent, terete, 1.5 - 4 mm ⌀; **Inf** racemose, to 60 cm, up to 35-flowered, erect; **Bra** to 2.5 mm, spurred, soon deciduous; **Fl** spreading, purplish; **Tep** reflexed, fused at the base, 1 - 1.6 cm; **Fil** linear-lanceolate, fused to the mouth of the **Per**, to 0.6 mm long; **Anth** to 2 mm; **Ov** to 3 mm, ovoid; **Sty** erect, 6 - 8 mm, 3-lobed; **Fr** oblong, 9 - 11 mm; **Se** 6 - 7 mm, flattened.

HYACINTHUS

U. Eggli

Hyacinthus Linné (Spec. Pl. [ed. 1], 316, 1753). **T:** *Hyacinthus orientalis* Linné [According to ING, designated by Hitchcock, Prop. Brit. Bot., 146, 1929.]. – **D:** Europe, SW Asia, Africa, Madagascar.

Dwarf geophytes with globose bulbs, innermost tunics exserted and forming a neck around the shoot; **L** up to 10, in a basal **Ros**, spreading to erect, linear to canaliculate, often quite fleshy, sheathing at the base; **Inf** terminal few- to many-flowered racemes positioned between the **L**; **Bra** small; **Ped** often diminishing in size from bottom to top; **Per** pale pink to lilac (numerous other colours in the cultivated hyacinths), with a short tube and erect to erect-spreading lobes; **St** 6 in 2 whorls, arising from the tube at or below the throat, exserted or not; **Anth** versatile, introrse; **Ov** ovoid, 3-locular with few ovules; **Sty** filiform, short to long; **Sti** terminal, small; **Fr** ovoid capsules, obtuse, trigonous, scarious, loculicidal; **Se** globose, black.

A small genus of ± 30 species. *H. orientalis* is the commonly cultivated Hyacinth, of which numerous cultivars are of considerable horticultural importance. A single, somewhat aberrant species from Madagascar can be considered succulent:

H. cryptopodus Baker (JLSB 20: 271, 1883). – **D:** C Madagascar (Ankaratra, Ibity); amongst grasses, ± 1800 m. **I:** Perrier (1938: 137, pl. 18: 1).

Dwarf glabrous geophytes; bulb underground, globose, 2 - 3 cm ⌀, outer tunics fleshy, green; **L** present at flowering time, 6 - 10 in a basal **Ros**, ascending, linear-lanceolate, 6 × 1 cm, elongating after flowering to 25 × 1 cm, somewhat fleshy, dark green, glossy; **Inf** to 5 cm, erect, sessile or up to 5 cm pedunculate; **Fl** densely arranged, almost sessile, ascending-spreading; **Per** ± 7 mm long, white with pink, free lobes oblong, 2.5 × 1.5 mm; **St** attached in the middle of the short **Per** tube; **Fil** filiform; **Ov** ovoid; **Sty** cylindrical, 2.5 mm; **Fr** trigonous, almost globose, 4 × 4.5 mm; **Se** black, irregularly angled.

LACHENALIA

E. van Jaarsveld

Lachenalia J. Jacquin *ex* Murray (Syst. Veg., ed. 14, 314, 1784). **T:** *Lachenalia tricolor* J. Jacquin. – **Lit:** Duncan (1988). **D:** RSA; Succulent Karoo, Renosterveld, Fynbos, mainly winter-rainfall regions. **Etym:** For Werner de [von] Lachenal (1736 - 1800); Swiss botanist in Basel.

Perennial bulbous geophytes, solitary or clustering; **L** synanthous, variable, occasionally spotted or with cross-bands, erect or horizontally spreading; **Inf** racemose, ascending, with broad membranous floral **Bra**; **Per** erect to pendulous; **Tep** 6, fused at the base, spreading or forming a tube, usually coloured; **ITep** occasionally longer than **OTep**; **St** 6, free; **Fil** terete, included or exserted; **Anth** versatile, introrse; **Ov** ovoid; **Sty** terete; **Sti** capitate; **Fr** ovoid loculicidal capsules; **Se** globose, black.

A genus of ± 70 species, of which only the following can be regarded as being (weakly) succulent. Many more taxa are, however, commonly encountered in cultivation, esp. *L. aloides* in its many colour variants.

L. patula Jacquin (Collectanea 4: 149, 1790). **T:**

[icono]: Ic. Pl. Rar., t. 384, 1791. – **D:** RSA (Western Cape); Succulent Karoo. **I:** Duncan (1988: t. 22d).

Incl. *Lachenalia succulenta* Masson *ex* Baker (1886).

Geophytes 6 - 15 cm tall, forming small groups; bulbs to 1.5 cm ∅ with dark tunics; **L** 2, opposite, linear-laneolate, 4.5 - 15 cm, subterete, fleshy, purple-tinged, apex obtuse, mucronate; **Inf** erect; 10 - 17 cm; **Bra** deltoid; **Ped** 5 mm; **Per** 15 mm, ascending, white or pale pink, **Tep** fused at the base, forming a tube with spreading tips; **ITep** occasionally longer than **OTep**; **Fil** terete, included or exserted; **Fr** ovoid; **Se** globose, black.

LEDEBOURIA

E. van Jaarsveld

Ledebouria Roth (Nov. Pl. Sp., 194, 1821). **T:** *Ledebouria hyacinthina* Roth. – **Lit:** Jessop (1970: 244-264). **D:** Tropical and S Africa, India. **Etym:** For Carl F. von Ledebour (1785 - 1851), German botanist widely travelling in Russia.

Perennial geophytes, bulbs with ± fleshy tunics, solitary or proliferating to form groups; **L** synanthous, variable, spreading, mostly spotted; **Inf** racemose, ascending with small filiform floral **Bra**; **Tep** 6, fused at the base, spreading to reflexed, papillate; **St** 6, free; **Fil** terete; **Anth** versatile, introrse; **Ov** conical with crenate base, stipitate; **Sty** terete; **Sti** capitate; **Fr** 3-angled obovoid loculicidal capsules; **Se** black, obovoid.

A small genus of ± 16 species, closely related to *Scilla*, but differing in its normally conspicuously spotted leaves, the axillary inflorescence and the free filaments. Only the following few species are ± succulent:

L. concolor (Baker) Jessop (JSAB 36(4): 254, 1970). **T:** RSA, Eastern Cape (*Cooper* s.n. [K, PRE [photo]]). – **D:** RSA (Eastern Cape); arid savanna, summer-rainfall regions.

≡ *Scilla concolor* Baker (1870); **incl.** *Drimia cooperi* Baker (1868).

Above-ground bulbous succulents; **R** succulent, 2 mm ∅; bulb conical, to 6 × 4.5 cm, purplish-green, proliferating from the base, tunics tight, withering grey, translucent; **L** 5 - 10, ovate-lanceolate to ovate, 8 - 11 × 4.5 - 6 cm, young ascending, old drooping, green, ± fleshy, unspotted, obscurely striate; **Inf** 30 - 45 cm, ascending; **Ped** 6 - 8 mm; **Bra** small, ovate; **Per** greenish, 8 mm.

L. ovatifolia (Baker) Jessop (JSAB 36(4): 262, 1970). **T:** RSA, KwaZulu-Natal (*Cooper* s.n. [K]). – **D:** RSA (Eastern Cape to Mpumalanga); Bushveld, autumn-flowering.

≡ *Scilla ovatifolia* Baker (1870); **incl.** *Scilla cicatricosa* C. A. Smith (1930); **incl.** *Scilla climacocarpha* C. A. Smith (1930); **incl.** *Scilla guttata* C.

A. Smith (1930); **incl.** *Scilla elevans* Van der Merwe (1944); **incl.** *Scilla collina* Hutchinson (1946).

Bulbs above- or underground, solitary or proliferating to form small clusters, globose, tapering, 2 - 3.5 × 2 - 3.5 cm, purplish-green with fleshy imbricate tunics in horizontal successive layers; **R** succulent, 1 mm ∅; **L** 3 - 5, ovate, 40 - 85 × 18 - 45 mm, spreading, ± appressed to the ground, attractively mottled with silvery sheen, fleshy; **Inf** 4 - 19 cm, ascending; **Bra** small, inconspicuous, truncate; **Ped** 1 mm; **Per** 7 mm, greenish-white, campanulate; **St** 2 mm; **Ov** ovate, stipitate, 1.5 mm; **Sty** 1.5 mm; **Sti** capitate.

The bulb scales produce copious threads when damaged.

L. socialis (Baker) Jessop (JSAB 36(4): 253, 1970). **T:** RSA, KwaZulu-Natal (*Cooper* 3635 [K]). – **D:** RSA (Eastern Cape, KwaZulu-Natal); arid savanna, summer-rainfall regions. **Fig. XXX.f**

≡ *Scilla socialis* Baker (1870); **incl.** *Scilla paucifolia* Baker (1870); **incl.** *Scilla violacea* Hutchinson (1932) ≡ *Ledebouria violacea* (Hutchinson) W. L. Tjaden (1989).

Bulbs above-ground, conical, tapering, 15 - 35 × 8 - 20 mm, green with peeling papery tunics, proliferating from the base to form dense clusters; **R** succulent, 1 mm ∅; **L** 2 - 4, elliptic, 4 - 7 × 1 - 1.3 cm, ascending to spreading, attractively mottled with silvery sheen, slightly fleshy; **Inf** 9 - 17 cm, spreading-ascending; **Ped** 6 mm, **Bra** small, ovate; **Per** greenish, campanulate, 3 mm; **Fil** purple, exserted; **Ov** globose, stipitate.

LITANTHUS

E. van Jaarsveld

Litanthus Harvey (J. Bot. 3: 314, t. 9, 1884). **T:** *Litanthus pusillus* Harvey. – **Lit:** Jessop (1977: 307-308). **D:** RSA. **Etym:** Gr. 'lithos', stone; and Gr. 'anthos', flower; probably alluding to the preference for rocky habitats; or barbaric combination of Lat. 'litus', coastline, coast; and as before, with uncertain application.

Perennials; bulbs dwarf, whitish, globose, clustering, above- to underground, to 13 mm ∅; **L** 1 - 3, hysteranthous, filiform, to 7 cm; **Inf** reduced to 1- to 2-flowered racemes; **Bra** 2, to 1 mm, spurred; **Per** pendulous, white to pink, tubular, to 5 mm; **Tep** fused in the lower ½; **St** fused to the **Per** tube; **Anth** dorsifixed; **Ov** sessile, ellipsoid; **Fr** loculicidal capsules to 5 mm, transparent; **Se** angled, 0.5 mm.

A monotypic genus related to *Drimia*. Its only species is perhaps the smallest of all bulbous species in RSA.

L. pusillus Harvey (J. Bot. 3: 315, t. 9, 1884). **T:** RSA, Cape Prov. (*Zeyher* s.n. [TCD]). – **D:** RSA; widespread in rock crevices.

Description as for the genus.

MASSONIA

E. van Jaarsveld

Massonia Thunberg *ex* Houttuyn (Nat. Hist. 12: 424, t. 85: fig. 1, 1780). **T:** *Massonia depressa* Houttuyn. – **Lit:** Müller-Doblies & Müller-Doblies (1997). **D:** RSA (Eastern Cape, Free State, Mpumalanga); mainly Succulent Karoo. **Etym:** For Francis Masson (1741 - 1805), British horticulturist collecting esp. in S Africa.

Incl. *Neobakeria* Schlechter (1924). **T:** not designated.

Soft succulent perennial geophytes with globose bulbs to 4.5 cm ⌀ with fleshy tunics; **L** 2, opposite, synanthous, large, appressed to the ground (or semi-ascending in cultivation), elliptic-oblong to lanceolate; **Inf** dense reduced terminal subcapitulate racemes with short peduncle; basal **Bra** large, surrounding the **Inf**; **Fl** pedicellate, **Tep** fused into a tube with the oblong tips free; **St** 6, fused to the mouth of the **Per** tube; **Anth** dorsifixed; **Ov** sessile, oblong to ovoid; **Sty** short; **Sti** capitate; **Fr** winged loculicidal lobed capsules; **Se** subglobose, black, to 2 mm ⌀.

A small genus of 12 species, closely related to *Whiteheadia*. Rarely cultivated, and only one species is moderately succulent:

M. depressa Houttuyn (Nat. Hist. 12: 424, t. 85: fig. 1, 1780). **T:** [icono]: l.c., t. 85: 1. – **D:** RSA (Eastern Cape to Free State); Succulent Karoo. **Fig. XXX.d, XXX.e**

Incl. *Massonia latifolia* Linné *fil.* (1781); **incl.** *Massonia sanguinea* Jacquin (1804); **incl.** *Massonia grandiflora* Lindley (1826); **incl.** *Massonia brachypus* Baker (1874); **incl.** *Massonia namaquensis* Baker (1897); **incl.** *Massonia triflora* Compton (1931).

Succulent geophytes with globose bulbs to 4.5 cm ⌀; tunics fleshy white; **L** 2, opposite, appressed to the ground, oblong to orbicular, to 26 × 15 cm (frequently less), fleshy, green with purple spots; **Inf** dense, capitate, to 30-flowered; basal **Bra** large, surrounding the **Inf**; **Fl** shortly pedicellate, greenish.

ORNITHOGALUM

E. van Jaarsveld

Ornithogalum Linné (Spec. Pl. [ed. 1], 306, 1753). **T:** *Ornithogalum umbellatum* Linné [Lectotype (Reneaulme, Specim. Hist. Pl., t. 87, 1611) designated by Stearn in Ann. Mus. Goulandris 6: 139-170, 1983, and proposed for conservation by Jarvis, Taxon 41: 566, 1992.]. – **Lit:** Obermeyer (1978); Müller-Doblies & Müller-Doblies (1996). **D:** Europe, W Asia, Africa. **Etym:** Lat. 'ornithogale' = Gr. 'ornithogalon', "Bird's Milk", a plant (from Gr. 'ornis, ornithos', bird; and Gr. 'gala', milk); refer-

ring to the egg-shale-coloured flowers of some European taxa; or going back to a Roman allusion of something rare or beautiful "as bird's milk".

Incl. *Honorius* Gray (1821). **T:** *Ornithogalum nutans* Linné.

Incl. *Caruelia* Parlatore (1854). **T:** *Ornithogalum arabicum* Linné.

Incl. *Ardernia* Salisbury (1866). **T:** not designated.

Incl. *Aspasia* Salisbury (1866) (*nom. illeg.*, Art. 53.1). **T:** not designated.

Incl. *Beryllis* Salisbury (1866). **T:** not designated.

Incl. *Brizophile* Salisbury (1866) (*nom. illeg.*, Art. 52.1). **T:** *Ornithogalum nutans* Linné.

Incl. *Cathissa* Salisbury (1866). **T:** not designated.

Incl. *Eustachys* Salisbury (1866). **T:** not designated.

Incl. *Monotassa* Salisbury (1866) (*nom. illeg.*, Art. 52.1). **T:** *Ornithogalum secundum* Jacquin.

Incl. *Myanthe* Salisbury (1866) (*nom. illeg.*, Art. 52.1). **T:** *Ornithogalum arabicum* Linné.

Incl. *Osmyne* Salisbury (1866). **T:** *Ornithogalum odoratum* J. Kenney.

Incl. *Phaeocles* Salisbury (1866). **T:** *Ornithogalum maculatum* Jacquin.

Incl. *Taeniola* Salisbury (1866) (*nom. illeg.*, Art. 53.1). **T:** *Albuca vittata* Ker Gawler.

Incl. *Urophyllon* Salisbury (1866). **T:** *Ornithogalum longibracteatum* Jacquin.

Incl. *Elsiea* F. M. Leighton (1944). **T:** not designated.

Perennial bulbous geophytes; bulbs variable, underground or above-ground, globose to pear- or drop-shaped, rarely asymmetrical; tunics fleshy; **L** very variable, withering in the dry season, rarely evergreen, terete to flattened, angular or clavate (*O. scabrocostatum, O. unifoliatum*), solitary to densely rosulate, rarely distichous; **Inf** racemose, subspicate to subcorymbose, erect, peduncle ebracteate and smooth; floriferous **Bra** small to large, membranous to foliaceous; **Fl** actinomorphic; **Per** stellate, **Tep** 6, free (rarely fused at the base), in 2 whorls of 3, mostly white (sometimes basally with black), rarely yellow, yellow-green or orange; **St** 6, the 3 outer with narrower **Fil**; **Fil** sometimes with appendages; **Anth** bilocular, versatile; **Ov** 3-locular; **Sty** erect or deflexed; **Sti** 3-lobed; **Fr** 3-locular loculicidal variable capsules; **Se** very variable in size and shape, black, shiny, papillate to echinulate.

A genus of ± 200 species with its main centre of diversity in S Africa. The S African taxa of the genus are classified into the following subgenera by Obermeyer (1978):

[1] Subgen. *Aspasia* (Salisbury) Obermeyer 1978: **L** mainly proteranthous; **Bra** cymbiform; **Fil** with outgrowths in the lower ½; **Fr** capsules fusiform to ellipsoid; **Se** variable in shape, minute, not flattened, papillate to echinate. Mainly from winter-rainfall regions.

[2] Subgen. *Urophyllon* (Salisbury) Obermeyer 1978: **L** synanthous or proteranthous; **Bra** with-

ering early; **Fil** terete; **Fr** capsules tricostate; **Se** large (to 9 mm), oblong, flat but angular and with ridges. Mainly from summer-rainfall regions.

[3] Subgen. *Osmyne* (Salisbury) Obermeyer 1978: **Tep** narrowly elliptic; **Fr** capsules ovoid to oblong-globose; **Se** D-shaped, flat, ridged. Mainly from winter-rainfall regions.

O. geniculatum Obermeyer (BT 12(3): 344, fig. 18, 1978). **T:** RSA, Northern Cape (*Marloth* 13249 [PRE]). – **D:** S Namib Desert of Namibia and RSA (Northern Cape); Succulent Karoo, spring-flowering.

[1] Dwarf solitary geophytes; bulbs globose-ovoid, to 1 cm ∅ and high, tunics white, fleshy, outer tunics dry; **L** synanthous, variable, oblong, succulent, lower **L** 4 - 5 × 2 cm, oblong, upper **L** 3 × 0.3 cm, tubular at the base and clasping the stem; **Inf** racemose, to 15 cm, up to 5-flowered; scape wiry, geniculate at the base; **Bra** linear-acuminate; **Ped** 1 cm; **Tep** white, 10 × 3 mm, linear-elliptic; **St** 5 mm, outer **Fil** filiform, inner ovate; **Ov** ovoid; **Sty** erect; **Sti** capitate; **Fr** capsules 1 × 1 cm ∅, transparent, oblong-ovoid; **Se** black, flat, 1 mm ∅.

O. juncifolium Jacquin (Pl. Hort. Schoenbr. 1: 46, t. 90, 1797). **T:** [icono]: l.c., t. 90. – **D:** RSA; widespread on rocky well-drained terrain.

Incl. *Ornithogalum setifolium* Kunth (1843); **incl.** *Ornithogalum compactum* Baker (1873); **incl.** *Ornithogalum griseum* Baker (1873); **incl.** *Ornithogalum subulatum* Baker (1874); **incl.** *Ornithogalum leptophyllum* Baker (1897); **incl.** *Ornithogalum oliganthum* Baker (1897); **incl.** *Ornithogalum tortuosum* Baker (1897); **incl.** *Ornithogalum stenostachyum* Baker (1901); **incl.** *Ornithogalum tenuipes* C. H. Wright (1901); **incl.** *Ornithogalum capillifolium* Fourcade (1934); **incl.** *Ornithogalum limosum* Fourcade (1934); **incl.** *Ornithogalum petraeum* Fourcade (1934); **incl.** *Ornithogalum brevifolium* F. M. Leighton (1943); **incl.** *Ornithogalum epigeum* F. M. Leighton (1943); **incl.** *Ornithogalum comptonii* F. M. Leighton (1944); **incl.** *Ornithogalum langebergense* F. M. Leighton (1945).

[1] Bulbs above-ground and cluster-forming, globose to 4 cm ∅ and tall, tunics grey-brown, leathery and exposing green living tissue; **L** synanthous, 10 - 20 cm × 2 - 3 mm, linear, filiform, green, base sheathing, tubular forming a membranous neck on top of the bulb; **Inf** racemose, to 40 cm, to 15-flowered; scape terete, erect; **Bra** deltoid-cuspidate, auriculate; **Ped** ± 2 mm (lengthening to 7 mm in fruit); **Per** stellate; **Tep** linear-lanceolate, 7 - 10 × 2.5 mm; **Fil** 5 mm, outer linear, inner shorter and ovate-acuminate; **Ov** ovate, 3 mm, shortly stipitate; **Sty** erect, 2.5 mm; **Fr** 5 mm, ovoid.

O. longibracteatum Jacquin (Hort. Bot. Vindob. 3: t. 29, 1776). **T:** [icono]: l.c., t. 29. – **D:** RSA (Eastern Cape to KwaZulu-Natal); rocky well-drained terrain, subtropical thicket. **Fig. XXXI.a**

Incl. *Ornithogalum caudatum* Aiton (1789); **incl.** *Ornithogalum bracteatum* Thunberg (1794); **incl.** *Ornithogalum scilloides* Jacquin (1797); **incl.** *Urginea mouretii* Battandier & Trabut (1921).

[2] Bulbs above-ground and cluster-forming, globose to 8 cm ∅, bulbiferous, tunics succulent, green, withering grey, exposing green live tissue; **R** white, terete, succulent; **L** synanthous, 20 - 100 × 2 - 5 cm, rosulate, ascending to curving, linear, flaccidly succulent, channelled, withering from the apex; **Inf** racemose, to 1 m, densely flowered; scape terete, erect; **Bra** to 4 cm, filiform, broadening at the base; **Ped** to 5 mm (lengthening to 15 mm in fruit); **Per** stellate; **Tep** linear-elliptic, 9 × 2.5 mm, green with white margins; **Ov** globose; **Fr** trigonous, 10 × 6 mm; **Se** oblong, angular, 4 × 1.5 mm.

Frequently cultivated and colloquially known as "Pregnant Onions".

O. maculatum Jacquin (Collectanea 2: 368, t. 18: fig. 3, 1789). **T:** [icono]: l.c., t. 18: fig. 3. – **D:** RSA (Northern Cape, Western Cape); Succulent Karoo, in shallow quartzitic sandstone rock pockets. **Fig. XXX.g**

Incl. *Ornithogalum maculatum* Thunberg (1794) (*nom. illeg.*, Art. 53.1); **incl.** *Ornithogalum speciosum* Baker (1891) ≡ *Ornithogalum maculatum* var. *speciosum* (Baker) F. M. Leighton (1944); **incl.** *Ornithogalum thunbergianum* var. *concolor* Baker (1897); **incl.** *Ornithogalum splendens* L. Bolus (1931); **incl.** *Ornithogalum insigne* F. M. Leighton (1943); **incl.** *Ornithogalum magnificum* von Poellnitz (1946).

[1] Bulbs shallowly underground, solitary, obturbinate to 3 cm ∅, tunics grey-brown when dry, white when alive; **L** 2 - 3 (-5), synanthous, 2.5 - 15 × 0.5 - 2 cm, erect, linear, filiform to flat, glaucous, tip acute or rarely cirrose; **Inf** erect, 7 - 60 cm, to 8-flowered, scape terete; **Bra** ovate-acuminate, to 7 mm; **Ped** to 3 cm; **Per** stellate, orange to yellow, **OTep** with a black spot towards the tips; **Tep** elliptic to obovate, 10 - 25 × 5 - 6 mm; **St** 7 - 8 mm, outer **Fil** subulate, inner flattened; **Ov** oblong-globose, 6 mm ∅; **Sty** 2 mm; **Fr** ellipsoid; **Se** minute.

O. multifolium Baker (JLSB 13: 271, 1873). **T:** RSA, Cape Prov. (*Whitehead* s.n. [TCD]). – **D:** RSA (Northern Cape, Western Cape); Succulent Karoo in shallow quartzitic sandstone rock pockets. **I:** Obermeyer (1978: 341).

Incl. *Ornithogalum aurantiacum* Baker (1878); **incl.** *Ornithogalum ranunculoides* L. Bolus (1933).

[1] Bulbs underground, solitary or forming clusters, ovoid to 4 cm ∅, tunics grey-brown, leathery and exposing green live tissue; **L** to ± 10, synanthous, 2 - 10 cm × 2 - 3 mm, linear-terete, glaucous, often spreading; **Inf** erect, 4 - 15 cm, 1- to 15-

flowered; scape terete; **Bra** cymbiform, acuminate, 4 - 25 mm; lower **Ped** to 2 cm; **Per** stellate, yellow to orange; **Tep** ovate-lanceolate to obovate, 6 - 15 × 4 mm; **St** ± 3 - 7 mm, linear-ovate, outer **Fil** narrower; **Ov** broadly ovate, 3 - 6 mm; **Sty** erect, 1 - 2 mm; **Fr** to 1 cm, ovoid.

O. naviculum W. F. Barker (BT 12(3): 348, fig. 23, 1978). **T:** RSA, Western Cape (*Hall* 2869 [NBG]). – **D:** RSA (N Western Cape); Succulent Karoo, flowering in mid-summer.

[1] Dwarf solitary geophytes; bulbs globose; **L** hysteranthous, obovoid to oblong-obovoid, 20 × 8 mm, cymbiform, succulent, canaliculate above, basally amplexicaul; **Inf** racemose, to 10 cm, erect, up to 12-flowered; **Bra** deltoid-acuminate, 3 mm; **Ped** 15 mm, asceding; **Tep** spreading, becoming reflexed, white, 7 × 2 mm, narrowly ovoid; **St** 5 mm, **Fil** filiform; **Ov** narrowly obovoid, yellow; **Sty** erect; **Sti** capitate; **Fr** and **Se** not seen.

O. sardienii van Jaarsveld (Bradleya 12: 32-34, ills., 1994). **T:** RSA, Western Cape (*van Jaarsveld & al.* 10854 [NBG]). – **D:** RSA (Western Cape: Little Karoo); Succulent Karoo.

[1] Dwarf, bulb globose, 8 - 20 × 5 - 18 mm ∅, above-ground, solitary or forming small clumps, tunics grey-white, succulent, slightly translucent; **R** fibrous, terete; **L** evergreen, numerous (20 - 50) in a dense basal **Ros**, ascending to erectly spreading, lamina linear-lanceolate, green with up to 6 rows of white translucent porrect cilia, 15 - 25 × basally 1 mm, keeled, triangular in cross-section, upper face flat, margin ciliate, lower face convex in the lower part, apex with a yellowish-green mucro; **Inf** 1 - 5, racemose, 15 - 24 cm, up to 30-flowered, erect; **Fl** pointing upwards, ± 12 mm ∅; lower **Bra** 3.5 mm, upper **Bra** smaller; **Ped** of lower **Fl** 20 mm; **Per** stellate; **Tep** 5 - 6 mm, white, obtuse; **OTep** 2 mm broad; **ITep** 1.5 mm broad; outer **St** 3 mm, inner 4 mm; **Fil** filiform; **Ov** 2 mm, obovoid-ellipsoid to globose, 6-grooved with light green tubercles along the middle part; **Sty** erect, 2 mm, 6-grooved; **Sti** obscurely 3-lobed, yellow; **Fr** ovoid, 3.5 mm; **Se** pear-shaped, 0.5 mm, many-angled.

O. scabrocostatum U. Müller-Doblies & D. Müller-Doblies (Feddes Repert. 107(5-6): 516-517, pl. IVc-f, 1996). **T:** RSA, Northern Cape (*Müller-Doblies* 79176k [PRE, B, BOL, BR, BTU, G, K, LI, M, MO, NBG, S, WIND, Z]). – **D:** RSA (Northern Cape: Richtersveld). **I:** Aloe 35: 25-26, 1998.

[3] Similar to *O. unifoliatum*, but the single **L** spreading at an angle (instead of erect), to 60 × 6 mm, subclavate, upper face flat or slightly canaliculate, lower face convex, margins hyaline, surface covered with ribs with rough raised rounded projections; **Inf** to 12 cm; **Tep** canary-yellow with a deep green midvein. – [U. Eggli]

O. unifoliatum (G. D. Rowley) Obermeyer (BT 12(3): 370, fig. 52, 1978). **T:** RSA, Northern Cape (*Hall & Rowley* 329 [K]). – **D:** RSA (Northern Cape); Succulent Karoo, spring-flowering. **Fig. XXXI.b, XXXI.c**

≡ *Albuca unifoliata* G. D. Rowley (1975).

[3] Dwarf solitary geophytes; bulbs globose, to 2 cm ∅, yellow; **L** solitary, proteranthous, terete, club-shaped, 4 × 1 cm; **Inf** racemose, to 12 cm, erect, 5- to 9-flowered; **Bra** deltoid-acuminate, 1 cm; **Ped** 1 cm, ascending; **Per** stellate, yellow, **Tep** 15 × 4 mm, elliptic; **St** 5 mm, **Fil** filiform; **Sty** erect; **Sti** capitate; **Se** flat, semi-orbicular, black.

O. zebrinum (Baker) Obermeyer (BT 12(3): 369-370, fig. 51, 1978). **T:** RSA, Northern Cape (*Schlechter* 11371 [GRA]). – **D:** RSA (Northern Cape); Succulent Karoo, spring-flowering.

≡ *Albuca zebrina* Baker (1904).

[3] Dwarf solitary geophytes; bulbs globose to 2 cm ∅ and high; tunics fleshy, outer tunics dry, dark; **L** synanthous, solitary, linear, terete, to 14 cm, with basal membranous sheaths with dark cross-bands; **Inf** racemose, to 20 cm, to 10-flowered; **Bra** linear-ovate, acuminate, 1 cm; **Ped** to 1 cm; **Tep** yellow, 8 × 3 mm, linear-elliptic; **Fil** subulate, minutely scabrid; **Ov** 4 mm, oblong; **Fr** and **Se** not known.

RHADAMANTHUS

E. van Jaarsveld

Rhadamanthus Salisbury (Gen. Pl., 37, 1866). **T:** *Hyacinthus convallarioides* Linné fil. – **Lit:** Nordenstam (1970). **D:** Namibia, RSA; Succulent Karoo and arid regions. **Etym:** Named for Rhadamanthus, the son of Zeus and Europa, and brother of Minos.

Dwarf geophytes, solitary or clustering and with fleshy or papery bulb **Sc**; **R** fibrous; **L** hysteranthous, erect or spreading, variable; **Inf** racemose; scape erect, scabrid; **Bra** membranous, small, lower spurred; **Ped** ascending with drooping **Fl**; **Per** campanulate, subglobose; **Tep** 6, fused at the base, spreading to semiconnivent; **St** 6, basally adnate to the **Tep**; **Anth** basifixed, introrse; **Ov** globose-triquetrous; **Sty** terete, subclavate; **Sti** capitate; **Fr** 3-angled capsules, subglobose; **Se** blackish, flattened.

8 of the ± 12 species of this small genus are ± succulent and are adapted to dry conditions, though rarely seen in cultivation; they are covered below. 2 subgenera are recognized:

[1] *Rhadamanthus*: **Anth** opening at the tips only.
[2] *Rhadamanthopsis* Obermeyer 1980: **Anth** opening along the whole length.

R. albiflorus B. Nordenstam (Bot. Not. 123: 177-178, ills., 1970). **T:** RSA, Western Cape (*Acocks*

23242 [PRE]). – **D:** RSA (Western Cape); Fynbos, midsummer-flowering.

[1] Bulb ovoid to pear-shaped, to 3 × 2.5 cm ∅, tunics papery, grey; **L** unknown; **Inf** to 25 cm, scape reddish-brown, erect, raceme lax to 6 cm, to 15-flowered; **Bra** membranous, ovate-deltoid to 2 mm, acuminate, spurred; **Ped** to 9 mm; **Per** campanulate, drooping, to 6 mm; **Tep** white, oblong, basally connate for 1.5 mm, obtuse; free part of **Fil** flat; **Anth** to 2.7 mm, introrse; **Ov** ovoid, to 2 mm.

R. arenicola B. Nordenstam (Bot. Not. 123: 166-169, ills., 1970). **T:** RSA, Northern Cape (*Pillans* s.n. [BOL 18253]). – **D:** RSA (Northern Cape); Succulent Karoo; late spring- and early summer-flowering.

[1] Geophytes with loose fleshy bulb **Sc**, these imbricate, ovate-lanceolate, 10 - 30 × 5 - 15 mm, tips acuminate; **L** erect, filiform; **Inf** 5 - 15 cm, erect, raceme lax; **Bra** membranous, ovate-deltoid to 2 × 1.5 mm, acuminate; **Per** campanulate, to 4 mm; **Tep** elliptic, connate, 4 × 2 mm, dirty white, obtuse; **Anth** to 1 mm; **Ov** ovoid to 2.5 mm; **Fr** ovoid-globose, to 5 mm; **Se** oblong to 4 mm, shiny black, rugose.

R. fasciatus B. Nordenstam (Bot. Not. 123: 174-177, ills., 1970). **T:** RSA, Northern Cape (*Leistner* 1983 [PRE, LD]). – **D:** RSA (Northern Cape); arid savanna, spring-flowering.

[1] Bulb ovoid to pear-shaped, to 4 × 3 cm ∅, tunics papery, grey-brown; basal **L** sheath cross-banded; **L** filiform, solitary; **Inf** to 25 cm, scape brownish, erect, raceme lax, to 11 cm, to 30-flowered; **Bra** ovate-deltoid to 1.5 mm, spurred; **Ped** to 11 mm; **Per** campanulate, drooping, to 5.5 mm; **Tep** white to pale yellow, oblong; **Fil** free; **Anth** green to yellowish-green, to 2.3 mm; **Ov** globose-ovoid, to 2 mm.

R. karooicus Obermeyer (BT 13(1-2): 138-139, ills., 1980). **T:** RSA, Western Cape (*Van Zanten* s.n. [PRE 45560]). – **D:** RSA (Western Cape: Little and S Great Karoo); in rock crevices, midsummer-flowering.

[2] Bulbs clustering, oblong-globose to 3 cm ∅, with exposed green fleshy bulb **Sc** withering grey at the tips; **R** terete, succulent, to 2 mm ∅; **L** to 6, lorate-lanceolate, spreading, to 8 × 2 cm, green, glabrous and shiny, apex acute, mucronate; **Inf** to 20 cm, scape erect, to 30-flowered in the upper part; **Bra** membranous; **Ped** to 8 mm; **Per** campanulate, mauve; **Fil** basally connate; **Anth** 1 mm; **Ov** ovoid.

R. montanus B. Nordenstam (Bot. Not. 123: 162-164, fig. (p. 161), 1970). **T:** RSA, Western Cape (*Esterhuysen* 11456 [BOL, K, NBG]). – **D:** RSA (Western Cape); Fynbos, quartzitic sandstone rock crevices, midsummer-flowering.

[1] Bulb ovoid to pear-shaped, to 5 cm ∅, tunics papery, grey-brown, translucent; **L** to 24, filiform, to 10 cm; **Inf** to 40 cm, scape reddish-brown, erect, raceme to 22 cm, to 50-flowered; **Bra** ovate-deltoid to 2 mm, spurred; **Ped** to 18 mm; **Per** campanulate, drooping, to 8 mm; **Tep** white to pale yellow, to 8 × 3 mm, oblong; **Fil** basally connate, free parts flat; **Anth** yellowish-green, to 1.6 mm; **Ov** oblong-ovoid, to 3 mm; **Fr** subglobose; **Se** numerous.

R. namibensis Obermeyer (BT 13(1-2): 137-138, ills., 1980). **T:** Namibia (*Giess* 13781 [PRE, WIND]). – **D:** S Namibia; Succulent Karoo, midsummer-flowering.

[2] Bulb ovoid-globose, to 5 cm ∅, with fleshy bulb **Sc**, **L** bases transversely banded, becoming white; **R** terete, succulent; **L** 2 - 4, hysteranthous, linear-lanceolate, ascending, to 24 × 2.5 cm, glaucous, canaliculate; **Inf** to 70 cm, erect, many-flowered in the upper part; **Bra** spurred; **Ped** to 8 mm; **Per** campanulate, pale mauve; **Fil** basally connate, 1.5 mm; **Ov** ovoid; **Fr** ovoid to 7 mm; **Se** flat, black, oblong, to 5 mm.

R. platyphyllus B. Nordenstam (Bot. Not. 123: 171-174, ills., 1970). **T:** RSA, Western Cape (*Esterhuysen* 18135 [BOL]). – **D:** RSA (Western Cape); Fynbos, quartzitic sandstone rock crevices, midsummer-flowering.

[1] Bulb ovoid-globose, to 5 × 4 cm, white, softly succulent; **L** 2, subopposite, flat, linear to ovate-elliptic, horizontally patent, to 4 × 2.5 cm, upper surface velutinous, apex obtuse; **Inf** to 30 cm, scape reddish-brown, erect, papillate-hirsute, raceme to 12 cm, to 50-flowered; **Bra** ovate-deltoid, acuminate, to 2 mm, spurred; **Ped** to 1 cm; **Per** urceolate, drooping, to 6 mm; **Tep** brownish, to 6 × 2.5 mm, oblong, fused in the lower part; **Fil** basally connate, free parts hairy; **Anth** yellowish, to 1.8 mm; **Ov** ovoid, to 2 mm, glabrous; **Fr** subglobose to 7 mm; **Se** flat, oblong, to 4 mm.

R. secundus B. Nordenstam (Bot. Not. 123: 168-171, ills., 1970). **T:** Namibia (*Giess* 2350 [M, WHK]). – **D:** SW Namibia; Succulent Karoo, winter-rainfall region, spring-flowering.

[1] Bulb lax and open, the **Sc** forming a **Ros**, clavate to fusiform, succulent, to 4 × 20 mm, apically contracted; **L** linear-filiform, 8 cm × 1.5 mm, apex obtuse; **Inf** to 5 cm, to 25-flowered, secund; **Bra** ovate, to 2 mm, spurred; **Ped** to 2 mm; **Per** campanulate, to 5 mm; **Tep** connate in the lower ½, brownish; **Fil** basally connate, free parts filiform; **Anth** light yellow, to 1.6 mm; **Ov** subglobose, to 2.5 mm.

RHODOCODON

U. Eggli

Rhodocodon Baker (JLSB 18: 280, 1881). **T:** *Rhodocodon madagascariensis* Baker [Typification by inference, only element included.]. – **Lit:** Perrier (1938: 113-124). **D:** Madagascar. **Etym:** Gr. 'rhodos' rose-red; and Gr. 'kodon', bell; for the bell-shaped rose-red flowers of the type species.

Perennials with bulbs underground or rarely at ground-level (often epiphytic in *R. urgineoides*), covered in thin papery tunics, or green; **L** few to numerous, synanthous or hysteranthous, thin or thickish and slightly succulent, 10 - 60 cm, narrowly lanceolate to linear, normally ± erect; **Inf** long-pedunculate lax unbranched spikes; **Bra** (esp. the lower) with a long spur-like appendage; **Fl** ± drooping, pedicellate; **Per** (broadly) globular or shortly tubular, white or rose-red (only *R. madagascariensis*), segments united for the greater part of their length, persistent at fruiting time; **St** 6, all equal; **Fil** basally united with the **Per** tube, glabrous; **Ov** sessile, with few to many ovules; **Fr** membranous capsules.

The genus is divided into 2 sections by Perrier (1938), and is endemic to Madagascar. Only *R. rotundatus* (Sect. *Rhodocodon*) is somewhat succulent:

R. rotundatus H. Perrier (Arch. Bot. Mém. 5: 13, 1931). **T:** Madagascar (*Perrier* 16583 [P ?]). – **D:** W Madagascar. **I:** Perrier (1938: 121, fig. 16: 7-8).

Bulb at ground-level or above-ground, broader than tall, greenish or covered with thin papery ochre tunics, often dividing and forming groups, succulent; **L** hysteranthous, numerous, semi-erect to flatly spreading, dark green, narrowly lanceolate, to 20 cm × 7 mm, slightly succulent with thickened midrib; **Inf** to 40 cm, peduncle to 20 cm; **Fl** laxly arranged; **Bra** narrowly lanceolate-linear, 5 - 8 mm, basally 3-veined, spur thin, 1.5 - 2 mm, variously shaped; **Ped** thin, 2 - 2.5 cm; **Per** globose, 7 - 8 mm ∅, greenish-white to greenish-brown; **Fil** subulate; **Anth** ovoid.

SCHIZOBASIS

E. van Jaarsveld

Schizobasis Baker (J. Bot. 11: 105, 1873). **T:** *Schizobasis macowanii* Baker. – **Lit:** Jessop (1977: 309-312). **D:** Namibia, RSA; well drained arid to semi-arid regions, in rock crevices, summer-flowering. **Etym:** Gr. 'schizein', to split; and Gr. 'basis', base; for the capsules.

With above-ground bulbs or underground geophytes; bulbs pear-shaped to globose with soft often loose imbricate fleshy tunics; juvenile **L** solitary, linear; **L** absent from adult plants; **Inf** 1 - 3 per plant, erect, thin, wiry, green, copiously branched to form roundish panicles; **Fl** solitary, small, nocturnal; **Per** white, persistent beyond flowering; **Tep** 6, lorate; **St** 6, slightly shorter than the **Tep** and fused to the base of the **Tep**; **Anth** dorsifixed, < 1 mm; **Ov** superior, sessile, 3-locular, ovoid, to 1.5 mm;

Sty terete, to 1.5 mm; **Sti** 3-lobed; **Fr** ellipsoid to globose capsules; **Se** angled, black.

Related to *Bowiea* but differing in the ± stiffly erect very thin and delicate inflorescences and minute nocturnal flowers. Rarely cultivated as an oddity.

S. cuscutoides (Burchell *ex* Baker) Bentham & Hooker (Gen. Pl. 3: 786, 1883). **T:** RSA, Cape Prov. (*Burchell* 2673 [WHK]). – **D:** S Namibia, RSA.

≡ *Asparagus cuscutoides* Burchell *ex* Baker (1875); **incl.** *Schizobasis buchubergensis* Dinter (1932).

Bulbs globose, tunics pale-coloured; juvenile **L** not recorded; **Inf** 1 - 3, erect, to 15 cm, tortuous at the nodes; **Bra** 2 - 3 mm, spurred at the base; **Ped** 0.5 - 1 cm, arcuate-ascending; **Tep** white, spreading, to 4 mm.

S. intricata (Baker) Baker (J. Bot. 1874: 140, 1874). **T** [syn]: RSA, Cape Prov. (*Burke* s.n. [SAM]). – **D:** Ethiopia, Zimbabwe, Zambia, Moçambique, Angola, Tanzania, Namibia, RSA; well drained rock-crevices, to 1500 m. **I:** Stedje (1996: 29). **Fig. XXXI.g, XXXI.h**

≡ *Anthericum intricatum* Baker (1872); **incl.** *Anthericum flagelliforme* Baker (1872) ≡ *Schizobasis flagelliformis* (Baker) Baker (1876); **incl.** *Schizobasis macowanii* Baker (1873); **incl.** *Schizobasis schlechteri* Baker (1901); **incl.** *Schizobasis dinteri* Krause (1912); **incl.** *Asparagus micranthus* Thunberg *ex* Jessop (1977) (*nom. inval.*, Art. 34.1a).

Bulbs globose, tunics white or pink; juvenile **L** filiform, to 6 cm; **Inf** 1 - 3, erect, to 50 cm, zigzag at the nodes; **Bra** 2 - 3 mm, spurred at the base; **Ped** 0.7 - 5 cm, arcuate-ascending; **Tep** white to pale yellow, spreading, to 4 mm; **St** fused to the base of the **Tep** and slightly shorter than the **Tep**.

URGINEA

E. van Jaarsveld

Urginea Steinheil (Ann. Sci. Nat. Bot., sér. 2, 1: 321, 1834). **T:** *Urginea fugax* Steinheil. – **D:** Mediterranean, Africa, India. **Etym:** For the Beni Urgen tribe in Algeria where the type was collected (Jackson 1990).

Incl. *Urgineopsis* Compton (1930). **T:** *Urgineopsis salteri* Compton.

Perennial, usually hysteranthous, geophytes, bulbous; bulbs variable, solitary or cluster-forming, under- or above-ground, globose with loose imbricate fleshy **Sc**; **L** rosulate, very variable, linear, terete, flattened or channelled; **Inf** erect, scape naked, **Fl** in cylindrical to subspicate racemes; **Bra** caudate, small, early deciduous; **Ped** ascending, rarely drooping; **Per** ascending to drooping, stellate; **Tep** 6, ascending, free (rarely fused at the base), green, white, brown with dark median stripe; **St** 6, some-what conivent to the base of the **Tep**; **Fil** terete to flattened; **Anth** introrse, versatile; **Ov** 3-locular, sessile, ovules axillary; **Sty** terete; **Sti** capitate; **Fr** 3-locular globose capsules; **Se** black, shiny, papillate.

A genus of ± 100 species with a few members with above-ground bulbs. Some ± succulent species are cultivated in subtropical Bushveld gardens in RSA and need to be kept dry during winter months.

U. delagoensis Baker (FC 6: 467, 1897). **T:** Moçambique (*Bolus* 7627 [K]). – **D:** Border areas of RSA (Mpumalanga, KwaZulu-Natal) with Moçambique, Swaziland; dry savanna on the Lebombo Mts., spring- and early summer-flowering. **I:** FPA 22: t. 858, 1942.

Bulbs above-ground, solitary or dividing to form small groups of up to 5 plants, globose, 7 - 10 cm ∅ with large imbricate loose succulent **Sc**; **Sc** globose, purplish-green, 3.5 × 3.5 cm, dry parts brown, older **Sc** becoming brownish and truncate at the top, ultimately withering; **R** fleshy, 3 mm ∅, terete; **L** 5 - 10, synanthous, green to glaucous, ascending, lamina linear-acuminate, 14 - 50 cm, upper face channelled, lower face convex; **Inf** racemose, 45 - 50 cm, up to 50-flowered (**Fl** in upper ⅓), erect; scape terete, 6 mm ∅ at the base; floral **Bra** inconspicuous, triangular-lanceolate, white, 2 mm, withering before **Fl** mature; **Ped** 5 - 6 mm; **Fl** 5 - 10 mm apart, spreading to drooping; **Per** stellate, ± 6 mm ∅; **Tep** linear-obovate, 6 × 2 mm, erectly spreading, white with purplish median stripes, tips obtuse; **St** 3 mm, **Fil** filiform; **Anth** 0.5 mm; **Ov** greenish, 2 mm, globose-obovoid; **Sty** erect, 3 mm; **Fr** oblong-ovoid, 10 - 12 × 4 - 5 mm; **Se** black, oblong, winged, 7 × 1.5 mm.

U. epigea R. A. Dyer (FPA 26: t. 1027 + text, 1947). **T:** RSA, Transvaal (*van der Merwe* 2203 [PRE]). – **D:** RSA, adjacent SE Africa; dry savanna, flowers in early spring. **Fig. XXXI.e**

Bulbs above-ground, solitary or dividing to form small groups of up to 4 bulbs, globose to 15 cm ∅ and high, bulb **Sc** fleshy, tightly imbricate, to 7 cm wide, withering from the tip: **L** hysteranthous, 15 - 30 × 2 - 3 cm, linear-lanceolate, tough, ascending, green and shiny; **Inf** racemose, to 1 m, up to 100-flowered, scape terete, to 15 mm ∅ at the base, erect, **Fl** in the upper ½ only; lower **Bra** linear-lanceolate, to 8 mm, basally spurred; **Ped** to 25 mm; **Per** stellate; **OTep** linear-lanceolate, 6 × 2.5 mm, **ITep** 6 × 3 mm, white with green keel; **St** 5 mm, **Fil** tapering; **Ov** greenish, 4 mm, oblong-ovoid; **Sty** erect, 3 mm; **Fr** 1 × 1 cm, 3-angled; **Se** black, winged.

Used medicinally by the inhabitants of Sekukuni-land (Mpumalanga, RSA).

U. lydenburgensis R. A. Dyer (FPA 22: t. 859 + text, 1942). **T:** RSA, Mpumalanga (*Swart* s.n.

[PRE 23303]). – **D:** RSA (Mpumalanga); savanna among rocks in crevices, flowers in early summer.

Bulbs above-ground, dividing to form small dense groups to 10 plants, almost pear-shaped, to 7 × 4 cm with imbricate, loose, succulent **Sc**, **Sc** purplish-green, clasping, membranous apically, dry parts grey-brown; **R** fleshy, 2 mm ∅, terete; **L** 1 - 4, hysteranthous, green, ascending, lamina linear, subterete, 20 - 50 cm × 1.5 - 5 mm, base channelled; **Inf** racemose, to 24 cm, up to 20-flowered, erect; scape 1 mm ∅ at the base; floral **Bra** inconspicuous, triangular-lanceolate, white, 1 - 1.5 mm, withering before **Fl** mature; **Fl** spreading to slightly drooping; **Per** stellate, ± 9 - 10 mm ∅; **Ped** 3 - 4 mm; **Tep** lorate-spatulate, 6 × 2 mm, white with brownish median stripes, apex obtuse; **St** 4 mm, **Fil** filiform; **Anth** 0.5 mm; **Ov** greenish, 2 mm, globose-obovoid; **Sty** erect, 3.5 mm; **Fr** 3.5 - 5 × 2 mm, oblong; **Se** pear-shaped, many-angled, 0.5 mm.

U. multifolia G. J. Lewis (Ann. South Afr. Mus. 40: 9-10, ill., 1952). **T:** RSA, Northern Cape (*Lewis* 60870 [SAM]). – **D:** RSA (Northern Cape: Namaqualand); Succulent Karoo. **Fig. XXXI.d**

Bulbs underground, solitary, to 4 cm long, globose to ellipsoid; **L** to 50, synanthous, 10 - 20 cm × 0.2 mm, green, terete, spirally twisted; **Inf** racemose, erect, to 20 cm, up to 15-flowered; scape naked; **Bra** triangular-lanceolate, to 4 mm, spurred at the base; **Ped** to 8 mm; **Per** stellate; **Tep** elliptic, 12 × 5 mm; white with brownish median stripes, apex obtuse; **St** 5 mm, **Fil** filiform; **Ov** 4 mm, oblong-ellipsoid; **Sty** erect, 3 mm; **Fr** and **Se** not seen.

WHITEHEADIA

E. van Jaarsveld

Whiteheadia Harvey (Gen. South Afr. Pl., ed. 2, 396, 1868). **T:** *Whiteheadia latifolia* Harvey. – **D:** Namibia, RSA. **Etym:** For Rev. Henry Whitehead (1817 - 1884), Anglican missionary from England who collected in Namaqualand (RSA).

Perennials; bulb small, to 25 mm ∅, tunics thin, fleshy; **L** 2, green, softly fleshy, opposite, to 13 × 9 cm, broadly ovate, horizontally spreading, fragile, longitudinally striate; **Fl** in dense terminal semipyramidal spikes to 10 cm (lengthening to 35 cm at **Fr** time) with short peduncle; **Bra** large, ovate-acuminate, to 3.5 cm; **Per** to 15 mm, cup-shaped, green, lobes ovate, fused below; **St** 6, to 8 mm, basally fused to form a short tube; **Anth** basifixed, introrse, to 4.5 mm; **Ov** triangular, 3-locular, green, to 5 mm; **Sty** short, to 5 mm; **Sti** capitate; **Fr** winged loculicidal capsules; **Se** globose, black, shiny, to 2 mm ∅.

The genus was for a long time monotypic. A second species, *W. etesionamibensis*, was recently described, together with the new, closely related, monotypic genus *Namophila* (Müller-Doblies &

Müller-Doblies 1997). Both *Whiteheadia* and *Namophila* are closely related to *Massonia*.

W. bifolia (Jacquin) Baker (JLSB 13: 226, 1873). **T:** [icono]: Jacquin, Icon. Pl. Rar. 2(16): t. 449, 1795. – **D:** S Namibia, RSA (Northern Cape); Succulent Karoo in rock crevices. **Fig. XXXI.f**

≡ *Eucomis bifolia* Jacquin (1791) ≡ *Basilaea bifolia* (Jacquin) Poiret (1810); **incl.** *Melanthium massoniifolium* Andrews (1804); **incl.** *Whiteheadia latifolia* Harvey (1868).

Description as for the genus.

Nolinaceae

Woody, generally large pachycaul arborescent plants with a stout, simple or sparingly branched stem up to a few metres tall, or stems occasionally underground rhizomes, or acaulescent and xerophytic shrubs; **L** in **Ros** at **Br** tips or in multiple **Ros** on the basal caudex, usually linear, narrow, sessile, parallel-veined, fibrous and hardly or not succulent, without terminal **Sp**; **Inf** panicles, often of considerable size with very numerous **Fl**; **Fl** small (< 1 cm ∅), actinomorphic, unisexual (plants polygamo-dioecious or dioecious), articulated on the **Ped**; **Tep** 6, equal and free; **St** 3 + 3; **Ov** superior, generally 3-locular (1-locular in *Beaucarnea* and *Dasylirion*); **Sty** relatively short; **Fr** ± dry, capsule-like and indehiscent, functioning like nutlets rather than berries; **Se** 1 - 3. – *Cytology:* n = 19.

Distribution: S USA, Mexico, C America (Guatemala, Belize, Honduras, Nicaragua?).

Literature: Trelease (1911); Standley (1920-1926: 94-101); Dahlgren & al. (1985); Delange (1990); García-Mendoza & Galván (1995); Rudall & al. (1995); Bogler (1998a).

A very small family of ± 50 species in 4 genera, and for long placed in the *Agavaceae*. However, recent molecular studies, e.g. Bogler & Simpson (1996), notably of cpDNA, have shown that the 4 genera of this family form a monophyletic group, but are distinct. Rowley (1990) considered *Beaucarnea* and *Calibanus* to be synonymous with *Nolina*. However, the cpDNA data strongly indicates that *Beaucarnea* and *Calibanus* are closely related but distinct from *Nolina*, the approach adopted here. *Nolina*, with 3-locular ovaries, is considered basal to the other genera. Like *Dasylirion*, *Beaucarnea* has 3-winged 1-locular nutlets. The fruit of *Calibanus* is uniquely fleshy and roundish, but its globular caudex is woody and swollen as in *Beaucarnea*. All the genera are poorly understood, notably the Mexican and Mesoamerican species, and are in need of revision.

Cytologically the family appears to be distantly related to the *Agavaceae* as presently constituted. Apparent relatives include *Yucca* (*Agavaceae*) and the *Dracaenaceae*. The *Nolinaceae* differ from the former in their small, (polygamo-) dioecious flowers, and small usually indehiscent fruits. The *Dracaenaceae* differ in the usually somewhat united perianth segments, bisexual flowers, and prevailingly fleshy fruit. Rudall & al. (1995) placed the *Nolinaceae* in the *Asteliales*, along with the families *Convallariaceae*, *Dracaenaceae*, *Asteliaceae* and *Dasypogonaceae*. *Nolinaceae*, like *Dracaenaceae* and *Convallariaceae*, have seeds lacking a phytomelan coat.

The leaves of *Nolina* and *Dasylirion* are used for

thatching, mats, baskets, hats etc. The pulp from the shoot tip of *Dasylirion* contains sugar and was used by the Indians as food or for preparing a beverage called "Sotol". Species of *Beaucarnea* and *Calibanus* are attractive ornamental caudiciforms or pachycauls, widely cultivated and treated here in full. As young plants they make attractive pot plants. *Nolina* and *Dasylirion* are less widely cultivated and hence only a selection of species is included. For information on outdoor cultivation in warm climates see Folsom & al. (1995).

Key to the genera

1 **L** margins with sharp curved prickles:
                                          Dasylirion
– **L** margins without sharp prickles: **2**
2 Caudex large, woody, globular, ± without **Br**; **Fr** fleshy and globular: **Calibanus**
– Caudex woody, sometimes globular when immature but often arborescent in old plants with few to numerous **Br**, sometimes acaulescent: **3**
3 Caudex prominent, often globular when immature but pachycaul and basally swollen in adult plants; **Ov** 1-celled; **Fr** 3-winged capsules:
                                          Beaucarnea
– Caudex never globular when immature, stem woody at maturity, not basally swollen, sometimes acaulescent; **Ov** 3-celled; **Fr** with 1 **Se** per locule, often only 1 locule developing: **Nolina**

[C. C. Walker]

BEAUCARNEA

C. C. Walker

Beaucarnea Lemaire (Ill. Hort. 8(Misc.): 59, 1861). **T:** *Beaucarnea recurvata* Lemaire. – **Lit:** Trelease (1911); Hernández Sandoval (1993). **D:** Mexico, C America (Guatemala, Belize, Honduras, Nicaragua?). **Etym:** For Monsieur Beaucarne, Belgian succulent plant grower and notary from Eename near Audenarde, who first collected flowers of *Beaucarnea recurvata*.

Incl. *Pincenectitia* Lemaire (1861). **T:** not typified.

Shrubs to trees; stem elongate, irregularly scarred by the remains of the **L** bases, basally swollen, tapering towards the tip; **L** in **Ros**, persistent, broadly linear, stiff, acuminate, grooved, bases broadened, glabrous, margin smooth or slightly rough; **Inf** paniculate; **Fl** unisexual, rarely a few bisexual, numerous, very small, pedicellate, white, slightly fragrant, **male Fl** short-lived; **Tep** 6; **Ov** 1-celled with 2 - 3 ovules; **Fr** 3-winged capsules.

A genus of ± 8 species, often previously included in *Nolina*. Poorly understood and in need of revision. All species are pachycaul, some with nearly globular caudices when immature, but only basally swollen when mature. *B. recurvata* is the most com-

monly cultivated species of the family, making an attractive pot plant when small.

Trelease (1911) divided the genus into 2 sections:
[1] Sect. *Beaucarnea*: Slender trees, basally moderately swollen; **Br** elongate; **L** with smooth grooves and nearly smooth margins, thin, nearly flat, recurved, green; **Fr** large, rather long-stalked.
[2] Sect. *Papillatae* Trelease 1911: Slender trees, basally massively swollen; **Br** short; **L** papillate-grooved, rather rough-margined, firm, ± concave, keeled or plicate, nearly straight, pale or glaucous; **Fr** small, very short-stalked.

B. goldmanii Rose (CUSNH 12: 261, pl. 20, 1909). **T:** Mexico, Chiapas (*Goldman* 887 [US]). – **D:** SW Mexico (Chiapas), Guatemala.

[1] Stem slender, branched above, size unknown; **L** linear, to 1 m × 15 mm, acuminate, margin entire; **Inf** compound-paniculate, nearly sessile, primary divisions 15 - 20 cm, **Br** few, to 10 cm; **Tep** ± 2 mm; **Fr** ellipsoid, glaucous, broadly 3-winged, very large, 18 - 20 × 12 - 15 mm.

Especially poorly known and probably not in cultivation. It resembles *B. guatemalensis*, but the leaves are large and the fruits are narrower and glaucous.

B. gracilis Lemaire (Ill. Hort. 8(Misc.): 61, 1861). – **D:** Mexico (Puebla, Oaxaca). **I:** CSJA 48: 64, 1976; Riha & Subik (1981: fig. 230); Rowley (1987: 89, as *Nolina*). **Fig. XXXII.a**
≡ *Nolina gracilis* (Lemaire) Ciferri & Giacomini (1950); **incl.** *Pincenectitia gracilis* Lemaire (1861); **incl.** *Beaucarnea aedipus* Rose (1906).

[2] Stem basally enormously swollen, circular in cross-section, variously and irregularly branched, to 12 m overall; **L** linear, erect, 30 - 60 cm × 4 - 7 mm, very glaucous, margins minutely but sharply serrulate-scabrous; **Inf** short-stalked, ovoid- or oblong-paniculate, primary divisions to 30 cm, **Br** rather weak, to 14 cm; **Ped** 3.2 - 3.5 mm; **Tep** 1.2 × 1 mm; **Fr** round-ellipsoid, openly notched at tip and base, 10 × 7 - 9 mm; **Se** smooth, 3.6 - 2.8 mm.

García-Franco & al. (1995) record a relatively high population density in the valley of Zapotitlán de las Salinas (Puebla) with a sex ratio close to 1:1. A rare occurrence of parasitism by *Psittacanthus calyculatus* (*Loranthaceae*) is also reported.

B. guatemalensis Rose (CUSNH 10: 88, fig. 1, 1906). **T:** Guatemala (*Kellermann* 4320 [US]). – **D:** Guatemala.
≡ *Nolina guatemalensis* (Rose) Ciferri & Giacomini (1950).

[1] Trees 6 - 12 m with thickened bulbous base, abruptly contracted into a slender stem often with slender multiple **Br**; **L** linear, to 1 m × 25 - 30 mm, margin entire; **Inf** short-stalked, broadly ovoid-

paniculate, primary divisions 30 cm, **Br** rather spreading, to 15 cm, branchlets to 6 cm; **Tep** 3 mm; **Fr** ellipsoid-obovate, openly notched at tip and base, 15 - 18 × 13 - 15 mm; **Se** irregularly 3-lobed, 5 mm ∅, smooth.

Related to *B. inermis* and *B. pliabilis*, but the fruit has broader wings than in the former, and the leaves are broader than in the latter.

B. hiriartiae L. Hernández S. (Acta Bot. Mex. 18: 25-29, ills., 1992). **T:** Mexico, Guerrero (*Hernández Sandoval & Martínez* 1629 [MEXU, TEX, UAT]). – **D:** Mexico (Guerrero).

[2] Trees to 8 m; stem basally swollen, oval in cross-section, often with slender multiple **Br**; **L** recurved, linear, to 70 - 90 cm × 10 - 15 mm, pale green, concave, margin minutely serrulate; **Inf** short-stalked, ovoid-elliptically paniculate, primary divisions 20 - 25 cm, **Br** rather spreading, 3 - 10 cm; **Ped** 5.5 - 6.5 mm; **Tep** ovate, 2.5 × 2 mm; **Fr** obovate-oblong, 8 - 11 × 7 - 10 mm; **Se** ellipsoid-obovate, 3.5 × 3.1 mm.

Closely related to *B. stricta* from the neighbouring States of Puebla and Oaxaca, from which it may differ only quantitatively in the length of organs such as leaves and pedicels. It differs though in that the stem base in cross section is oval and the leaves are recurved. Probably not in cultivation.

B. inermis (S. Watson) Rose (CUSNH 10: 88, fig. 2, 1906). **T:** Mexico, San Luis Potosí (*Palmer* 3108 [not located]). – **D:** E-C Mexico (San Luis Potosí, Vera Cruz).

≡ *Dasylirion inerme* S. Watson (1891).

[1] Stem to 13 m, 1.5 m ∅, closely few-branched at top; **L** linear, to 1 m × 12 - 15 mm, margin smooth; **Inf** long-stalked, narrowly pyramidal-paniculate, primary divisions to 30 cm, lower **Br** slender, to 15 cm, and few branchlets 3 - 4 cm; **Tep** scarcely 2 mm; **Fr** elongated-ellipsoid, 14 × 10 mm; **Se** 3 × 2 mm.

A very poorly understood taxon and possibly a short-leaved form of *B. recurvata*.

B. pliabilis (Baker) Rose (CUSNH 10: 89, 1906). **T:** Mexico, Yucatán (*Schott* 892 [BM]). – **D:** SE Mexico (Yucatán, Quintana Róo), Guatemala, Belize. **I:** Trelease (1911: pl. 10); Lundell (1939: fig. 1). **Fig. XXXII.c**

≡ *Dasylirion pliabile* Baker (1880) ≡ *Nolina pliabilis* (Baker) Lundell (1939); **incl.** *Dracaena petenensis* Lundell (1935) ≡ *Beaucarnea petenensis* (Lundell) Lundell (1939); **incl.** *Beaucarnea ameliae* Lundell (1939).

[1] Trees 4 - 12 m, base swollen to 90 cm ∅, stem openly branched; **Br** slender; **L** linear, to 1 m × 15 mm, smooth, acuminate, pendent, margin serrulate, tip entire, base amplexicaul; **Inf** compound-paniculate, primary divisions 30 cm, with few rather short spreading **Br** to 58 cm; **Tep** ovate-oblong or ob-

long-elliptic, pale yellow-white, 4 × 3 mm; **Fr** somewhat obovately round-ellipsoid, 13 - 18 × 11 - 12 mm; **Se** irregularly 3-lobed, transversely wrinkled, 4 × 3 mm.

Resembling *B. guatemalensis*, but the latter has a finely scabrous leaf surface, whereas *B. pliabilis* has entirely smooth blades. Vernacular name: "Tzipil". Only rarely cultivated.

B. recurvata Lemaire (Ill. Hort. 8(Misc.): 61, 1861). – **D:** SE Mexico (Vera Cruz). **I:** CSJA 41: 52, 1969; Rowley (1987: 89, as *Nolina*). **Fig. XXXII.d**

≡ *Nolina recurvata* (Lemaire) Hemsley (1884); **incl.** *Pincenectitia tuberculata* Lemaire (1861) ≡ *Beaucarnea tuberculata* (Lemaire) Roezl (1883).

[1] Trees to 9 m, stem basally moderately swollen, slender and few-branched above, caudex almost globular when immature, later 4 - 6 m, 50 cm ∅ and more at the base, bark smooth; **L** linear, 90 - 180 cm × 15 - 20 mm, slightly tapering, recurved, thin, flat or slightly grooved, green, margin smooth; **Inf** to > 1 m, nearly sessile, broadly ovoid-paniculate, primary divisions to 30 cm, lower **Br** to 15 cm, branchlets 5 cm; **Tep** 3 mm; **Fr** orbicular, tip and base emarginate, 3 - 4 mm ∅.

Readily grown from seed and therefore now extremely common in cultivation, and widely distributed through the commercial nursery trade. *B. recurvata* retains its immature form of an almost globular caudex for many years before the stem tip starts to elongate. As a young plant, therefore, with the long narrow leaves emerging from the globose base it makes an unusual and attractive pot plant. First flowering of 15 year old pot-grown specimens has been reported at ± 1 m tall and ± 20 - 25 cm basal ∅. At maturity it has a rather different form, basally swollen with a tapering main stem and a few terminal branches. Branching presumably occurs after flowering. Vernacular name (USA): "Ponytail Palm".

B. stricta Lemaire (Ill. Hort. 8(Misc.): 61, 1861). – **Lit:** Matuda (1960). **D:** Mexico (Puebla, Oaxaca). **I:** CSJA 41: 53-54, 1969. **Fig. XXXII.e**

≡ *Beaucarnea recurvata* var. *stricta* (Lemaire) Baker (1880) ≡ *Dasylirion strictum* (Lemaire) Macbride (1918) ≡ *Nolina stricta* (Lemaire) Ciferri & Giacomini (1950); **incl.** *Pincenectitia glauca* Lemaire (1861) ≡ *Beaucarnea glauca* (Lemaire) Hereman (1868); **incl.** *Beaucarnea purpusii* Rose (1906).

[2] Trees to 10 m; stem greatly swollen, circular in cross-section at the base, irregularly and modestly branched, bark corky, fissured; **L** linear, erect, to 55 - 80 cm × 8 - 15 mm, stiffly spreading, keeled, pale or glaucous grey, margin yellowish, usually minutely serrulate-scabrous; **Inf** short-stalked, ovoid-paniculate, primary divisions 20 cm, **Br** short, the lower with branchlets to 3 cm; **Ped** 2.8 - 3 mm;

Tep 2.5 × 2 mm; **Fr** broadly ellipsoid, openly notched at tip and base, 12 × 8 - 10 mm; **Se** irregularly 3-lobed, smooth, 3.6 × 3.3 mm.

CALIBANUS

C. C. Walker

Calibanus Rose (CUSNH 10: 90, 1906). **T:** *Dasylirion hookeri* Lemaire. – **D:** Mexico. **Etym:** Named for Shakespeare's monster Caliban from the play 'The Tempest'; perhaps referring to the massive caudex.

Caudex globose, slightly flattened, to 50 cm ∅, bark thick, corky and fissured, later woody, with numerous crowns of **L** from very short monocarpic **Br** not elevated over the surface of the caudex; **L** thin, grass-like, wiry, narrowly linear, somewhat concave and keeled, to 30 cm × 2 - 3 mm, margins serrulate-scabrous, tip entire; **Inf** simple panicles, normally unisexual or rarely with some hermaphrodite **Fl**, to 25 cm, shorter than the **L**, shortly pedunculate, lax with thin spreading **Br** 6 - 8 cm; **Bra** scarious, ovate or lanceolate, much shorter than the subtended **Br**; **Tep** 3 + 3, elliptic-obovate, 2 - 3 × 1.5 - 2 mm, **Sc**-like, dirty whitish-yellow, translucent; **St** 6, subequal, 2.5 - 3 mm, **Anth** medifixed, 1.2 - 1.5 mm; **Ov** ovoid, 2.5 - 3 × 2 mm, 3-locular with 1 ovule in each locule; **Sti** sessile, 3-lobed; **Fr** globose-ovoid, 3-angled, 8 - 9 × 6 - 7 mm, pale straw-brown, indehiscent, with a single **Se**; **Se** melon-shaped, 3 - 4 × 3 mm. – *Cytology:* 2n = 38.

The genus *Calibanus* is a close ally of *Nolina* and *Beaucarnea*, resembling the former in leaf anatomy and pollen, and the latter in cytology (Johnson & Gale 1983). At present monotypic, there is a recent brief report of the discovery of a second species (Fitz Maurice & al. 1997). The putative new species differs from *C. hookeri* in its "light-blue coloured and wider leaves, longer inflorescences and somewhat triangular seeds" but its formal status has yet to be determined.

C. hookeri (Lemaire) Trelease (Proc. Amer. Philos. Soc. 50: 426, 1911). – **Lit:** Johnson & Gale (1983). **D:** Mexico (Hidalgo, San Luis Potosí). **I:** CSJA 42: 269, 1970; Rowley (1987: 32, 90). **Fig. XXXII.b**
≡ *Dasylirion hookeri* Lemaire (1859) ≡ *Beaucarnea hookeri* (Lemaire) Baker (1872) ≡ *Nolina hookeri* (Lemaire) G. D. Rowley (1990); incl. *Dasylirion hartwegianum* Hooker (1859) (*nom. illeg.*, Art. 53.1); **incl.** *Dasylirion caespitosum* Scheidweiler (1861) ≡ *Calibanus caespitosus* (Scheidweiler) Rose (1906).

Description as for the genus.

C. hookeri was first described around 1845. It was not described until 1859 when it was misidentified as *Dasylirion hartwegianum* by Sir William Hooker at Kew. A month later Lemaire corrected the mistake, and in 1906 Rose established the monotypic genus *Calibanus* but used an incorrect

specific epithet. Finally W. Trelease published its current name in 1911. Despite its long and complicated nomenclatural history, the taxon is a relatively recent introduction into cultivation. It was rediscovered by C. Glass and B. Foster in 1968 in San Luis Potosí, where it appears to be well camouflaged in habitat. Most of the material now widespread in cultivation originates from this introduction, since the plant has taken well to cultivation. It has proved to be almost hardy, and easily propagated from seed; Johnson & Gale (1983) report successful artificial pollination with one-year old refrigerated pollen. *C. hookeri* is a remarkable caudiciform, eventually developing a large woody caudex, curious rather than attractive. It is known locally in Mexico as "Sacamecate" and its leaves are used for thatching and for scouring dishes.

DASYLIRION

C. C. Walker

Dasylirion Zuccarini (Allg. Gartenzeitung 6: 258, 1838). **T:** *Dasylirion graminifolium* Zuccarini. – **Lit:** Trelease (1911); Bogler (1994); Bogler (1995). **D:** S USA to Mexico. **Etym:** Gr. 'dasys', dense, rough, shaggy; and Gr. 'leirion', lily; presumably for the long and untidy appearance of the leaves.
Incl. *Dasylirium* Lemaire (1865) (*nom. inval.*, Art. 61.1).

Short arborescent perennial shrubs with thick unbranched stems crowned with dense **Ros**, sometimes **Ros** ± stemless; **L** serrulate or prickly, or unarmed, linear, hard and fibrous; **Inf** paniculate, elongate, conspicuously bracteate, to 6 m; **Ped** jointed at base of **Fl**; **Per** small, whitish, persistent; **St** 6, vestigial in female **Fl**; **Ov** 1-celled, ovules 2 or 3 but only 1 developing; **Sty** very short, erect; **Sti** 3; **Fr** 1-seeded small capsules, 3-winged, thin-walled.

A genus of about 15 species, many poorly understood. All are xerophytic shrubs and not truly succulent, but some are sometimes cultivated in succulent plant collections. Several species, commonly known by the Indian vernacular name "Sotol", have various uses. Their stems were used for building and for fuel. The leaves are trimmed off and the remaining stump is roasted or boiled. The heads are often baked in pits dug in the ground. The roasted stems are also allowed to ferment to produce an alcoholic beverage. The leaves are much used for thatching, mats, baskets, rough hats etc., and their fibres for rough cordage. Dried and varnished expanded leaf bases, called "Desert Spoons", are widely used in flower arrangements etc.

Trelease (1911) divided the genus into 2 sections as follows:
[1] Sect. *Dasylirion*: **L** 2-edged, usually somewhat concave and irregularly keeled, margins prickly and usually rough with minute intervening denticles.

[2] Sect. *Quadrangulatae* Trelease 1911: **L** 4-sided, unarmed.

D. acrotrichum (Schiede) Zuccarini (Abh. math.-phys. Cl. König. Bayer. Akad. Wiss. 3: 226, 1840). – **D:** Mexico (Hidalgo, Querétaro, San Luis Potosí). **I:** CBM 84: t. 5030, 1858; Jacobsen (1960: 335-336).

≡ *Yucca acrotricha* Schiede (1829); **incl.** *Roulinia gracilis* Brongniart (1840) ≡ *Barbacenia gracilis* (Brongniart) Brongniart (1840) ≡ *Bonapartea gracilis* (Brongniart) Otto (1841) ≡ *Yucca gracilis* (Brongniart) Otto (1841) (*nom. illeg.*, Art. 53.1) ≡ *Dasylirion gracile* (Brongniart) Zuccarini (1845) ≡ *Littaea gracilis* (Brongniart) Verschaffelt (1864).

[1] Stem to 1 m; outer **L** recurved, linear, green and glossy or somewhat glaucous and dull, 60 - 100 × 0.9 - 1.8 cm, tip split into 20 - 30 spreading fibres, margin distinctly and finely toothed between the prickles, prickles 5 - 10 (-15) mm apart, to 2 mm long, rather straight, pale yellowish with slightly brown tips; **Inf** paniculate, 3 - 5 m; **Bra** ovate, entire; **Tep** 2 - 3 mm; **Fr** round-cordate, shallowly notched, wings broadening upwards, 8 - 9 × 6 - 7 mm; **Se** 3.5 × 3 mm.

D. glaucophyllum Hooker (CBM 84: t. 5041 + text, 1858). – **D:** E-C Mexico. **I:** Riha & Subik (1981: fig. 225).

Incl. *Dasylirion glaucum* Carrière (1872).

[1] Stem short; **L** linear, acuminate, intensely glaucous, bluish-green, 60 - 120 × 1.2 - 1.8 cm, tip entire, marginal prickles 1 - 2 mm long, horny, deep yellow, margin finely toothed between the prickles; **Inf** densely paniculate, to 1.2 m; **Tep** greenish-white, red-tipped, 2 mm; **Fr** subellipsoid, 9 - 10 × 6 mm; **Se** 4 × 2.5 mm.

D. graminifolium Zuccarini (Allg. Gartenzeitung 6: 259, pl. 1, 1833). – **D:** Mexico (San Luis Potosí).

≡ *Yucca graminifolia* (Zuccarini) Zuccarini (1837).

[1] Stem short, < 80 cm; **L** linear, long-acuminate, bright green, smooth, glossy, 90 - 120 × 1.2 - 1.4 cm, tip with 6 - 8 spreading fibres, marginal prickles 5 - 10 mm apart, 1 - 2 mm long, horny, yellow or tips slightly darkened; **Inf** paniculate, 2.4 - 2.7 m; **Tep** 2 mm; **Fr** ellipsoid, 8 - 9 × 6 mm.

D. leiophyllum Engelmann *ex* Trelease (Proc. Amer. Philos. Soc. 50: 433, 1911). **T:** USA, Texas (*Havard* s.n. [not located]). – **D:** USA (Texas, New Mexico), Mexico (Chihuahua). **I:** Lamb & Lamb (1974: pl. 17).

[1] Stem short, < 80 cm, **Ros** to 1.3 m ∅; **L** linear, smooth, glossy green or somewhat glaucous, 100 × 2.5 - 3 cm, tip fibrous, marginal prickles 1 - 1.5 cm apart, 3 - 4 mm long, recurved, becoming orange or reddish; **Inf** paniculate, to 3 m; **Tep** greenish, 2 mm; **Fr** broadly ellipsoid, openly and deeply notched at the tip, 6 - 9 × 2 - 6 mm; **Se** 3 × 2 mm.

D. longissimum Lemaire (Ill. Hort. 3(Misc.): 91, 1856). **T** [neo]: Mexico, Hidalgo (*Quintero* 3329 [MEXU, ARIZ, CAS, F, GH, MO, NY, RSA, TEX, US]). – **Lit:** Alanis Flores & al. (1994). **D:** Mexico (Tamaulipas, San Luis Potosí, Hidalgo). **I:** CBM 126: t. 7749, 1900; CSJA 56: 19, 1984. **Fig. XXXII.f, XXXII.g**

Incl. *Dasylirion quadrangulatum* S. Watson (1879); **incl.** *Dasylirion juncifolium* Rehnelt (1906).

[2] Stem solitary, 1 - 2 m; **L** very numerous, spreading in all directions, lower ones recurved against the stem, all **L** narrowly linear, green, dull, smooth, rhombic or square in cross-section, upper and lower surfaces raised to low keels, 2 m × 3 - 8 mm, tip entire, margin minutely granular-roughened or smooth; **Inf** paniculate, 2 - 6 m; **Tep** 3 - 4 mm; **Fr** broadly obovate or ellipsoid, scarcely notched, 7 - 10 × 5 - 8 mm; **Se** 3 - 4 × 3 mm.

Bogler (1998b) studied the variability of this species and recognizes a variety, var. *treleasei*.

D. serratifolium (Schultes) Zuccarini (Abh. math.-phys. Cl. König. Bayer. Akad. Wiss. 3: 225, 1840). – **D:** Mexico (Oaxaca). **I:** Gómez-Pompa & al. (1971: fig. 8).

≡ *Yucca serratifolia* Karwinsky *ex* Schultes *fil.* (1830) ≡ *Roulinia serratifolia* (Schultes) Brongniart (1840); **incl.** *Dasylirion laxiflorum* Baker (1872).

[1] Subacaulescent; **L** glaucous, finely roughened on one or both faces, 60 - 100 × 1.5 - 3 cm, tip fibrous, marginal prickles hooked, 2 - 3 cm apart, 1 - 3 mm long, horny, deep yellow; **Inf** loosely paniculate; **Tep** 2 mm; **Fr** ± globose, broadly winged, tip deeply notched, 6 - 8 × 6 - 8 mm; **Se** 4 × 3 mm.

D. wheeleri S. Watson (in Rothrock, Rep. US Geogr. Surv. Wheeler, 6: 378, 1878). **T** [syn]: USA, Arizona (*Rothrock* 329 [MO]). – **Lit:** Laferrière (1991). **D:** SW USA, Mexico.

D. wheeleri var. **durangense** (Trelease) J. E. Laferrière (Ann. Missouri Bot. Gard. 78(2): 519, 1991). **T:** Mexico, Durango (*Palmer* 557 p.p. [MO, MEXU]). – **D:** Mexico (Sonora, Chihuahua, Durango, Zacatecas). **I:** Trelease (1911: pl. 11-12).

≡ *Dasylirion durangense* Trelease (1911).

[1] Differs from var. *wheeleri*: **Fr** broadly elliptic-cordate, 9 × 7 - 8 mm, distal wing tips extending > 1.5 mm above the base of the **Sty** remains.

Laferrière (1991) reports a great deal of phenotypic variation within individual populations of *D. wheeleri*, and indeed records one population from W Chihuahua where both varieties occur. However, he recognizes var. *durangense* as distinct, based on a statistical analysis, which shows a consistent difference in notch depth between specimens from the S part of the range and those from further north.

D. wheeleri var. **wheeleri** – **D:** USA (Arizona, New Mexico, Texas), Mexico (Chihuahua). **I:** Gentry (1972: 176, fig. 69); Folsom & al. (1995: 67).

Incl. Dasylirion wheeleri var. *wislizeni* Trelease (1911).

[1] Perennial bushy shrubs, 0.5 - 1.5 m tall, **Ros** in shrubby groups, skirted by persistent dry recurved **L**; **L** numerous, glaucous, nearly smooth, 60 - 100 × 1.5 - 2 (-2.5) cm, tip attenuate, margin armed with sharp slender straight to antrorse teeth, teeth yellow to brownish, 2 - 5 mm; **Inf** paniculate, slender, 3 - 4 m, peduncle extending above the **L**, with numerous short ascending lateral **Br** subtended by broad scarious fimbriate bractlets; **Fr** round-obovate, 7 - 9 × 6 - 7 mm, remains of **Sty** normally about equalling the apical notch which is ± 1 mm deep, tips of the wings acute to obtuse, distal wing tips extending < 1.5 mm above the base of the **Sty**; **Se** brownish, 3.5 - 4 mm. – *Cytology:* 2n = 38.

D. wheeleri is commonly known by the Indian name "Sotol". Native Americans have many uses for it – after removal of the teeth! – including thatching, mats, baskets. Stems are used for posts in houses and corals. The alcoholic beverage "Sotol" is distilled from the fermented soft meristematic tissue.

NOLINA

C. C. Walker

Nolina Michaux (Fl. Bor.-Amer. 1: 207, 1803). **T:** *Nolina georgiana* Michaux. – **Lit:** Trelease (1911). **D:** S USA (California, Arizona, Texas to Florida), Mexico. **Etym:** For P. C. Nolin, French agriculturalist and horticultural author.

Acaulescent or arborescent perennial shrubs, stems sometimes basally swollen, occasionally with extensive underground rhizomes; **L** linear, hard, fibrous, margins rough or serrulate; **Inf** paniculate, diffuse and racemose, **Br** subtended by **Bra**; **Ped** jointed; **Tep** 6, small, persistent, spreading; **St** 6, usually abortive in fertile female **Fl**; **Fil** short, slender; **Ov** deeply 3-lobed, sessile or shortly stipitate, abortive in male **Fl**; **Fr** papery 3-celled capsules, dehiscent; **Se** 1 - 3, globose to oblong.

A genus of ± 20 species. As the genus is here circumscribed, and separated from *Beaucarnea* and *Calibanus*, it consists of xerophytic shrubs, not truly succulent. Most taxa occur in Mexico, but several species grow across the S USA from California to Florida. Many are poorly known. Gentry (1946) notes that the genus possesses few striking morphological characters, with flowers being monotonously similar and fruits varying only in size and dehiscence. He suggests leaf characters appear to be the most useful for determining the nature of closely related entities.

In Mexico species have several common names, including "Zacate Cortador", "Zacate de Armazón" and "Palmilla". In the USA, species are commonly known as "Beargrass", and as the name suggests, plants resemble coarse grass when not in flower. Leaves are tough and have been used for thatching, brooms, baskets, hats, mats etc.

Trelease (1911) divided the genus into 4 sections as follows:

[1] Sect. *Nolina* (= Sect. *Graminifoliae* Trelease 1911): Acaulescent, **L** thin and grass-like but hard and fibrous, linear, rarely > 5 mm wide, rather flat, usually not brush-like at the tip (no taxa treated here).

[2] Sect. *Erumpentes* Trelease 1911: **L** rather thick, linear or narrowly oblong-triangular, to 12 mm wide, green, ± concave, tip often fibrous-lacerate; **Fr** small, not inflated.

[3] Sect. *Microcarpae* Trelease 1911: **L** rather thick, linear or narrowly oblong-triangular, to 12 mm wide, green, ± concave, tip often fibrous-lacerate; **Fr** moderately sized, somewhat inflated (no taxa treated here).

[4] Sect. *Arborescentes* Trelease 1911: Trees; **L** relatively thin, 15 - 40 mm wide, tip usually not brush-like; **Fr** large, inflated.

N. bigelovii (Torrey) S. Watson (Proc. Amer. Acad. Arts 14: 247, 1879). **T:** USA, Arizona (*Bigelow* s.n. [not located]). – **D:** USA (Arizona). **I:** Gentry (1972: 181, fig. 71).

≡ *Dasylirion bigelovii* Torrey (1857) ≡ *Beaucarnea bigelovii* (Torrey) Baker (1872).

[4] Stem 1 - 3 m with a large crown of stiff **L** persisting dry and reflexed on the trunk; **L** linear, 80 - 120 × 1.5 - 3.5 cm, tip entire, margins at first serrulate then filiferous; **Inf** paniculate, 60 - 100 cm; **Bra** deltoid-lanceolate, thin, 4 - 10 × 1 - 2 cm, attenuate, soon deciduous, primary **Br** slender, 10 - 20 cm, ascending, glabrous, smooth; **Tep** oblong-linear, 2.5 - 3 mm, **OTep** introrsely shortly apiculate, reflexed in female **Fl**, **ITep** erect or ascending; **Fr** narrowly emarginate at tip and base, 8 - 12 × 9 - 12 mm; **Se** ovate to oblong, whitish, wrinkled, 2.5 - 3.5 mm.

Gentry (1972) anticipates that the taxon should also occur in adjacent Mexico (Sonora).

N. cismontana Dice (Novon 5(2): 162-164, 1995). **T:** USA, California (*Dice & Oberbauer* 650 [SD 121705, ARIZ, NY, RSA, UC]). – **D:** USA (S California).

[4] Stem 0.5 - 1.5 m, branching above and below ground, mature **Ros** with 30 - 90 **L**; **L** lanceolate-linear, 50 - 140 × 1.2 - 3 cm, base expanded, margin serrulate; **Inf** 1 - 3 m, with rather narrow **Br** 13 - 35 cm long and spreading; **Bra** large, papery, persistent; **Tep** cream-white, ovate, 3 - 5 × 1.5 - 2.5 mm; **Fr** orbicular, papery, emarginate at base and tip, 8 - 12 mm tall and slightly broader; **Se** ovoid to oblong, reddish-brown, 4 - 5 × 3 - 4 mm. – *Cytology:* n = 19.

N. interrata Gentry (Madroño 8: 181, fig. 1, pl. 1, 1946). **T:** USA, California (*Gentry* 7330 [SD]). – **D:** USA (S California: San Diego County), Mexico (N Baja California). **I:** CSJA 44: 177, 1972; 48: 155, 1976.

[4] Plants with underground branching rhizome to 3 m × 30 cm ∅, above-ground stems not obvious, mature **Ros** with < 45 **L**; **L** linear, glaucous, 70 - 100 × 0.8 - 1.5 cm, base barely expanded, tip dry, slender, not filiferous, margin minutely and persistently serrate, armed with denticles of 2 sizes; **Inf** open compound panicles 1.5 - 2 m, base of peduncle 0.5 - 1.8 cm ∅; **Bra** persistent, 20 - 40 cm; **female Fl** with staminodes inserted on the **Tep**; **Fr** large, broader than long, 12 - 15 mm wide; **Se** reddish-brown to yellowish, wrinkled, 5 × 4 mm.

N. interrata is distinguished by the glaucous leaves with coarse armature. The horizontal underground stem or rhizome is esp. noteworthy, but this structure may be present in other species assumed to be acaulescent, since this feature may be readily overlooked. Known locally in San Diego County as "Dehesa Nolina". *N. interrata* is one of the most endangered plant species in California, occurring in only 2 small stands. The populations are stable but are slowly being affected by land clearance for housing development. 3 additional populations are known in Baja California.

N. longifolia (Karwinsky *ex* Schultes *fil.*) Hemsley (Biol. Centr.-Amer., Bot. 3: 373, 1884). – **D:** Mexico (Puebla, Oaxaca). **I:** Rowley (1990: figs. 1,2,4); Folsom & al. (1995: 104-105).
    ≡ *Yucca longifolia* Karwinsky *ex* Schultes *fil.* (1830) ≡ *Dasylirion longifolium* (Karwinsky *ex* Schultes *fil.*) Zuccarini (1840) ≡ *Beaucarnea longifolia* (Schultes) Baker (1872); **incl.** *Roulinia karwinskiana* Brongniart (1840).

[4] Stem 2 - 3 m, basally swollen, narrowing towards the tip, bark rough; **Br** few, short, with a dense crown of **L**; **L** > 2 m × 2.5 cm, long-tapering, pendent, thin, firm, tip frayed and brush-like, margins minutely rough; **Inf** paniculate-ramose, to 2 m, primary **Br** to 30 cm; **Tep** of **female Fl** ± 1.5 mm, white; **Fr** inflated, suborbicular or rather depressed, 10 - 12 × 8 mm; **Se** 4 × 3 mm.

N. matapensis Wiggins (Contr. Dudley Herb. 3: 65, 1940). **T:** Mexico, Sonora (*Wiggins* 7515 [Dudley Herb.]). – **D:** Mexico (Sonora). **I:** CSJA 48: 92, fig. 12, 1976.

[4?] Small trees, stem to ± 6 m, basally swollen, simple or modestly branched, bark fissured; **Ros** small; **L** linear, 70 - 120 × 1 - 1.5 cm, recurved, hard, striate, greenish-yellow, somewhat glaucous, margin finely serrate; **Inf** paniculate, 1.5 - 3× longer than the **L**, primary **Br** 15 - 35 cm, branchlets 10 - 15 cm; **Tep** oblong-linear, 2 mm; **Fr** broadly ellipsoid, depressed, deeply notched at both ends, tardily dehiscent, 8 - 9 × 5 mm; **Se** pale brown,

nearly round, 2.5 - 3 mm ∅.

The leaves are used for basket-making, the stems for posts. A relatively unknown species, first introduced into cultivation in 1976 as ISI 982. Plants from this introduction are now multi-headed specimens at the Huntington Botanical Garden. The species is recommended as an attractive landscape plant for subtropical areas such as S California. Local vernacular names: "Palmito", "Tuya".

N. parryi S. Watson (Proc. Amer. Acad. Arts 14: 247, 1879). **T:** USA, California (*Parry* s.n. [GH, MO]). – **Lit:** Mitich (1982); Hess & Dice (1995). **D:** USA (California, Arizona), Mexico (Baja California). **I:** Lamb & Lamb (1974: pl. 134). **Fig. XXXII.h**
    ≡ *Nolina bigelovii* var. *parryi* (S. Watson) L. D. Benson (1945) ≡ *Nolina bigelovii* ssp. *parryi* (S. Watson) E. Murray (1983); **incl.** *Nolina parryi* ssp. *parryi*; **incl.** *Nolina parryi* ssp. *wolfii* Munz (1950) ≡ *Nolina bigelovii* var. *wolfii* (Munz) L. D. Benson (1954) ≡ *Nolina wolfii* (Munz) Munz (1974) ≡ *Nolina bigelovii* ssp. *wolfii* (Munz) E. Murray (1983).

[4] Stem 1 - 2 m, basally branching, mature **Ros** with up to 200 **L**; **L** linear, 50 - 150 × 2 - 4 cm, almost pungent, rather thick, concave, keeled, margin serrulate-scabrous, base strongly expanded; **Inf** to 4 m, with rather narrow **Br** 15 - 30 cm long and spreading, densely flowered branchlets < 4 cm long; **Bra** large, papery, persistent; **Fl** large; **Tep** 4 mm; **Fr** very large, orbicular, deeply notched at both ends, 12 - 15 mm ∅; **Se** reddish-brown, 4 × 3 mm.

Mitich (1982) records spectacular specimens of this species in the Kingston Mountains, Mojave Desert (California). Here some individuals reach a height of 5 m with stems 1 m ∅, and with leaves 1.5 m long; the plants are 4 m across. Inflorescences are enormous too, being 4 m long with the fertile portion 2 m in length. He reports that *N. parryi* grows readily from seed, requiring a mild climate for outdoor cultivation. In such conditions plants flower after 7 - 8 years.

N. texana S. Watson (Proc. Amer. Acad. Arts 14: 248, 1879). – **D:** USA (Arizona, Texas). **I:** Small (1916); Gentry (1972: 183, fig. 72).
    ≡ *Beaucarnea texana* (S. Watson) Baker (1880).

[2] Low-growing acaulescent spreading shrubs; **L** narrowly linear, 70 - 120 × 0.3 - 0.4 cm, deeply rounded below, margins somewhat serrulate or smooth; **Inf** paniculate, slender, 40 - 50 cm, primary **Br** 10 - 15 cm; **Bra** caudate-attenuate, surpassing the **Br**, dry, yellowish, persisting; **Fr** 5 - 7 × 4 - 5 mm; **Se** globose, 3 mm ∅. – *Cytology:* 2n = 38.

Gentry (1972) anticipates localities being found in NW Mexico (Sonora).

Orchidaceae

Perennial terrestrial or predominantly epiphytic herbs, rarely shrubs or lianas, very rarely saprophytic and without chlorophyll or completely subterranean; **R** often tuberous in terrestrial taxa, with multi-layered velamen in epiphytic taxa; stems virtually absent, or forming corms or rhizomes, often distinctly swollen to form pseudobulbs, often rooting at the nodes; **L** entire, spirally arranged (sometimes distichous, rarely opposite or whorled), sometimes **Sc**-like or absent, often fleshy; **Inf** racemes or panicles, or **Fl** solitary; **Fl** mostly bisexual (rarely unisexual and plants dioecious), 3-merous, often resupinate; **Per** of 3 + 3 usually petaloid **Tep**, one of the inner **Tep** (labellum) usually much larger and / or differently coloured; **St** 1 (-3), mostly united with the **Sty** to form a **Gy**; pollen normally united into complex pollinia (rarely pollen dustlike); **Ov** of 3 united **Ca**, inferior, normally 1-locular; **Fr** capsules opening with 3 (or 6) longitudinaly slits but remaining closed at either end; **Se** numerous, dustlike with undeveloped minute embryo.

Distribution: Worldwide but concentrated in the tropics and subtropics.

Literature: Pridgeon (1992); Dressler (1993); Bechtel & al. (1993).

With ± 800 genera and (depending on the source consulted) 17'500 to 25'000 species, the orchids are the largest plant family. Orchids show a number of special developments. Like in the *Asclepiadaceae*, the pollen is united into complex pollinia in most orchids. Unlike the pollinia of the asclepiads with their clasping mechanism of the translator, orchid pollinia have a sticky 'plate' (viscidium) at the basal end of the caudicle and are glued on the body of the pollinator. The minute orchid seeds in nature normally germinate only in the presence of certain mycorrhizal fungi (but can be germinated on appropriate laboratory media in cultivation). Mycorrhiza are probably universally present throughout the family also in adult plants.

The classification of this vast family offers several problems, and a number of genera has not been properly classified up to the present (Dressler 1993). The following outline classification is undisputed:

Apostasioideae: **L** spiral, plicate; **Fl** sometimes resupinate, indistinctly irregular; **St** 2 - 3, pollen never in pollinia. – This small group is regarded as evolutionary primitive and is sometimes recognized as separate family *Apostasiaceae*. It embraces only the 2 genera *Apostasia* and *Neuwiedia* from Indomalesia.

Cypripedioideae: **L** spiral or distichous, sometimes plicate; **Fl** resupinate, labellum slippershaped; **St** 2 (of the inner whorl) + 1 staminode (of the outer whorl); pollen rarely in true pollinia. – This group, sometimes segregated as separate family *Cypripediaceae*, is widely distributed in the N hemisphere from temperate to tropical areas (but excl. Africa) and is also regarded as ± primitive. It embraces only 4 genera: *Cypripedium, Paphiopedilum, Phragmipedium* and *Selenipedium*.

Orchidoideae: **L** normally alternate, often fleshy; **Fl** various; **St** 1 (very rarely 2); pollen usually in pollinia. – This largest subfamily embraces the remainder of the family and is regarded as evolutionary advanced. It is further divided by Dressler (1993), who in addition recognizes the subfamilies *Spiranthoideae* and *Epidendroideae* plus the informal groups Cymbidioids, Epidendroids and Dendrobioids.

Many orchids can be considered true succulents and occur widely in semi-arid climates. Succulence occurs either as stem succulence (fleshy pseudobulbs) or as leaf succulence (esp. in epiphytic taxa). Orchids rival with bromeliads and cacti as *the* most important horticultural group, and numerous taxa are more or less often cultivated. Like for succulents, numerous hobby associations exist throughout the world. Despite their undisputed claim to succulence, they are not further considered here because of the vast literature already available. [U. Eggli]

Xanthorrhoeaceae

Perennial pachycaul small trees to stemless herbs with massive underground **R**stock or rhizome; stems sometimes with secondary growth; **L** mostly numerous in dense **Ros**, narrowly oblong to filiform-linear, xeromorphic and wiry, tough, tips often spiny, basally often broadened and persistent; **Inf** spikes, heads, or thyrses, or head- or umbel-like, or **Fl** solitary; **Fl** bisexual or unisexual (plants monoecious or dioecious and male and female **Fl** often dissimilar), 3-merous, normally numerous and small; **Per** with 6 dry and chaffy persistent **Tep** in 2 series, free or basally united; **St** 3 + 3, the inner often basally united with the **Tep**; **Anth** basifixed or versatile, opening longitudinally; **Ov** superior, of 3 united **Ca**, with axile or axile-basal placentation, 3- or 1-locular; **Sty** 3 or 1 with 3 **Sti**, ovules 1 to several per locule; **Fr** loculicidal capsules or indehiscent, with numerous **Se** or 1-seeded nuts; endosperm copious.

Distribution: Australia except few species in New Guinea and New Caldeonia.

Literature: Bedford & al. (1986); Clifford (1998b).

This small family (10 genera with ± 100 species) with its restricted distribution is probably heterogeneous and some authors restrict it to the genus *Xanthorrhoea*, removing the remaining genera to the families *Calectasiaceae* and *Dasypogonaceae* (Clifford 1998b).

Species of *Xanthorrhoea* (28 species in Australia) are sometimes included amongst succulent plant collections but are more properly regarded as being merely xerophytic with woody stems.

[U. Eggli]

References

APG [Angiosperm Phylogeny Group] (1998) An ordinal classification for the families of Flowering Plants. Ann. Missouri Bot. Gard. 85: 531-553, diags.

Adams, C. D. (1972) Flowering plants of Jamaica. Mona (Jamaica): University of the West Indies.

Alanis Flores, G. J. & al. (1994) Datos fenológicos de *Dasylirion longissimum* en un jardín botánico. Cact. Suc. Mex. 39(2): 43-47, ill.

Álvarez de Zayas, A. (1985) Los Agaves de Cuba occidental. Revista Jard. Bot. Nac. Univ. Habana 5(3): 3-16, ills.

Álvarez de Zayas, A. (1996a) El género *Furcraea* (*Agavaceae*) en Cuba. Anales Inst. Biol. UNAM, Ser. Bot. 67(2): 329-346, ills., maps, key.

Álvarez de Zayas, A. (1996b) Los Agaves de Cuba central. Fontqueria 44: 117-128, ills., key, maps.

Anderson, E. S. & Woodson, R. E. (1935) The species of *Tradescantia* indigenous to the United States. Contr. Arnold Arbor. 9: 1-132, key, ills.

Archibald, E. E. A. (1967) The genus *Dioscorea* in the Cape Province west of East London. J. South Afr. Bot. 33: 1-46, ills.

Audissou, J.-A. (1999) The Dragon Tree in Morocco. *Dracaena draco* ssp. *ajgal* Benabid & Cuzin. Cact. Aventures No. 41: 23-25, ills.

Baensch, U. (1994) Blühende Bromelien. Nassau (Bahamas): Tropic Beauty Publishers. 269 pp., maps, ills.

Bally, P. R. O. (1967) Miscellaneous notes on the flora of Tropical East Africa, including description of new taxa, 35-37. Candollea 22(2): 255-263, ills., maps.

Bayer, M. B. (1976) *Haworthia* handbook. A guide to the species, with identification keys and illustrations. Kirstenbosch (RSA): National Botanic Gardens of South Africa. 184 pp., ills.

Bayer, M. B. (1982) The new *Haworthia* handbook. A revised guide to the literature of the genus, with discussion of the species, identification keys and colour illustrations. Kirstenbosch (RSA): National Botanic Gardens of South Africa. 124 pp., ills., key.

Bayer, M. B. (1999) *Haworthia* revisited. A revision of the genus. Hatfield (RSA): Umdaus Press. 250 pp., ills., maps.

Bechtel, H. & al. (1993) Orchideen-Atlas. Lexikon der Kulturorchideen. Stuttgart (D): Verlag Eugen Ulmer. 3. Ed.; 590 pp., ills.

Bedford, D. J. & al. (1986) *Xanthorrhoeaceae*. In: Flora of Australia, 46: 88-171, ills., keys. Canberra (AUS): Australian Government Publishing Service.

Benítez B., G. (1986) Arboles y flores del Ajusco. México D.F. (MEX): Instituto de Ecología, Museo de Historia Natural de la Ciudad de México. 183 pp., ills.

Benson, L. (1943) Revisions of status of southwestern desert trees and shrubs. Amer. J. Bot. 30: 230-240.

Benson, L. & Darrow, R. A. (1981) Trees and shrubs of the southwestern deserts. Tucson (US: AZ): University of Arizona Press. Ed. 3; 416 pp., maps, ills.

Berger, A. (1908) *Liliaceae − Asphodeloideae − Aloineae*. In: Engler, A. (ed.): Das Pflanzenreich IV.38 (Heft 33). Leipzig (D): Wilhelm Engelmann. 347 pp., ills.

Berger, A. (1915) Die Agaven. Beiträge zu einer Monographie. Jena (D): Verlag Gustav Fischer. 288 pp., ills., maps.

Berry, P. E. (1995) *Agavaceae*. In: Berry, P. E. & al. (eds.): Flora of the Venezuelan Guayana, 2: 374-375, ill. St. Louis (US: MO): Missouri Botanical Garden / Portland (US: OR): Timber Press.

Bogler, D. J. (1994) Systematics of *Dasylirion* (*Nolinaceae*). Amer. J. Bot. 81(6: Suppl.): 142 [abstract 399].

Bogler, D. J. (1995) Systematics of *Dasylirion*. Taxonomy and molecular phylogeny. Bol. Soc. Bot. México 56: 69-76.

Bogler, D. J. (1998a) *Nolinaceae*. In: Kubitzki, K. (ed.): The families and genera of vascular plants; 3: 392-397, ills., key. Berlin (D) etc.: Springer Verlag.

Bogler, D. J. (1998b) Three new species of *Dasylirion* (*Nolinaceae*) from Mexico and a clarification of the *D. longissimum* complex. Brittonia 50(1): 71-86, ills.

Bogler, D. J. & Simpson, B. B. (1995) A chloroplast DNA study of the *Agavaceae*. Syst. Bot. 20(2): 191-205, diags.

Bogler, D. J. & Simpson, B. B. (1996) Phylogeny of *Agavaceae* based on ITS rDNA sequence variation. Amer. J. Bot. 83(9): 1225-1235, diags.

Bogler, D. J. & al. (1995) Multiple origins of the *Yucca* − Yucca Moth association. Proc. Nation. Acad. Sci. USA 92(15): 6864-6867.

Bolliger, T. (1998) Nach 13 Jahren kam die Blüte: *Yucca whipplei* Torrey ssp. *parishii* in Mitteleuropa im Freiland gehalten. Kakt. and. Sukk. 49(8): 187-189, ills.

Bos, J. J. (1984) *Dracaena* in West Africa. Agric. Univ. Wageningen Pap. 84-1: 126 pp., ills., maps, key.

Bos, J. J. (1998) *Dracaenaceae*. In: Kubitzki, K. (ed.): The families and genera of vascular plants; 3: 238-241, ills., key. Berlin (D) etc.: Springer Verlag.

Bosser, J. (1968) Espèces et hybride nouveaux

d'Aloes de Madagascar. Adansonia, n.s., 8(4): 505-512, ills.

Brako, L. & Zarucchi, J. L. (1993) Catalogue of the Flowering Plants and Gymnosperms of Peru. Catálogo de las Angiospermas y Gimnospermas del Perú. St. Louis (US: MO): Missouri Botanical Garden. 1326 pp., maps.

Bramwell, D. & Bramwell, Z. (1974) Wild flowers of the Canary Islands. London (GB): Stanley Thornes Publ. Ltd. 261 pp., ill., maps.

Brandham, P. E. (1969) Chromosome behaviour in the *Aloineae*. I. The nature and significance of E-type bridges. Chromosoma 27: 201-215.

Brandham, P. E. (1971) The chromosomes of the *Liliaceae*: II. Polyploidy and karyotype variation in the *Aloineae*. Kew Bull. 25: 381-399, ills.

Brandham, P. E. (1973) New hybrids in the *Aloineae*. Nation. Cact. Succ. J. 28(1): 16-19, ills.

Brandham, P. E. & Carter, S. (1990) A revision of the *Aloe tidmarshii / A. ciliaris* complex in South Africa. Kew Bull. 45(4): 637-645.

Brandham, P. E. & al. (1994) A multidisciplinary study of relationships among the cremnophilous Aloes of northeastern Africa. Kew Bull. 49(3): 415-428, maps, keys.

Breedlove, D. E. (1986) Flora de Chiapas. Listados florísticos de México IV. Mexico City (MEX): Instituto de Biología, UNAM. v + 246 pp.

Breuer, I. (1998a) The world of Haworthias. Volume 1. Bibliography and annotated index. Niederzier / Homburg (D): Ingo Breuer / Arbeitskreis für Mammillarienfreunde. xii + pp. 1-340, pl. A-X, ills.

Breuer, I. (1998b) Die Gattung *Haworthia* im Überblick. Schumannia 2: 3-74, ills., maps.

Breuer, I. & Metzing, D. (1997) Types of names accepted in *Haworthia* (*Aloaceae*). Taxon 46(1): 3-14.

Bridson, G. D. R. & Smith, E. R. (1991) Botanico-Periodicum-Huntianum / Supplementum. Pittsburgh (USA: PA): Hunt Institute for Botanical Documentation, Carnegie Mellon University. 1068 pp.

Brown, N. E. (1915) *Sansevieria*. A monograph of all known species. Bull. Misc. Inform. [Kew] 1915(5): 1-81, ills., key.

Brummitt, R. K. & Powell, C. E. (eds.) (1992) Authors of plant names. A list of authors of scientific names of plants, with recommended standard forms of their names, including abbreviations. Richmond (GB): The Board of Trustees of The Royal Botanic Gardens, Kew. 732 pp.

Burkill, I. H. (1952) *Testudinaria* as a section of the genus *Dioscorea*. J. South Afr. Bot. 18: 177-191, maps, ills.

Carter, S. (1994) *Aloaceae*. In: Polhill, R. M. (ed.): Flora of Tropical East Africa. Rotterdam

(NL) / Brookfield (US): A. A. Balkema. 60 pp., ills., map, 4 pl., key.

Carter, S. (1996) New *Aloe* taxa in the Flora Zambesiaca area. Kew Bull. 51(4): 777-785, ills.

Carter, S. & Brandham, P. E. (1983) New species of *Aloe* from Somalia. Bradleya 1: 17-24, ills.

Carter, S. & Reynolds, T. (1990) *Aloe penduliflora* and *Aloe confusa*. Kew Bull. 45(4): 647-651.

Carter, S. & al. (1984) A multidisciplinary approach to a revision of the *Aloe somaliensis* complex (*Liliaceae*). Kew Bull. 39: 611-633, ills.

Cave, M. S. (1948) Sporogenesis and embryo sac development of *Hesperocallis* and *Leucocrinum* in relation to their systematic position. Amer. J. Bot. 35: 343-349.

Cedano M., M. & al. (1993) Una nueva espécie de *Polianthes* (*Agavaceae*) del Estado de Michoacán y nota complementaria sobre *Polianthes longiflora* Rose. Bol. Inst. Bot. (Guadalajara) 1(7): 521-530, ills., map.

Chahinian, B. J. (1986) The *Sansevieria trifasciata* varieties. A presentation of all cultivated varieties. Reseda (US: CA): Trans Terra Publishing. 109 pp., ills.

Chahinian, B. J. (1993) *Sansevieria ehrenbergii* & *Sansevieria robusta*. Sansevieria J. 2(1): 7-8, ill.

Cházaro Basañez, M. J. (1981) Nota sobre la tipificación de *Agave obscura* Schiede y su confusión con *Agave xalapensis* Roezl. Biotica 6(4): 435-446, ills.

Cházaro Basáñez, M. J. & Machuca Núñez, J. A. (1995) Nota sobre *Polianthes longiflora* Rose (*Agavaceae*). Cact. Suc. Mex. 30(1): 20-22.

Chupov, V. S. & Kutiavina, N. G. (1981) [Russian:] Serological studies in the order *Liliales*. Bot. Zhurn. (Moscow & Leningrad) 66: 75-81.

Clarke, C. B. (1881) *Commelinaceae*. Monogr. Phan. 3: 113-324.

Clary, K. H. & Simpson, B. B. (1995) Systematics and character evolution of the genus *Yucca* L. (*Agavaceae*): Evidence from morphology and molecular analyses. Bol. Soc. Bot. México 56: 77-88.

Clifford, H. T. (1998a) *Doryanthaceae*. In: Kubitzki, K. (ed.): The families and genera of vascular plants; 3: 236-238, ills., key. Berlin (D) etc.: Springer Verlag.

Clifford, H. T. (1998b) *Xanthorrhoeaceae*. In: Kubitzki, K. (ed.): The families and genera of vascular plants; 3: 467-470, ills. Berlin (D) etc.: Springer Verlag.

Collenette, S. (1999) Wildflowers of Saudi Arabia. Riad (Saudi Arabia): National Commission for Wildlife Conservation and Development.

Colunga-García Marín, P. & al. (1999) Isozymatic variation and phylogenetic relationships between Henequen (*Agave fourcroydes*) and its wild ancestor *A. angustifolia* (*Agavaceae*). Amer. J. Bot. 86(1): 115-123.

Colunga-García Marín, P. & al. (1996) Patterns of morphological variation, diversity, and domestication of wild and cultivated populations of *Agave* in Yucatán, Mexico. Amer. J. Bot. 83(8): 1069-1082.

Conran, J. G. (1998) *Anthericaceae*. In: Kubitzki, K. (ed.): The families and genera of vascular plants; 3: 114-121, ills., key. Berlin (D) etc.: Springer Verlag.

Correll, D. S. & Correll, H. B. (1982) Flora of the Bahama Archipelago (including the Turks and Caicos islands). Vaduz (FL): J. Cramer. 50 + 1692 pp., ills., keys.

Cronquist, A. (1981) An integrated system of classification of Flowering Plants. New York (US): Columbia University Press. 1262 pp., ills.

Cseh, T. A. (1993) In search of *Agave panamana* Trel. Haseltonia 1: 34-44, ills., map.

Cullen, J. (1986) *Yucca* (*Agavaceae*). In: Walters, S. M. & al. (eds.): The European Garden Flora, Vol. 1: 273-276. Cambridge (GB): Cambridge University Press.

Cumming, D. M. (1999) New nothogenera proposed for four trigeneric hybrids within *Aloaceae*. Haworthiad 13(1): 20-21.

Cunningham, G. M. & al. (1981) Plants of Western New South Wales. Sydney (AUS): New South Wales Government Printer.

Cutler, D. F. & Brandham, P. E. (1977) Experimental evidence for the genetic control of leaf surface characters in hybrid *Aloineae* (*Liliaceae*). Kew Bull. 32: 23-32, ills., chrom.nos.

Cutler, D. F. & al. (1980) Morphological, anatomical, cytological and biochemical aspects of evolution in East African shrubby species of *Aloe* L. Bot. J. Linn. Soc. 80(4): 293-317, ills., maps.

Dahlgren, R. M. T. & al. (1985) The families of the Monocotyledons. Structure, evolution, and taxonomy. Berlin, Heidelberg (D) etc.: Springer-Verlag. 520 pp., ills.

Darlington, C. D. & Wylie, A. P. (1955) Chromosome atlas of flowering plants. London (GB): Allen & Unwin. xix + 519 pp., ills.

DeMason, D. A. (1984) Offshoot variability in *Yucca whipplei* ssp. *percursa* (*Agavaceae*). Madroño 31: 197-202.

Delange, Y. (1990) *Nolina* Michaux, *Beaucarnea* Lemaire et *Dasylirion* Zucc. Succulentes 13(3): 18-22, ills.

Desmet, P. (2000) Namaqualand. Journey of discovery. Veld. Fl. (1975+) 86(2): 62-66, ills.

Diggs, G. M. jr. & al. (1999) Shinner's & Mahler's Illustrated Flora of North Central Texas. Sida Bot. Misc. keys, ills.

Dressler, R. L. (1993) Phylogeny and classification of the orchid family. Cambridge (GB) etc.: Cambridge University Press. 314 pp., 91 figs., 16 pl.

Drummond, J. R. (1907) The literature of *Furcraea* with a synopsis of the known species. Annual Rep. Missouri Bot. Gard. 18: 25-75, 4 pl.

Duncan, G. D. (1988) The *Lachenalia* handbook. Ann. Kirstenbosch Bot. Gard. 17: 71 pp., ills.

Dyer, R. A. (1950) A revision of the genus *Brunsvigia*. Pl. Life 6: 63-83.

Dyer, R. A. (1975-76) The genera of Southern African flowering plants. Vol. I: Dicotyledons; Vol. II: Gymnosperms and Monocotyledons. Pretoria (RSA): Department of Agricultural Technical Services. 1040 pp., maps.

Dyer, R. A. & Hardy, D. (1969) Addendum. In: Reynolds, G. W.: Aloes of South Africa, ed. 2, pp. 509-516. Rotterdam (NL): A. A. Balkema.

Eggli, U. (1985) A bibliography of succulent plant periodicals. Bradleya 3: 103-119.

Eggli, U. (1994) Sukkulenten. Stuttgart (D): Ulmer Verlag. 336 pp., ills., maps, keys.

Eggli, U. (1998a) Bibliography of succulent plant periodicals. Bibliografie casopisu o sukulentních rostlinách. Friciana 60: 139 pp.

Eguiarte, L. E. (1988) Reducción en la fecundidad en *Manfreda brachystachya* (Cav.) Rose, una Agavácea polinizado por murciélagos: Los riesgos de la especialización en la polinización. Bol. Soc. Bot. México 48: 147-149.

Eguiarte, L. E. (1995) Hutchinson (*Agavales*) vs. Huber y Dahlgren (*Asparagales*). Análisis moleculares sobre filogenia y evolución de la familia *Agavaceae* sensu Hutchinson dentro de las monocotiledóneas. Bol. Soc. Bot. México 56: 45-56.

Eguiarte, L. E. & Búrquez, A. (1987) Reproductive biology of *Manfreda brachystachya*, an iteroparous species of *Agavaceae*. Southw. Naturalist 32(2): 169-178.

Eguiarte, L. E. & al. (1994) The systematic status of the *Agavaceae* and *Nolinaceae* and related *Asparagales* in the monocotyledons: An analysis based on the rbcL gene sequence. Bol. Soc. Bot. México 54: 36-56.

Elliot, W. R. & Jones, D. L. (1984) Encyclopaedia of Australian plants suitable for cultivation. Vol. 3. Melbourne etc. (AUS): Lothian Publ. Co.

Engelmann, G. (1875) Notes on *Agave*. Trans. Acad. Sci. St. Louis 3: 291-322.

Engler, A. (1905) *Araceae – Pothoideae*. In: Engler, A. (ed.): Das Pflanzenreich, IV.23B. Leipzig (D): W. Engelmann.

Everett, T. H. (1981-1982) New York Botanical Garden illustrated encyclopedia of horticulture. New York (US: NY): Garland Publishing. 10 vols., ills.

Faden, R. B. (1985) *Commelinaceae*. In: Dahlgren, R. M. T. & al. (eds.): The families of Monocotyledons; pp. 381-387. Berlin / Heidelberg (D) etc.: Springer-Verlag.

Faden, R. B. (1991) The morphology and taxonomy of *Aneilema* R. Brown (*Commelinaceae*). Smithsonian Contr. Bot. 76: 166 pp., ills., keys, maps.

Faden, R. B. (1998) *Commelinaceae*. In: Kubitzki, K. (ed.): The families and genera of vascular plants; 4: 109-128, ills., key. Berlin (D) etc.: Springer-Verlag.

Faden, R. B. & Hunt, D. R. (1991) The classification of the *Commelinaceae*. Taxon 40: 19-32.

Favell, P. & al. (1999) Notes on two Aloes from Yemen, including the description of a new species, *A. ahmarensis*. Cact. Succ. J. (US) 71(5): 257-261, ills.

Fellingham, A. C. & Meyer, N. L. (1995) New combinations and a complete list of *Asparagus* species in southern Africa (*Asparagaceae*). Bothalia 25(2): 205-209.

Ferguson, N. (1926) The *Aloineae*: A cytological study, with special reference to the form and size of the chromosomes. Philos. Trans., Ser. B, 215: 225-253.

Fernandes, A. (1930) Etudes sur les chromosomes. 4. Sur le nombre et la morphologie des chromosomes de quelques espèces du genre *Aloe* L. Bol. Soc. Brot., sér. 2, 6: 294-308.

Fitz Maurice, W. A. & al. (1997) *Mammillaria marcosii*, a new species of Series *Stylothelae* from northeastern Guanajuato, Mexico. Cact. Succ. J. (US) 69(1): 10-14, ills.

Folsom, J. & al. (1995) Dry climate gardening with succulents. New York (US: NY): Pantheon Books, Knopf Publishing Group. 224 pp., ills.

Forster, P. I. (1993) *Bulbine vagans* E. M. Watson (*Asphodelaceae*), a restricted Australian endemic. Aloe 30(1): 23-24, ills.

Friis, I. (1995) *Sansevieria forskaoliana*, a new name for *Sansevieria abyssinica*. Sansevieria J. 4(1): 3-8, ills.

Galván Villanueva, R. (1990) *Amaryllidaceae*. In: Rzedowski, J. & Calderón de Rzedowski, G. C. (eds.): Flora Fanerogámica del Valle de México. Vol. 3: 305-320, key. Patzcuaro (MEX): Instituto de Ecología, Centro Regional del Bajío.

García-Franco, J. & al. (1995) Parasitismo de *Psittacanthus calyculatus* (*Loranthaceae*) sobre *Beaucarnea gracilis* (*Nolinaceae*) en el Valle de Zapotitlán de las Salinas, Puebla, México. Cact. Suc. Mex. 40(3): 62-65, ill.

García-Mendoza, A. (1987) Monografia del género *Beschorneria* Kunth (*Agavaceae*). México D.F. (MEX): M.Sc. thesis, UNAM. 131 pp., ills., maps.

García-Mendoza, A. (1999) Una especie nueva de *Furcraea* (*Agavaceae*) de Chiapas, México. Novon 9(1): 42-45, ills.

García-Mendoza, A. & Castañeda Rojas, A. (2000) *Manfreda littoralis* (*Agavaceae*), nueva especie de Guerrero y Oaxaca, México. Acta Bot. Mex. 50: 39-45, ills., key.

García-Mendoza, A. & Galván, R. (1995) Riqueza de las familias *Agavaceae* y *Nolinaceae* en México. Bol. Soc. Bot. México 56: 7-24, maps.

García-Mendoza, A. & Martínez Salas, E. (1998) Una nueva especia de *Agave*, subgenero *Littaea* (*Agavaceae*) de Guerrero y Oaxaca, México. Sida 18(1): 227-230, ills.

Geesink, R. & al. (1981) Thonner's analytical key to the families of flowering plants. Den Haag (NL): Leiden University Press. 231 pp.

Gentry, H. S. (1946) A new *Nolina* from southern California. Madroño 8: 179-184.

Gentry, H. S. (1972) The *Agave* family in Sonora. Agricultural handbook No. 399. Washington D.C. (US): U.S. Department of Agriculture. 195 pp., ills., maps.

Gentry, H. S. (1982) Agaves of Continental North America. Tucson (US: AZ): University of Arizona Press. xiv + 670 pp., ills., maps, keys.

Gilbert, M. G. & Sebsebe, Demissew (1992) Notes on the genus *Aloe* in Ethiopia: Misinterpreted taxa. Kew Bull. 47(4): 647-653.

Gilbert, M. G. & Sebsebe, Demissew (1997) Further notes on the genus *Aloe* in Ethiopia and Eritrea. Kew Bull. 52(1): 139-152.

Glen, H. F. & Hardy, D. S. (1987) Nomenclatural notes on three southern African representatives of the genus *Aloe*. South Afr. J. Bot. 53(6): 489-492, maps, keys.

Glen, H. F. & Hardy, D. S. (1988) The identity of *Aloe penduliflora* Bak. Kew Bull. 43: 523-529, ills.

Glen, H. F. & Hardy, D. S. (1995) *Aloe* section *Anguialoe* and the problem of *Aloe spicata* L.f. (*Aloaceae*). Haseltonia 3: 92-103, ills., SEM-ills., maps, diag.

Glen, H. F. & Hardy, D. S. (2000) *Aloaceae* (first part): *Aloe*. In: Germishuizen, G. (ed.): Flora of Southern Africa, Vol. 5, part 1, fascicle 1. Pretoria (RSA): National Botanical Institute. vi + 167 pp., ills., keys, maps.

Gómez-Pompa, A. & al. (1971) Studies in the *Agavaceae* I. Chromosome morphology and number of seven species. Madroño 21: 208-221, ills.

González Medrano, F. (1991) Nota sobre la tipificación de *Manfreda guerrerensis* Matuda (*Agavaceae*). Cact. Suc. Mex. 36(1): 16.

Groen, L. E. (1986) *Astroloba* Uitew. Succulenta 65(1): 19-23, ills.

Groen, L. E. (1987) *Astroloba* Uitew. Succulenta 66(3): 51-55, (4): 82-87, (5): 110-113, (7/8): 162-167, (9): 171-174, (12): 261-263, key, maps, ills., SEM-ills.

Gunn, M. & Codd, L. E. (1981) Botanical exploration of Southern Africa. Cape Town (RSA): A. A. Balkema for the Botanical Research Institute. 400 pp., ill., maps.

Haines, L. (1941) Variation in *Yucca whipplei*. Madroño 6: 33-64.

Handlos, W. L. (1975) The taxonomy of *Tripogandra* (*Commelinaceae*). Rhodora 77: 213-333, key, ills.

Heath, P. V. (1994) New nothogeneric names in the *Asphodelaceae*. Calyx 4(4): 146-147.

Hepper, F. N. & Friis, I. (1994) The plants of Pehr Forsskål's Flora Aegyptiaco-Arabica. Richmond (GB) etc.: Royal Botanic Gardens Kew in association with the Botanical Museum Copenhagen. 412 pp., map, ills.

Hernández Sandoval, L. G. (1993) *Beaucarnea*, un genero amenazado ? Cact. Suc. Mex. 38(1): 11-13, ill.

Hernández Sandoval, L. G. (1995) Análisis cladístico de la familia *Agavaceae*. Bol. Soc. Bot. México 56: 57-68.

Hess, W. J. & Dice, J. C. (1995) *Nolina cismontana* (*Nolinaceae*), a new species name for an old taxon. Novon 5(2): 162-164, key.

Hill, S. R. & James, A. (1998) New plant records for Dominica, Lesser Antilles. Sida 18(1): 297-305.

Hochstätter, F. (1998) Het geslacht *Yucca* (*Agavaceae*). Deel 1 / Deel 2. Succulenta 77(2): 69-84, (5): 220-229, ills., map, key.

Hochstätter, F. (1999a) Het geslacht *Yucca* (*Agavaceae*). Deel 2 (vervolg) / Deel 3 / Deel 4. Succulenta 78(1): 24-34, (3): 120-129, (5): 207-218, ills., maps, keys.

Hochstätter, F. (1999b) Redefinition of *Yucca*-varieties to subspecies. Überführung von *Yucca*-Varietäten in Unterarten. Cactaceae Rev. 1(2): 21.

Hochstätter, F. (1999c) *Yucca*. Mannheim (D): Published by the author. CD-ROM edition.

Hochstätter, F. (2000a) Het geslacht *Yucca* (*Agavaceae*). Deel 5. Sectie *Hesperoyucca*, *Yucca whipplei*. Succulenta 79(1): 32-43, ills., key, map.

Hochstätter, F. (2000b) *Yucca* I (*Agavaceae*) in the Southwest and Midwest of the USA and Canada. Mannheim (D): Published by the author. 256 pp., ills., maps, key.

Hodgson, W. C. (1999) A new Flora for Arizona in preparation. *Agavaceae*. Agave Family. Part 1. *Agave* L. Century Plant, Maguey. J. Arizona-Nevada Acad. Sci. 32(1): 1-21, keys, maps, ills.

Horich, C. K. (1973) *Agave wercklei* Weber *ex* Trelease. Kakt. and. Sukk. 24(7): 158-160, ills.

Howard, R. A. & al. (1979) Flora of the Lesser Antilles. Vol. 3. *Monocotyledoneae*. Jamaica Plain (US: MA): Arnold Arboretum. 586 pp., ills.

Huber, H. (1969) Die Samenmerkmale und Verwandtschaftsverhältnisse der *Liliiflorae*. Mitt. Bot. Staatssamml. München 8: 219-538.

Huber, H. (1998) *Dioscoreaceae*. In: Kubitzki, K. (ed.): The families and genera of vascular plants; 3: 216-235, ills., key. Berlin (D) etc.: Springer-Verlag.

Hummelinck, P. W. (1936) Notes on *Agave* in Aruba, Curaçao, Bonaire and some parts of the South American continent. Recueil Trav. Bot. Néerl. 33: 223-249, pl. 1-8.

Hummelinck, P. W. (1938) Notes on *Agave* in the Netherlands West Indies and North Venezuela. Recueil Trav. Bot. Néerl. 35: 14-28, pl. 1-4, key.

Hummelinck, P. W. (1987) Agavenproblemen op de Bovenwindse Eilanden der Kleine Antillen. Succulenta 66(1): 10-13, (3): 65-69, (6): 127-132, (9): 187-189, (10): 205-211, (12): 265-270, key, ills., tab.

Hummelinck, P. W. (1993) Agaven op Curaçao, Aruba en Bonaire. 1-3. Succulenta 72(1): 5-11, (3): 104-109, (5): 214-223, ills., key.

Hunt, D. R. (1980) Sections and series in *Tradescantia*. American *Commelinaceae* IX. Kew Bull. 35: 437-442.

Hunt, D. R. (1986a) *Campelia*, *Rhoeo* and *Zebrina* united with *Tradescantia*. American *Commelinaceae* XIII. Kew Bull. 41: 400-405.

Hunt, D. R. (1986b) Amplification of *Callisia* Loefl. American *Commelinaceae* XV. Kew Bull. 41: 407-412.

Hunt, D. R. (1993a) *Commelinaceae*. In: McVaugh, R. (ed.): Flora Nova-Galiciana; 13: 130-201, ills. Ann Arbor (US: MI): University of Michigan Herbarium

Hunt, D. R. (1993b) The *Commelinaceae* of Mexico. In: Ramamoorthy, T. P. & al. (eds.): Biological diversity of Mexico: Origins and distribution; pp. 421-446. Oxford (GB) / New York (US): Oxford University Press.

Hutchinson, J. (1934) The families of flowering plants. Vol. 2. London (GB): MacMillan.

Jaarsveld, E. J. van (1992a) The genus *Gasteria*, a synoptic review. Aloe 29(1): 5-30, ills., maps, key, (2): 49 [erratum].

Jaarsveld, E. J. van (1992b) *Bowiea gariepensis* and *Bowiea volubilis*. Brit. Cact. Succ. J. 10(4): 96-98, ills., key.

Jaarsveld, E. J. van (1994) The *Sansevieria* species of South Africa and Namibia. Aloe 31(1): 11-15, ills., key.

Jaarsveld, E. J. van (1998) A new taxon and new combinations in the *Gasteria carinata* complex. Cact. Succ. J. (US) 70(2): 65-71, ills., map, key.

Jaarsveld, E. J. van & Ward-Hilhorst, E. (1994)

Gasterias of South Africa. A new revision of a major succulent group. Vlaeberg (RSA): Fernwood Press in association with the National Botanical Institute. 96 pp., ills., maps, keys.

Jacobsen, H. (1960) A handbook of succulent plants. London (GB): Blandford Press. 3 vols., 1441 pp., ills.

Jacobsen, H. (1981) Das Sukkulentenlexikon. Kurze Beschreibung, Herkunftsangaben und Synonymie der sukkulenten Pflanzen mit Ausnahme der *Cactaceae*. Stuttgart (D): Gustav Fischer Verlag. 2. Ed.; 645 pp., 216 pl.

Janaki-Ammal, E. K. (1945) In: Darlington, C. D. & Janaki-Ammal, E. K.: Chromosome atlas of cultivated plants. London (GB): Allen & Unwin. 397 pp., maps.

Jankowitz, W. J. (1975) Aloen von Südwestafrika. Windhoek (Namibia): Abteilung Naturschutz und Fremdenverkehr. 61 pp., ill.

Jessop, J. P. (1970) Studies in the bulbous *Liliaceae*: 1. *Scilla, Schizocarphus* and *Ledebouria*. J. South Afr. Bot. 36(4): 233-266, keys.

Jessop, J. P. (1977) Studies in the bulbous *Liliaceae* in South Africa: 7. The taxonomy of *Drimia* and certain allied genera. J. South Afr. Bot. 43(4): 265-319, keys.

Johnson, M. A. T. & Brandham, P. E. (1997) New chromosome numbers in petaloid monocotyledons and in other miscellaneous angiosperms. Kew Bull. 52(1): 121-138, chrom. nos.

Johnson, M. A. T. & Gale, R. M. O. (1983) Observations on the leaf-anatomy, pollen, cytology and propagation of *Calibanus hookeri* (Lem.) Trelease. Bradleya 1: 25-32, ills.

Joyner, J. F. & al. (1951) The potentialities of *Sansevieria* for fiber production in South Florida. Proc. Soil Soc. Florida 11: 138-139.

Jumelle, H. (1923) Le *Sansevieria canaliculata* et le *Sansevieria stuckyi*. Bull. Mus. Nation. Hist. Nat. 29(8): 607-612.

Kartesz, J. T. (1996) A synonymized checklist for the vascular flora of the United States, Canada, and Greenland. Internet version. Electronically published by the Biota of North America Program of the North Carolina Botanical Garden; http: //ils.unc.edu/botanical/gardens.html.

Kativu, S. & Nordal, I. (1993) New combinations of African species in the genus *Chlorophytum* (*Anthericaceae*). Nordic J. Bot. 13(1): 59-65, ills.

Keay, R. W. J. (1963) The Nigerian species of *Aloe*. Kew Bull. 17: 65-69, ill.

Keeley, J. E. & Tufenkian, D. A. (1983) Garden comparision of germinability and seedling growth of *Yucca whipplei* subspecies (*Agavaceae*). Madroño 31(1): 24-29.

Knuth, R. (1924) *Dioscoreaceae*. In: Engler, A. (ed.): Das Pflanzenreich, IV. 43 (Heft 87). Leipzig (D): W. Engelmann. 387 pp., ills., keys.

Kondo, N. & Megata, M. (1943) Chromosome studies in *Aloineae*. Rep. Kihara Inst. Biol. Res. 2: 69-82.

Koshy, T. K. (1937) Number and behaviour of chromosomes in *Aloe litoralis*. Ann. Bot. (London), n.s. 1: 43-58.

Krause, K. (1930) *Liliaceae*. In: Engler, A. & Prantl, K. (eds.): Die natürlichen Pflanzenfamilien, ed. 2, 15a: 227-385, key, ills. Leipzig (D): W. Engelmann.

Kubitzki, K. & Rudall, P. J. (1998) *Asparagaceae*. In: Kubitzki, K. (ed.): The families and genera of vascular plants; 3: 125-129, ills., key. Berlin (D) etc.: Springer-Verlag.

Laferrière, J. E. (1991) *Dasylirion wheeleri* var. *durangense*: A new combination in the *Nolinaceae*. Ann. Missouri Bot. Gard. 78: 516-520, ills., map.

Lamb, E. & Lamb, B. (1974) Colourful cacti and other succulents of the desert. London (GB): Blandford Press. 236 pp., ill.

Lavranos, J. J. (1965) Notes on the Aloes of Arabia, with descriptions of six new species. J. South Afr. Bot. 31: 55-81, ills.

Lavranos, J. J. (1969) The genus *Aloe* L. in the Socotra Archipelago, Indian Ocean. A revision. Cact. Succ. J. (US) 41(5): 202-207, ills.

Lavranos, J. J. (1992) *Aloe inermis* Forsk. and its relatives. With description of a new species. Cact. Succ. J. (US) 64(4): 206-208, ills.

Lavranos, J. J. (1995) *Aloaceae*. In: Thulin, M. (ed.): Flora of Somalia, 4: 35-42. Richmond (GB): Royal Botanic Gardens Kew.

Lawrence, G. H. M. & al. (eds.) (1968) Botanico-Periodicum-Huntianum. Pittsburgh (USA: PA): Hunt Botanical Library. 1063 pp.

Leach, L. C. (1970) The identity of *Aloe vituensis* Baker. J. South Afr. Bot. 36(2): 57-62, ills.

Leach, L. C. (1971) Two new species of *Aloe (Liliaceae)* from South Tropical Africa. J. South Afr. Bot. 37(4): 249-266, ills.

Leach, L. C. (1977) Notes on *Aloe* (*Liliaceae*) species of the Flora Zambesiaca area. Kirkia 10(2): 385-389, ill.

Lenz, L. W. (1992) An annotated catalogue of the plants of the Cape region, Baja California Sur, Mexico. Claremont (US: CA): The Cape Press. 114 pp.

Lenz, L. W. & Hanson, M. A. (2000) Typification and change in status of *Yucca schottii* (*Agavaceae*). Aliso 19(1): 93-98, ills.

León, H. (1946) Flora de Cuba. Vol. 1: Gimnospermas, Monocotiledoneas. Contr. Ocas. Mus. Hist. Nat. Colegio "De La Salle" No. 8.

Little, E. L. (1981) *Agave*. In: Benson, L. & Darrow, R. A.: Trees and shrubs of the Southwestern Deserts; pp. 63-76. Tucson (US: AZ): University of Arizona Press.

Lott, E. J. & García-Mendoza, A. (1994) *Agavaceae*. In: Davidse, G. & al. (eds.): Flora Mesoamericana, 6: 35-47, keys. México

(MEX): UNAM, Instituto de Biología / St. Louis (US: MO): Missouri Botanical Garden / London (GB): Natural History Museum.

Lott, E. J. & Verhoek-Williams, S. E. (1991) *Manfreda chamelensis* (*Agavaceae: Poliantheae*), a new species from western Mexico. Phytologia 70(5): 366-370.

Lundell, C. L. (1939) Studies of Mexican and Central American plants. VIII. Bull. Torrey Bot. Club 66: 583-604.

Mabberley, D. J. (1987) The plant-book. A portable dictionary of the higher plants. Cambridge (GB): Cambridge University Press. 706 pp.

Macbride, J. F. (1936) *Agavaceae*. In: Flora of Peru. Field Mus. Nat. Hist., Bot. Ser. 13(1,3): 666-667.

Mägdefrau, K. (1975) Das Alter der Drachenbäume auf Tenerife. Flora 164: 347-357.

Marais, W. (1975) A further note on *Lomatophyllum* (*Liliaceae*). Kew Bull. 30(4): 601-602.

Marais, W. (1978) Famille 183: Liliacées: *Aloe, Lomatophyllum*. In: Bosser, J. & al.: Flore des Mascareignes, 177-188. Richmond (GB): Royal Botanic Gardens Kew etc.

Marrero, A. & al. (1998) A new species of the wild dragon tree, *Dracaena* (*Dracaenaceae*) from Gran Canaria and its taxonomic and biogeographic implications. Bot. J. Linn. Soc. 128: 291-314, ills., map.

Matuda, E. (1960) *Beaucarnea stricta*. Cact. Suc. Mex. 5: 93-94, ill.

Matuda, E. (1961) Las Amarilidáceas y Liliáceas del Valle de Mexico y sus alrededores. Anales Inst. Biol. UNAM 31(1-2): 53-118, ills., keys.

Matuda, E. (1967) *Beschorneria hidalgorupicola* Matuda, sp.nov. Anales Inst. Biol. UNAM 37: 79-80, ills.

Matuda, E. & Piña Lujan, I. (1980) Las plantas mexicanas del género *Yucca*. Toluca (MEX): Miscelanea Estado de México. 154 pp., ills., maps, key.

Mayo, S. & al. (1997) The genera of *Araceae*. Richmond (GB): Royal Botanic Gardens Kew.

Mbugua, P. K. (1994) Three *Sansevieria* species of Kenya. 1. *Sansevieria robusta* N. E. Br. Cact. Succ. J. (US) 66(2): 87-88, ills., map.

Mbugua, P. K. (1998) Three *Sansevieria* species of Kenya. II. *S. braunii*. Cact. Succ. J. (US) 70(6): 311-312, ills.

McKelvey, S. D. (1933) Taxonomic and cytological relationships of *Yucca* and *Agave*. J. Arnold Arbor. 14: 76-81.

McKelvey, S. D. (1938) Yuccas of the Southwestern United States. Part I. Jamaica Plain (US: MA): Arnold Arboretum of Harvard University. 150 pp., ills., keys.

McKelvey, S. D. (1947) Yuccas of the Southwestern United States. Part II. Jamaica Plain (US: MA): Arnold Arboretum of Harvard University. 192 pp., ills., keys.

McKinney, K. K. (1993) *Agave*. In: Hickman, J. C. (ed.): The Jepson Manual. Higher plants of California; p. 1172. Berkeley etc. (US: CA): University of California Press.

McKinney, K. K. & Hickman, J. C. (1993) *Yucca*. In: Hickman, J. C. (ed.): The Jepson Manual. Higher plants of California; p. 1210. Berkeley etc. (US: CA): University of California Press.

McVaugh, R. (1989) Flora Novo-Galiciana. A descriptive account of the vascular plants of Western Mexico. Vol. 15: *Bromeliaceae* to *Dioscoreaceae*. Ann Arbor (US: MI): University of Michigan Herbarium.

Meerow, A. W. & Snijman, D. A. (1998) *Amaryllidaceae*. In: Kubitzki, K. (ed.): The families and genera of vascular plants; 3: 83-110, ills., key. Berlin (D) etc.: Springer-Verlag.

Miller, A. G. & Morris, M. (1988) Plants of Dhofar. The southern region of Oman. Traditional, economic and medicinal uses. Sultanate of Oman: Government of Oman. 388 pp., ills., maps.

Mitich, L. W. (1982) Parry's *Nolina* (*Nolina parryi*). Cact. Succ. J. Gr. Brit. 44: 67-68, ill.

Moore, H. E. (1958) *Callisia elegans*, a new species, with notes on the genus. Baileya 6: 135-147.

Moore, H. E. (1963) *Hadrodemas*, a new genus of *Commelinaceae*. Baileya 10: 131-136.

Müller, F. S. (1941) 'n sitologiese studie van 'n aantal *Aloe*-soorte. I. Gromosoomgetalle. Tydskr. Wetensk. Kuns, n.s., 2: 99-104.

Müller, F. S. (1945) 'n chromosoomstudie van 'n aantal species van die genus *Aloe* Linn. Met spesiale verwysing na die morfologie en betekenis van die somatiese chromosome. Publ. Univ. Pretoria 2: 1-157.

Müller-Doblies, U. & Müller-Doblies, D. (1996) Revisionula incompleta Ornithogalorum austro-africanorum (*Hyacinthaceae*). Feddes Repert. 107(5-6): 361-548, ills. key.

Müller-Doblies, U. & Müller-Doblies, D. (1997) A partial revision of the tribe *Massonieae* (*Hyacinthaceae*). 1. Survey, including three novelties from Namibia: A new genus, a second species in the monotypic *Whiteheadia* and a new combination in *Massonia*. Feddes Repert. 108(1-2): 49-96, ills., keys.

Navas B., E. & Erba V., G. (1968) El género *Dioscorea* en Chile. Revista Univ. (Santiago) 53: 41-60, ills., key.

Newton, L. E. (1970) Chromosome numbers of West African Aloes. J. South Afr. Bot. 36(2): 69-72, ill.

Newton, L. E. (1976) The taxonomic status of *Aloe keayi* Reynolds. Nation. Cact. Succ. J. 31(3): 49-53, ills., maps.

Newton, L. E. (1979) In defence of the name *Aloe vera*. Cact. Succ. J. Gr. Brit. 41(2): 29-30.

Newton, L. E. (1991) On the identity of *Aloe penduliflora* Baker (*Liliaceae / Aloaceae*). Taxon 40(1): 53-60, ills.

Newton, L. E. (1992) The identity of *Aloe archeri* Lavranos (*Liliaceae / Aloaceae*). Taxon 41(1): 25-34, ills.

Newton, L. E. (1994) Observations on flowering of *Sansevieria robusta* in Kenya. EANHS Bull. 24(1): 8-11.

Newton, L. E. (1996) *Aloe chrysostachys* and *Aloe meruana* (*Aloaceae*). Cact. Succ. J. (US) 68(3): 126-129, ills.

Newton, L. E. (1998a) Natural hybrids in the genus *Aloe* (*Aloaceae*) in East Africa. J. East Afr. Nat. Hist. Soc. 84: 141-145, ill.

Newton, L. E. (1998b) Two new succulent plant records for East Africa. Bradleya 16: 109-114, ills.

Newton, L. E. (1998c) Unnamed hybrids of ×*Gasteraloe* alliance. Bradleya 16: 115-118, ills.

Newton, L. E. & Lavranos, J. J. (1990) Two new Aloes from Kenya, with notes on the identity of *Aloe turkanensis*. Cact. Succ. J. (US) 62(5): 215-221, ills., map.

Nobel, P. S. (1988) Environmental biology of agaves and cacti. Cambridge (GB) etc.: Cambridge University Press. x + 270 pp., ills., diags.

Nordal, I. & al. (1997) *Anthericaceae*. In: Polhill, R. M. (ed.): Flora of Tropical East Africa. Rotterdam (NL) / Brookfield (US): A. A. Balkema. 67 pp., ills., keys, map.

Nordenstam, B. (1970) Studies in South African *Liliaceae*. III. The genus *Rhadamanthus*. Bot. Not. 123: 155-182, key, ills.

Obermeyer, A. A. (1962) A revision of the South African species of *Anthericum, Chlorophytum* and *Trachyandra*. Bothalia 7(4): 669-767, keys, ill.

Obermeyer, A. A. (1978) *Ornithogalum*: A revision of the southern African species. Bothalia 12(3): 323-376, ills., key.

Obermeyer, A. A. & al. (1992) *Dracaenaceae, Asparagaceae, Luzuriagaceae* and *Smilacaceae*. In: Leistner, O. A. (ed.): Flora of Southern Africa, Vol. 5, part 3. Pretoria (RSA): National Botanic Institute. 98 pp., ills., maps, keys.

Oliver, G. W. & Bailey, L. H. (1927) *Beschorneria*. In: Bailey, L. H. (ed.): The Standard Cyclopedia of Horticulture; pp. 495-496. New York (US: NY): MacMillan Co.

Riley, H. P. (1959) Chromosome numbers in *Aloe*. J. South Afr. Bot. 25: 237-246.

Pax, F. & Hoffmann, K. (1930) *Amaryllidaceae*. In: Engler, A. & Prantl, K. (eds.): Die natürlichen Pflanzenfamilien, Ed. 2, 15a: 391-430, key, ills. Leipzig (D): W. Engelmann.

Pedley, L. (1986) *Cordyline. Doryanthes*. In: George, A. S. (ed.): Flora of Australia, 46: 81-88. Canberra (AUS): Australian Government Publishing Service.

Perrier, H. (1926) Les *Lomatophyllum* et les *Aloe* de Madagascar. Mém. Soc. Linn. Normandie, Bot. 1(1): 1-58, 8 pl., map.

Perrier, H. (1938) 40e famille. Liliacées (*Liliaceae*). In: Humbert, H. (ed.): Flore de Madagascar. Tananarive (Madagascar): Imprimerie Officielle. 147 pp., ills., keys.

Perry, P. L. (1994) A revision of the genus *Eriospermum* (*Eriospermaceae*). Contr. Bolus Herb. 17: 320 pp., maps, ills., key.

Pfennig, H. (1977a) Rasenbildend bis baumartig: Die Sansevierien. Gartenpraxis 1977: 506-511, ills.

Pfennig, H. (1977b) Die Kultur der Sansevierien. Gartenpraxis 1977: 550-553, ills.

Pfennig, H. (1981) Zur Systematik und Kultur einiger ostafrikanischer *Sansevieria*-Arten (*Agavaceae*). Bot. Jahrb. Syst. 102(1-4): 169-179, map.

Pilbeam, J. (1983) *Haworthia* and *Astroloba* – a collector's guide. London (GB): Batsford Ltd. 167 pp., ills., maps.

Piña Luján, I. (1985) Consideraciones sobre el género *Manfreda*. I - III. Cact. Suc. Mex. 30(2): 27-32, (3): 56-64, (4): 84-90.

Piña Luján, I. (1986) Consideraciones sobre el género *Manfreda*. IV. Cact. Suc. Mex. 31(1): 12-18.

Piña Luján, I. (1994) Consideraciones sobre la localidad de *Agave atrovirens*. Cact. Suc. Mex. 39(1): 8-12, ill.

Pinkava, D. J. & Baker, M. A. (1985) Chromosome and hybridization studies of Agaves. Desert Pl. 7(2): 93-100, ills.

Poindexter, R. W. (1936) New garden species. VII. *Gastrolea sculptilis*. Cact. Succ. J. (US) 7(9): 136, ill.

Potter-Bassano, G. (1991) *Yucca baccata*: From the beginning. Arid Lands Newslett. 31(fall): 24-29, ills.

Powell, J. A. (1984) Biological interrelationships of moths and *Yucca schottii*. Los Angeles (US): University of California Press. 93 pp., ill.

Pridgeon, A. (ed.) (1992) The illustrated encyclopedia of orchids. Sydney (AUS): Kevin Weldon / Portland (US: OR): Timber Press. 304 pp., ills.

Proctor, G. R. (1984) Flora of the Cayman Islands. Kew Bull., Add. Ser. 11.

Rauh, W. (1958) Beitrag zur Kenntnis der peruanischen Kakteenvegetation. Sitzungsber. Heidelberger Akad. Wiss., Math.-Naturwiss. Kl. 1: 542 pp., ill.

Rauh, W. (1963) Über einige interessante Sukkulenten aus Kenia. Sukkulentenkunde 7/8: 108-127, ills.

Rauh, W. (1990) Bromelien. Stuttgart (D): Verlag Eugen Ulmer. 3. ed., 458 pp., ill.

Rauh, W. (1995a) Succulent and xerophytic plants of Madagascar. Vol. 1. Mill Valley (US: CA): Strawberry Press. 343 pp., ills., maps.

Rauh, W. (1998) Three new species of *Lomatophyllum* and one new *Aloe* from Madagascar. Bradleya 16: 92-100, ills., key.

Ravenna, P. (1969) Contribution to South American *Amaryllidaceae*. II. Pl. Life 25: 55-76.

Rebelo, A. G. & al. (1989) Branching patterns in *Aloe dichotoma* – is *Aloe ramosissima* a separate species ? Madoqua 16(1): 23-26, maps.

Reichenbacher, F. W. (1985) Conservation of southwestern Agaves. Desert Pl. 7(2): 103-106, 88, maps.

Resende, F. (1937) Über die Ubiquität der SAT-Chromosomen bei den Blütenpflanzen. Planta 26: 757-807.

Reveal, J. L. (1977) *Agavaceae*. In: Cronquist, A. & al. (eds.): Intermountain Flora, 6: 526-538, ills., keys. New York (USA): The New York Botanical Gardens / Columbia University Press.

Reynolds, G. W. (1950) The Aloes of South Africa. Johannesburg (RSA): Aloes of South Africa Book Fund. 520 pp., ills., keys.

Reynolds, G. W. (1958) Les Aloés de Madagascar. Revision. Tananarive (Madagascar): Institut de Recherche Scientifique de Madagascar. 156 pp., ills., keys.

Reynolds, G. W. (1966) The Aloes of tropical Africa and Madagascar. Mbabane (Swaziland): Aloes Book Fund. 537 pp., ills., maps, keys.

Riha, J. & Subik, R. (1981) Illustrated encyclopedia of cacti and other succulents. London (GB): Octopus Books. 352 pp., ills.

Riley, H. P. & Majumdar, S. K. (1979) The *Aloineae*. A biosystematic survey. Lexington (US: KT): The University Press of Kentucky. x + 180 pp., ills.

Riley, S. (1993) New *Aloe* hybrids from John Bleck. Cact. File 1(11): 22-23, ills.

Roberts Reinecke, P. (1965) The genus *Astroloba* Uitewaal (*Liliaceae*). Cape Town (RSA): M.Sc. thesis, unpublished.

Rogers, G. K. (2000) A taxonomic revision of the genus *Agave* (*Agavaceae*) in the Lesser Antilles, with an ethnobotanical hypothesis. Brittonia 52(3): 218-233.

Rose, J. N. (1903) Revision of *Polianthes* with new species. Contr. US Nation. Herb. 8: 8-13.

Rowley, G. D. (1968) Rowley Reporting. Gastroleas. Nation. Cact. Succ. J. 23(3): 57.

Rowley, G. D. (1980) Name that succulent. Keys to the families and genera of succulent plants in cultivation. Cheltenham (GB): Stanley Thornes (Publishers) Ltd. 268 pp., ill., keys.

Rowley, G. D. (1982) Intergeneric hybrids in succulents. Nation. Cact. Succ. J. 37(1): 2-6, (2): 45-49, (3): 76-80, (4): 119, ills.

Rowley, G. D. (1987) Caudiciform and pachycaul succulents. Pachycauls, bottle-, barrel- and ele-phant-trees and their kin: A collector's miscellany. Mill Valley (US: CA): Strawberry Press. 282 pp., ills.

Rowley, G. D. (1990) *Nolina* en culture. Cact. Aventures No. 5: 2-4.

Rowley, G. D. (1997) The berried Aloes: *Aloe* section *Lomatophyllum*. Excelsa 17: 59-62, ill., map.

Roy, M. (1956) A cytological investigation of the different species of *Sansevieria* with the aid of the improved technique. Caryologia 8: 221-230.

Rudall, P. J. & al. (1995) Monocotyledons. Systematics and evolution. Kew (GB): Royal Botanic Gardens. 2 vols., 750 pp., ills.

Satô, D. (1937) Analysis of karyotypes in *Aloinae* with special reference to the SAT-chromosomes. Cytologia Fujii Vol.: 80-95.

Scott, C. L. (1985) The genus *Haworthia* (*Liliaceae*), a taxonomic revision. Johannesburg (RSA): Aloe Books. 190 pp., ills., maps.

Sebsebe, Demissew & Gilbert, M. G. (1997) *Aloaceae*. In: Edwards, S. & al. (eds.): Flora of Ethiopia and Eritrea 6: 117-135, ills., keys. Addis Ababa (ETH): Addis Ababa University / Uppsala (S): Uppsala University.

Sharma, A. K. & Chaudhuri, M. (1964) Cytological studies as an aid in assessing the status of *Sansevieria, Ophiopogon* and *Curculigo*. Nucleus (Calcutta) 7(1): 43-58.

Small, J. K. (1916) *Nolina texana*. Addisonia 2: 1-2, t. 41.

Smith, G. F. (1990) Nomenclatural notes on the subsection *Bowiea* in *Aloe* (*Asphodelaceae*). South Afr. J. Bot. 56(3): 303-308.

Smith, G. F. (1991) Historical review of the taxonomy of *Chortolirion* Berger (*Asphodelaceae: Alooideae*). Aloe 28: 90-95, ills.

Smith, G. F. (1994) Taxonomic history of *Poellnitzia* Uitewaal, a unispecific genus of *Alooideae* (*Asphodelaceae*). Haseltonia 2: 74-78, ills.

Smith, G. F. (1995a) FSA contributions 2: *Asphodelaceae / Aloaceae*, 1029010 *Chortolirion*. Bothalia 25(1): 31-33, ill., map.

Smith, G. F. (1995b) FSA contributions 3: *Asphodelaceae / Aloaceae*, 1028000 *Poellnitzia*. Bothalia 25(1): 35-36, ill., map.

Smith, G. F. & Crouch, N. R. (1995) Notes on *Aloe pruinosa* Reyn. (*Aloaceae*), a rare and little-known maculate from the midlands of KwaZulu-Natal, South Africa. Aloe 32(3-4): 66-69, ills., map.

Smith, G. F. & Mössmer, M. (1995) The correct orthography of *Aloe micracantha* (*Aloaceae*). Haseltonia 3: 9.

Smith, G. F. & Wyk, B-E. van (1991) Generic relationships in the *Alooideae* (*Asphodelaceae*). Taxon 40(4): 557-581, ills., maps.

Smith, G. F. & Wyk, B-E. van (1998) *Asphodelaceae*. In: Kubitzki, K. (ed.): The families and genera of vascular plants; 3: 130-140, ills., key. Berlin (D) etc.: Springer-Verlag.

Smith, G. F. & al. (1994) *Aloe barberae* to replace *A. bainesii*. Bothalia 24(1): 34-35.

Smith, L. B. & Downs, R. J. (1974-1979) Flora Neotropica. Monograph No. 14 [*Bromeliaceae*]. Part 1: *Pitcairnioideae*; Part 2: *Tillandsioideae*; Part 3: *Bromelioideae*. New York (US: NY): Hafner Press & New York Botanical Garden. 2142 pp., ills., keys, maps.

Smith, L. B. & Till, W. (1998) *Bromeliaceae*. In: Kubitzki, K. (ed.): The families and genera of vascular plants; 4: 75-99, ills., key. Berlin (D) etc.: Springer-Verlag.

Snijman, D. A. (1984) A revision of the genus *Haemanthus* L. (*Amaryllidaceae*). J. South Afr. Bot. 12(Suppl.): 138 pp., ills., key, maps.

Snoad, B. (1951) Chromosome numbers of succulent plants. Heredity 5(2): 279-283, ills.

Stafleu, F. A. & Cowan, R. S. (1976-1988) Taxonomic literature. Utrecht (NL): Bohn, Scheltema & Holkema, etc. 2. Ed.; 7 vols.

Stafleu, F. A. & Mennega, E. A. (1992-2000) Taxonomic literature. Königstein (D): Koeltz Scientific Books. Supplements to Ed. 2; 6 vols.

Standley, P. C. (1920-1926) Trees and shrubs of Mexico. Contr. US Nation. Herb. 23(1-5): 1721 pp.

Standley, P. C. & Steyermark, J. A. (1952) Flora of Guatemala. *Commelinaceae*. Fieldiana, Bot. 24(3): 1-42, key, ills.

Starr, G. (1995) *Hesperaloe*: Aloes of the West. Desert Pl. 11(4): 3-8, ills.

Starr, G. (1997) A revision of the genus *Hesperaloe* (*Agavaceae*). Madroño 44(3): 282-296, map, ills., key.

Stearn, W. T. (1957) The Boat Lily (*Rhoeo spathacea*). Baileya 5: 195-198.

Stearn, W. T. (1992) Botanical Latin. Newton Abbot (GB): David & Charles Publishers. Ed. 4; 560 pp.

Stedje, B. (1996) *Hyacinthaceae*. In: Polhill, R. M. (ed.): Flora of Tropical East Africa. Rotterdam (NL) / Brookfield (US): A. A. Balkema. 32 pp., ills., keys.

Sutaria, R. N. (1932) Somatic cell division in *Aloe vera* L. J. Indian Bot. Soc. 11: 132-136.

Tamura, M. N. (1995) A karyological review of the orders *Asparagales* and *Liliales* (*Monocotyledoneae*). Feddes Repert. 106(1-2): 83-111.

Taylor, W. R. (1925) Cytological studies on *Gasteria*. II. A comparison of the chromosomes of *Gasteria, Aloe,* and *Haworthia*. Amer. J. Bot. 12: 219-223, ills., chrom. nos.

Teketay, D. (1995) The genus *Sansevieria* Thunb. in Ethiopia. A contribution to the flora of Ethiopia. Sansevieria J. 4(2): 43-58, ills., map, key.

Thiede, J. (1993) Notes on the *Sansevieria* species of Malawi. Sansevieria J. 2(2): 27-34, (3): 51-52, ills., map, key.

Thiede, J. & Eggli, U. (1999) Einbeziehung von *Manfreda* Salisbury, *Polianthes* Linné und *Prochnyanthes* in *Agave* (*Agavaceae*). Kakt. and. Sukk. 50(5): 109-113, ill.

Thulin, M. (ed.) (1995) Flora of Somalia. Volume 4. *Angiospermae* (*Hydrocharitaceae - Pandanaceae*). Richmond (GB): Royal Botanic Gardens Kew. ii + 298 pp., ills., keys.

Trelease, W. (1902) The *Yucceae*. Annual Rep. Missouri Bot. Gard. 13: 27-133, ills.

Trelease, W. (1907) *Agave macroacantha* and allied Euagaves. Annual Rep. Missouri Bot. Gard. 18: 231-256, 16 pl.

Trelease, W. (1910) Observations on *Furcraea*. Ann. Jard. Bot. Buitenzorg 3(Suppl. 2): 905-916.

Trelease, W. (1911) The desert group *Nolineae*. Proc. Amer. Philos. Soc. 50: 404-443, pl. 1-17.

Trelease, W. (1913) Agave in the West Indies. Mem. Nation. Acad. Sci. 11: 55 pp., 116 pl., maps, key.

Trelease, W. (1915a) *Furcraea*. In: Bailey, L. H. (ed.): Standard Cyclopedia of Horticulture; 3: 1305-1306. New York (US): Macmillan and Co.

Trelease, W. (1915b) The *Agaveae* of Guatemala. Trans. Acad. Sci. St. Louis 23(3): 129-152, pl. 6 - 35.

Trelease, W. (1920) *Furcraea*. In: Standley, P. C.: Trees and shrubs of Mexico. Contr. US Nation. Herb. 23(1): 105-107.

Turner, R. M. & al. (1995) Sonoran Desert plants. An ecological atlas. Tucson (US: AZ): University of Arizona Press. xvi + 504 pp., ills., maps.

USDA (2001) The PLANTS database. Baton Rouge (US: LA): National Plant Data Center. Electronically published at http://plants.usda.gov.

Ullrich, B. (1989) *Manfreda nanchititlensis* Matuda. Kakt. and. Sukk. 40(9): centre page pullout 1989/28, ills.

Ullrich, B. (1990a) *Hesperaloe funifera* (Koch) Trelease. Kakt. and. Sukk. 41(3): centre page pullout 1990/7, ills.

Ullrich, B. (1990b) *Agave bracteosa* S. Watson *ex* Engelmann. Kakt. and. Sukk. 41(5): centre page pullout 1990/13, ills.

Ullrich, B. (1990c) *Agave striata* Zuccarini ssp. *stricta* (Salm-Dyck) Ullrich *stat. nov.* Kakt. and. Sukk. 41(5): centre page pullout 1990/14, ills.

Ullrich, B. (1990d) *Agave grijalvensis* – Eine neue Art aus Chiapas. Kakt. and. Sukk. 41(6): 102-108, ills.

Ullrich, B. (1990e) *Agave victoriae-reginae* T. Moore. Kakt. and. Sukk. 41(7): centre page pullout 1990/22, ills.

Ullrich, B. (1990f) Anmerkungen zu *Agave parviflora* Torrey und *Agave hartmanii* S. Watson. Kakt. and. Sukk. 41(7): 133-136, ills.

Ullrich, B. (1990g) Ein neuer Standort für *Agave dasylirioides* Jacobi & Bouché in Oaxaca. Kakt. and. Sukk. 41(8): 164-166, ills., map.

Ullrich, B. (1990h) *Agave attenuata* Salm-Dyck op haar natuurlijke groeiplaatsen. Succulenta 69(6): 121-127, ills., map.

Ullrich, B. (1990i) *Agave macroculmis* Todaro en *Agave gentryi* Ullrich spec. nov. Succulenta 69(9): 190-193, (10): 210-214, ills.

Ullrich, B. (1990j) *Agave obscura* Schiede y *Agave horrida* Lemaire *ex* Jacobi ssp. *perotensis* Ullrich ssp. nov. Cact. Suc. Mex. 35(4): 75-82, ills.

Ullrich, B. (1991a) *Agave colimana* Gentry oder *Agave ortgiesiana* Roezl? Kakt. and. Sukk. 42(1): 24-28, ills., map.

Ullrich, B. (1991b) *Agave gypsophila*. Kakt. and. Sukk. 42(3): centre page pullout 1991/8, ills.

Ullrich, B. (1991c) Notiz zu *Agave breviscapa* Berger *ex* Roster und *Agave vernae* Berger. Kakt. and. Sukk. 42(10): 242-243, ill.

Ullrich, B. (1991d) Zum Verbreitungsgebiet von *Agave victoriae-reginae* T. Moore. Kakt. and. Sukk. 42(11): 262-263, ill., map.

Ullrich, B. (1991e) *Beschorneria yuccoides* W. J. Hooker. Kakt. and. Sukk. 42(12): centre page pullout 1991/36, ills.

Ullrich, B. (1991f) De systematische positie van *Agave felgeri* Gentry. Succulenta 70(7/8): 147-150, ills., map.

Ullrich, B. (1991g) Aantekeningen bij *Agave guiengola* Gentry. Succulenta 70(12): 257-260, ills.

Ullrich, B. (1991h) Notas sobre la publicación del género *Furcraea* Ventenat (*Agavaceae*). Cact. Suc. Mex. 36(1): 8-9.

Ullrich, B. (1991i) El complejo *Furcraea longaeva* Karwinski et Zuccarini. Cact. Suc. Mex. 36(2): 30-36, (3): 56-61, (4): 79-83, ills., map.

Ullrich, B. (1991j) De taxonomische rangschikking van *Agave nizandensis* Cutak. Succulenta 70(4): 89-92, ills.

Ullrich, B. (1992a) Anmerkungen zu drei Taxa der Gattung *Agave* aus El Salvador. Kakt. and. Sukk. 43(3): 50-53, ills.

Ullrich, B. (1992b) Wer ist der Autor von *Agave wercklei*? Kakt. and. Sukk. 43(8): 181-183, (9): 212-216, ills.

Ullrich, B. (1992c) The discovery of *Agave seemanniana* Jacobi. Brit. Cact. Succ. J. 10(1): 24-28, ills., map.

Ullrich, B. (1992d) On the discovery of *Agave schidigera* Lemaire and status of certain taxa of section *Xysmagave* Berger. Brit. Cact. Succ. J. 10(3): 61-70, ills., map.

Ullrich, B. (1992e) Zum Status von *Agave chrysantha* Peebles. And. Sukk. No. 18: 3-5, map.

Ullrich, B. (1992f) On the history of *Agave asperrima* and *A. scabra* (*Agavaceae*) as well as some taxa of the *Parryanae*. Sida 15(2): 241-261, ills.

Ullrich, B. (1992g) Sobre *Agave langlassei* André y la tipificación de *Manfreda brachystachys* (Cav.) Rose. Cact. Suc. Mex. 37(3): 60-63, ill.

Ullrich, B. (1992h) *Furcraea* (*Agavaceae*) en Sudamerica. Quepo 6: 67-75, ills.

Ullrich, B. (1993a) Observations sur *Agave mitis* Martius, *A. celsii* Hooker et *A. albicans* Jacobi. Succulentes 16(1): 26-32, ills.

Ullrich, B. (1993b) Early illustrations of *Polianthes tuberosa* L. (*Agavaceae*). Herbertia, ser. 3, 49(1-2): 50-57.

Ullrich, B. (1995) La questione della priorità di *Agave spicata* Cavanilles su *Agave yuccaefolia* F. Delaroche in Redouté (Parte I). Zur Priorität von *Agave spicata* Cavanilles über *Agave yuccaefolia* F. Delaroche in Redouté. Piante Grasse 15(4): 116-123, ills., diag.

Ullrich, B. (1996) La questione della priorità di *Agave spicata* Cavanilles su *Agave yuccaefolia* F. Delaroche in Redouté (Parte II). Zur Priorität von *Agave spicata* Cavanilles über *Agave yuccaefolia* F. Delaroche in Redouté. Piante Grasse 16(1): 23-32, ills.

Valverde, P. L. & al. (1996) A morphometric analysis of a putative hybrid between *Agave marmorata* Roezl and *Agave kerchovei* Lem.: *Agave peacockii* Croucher. Bot. J. Linn. Soc. 122(2): 155-161, ills.

Verhoek-Williams, S. (1975) A study of the tribe *Poliantheae* (including *Manfreda*) and revision of *Manfreda* and *Prochnyanthes* (*Agavaceae*). Ithaca (US: NY): Ph.D. Thesis, Cornell University, unpublished. x + 405 pp., key, ills.

Verhoek-Williams, S. (1976) *Polianthes howardii* (*Agavaceae*): A new species from Colima. Phytologia 34(4): 365-368, ills.

Verhoek-Williams, S. (1978) Two new species and a new combination in *Manfreda* (*Agavaceae*). Brittonia 30(2): 165-171.

Verhoek-Williams, S. (1998) *Agavaceae*. In: Kubitzki, K. (ed.): The families and genera of vascular plants; 3: 60-70, ills., key. Berlin (D) etc.: Springer-Verlag.

Viljoen, A. M. & al. (1996) A chemotaxonomic and biochemical evaluation of the identity of *Aloe candelabrum* (*Aloaceae*). Taxon 45(3): 461-471, map, diags.

Vorster, P. (1983) *Aloe meyeri* van Jaarsveld and *A. richtersveldensis* Venter & Beukes. J. South Afr. Bot. 49(2): 175.

Vosa, C. G. (1982) Chromosome studies in the southern African flora: 30-37. J. South Afr. Bot. 48(3): 409-424, ills., chrom. nos.

Waidhofer, A. (1996) *Sansevieria deserti* (*Agavaceae*). Cact. Succ. J. (US) 68(1): 9-11, ills.

Walker, C. C. (1999) A tale of dragons – the pachycaul species of *Dracaena*. Brit. Cact. Succ. J. 17(4): 171-177, ills., map.

Walther, E. (1930) *Gastrolea*. Cact. Succ. J. (US) 2(3): 303-307, ills.

Watson, E. M. (1987) *Bulbine*. In: George, A. S. (ed.): Flora of Australia, 45: 236-241. Canberra (AUS): Australian Government Publishing Service.

Webber, J. M. (1953) Yuccas of the Southwest. Washington D.C. (US): US Dept. of Agriculture. 97 pp., 72 pl.

Weberbauer, A. (1911) Die Pflanzenwelt der peruanischen Anden. Die Vegetation der Erde, Band 12 (ed. A. Engler & O. Drude). Leipzig (D): W. Engelmann. 355 pp., ills.

Wijnands, D. O. (1973) Typification and nomenclature of two species of *Sansevieria* (*Agavaceae*). Taxon 22(1): 109-114.

Wijnands, D. O. (1983) The Botany of the Commelins. A taxonomical, nomenclatural and historical account of the plants depicted in the Moninckx Atlas and in the four books by Jan and Caspar Commelin on the plants in the Hortus Medicus Amstelodamensis 1682-1710. Rotterdam (NL): A. A. Balkema. 238 pp., ills., + 64 col. plates.

Williamson, G. (1998) The ecological status of *Aloe pillansii* (*Aloaceae*) in the Richtersveld with particular reference to Cornelskop. Bradleya 16: 1-8, maps, ills.

Wilson, F. D. & al. (1962) Florida H-13, a promising *Sansevieria* hybrid as a source of cordage fiber. Gainesville (US: FL): Everglades Experimental Station. 7 pp., ills.

Wood, J. R. I. (1983) The Aloes of the Yemen Arab Republic. Kew Bull. 38(1): 13-31, ills., maps, keys.

Woodson, R. E. (1942) Commentary on the North American genera of *Commelinaceae*. Ann. Missouri Bot. Gard. 29: 141-154.

Worsley, A. (1911) The genus *Polianthes* (including *Prochnyanthes* and *Bravoa*). J. Roy. Hort. Soc. 36(3): 603-605.

Wyk, B-E. van & Smith, G. F. (1996) Guide to the Aloes of South Africa. Pretoria (RSA): Briza Publications. 302 pp., ills., maps.

Wyk, B-E. van & al. (1997) Medicinal plants of South Africa. Pretoria (RSA): Briza Publications. 304 pp., ills.

Zamudio Ruiz, S. & Sánchez Martinez, E. (1995) Una nueva especie de *Agave* del subgenero *Littaea* (*Agavaceae*) de la Sierra Madre Oriental, México. Acta Bot. Mex. 32: 47-52, ills.

Zomlefer, W. B. (1994) Guide to flowering plant families. Chapel Hill (US: NC) / London (GB): University of North Carolina Press. 430 pp., ills.

Taxonomic Cross-Reference Index

[Aloe verrucosa]
– – var. latifolia → Gasteria carinata var.
    verrucosa: 196
– – – striata → Gasteria carinata var. verrucosa:
    196
– **versicolor**: 183
– **veseyi**: 183
– **viguieri**: 183
– virens → A. sp.: 105
– virescens → Haworthia marginata: 212
– **viridiflora**: 183
– viscosa → Haworthia viscosa: 222
– – var. indurata → Haworthia viscosa: 222
– – – major → Haworthia viscosa: 222
– – – minor → Haworthia viscosa: 222
– – – parvifolia → Haworthia viscosa: 222
– **vituensis**: 184
– vivipara → Agave vivipara: 72
– **vogtsii**: 184
– **volkensii**: 184
– – ssp. **multicaulis**: 184
– – – **volkensii**: 184
– **vossii**: 184
– **vryheidensis**: 184
– vulcanica → Aloe: 105
– vulgaris → A. vera: 182
– weingartii → Aloe: 104
– **whitcombei**: 185
– wickensii → A. cryptopoda: 126
– – var. lutea → A. cryptopoda: 126
– – – wickensii → A. cryptopoda: 126
– **wildii**: 185
– **wilsonii**: 185
– winteri → Aloe: 104
– **wollastonii**: 185
– **woodii**: 185
– woolliana → A. chortolirioides var. woolliana:
    122
– **wrefordii**: 185
– xanthacantha → A. mitriformis: 156
– – → A. sp.: 105
– **yavellana**: 186
– **yemenica**: 186
– yuccaefolia → Hesperaloe parviflora: 85
– zanzibarica → A. squarrosa: 175
– **zebrina**: 186
– zeyheri → Gasteria bicolor var. bicolor: 194
– zeylanica → Sansevieria zeylanica: 272
– **zombitsiensis**: 186
Aloinella haworthioides → Aloe haworthioides:
    140
Alolirion : 186
Aloloba : 187
Alworthia : 187
Amaryllidaceae : 224
Amaryllis disticha → Boophane disticha: 224
– radula → Brunsvigia radula: 225
Ananas → Bromeliaceae : 247
Aneilema : 248

[Aneilema]
– **zebrinum**: 248
Anoiganthus → Cyrtanthus: 225
Anthericaceae : 228
Anthericum acuminatum → Chlorophytum
    suffruticosum: 229
– alooides → Bulbine alooides: 233
– annuum → Bulbine annua: 234
– asphodeloides → Bulbine asphodeloides: 234
– blepharophoron → Trachyandra ciliata: 245
– bulbosum → Bulbine bulbosa: 234
– campestre → Chlorophytum suffruticosum: 229
– canaliculatum → Trachyandra ciliata: 245
– ciliatum → Trachyandra ciliata: 245
– comosum → Chlorophytum comosum: 228
– favosum → Bulbine favosa: 236
– flagelliforme → Schizobasis intricata: 284
– frutescens → Bulbine frutescens: 238
– hamatum → Trachyandra ciliata: 245
– inexpectatum → Chlorophytum suffruticosum:
    229
– intricatum → Schizobasis intricata: 284
– involucratum → Trachyandra involucrata: 246
– lagopus → Bulbine lagopus: 239
– latifolium → Bulbine latifolia: 239
– longifolium → Trachyandra ciliata: 245
– longiscapum → Bulbine longiscapa: 239
– longisetosum → Chlorophytum suffruticosum:
    229
– maculatum → Trachyandra ciliata: 245
– nutans → Bulbine nutans: 241
– oocarpum → Trachyandra tortilis: 246
– pilosiflorum → Trachyandra ciliata: 245
– – var. subpapillosum → Trachyandra ciliata:
    245
– planifolium → Chlorophytum comosum: 228
– praemorsum → Bulbine alooides: 233
– pugioniforme → Bulbine cepacea: 235
– recurvatum → Trachyandra ciliata: 245
– rostratum → Bulbine frutescens: 238
– salteri → Trachyandra tortilis: 246
– semibarbatum → Bulbine semibarbata: 243
– spongiosum → Trachyandra ciliata: 245
– sternbergianum → Chlorophytum comosum: 228
– suffruticosum → Chlorophytum suffruticosum:
    229
– tortile → Trachyandra tortilis: 246
– vespertinum → Trachyandra ciliata: 245
Apicra → Haworthia: 200
– – → Astroloba: 187
– albicans → Haworthia marginata: 212
– anomala → Haworthia sp.: 200
– arachnoidea → Haworthia arachnoidea: 202
– aspera → Haworthia sp.: 200
– – var. major → Haworthia sp.: 201
– atrovirens → Haworthia herbacea: 210
– attenuata → Haworthia attenuata: 203
– bicarinata → Astroworthia bicarinata: 189
– bullulata → Astroloba bullulata: 187

[Yucca angustissima]
– – var. **angustissima**: 89
– – – **avia**: 89
– – – **kanabensis**: 89
– – – **toftiae**: 89
– antwerpensis → Y. filamentosa: 94
– arborescens → Y. brevifolia var. brevifolia: 90
– arcuata → Y. aloifolia: 88
– argospatha → Y. treculiana: 101
– argyrophylla → Furcraea bedinghausii: 79
– arizonica → Y. schottii: 100
– **arkansana**: 89
– – ssp. freemanii → Y. louisianensis: 97
– – – louisianensis → Y. louisianensis: 97
– – var. paniculata → Y. arkansana: 89
– armata → Y. aloifolia: 88
– aspera → Y. treculiana: 101
– atkinsii → Y. sp.: 88
– australis → Y. faxoniana: 93
– – → Y. filifera: 94
– – var. valida → Y. faxoniana: 93
– **baccata**: 89
– – fa. parviflora → Y. baccata var. baccata: 90
– – var. australis → Y. filifera: 94
– – – **baccata**: 90
– – – brevifolia → Y. schottii: 100
– – – circinata → Y. periculosa: 98
– – – macrocarpa → Y. torreyi: 101
– – – periculosa → Y. periculosa: 98
– – – **vespertina**: 90
– **baileyi**: 90
– – ssp. intermedia → Y. baileyi var. intermedia: 90
– – var. **baileyi**: 90
– – – **intermedia**: 90
– – – **navajoa**: 90
– barrancasecca → Y. sp.: 88
– boerhaavii → Y. recurvifolia: 99
– boscii → Agave geminiflora: 34
– **brevifolia**: 90
– – → Y. schottii: 100
– – fa. herbertii → Y. brevifolia var. brevifolia: 90
– – – kernensis → Y. brevifolia: 90
– – var. **brevifolia**: 90
– – – herbertii → Y. brevifolia var. brevifolia: 91
– – – **jaegeriana**: 91
– – – wolfei → Y. brevifolia var. jaegeriana: 91
– californica → Hesperoyucca whipplei: 86
– – → Y. schidigera: 100
– **campestris**: 91
– canaliculata → Y. treculiana: 101
– – var. filifera → Y. filifera: 94
– **capensis**: 91
– **carnerosana**: 91
– carrierei → Y. sp.: 88
– circinata → Y. periculosa: 98
– **coahuilensis**: 91
– coloma → Y. harrimaniae: 96

[Yucca]
– concava → Y. flaccida: 94
– – → Y. treculiana: 101
– confinis → Y. schottii: 100
– conspicua → Y. aloifolia: 88
– – → Y. sp.: 88
– **constricta**: 92
– contorta → Y. sp.: 88
– cornuta → Y. treculiana: 101
– crassifila → Y. torreyi: 101
– crenulata → Y. aloifolia: 88
– crinifera → Y. sp.: 88
– **decipiens**: 92
– **declinata**: 92
– desmetiana → Y. sp.: 88
– draco → Dracaena draco: 260
– draconis → Y. aloifolia: 88
– – var. arborescens → Y. brevifolia var. brevifolia: 90
– ehrenbergii → Y. sp.: 88
– **elata**: 92
– – ssp. utahensis → Y. elata var. utahensis: 93
– – – verdiensis → Y. elata var. verdiensis: 93
– – var. **elata**: 92
– – – magdalenae → Y. elata var. elata: 92
– – – **utahensis**: 93
– – – **verdiensis**: 93
– **elephantipes**: 93
– ellacombei → Y. gloriosa: 95
– **endlichiana**: 93
– ensifera → Y. sp.: 88
– ensifolia → Y. sp.: 88
– exigua → Y. filamentosa: 94
– eylesii → Y. recurvifolia: 99
– falcata → Y. recurvifolia: 99
– **faxoniana**: 93
– **filamentosa**: 93
– – → Y. baccata: 89
– – fa. concava → Y. flaccida: 94
– – – flaccida → Y. flaccida: 94
– – – genuina → Y. filamentosa: 94
– – – glaucescens → Y. flaccida: 94
– – – orchioides → Y. flaccida: 94
– – – puberula → Y. flaccida: 94
– – var. angustifolia → Y. filamentosa: 94
– – – bracteata → Y. filamentosa: 94
– – – concava → Y. flaccida: 94
– – – flaccida → Y. flaccida: 94
– – – glaucescens → Y. flaccida: 94
– – – grandiflora → Y. filamentosa: 94
– – – laevigata → Y. filamentosa: 94
– – – latifolia → Y. filamentosa: 94
– – – mexicana → Y. filamentosa: 94
– – – puberula → Y. flaccida: 94
– – – recurvifolia → Y. filamentosa: 94
– – – smalliana → Y. flaccida: 94
– **filifera**: 94
– – → Y. baccata: 89
– **flaccida**: 94

a   *Agave attenuata*

b   *Agave avellanidens*

c   *Agave bovicornuta*

d   *Agave bracteosa*

e   *Agave bovicornuta*

f   *Agave colorata*

a   *Agave deserti* ssp. *deserti*

b   *Agave cerulata* ssp. *nelsonii*

c   *Agave cerulata* ssp. *subcerulata*

d   *Agave chrysoglossa*

e   *Agave deserti* ssp. *deserti*

f   *Agave cupreata*

g   *Agave deserti* ssp. *simplex*

a   *Agave leopoldii*

b   *Agave filifera* ssp. *multifilifera*

c   *Agave ghiesbreghtii*

d   *Agave guiengola*

e   *Agave havardiana*

f   *Agave horrida* ssp. *perotensis*

g   *Agave leopoldii*

a *Agave marmorata*

b *Agave marmorata*

c *Agave kerchovei*

d *Agave lechuguilla*

e *Agave lophantha*

f *Agave macroacantha*

g *Agave mckelveyana*

a   *Agave polianthiflora*

b   *Agave polianthiflora*

c   *Agave mitis* var. *mitis*

d   *Agave neomexicana*

e   *Agave nizandensis*

f   *Agave ocahui* var. *ocahui*

g   *Agave palmeri*

a   *Agave stricta*

b   *Agave striata* ssp. *striata*

c   *Agave striata* ssp. *striata*

d   *Agave parryi* var. *parryi*

e   *Agave schottii* var. *schottii*

f   *Agave shrevei* ssp. *shrevei*

g   *Agave titanota*

a   *Agave vivipara* var. *vivipara*

b   *Agave toumeyana* ssp. *bella*

c   *Agave toumeyana* ssp. *toumeyana*

d   *Agave utahensis* ssp. *kaibabensis*

e   *Agave victoriae-reginae*

f   *Agave triangularis*

g   *Agave zebra*

b   *Beschorneria yuccoides* ssp. *yuccoides*

a   *Beschorneria yuccoides* ssp. *yuccoides*

c   *Furcraea bedinghausii*

e   *Hesperoyucca whipplei*

d   *Hesperoyucca whipplei*

f   *Furcraea guerrerensis*

a   *Hesperaloe parviflora*

b   *Yucca baccata* var. *vespertina*

c   *Yucca brevifolia* var. *brevifolia*

d   *Yucca endlichiana*

e   *Yucca brevifolia* var. *brevifolia*

f   *Yucca faxoniana*

a *Yucca faxoniana*

b *Yucca rigida*

c *Yucca schidigera*

d *Aloe acutissima* var. *acutissima*

e *Aloe africana*

f *Aloe albida*

g *Yucca × schottii* ('*Y. thornberi*')

h *Aloe aageodonta*

a   *Aloe alooides*

b   *Aloe ballii* var. *ballii*

c   *Aloe ballyi*

d   *Aloe belavenokensis*

e   *Aloe bellatula*

f   *Aloe castanea*

g   *Aloe asperifolia*

h   *Aloe bakeri*

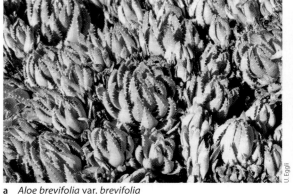

a   *Aloe brevifolia* var. *brevifolia*

b   *Aloe bussei*

c   *Aloe debrana*

d   *Aloe capitata* var. *capitata*

e   *Aloe chabaudii* var. *chabaudii*

f   *Aloe claviflora*

g   *Aloe diolii*

h   *Aloe divaricata* var. *divaricata*

a   *Aloe elata*

b   *Aloe fragilis*

c   *Aloe fragilis*

d   *Aloe francombei*

e   *Aloe gariepensis*

f   *Aloe helenae*

g   *Aloe calcairophila*

h   *Aloe cryptopoda*

a   *Aloe ferox*

b   *Aloe framesii*

c   *Aloe gilbertii* ssp. *gilbertii*

d   *Aloe globuligemma*

e   *Aloe grandidentata*

f   *Aloe guillaumetii*

g   *Aloe grandidentata*

h   *Aloe humilis*

a   *Aloe ketabrowniorum*

b   *Aloe kilifiensis*

c   *Aloe rauhii*

d   *Aloe kulalensis*

e   *Aloe lateritia* var. *graminicola*

f   *Aloe massawana*

g   *Aloe mitriformis*

a   *Aloe isaloensis*

b   *Aloe isaloensis*

c   *Aloe krapohliana* var. *krapohliana*

d   *Aloe leptosiphon*

e   *Aloe maculata*

f   *Aloe marlothii* var. *marlothii*

g   *Aloe melanacantha* var. *melanacantha*

h   *Aloe ngongensis*

a   *Aloe mawii*

b   *Aloe mzimbana*

c   *Aloe nuttii*

d   *Aloe pillansii*

e   *Aloe plicatilis*

f   *Aloe pustuligemma*

g   *Aloe pearsonii*

h   *Aloe ortholopha*

a   *Aloe porphyrostachys*

b   *Aloe pretoriensis*

c   *Aloe powysiorum*

d   *Aloe striata* ssp. *karasbergensis*

e   *Aloe striatula* var. *striatula*

a   *Aloe propagulifera*

b   *Aloe soutpansbergensis*

c   *Aloe suarezensis*

d   *Aloe secundiflora* var. *secundiflora*

e   *Aloe variegata*

f   *Aloe vera*

a   *Aloe suzannae*

b   *Aloe tomentosa*

c   *Aloe trichosantha* ssp. *trichosantha*

d   *Astroloba corrugata*

e   *Chortolirion angolense*

f   *Gasteria pillansii* var. *ernesti-ruschii*

g   *Astroloba spiralis*

h   *Gasteria pillansii* var. *pillansii*

a *Gasteria batesiana*

b *Gasteria bicolor* var. *bicolor*

c *Gasteria bicolor* var. *bicolor*

d *Gasteria bicolor* var. *liliputana*

e *Gasteria pillansii* var. *pillansii*

f *Gasteria carinata* var. *retusa*

g *Gasteria carinata* var. *glabra*

a   *Haworthia cymbiformis* var. *cymbiformis*

b   *Haworthia arachnoidea* var. *arachnoidea*

c   *Haworthia bolusii* var. *blackbeardiana*

d   *Haworthia attenuata* var. *attenuata*

e   *Haworthia emelyae* var. *comptoniana*

f   *Haworthia herbacea* var. *herbacea*

a　*Haworthia mucronata* var. *mucronata*

b　*Haworthia mucronata* var. *habdomadis*

c　*Haworthia pungens*

d　*Haworthia reinwardtii* var. *reinwardtii*

e　*Haworthia truncata* var. *truncata*

f　*Haworthia venosa* ssp. *tessellata*

g　*Haworthia viscosa*

h　*Haworthia wittebergensis*

a   *Boophane disticha*

b   *Poellnitzia rubriflora*

c   *Poellnitzia rubriflora*

d   *Cyrtanthus herrei*

e   *Cyrtanthus flammosus*

f   *Haemanthus albiflos*

a　*Chlorophytum suffruticosum*

b　*Rauhia multiflora*

c　*Zamioculcas zamiifolia*

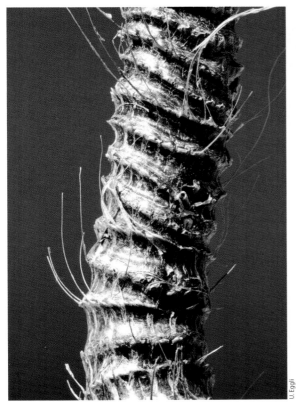

d　*Chlorophytum suffruticosum*

e　*Trachyandra ciliata*

f　*Bulbine fallax*

a *Bulbine frutescens* 'Hallmark'

b *Bulbine haworthioides*

c *Bulbine latifolia*

d *Bulbine margarethae*

e *Aneilema zebrinum*

f *Cyanotis somaliensis*

g *Callisia navicularis*

h *Callisia repens*

a *Dioscorea elephantipes*

b *Dioscorea fastigiata*

c *Tradescantia sillamontana*

d *Dioscorea elephantipes*

e *Dioscorea hemicrypta*

f *Dioscorea basiclavicaulis*

g *Dioscorea sylvatica*

h *Doryanthes palmeri*

a   *Dracaena serrulata*

b   *Sansevieria arborescens*

c   *Dracaena cinnabari*

d   *Cordyline australis*

e   *Sansevieria aethiopica*

f   *Sansevieria conspicua*

g   *Sansevieria eilensis*

h   *Sansevieria kirkii* var. *pulchra*

a   *Sansevieria powellii*

Bowiea gariepensis

b   *Bowiea gariepensis*

c   *Albuca juncifolia*

d   *Sansevieria pinguicula*

e   *Sansevieria stuckyi*

a *Drimia haworthioides*

b *Drimia haworthioides*

c *Drimia haworthioides*

d *Massonia depressa*

e *Massonia depressa*

f *Ledebouria socialis*

g *Ornithogalum maculatum*

a   *Ornithogalum longibracteatum*

b   *Ornithogalum unifoliatum*

c   *Ornithogalum unifoliatum*

d   *Urginea multifolia*

e   *Urginea epigea*

f   *Whiteheadia bifolia*

g   *Schizobasis intricata*

h   *Schizobasis intricata*

a *Beaucarnea gracilis*

b *Calibanus hookeri*

c *Beaucarnea pliabilis*

d *Beaucarnea recurvata*

e *Beaucarnea stricta*

f *Dasylirion longissimum*

g *Dasylirion longissimum*

h *Nolina parryi*